Physical Metallurgy Principles

Fifth Edition

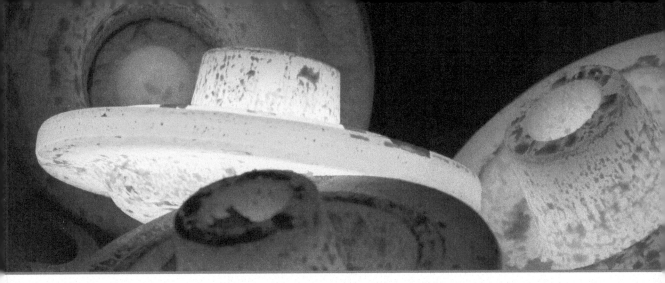

Physical Metallurgy Principles

Fifth Edition

Reza Abbaschian
Lara Abbaschian

Australia • Brazil • Canada • Mexico • Singapore • United Kingdom • United States

Physical Metallurgy Principles, Fifth Edition
Reza Abbaschian, Lara Abbaschian

SVP, Product: Cheryl Costantini

VP, Product: Thais Alencar

Portfolio Product Director: Rita Lombard

Senior Portfolio Product Manager: Timothy Anderson

Product Assistant: Emily Smith

Learning Designer: MariCarmen Constable

Content Manager: Samantha Enders

Digital Project Manager: Nikkita Kendrick

VP, Product Marketing: Jason Sakos

Senior Director, Product Marketing: Danae April

Product Marketing Manager: Mackenzie Paine

Content Acquisition Analyst: Deanna Ettinger

Production Service: MPS Limited, RPK Editorial Services

Designer: Chris Doughman

Cover Image Source: Jackfoto/Shutterstock.com

Copyright © 2025 Cengage Learning, Inc. ALL RIGHTS RESERVED.
WCN: 01-100-371

No part of this work covered by the copyright herein may be reproduced or distributed in any form or by any means, except as permitted by U.S. copyright law, without the prior written permission of the copyright owner.

The names of all products mentioned herein are used for identification purposes only and may be trademarks or registered trademarks of their respective owners. Cengage Learning disclaims any affiliation, association, connection with, sponsorship, or endorsement by such owners.

Previous Editions: © 2009, © 1992

> For product information and technology assistance, contact us at
> **Cengage Customer & Sales Support, 1-800-354-9706 or support.cengage.com.**
>
> For permission to use material from this text or product, submit all requests online at **www.copyright.com**.

Library of Congress Control Number: 2024931732

ISBN: 979-8-214-00166-1

Cengage
5191 Natorp Boulevard
Mason, OH 45040
USA

Cengage is a leading provider of customized learning solutions. Our employees reside in nearly 40 different countries and serve digital learners in 165 countries around the world. Find your local representative at **www.cengage.com**.

To learn more about Cengage platforms and services, register or access your online learning solution, or purchase materials for your course, visit **www.cengage.com**.

Notice to the Reader

Publisher does not warrant or guarantee any of the products described herein or perform any independent analysis in connection with any of the product information contained herein. Publisher does not assume, and expressly disclaims, any obligation to obtain and include information other than that provided to it by the manufacturer. The reader is expressly warned to consider and adopt all safety precautions that might be indicated by the activities described herein and to avoid all potential hazards. By following the instructions contained herein, the reader willingly assumes all risks in connection with such instructions. The publisher makes no representations or warranties of any kind, including but not limited to, the warranties of fitness for particular purpose or merchantability, nor are any such representations implied with respect to the material set forth herein, and the publisher takes no responsibility with respect to such material. The publisher shall not be liable for any special, consequential, or exemplary damages resulting, in whole or in part, from the readers' use of, or reliance upon, this material.

Printed in the United States of America
Print Number: 01   Print Year: 2024

Dedication

We dedicate this edition in honor of Professor Robert E. Reed-Hill, who spent a lifetime to advance physical metallurgy.
We also give special thanks to Janette, Cyrus, Gwen, David and Emily.

Contents

Preface xix
About the Authors xxi
Digital Resources xxii

Chapter 1 The Structure of Metals 1

1.1	The Structure of Metals	2
1.2	Unit Cells	2
1.3	The Body-Centered Cubic Structure (BCC)	4
1.4	Coordination Number of the Body-Centered Cubic Lattice	4
1.5	The Face-Centered Cubic Lattice (FCC)	5
1.6	The Unit Cell of the Hexagonal Closed-Packed (HCP) Lattice	6
1.7	Comparison of the Face-Centered Cubic and Close-Packed Hexagonal Structures	7
1.8	Coordination Number of the Systems of Closest Packing	8
1.9	Anisotropy	8
1.10	Textures or Preferred Orientations	9
1.11	Miller Indices	10
	Direction Indices in the Cubic Lattice	11
	Cubic Indices for Planes	12
	Miller Indices for Hexagonal Crystals	14
1.12	Crystal Structures of the Metallic Elements	15
1.13	The Stereographic Projection	16
1.14	Directions that Lie in a Plane	18
1.15	Planes of a Zone	18
1.16	The Wulff Net	20
	Rotation About an Axis in the Line of Sight	21
	Rotation About the North–South Axis of the Wulff Net	21
1.17	Standard Projections	23
1.18	The Standard Stereographic Triangle for Cubic Crystals	24
	Problems	27
	References	30

Chapter 2 Characterization Techniques 31

2.1	The Bragg Law	32
2.2	Laue Techniques	36

2.3	The Rotating-Crystal Method	38
2.4	The Debye-Scherrer or Powder Method	38
2.5	The X-Ray Diffractometer	42
2.6	The Transmission Electron Microscope	43
2.7	Interactions between the Electrons in an Electron Beam and a Metallic Specimen	49
2.8	Elastic Scattering	49
2.9	Inelastic Scattering	49
2.10	Electron Spectrum	51
2.11	The Scanning Electron Microscope	51
2.12	Topographic Contrast	53
2.13	The Picture Element Size	56
2.14	The Depth of Focus	57
2.15	Microanalysis of Specimens	58
2.16	Electron Probe X-Ray Microanalysis	58
2.17	The Characteristic X-Rays	59
2.18	Auger Electron Spectroscopy (AES)	61
2.19	The Scanning Transmission Electron Microscope (STEM)	63
	Problems	64
	References	65

Chapter 3 Crystal Binding 66

3.1	The Internal Energy of a Crystal	66
3.2	Ionic Crystals	67
3.3	The Born Theory of Ionic Crystals	68
3.4	Van Der Waals Crystals	72
3.5	Dipoles	72
3.6	Inert Cases	74
3.7	Induced Dipoles	74
3.8	The Lattice Energy of an Inert-Gas Solid	76
3.9	The Debye Frequency	76
3.10	The Zero-Point Energy	78
3.11	Dipole-Quadrupole and Quadrupole-Quadrupole Terms	79
3.12	Molecular Crystals	80
3.13	Refinements to the Born Theory of Ionic Crystals	80
3.14	Covalent and Metallic Bonding	81
	Problems	84
	References	85

Chapter 4 Introduction to Dislocations 86

4.1	The Discrepancy Between the Theoretical and Observed Yield Stresses of Crystals	86
4.2	Dislocations	89
4.3	The Burgers Vector	97
4.4	Vector Notation for Dislocations	100
4.5	Dislocations in the Face-Centered Cubic Lattice	101
4.6	Intrinsic and Extrinsic Stacking Faults in Face-Centered Cubic Metals	105
4.7	Extended Dislocations in Hexagonal Metals	106
4.8	Climb of Edge Dislocations	107
4.9	Dislocation Intersections	108
4.10	The Stress Field of a Screw Dislocation	111
4.11	The Stress Field of an Edge Dislocation	112
4.12	The Force on a Dislocation	115
4.13	The Strain Energy of a Screw Dislocation	118
4.14	The Strain Energy of an Edge Dislocation	119
	Problems	119
	References	122

Chapter 5 Dislocations and Plastic Deformation 123

5.1	The Frank-Read Source	124
5.2	Nucleation of Dislocations	125
5.3	Bend Gliding	128
5.4	Rotational Slip	130
5.5	Slip Planes and Slip Directions	133
5.6	Slip Systems	134
5.7	Critical Resolved Shear Stress	134
5.8	Slip on Equivalent Slip Systems	138
5.9	The Dislocation Density	138
5.10	Slip Systems in Different Crystal Forms	138
	Face-Centered Cubic Metals	138
	Hexagonal Metals	140
	Easy Glide in Hexagonal Metals	142
	Body-Centered Cubic Crystals	142
5.11	Cross-Slip	143
5.12	Slip Bands	145
5.13	Double Cross-Slip	145
5.14	Extended Dislocations and Cross-Slip	147
5.15	Crystal Structure Rotation during Tensile and Compressive Deformation	149
5.16	The Notation for the Slip Systems in the Deformation of fcc Crystals	151
5.17	Work Hardening	153

5.18	Considère's Criterion	155
5.19	The Relation Between Dislocation Density and the Stress	156
5.20	Taylor's Relation	157
5.21	The Orowan Equation	158
	Problems	159
	References	161

Chapter 6 Elements of Grain Boundaries   163

6.1	Grain Boundaries	163
6.2	Dislocation Model of a Small-Angle Grain Boundary	164
6.3	The Five Degrees of Freedom of a Grain Boundary	167
6.4	The Stress Field of a Grain Boundary	168
6.5	Grain-Boundary Energy	169
6.6	Low-Energy Dislocation Structures, LEDS	172
6.7	Dynamic Recovery	177
6.8	Surface Tension of the Grain Boundary	179
6.9	Boundaries between Crystals of Different Phases	180
6.10	The Grain Size	183
6.11	The Effect of Grain Boundaries on Mechanical Properties: Hall-Petch Relation	185
6.12	Grain Size Effects in Nanocrystalline Materials	187
6.13	Coincidence Site Boundaries	190
6.14	The Density of Coincidence Sites	191
6.15	The Ranganathan Relations	191
6.16	Examples Involving Twist Boundaries	192
6.17	Tilt Boundaries	194
	Problems	197
	References	198

Chapter 7 Vacancies and Thermodynamics   200

7.1	Thermal Behavior of Metals	200
7.2	Internal Energy	202
7.3	Entropy	202
7.4	Spontaneous Reactions	203
7.5	Gibbs Free Energy	203
7.6	Statistical Mechanical Definition of Entropy	205
7.7	Vacancies	209
7.8	Vacancy Motion	215
7.9	Interstitial Atoms and Divacancies	217
	Problems	220
	References	222

Chapter 8 Annealing 223

- 8.1 Stored Energy of Cold Work 223
- 8.2 The Relationship of Free Energy to Strain Energy 225
- 8.3 The Release of Stored Energy 225
- 8.4 Recovery 227
- 8.5 Recovery in Single Crystals 228
- 8.6 Polygonization 231
- 8.7 Dislocation Movements in Polygonization 232
- 8.8 Recovery Processes at High and Low Temperatures 236
- 8.9 Recrystallization 236
- 8.10 The Effect of Time and Temperature on Recrystallization 237
- 8.11 Recrystallization Temperature 239
- 8.12 The Effect of Strain on Recrystallization 239
- 8.13 The Rate of Nucleation and the Rate of Nucleus Growth 240
- 8.14 Formation of Nuclei 241
- 8.15 Driving Force for Recrystallization 243
- 8.16 The Recrystallized Grain Size 243
- 8.17 Other Variables in Recrystallization 245
- 8.18 Purity of the Metal 245
- 8.19 Initial Grain Size 247
- 8.20 Grain Growth 247
- 8.21 Geometrical Coalescence 249
- 8.22 Three-Dimensional Changes in Grain Geometry 251
- 8.23 The Grain Growth Law 252
- 8.24 Impurity Atoms in Solid Solution 256
- 8.25 Impurities in the Form of Inclusions 256
- 8.26 The Free-Surface Effects 259
- 8.27 The Limiting Grain Size 260
- 8.28 Preferred Orientation 261
- 8.29 Secondary Recrystallization 262
- 8.30 Strain-Induced Boundary Migration 263
  - Problems 264
  - References 265

Chapter 9 Solid Solutions 267

- 9.1 Solid Solutions 267
- 9.2 Intermediate Phases 268
- 9.3 Interstitial Solid Solutions 269
- 9.4 Solubility of Carbon in Body-Centered Cubic Iron 269
- 9.5 Substitutional Solid Solutions and the Hume-Rothery Rules 273

9.6 Interaction of Dislocations and Solute Atoms 274
9.7 Dislocation Atmospheres 274
9.8 The Formation of a Dislocation Atmosphere 275
9.9 The Evaluation of A 277
9.10 The Drag of Atmospheres on Moving Dislocations 277
9.11 The Sharp Yield Point and Lüders Bands 279
9.12 The Theory of the Sharp Yield Point 281
9.13 Strain Aging 282
9.14 The Cottrell-Bilby Theory of Strain Aging 283
9.15 Dynamic Strain Aging 287
Problems 291
References 292

Chapter 10 Phases 293

10.1 Basic Definitions 293
10.2 The Physical Nature of Phase Mixtures 295
10.3 Thermodynamics of Solutions 295
10.4 Equilibrium between Two Phases 298
10.5 The Number of Phases in an Alloy System 299
One-Component Systems 299
Two-Component Systems 304
Ideal Solutions 304
Nonideal Solutions 305
10.6 Two-Component Systems Containing Two Phases 308
10.7 Graphical Determinations of Partial-Molar Free Energies 310
10.8 Two-Component Systems with Three Phases in Equilibrium 312
10.9 The Gibbs Phase Rule 313
10.10 Ternary Systems 315
Problems 316
References 317

Chapter 11 Binary Phase Diagrams 318

11.1 Phase Diagrams 318
11.2 Isomorphous Alloy Systems 319
11.3 The Lever Rule 320
11.4 Equilibrium Heating or Cooling of an Isomorphous Alloy 323
11.5 The Isomorphous Alloy System from the Point of View of Free Energy 325
11.6 Maxima and Minima 327
11.7 Superlattices 329
11.8 Miscibility Gaps 333
11.9 Eutectic Systems 334

- 11.10 The Microstructures of Eutectic Systems 335
- 11.11 The Peritectic Transformation 340
- 11.12 Monotectics 343
- 11.13 Other Three-Phase Reactions 347
- 11.14 Intermediate Phases 348
- 11.15 The Copper-Zinc Phase Diagram 350
- 11.16 Ternary Phase Diagrams 353
  - Problems 356
  - References 357

Chapter 12 Diffusion in Substitutional Solid Solutions 358

- 12.1 Diffusion in an Ideal Solution 359
- 12.2 The Kirkendall Effect 362
- 12.3 Pore Formation 366
- 12.4 Darken's Equations 367
- 12.5 Fick's Second Law 371
- 12.6 The Matano Method 373
- 12.7 Determination of the Intrinsic Diffusivities 377
- 12.8 Self-Diffusion in Pure Metals 378
- 12.9 Temperature Dependence of the Diffusion Coefficient 380
- 12.10 Chemical Diffusion at Low-Solute Concentration 383
- 12.11 The Study of Chemical Diffusion Using Radioactive Tracers 384
- 12.12 Diffusion along Grain Boundaries and Free Surfaces 388
- 12.13 Fick's First Law in Terms of a Mobility and an Effective Force 391
- 12.14 Diffusion in Non-Isomorphic Alloy Systems 392
  - Problems 397
  - References 399

Chapter 13 Interstitial Diffusion 400

- 13.1 Measurement of Interstitial Diffusivities 401
- 13.2 The Snoek Effect 402
- 13.3 Experimental Determination of the Relaxation Time 409
- 13.4 Experimental Data 415
- 13.5 Anelastic Measurements at Constant Strain 416
  - Problems 417
  - References 418

Chapter 14 Solidification of Metals 419

- 14.1 The Liquid Phase 420
- 14.2 Nucleation 423

14.3	Metallic Glasses 425
14.4	Atomic Movement at S/L Interface 431
14.5	The Heats of Fusion and Vaporization 432
14.6	The Nature of the Liquid-Solid Interface 434
14.7	Continuous Growth 436
14.8	Lateral Growth 438
14.9	Stable Interface Freezing 439
14.10	Dendritic Growth in Pure Metals 441
14.11	Freezing in Alloys with Planar Interface 444
14.12	The Scheil Equation 446
14.13	Dendritic Freezing in Alloys 449
14.14	Freezing of Ingots 451
14.15	The Grain Size of Castings 454
14.16	Segregation 455
14.17	Homogenization 457
14.18	Inverse Segregation 461
14.19	Porosity 462
14.20	Eutectic Freezing 466
	Problems 471
	References 473

Chapter 15 Nucleation and Growth Kinetics 475

15.1	Nucleation of a Liquid from the Vapor 476
15.2	The Becker-Döring Theory 483
15.3	Freezing 485
15.4	Solid-State Reactions 487
15.5	Heterogeneous Nucleation 490
15.6	Growth Kinetics 493
15.7	Diffusion Controlled Growth 496
15.8	Interference of Growing Precipitate Particles 500
15.9	Interface Controlled Growth 501
15.10	Transformations That Occur on Heating 504
15.11	Dissolution of a Precipitate 505
	Problems 508
	References 509

Chapter 16 Precipitation Hardening 511

16.1	The Significance of the Solvus Curve 512
16.2	The Solution Treatment 513
16.3	The Aging Treatment 514

16.4 Development of Precipitates 517
16.5 Aging of Al-Cu Alloys at Temperatures above 100°C (373 K) 519
16.6 Precipitation Sequences in Other Aluminum Alloys 522
16.7 Homogeneous Versus Heterogeneous Nucleation of Precipitates 523
16.8 Interphase Precipitation 525
16.9 Theories of Hardening 527
16.10 Additional Factors in Precipitation Hardening 529
Problems 531
References 532

Chapter 17 Deformation Twinning and Martensite Reactions 533

17.1 Deformation Twinning 534
17.2 Formal Crystallographic Theory of Twinning 536
17.3 Twin Boundaries 542
17.4 Twin Nucleation and Growth 543
17.5 Accommodation of the Twinning Shear 546
17.6 The Significance of Twinning in Plastic Deformation 547
17.7 The Effect of Twinning on Face-Centered Cubic Stress-Strain Curves 548
17.8 Martensite 550
17.9 The Bain Distortion 551
17.10 The Martensite Transformation in an Indium-Thallium Alloy 553
17.11 Reversibility of the Martensite Transformation 554
17.12 Athermal Transformation 554
17.13 Phenomenological Crystallographic Theory of Martensite Formation 555
17.14 Irrational Nature of the Habit Plane 561
17.15 The Iron-Nickel Martensitic Transformation 562
17.16 Isothermal Formation of Martensite 564
17.17 Stabilization 564
17.18 Nucleation of Martensite Plates 565
17.19 Growth of Martensite Plates 566
17.20 The Effect of Stress 566
17.21 The Effect of Plastic Deformation 567
17.22 Thermoelastic Martensite Transformations 567
17.23 Elastic Deformation of Thermoelastic Alloys 569
17.24 Stress-Induced Martensite (SIM) 569
17.25 The Shape-Memory Effect 571
Problems 572
References 574

Chapter 18 The Iron-Carbon Alloy System 576

- 18.1 The Iron-Carbon Diagram 576
- 18.2 The Proeutectoid Transformations of Austenite 579
- 18.3 The Transformation of Austenite to Pearlite 581
- 18.4 The Growth of Pearlite 586
- 18.5 The Effect of Temperature on the Pearlite Transformation 587
  - The Interlamellar Spacing and the Rate of Growth 587
- 18.6 Forced-Velocity Growth of Pearlite 589
- 18.7 The Effects of Alloying Elements on the Growth of Pearlite 592
- 18.8 The Rate of Nucleation of Pearlite 595
- 18.9 Time-Temperature-Transformation Curves 597
- 18.10 The Bainite Reaction 598
- 18.11 The Complete *T-T-T* Diagram of an Eutectoid Steel 605
  - Path 1 606
  - Path 2 607
  - Path 3 607
  - Path 4 607
- 18.12 Slowly Cooled Hypoeutectoid Steels 607
- 18.13 Slowly Cooled Hypereutectoid Steels 609
- 18.14 Isothermal Transformation Diagrams for Noneutectoid Steels 611
  - Problems 615
  - References 616

Chapter 19 The Hardening of Steel 618

- 19.1 Continuous Cooling Transformations (CCT) 618
- 19.2 Hardenability 622
- 19.3 The Variables That Determine the Hardenability of a Steel 628
- 19.4 Austenitic Grain Size 629
- 19.5 The Effect of Austenitic Grain Size on Hardenability 630
- 19.6 The Influence of Carbon Content on Hardenability 630
- 19.7 The Influence of Alloying Elements on Hardenability 631
- 19.8 The Significance of Hardenability 636
- 19.9 The Martensite Transformation in Steel 637
- 19.10 The Hardness of Iron-Carbon Martensite 642
- 19.11 Dimensional Changes Associated with Transformation of Martensite 646
- 19.12 Quench Cracks 647
- 19.13 Tempering 648
- 19.14 Tempering of a Low-Carbon Steel 654
- 19.15 Spheroidized Cementite 656
- 19.16 The Effect of Tempering on Physical Properties 658

19.17　The Interrelation Between Time and Temperature in Tempering　661
19.18　Secondary Hardening　661
　　　　Problems　663
　　　　References　664

Chapter 20　Selected Nonferrous Alloy Systems　666

20.1　Commercially Pure Copper　666
20.2　Copper Alloys　669
20.3　Copper Beryllium　673
20.4　Other Copper Alloys　675
20.5　Aluminum Alloys　675
20.6　Aluminum-Lithium Alloys　676
20.7　Titanium Alloys　683
20.8　Classification of Titanium Alloys　685
20.9　The Alpha Alloys　691
20.10　The Beta Alloys　692
20.11　The Alpha-Beta Alloys　692
20.12　Superalloys　694
20.13　Creep Strength　697
　　　　Problems　698
　　　　References　700

Chapter 21　Failure of Metals　702

21.1　Failure by Easy Glide　703
21.2　Rupture by Necking (Multiple Glide)　704
21.3　The Effect of Twinning　705
21.4　Cleavage　706
21.5　The Nucleation of Cleavage Cracks　707
21.6　Propagation of Cleavage Cracks　709
21.7　The Effect of Grain Boundaries　712
21.8　The Effect of the State of Stress　713
21.9　Ductile Fractures　715
21.10　Intercrystalline Brittle Fracture　721
21.11　Blue Brittleness　721
21.12　Fatigue Failures　722
21.13　The Macroscopic Character of Fatigue Failure　723
21.14　The Rotating-Beam Fatigue Test　724
21.15　Alternating Stress Parameters　726
21.16　The Microscopic Aspects of Fatigue Failure　729
21.17　Fatigue Crack Growth　731
21.18　The Effect of Nonmetallic Inclusions　735

21.19 The Effect of Steel Microstructure on Fatigue   737
21.20 Low-Cycle Fatigue   737
21.21 The Coffin-Manson Equation   741
21.22 Certain Practical Aspects of Fatigue   743
   Problems   743
   References   744

Appendices   747
   A: Angles Between Crystallographic Planes in the Cubic System (In Degrees)   748
   B: Angles Between Crystallographic Planes for Hexagonal Elements   749
   C: Indices of the Reflecting Planes for Cubic Structures   750
   D: Conversion Factors and Constants   751
   E: Twinning Elements of Several of the More Important Twinning Modes   752
   F: Selected Values of Intrinsic Stacking-Fault Energy γ_I, Twin-Boundary Energy γ_T, Grain-Boundary Energy γ_G, and Crystal-Vapor Surface Energy γ for Various Materials in ergs/cm^2 752

Index   753

Preface

The original philosophy of the text has been preserved in this Fifth Edition. The theoretical approach to physical metallurgy is premised on the belief that the properties of metals and alloys are determined by simple physical laws. The conceptional framework used throughout is based on the fundamentals of the materials structure-properties-processing-performance relationships. As such, it is not necessary to consider each alloy as a separate entity nor to spend time learning large numbers of apparently unrelated facts that are easily forgotten. Today's alloys and methods have evolved since the publication of the First Edition and will continue to do so. However, the approach embodied by this text ensures a foundational understanding that translates into durable utility and relevance.

We have retained the easy-to-read format so that the essence of the information is most successfully communicated. Throughout the years, we have found this format to be very useful not only to current students, but also to practitioners who would like to refresh their knowledge of the field of physical metallurgy principles.

This book is intended for use as a comprehensive introduction to metallurgy and materials science and engineering. It is appropriate for all engineering students at the junior or senior level. Graduate students seeking to engage with the materials structure-properties-processing-performance relationship as they develop their research plans will also find the book valuable. Recommended prerequisites to this text are college physics, chemistry, and strength of materials. An engineering course in thermodynamics or physical chemistry is also considered desirable, but not essential since relevant core concepts are explained in Chapters 7, 10, and 11. When this textbook is used for a one semester course, a number of chapters dealing with advanced topics, such as Chapters 10, 11, and 15, may be omitted. Alternatively, the newly added Learning Objectives can serve as a guide for topic selection and syllabus creation.

Features of the Book

Following the theme of this text's pedagogy, properties-processing and performance as well as the strong interrelationship among them are discussed in the early chapters. Chapters 1 through 3 cover the structure of metals, characterization, and bonding of atoms, in that order. Because of the importance of sophisticated techniques that reach atomic-level resolution, Chapter 2 covers x-ray techniques, scanning electron microscopy, transmission and scanning transmission electron microscopy, as well as Auger electron spectroscopy. Chapters 4 and 5 cover geometrical aspects of dislocations, slip planes, and directions during plastic deformation, followed by Chapter 6 with discussion of grain boundaries and their important effects on mechanical behavior. Chapter 7 introduces a broad coverage of thermodynamic concepts, vacancies, their motion, and interstitial impurity atoms. Chapter 8 introduces annealing of cold-worked metals, recrystallization, and the release of stored energy of deformation. Chapters 9 through 11 provide background information on solid solutions, phases, and phase diagrams, which form the foundations for alloy development. Chapters 12 and 13 deal with substitutional and interstitial diffusion and kinetic concepts. Chapters 14 and 15 cover liquid-solid and solid-solid phase transformation, microstructural developments and compositions, and growth kinetics. Theories of precipitation hardening and development of precipitates during aging of Al-Cu alloys are covered in Chapter 16. Twinning and martensitic transformations are discussed in Chapter 17, together with phenomenological crystallographic theory of martensite formation followed by its reversibility in shape-memory alloys. Chapter 18 presents an in-depth look at isothermal transformations of austenite to pearlite or bainite, their microstructures and kinetics, and the TTT diagrams of eutectoid and non-eutectoid steels. In Chapter 19, martensite formation during continuous cooling is explained with a specific focus on atomic rearrangement, hardenability, influence of composition, and grain size. The last half of the chapter discusses tempering and the effect of temperature and time on physical properties. Chapter 20 covers the important topic of non-ferrous metals. In the final chapter, Chapter 21, failure of metals and alloys is covered including cleavage, brittle and ductile failures, and fatigue.

New to This Edition

The subject matter of *Physical Metallurgy Principles* is fundamental and borne out of decades of research by pioneers that does not disappear at the introduction of this Fifth Edition. Recent references have been incorporated to complement the remaining original references, which provide historic context. Indeed, with today's electronic search engines limited to more recently published literature, such identification and acknowledgment of the pioneers of the field often get overlooked. Other changes introduced in the Fifth Edition are the addition of chapter-by-chapter Learning Objectives to improve learning outcomes, new figures to augment understanding of the text, and the inclusion of color and enhancement to figures to increase usability.

Acknowledgments

The authors would like to acknowledge the critical inputs and contributions of all who are listed in the prefaces of the earlier editions. Our gratitude extends to A. S. Nowick, W. D. Robertson, F. N. Rhines, Richard W. Heckel, Walter S. Owen, Marvin Metzger, John Kronsbein, John Hren, Robert T. DeHoff, Derek Dove, Ellis Verink, William C. Leslie, Daniel N. Beshers, Paul C. Holloway, Rolf N. Hummel, William A. Jesser, William G. Ovens, Dale E. Wittmer, James C. M. Li, Alan R. Pelton, Samuel J. Hruska, Richard B. Griffin, Dong-Joo (Daniel) Kim, Anthony P. Reynolds, Christopher A. Schuh, Jiahong Zhu, Abraham Munitz, and Gerald Bourne. Special thanks to the publishing team for this edition: Tim Anderson, Samantha Enders Rose Kernan, and John Kronsbein. We also thank Shahin Amini and Steven Herrera for contributing some pictures incorporated in this edition.

Last, and most important, we would like to acknowledge that this text would not exist were it not for the late Professor Reed-Hill and his vision of a comprehensive tome dedicated to the field of metallurgy and materials science. One of the authors (R.A.) remembers many hours of intellectually stimulating discussions with Professor Reed-Hill and other colleagues at the University of Florida.

Reza Abbaschian
Lara Abbaschian

About the Authors

Reza Abbaschian

Reza Abbaschian is Director of Winston Chung Global Energy Center, Distinguished Professor, and Winston Chung Endowed Professor in Sustainability at University of California Riverside (UCR). He began his tenure at UCR in September 2005 as Dean of the Bourns College of Engineering (BCOE), a position that he held until July 2016. Prior to his appointment as dean at UCR, Dr. Abbaschian was the Vladimir A. Grodsky Professor of Materials Science and Engineering Department at the University of Florida, where he also served as Chair of the Department for 16 years. During his tenure, the Department rose to a *U.S. News & World Report* top-ten ranking for both undergraduate and graduate education.

Dr. Abbaschian received his PhD in Materials Science and Engineering from the University of California, Berkeley, his MS in Metallurgical Engineering from Michigan Technological University, and his BSc in Mining and Metallurgy from Tehran University. He has published more than 250 scientific articles including eight books on subjects ranging from metal processing to composites and high temperature-high pressure growth of diamonds. His research led to the introduction of man-made diamonds to the market by Gemesis Diamond Company and he holds five patents and eight trade secrets held by Gemesis.

Dr. Abbaschian is a past President of ASM International. His awards include the TMS Educator Award, Structural Material Division's Distinguished Scientist/Engineer Award, TMS Leadership Award, ASEE Donald E. Marlowe Award, Davis Productivity Award of the State of Florida, and the ASM Albert Sauveur Achievement Award. He has been elected a Fellow of ASM, TMS, and AAAS. In 2017, he was bestowed the AIME Honorary Membership, an honor bestowed on only 1/10th of 1% of its membership, in recognition of being an "outstanding scientist and researcher in solidification fundamentals and materials processing, educator and leader in advancing the materials profession."

Lara Abbaschian

Lara Abbaschian received her M.Eng. and S.B. degrees in Materials Science and Engineering from Massachusetts Institute of Technology. As a graduate student at MIT, Lara was a member of the Langer Lab, a Whitaker Fellow, and a recipient of the John Wulff Award for Excellence in Teaching. Lara holds an MBA from the Tuck School of Business at Dartmouth where she was a Forté Fellow. In her professional capacity, Lara has been awarded the Pinnacle Award and the Brain Award.

Lara has built her career at the forefront of medical device and materials innovation beyond the university setting. Her professional experience encompasses more than 20 years in the multidisciplinary healthcare space across a range of therapeutic areas and cutting-edge technologies, including cardiovascular implants, neurovascular interventional devices, stents, drug-device interfaces, and novel materials platforms. Lara has contributed to all facets of the medical device lifecycle—from inception through commercialization. Her deep experience involves bridging research and early-stage innovation to real-world realization in the highly regulated industry of medical technology. Her expertise in company venture funding and formation, product strategy and growth, and in-house R&D have allowed her to bring numerous life-saving technologies from early ideation to market and into the hands of clinicians.

Digital Resources

MindTap Reader

Available via our digital subscription service, Cengage Unlimited, **MindTap Reader** is Cengage's next-generation eBook for engineering students.

The MindTap Reader provides more than just text learning for the student. It offers a variety of tools to help our future engineers learn chapter concepts in a way that resonates with their workflow and learning styles.

- **Personalize their experience**

Within the MindTap Reader, students can highlight key concepts, add notes, and bookmark pages. These are collected in My Notes, ensuring they will have their own study guide when it comes time to study for exams.

- **Flexibility at their fingertips**

The ReadSpeaker feature reads text aloud to students, so they can learn on the go—wherever they are.

Cengage Read

Available on iOS and Android smartphones, the MindTap Mobile App provides convenience. Students can access their entire textbook anyplace and anytime. They can take notes, highlight important passages, and have their text read aloud whether they are online or off.

To download the mobile app, visit https://www.cengage.com/mindtap/mobileapp.

Cengage Unlimited

All-You-Can-Learn Access with Cengage Unlimited

Cengage Unlimited is the first-of-its-kind digital subscription that gives students total and on-demand access to all the digital learning platforms, eBooks, online homework, and study tools Cengage has to offer—in one place, for one price. With Cengage Unlimited, students get access to their WebAssign courseware, as well as content in other Cengage platforms and course areas from day one. That's 70 disciplines and 675 courses worth of material, including engineering.

With Cengage Unlimited, students get **unlimited access** to a library of more than 22,000 products. To learn more, visit https://www.cengage.com/unlimited.

Chapter 1

The Structure of Metals

Learning Objectives
Upon completion of Chapter 1, you will be able to:
1. Describe, via examples, how the structure of a material impacts its properties and, thus, its applications
2. Define the difference between microstructure and macrostructure
3. Draw the unit cell structure for face-centered cubic, body-centered cubic, and hexagonal close-packed lattices
4. Diagram how planes of atoms come together to form close-packed structures
5. Explain coordination numbers, anisotropy, and how manufacturing processes can impact both
6. Create Miller indices for crystallographic directions and crystallographic planes
7. Use a stereographic projection to map three-dimensional structure in two dimensions
8. Understand the use of Wulff nets to identify planes, directions and poles

The most important aspect of any engineering material is its structure, because its properties are closely related to this feature. To be successful, a materials engineer must have a good understanding of this relationship between structure and properties. By way of illustration, wood is a very easy material in which to see the close interaction between structure and properties. A typical structural wood, such as southern yellow pine, is essentially an array of long hollow cells or fibers. These fibers, which are formed largely from cellulose, are aligned with the grain of the wood and are cemented together by another weaker organic material called lignin. The structure of wood is thus roughly analogous to that of a bundle of drinking straws. It can be split easily along its grain; that is, parallel to the cells. Wood is also much stronger in compression (or tension) parallel to its grain than it is in compression (or tension) perpendicular to the grain. It makes excellent columns and beams, but it is not really suitable for tension members required to carry large loads, because the low resistance of wood to shear parallel to its grain makes it difficult to attach end fastenings that will not pull out. As a result, wooden bridges and other large wooden structures are often constructed containing steel tie rods to support the tensile loads.

1.1 The Structure of Metals

The structure in metals is of similar importance to that in wood, although often in a more subtle manner. Metals are usually crystalline when in the solid form. A *crystal* is defined as an orderly array of atoms in space. While very large single crystals can be prepared, the normal metallic object consists of an aggregate of many very small crystals. Metals are therefore *polycrystalline*. The crystals in these materials are normally referred to as its grains. Because of their very small sizes, an optical microscope, operating at magnifications between about 100 and 1000 times, is usually used to examine the structural features associated with the grains in a metal. Structures requiring this range of magnification for their examination fall into the class known as *microstructures*. Occasionally, metallic objects, such as castings, may have very large crystals that are discernible to the naked eye or are easily resolved under a low-power microscope. Structure in this category is called *macrostructure*. On the other hand, there are materials whose grains or sizes are much finer and in the nanoscale range. These microstructures are commonly referred to as nanostructure, with scales on the order of one billionth of a meter. It should be noted that nanoscale features can be in one dimension, as in nanosurfaces or nanofilms; in two dimensions, as in nanotubes or whiskers; or in three dimensions, as in nanoparticles. Nanoprecipitates such as Guinier and Preston (GP) Zones have been used for decades for precipitation hardening of aluminum alloys, as discussed in Chapter 16. Finally, there is the basic structure inside the grains themselves: that is, the atomic arrangements inside the crystals. This form of structure is logically called the *crystal structure*.

Of the various forms of structure, microstructure (that visible under the optical microscope) has been historically of the greatest use and interest to the metallurgist. Because the metallurgical microscope is normally operated at magnifications where its depth of field is extremely shallow, the metallic surface to be observed must be very flat. At the same time, it must reveal accurately the nature of the structure inside the metal. One is therefore presented with the problem of preparing a very smooth flat and undistorted surface, which is by no means an easy task. The procedures required to obtain the desired goal fall under the general heading of metallographic specimen preparation. Examination of the polished surface is often aided by slight chemical attacking of the surface, called "etching." During etching, grain boundaries and certain orientations are attacked more than others. Detailed description of metallographic sample preparation techniques and examples of microstructures can be found in Reference 1.

There are many different types of crystal structures, some of which are quite complicated. Fortunately, most metals crystallize in one of three relatively simple structures: the face-centered cubic, the body-centered cubic, and the close-packed hexagonal.

1.2 Unit Cells

The *unit cell* of a crystal structure is the smallest group of atoms possessing the symmetry of the crystal which, when repeated in all directions, will develop the crystal lattice. Figure 1.1A shows the unit cell of the body-centered cubic lattice. It is evident that its name is derived from the shape of the unit cell. Eight unit cells are combined in Fig. 1.1B in order to show how the unit cell fits into the complete lattice. Note that atom *a* of Fig. 1.1B does not belong uniquely to one

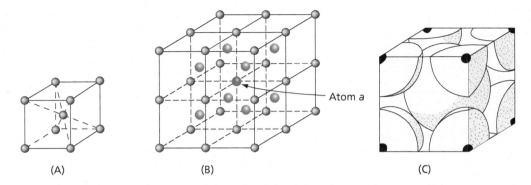

Fig. 1.1 **(A)** Body-centered cubic unit cell. **(B)** Eight unit cells of the body centered cubic lattice. **(C)** Cut view of a unit cell. The coloring scheme is used for ease of visualization. Otherwise all atoms are the same.

unit cell, but is a part of all eight unit cells that surround it. Therefore, it can be said that only one-eighth of this corner atom belongs to any one-unit cell. This fact may be used to compute the number of atoms per unit cell in a body-centered cubic crystal. Even a small crystal will contain billions of unit cells, and the cells in the interior of the crystal must greatly exceed in number than those lying on the surface. Therefore, surface cells may be neglected in our computations. In the interior of a crystal, each corner atom of a unit cell is equivalent to atom *a* of Fig. 1.1B and contributes one-eighth of an atom to a unit cell. In addition, each cell also possesses an atom located at its center that is not shared with other unit cells. The body-centered cubic lattice thus has two atoms per unit cell; one contributed by the corner atoms, and one located at the center of the cell, as shown in Fig. 1.1C.

The unit cell of the face-centered cubic lattice is shown in Fig. 1.2. In this case, the unit cell has an atom in the center of each face. The number of atoms per unit cell in the face-centered cubic lattice can be computed in the same manner as in the body-centered cubic lattice. The eight corner atoms again contribute one atom to the cell, as shown in Fig. 1.2B. There are also six face-centered

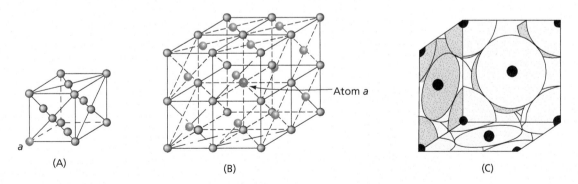

Fig. 1.2 **(A)** Face-centered cubic unit cell. **(B)** Eight unit cells of the face-centered cubic lattice. The face-centered atoms of the top, front and right side faces are shown. **(C)** Cut view of a unit cell

atoms to be considered, each a part of two unit cells. These contribute six times one-half an atom, or three atoms. The face-centered cubic lattice has a total of four atoms per unit cell, or twice as many as the body-centered cubic lattice.

1.3 The Body-Centered Cubic Structure (BCC)

It is frequently convenient to consider metal crystals as structures formed by stacking together hard spheres. This leads to the so-called *hard-ball model* of a crystalline lattice, where the radius of the spheres is taken as half the distance between the centers of the most closely spaced atoms.

Figure 1.3 shows the hard-ball model of the body-centered cubic (bcc) unit cell. A study of the figure shows that the atom at the center of the cube is colinear with each corner atom; that is, the atoms connecting diagonally opposite corners of the cube form straight lines, each atom touching the next in sequence. These linear arrays do not end at the corners of the unit cell, but continue on through the crystal much like a row of beads strung on a wire (see Fig. 1.1B). These four cube diagonals constitute the close-packed directions of the body-centered cubic crystal, directions that run continuously through the lattice on which the atoms are as closely spaced as possible.

Further consideration of Figs. 1.3 and 1.1B reveals that all atoms in the body-centered cubic lattice are equivalent. Thus, the atom at the center of the cube of Fig. 1.3 has no special significance over those occupying corner positions. Each of the latter could have been chosen as the center of a unit cell, making all corner atoms of Fig. 1.1B centers of cells, and all centers of cells corners.

1.4 Coordination Number of the Body-Centered Cubic Lattice

The coordination number (CN) of a crystal structure equals the number of nearest neighbors that an atom possesses in the lattice. In the body-centered cubic unit cell, the center atom has eight neighbors touching it (see Fig. 1.3). We have already seen that all atoms in this lattice

Fig. 1.3 Hard-ball model of the body-centered cubic unit cell

1.5 The Face-Centered Cubic Lattice (FCC)

are equivalent. Therefore, every atom of the body-centered cubic structure not lying at the exterior surface possesses eight nearest neighbors, and the coordination number of the lattice is eight.

The hard-ball model assumes special significance in the face-centered cubic crystal, for in this structure the atoms or spheres are packed together as closely as possible. The CN for fcc is 12. Figure 1.4A shows a complete face-centered cubic (fcc) cell, and Fig. 1.4B shows the same unit cell with a corner atom removed to reveal a close-packed plane (octahedral plane) in which the atoms are spaced as tightly as possible. A larger area from one of these close-packed planes is shown in Fig. 1.5. Three close-packed directions lie in the octahedral plane (the directions *aa*, *bb*, and *cc*). Along these directions the spheres touch and are colinear.

Returning to Fig. 1.4A, we see that the close-packed directions of Fig. 1.5 correspond to diagonals that cross the faces of the cube. There are six of these close-packed directions in the face-centered cubic lattice, as shown in Fig. 1.4C. Face diagonals lying on the reverse faces of the cube are not counted in this total because each is parallel to a direction lying on a visible face, and, in considering crystallographically significant directions, parallel directions are the same. It should also be pointed out that the face-centered cubic structure has four close-packed or octahedral planes. This can be verified as follows. If an atom is removed from each of the corners of a unit cell in a manner similar to that of Fig. 1.4B, an octahedral plane will be revealed in each instance. There are eight of these planes, but since diagonally opposite planes are parallel, they are crystallographically equal. This reduces the number of different octahedral planes to four. The face-centered cubic lattice, however, is unique in that it contains as many as four planes of closest packing, each containing three close-packed directions. No other lattice possesses such a large number of close-packed planes and closed-packed directions. This fact is important, since it gives face-centered cubic metals physical properties different from those of other metals, one of which is the ability to undergo severe plastic deformation.

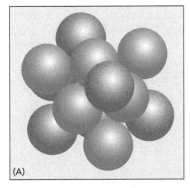
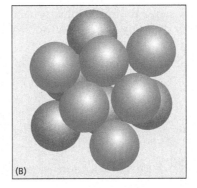
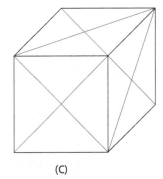

Fig. 1.4 **(A)** Face-centered cubic unit cell (hard-ball model). **(B)** Same cell with a corner atom removed to show an octahedral plane. **(C)** The six-face diagonal directions

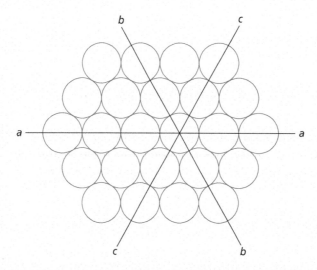

Fig. 1.5 Atomic arrangement in the octahedral plane of a face-centered cubic metal. Notice that the atoms have the closest possible packing. This same configuration of atoms is also observed in the basal plane of close-packed hexagonal crystals. The close-packed directions are *aa*, *bb*, and *cc*

1.6 The Unit Cell of the Hexagonal Closed-Packed (HCP) Lattice

The configuration of atoms most frequently used to represent the hexagonal close-packed structure is shown in Fig. 1.6. This group of atoms contains more than the minimum number of atoms needed to form an elementary building block for the lattice; therefore it is not a true unit cell. However, because the arrangement of Fig. 1.6, which contains three primitive cells, brings out important crystallographic features, including the sixfold symmetry of the lattice, it is commonly used as the unit cell of the close-packed hexagonal structure. A comparison of Fig. 1.6 with Fig. 1.5 shows that the atoms in the planes at the top, bottom, and center of the unit cell belong to a plane of closest packing, the basal plane of the crystal. The figure also shows that the atoms in these basal planes have the proper stacking sequence for the hexagonal close-packed lattice (*ABA* ...); atoms at the top of the cell are directly over those at the bottom, while atoms in the central plane have a different set of positions.

The basal plane of a hexagonal metal, like the octahedral plane of a face-centered cubic metal, has three close-packed directions. These directions correspond to the lines *aa*, *bb*, *cc* of Fig. 1.5.

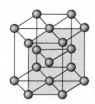

Fig. 1.6 The close-packed hexagonal unit cell

1.7 Comparison of the Face-Centered Cubic and Close-Packed Hexagonal Structures

The face-centered cubic lattice can be constructed by first arranging atoms into a number of close-packed planes, similar to that shown in Fig. 1.5, and then by stacking these planes over each other in the proper sequence. There are a number of ways in which planes of closest packing can be stacked. One sequence gives the close-packed hexagonal lattice, another the face-centered cubic lattice. The reason that there is more than one way of stacking close-packed planes is because any one plane can be set down on a previous one in two different ways. For example, consider the close-packed plane of atoms in Fig. 1.7. The center of each atom in the figure is indicated by the symbol A. Now, if a single atom is placed on top of the configuration of Fig. 1.7, it will be attracted by interatomic forces into one of the natural pockets that occur between any three contiguous atoms. Suppose that it falls into the pocket marked B_1 at the upper left of the figure; then a second atom cannot be dropped into either C_1 or C_2 because the atom at B_1 overlaps the pocket at these two points. However, the second atom can fall into B_2 or B_3 and start the formation of a second close-packed plane consisting of atoms occupying all B positions. Alternatively, the second plane could have been set down in such a way as to fill only C positions. Thus if the first close-packed plane occupies A positions, the second plane may occupy B or C positions. Let us assume that the second plane has the B configuration. Then the pockets of the second plane fall half over the centers of the atoms in the first plane (A positions) and half over the C pockets in the first plane. The third plane may now be set down over the second plane into either A or C positions. If set down into A positions, the atoms in the third layer fall directly over atoms in the first layer. This is not the face-centered cubic order, but that of the close-packed hexagonal structure. The face-centered cubic stacking order is: A for the first plane, B for the second plane, and C for the third plane, which may be written as ABC. The fourth plane in the face-centered cubic lattice, however, does fall on the A position, the fifth on B, and the sixth on C, so that the stacking order for face-centered cubic crystals is $ABCABCABC$, etc. In the close-packed hexagonal structure, the atoms in every other plane fall directly over one another, corresponding to the stacking order $ABABAB$....

There is no basic difference in the packing obtained by the stacking of spheres in the face-centered cubic or the close-packed hexagonal arrangement, since both give an ideal close-packed structure. There is, however, a marked difference between the physical properties of hexagonal close-packed metals (such as cadmium, zinc, and magnesium) and the face-centered

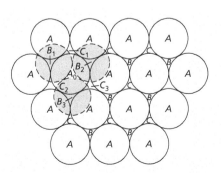

Fig. 1.7 Stacking sequences in close-packed crystal structures

cubic metals, (such as aluminum, copper, and nickel), which is related directly to the difference in their crystalline structure. The most striking difference is in the number of close-packed planes. In the face-centered cubic lattice there are four planes of closest packing, the *octahedral planes*; but in the close-packed hexagonal lattice only one plane, the *basal plane*, is equivalent to the octahedral plane. The single close-packed plane of the hexagonal lattice engenders, among other things, plastic deformation properties that are much more directional than those found in cubic crystals.

1.8 Coordination Number of the Systems of Closest Packing

The *coordination number* of an atom in a crystal has been defined as the number of nearest neighbors that it possesses. This number is 12 for both face-centered cubic and close-packed hexagonal crystals, as may be verified with the aid of Fig. 1.7. Thus, consider atom A_0 lying in the plane of atoms shown as circles drawn with continuous lines. Six other atoms lying in the same close-packed plane as A_0 are in nearest neighbor positions. Atom A_0 also touches three atoms in the plane directly above. These three atoms could occupy B positions, as is indicated by the dashed lines around pockets B_1, B_2, and B_3, or they could occupy positions C_1, C_2, and C_3. In either case, the number of nearest neighbors in the plane just above A_0 is limited to three. In the same manner, it may be shown that A_0 has three nearest neighbors in the next plane below the close-packed plane containing A_0. The number of nearest neighbors of atom A_0 is thus twelve: six in its own plane, three in the plane above it, and three in the plane below it. Since the argument is valid no matter whether the atoms in the close-packed planes just above or below atom A_0 are in B or C positions, it holds for both face-centered cubic and close-packed hexagonal stacking sequences. We conclude, therefore, that the coordination number in these lattices is 12.

1.9 Anisotropy

When the properties of a substance are independent of direction, the material is said to be isotropic. Thus, in an isotropic material, one should expect to find that it has the same strength in all directions. Or, if its electrical resistivity were measured, the same value of this property would be obtained irrespective of how a resistivity specimen was cut from a quantity of the material. The physical properties of crystals normally depend strongly on the direction along which they are measured. This means that, basically, crystals are not isotropic, but anisotropic. In this regard, consider a body-centered cubic crystal of iron. The three most important directions in this crystal are the directions labeled a, b, and c in Fig. 1.8. That these directions are not equivalent can be recognized from the fact that the spacings of the atoms along the three directions are different, being equal respectively, in terms of the lattice parameter a (the length of one edge of the unit cell), to a, $\sqrt{2}a$, and $\sqrt{\frac{3}{2}}a$. The physical properties of iron, measured along these three directions, also tend to be different. As an example, consider the I-H curve for the magnetization of iron crystals. As may be seen in Fig. 1.9, the intensity of magnetization I rises most rapidly with the magnetic field intensity H along the direction a, at an intermediate rate along b, and least rapidly along c. Interpreted in another way, we may say that a is the direction of easiest magnetization, while c is correspondingly the most difficult. Another example is shown for nickel single crystals. Here the intensity of magnetization rises rapidly in the direction c and least rapidly in direction a.

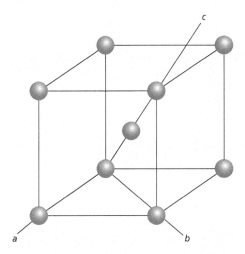
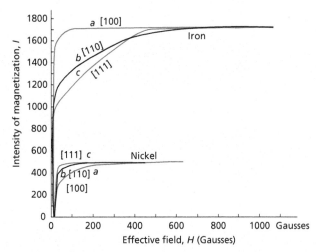

Fig. 1.8 The most important directions in a body-centered cubic crystal

Fig. 1.9 An iron crystal is much easier to magnetize along an *a* direction of Fig. 1.8 than along a *b* or *c* direction. The opposite is the case for nickel (Refs. 2 and 3)

It should be noted that the above-mentioned bulk anisotropies relate to the crystal structures. When these materials are used in thin layers having a thickness on the order of a few atomic layers, additional anistropies related to surfaces and interfaces may also appear (Ref. 4).

Ideally, a polycrystalline specimen might be expected to be isotropic if its crystals were randomly oriented, for then, from a macroscopic point of view, the anisotropy of the crystals should be averaged out. However, a truly random arrangement of the crystals is seldom achieved, because manufacturing processes tend to align the grains in a metal so that their orientations are not uniformly distributed. The result is known as a texture or a preferred orientation. Because most polycrystalline metals have a preferred orientation, they tend to be anisotropic, the degree of this anisotropy depending on the degree of crystal alignment.

1.10 Textures or Preferred Orientations

Wires are formed by pulling rods through successively smaller and smaller dies. In the case of iron, this kind of deformation tends to align a *b* direction of each crystal parallel to the wire axis. About this direction the crystals are normally considered to be randomly arranged. This type of preferred arrangement of the crystals in an iron or steel wire is quite persistent. Even if the metal is given a heat treatment* that completely reforms the crystal structure, the crystals tend to keep a *b* direction parallel to the wire axis. Because the deformation used in forming sheets and plates is basically two-dimensional in character, the preferred orientation found in them is more restrictive than that observed in wires. As indicated in Fig. 1.10A, not only does one tend to find a *b* direction parallel

---

*Recrystallization following cold work is discussed in Chapter 8.

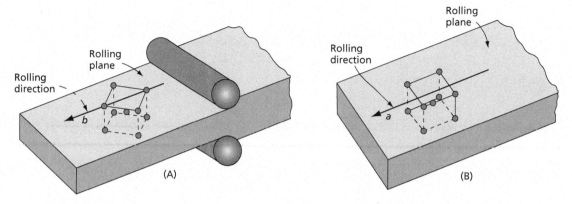

Fig. 1.10 Two basic crystalline orientations that can be obtained in rolled plates of body-centered cubic metals

to the rolling direction or length of the plate, but there is also a strong tendency for a cube plane, or face of the unit cell, to be aligned parallel to the rolling plane or surface of the sheet or plate.

There are a number of reasons why an understanding of crystal properties is important to engineers. One of these is that the basic anisotropy of crystalline materials is reflected in the polycrystalline objects of commerce. It should be noted also that this is not always undesirable. Preferred orientations can often result in materials with superior properties. An interesting example of this sort is found in the alloy of iron with 4 percent of silicon, used for making transformer coils. In this case, by a complicated combination of rolling procedures and heat treatments, it is possible to obtain a very strong preferred orientation in which an *a* direction of the crystals is aligned parallel to the rolling direction, while a cube plane, or face of the unit cell, remains parallel to the rolling plane. This average orientation is shown schematically in Fig. 1.10B. The significant feature of this texture is that it place the direction of easy magnetization parallel to the length of the sheet. In manufacturing transformers, it is then a rather simple matter to align the plates in the core so that this direction is parallel to the direction along which the magnetic flux runs. When this occurs, the resultant hysteresis loss can be made very small.

1.11 Miller Indices

As one becomes more and more involved in the study of crystals, the need for symbols to describe the orientation in space of important crystallographic directions and planes becomes evident. Thus, while the directions of closest packing in the body-entered cubic lattice may be described as the diagonals that traverse the unit cell, and the corresponding directions in the face-centered cubic lattice as the diagonals that cross the faces of a cube, it is much easier to define these directions in terms of several simple integers. The Miller system of designating indices for crystallographic planes and directions is universally accepted for this purpose. In the discussion that follows, the Miller indices for cubic and hexagonal crystals will be considered. The indices for other crystal structures are not difficult to develop.

Direction Indices in the Cubic Lattice Let us take a cartesian coordinate system with axes parallel to the edge of the unit cell of a cubic crystal. (See Fig. 1.11.) In this coordinate system, the unit of measurement along all three axes is the length of the edge of a unit cell, designated by the symbol a in the figure. The Miller indices of directions are introduced with the aid of several simple examples. Thus the cube diagonal m of Fig. 1.11 has the same direction as a vector t with a length that equals the diagonal distance across the cell. The component of the vector t on each of the three coordinate axes is equal to a. Since the unit of measurement along each axis equals a, the vector has components 1, 1 and 1 on the x, y, and z axes, respectively. The Miller indices of the direction m are now written [111]. In the same manner, the direction n, which crosses a face of the unit cell diagonally, has the same direction as a vector s the length of which is the face diagonal of the unit cell. The x, y, and z components of this vector are 1, 0 and 1 respectively; the corresponding Miller indices are [101]. The indices of the x axis are [100], the y axis [010], and the z axis [001].

A general rule for finding the Miller indices of a crystallographic direction can now be stated. Draw a vector from the origin parallel to the direction whose indices are desired. Make the magnitude of the vector such that its components on the three coordinate axes have lengths that are simple integers. These integers must be the smallest numbers that will give the desired direction. Thus, the integers 1, 1, and 1, and 2, 2, and 2 represent the same direction in space, but, by convention, the Miller indices are [111] and not [222].

Let us apply the above rule to the determination of the Miller indices of a second cube diagonal; that indicated by the symbol p in Fig. 1.12. The vector (which starts at the origin in Fig. 1.12B) is parallel to the direction p. The components of q are 1, -1, and 1, and, by the above definition, the corresponding Miller indices of p are [1$\bar{1}$1], where the negative sign of the y index is indicated by a bar over the corresponding integer. The indices of the diagonal m in Fig. 1.12A have already been shown to be [111], and it may also be shown that the indices of the diagonals u and v are [11$\bar{1}$] and [$\bar{1}$11]. The four cube diagonals thus have indices [111], [$\bar{1}$11], [1$\bar{1}$1], and [11$\bar{1}$].

When a specific crystallographic direction is referred to, the Miller indices are enclosed in square brackets as shown here. However, it is sometimes desirable to refer to all of the directions of the same form. In this case, the indices of one of these directions are enclosed in carets $\langle 111 \rangle$, and the symbol is read to signify all four directions ([111], [$\bar{1}$11], [1$\bar{1}$1], and [11$\bar{1}$]), which are considered as a class. Thus, one might say that the close-packed directions in the body-centered cubic lattice are $\langle 111 \rangle$ directions, whereas a specific crystal might be stressed in tension along its [111] direction and simultaneously compressed along its [1$\bar{1}$1] direction.

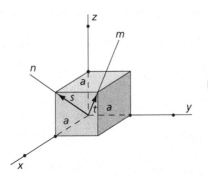

Fig. 1.11 The [111] and [101] directions in a cubic crystal; directions m and n, respectively

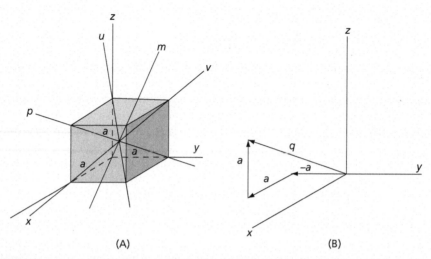

Fig. 1.12 **(A)** The four cube diagonals of a cubic lattice, m, n, u, and v. **(B)** The components of the vector q that parallels the cube diagonal p are a, −a, and a. Therefore, the indices of q are [111]

Cubic Indices for Planes Crystallographic planes are also identified by sets of integers. These are obtained from the intercepts that the planes make with the coordinate axes. Thus, in Fig. 1.13 the indicated plane intercepts the x, y, and z axes at 1, 3, and 2 unit-cell distances, respectively. The Miller indices are proportional not to these intercepts, but to their reciprocals $\frac{1}{1}, \frac{1}{3}, \frac{1}{2}$ and, by definition, the Miller indices are the smallest integers having the same ratios as these reciprocals. The desired integers are, therefore, 6, 2, 3. The Miller indices of a plane are enclosed in parentheses, for example (623), instead of brackets, thus making it possible to differentiate between planes and directions.

Let us now determine the Miller indices of several important planes of cubic crystals. The plane of the face a of the cube shown in Fig. 1.14A is parallel to both the y and z axes and, therefore, may be said to intercept these axes at infinity. The x intercept, however, equals 1, and the reciprocals of the three intercepts are $\frac{1}{1}, \frac{1}{\infty}, \frac{1}{\infty}$. The corresponding Miller indices are (100). The indices of the face b are (010), while those of c are (001). The indicated plane of Fig. 1.14B has indices (011), and that of Fig. 1.14C (111). The latter plane is an octahedral plane, as may be seen by referring

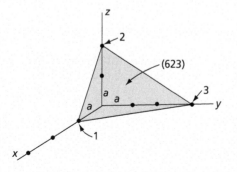

Fig. 1.13 The intercepts of the (623) plane with the coordinate axes

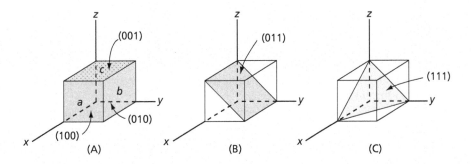

Fig. 1.14 **(A)** Cube planes of a cubic crystal: *a* (100); *b* (010); *c* (001). **(B)** The (011) plane. **(C)** The (111) plane

to Fig. 1.15. Other octahedral planes have the indices ($\bar{1}11$), ($1\bar{1}1$), and ($11\bar{1}$), where the bar over a digit represents a negative intercept. By way of illustration, the ($\bar{1}11$) plane is shown in Fig. 1.15, where it may be seen that the *x* intercept is negative, whereas the *y* and *z* intercepts are positive. This figure also shows that the ($1\bar{1}\bar{1}$) plane is parallel to the ($\bar{1}11$) plane and is, therefore, the same crystallographic plane. Similarly, the indices ($\bar{1}1\bar{1}$) and ($\bar{1}\bar{1}1$) represent the same planes as ($1\bar{1}1$) and ($11\bar{1}$), respectively.

The set of planes of a given form, such as the four octahedral planes (111), ($1\bar{1}1$), ($\bar{1}11$), and ($11\bar{1}$), are represented as a group with the aid of braces enclosing one of the indices, that is, {111}. Thus, if one wishes to refer to a specific plane in a crystal of known orientation, parentheses are used, but if the class of planes is to be referred to, braces are used.

An important feature of the Miller indices of cubic crystals is that the integers of the indices of a plane and of the direction normal to the plane are identical. Thus, face *a* of the cube in Fig. 1.14A has indices (100), and the *x* axis, perpendicular to this plane, has indices [100]. In the same manner, the octahedral plane of Fig. 1.14C and its normal, the cube diagonal, have indices (111) and [111], respectively. Noncubic crystals do not, in general, possess this equivalence between the indices of planes and normals to the planes. The spacing between crystallographic planes is covered later in Section 2.4.

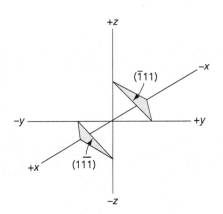

Fig. 1.15 The ($\bar{1}11$) and ($1\bar{1}\bar{1}$) planes are parallel to each other and therefore represent the same crystallographic plane

Miller Indices for Hexagonal Crystals

Miller Indices for Hexagonal Crystals Planes and directions in hexagonal metals are defined almost universally in terms of Miller indices containing four digits instead of three. The use of a four-digit system gives planes of the same form similar indices. Thus, in a four-digit system, the planes $(11\bar{2}0)$ and $(\bar{1}2\bar{1}0)$ are equivalent planes. The three-digit system, on the other hand, gives equivalent planes indices that are not similar. Thus, the previously mentioned two planes would have indices (110) and $(1\bar{2}0)$ in the three-digit system.

Four-digit hexagonal indices are based on a coordinate system containing four axes. Three axes correspond to close-packed directions and lie in the basal plane of the crystal, making 120° angles with each other. The fourth axis is normal to the basal plane and is called the c axis, where as the three axes that lie in the basal plane are designated the a_1, a_2, and a_3 axes. Figure 1.16 shows the hexagonal unit cell superimposed upon the four-axis coordinate system. It is customary to take the unit of measurement along the a_1, a_2, and a_3 axes as the distance between atoms in a close-packed direction. The magnitude of this unit is indicated by the symbol a. The unit of measurement for the c axis is the height of the unit cell that is designated as c.

Let us now determine the Miller indices of several important close-packed hexagonal lattice planes. The uppermost surface of the unit cell in Fig. 1.16 corresponds to the basal plane of the crystal. Since it is parallel to the axes a_1, a_2, and a_3, it must intercept them at infinity. Its c axis intercept, however, is equal to 1. The reciprocals of these intercepts are $\frac{1}{\infty}, \frac{1}{\infty}, \frac{1}{\infty}, \frac{1}{1}$. The Miller indices of the basal plane are, therefore, (0001). The six vertical surfaces of the unit cell are known as *prism planes* of Type 1. Consider now the prism plane that forms the front face of the cell, which has intercepts as follows: a_1 at 1, a_2 at ∞, a_3 at -1, and c at ∞. Its Miller indices are, therefore, $(10\bar{1}0)$. Another important type of plane in the hexagonal lattice is shown in Fig. 1.17. The intercepts are a_1 at 1, a_2 at ∞, a_3 at -1, and c at $\frac{1}{2}$, and the Miller indices are accordingly $(10\bar{1}2)$.

Miller indices of directions are also expressed in terms of four digits. In writing direction indices, the third digit must always equal the negative sum of the first two digits. Thus, if the first two digits are 3 and 1, the third must be -4, that is, $[31\bar{4}0]$.

Let us investigate directions lying only in the basal plane, since this will simplify the presentation. If a direction lies in the basal plane, then it has no component along the c axis, and the fourth digit of the Miller indices will be zero.

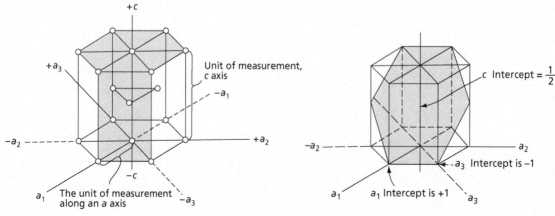

Fig. 1.16 The four coordinate axes of a hexagonal crystal

Fig. 1.17 The $(10\bar{1}2)$ plane of a hexagonal metal

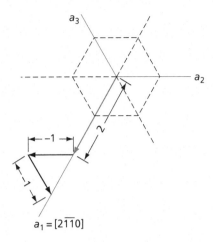

Fig. 1.18 Determination of indices of a digonal axis of Type I—$[2\bar{1}\bar{1}0]$

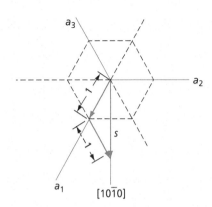

Fig. 1.19 Determination of indices of a digonal axis of Type II—$[10\bar{1}0]$

As our first example, let us find the Miller indices of the a_1 axis. This axis has the same direction as the vector sum of three vectors (Fig. 1.18), one of length $+2$ along the a_1 axis, another of length -1 parallel to the a_2 axis, and the third of length -1 parallel to the a_3 axis. The indices of this direction are, accordingly $[2\bar{1}\bar{1}0]$. This unwieldy method of obtaining the direction indices is necessary in order that the relationship mentioned above be maintained between the first two digits and the third. The corresponding indices of the a_2 and a_3 axes are $[\bar{1}2\bar{1}0]$ and $[\bar{1}\bar{1}20]$. These three directions are known as the digonal axes of Type I. Another important set of directions lying in the basal plane are the digonal axes of Type II; a set of axes perpendicular to the digonal axes of Type I. Figure 1.19 shows one of the axes of Type II and indicates how its direction indices are determined. The vector s in the figure determines the desired direction and equals the vector sum of a unit vector lying on a_1, and another parallel to a_3, but measured in a negative sense. The indices of the digonal axis of Type II are thus $[10\bar{1}0]$. In this case, the second digit is zero because the projection of the vector s on the a_2 axis is zero.

1.12 Crystal Structures of the Metallic Elements

Crystalline structures can form in two-dimensional or three-dimensional space. A good example of the former is graphene, which is a single layer of carbon atoms arranged in a hexagonal lattice, with each atom tightly bonded to three other carbon atoms. A review of graphene, its properties, and applications can be found in a recent review by Urade et al. (Ref. 5). In three-dimensional space, atoms can be arranged in seven lattice systems, consisting of Cubic, Hexagonal, Tetragonal, Orthorhombic, Rhombohedral, Monoclinic and Triclinic (Ref. 6). These primitive unit cells may also contain additional atoms as body-centered, face-centered, or base-centered lattices, leading to total of 14 lattice systems often referred to as Bravais lattices.

Some of the most important metals are classified according to their crystal structures in Table 1.1.

Table 1.1 Crystal Structure of Some of the More Important Metallic Elements

Face-Centered Cubic	Closed-Packed Hexagonal	Body-Centered Cubic
Iron (911.5 to 1396 °C)	Magnesium	Iron (below 911.5 and from 1396 to 1538 °C)
Copper	Zinc	Titanium (882 to 1670 °C)
Silver	Titanium (below 882 °C)	Zirconium (863 to 1855 °C)
Gold	Zirconium (below 863 °C)	Tungsten
Aluminum	Beryllium	Vanadium
Nickel	Cadmium	Molybdenum
Lead		Alkali Metals (Li, Na, K, Rb, Ca)
Platinum		

A number of metals are polymorphic, that is, they crystallize in more than one structure. The most important of these is iron, which crystallizes as either body-centered cubic or face-centered cubic, with each structure stable in separate temperature ranges. Thus, at all temperatures below 911.5 °C and above 1396 °C to the melting point, the preferred crystal structure is body-centered cubic, whereas between 911.5 °C and 1396 °C the metal is stable in the face-centered cubic structure. It may also be seen in Table 1.1 that titanium and zirconium are also polymorphic, being body-centered cubic at higher temperatures and close-packed hexagonal at lower temperatures.

1.13 The Stereographic Projection

The stereographic projection is a useful metallurgical tool, for it permits the mapping in two dimensions of crystallographic planes and directions in a convenient and straightforward manner.[*] The real value of the method is attained when it is possible to visualize crystallographic features directly in terms of their stereographic projections. The purpose of this section is to concentrate on the geometrical correspondence between crystallographic planes and directions and their stereographic projections. In each case, a sketch of a certain crystallographic feature, in terms of its location in the unit cell, is compared with its corresponding stereographic projection.

Several simple examples will be considered, but before this is done, attention will be called to several pertinent facts. The stereographic projection is a two-dimensional drawing of three-dimensional data. The geometry of all crystallographic planes and directions is accordingly reduced by one dimension. Planes are plotted as great circle lines, and directions are plotted as points. Also, the normal to a plane completely describes the orientation of a plane.

As our first example, consider several of the more important planes of a cubic lattice: specifically the (100), (110), and (111) planes. All three planes are treated in the three parts of Fig. 1.20. Notice that the stereographic projection of each plane can be represented either by a great circle or by a point showing the direction in space that is normal to the plane.

---

[*]Stereographic projection is used for analysis of diffraction patterns discussed in Chapter 2.

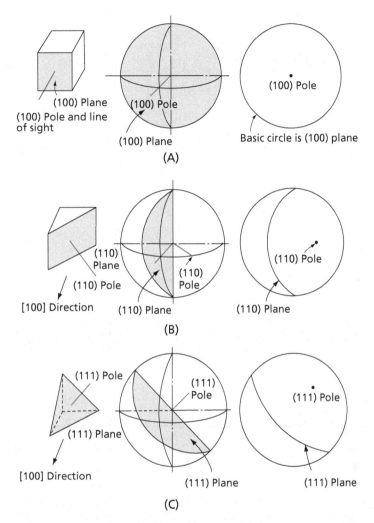

Fig. 1.20 Stereographic projections of several important planes of a cubic crystal. **(A)** The (100) plane, line of sight along the [100] direction. **(B)** The (110) plane, line of sight along the [100] direction. **(C)** The (111) plane, line of sight along the [100] direction

Many crystallographic problems can be solved by considering the stereographic projections of planes and directions in a single hemisphere, that is, normally the one in front of the plane of the paper. The three examples given in Fig. 1.20 have all been plotted in this manner. If the need arises, the stereographic projections in the rear hemisphere can also be plotted in the same diagram. However, it is necessary that the projections in the two hemispheres be distinguishable from each other. This may be accomplished if the stereographic projections of planes and directions in the forward hemisphere are drawn as solid lines and dots, respectively, while those in the rear hemisphere are plotted as dotted lines and circled dots, respectively. As an illustration, consider Fig. 1.21, in which the projections in both hemispheres of a single plane are shown. The (120) plane of a cubic lattice is used in this example.

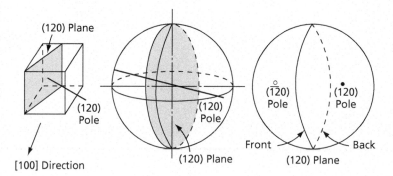

Fig. 1.21 Cubic system, the (120) plane, showing the stereographic projections from both hemispheres, line of sight the [100] direction

1.14 Directions that Lie in a Plane

Frequently one desires to show the positions of certain important crystallographic directions that lie in a particular plane of a crystal. Thus, in a body-centered cubic crystal one of the more important planes is {110}, and in each of these planes one finds two close-packed ⟨111⟩ directions. The two that lie in the (101) plane are shown in Fig. 1.22, where they appear as dots lying on the great circle representing the (101) plane.

1.15 Planes of a Zone

Those planes that mutually intersect along a common direction form the planes of a zone, and the line of intersection is called the *zone axis*. In this regard, consider the [111] direction as a zone axis. Figure 1.23 shows that there are three {110} planes that pass through the [111] direction. There are also three {112} planes and six {123} planes, as well as a number of higher indice planes that have the same zone axis. The pertinent {112} and {123} planes are shown in Fig. 1.24,

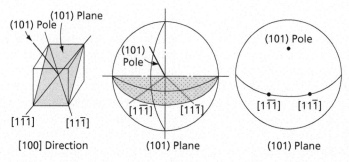

Fig. 1.22 Cubic system, the (101) plane and the two ⟨111⟩ directions that lie in this plane, line of sight [100]

1.15 Planes of a Zone

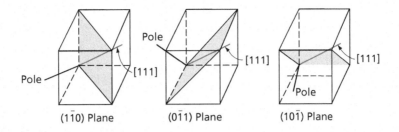

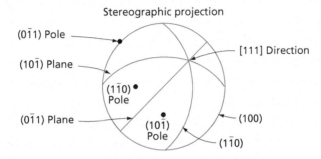

Fig. 1.23 Cubic system, zone of planes the zone axis of which is the [111] direction. The three {110} planes that belong to this zone are illustrated in the figures

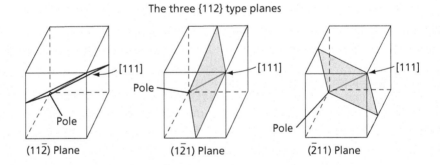

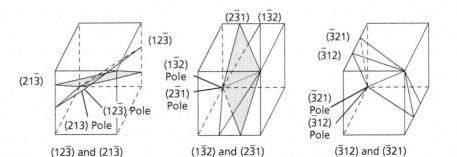

Fig. 1.24 The {112} and {123} planes that have [111] as their zone axis

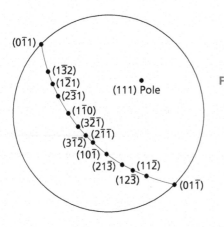

Fig. 1.25 Stereographic projection of the zone containing the 12 planes shown in Figs. 1.23 and 1.24. Only the poles of the planes are plotted. Notice that all of the planar poles lie in the (111) plane

whereas the stereographic projection of these and the previously mentioned {110} planes are shown in Fig. 1.25. Notice that in this latter figure only the poles of the planes are plotted, and it is significant that all of the poles fall on the great circle representing the stereographic projection of the (111) plane.

1.16 The Wulff Net

The *Wulff net* is a stereographic projection of latitude and longitude lines in which the north–south axis is parallel to the plane of the paper. The latitude and longitude lines of the Wulff net serve the same function as the corresponding lines on a geographical map or projection; that is, they make possible graphical measurements. However, in the stereographic projection we are primarily interested in measuring angles, whereas in the geographical sense distance is usually more important. A typical Wulff, or meridional, net drawn to 2° intervals is shown in Fig. 1.26.

Several facts about the Wulff net should be noted. First, all meridians (longitude lines), including the basic circle, are great circles. Second, the equator is a great circle. All other latitude lines are small circles. Third, angular distances between points representing directions in space can be measured on the Wulff net only if the points are made to coincide with a great circle of the net.

In the handling of many crystallographic problems, it is frequently necessary to rotate a stereographic projection corresponding to a given crystal orientation into a different orientation. This is done for a number of reasons. One of the most important is to bring experimentally measured data into a standard projection where the basic circle is a simple close-packed plane such as (100) or (111). Deformation markings, or other experimentally observed crystallographic phenomena, usually can be more readily interpreted when studied in terms of standard projections.

In solving problems with the aid of the Wulff net, it is customary to cover it with a piece of tracing paper. A common pin is then driven through the paper and into the exact center of the net. The paper, thus mounted, serves as a work sheet on which crystallographic data are plotted. The following two types of rotation of the plotted data are possible.

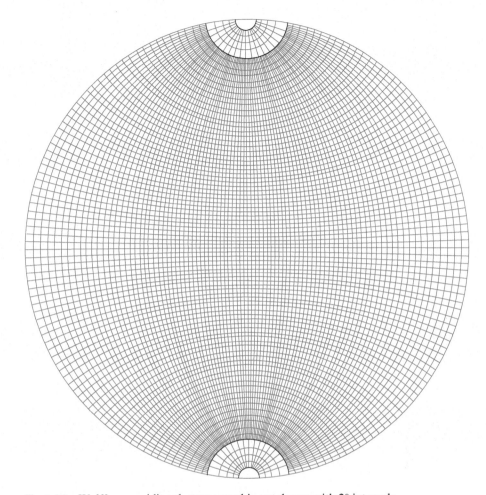

Fig. 1.26 Wulff, or meridional, stereographic net drawn with 2° intervals

Rotation About an Axis in the Line of Sight This rotation is easily performed by merely rotating the tracing paper, relative to the net, about the pin. As an example, let us rotate a cubic lattice 45° clockwise around the [100] direction as an axis. This rotation has the effect of placing the pole of the (111) plane, as plotted in Fig. 1.20C, on the equator of the Wulff net. Figure 1.27A shows the effect of the desired rotation on the orientation of the cubic unit cell when the cell is viewed along the [100] direction. Note that because the basic circle represents the (100) plane, a simple rotation of the paper by 45° about the pin produces the desired rotation in the stereographic projection.

Rotation About the North–South Axis of the Wulff Net This rotation is not as simple to perform as that given previously, which can be accomplished by merely rotating the work

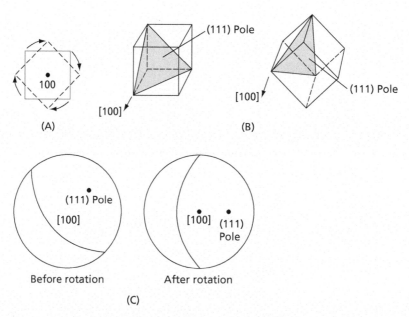

Fig. 1.27 Rotation about the center of the Wulff net. **(A)** The effect of the desired rotation on the cubic unit cell. Line of sight [100]. **(B)** Perspective view of the (111) plane before and after the rotation. **(C)** Stereographic projection of the (111) plane and its pole before and after rotation. Rotation clockwise 45° about the [100] direction

sheet about the pin. Rotations of this second type are accomplished by a graphical method. The data are first plotted stereographically and then rotated along latitude lines and replotted in such a manner that each point undergoes the same change in longitude. The method will be quite evident if one considers the drawings of Fig. 1.28. In the present example, it is assumed that the forward face (100) of

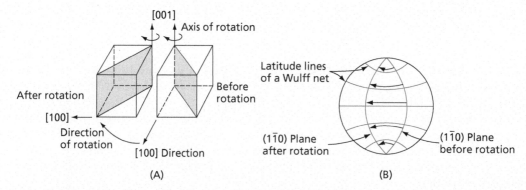

Fig. 1.28 Rotation about the north-south axis of the Wulff net. **(A)** Perspective views of the unit cell before and after the rotation showing the orientation of the (1$\bar{1}$0) plane. **(B)** Stereographic projection showing the preceding rotation. For the sake of clarity of presentation, only the (1$\bar{1}$0) plane is shown. The rotation of the pole is not shown. Also, the meridians of the Wulff net are omitted

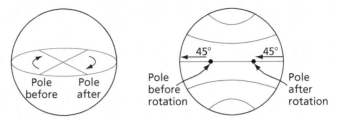

Fig. 1.29 The rotation of the pole of the (1$\bar{1}$0) plane is given here. The diagram on the left shows the rotation in a perspective figure, whereas that on the right shows the motion of the pole along a latitude line of a stereographic projection which, in this case, is the equator

the unit cell is rotated to the left about the [001] direction as an axis. Consider now the effect of this rotation on the spatial orientation and stereographic projection of the (1$\bar{1}$0) plane. In Fig. 1.28A, the right- and left-hand drawings represent the cubic unit cell before and after the rotation respectively. The effect of the rotation on the stereographic projection of the (1$\bar{1}$0) plane is shown in Fig. 1.28B. Each of the curved arrows shown in this drawing represents a change in longitude of 90°. In these drawings, the pole of the (1$\bar{1}$0) plane is not shown in order to simplify the presentation. The rotation that the (1$\bar{1}$0) pole undergoes is shown, however, in Fig. 1.29.

By using simple examples, the two basic rotations that can be made when the Wulff net is used have been pointed out here. All possible rotations of a crystal in three dimensions can be duplicated by using these rotations on a stereographic projection.

1.17 Standard Projections

A stereographic projection, in which a prominent crystallographic direction or pole of an important plane lies at the center of the projection, is known as a standard stereographic projection. Such a projection for a cubic crystal is shown in Fig. 1.30, where the (100) pole is assumed to be normal to the plane of the paper. This figure is properly called a standard 100 projection of a cubic crystal. In this diagram, note that the poles of all the {100}, {110}, and {111} planes have been plotted at their proper orientations. Each of these basic crystallographic directions is represented by a characteristic symbol. For the {100} poles, this is a square, signifying that these poles correspond to four-fold symmetry axes. If the crystal is rotated 90° about any one of these directions, it will be returned to an orientation exactly equivalent to its original orientation. In a 360° rotation about a {100} pole, the crystal reproduces its original orientation four times. In the same fashion, a ⟨111⟩ direction corresponds to a three-fold symmetry axis, and these directions are indicated in the stereographic projection by triangles. Finally, the two-fold symmetry of the ⟨110⟩ directions is indicated by the use of small ellipses to designate their positions in the stereographic projection.

A more complete 100 standard projection of a cubic crystal is shown in Fig. 1.31. This includes the poles of other planes of somewhat higher Miller indices. Figure 1.31 can be considered to

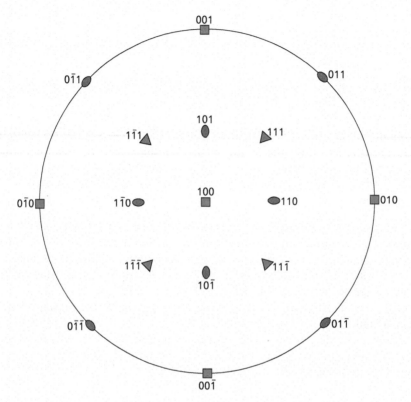

Fig. 1.30 A 100 standard stereographic projection of a cubic crystal

be either a projection showing the directions in a cubic crystal or the poles of its planes. This is because in a cubic crystal, a plane is always normal to the direction with the same Miller indices. In a hexagonal close-packed crystal, however, the projection showing the poles of the planes is not the same as one showing crystallographic directions.

Figure 1.32 shows a 111 standard projection that contains the same poles as the 100 projection in Fig. 1.31. The three-fold symmetry of the crystal structure about the pole of a {111} plane is clearly evident in this figure. At the same time, attention is called to the fact that the 100 projection of Fig. 1.31 also plainly reveals the four-fold symmetry about a {100} pole.

1.18 The Standard Stereographic Triangle for Cubic Crystals

The great circles corresponding to the {100} and {110} planes of a cubic crystal are also shown in the standard projections of Figs. 1.31 and 1.32. These great circles pass through all of the poles shown on the diagram except those of the {123} planes. At the same time, they divide the

1.18 The Standard Stereographic Triangle for Cubic Crystals

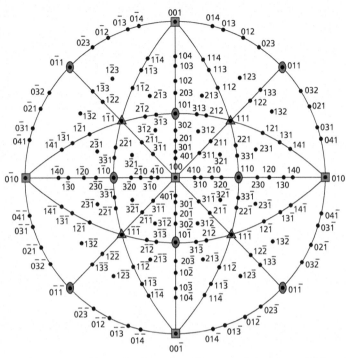

Fig. 1.31 A 100 standard stereographic projection of a cubic crystal showing additional poles

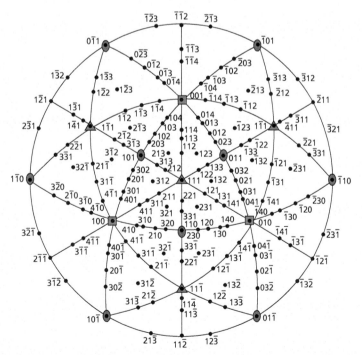

Fig. 1.32 A 111 standard projection of a cubic crystal

standard projection into 24 spherical triangles. These spherical triangles all lie in the forward hemisphere of the projection. There are, of course, 24 similar triangles in the rear hemisphere. An examination of the triangles outlined in Figs. 1.31 and 1.32 shows an interesting fact: in every case, the three corners of the triangles are formed by a $\langle 111 \rangle$ direction, a $\langle 110 \rangle$ direction, and a $\langle 100 \rangle$ direction. This is a highly significant observation, since it means that each triangle corresponds to a region of the crystal that is equivalent. In effect, this signifies that the three lattice directions, marked a_1, a_2, and a_3, and shown in Fig. 1.33, are crystallographically equivalent, because they are located at the same relative positions inside three stereographic triangles. To illustrate this point, let us assume that it is possible to cut three tensile specimens, with axes parallel to a_1, a_2, and a_3, out of a very large single crystal. If tensile tests were to be performed on these three smaller crystals now, one would expect to get identical stress-strain curves for the three specimens. A similar result should be obtained if some other physical property, such as the electrical resistivity, were to be measured along these three directions. The plotting of crystallographic data is often simplified because of the equivalence of the stereographic triangles. For example, if one has a large number of long, cylindrical crystals and wishes to plot the orientations of the individual crystal axes, this can be done conveniently in a single stereographic triangle, as shown in Fig. 1.34.

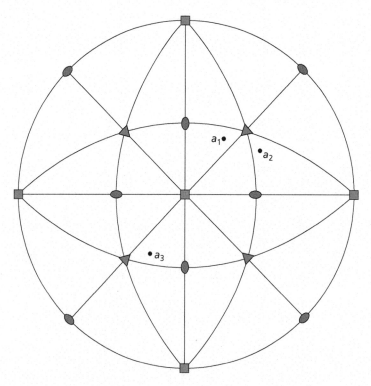

Fig. 1.33 The crystallographic directions a_1, a_2, and a_3 shown in this standard projection are equivalent because they lie in similar positions inside their respective standard stereographic triangles

1.18 The Standard Stereographic Triangle for Cubic Crystals

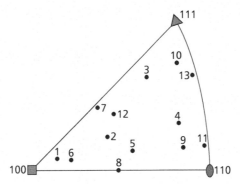

Fig. 1.34 When it is necessary to compare the orientations of a number of crystals, this often can be done conveniently by plotting the crystal axes in a single stereographic triangle, as indicated in this figure

Problems

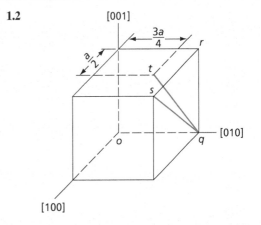

1.1 Determine the direction indices for (**a**) line *om*, (**b**) line *on*, and (**c**) line *op* in the accompanying drawing of a cubic unit cell.

1.2 Determine direction indices for lines (**a**) *qr*, (**b**) *qs*, and (**c**) *qt*.

1.3

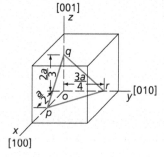

In this figure, plane *pqr* intercepts the *x*, *y*, and *z* axes as indicated. What are the Miller indices of this plane?

1.4

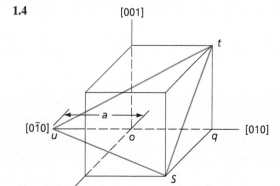

What are the Miller indices of plane *stu*?

1.5

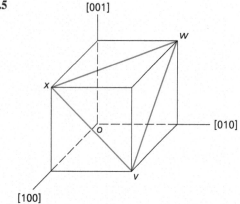

Write the Miller indices for plane *vwx*.

1.6 Linear density in a given crystallographic direction represents the fraction of a line length that is occupied by atoms whereas linear mass density is mass per unit length. Similarly, planar density is the fraction of a crystallographic plane occupied by atoms. The fraction of the volume occupied in a unit cell, on the other hand, is called the atomic packing factor. The latter should not be confused with bulk density, which represents weight per unit volume.

(a) Calculate the linear density in the [100], [110], and [111] directions in body-centered cubic (BCC) and face-centered cubic (FCC) structures.

(b) Calculate planar densities in (100) and (110) planes in bcc and fcc structures.

(c) Show that atomic packing factors for BCC, FCC, and hexagonal close-packed (HCP) structures are 0.68, 0.74, and 0.74, respectively.

1.7 Show that the c/a ratio (see Fig. 1.16) in an ideal hexagonal close-packed (HCP) structure is 1.63. (*Hint*: Consider an equilateral tetrahedron of four atoms which touch each other along the edges.)

1.8 Iron has a BCC structure at room temperature. When heated, it transforms from BCC to FCC at 1185 K. The atomic radii of iron atoms at this temperature are 0.126 and 0.129 nm for bcc and fcc, respectively. What is the percentage volume change upon transformation from BCC to FCC?

1.9

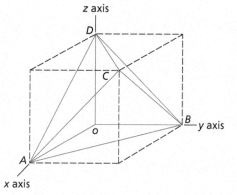

This diagram shows the Thompson Tetrahedron, which is a geometrical figure formed by the four cubic {111} planes. It has special significance with regard to plastic deformation in face-centered cubic metals. The corners of the tetrahedron are marked with the letters *A, B, C,* and *D*. The four surfaces of the tetrahedron are defined by the triangles *ABC, ABD, ACD,* and *BCD*. Assume that the cube in the above figure corresponds to a face-centered cubic unit cell and identify, with their proper Miller indices, the four surfaces of the tetrahedron.

1.10

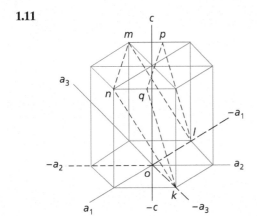

The figure accompanying this problem is normally used to represent the unit cell of a close-packed hexagonal metal. Determine Miller indices for the two planes, *defg* and *dehj*, that are outlined in this drawing.

1.11

Two other hexagonal closed-packed planes, *klmn* and *klpq*, are indicated in this sketch. What are their Miller indices?

1.12

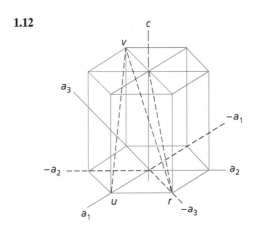

Determine the hexagonal close-packed lattice directions of the lines *rt, ut,* and *uv* in the figure for this problem. To do this, first determine the vector projection of a line in the basal plane and then add it to the *c* axis projection of the line. Note that the direction indices of the *c* axis are [0001], and that if [0001] is considered a vector its magnitude will equal the height of the unit cell. A unit distance along a digonal axis of Type I, such as the distance *or*, equals one-third of the length of $[2\bar{1}\bar{1}0]$ in Fig. 1.18. The magnitude of this unit distance is thus equal to $\frac{1}{3}[2\bar{1}\bar{1}0]$. Combine these two quantities to obtain the direction indices of each of the lines.

Stereographic Projection

The following problems involve the plotting of stereographic projections and require the use of the Wulff net, shown in Fig. 1.26, and a sheet of tracing paper. In each case, first place the tracing paper over the Wulff net and then pass a pin through the tracing paper and the center of the net so that the tracing paper may be rotated about the center of the net. Next, trace the outline of the basic circle on the tracing paper and place a small vertical mark at the top of this traced circle to serve as an index.

1.13 Place a piece of tracing paper over the Wulff net as described above, and draw an index mark on the tracing paper over the north pole of the Wulff net. Then draw on the tracing paper the proper symbols that identify the three $\langle 100 \rangle$ cube poles, the six $\langle 110 \rangle$ poles, and the four $\langle 111 \rangle$ octahedral poles as in Fig. 1.33. On the assumption that the basic circle is the (010) plane and that the north pole is [100], mark on tracing paper the correct Miller indices of all of the $\langle 100 \rangle$, $\langle 110 \rangle$, and $\langle 111 \rangle$ poles. Draw in the great circles corresponding to the planes of the plotted poles (see Fig. 1.33). Finally, identify these planes with their Miller Indices.

1.14 Place a piece of tracing paper over the Wulff net and draw on it the index mark at the north pole of the net as well as the basic circle. Mark on this tracing paper all of the poles shown in Fig. 1.30 in order to obtain a 100 standard projection. Now rotate this standard projection about the north–south polar axis by 45° so that the (110) pole moves to the center of the stereographic projection. In this rotation all of the other poles should also be moved through 45° along the small circles of the Wulff net on which they lie. This type of rotation is facilitated by placing a second sheet of tracing paper over the first and by plotting the rotated data on this sheet. This exercise shows one of the basic rotations that can be made with a stereographic projection. The other primary rotation involves a simple rotation of the tracing paper around the pin passing through the centers of both the tracing paper and the Wulff net.

References

1. "Metallography and Microstructures," *ASM Handbook*, Volume 9, ASM International, pp. 23–69.
2. K. Honda and S. Kaya, *Sci. Rep.* Tohoku Univ., 15, pp. 721–754 (1926).
3. S. Kaya, *Sci. Rep.* Tohoku Univ., 17, 639 (1928).
4. "Magnetic Anisotropy of Epitaxial Iron Films on Single-Crystal MgO.001. and Al2O3(112–0) Substrates," Yu. V. Goryunov and I. A. Garifullin, *Journal of Experimental and Theoretical Physics*, 88, 2 (1999), pp. 377–384.
5. A. R. Urade, I. Lahiri, and K. S. Suresh, *JOM* 75, 614–630 (2023).
6. *Smithells Metals Reference Book*, 6th Edition, edited by Eric A. Brandes, Butterworths, P 5.2.

Chapter 2

Characterization Techniques

Learning Objectives

Upon completion of Chapter 2, you will be able to:

1. Derive Bragg's Law to understand the relationship between incident wavelengths, interplanar distances, angles of reflection, and the constraints between them
2. Explain how Bragg's Law is employed in the Laue methods, rotating-crystal method, and Debye-Scherrer method to understand the crystalline phase of a metal
3. Describe how X-Ray diffractometers relates intensity to Bragg angles and thus multiphase mixtures
4. Compare and contrast the transmission electron microscope (TEM) to a Scanning Electron Microscope (SEM) to an optical microscope
5. Outline some key considerations in sample preparation for TEM and for SEM, taking into account the principles of the measurement technique
6. Relate image formation via TEM to Bragg's Law, Miller indices/crystallographic planes, diffraction patterns, and defects
7. Outline how the electrons in an electron beam (TEM or SEM) will interact with a metallic specimen, including the different kinds of scattering as well as the role of secondary electrons
8. Summarize how X-ray spectra can be used to identify the composition of a sample

The most commonly used characterization technique in metallurgy is through optical characterization called metallography. The technique involved preparation of metallographic specimens with a suitable undistorted flat surface by grinding, polishing and appropriately etching the surface. The details of the procedure can be found in standard texts (Ref. 1). The maximum magnification of light microscope is around 2,000X, with spatial resolution of around 200 nm. As such, the technique can provide qualitative and quantitative information for most metallurgical purposes. However, because of the use of optical waves, the depth of the characterization is extremely shallow. As such, metallography is often complemented by other characterization technique described in this chapter.

Because crystals are symmetrical arrays of atoms containing rows and planes of high atomic density, they are able to act as three-dimensional diffraction gratings. If light rays are to be efficiently diffracted by a grating, then the spacing of the grating (distance between ruled lines on a grating) must be approximately equal to the wavelength of the light waves. In the case of visible light, gratings with line separations between 1000 to 2000 nm are used to diffract wavelengths in the range from 400 to 800 nm. In crystals, however, the separation between equally spaced parallel rows of atoms or atomic planes is much smaller and of the order of a few tenths of a nanometer. Fortunately, low-voltage X-rays have wavelengths of the proper magnitudes to be diffracted by crystals; that is, X-rays produced by tubes operated in the range between 20,000 and 50,000 volts, as contrasted to those used in medical applications, where voltages exceed 100,000 volts.

When X-rays of a given frequency strike an atom, they interact with its electrons, causing them to vibrate with the frequency of the X-ray beam. Since the electrons become vibrating electrical charges, they reradiate the X-rays with no change in frequency. These reflected rays come off the atoms in any direction. In other words, the electrons of an atom "scatter" the X-ray beam in all directions.

When atoms spaced at regular intervals are irradiated by a beam of X-rays, the scattered radiation undergoes interference. In certain directions constructive interference occurs; in other directions destructive interference occurs. For instance, if a single atomic plane is struck by a parallel beam of X-rays, the beam undergoes constructive interference when the angle of incidence equals the angle of reflection. Thus, in Fig. 2.1, the rays marked a_1 to a_3 represent a parallel beam of X-rays. A wave front of this beam, where all rays are in phase, is shown by the line AA. The line BB is drawn perpendicular to rays reflected by the atoms in a direction such that the angle of incidence equals the angle of reflection. Since BB lies at the same distance from the wave front AA when measured along any ray, all points on BB must be in phase. It is, therefore, a wave front, and the direction of the reflected rays is a direction of constructive interference.

2.1 The Bragg Law

The preceding discussion does not depend on the frequency of the radiation. However, when the X-rays are reflected, not from an array of atoms arranged in a solitary plane, but from atoms on a number of equally spaced parallel planes, such as exist in crystals, then constructive interference can occur only under highly restricted conditions. The law that governs the latter case is known as

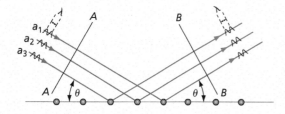

Fig. 2.1 An X-ray beam, with wave length λ, is reflected with constructive interference when the angle of incidence equals the angle of reflection

2.1 The Bragg Law

Bragg's law. Let us now derive an expression for this important relationship. For this purpose, let us consider each plane of atoms in a crystal as a semitransparent mirror; that is, each plane reflects a part of the X-ray beam and also permits part of it to pass through. When X-rays strike a crystal, the beam is reflected not only from the surface layer of atoms, but also from atoms underneath the surface to a considerable depth. Figure 2.2 shows an X-ray beam that is being reflected simultaneously from two parallel lattice planes. In an actual case, the beam would be reflected not from just two lattice planes, but from a large number of parallel planes. The lattice spacing, or distance between planes, is represented by the symbol d in Fig. 2.2. The line oA_i is drawn perpendicular to the incident rays and is therefore a wave front. Points o and m, which lie on this wave front, must be in phase. The line oA_r is drawn perpendicular to the reflected rays a_1 and a_2, and the condition for OA_r to be a wave front is that the reflected rays must be in phase at points o and n. This condition can only be satisfied if the distance mpn equals a multiple of a complete wavelength; that is, it must equal λ or 2λ or 3λ or $n\lambda$, where λ is the wavelength of X-rays and n is an arbitrary integer.

An examination of Fig. 2.2 shows that both the distances mp and pn equal $d \sin \theta$. The distance mpn is, accordingly, $2d \sin \theta$. If this quantity is equated to $n\lambda$, we have Bragg's law:

$$n\lambda = 2d \sin \theta \qquad \qquad \textbf{2.1}$$

where $n = 1, 2, 3, \ldots$
    λ = wavelength in nm
    d = interplanar distance in nm
    θ = angle of incidence or reflection of X-ray beam

When this relationship is satisfied, the reflected rays a_1 and a_2 will be in phase and constructive interference will result. Furthermore, the angles at which constructive interference occur when a narrow beam of X-rays strikes an undistorted crystal are very sharply defined, because the reflections originate on many thousands of parallel lattice planes. Under this condition, even a very small deviation from the angle θ satisfying the above relationship causes destructive interference of the reflected rays. As a consequence, the reflected beam leaves the crystal as a narrow pencil of rays capable of producing sharp images of the source on a photographic plate.

Let us now consider a simple example of an application of the Bragg equation. The {110} planes of a body-centered cubic crystal have a separation of 0.1181 nm. If these planes are irradiated with X-rays

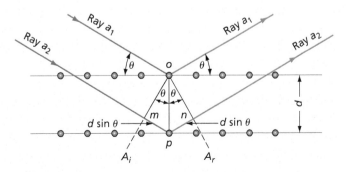

Fig. 2.2 The Bragg law

from a tube with a copper target, the strongest line of which, the K_{α_1}, has a wavelength 0.1541 nm, first-order ($n = 1$) reflection will occur at an angle of

$$\theta = \sin^{-1}\left(\frac{n\lambda}{2d}\right) = \sin^{-1}\frac{(1)0.1541}{2(0.1181)} = 40.7°$$

A second-order reflection from these {110} planes is not possible with radiation of this wavelength because the argument of the arc sin {$n\lambda/2d$} is

$$\frac{2(0.1541)}{2(0.1181)} = 1.305$$

a number greater than unity, and therefore the solution is impossible. On the other hand, a tungsten target in an X-ray tube gives a K_{α_1} line with a wavelength of 0.02090 nm. Eleven orders of reflections are now possible. The angle θ, corresponding to several of these reflections, is shown in Table 2.1, and Fig. 2.3 shows a schematic representation of the same reflections.

In considering the preceding example, it is important to notice that, although there are eleven angles at which a beam of wavelength 0.02090 nm will be reflected with constructive interference from the {110} planes, only a very slight deviation in the angle θ away from any of these eleven values causes destructive interference and cancellation of the reflected beam. Whether a beam of X-rays is reflected from a set of crystallographic planes is thus a sensitive function of the angle of inclination of the X-ray beam to the plane, and a constructive reflection should not be expected to occur every time a monochromatic beam impinges on a crystal.

Suppose that a crystal is maintained in a fixed orientation with respect to a beam of X-rays and that this beam is not monochromatic but contains all wavelengths longer than a given minimum value λ_0. This type of X-ray beam is called a *white* X-ray beam, since it is analogous to white light, which contains all the wavelengths in the visible spectrum. Although the angle of the beam is fixed with respect to any given set of planes in the crystal, and the angle θ of Bragg's law is therefore a constant, reflections from all planes can now occur as a result of the fact that the X-ray beam is continuous. The point in question can be illustrated with the aid of a simple cubic lattice.

Let the X-ray beam have a minimum wavelength of 0.05 nm, and let it make a 60° angle with the surface of the crystal, which, in turn, is assumed to be parallel to a set of {100} planes. In addition, let the {100} planes have a spacing of 0.1 nm. Substituting these values in the Bragg equation gives

$$n\lambda = 2d \sin \theta$$

or

$$n\lambda = 2(0.1) \sin 60° = 0.1732$$

Table 2.1

Order of Reflection	θ, Angle of Incidence or Reflection
1	5° 5′
2	10° 20′
5	26° 40′
11	80°

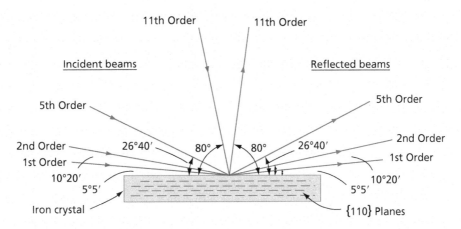

Fig. 2.3 Four of the eleven angles at which Bragg reflections occur using a crystal with an interplanar spacing of 0.1181 nm and X-rays of wavelength 0.02090 nm (WK_{α_1})

Thus, the rays reflected from {100} planes will contain the following wavelengths

0.1732 nm   for first-order reflection
0.0866 nm   for second-order reflection
0.0546 nm   for third-order reflection

All other wavelengths will suffer destructive interference.

In the preceding examples, the reflecting planes were assumed parallel to the crystal surface. This is not a necessary requirement for reflection; it is quite possible to obtain reflections from planes that make all angles with the surface. Thus, in Fig. 2.4, the incident beam is shown normal to the surface and a (001) plane, while making an angle θ with two {210} planes—(012) and (0$\bar{1}$2). The reflections from these two planes are shown schematically in Fig. 2.4. It may be concluded that when

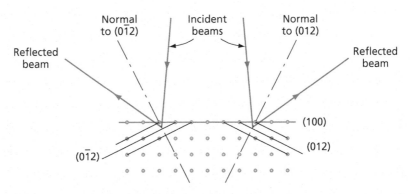

Fig. 2.4 X-ray reflections from planes not parallel to the surface of the specimen

a beam of white X-rays strikes a crystal, many reflected beams will emerge from the crystal, each reflected beam corresponding to a reflection from a different crystallographic plane. Furthermore, in contrast to the incident beam that is continuous in wavelength, each reflected beam will contain only discrete wavelengths as prescribed by the Bragg equation.

2.2 Laue Techniques

The Laue X-ray diffraction methods make use of a crystal with an orientation that is fixed with respect to a beam of continuous X-rays, as described in the preceding section. There are two basic Laue techniques: in one, the beams reflected back in directions close to that of the incident X-ray beam are studied; in the other, the reflected beams that pass through the crystal are studied. Clearly the latter method cannot be applied to crystals of appreciable thickness (1 mm or more) because of the loss in intensity of the X-rays by their absorption in the metal. The first method is known as the *back-reflection Laue technique*; the latter as the *transmission Laue technique*.

The back-reflection Laue method is especially valuable for determing the orientation of the lattice inside crystals when the crystals are large and therefore opaque to X-rays. Many physical and mechanical properties vary with direction inside crystals. The study of these anisotropic properties of crystals requires a knowledge of the lattice orientation in the crystals.

Figure 2.5 shows the arrangement of a typical Laue back-reflection camera. X-rays from the target of an X-ray tube are collimated into a narrow beam by a tube several inches long with an internal diameter of about 1 mm. The narrow beam of X-rays impinges on the crystal at the right of the figure, where it is diffracted into a number of reflected beams that strike the cassette containing a photographic film. The front of the cassette is covered with a thin sheet of material, for example, black paper, opaque to visible light, but transparent to the reflected X-ray beams. In this way, the positions of the reflected beams are recorded on the photographic film as an array of small dark spots. More modern X-ray instruments use Charge-Coupled Device (CCD) cameras or flat plate detectors in place of photographic films. The CCD detector not only records the angles of diffracted beams, but also their intensities.

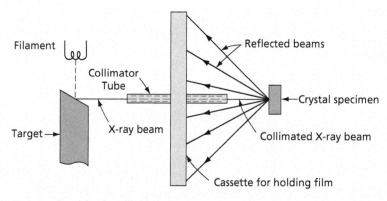

Fig. 2.5 Laue back-reflection camera

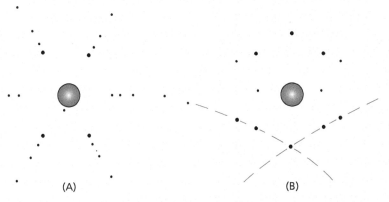

Fig. 2.6 Laue back-reflection photographs. **(A)** Photograph with X-ray beam perpendicular to the basal plane (0001). **(B)** Photograph with X-ray beam perpendicular to a prism plane (11$\bar{2}$0). Dashed lines on the photograph are drawn to show that the back-reflection spots lie on hyperbolas

Figure 2.6A shows the back-reflection X-ray pattern of a magnesium crystal oriented so that the incident X-ray beam was perpendicular to the basal plane of the crystal. Each spot corresponds to a reflection from a single crystallographic plane, and the sixfold symmetry of the lattice, when viewed in a direction perpendicular to the basal plane, is quite apparent. If the crystal is rotated in a direction away from the one that gives the pattern of Fig. 2.6A, the pattern of spots changes (Fig. 2.6B); nevertheless it still defines the orientation of the lattice in space. Therefore, the orientation of the crystal can be determined in terms of a Laue photograph.

Transmission Laue patterns can be obtained with an experimental arrangement similar to that for back-reflection patterns, but the film is placed on the opposite side of the specimen from the X-ray tube. Specimens may have the shape of small rods or plates, but must be small in their thickness perpendicular to the X-ray beam. While the back-reflection technique reflects the X-ray beam from planes nearly perpendicular to the beam itself, the transmission technique records the reflections from planes nearly parallel to the beam, as can be seen in Fig. 2.7.

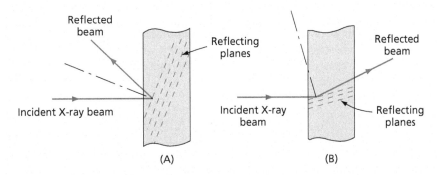

Fig. 2.7 **(A)** Laue back-reflection photographs record the reflections from planes nearly perpendicular to the incident X-ray beam. **(B)** Laue transmission photographs record the reflections from planes nearly parallel to the incident X-ray beam

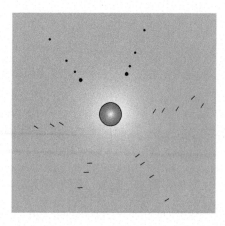

Fig. 2.8 Asterism in a Laue back-reflection photograph. The reflections from distorted- or curved-crystal planes form elongated spots

Laue transmission photographs, like back-reflection photographs, consist of arrays of spots. However, the arrangements of the spots differ in the two methods: transmission patterns usually have spots arranged on ellipses, back-reflection on hyperbolas. (See Fig. 2.6B.)

The Laue transmission technique, like the back-reflection technique, is also used to find the orientation of crystal lattices. Both methods can be used to study a phenomenon called *asterism*. A crystal that has been bent, or otherwise distorted, will have curved lattice planes that act in the manner of curved mirrors and form distorted or elongated spots instead of small circular images of the X-ray beam. A typical Laue pattern of a distorted crystal is shown in Fig. 2.8. In many cases, analysis of the asterism, or distortion, of the spots in Laue photographs leads to valuable information concerning the mechanisms of plastic deformation.

In the preceding examples (Laue methods), a crystal is maintained in a fixed orientation with respect to the X-ray beam. Reflections are obtained because the beam is continuous; that is, the wavelength is the variable. Several important X-ray diffraction techniques that use X-rays of a single frequency or wavelength will now be considered. In these methods, since λ is no longer a variable, it is necessary to vary the angle θ in order to obtain reflections.

2.3 The Rotating-Crystal Method

In the rotating-crystal method, crystallographic planes are brought into reflecting positions by rotating a crystal about one of its axes while simultaneously radiating it with a beam of monochromatic X-rays. The reflections are usually recorded on a photographic film that surrounds the specimen. (See Fig. 2.9 for a schematic view of the method.)

2.4 The Debye-Scherrer or Powder Method

In this method care is taken to see that the specimen contains not one crystal, but more than several hundred randomly oriented crystals. The specimen may be either a small polycrystalline metal wire, or a finely ground powder of the metal contained in a plastic, cellulose, or glass tube. In either case, the crystalline aggregate consists of a cylinder about 0.5 mm in diameter with

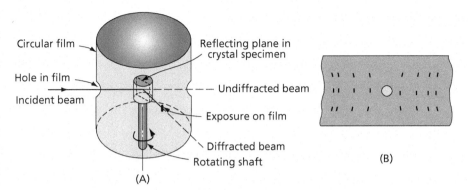

Fig. 2.9 **(A)** Schematic diagram of a rotating single-crystal camera. **(B)** Schematic representation of the diffraction pattern obtained with a rotating crystal camera. Reflected beams make spots lying in horizontal rows

crystals approximately 0.1 mm in diameter or smaller. In the Debye-Scherrer method, as in the rotating single-crystal method, the angle θ is the variable; the wavelength λ remains constant. In the powder method, a variation of θ is obtained, not by rotating a single crystal about one of its axes, but through the presence of many small crystals randomly oriented in space in the specimen. The principles involved in the Debye-Scherrer method can be explained with the aid of an example.

For the sake of simplicity, let us assume a crystalline structure with the simple cubic lattice shown in Fig. 2.10, and that the spacing between {100} planes equals 0.1 nm. It can easily be shown that the interplanar spacing for planes of the {110} type equals that of {100} planes divided by the square root of two, and is, therefore, 0.0707 nm. (See Fig. 2.10.) The {110} spacing is, therefore, smaller than the {100} spacing. In fact, all other planes in the simple cubic lattice have a smaller spacing than that of the cube, or {100}, planes, as is shown by the following equation for the spacing of crystallographic planes in a cubic lattice, where h, k, and l are the three Miller indices of a plane in the crystal; d_{hkl} the interplanar spacing of the plane; and a the length of the unit-cell edge

$$d_{hkl} = \frac{a}{\sqrt{h^2 + k^2 + l^2}} \qquad 2.2$$

In the simple cubic structure, the distance between cubic planes, d_{100}, equals a. Therefore the preceding expression is written

$$d_{hkl} = \frac{d_{100}}{\sqrt{h^2 + k^2 + l^2}} \qquad 2.3$$

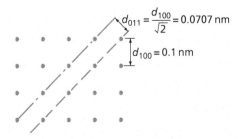

Fig. 2.10 Simple cubic lattice. Relative interplanar spacing for {100} and {110} planes. Note the unit cell of this lattice is a cube with atoms at its eight corners

Now, according to the Bragg equation (Eq. 2.1)

$$n\lambda = 2d \sin \theta$$

and for first-order reflections, where n equals one, we have

$$\theta = \sin^{-1}\left(\frac{\lambda}{2d}\right) \qquad \textbf{2.4}$$

This equation tells us that planes with the largest spacing will reflect at the smallest angle θ. If now it is arbitrarily assumed that the wavelength of the X-ray beam is 0.04 nm, first-order reflections will occur from {100} planes (with the assumed 0.1 nm spacing) when

$$\theta = \sin^{-1} \frac{0.04}{2(0.1)} = \sin^{-1} \frac{1}{5} = 11° 30'$$

On the other hand, {110} planes with a spacing 0.0707 nm reflect when

$$\theta = \sin^{-1} \frac{0.04}{2(0.0707)} = 16° 28'$$

Per equation 2.2, all other planes with larger indices (that is, {111}, {234}, etc.) reflect at still larger angles.

Figure 2.11 shows how the Debye-Scherrer reflections are found. A parallel beam of monochromatic X-rays coming from the left of the figure is shown striking the crystalline aggregate. Since the specimen contains hundreds of randomly oriented crystals in the region illuminated by the incident X-ray beam, many of these will have {100} planes at the correct Bragg angle of 11°30'. Each of these crystals will therefore reflect a part of the incident radiation in a direction that makes an angle twice 11°30' with the original beam. However, because the crystals are randomly oriented in space, not all the reflections will lie in the same direction. Instead, the reflections will form a cone with an apex angle of 23°. This cone is symmetric about the line of the incident beam. In the same manner, it can be shown that first-order reflections from {110} planes form a cone, the surface of which makes an angle of twice 16°28', or 32°56' with the original direction of the beam, and that the planes of still higher indices form cones of reflected rays, making larger and larger angles with the original direction of the beam.

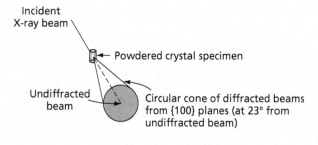

Fig. 2.11 First-order reflections from {100} planes of a hypothetical simple cubic lattice. Powdered crystal specimen

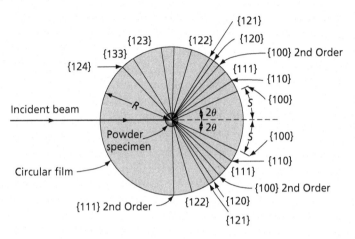

Fig. 2.12 Schematic representation of the Debye or powder camera. Specimen is assumed to be simple cubic. Not all reflections are shown

The most commonly used powder cameras employ a long strip film that forms a circular cylinder surrounding the specimen, as shown in Fig. 2.12. A schematic view of a Debye-Scherrer film after exposure and development is shown in Fig. 2.13. It should be noted that, instead of X-ray film that was originally used, flat-bed or cylindrical cameras are commonly used presently.

On a Debye-Scherrer film, the distance $2S$ between the two circular segments of the {100} cone is related to the angular opening of the cone and therefore to the Bragg angle θ between the reflecting plane and the incident beam. Thus, the angle in radians between the surface of the cone and the X-ray beam equals S/R, where R is the radius of the circle formed by the film. However, this same angle also equals 2θ, and, therefore,

$$2\theta = \frac{S}{R} \qquad 2.5$$

or

$$\theta = \frac{S}{2R} \qquad 2.6$$

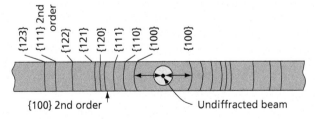

Fig. 2.13 Powder camera photograph. Diffraction lines correspond to the reflections shown in Fig. 2.12

This last relationship is important because it is possible to measure the Bragg angle θ with it. In the preceding example, the spacing between parallel lattice planes was assumed to be known. This assumption was made in order to explain the principles of the Debye-Scherrer method. In many cases, however, one may not know the interplanar spacings of a crystal, and measurements of the Bragg angles can then be used to determine these quantities. The powder method is, accordingly, a powerful tool for determining the crystal structure of a metal. In complicated crystals, other methods may have to be used in combination with the powder method in order to complete an identification. In any case, the Debye-Scherrer method is probably the most important of all methods used in the determination of crystal structures. Another very important application of the powder method is based upon the fact that each crystalline material has its own characteristic set of interplanar spacings. Thus, while copper, silver, and gold all have the same crystal structure (face-centered cubic), the unit cells of these three metals are different in size, and, as a result, the interplanar spacings and Bragg angles are different in each case. Since each crystalline material has its own characteristic Bragg angles, it is possible to identify unknown crystalline phases in metals with the aid of their Bragg reflections. For this purpose, a card file system (X-ray Diffraction Data Index) has been published listing for approximately a thousand elements and crystalline compounds, not only the Bragg angle of each important Debye-Scherrer diffraction line, but also its relative strength or intensity. The identification of an unknown crystalline phase in a metal can be made by matching powder pattern Bragg angles and reflected intensities of the unknown substance with the proper card of the index. The method is analogous to a fingerprint identification system and constitutes an important method of qualitative chemical analysis.

2.5 The X-Ray Diffractometer

The X-ray diffractometer is a device that measures the intensity of the X-ray reflections from a crystal with an electronic detector, such as a Geiger counter or ionization chamber, instead of a photographic film. Figure 2.14 shows the elementary parts of a diffractometer—a crystalline specimen, a parallel beam of X-rays, and a detector. The apparatus is so arranged that both the crystal and the intensity measuring device (Geiger counter) rotate. The detector, however, always moves at twice the speed of the specimen, which keeps the intensity recording device at the proper angle during the rotation of the crystal so that it can pick up each Bragg reflection as it occurs. In modern instruments of this type, the intensity measuring device is connected through a suitable amplification system to a chart recorder or

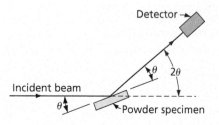

Fig. 2.14 X-ray diffractometer—the detector rotates at twice the speed of the specimen

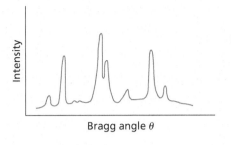

Fig. 2.15 The X-ray diffractometer records on a chart the reflected intensity as a function of Bragg angle. Each intensity peak corresponds to a crystallographic plane in a reflecting position

a digital storage device. In this manner, one obtains an accurate plot of intensity against Bragg angles. A typical X-ray diffractometer plot is shown in Fig. 2.15. Crystal structure and phase identification in multiphase materials is commonly performed by comparison of the stored diffraction pattern to databases containing powder diffraction files (PDFs) of known standards.

The X-ray diffractometer is most commonly used with a powder specimen in the form of a rectangular plate with dimensions about 25 mm long and 12.5 mm wide. The specimen may be a sample of a polycrystalline metal, and it should be noted that, in contrast to the Debye-Scherrer method where the specimen is a fine wire (approximately 0.5 mm diameter), the diffractometer sample has a finite size, which makes the specimen much easier to prepare and is therefore advantageous. Because the X-ray diffractometer is capable of measuring the intensities of Bragg reflections with great accuracy, both qualitative and quantitative chemical analyses can be made by this method. For a multiphase mixture, the technique will allow for determination of relative concentration of the constituent phases as well as identification of each phase. The details of the procedure can be found in standard texts dealing with X-rays (Refs. 2–4).

2.6 The Transmission Electron Microscope

Since the 1960s a very powerful technique has become available to metallurgists. It involves the use of the electron microscope to study the internal structure of thin crystalline films or foils. These foils, which can be removed from bulk samples, are normally only several hundred nm thick. The thickness is dictated by the voltage at which the microscope is operated. The common commercial instrument is rated between 100 and 400 kV, and electrons accelerated by this voltage can give an acceptable image if the foil is not made thicker than the indicated value. Thinner foils, on the other hand, tend to be less useful in revealing the nature of the structure in the metal. Some instruments have been developed that operate at much higher voltages (of the order of a million volts), and in them the foils can be proportionally thicker. However, equipment costs are also much greater, and few of these instruments are available.

A word should be said about the technique involved in preparing a foil specimen for examination in the transmission electron microscope (TEM). This involves cutting a thin slice of the metal that is to be examined. A great deal of care must be taken so that the specimen is not deformed during its preparation. This is because plastic deformation can introduce unwanted structural defects in the microstructure that are visible in the TEM image. A convenient machine that is useful for preparing specimens with a minimum of distortion is a spark cutter designed to cut thin slices of a metal. In this case, electrical discharges between a wire and the specimen are used to cut the metal by removing small particles of metal from the surface of the specimen as the wire slices through the specimen. This leaves a thin, highly distorted layer near the surfaces of the cut that is later polished away by chemical or electrochemical means. A typical section obtained by this type of spark machining may be about 200 μm thick. This is much too thick to be transparent in the TEM. Therefore, it is necessary to thin the specimen to its final desired thickness of several hundred nm using a chemical, electrochemical, or ion miller technique. The details of the procedure may be found in standard texts dealing with electron microscopy (Refs. 5 and 6). The most recent technique for TEM sample preparation is by using a focused ion beam (FIB) instrument (Ref. 4). A typical FIB utilizes a liquid metal source to form a highly focused beam with high current density. When the ion beam strikes a specimen, it causes physical sputtering and material removal. The main advantage of the technique is that it allows for selective thinning at desired locations by cutting trenches in the sample. However, the technique also has the disadvantage of causing ion implantation and severe damage to the material (Ref. 7).

In the transmission electron microscope, the detail in the image is formed by the diffraction of electrons from the crystallographic planes of the object being investigated. The electron microscope is, in many respects, analogous to an optical microscope. The source is an electron gun instead of a light filament. The lenses are magnetic, being composed normally of a current-carrying coil surrounded by a soft iron case. The lenses are energized by direct current. An excellent, easy-to-read description of the electron microscope is given in Volume 9 of the *Metals Handbook* (Ref. 5). For our present purposes we shall concentrate on that part of the microscope containing the specimen and the objective lens. This region is indicated schematically in Fig. 2.16. In this diagram, the electron beam is shown entering the specimen from above. This beam originated in the electron gun and has passed through a set of condensing lenses before it reaches the specimen. On emerging from the specimen, the beam passes through the rear element of the instrument's objective lens. Shortly beyond this lens element, the rays converge to form a spot at point a in plane I_1. This spot is equivalent to an image of the source. Somewhat beyond this point, the image of the specimen is formed at plane I_2. Similar double image effects are observable in simple optical instruments where it is possible to form images of the light source at one position and images of a lantern slide or other object at other positions.

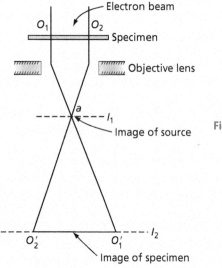

Fig. 2.16 Schematic drawing of a transmission electron microscope

2.6 The Transmission Electron Microscope

Because the image formation in the transmission electron microscope depends on the diffraction of electrons, it is necessary to consider some elementary facts about this type of diffraction. As demonstrated in Chapter 3, electrons not only have many of the attributes of particles, but they also possess wavelike properties. It will also be shown that the wavelength of an electron is related to its velocity v by the relation

$$\lambda = \frac{h}{mv} \qquad \qquad 2.7$$

where λ is the wavelength of the electron, m is its mass, and h is Planck's constant equal to 6.626×10^{-34} J/Hz. This expression shows that the wavelength of an electron varies inversely as its velocity. The higher the velocity, the shorter its wavelength.

Now let us assume that an electron is accelerated by a potential of 100 kV. This will give the electron a velocity of about 2×10^8 m/s and, by the above equation, a wavelength of about 4×10^{-12} m, or about 4×10^{-3} nm. This is about two orders of magnitude smaller than the average wavelength used in X-ray diffraction studies of metallic crystals. This causes a corresponding difference in the nature of the diffraction, as can be deduced by considering Bragg's law. Suppose that we are concerned with first-order diffraction, where $n = 1$. Then, by Bragg's law, Eq. 2.1, we have

$$\lambda = 2d \sin \theta$$

If d, the spacing of the parallel planes from which the electrons are reflected, is assumed to be about 0.2 nm, we have

$$\theta \approx \sin \theta = 0.01$$

The angle of incidence or reflection of a diffracted beam is thus only of the order of 10^{-2} radians, or about 30'. This means that when a beam of electrons is passed through a thin layer of crystalline material, only those planes nearly parallel to the beam can be expected to contribute to the resulting diffraction pattern.

Let us now consider how the image is formed in the electron microscope as the result of diffraction. In this regard, consider Fig. 2.17. Here it is assumed that some of the electrons, in passing through the specimen, are diffracted by one of the sets of planes in the specimen. In general, only part of the electrons will be diffracted, and the remainder will pass directly through the specimen without being diffracted. These latter electrons will form a spot at position a and an image of the specimen $(O'_2 - O'_1)$ at the plane I_2, as indicated in Fig. 2.16. On the other hand, the diffracted electrons will enter the objective lens at a slightly different angle and will converge to form a spot at point b. These rays that pass through point b will also form an image of the specimen at I_2 that is superimposed over that from the direct beam. In the above, it has been assumed that the crystal is so oriented that the electrons are reflected primarily from a single crystallographic plane. This should cause the formation of a single pronounced spot at point b as a result of the diffraction. It is also possible to have simultaneous reflections from a number of planes. In this case, instead of a single spot appearing in I_1 at point b, a typical array of spots or a diffraction pattern will form on plane I. A characteristic diffraction pattern will be described presently.

The electron microscope is so constructed that either the image of the diffraction pattern (formed at I_1) or the image of the detail in the specimen (formed at I_2) can be viewed on the fluorescent screen of the instrument. Alternatively, both of these images can be photographed on a plate or film. This is possible because a projection lens system (not shown) is located in the microscope below that part

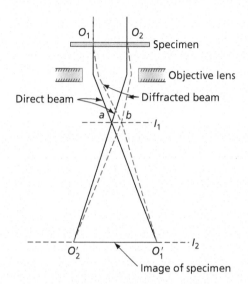

Fig. 2.17 Images can be formed in the transmission electron microscope corresponding to the direct beam or to a diffracted beam. (Images from more than one diffracted beam are also possible.)

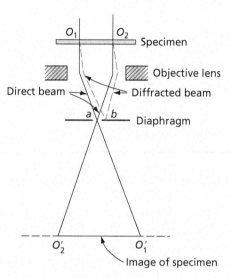

Fig. 2.18 The use of a diaphragm to select the desired image

of the instrument shown in Fig. 2.16. This lens system can be adjusted to project either the image of the diffraction pattern at plane I_1, or that of the specimen at plane I_2, onto the fluorescent screen or photographic emulsion.

In operating the instrument as a microscope, one has the choice of using either the image formed by the direct beam or the image formed by diffraction from a particular set of planes. The elimination of the beam causing either of these two types of images is made possible by the insertion of an aperture diaphragm at plane I_1 that allows only one of the corresponding beams to pass through it, as shown in Fig. 2.18. In this diagram, the diffracted beam is shown intercepted by the diaphragm, while the direct beam is allowed to pass through the aperture. When the specimen is viewed in this manner, a *bright-field image* is formed. Imperfections in the crystal will normally appear as dark areas in this image. These imperfections could be small inclusions of different transparency from the matrix crystal, and therefore visible in the image as a result of a loss in intensity of the beam where it passes through the more opaque particles. Of more general interest, however, is the case where the imperfections are faults in the crystal lattice itself. A very important defect of this type, that will be of considerable concern to us in subsequent chapters, is a dislocation. Without delving too deeply into the nature of dislocations at this time, it is necessary to point out that dislocations involve distortions in the arrangement of the crystal planes. Such local distortions will have effects on the diffraction of electrons, because the angle of incidence between the electron beam and the lattice planes around the dislocation are altered. In some cases this may cause an increase in the number of diffracted electrons, and in others a decrease. Since the direct beam can be considered to be the difference between the incident beam and the diffracted beam, a local change in the diffracting conditions in the specimen will be reflected by a corresponding alteration in the intensity recorded in the specimen image. Dislocations are thus visible in the image because they affect the diffraction of electrons. In a bright-field image, dislocations normally appear as dark lines. A typical bright-field photograph is shown in Fig. 4.6 (see p. 91).

The alternate method of using the electron microscope is to place the aperture so that a diffracted ray is allowed to pass, while the direct beam is cut off. The image of the specimen formed in this case is of the *dark-field image*. Here, dislocations appear as white lines lying on a dark background.

An important feature of the transmission electron microscope is the stage that holds the specimen. As indicated earlier, diffraction plays a very important role in making the defects in the crystal structure visible in the image. In order that the specimen may be capable of being aligned so that a suitable crystallographic plane can be brought into a reflecting condition, it is usually necessary for the specimen to be tilted with respect to the electron beam. The stage of an electron microscope is normally constructed so that the specimen may be rotated or tilted.

With regard to the diffraction patterns observable in the microscope, an interesting diffraction pattern is obtained when the specimen is tilted in its stage so that an important zone axis is placed parallel to the microscope axis. When this is done, a pattern is obtained whose spots correspond to the planes of the zone whose axis parallels the electron beam. By way of illustration, let us assume that the specimen is so oriented that a ⟨100⟩ direction is parallel to the instrument axis. Figure 2.19 shows a stereographic projection in which the zone axis is located at the center of the projection. The poles of the planes belonging to this zone should therefore lie on the basic circle of the stereographic projection. In the figure, only the planes of low indices are shown. The diffraction pattern corresponding to this zone is shown in Fig. 2.20. The Miller indices of the planes responsible for each spot are indicated alongside the corresponding spots.

The most significant feature of the diffraction pattern in Fig. 2.20 is that all spots correspond to planes parallel to the electron beam. Furthermore, as may be seen in Fig. 2.20, spots are indexed at both 100 and $\bar{1}$00. This implies that the electrons are reflected from both sides of the same planes. Obviously, the simple Bragg picture shown in Fig. 2.2, where the angle of incidence equals the angle

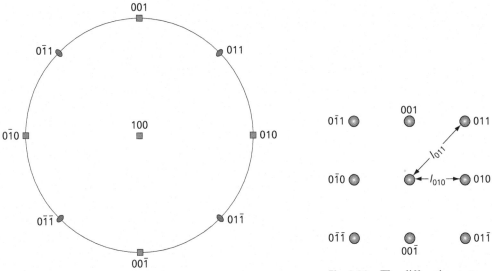

Fig. 2.19 This stereographic projection of a cubic crystal shows the principle planes whose zone axis is [100]

Fig. 2.20 The diffraction pattern corresponding to a beam directed along [100] in a cubic crystal

of reflection, does not apply in this case. The reasons for this are not easy to understand. However, several factors are undoubtedly involved. Of some importance is the small value of the Bragg angle θ, which is about 10^{-2} radians. Another is the fact that the transmission specimen is very thin, so that the electron beam, in traversing the specimen, sees a lattice that is nearly two-dimensional. This tends to relax the diffraction conditions. Finally, the electron microscope, with the specimen located inside a system of lenses, is not a simple diffracting device. For our present purposes, however, it is more important to note the nature of the diffraction pattern than the causes for it.

Attention is now called to the spacing of the spots in the diffraction pattern in Fig. 2.20. In this figure, the distance from the spot corresponding to the direct beam to that of a reflection from a {100} plane is indicated as 1_{010}, while the corresponding distance to a {110} reflection is 1_{011}. As can be deduced from the figure, $1_{011} = \sqrt{2} 1_{010}$. Attention is now called to Fig. 2.10, where it is shown that the spacing between the two respective planes varies inversely as $\sqrt{2}$. This indicates that the spacing of the spots in the diffraction pattern is inversely proportional to the interplanar spacing. This result, unlike the relationship of the incident beam to the planes from which it is reflected, is in good agreement with the Bragg law. In the present case, where the angle θ is small, $\sin \theta \approx \theta$ and Bragg's law reduces to

$$n\lambda = 2d\theta \qquad \text{2.8}$$

or

$$\theta = \frac{n\lambda}{2d} \qquad \text{2.9}$$

Since the angle θ is small, $\tan \theta$ is also approximately equal to θ (in radians) and we should expect the diffracted spots to be deviated through distances that are inversely proportional to the interplanar spacing d.

It is clear from the above that with the electron microscope it is possible both to investigate the internal defect structure of a crystalline specimen using the instrument as a microscope, and to determine a considerable degree of information about the crystallographic features of the specimen using it as a diffraction instrument. With regard to the latter application, the diffraction patterns can yield information both about the nature of the crystal structure and about the orientation of the crystals in a specimen. Furthermore, the electron microscope has a diaphragm in its optical path that controls the size of the area that is able to contribute to the diffraction pattern. As a result, it is possible to obtain information about an area of the specimen that has a radius as small as 0.5μ. The diffraction patterns are therefore called *selected area diffraction patterns*.

Examine Fig. 2.20 again. Note that each of the points, except for that at the center of the diffraction pattern, represents a specific set of crystal planes. Thus, the point at the upper-left-hand corner corresponds to the set of parallel planes represented by the Miller indices (011). Furthermore, the vector, 1_{011}, from the center of the figure to this point has a direction normal to these (011) planes and a length that is proportional to the interplanar spacing of the (011) planes. In addition, in the preceding text, it was deduced that $1_{001} = \sqrt{2}\, 1_{011}$. However, in a simple cubic lattice, the lattice parameter, a, equals 1_{100} so that it follows that $1_{011} = a/\sqrt{2}$. This can also be verified with the aid of Eq. 2.2. Actually, the diffraction pattern in Fig. 2.20 corresponds to a two-dimensional section taken through what is known as a *reciprocal lattice*. As defined in the *ASM Handbook*, "The reciprocal lattice is a lattice of points, each representing a set of planes in the crystal lattice, such that a vector from the origin of the reciprocal lattice to any point is normal to the crystal planes and has a length that is the reciprocal of the planar spacing." Reciprocal lattice can also be viewed as the Fourier transform of the regular lattice. For more information about the reciprocal lattice, see the book by Barrett and Massalski (Ref. 8).

2.7 Interactions between the Electrons in an Electron Beam and a Metallic Specimen

In the discussion about the transmission electron microscope, it was pointed out that electrons might penetrate the foil specimen to form the direct beam or be diffracted by certain planes of the crystals through which they passed. These are only two of the possibilities that may occur when a high-speed electron strikes a crystal surface. The others are also important. First, it is necessary to mention that the interaction between an electron and the atoms in a specimen may occur either elastically or inelastically.

2.8 Elastic Scattering

In this case, the path or trajectory of the moving electron is changed, but its energy or velocity is not altered significantly. For example, the diffraction of electrons by the planes of a TEM foil specimen can be classified as a form of elastic scattering. Elastic scattering can also occur more randomly due to collisions either between beam electrons and atomic nuclei of the specimen that are not completely screened by their bound electrons or between the beam electrons and tightly bound core electrons. These scattering events can alter the trajectories of the electrons through any angle up to and including 180°. However, the change in path direction is normally less than about 5° for a single collision. If the direction change is more than 90° and if the electron exits the specimen, it is said to have been *elastically backscattered*.

2.9 Inelastic Scattering

Inelastic scattering occurs when the moving electron loses some of its kinetic energy as a result of an interaction with the specimen. This energy loss can occur in a number of different ways. Often, several successive interactions may occur, with each resulting in a loss of a fraction of the electron's original energy.

A large fraction of the incident electron energy is spent creating phonons or atomic lattice vibrations; that is, in heating the specimens. The electrons in the beam can also lose a significant fraction of their energy by creating oscillations in a metal's electron gas. This is known as *plasmon excitation*. Another possibility is the energy loss associated with creating the bremsstrahlung or continuous X-ray radiation, which results from the deceleration of the beam electrons in the coulombic field of an atom.

There are, however, several other types of interactions between beam electrons and a specimen that are very useful for the microstructural analysis of specimens. Thus, beam electrons can cause loosely bound conduction band electrons to be ejected from a specimen's surface. These ejected electrons are known as *secondary electrons*, and they normally possess a relatively low energy; that is, <50 eV. If one plots the intensity of the secondary electrons as a function of their kinetic energy, as shown schematically in Fig. 2.21A, it is usually found that the intensity of the secondary electrons peaks between 3 and 5 eV. Additional details on the energy distribution curve of secondary electrons can be found in Reference 9.

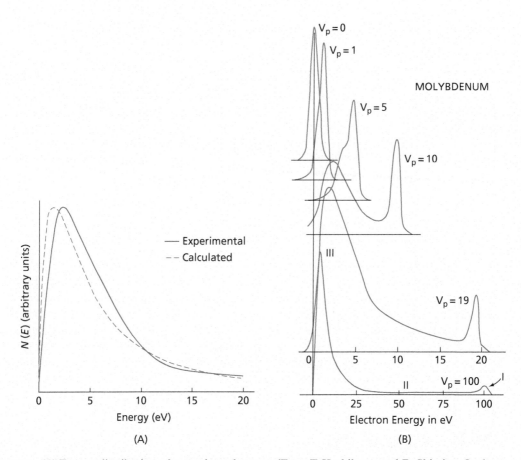

Fig. 2.21 **(A)** Energy distribution of secondary electrons (From T. Koshikawa and R. Shimizu, Osaka University, Department of Applied Physics, A Monte Carlo Calculation of low-energy secondary electron emission from metals, *Journal of Phys D: Applied Physics*, Vol **7** No 9, June 1974, p. 1309.) **(B)** The flux of electrons measured normal to the specimen surface, with a 45° incident beam, as a function of the energy of the emitted electrons (Reprinted Figure 3 with permission from American Physical Society: G. A. Harrower, *Physical Review*, Vol. **104**, pp. 52, 1956. © 1956 by the American Physical Society. http://prola.aps.org/abstract/PR/v104/i1/p52_1)

If the electrons in the incident beam are sufficiently energetic, they may also cause the ejection of electrons from the inner shells of the atoms. This ionizes the atoms and leaves them in an energetic state where electrons from outer shells are able to drop into the vacancies in the inner shells. Each of these events is accompanied by an energy release either in the form of (1) an emission of a characteristic X-ray photon or (2) the ejection of an electron from an outer shell of the atom. The electrons in the second type of emission are known as *Auger electrons*. Both types of emissions are associated with fixed amounts of energy, which are unique to the atomic elements from which they come and thus characteristic of the chemical elements in a specimen. Consequently, it is possible to obtain useful information about the chemical nature of a specimen in an electron microscope specimen by measuring either the characteristic X-ray wavelengths or the Auger electron energies. This will be discussed in more detail later.

2.10 Electron Spectrum

An example of an electron spectrum obtained using a pure metal, molybdenum (atomic number 47), bombarded by 100 keV electrons, is shown in Fig. 2.21B. This figure shows a plot in which the ordinate is proportional to the number (f) of backscattered electrons per unit energy interval, while the abscissa represents the energy (E) of the respective backscattered electrons normalized to the energy of the beam electrons (E_0). The data for this plot was obtained using a 45° angle of incidence for the incident beam and a detector for measuring the emitted electrons oriented normal to the surface. Note that the backscattered electron flux passes through a very small maximum, identified by the symbol I, at an energy close to E_0, the energy of the electrons in the incident beam. A backscattered electron with an energy E_0 corresponds to one that has been elastically backscattered. With increasing atomic number of the specimen, this peak in the emitted electron flux increases in height, becomes narrower, and moves closer to E_0. At lower energies, in the region marked II, the ordinate represents the flux of the backscattered electrons that have lost some of their incident energy as a result of various inelastic processes in the solid. In region III, which is very close to the zero end of the energy spectrum, the measured flux is almost entirely caused by secondary electrons, which are normally considered to have a maximum energy of about 50 eV. Note that the flux of the secondary electron peaks between 3 and 5 eV.

2.11 The Scanning Electron Microscope

It is useful to consider the scanning electron microscope (SEM) as an instrument that greatly extends the usefulness of the optical microscope for studying specimens that require higher magnifications and greater depths of field than can be attained optically. Many SEM specimens are normally polished and etched in the same manner as would be done for examination in an optical microscope. Thus, the lengthy and tedious procedures required for the preparation of TEM foil specimens are not needed for SEM specimens. The scanning electron microscope is capable of greatly extending the limited magnification range of the optical microscope, which normally extends to only about 1500×, to over 50,000×. In addition, with the SEM it is possible to obtain useful images of specimens that have a great deal of surface relief such as are found on deeply etched specimens or on fracture surfaces. The depth of field of the SEM can be as great as 300 times that of the optical microscope. This feature makes the SEM especially valuable for analyzing fractures.

On the other hand, at low magnifications, that is, below 300 to 400×, the image formed by the scanning electron microscope is normally inferior to that of an optical microscope. Thus, the optical and scanning microscopes can be viewed as complementing each other. The optical microscope is the superior instrument at low magnifications with relatively flat surfaces and the scanning microscope is superior at higher magnifications and with surfaces having strong relief.

The scanning electron microscope differs significantly from the transmission electron microscope in the way an image of the specimen is formed. First, the field of view in the TEM specimen is uniformly "illuminated" by the high-speed electrons of the incident beam. After passing through the foil specimen, these electrons are focused by a magnetic objective lens to form an image of the specimen that is analogous to an optical shadow picture of the structure in the foil. The contrast in this image is produced by the varying degree to which the electrons are diffracted as they pass through the specimen. The image is thus roughly similar to that formed by an optical slide projector. In the scanning electron microscope, on the other hand, the image is developed as in a television set. The specimen surface is scanned by a pointed electron beam over an area known as the *raster*. The interaction of this sharply pointed beam with the specimen surface causes several types of energetic emissions, including backscattered electrons, secondary electrons, Auger electrons (a special form of secondary electrons), continuous X-rays, and characteristic X-rays. Most of these emissions can furnish useful information about the nature of the specimen at the spot under the beam. In a standard scanning electron microscope, one normally uses the secondary electrons to develop an image. The reason for this is that the secondary electron signal comes primarily from the area directly under the beam and thus furnishes an image with a very high resolution or one in which the detail is better resolved. The secondary electron detector is shown to the right of the electron beam in the schematic drawing of Fig. 2.22. The front of this detector contains a screen biased at +200 V. Since most of the secondary electrons have energies only of the order of 3 to 5 eV, these low-energy electrons tend to be easily drawn into the detector by its 200 V bias.

In the part of the specimen surface used to form the image, that is, the raster, the electron beam is swept along a straight line over the entire width of the raster, as indicated in Fig. 2.23. As the beam moves across the line, the strength of the secondary electron emission from the surface is measured by the detector and is used to control the brightness of the synchronized spot on the cathode ray tube used to view or record the image. When the electron beam completes its line scan at the far end of the raster, it is returned quickly to the other side of the raster and to a point just below the start of the first line. During the time of its return, the beam of the cathode ray

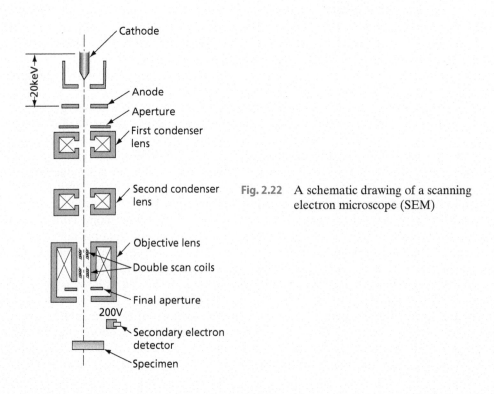

Fig. 2.22 A schematic drawing of a scanning electron microscope (SEM)

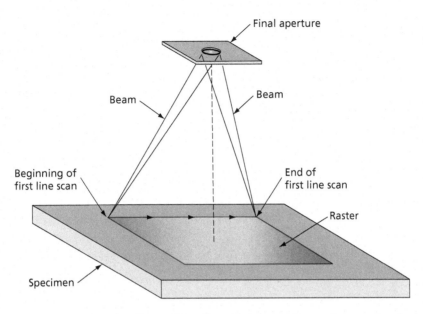

Fig. 2.23 The way the electron beam is moved across the specimen surface in an SEM

tube is turned off. By repeating this line scanning process, the entire surface of the raster can be surveyed. The typical SEM uses 1000 line scans to form a 10×10 cm image. A CRT screen with a long persistence phosphor is used so that the image will last long enough for the eye to be able to see a complete picture without problems of fading. The complete scanning process is repeated every thirtieth of a second, which conforms well to the one-twenty-fourth of a second frame time of a motion picture. To obtain a permanent photographic record of the image, on the other hand, a cathode ray tube with a short persistence phosphor is used. This avoids overlapping of images from adjacent lines.

2.12 Topographic Contrast

The means by which the SEM is able to interpret the topographic features of a metal surface will now be discussed very briefly. In Fig. 2.24, it is assumed that secondary electrons are being used to study the surface of a fractured metal specimen and that the incident beam of high-speed electrons (20 KeV) comes vertically from above. The energetic electrons in this beam cause the emission of low-energy secondary electrons from the specimen surface. Because of the 200 V positive bias on the detector, the secondary electrons are attracted to the detector, which lies to the right of the incident beam. The number of these electrons that reach the detector depends on several factors, one of which is the relative inclination of the surface with respect to the location of the detector. Thus, in general, for two different surfaces—for example, one inclined downwards to the right as in Fig. 2.24A and the other downwards to the left as in Fig. 2.24B—more secondary electrons will reach the detector from the former surface than from the latter as the incident beam moves

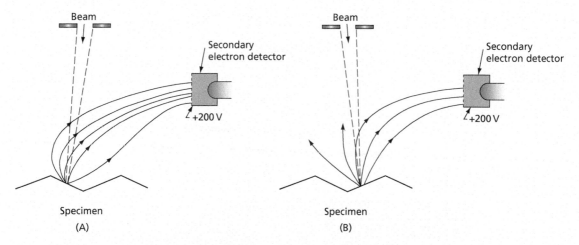

Fig. 2.24 An illustration of how the scanning electron microscope can reveal surface relief when used with a secondary electron detector

across each of these facets in turn. The result is that on the CRT screen, surface A will appear brighter than surface B. Figure 2.25 shows an example of the ability of the SEM to yield images with good contrast and excellent depth of focus. This photomicrograph shows the fracture surface of a Copper–4.9 at. % Tin specimen. This specific specimen was fractured under conditions that produced a failure largely along the grain boundaries, although there are a number of grains that failed transgranularly.

Backscattered electrons may also be used as a source of the signal sent from the specimen surface to the CRT screen. This is considered in Fig. 2.26, where the detector screen is assumed to be biased negatively (-50 V). The negative bias is sufficient so that secondary electrons are not able to reach the detector, but not nearly high enough to reject the energetic (16 to 18 keV) backscattered electrons. The backscattered electrons tend to travel in straight lines so that only those lying on a "line of sight" between the point at which the incident beam strikes the specimen and the front of the detector are effectively collected by the detector. The detector in this case is more sensitive to its orientation with respect to the surface than the detector used for secondary electrons. This means that by using backscattered electrons, it is possible to attain an increased topographic contrast over what would be achieved using secondary electrons. To illustrate, consider Fig. 2.26. Note that the detector is aligned to receive some backscattered electrons from surface A, while for surface B the alignment is much poorer. On the CRT screen, surface B should be very dark in comparison to surface A. The resulting contrast may actually be too great, so a detector capable of collecting both backscattered and secondary electrons may be a good compromise.

Another important feature of the backscattering is that the scattering coefficient, η, increases with increasing atomic number, as (Ref. 6)

$$\eta = -0.0254 + 0.016Z - 1.86 \times 10^{-4}Z^2 + 8.3 \times 10^{-7}Z^3 \qquad 2.10$$

where Z is the atomic number and η is defined as the ratio of the number of backscattered electrons to the number of electrons incident on the target. Thus, backscattering can be used to image

2.12 Topographic Contrast

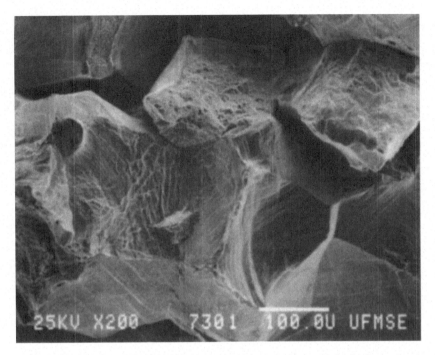

Fig. 2.25 An SEM micrograph of a fractured Cu–4.9 at. % Sn specimen. Note the large depth of field exhibited in the picture, which shows that the specimen failed by both transgranular (across the grains or crystals) and intergranular (along the grain boundaries) fracture

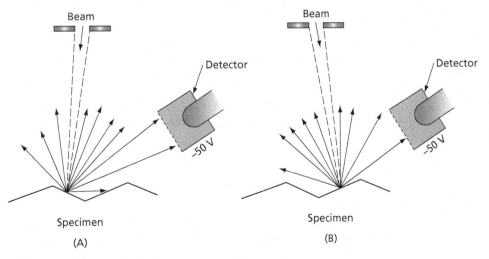

Fig. 2.26 Backscattered electrons are also able to reveal surface relief

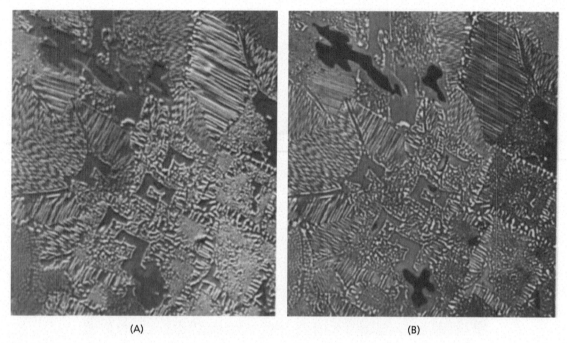

Fig. 2.27 A comparison of images from the same area of a specimen surface, obtained with **(A)** secondary electrons and **(B)** backscattered electrons

the atomic number difference between various regions of a specimen. Note that when the target is a homogeneous mixture of several elements, the coefficient η will depend on the weight fractions of the components.

A comparison between images produced by the secondary and backscattered electrons is shown in parts A and B of Fig. 2.27. The sample, with an overall composition Nb–20 at. % Si, contained three phases: Nb_5Si_3, Nb_3Si, and an Nb-rich solid solution. The first phase, which has a lower atomic number than the other two, appears as dark islands in the backscattered image, Fig. 2.27B. In contrast, the Nb solid solution appears white and the Nb_3Si solution appears gray. As will be discussed in Chapter 14 on the solidification of metals, the first phase to solidify is Nb_5Si, followed by the (peritectic) phase Nb_3Si, and then by an (eutectic) mixture of Nb_3Si and the Nb-rich solid solution. Note the terms *peritectic* and *eutectic* are defined in Chapter 11, on binary phase diagrams.

2.13 The Picture Element Size

The size of the picture element, also known as the *picture point*, is an important parameter for understanding the SEM. It is the area of the raster surface that supplies the information to a single spot on the CRT screen. For convenience, it will be considered that a line scan occurs digitally; that is, the beam advances in equally spaced steps as it scans along a line. Some scanning microscopes actually do this. Thus, if there are 1000 line scans on a 10×10 cm CRT screen, there normally would be 1000 steps along a 10 cm line. Each spot should then have a diameter of 100 μm. Dividing this diameter by the magnification of the instrument, M, gives the picture element size (PES), which is also equivalent to the distance that the beam moves along the raster in a single step.

$$PES = 100 \ \mu m / M \qquad \textbf{2.11}$$

The ratio of the picture element size to the area under the beam supplying information to a spot on the CRT screen is important. The latter area is normally larger than the beam area at the surface, because of scattering of the incident beam electrons after they penetrate the surface. It should also be pointed out that at the surface, the electron beam covers a finite dimension; it is not a point of zero diameter. Using extreme care and special equipment, such as a scanning microscope in which the specimen is placed within the objective lens, a probe diameter of the order of 2 nm may be obtained. However, in the conventional SEM the smallest beam diameter is about 5 nm, but the instrument is more often used with a beam diameter of 10 nm. For the sake of the present argument it will be assumed that the effective area supplying information has a diameter of 12.5 nm. If this area is smaller than the picture element, the image of the specimen surface on the CRT screen should be in sharp focus. With the aid of Eq. 2.11 it can be deduced that a magnification of 8000× will yield a picture element with a 12.5 nm diameter. All magnifications greater than 8000× will of course produce a picture element smaller than the area supplying information to the spot on the CRT screen. Thus, there will be an overlapping of the information sent to the spots on the screen. In brief, for magnifications greater than 8000×, the images will not contain more information than there is in the 8000× image. On the other hand, all magnifications less than 8000× should yield images in sharp focus.

2.14 The Depth of Focus

The very large depth of focus characteristic of the SEM can be easily understood with the aid of the picture element size concept. The incident electron beam, which scans the specimen surface, normally has a very small angle of divergence. This angle may be estimated using Fig. 2.28. In this sketch, the distance between the final aperture and the plane of optimum focus is designated by WD, which signifies the working distance of the microscope. If the radius, R, of the circular opening of the aperture is divided by the working distance WD, the beam divergence, α, which is the incident beam semicone angle, is obtained. Thus, $\alpha = R/WD$. In a typical case, R could be 100 μm and WD 10 mm, making α equal 0.01 rad.

Scanning electron microscope images with a very large depth of field are obtained when the area supplying information, which we will assume equals the beam size, is significantly smaller than the picture element size at the working distance, WD, of the microscope. Thus, because, in general, the divergence of the beam is small, there is a region below and a region above the plane of optimum focus in which the beam size is smaller than the picture element size. Objects will be in sharp focus within this space. This is illustrated in Fig. 2.29, where two parallel vertical lines are drawn on a view of the beam cross-section to represent the PES diameter. In this figure the depth of focus equals the vertical distance, D, or the total distance above and below the plane of optimum focus within which the beam diameter is smaller than the diameter of the picture element. Above and below the region defined by D, the beam diameter is larger than the diameter of the picture element and the focus is no longer sharp.

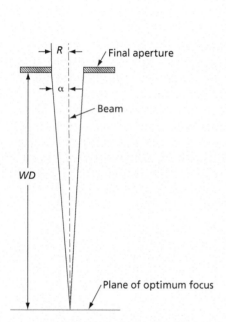

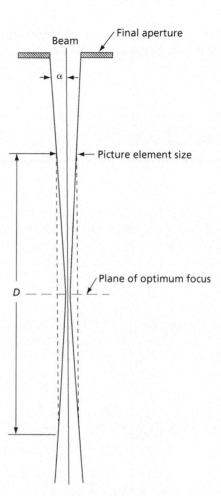

Fig. 2.28 The parameters used to compute the angle of divergence of the electron beam

Fig. 2.29 The important parameters associated with determining the depth of field in an SEM

2.15 Microanalysis of Specimens

The scanning electron microscope can be easily converted into an instrument capable of chemically microanalyzing specimens. Two important microanalyzing techniques will now be considered. These are electron probe X-ray microanalysis and Auger electron spectroscopy.

2.16 Electron Probe X-Ray Microanalysis

The electron probe microanalyzer uses the characteristic peaks of the X-ray spectrum resulting from the bombardment of the specimen by the beam electrons. The wavelengths and strengths of these peaks can yield valuable information about the chemical composition of the specimen.

An electron probe microanalyzer is thus basically an SEM equipped with X-ray detectors. Two basic types of detectors are used: 1) Energy Dispersive Spectrometers (EDS), and 2) Wavelength Dispersive Spectrometers (WDS). In the energy-dispersive X-ray spectrometer, a solid-state detector develops a histogram showing the relative frequency of the X-ray photons as a function of their energy. The wavelength-dispersive spectrometer uses X-ray diffraction to separate the X-ray radiation into its component wavelengths. Time does not permit further discussion of these devices; detailed descriptions can be found in books on microanalysis (Ref. 5).

2.17 The Characteristic X-Rays

The nature of the charactistic X-ray lines may be seen by examining the X-ray spectrum of a typical metallic element such as molybdenum. Normally, the most important characteristic lines of an element are the K_α and K_β lines. If the surface of a molybdenum specimen is bombarded by electrons accelerated through a potential of 20,000 V, the X-rays emitted form only the continuous spectrum, within the wavelength limits, of Fig. 2.30A, where the X-ray intensity is plotted against the wavelength. Note the sharp cutoff at the short wavelength side of the spectrum as well as the maximum in the X-ray intensity close to the short wavelength limit. Increasing the applied potential to 25,000 V causes several things to happen, as is shown in Fig. 2.30B. The short wavelength

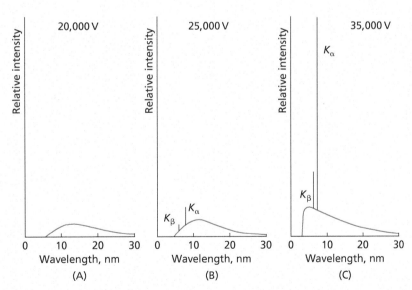

Fig. 2.30   The effect of the accelerating potential, applied to the electrons in the electron beam, on the X-ray spectrum of a molybdenum specimen. **(A)** When the electrons are accelerated by a potential of 20,000 volts, only a continuous X-ray spectrum is obtained. **(B)** When the potential is raised to 25,000 volts, two small characteristic peaks are superimposed upon the continuous spectrum. **(C)** Further increasing the potential applied to the electrons to 35,000 volts greatly increases the magnitudes of the characteristic lines

limit and the maximum move to shorter wavelengths, the X-ray intensity increases at all wavelengths, and two sharp peaks now appear superimposed on the continuous X-ray spectrum. These correspond to the K_α and K_β characteristic X-ray lines. Finally, increasing the accelerating voltage applied to the electrons to 35,000 V, as in Fig. 2.30C, again increases the intensity at all wavelengths, including those of the K_α and K_β peaks, which now become much more pronounced. Note that since the K_α and K_β wavelengths are fixed, these lines now lie to the right of the maximum in the continuous part of the spectrum instead of to the left of it as in Fig. 2.30B.

The continuous part of the X-ray spectrum in Fig. 2.30 is a result of interactions between incident beam electrons and atomic ions of the specimen in which electrons are decelerated as they pass through the coulomb force fields of the ions. The energy losses of the electrons in these inelastic collisions are converted into the energies of the X-ray photons. Since these collisions can occur in a multitude of different ways, the result is the continuous X-ray band or bremsstrahlung radiation. On the other hand, the characteristic lines are formed by collisions between beam electrons and the atoms of the sample in which inner shell electrons are ejected from their atoms. When one of these events occurs, an electron from one of the atom's outer shells falls almost instantly into the inner shell hole, accompanied by the emission of a characteristic X-ray photon or an Auger electron. In the present case, it is assumed that the emissions involve X-ray photons associated with the K_α line. This line actually consists of two lines of only slightly different wavelengths known as the K_{α_1} and the K_{α_2} lines. In Fig. 2.31A, the K and L shell energies of the molybdenum atom are plotted

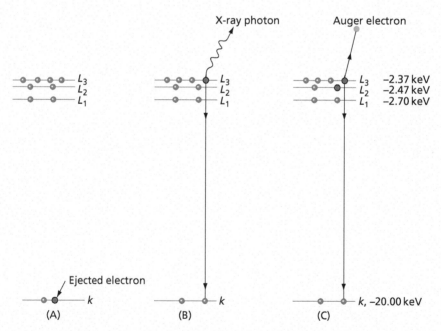

Fig. 2.31 Illustration of how K_α radiation and Auger electrons are formed.
(A) The ejection of an electron from the K shell is the first step.
(B) If an atom from the L shell falls into the hole in the K shell, it is possible for a K_α photon to be released. (C) Alternatively, the drop of an L shell electron into the K shell hole may result in the ejection of an Auger electron from the L shell

as horizontal lines. While the molybdenum atom also contains electrons in the *M* and *N* shells, the energy levels of these shells are not shown so as to simplify the presentation. In Fig. 2.31A, it is assumed that an inelastic collision between an electron in the beam and a molybdenum atom has resulted in the removal of an electron from the *K* level of the atom and then, as shown in Fig. 2.31B, the hole in the *K* shell is filled by an electron from the L_3 shell. This jump of the L_3 electron produces an X-ray photon whose energy equals

$$h\nu = E_K - E_{L_3} \qquad 2.12$$

where h is Planck's constant, ν is the frequency of the X-ray photon, and E_K and E_L are the ionization energies of the *K* and L_3 levels. Substituting for ν the quotient λ/c, where λ is the X-ray wavelength and c is the velocity of light, and rearranging Eq. 2.12 yields

$$\lambda_{K\alpha_1} = 1.2398/(E_K - E_{L_3}) \qquad 2.13$$

where λ is in nm and E_K and E_{L_3} are expressed in keV. In molybdenum, E_K is 19.9995 keV and E_{L_3} is 2.6251 keV so that $\lambda K_{\alpha_1} = 0.07093$ nm. The K_{α_2} line, which conforms to electron jumps between the L_2 and *K* energy levels, has a wavelength of 0.07136 nm. Since the K_{α_1} line is normally about twice as strong as the K_{α_2} line, a weighted average can be used to estimate the wavelength of an unresolved K_α line. This gives $\lambda_{k_\alpha} = 0.07107$ nm. For the purposes of making chemical analyses with the electron probe, it is normally not necessary to resolve the K_α line into its two components; the wavelength (or energy) corresponding to the K_α doublet is generally used. It is important to note that jumps from the L_1 energy level to the *K* level do not occur.

An important consideration in the use of the electron probe X-ray microanalyzer is that, while the electron beam striking the specimen surface may have a diameter as small as 10 nm, the area of the specimen surface from which the X-rays are emitted usually has a diameter of the order of 1 μm or 100 times larger than that of the beam. Furthermore, the X-rays may also evolve from depths below the surface of $\simeq 1\mu$m. In effect, this means that with the Electron Microprobe Analysis (EMPA) it is possible to chemically map a metal specimen over dimensions of the order of mm with a spatial resolution of about 1 μm and a detection limit of about 100 ppm. This technique cannot be applied to analyze elements below Boron (atomic number 5) and is only suitable for qualitative analyses for elements between Boron and Neon (atomic number 10). However, quantitative measurements can be made for Na (atomic number 11) and elements with higher atomic numbers. The electron microprobe is a useful instrument for the identification of the various phases in a metal specimen, including the nonmetallic inclusions found in almost all commercial metals. Another area where this instrument has proven valuable is in diffusion studies where information about composition gradients is required. It can also be used to prove whether or not a metal alloy has a homogeneous composition.

2.18 Auger[*] Electron Spectroscopy (AES)

In the previous section, it was assumed that after a beam electron knocks an electron out of an atom's inner shell (the *K* shell), an electron from an outer shell (an *L* shell) jumps into the inner shell hole with the simultaneous release of an X-ray photon. An alternative possibility is for

---

[*]Pronounced as (Ozha) after Pierre V. Auger, a French physicist.

the jump of the outer shell electron to be accompanied by the ejection of a second outer shell electron (an Auger electron) as illustrated in Fig. 2.31C. Thus, when an electron from an outer shell drops into a hole in an inner shell, one or the other of these two events (X-ray photon or Auger electron) occurs. The release of an Auger electron is preferred for elements with an atomic number less than 15 and for the study of low-energy transitions ($\Delta E \lesssim 2000$ eV) in elements of higher atomic number. With rising atomic number or $\Delta E \gtrsim 2000$ eV, the release of an X-ray photon becomes more and more probable. Consequently, although the use of Auger electron spectroscopy is favored for elements of low atomic number, it is useful for all elements, even through gold and higher atomic numbers.

A significant feature of the Auger reaction is that it involves three electrons: the electron knocked out of an inner shell, the outer-shell electron that jumps into the inner-shell hole, and the ejected outer-shell electron. It is common practice to describe a particular Auger reaction with a three-letter symbol identifying the three basic types of electron energy levels involved in the transition. The present example would be represented by the notation *KLL*.

When an Auger electron is ejected from an atom, it leaves with a fixed amount of kinetic energy. A large part of this energy can easily be lost if the Auger electron comes from an atom situated at any sizeable distance below the surface. However, if the atom lies very close to the specimen surface (about 2–3 nm), the electron may be able to leave the surface still possessing its characteristic energy. With a suitable detector, this energy may be measured. The energy of the Auger electron that the detector sees may be computed with the aid of the following equation:

$$E_{ke} = E_K - 2E_L - \phi \qquad 2.14$$

where E_{ke} is the kinetic energy of the Auger electron as seen by the detector, E_K is the ionization energy for a K level electron, E_L is the ionization energy of the L electrons, and ϕ is the work function of the detector (the work required for an electron to penetrate an electrostatic or electromagnetic deflection analyzer). For example, if an aluminum specimen is used where $E_K = 1559$ eV, $E_L = 73$ eV, and it is assumed that $\phi = 5$ eV, $E_{ke} = 1408$ eV.

One cannot perform an Auger analysis for an element having less than three electrons because three electrons are needed for an Auger transition. This eliminates hydrogen and helium from consideration. An Auger analysis is normally based on measurements of the strengths of the Auger peaks in a plot of ξ, the backscattered electron energy per unit energy interval, versus E, the energy of the electrons. This energy flux is simply equal to the electron energy times the number of electrons per unit energy interval; that is, $\xi = E \times f$. Such a spectrum is depicted for a pure silver specimen bombarded by 1000 V electrons in the lower curve of Fig. 2.32. Note that it is very difficult to resolve the Auger peaks on this curve because they are small and superimposed on a strong background signal due to backscattered electrons. If ξ is multiplied by 10 and the results replotted, the intermediate curve in this figure is obtained. The Auger peaks are now more pronounced, but still difficult to resolve. This problem can be solved by plotting the derivative of ξ with respect of E as in the uppermost curve of Fig. 2.32. The resulting $d\xi/dE$ curve clearly shows evidence of several peaks that fall within the range of energies between approximately 240 and 360 eV. These peaks are associated with *MNN* Auger transitions. The *KLL* and *LMM* peaks do not appear because the 1,000 eV beam electrons are not energetic enough to remove electrons from the K and L shells of silver, where the ionization potentials are approximately 25,000 and 3500 V, respectively.

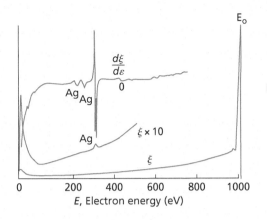

Fig. 2.32 Auger electron spectra of silver with an incident beam energy of 1 keV. Derivative and integral spectra are compared (After N. C. MacDonald)

In summary, Auger electron spectroscopy is useful for determining the compositions of surface layers to a depth of about 2 nm for elements above He. It also has a spatial resolution ≥ 100 nm, which is about a tenth of that of the electron probe X-ray microanalyzer. This makes this technique well suited to the studies of grain boundaries in metals and alloys, especially with specimens susceptible to brittle grain boundary fractures. It is also useful for surface segregation studies as in the solving of stress-corrosion problems.

2.19 The Scanning Transmission Electron Microscope (STEM)

If the beam in the transmission microscope is focused to form a probe, as in the SEM, and then scanned across the surface of a TEM thin foil specimen, it is possible to make local, small-area analyses of the material in the foil. Alternatively, small-area diffraction studies may be made. Accordingly, chemical analyses may be made of small particles or inclusions observed in the specimen. This technique also makes use of the fact that the volume from the signal that is sent to the detector is very small. In other words, the STEM is an instrument with a very high resolution. This is because the foil specimen is normally so thin that there is not much chance for the electrons to scatter out from the center of the beam in the specimen. Thus, in the electron probe X-ray microanalyzer, the signal comes from a bulbous volume roughly 1 μm deep by 1 μm in diameter. In the STEM, the corresponding volume is roughly the thickness of the foil in depth and 2 to 3 nm in diameter when the specimen is mounted in the lens, and 5–30 nm in diameter when it is placed in the conventional position below the lens.

Tremendous progress has been made in both TEM and STEM techniques, in their resolutions and analysis since its invention in the middle of the 20th century. This progress is attributed mostly to improvements in the microscope hardware, incident energy sources, detectors, as well as software and analysis capabilities. In recent years, additional improvements have been made in image resolution by incorporating what is referred to as aberration correction. A good review of these advances can be found in a review article by Mark P. Oxley et al (Ref. 10). Figure 2.33 shows that the current resolution utilizing these methods is on the order of 0.05 nm (0.5 Angstrom).

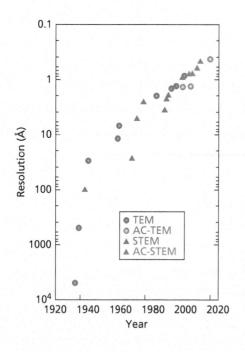

Fig. 2.33 Progress in TEM and STEM resolution—note the accelerated progress in the past two decades with the introduction of aberration correction. (Reprinted with permission Figure 1 of "Ultra-high resolution electron microscopy, "Mark P. Oxley, Andrew R. Lupini and Stephen J. Pennycook, *Rep. Prog. Phys.* 80 (2017) 026101. doi: 10.1088/1361-6633/80/2/026101

Problems

2.1 If a Bragg of $\theta = 41.31°$ angle is observed for first-order diffraction from the {110} plane of body centered cubic niobium using copper K_{α_1} radiation ($\lambda = 0.1541$ nm), what is the interplanar spacing of the {110} planes in this metal? Compare the calculated spacing with that determined from niobium lattice parameter of $a = 0.3301$ nm. Explain if the results are different.

2.2 With the aid of a sketch similar to that for a simple cubic lattice in Fig. 2.10, demonstrate that the interplanar spacing of the {100} planes in the body-centered cubic lattice also equals $a/\sqrt{2}$.

2.3 Using a geometrical argument similar to that in Prob. 2.2, show that the {100} interplanar spacing in the bcc lattice is $a/2$ and not a.

2.4 Consider the Bragg equation with respect to first-order reflections from {100} planes of a bcc metal. By how much of a wavelength do the reflected pathlengths differ for two adjacent parallel (100) planes if the interplanar spacing, d, is taken as a instead of $a/2$? Does this explain why the {100} plane is not listed as a reflecting bcc plane in Appendix C? Explain.

2.5 With the aid of Eq. 2.2, calculate d_{hkl} for all of the fcc reflecting planes listed in Appendix C using copper with $a = 0.3615$ nm.

2.6 Given a Laue back-reflection camera (see Figs. 2.5 and 2.7) with a 5 cm film-to-camera distance, a 10 cm in diameter circular film, and the (100) plane of a copper crystal specimen normal to the X-ray beam,

(a) What is the maximum angle that some other plane of the copper crystal can make with the (100) plane and still reflect a spot onto the film of the camera?

(b) With the aid of the data listed in Appendix A, now determine which of the planes whose poles are plotted in the 100 standard projection shown in Fig. 1.31 will produce spots on the camera film.

2.7 Determine powder pattern, S, values for the first four reflecting planes, listed in Appendix C, if the specimen is a gold powder. Assume that Cu K_{α_1} radiation ($\lambda = 0.1541$ nm) is used and that the lattice parameter of the gold crystal is 0.4078 nm.

2.8 Assume that the electrons in a transmission electron microscope are accelerated through a potential of 80,000 volts. Determine

(a) The electron velocity given by this potential by assuming that the energy the electrons gain falling through the potential equals the gain in their kinetic energy.

(b) The effective wavelength of the electrons.

(c) The Bragg angle, if the electrons undergo a first-order reflection from a {100} plane of a bcc vanadium crystal. Take the lattice parameter of vanadium as 0.3039 nm. *Note.* See Appendix D for values of the constants needed in the solution of this problem.

2.9 Rotate the 100 cubic standard projection of Fig. 1.30 about its north–south polar axis to obtain a 110 standard projection, and draw a figure similar to Fig. 2.19 showing the poles of the major planes of the zone whose axis is [110]. On the assumption that the electron beam of an electron microscope is parallel to [110], make a sketch, drawn to scale, of the electron diffraction pattern for this case that is similar to the one in Fig. 2.20, where the beam was parallel to [100].

2.10 If the effective area of the raster of an electron microscope specimen supplying information to the CRT of an electron microscope has a 5 nm diameter and the semicone angle, α, of the electron beam is 0.01 rad,

(a) What would be the maximum useable magnification of the microscope if a digital spot on the screen of the CRT has a 100 μm diameter?

(b) What would be the depth of field at this magnification?

(c) What would be the depth of field if this microscope were to be operated at a magnification of 2000×?

References

1. *The Principles of Metallographic Laboratory Practice* (Third Edition), George L. Kehl, McGraw-Hill, 1949.
2. *Elements of X-Ray Crystallography,* L.V. Azaroff, McGraw-Hill, 1968.
3. *X-Ray Multiple-Wave Diffraction*, Shih-Lin Chang, Springer, 2004.
4. *Elements of X-Ray Diffraction* (3rd ed.), B. D. Cullity and S. R. Stock, Addison-Wesley, 2001.
5. "Materials Characterization," *ASM Handbook*, Volume 10, ASM International, 1992 (third printing), pp. 450–453.
6. *Scanning Electron Microcopy and X-Ray Microanalysis*, Joseph I. Goldstein, D. E. Newburry, D. Echlin, D. C. Joy, C. Fiori, and E. Lifshin, Plenum Press, 1981.
7. "TEM Sample Preparation and FIB-Induced Damage," Joachim Mayor, L. A. Giannuzzi, T. Kamino, and J. Michael, *MRS Bulletin*, vol. 32, pp. 400–407, May 2007.
8. *Structure of Metals*, C. S. Barrett and T. B. Massalski, McGraw-Hill, New York, 1980, p. 84.
9. "Low-Energy Secondary-Electron Spectroscopy of Molybdenum," O. F. Panchenko, *Phys. Solid State*, 39 (10), October 1997, pp. 1537–1541.
10. "Ultra-high resolution electron microscopy," Mark P. Oxley, Andrew R. Lupini and Stephen J. Pennycook, *Rep. Prog. Phys.* 80 (2017) 026101.

Chapter 3

Crystal Binding

> **Learning Objectives**
>
> Upon completion of Chapter 3, you will be able to:
> 1. Understand the origins of the internal energy of a crystal
> 2. Explain attractive and repulsive forces in an ionic crystal
> 3. Become familiar with the four classifications of bonding in crystals
> 4. Qualitatively and quantitatively develop understanding of van der Waals bonding in inert gas-solid
> 5. Become familiar with bonding and electron contribution in a material with covalent bonding
> 6. Understand the nature of metallic bonding and electron gas
> 7. Develop general understanding of the influence of bonding and its strength on materials properties, such as melting temperatures, sublimation energy, electrical and optical properties

Crystalline solids are empirically grouped into four classifications: (*a*) ionic, (*b*) van der Waals, (*c*) covalent, and (*d*) metallic. This is not a rigid classification, for many solids are of an intermediate character and not capable of being placed in a specific class. Nevertheless, this grouping is very convenient and greatly used in practice to indicate the general nature of the various solids.

3.1 The Internal Energy of a Crystal

The internal energy of a perfect crystal is considered to be composed of two parts. First, there is the lattice energy U that is defined as the potential energy due to the electrostatic attractions and repulsions that atoms exert on one another. Second, there is the thermal energy of the crystal, associated with the vibrations of the atoms about their equilibrium lattice positions. It consists of the sum of all the individual vibrational energies (kinetic and potential) of the atoms. In order to most conveniently study the nature of the binding forces that hold crystals together (the lattice energy U), it is desirable to eliminate from our considerations, as far as possible, complicating considerations of the thermal energy. This can be done most conveniently by assuming that all cohesive calculations refer to zero degrees absolute. Quantum theory tells us that at this temperature the atoms will be in their lowest vibrational energy states and that the zero-point energy associated with these states is small. For the present, it will be assumed that the zero-point energy can be neglected and that all calculations refer to 0 K.

In setting up quantitative relationships to express the cohesion of solids, it is customary to work with cohesive energies rather than cohesive forces. The energy concept is preferred because it is more conveniently compared with experimental data. Thus, the heat of sublimation and/or the heat of formation of a compound are both related to cohesive energies. In fact, the heat of sublimation at 0 K, which is the energy required to dissociate a mole of a substance into free atoms at absolute zero, is a particularly convenient measure of the cohesive energy of a simple solid such as a metal.

3.2 Ionic Crystals

The sodium chloride crystal serves as a good example of an ionic solid. Figure 3.1 shows the lattice structure of this salt, which is simple cubic with alternating lattice positions occupied by positive and negative ions. This lattice can also be pictured as two interpenetrating face-centered cubic structures made up of positive and negative ions respectively. Figure 3.2 shows another form of ionic lattice, that of cesium chloride. Here each ion of a given sign is surrounded by eight neighbors of the opposite sign. In the sodium chloride lattice, the corresponding coordination number is 6. An example in which this number is 4 is shown in Fig. 3.3 for ZnS as zincblende, also known as sphalerite.*
In general, all three of the above structures are characteristic of two-atom ionic crystals in which

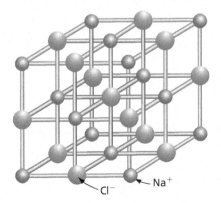

Fig. 3.1 The sodium chloride lattice

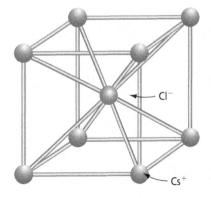

Fig. 3.2 The cesium chloride lattice

---

*ZnS as Wurtzite has an HCP structure.

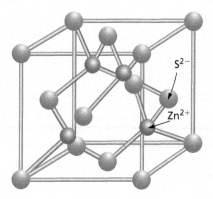

Fig. 3.3 The zincblende lattice, ZnS

the positive and negative ions both carry charges of the same size. This is equivalent to saying that the atoms that make up the crystal have the same valence. Ionic crystals can also be formed from atoms that have different valences, for example, CaF_2 and TiO_2. These, of course, form still different types of crystal lattices (Ref. 1) that will not be discussed.

Ideally, ionic crystals are formed by combining a highly electropositive metallic element with a highly electronegative element such as one of the halogens, oxygen, or sulfur. Certain of these solids have very interesting physical properties that are of considerable importance to the metallurgist. In particular, the study of plastic deformation mechanisms in ionic crystals, for example, LiF, AgCl, and MgO, has greatly added to our understanding of similar processes in metals.

3.3 The Born Theory of Ionic Crystals

The classical theory developed by Born and Madelung gives a simple and rather understandable picture of the nature of the cohesive forces in ionic crystals. It is first assumed that the ions are electrical charges with spherical symmetry and that they interact with each other according to simple central-force laws. In ionic crystals, these interactions take two basic forms, one long range and the other short range. The first is the well-known electrostatic, or coulomb, force that varies inversely with the square of the distance between a pair of ions, or

$$f = \frac{ke_1 e_2}{(r_{12})^2} \qquad 3.1$$

where e_1 and e_2 are the charges on the ions, r_{12} is the center-to-center distance between the ions, and k is a constant. If cgs units are used, $k = 1$ dyne (cm^2)/(statcoulombs)^2, and if mks units are employed, $k = 9 \times 10^9$ N m^2/C^2. The corresponding coulomb potential energy for a pair of ions is

$$\phi = \frac{ke_1 e_2}{r_{12}} \qquad 3.2$$

The other type of interaction is a short-range repulsion that occurs when ions are brought so close together that their outer electron shells begin to overlap. When this happens, very large forces are brought into play that force the ions away from each other. In a typical ionic crystal, such as NaCl, both the positive and negative ions have filled electron shells characteristic of inert gases. Sodium, in giving up an electron, becomes a positive ion with the electron configuration of neon ($1s^2, 2s^2, 2p^6$), while chlorine, in gaining an electron, assumes that of argon ($1s^2, 2s^2, 2p^6, 3s^2, 3p^6$). On a time-average basis, an atom with an inert-gas arrangement of electrons may be considered as a positively charged nucleus surrounded by a spherical volume of negative charge (corresponding to the electrons). Inside the outer limits of this negatively charged region all the available electronic energy states are filled. It is not possible to introduce another electron into this volume without drastically changing the energy of the atom. When two closed-shell ions are brought together so that their electron shells begin to overlap, the energy of the system (the two ions taken together) rises very rapidly, or, as it might otherwise be stated, the atoms begin to repel each other with large forces.

According to the Born theory, the total potential energy of a single ion in an ionic crystal of the NaCl type, due to the presence of all the other ions, may be expressed in the form

$$\phi = \phi_M + \phi_R \qquad 3.3$$

In this expression, ϕ is the total potential energy of the ion, ϕ_M is its energy due to coulomb interactions with all the other ions in the crystal, and ϕ_R is the repulsive energy. If one uses the cgs system of units where, in Eqs. 3.1 and 3.2, $k = 1$ this expression may also be written

$$\phi = -\frac{Az^2e^2}{r} + \frac{Be^2}{r^n} \qquad 3.4$$

Here e is the electron charge, z is the number of electronic charges on the ions, r is the distance between centers of an adjacent pair of negative and positive ions (Fig. 3.4), n is a large exponent, usually of the order of 9, and A and B are constants. If we now think in terms of the potential energy of a crystal containing one mole of NaCl, rather than in terms of a single ion, the preceding equation becomes

$$U = -\frac{ANz^2e^2}{r} + \frac{NBe^2}{r^n} \qquad 3.5$$

where N is Avogadro's number (6.022×10^{23}) and U is the total lattice potential energy. The first term on the right-hand side of this equation represents the electrostatic energy due to simple coulomb forces between ions, whereas the second term is that due to the repulsive interactions that arise when ions closely approach each other. It is a basic assumption of the Born

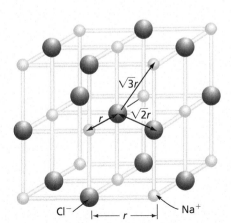

Fig. 3.4 Interionic distances in the sodium chloride lattice

theory that the repulsive energy can be expressed as a simple inverse power of the interionic distance. While quantum theory tells us that a repulsive term of the type Be^2/r^n is not rigorously correct, it is still a reasonable approximation for small variations of r from the equilibrium separation between atoms r_0.

We shall presently consider in more detail the individual terms of the Born equation, but before doing this let us look at the variation of the lattice energy with respect to the interionic distance r. This can be done conveniently by plotting each of the two terms on the right of the equation separately. The cohesive energy U is then obtained as a function of r by summing the curves of the individual terms. This is done in Fig. 3.5 for an assumed value of 9 for the exponent n. Note that the repulsive term, because of the large value of the exponent, determines the shape of the total energy curve at small distances, whereas the coulomb energy, with its smaller dependence on r, is the controlling factor at large values of r. The important factor in this addition is that the cohesive energy shows a minimum, U_0, at the interionic distance r_0, where r_0 is the equilibrium separation between ions at 0 K. If the separation between ions is either increased or decreased from r_0, the total energy of the crystal rises. Corresponding to this increase in energy is the development of restoring forces acting to return the ions to their equilibrium separation r_0.

Let us now consider the coulomb energy term of the Born equation, which for a single ion is

$$\phi_M = -\frac{z^2 e^2 A}{r} \qquad 3.6$$

or

$$\phi_M = -\frac{e^2 A}{r} \qquad 3.7$$

assuming that we are specifically interested in a sodium chloride crystal where there is a unit charge on each ion and $z^2 = 1$. Because the coulomb energy varies inversely as the first power

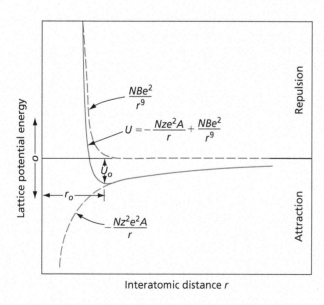

Fig. 3.5 Variation of the lattice energy of an ionic crystal with the spacing between ions

of the distance between charged ions, coulomb interactions act over large distances, and it is not sufficient to consider only the coulomb energy between a given ion and its immediate neighbors. That this is true may be seen in the following. Around each negative chlorine ion there are 6 positive sodium atoms at a distance of r. This may be confirmed by studying Fig. 3.4. There is an attractive energy between each of the six sodium ions and the chlorine ions equal to $-e^2/r$, or in total $-6e^2/r$. The next closest ions to a given chlorine ion are 12 other negatively charged chlorine ions at a distance $\sqrt{2}r$. The interaction energy between these ions and the given ion is accordingly $12e^2/\sqrt{2}r$. Following this there are 8 sodium ions at a distance of $\sqrt{3}r$, 6 chlorine ions at $\sqrt{4}r$, 24 sodium atoms at $\sqrt{5}r$, etc. It is evident, therefore, that the coulomb energy of a single ion equals a series of terms of the form

$$\phi_M = -\frac{6e^2}{\sqrt{1}r} + \frac{12e^2}{\sqrt{2}r} - \frac{8e^2}{\sqrt{3}r} + \frac{6e^2}{\sqrt{4}r} - \frac{24e^2}{\sqrt{5}r} \ldots \qquad 3.8$$

or

$$\phi_M = -\frac{Ae^2}{r} = -\frac{e^2}{r}[6 - 8.45 + 4.62 - 3.00 + 10.7\ldots] \qquad 3.9$$

The constant A of the coulomb energy, of course, equals the sum of the terms inside the brackets in the preceding expression. This series, as expressed earlier, does not converge because the terms do not decrease in size as the distance between ions is made larger. There are other mathematical methods of summing the ionic interactions (Refs. 2 and 3) and it is quite possible to evaluate the constant A, which is called the *Madelung number*. For sodium chloride, the Madelung number is 1.7476 and the coulomb, or Madelung energy, for one ion in the crystal is, accordingly,

$$\phi_M = -\frac{1.7476 e^2}{r} \qquad 3.10$$

For a single ion, the repulsive energy term is

$$\phi_R = \frac{Be^2}{r^n} \qquad 3.11$$

In this term the two quantities, B and n, must be evaluated. This can be accomplished with the aid of two experimentally determined quantities: r_0, the equilibrium interionic separation at 0 K; and K_0, the compressibility of the solid at 0 K. At r_0 the net force on an ion due to the other ions is zero so that the first derivative of the total potential energy with respect to the distance, which equals the force on the ion, is also zero or

$$\left(\frac{d\phi}{dr}\right)_{r=r_0} = \frac{d}{dr}\left(-\frac{Ae^2}{r} + \frac{Be^2}{r^n}\right) = 0 \qquad 3.12$$

Since the quantity A is already known, the preceding expression produces an equation relating n and B. A second equation is obtained from the fact that the compressibility is a function of the second derivative of the cohesive energy $(d^2\phi/dr^2)_{r=r_0}$ at $r = r_0$. In making these computations, the equilibrium separation of ions r_0 can be experimentally obtained from X-ray diffraction measurements of the lattice constant extrapolated to 0 K. In the NaCl crystal, this quantity equals 0.282 nm.

The compressibility is defined by the expression

$$K_0 = \frac{1}{V}\left(\frac{\partial V}{\partial p}\right)_T \qquad 3.13$$

where K_0 is the compressibility, V is the volume of the crystal, and $(\partial V/\partial p)_T$ is the rate of change of the volume of the crystal with respect to pressure at constant temperature. The compressibility is thus the relative rate of change of volume with pressure at constant temperature, and is a quantity capable of experimental evaluation and extrapolation to 0 K.

When the calculations outlined above are made, (Ref. 4) it is found that the Born exponent for the sodium chloride lattice is 8.0. In terms of this exponent, the computed cohesive energy is 180.4 K cal (7.56×10^5 J) per mole. The latter is actually the energy of formation of a mole of solid NaCl from a mole of Na^+ ions, in the vapor form, and a mole of gaseous Cl^- ions. It is not possible experimentally to measure this quantity directly, but it can be evaluated from measured values of the heat of formation of NaCl from metallic sodium and gaseous Cl_2 in combination with measured values of the energy to sublime sodium, the energy to ionize sodium, the energy to dissociate molecular chlorine to atomic chlorine, and the energy to ionize chlorine. When all of these values are considered, the experimental value of U for the NaCl lattice turns out to be 188 K cal (7.88×10^5 J) per mole.

The rather good correspondence between the measured value of the cohesive energy for NaCl and the value computed with the Born equation shows that the latter gives a good first approximation of the cohesive energy for a typical ionic solid.

3.4 Van Der Waals Crystals

In the final analysis, the cohesion of an ionic crystal is the result of its being composed of ions: atoms carrying electrical charges. In Figure 1.1, sodium, with one 3s electron, can give up that electron to become positively charged cation. Chlorine, on the other hand, can accept that electron to become anion with six 3p electrons in its outer shell. In forming the crystal, the atoms arrange themselves in such a way that the attractive energies between ions with unlike charges is greater than all of the repulsive energies between ions with charges of the same signs. We shall now consider another type of bonding that makes possible the formation of crystals from atoms or even molecules that are electrically neutral and possess electron configurations characteristic of inert gases. The forces that hold this type of solid together are usually quite small and of short range. They are called *van der Waals forces* and arise from nonsymmetrical charge distributions. The most important component of these forces can be ascribed to the interactions of electrical dipoles.

3.5 Dipoles

An electrical dipole is formed by a pair of oppositely charged particles ($+e_1$ and $-e_1$) separated by a small distance. Let us call this distance a. Because the charges are not concentric, they produce an electrostatic field that is capable of exerting a force on other electrical charges. Thus, in Fig. 3.6,

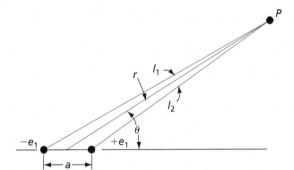

Fig. 3.6 An electrical dipole

let l_1 and l_2 be the respective distances from the two charges of the dipole to a point in space at a distance r from the midpoint of the dipole. If r is large compared to a, then the potential at the point p, using cgs units, is given by

$$V = \frac{e_1}{l_1} - \frac{e_2}{l_2} = \frac{e_1}{r - (a/2)\cos\theta} - \frac{e_1}{r + (a/2)\cos\theta} \qquad 3.14$$

or

$$V = \frac{e_1}{r}\left(\frac{1 + (a/2r)\cos\theta - 1 + (a/2r)\cos\theta}{1 - [(a/2r)\cos\theta]^2}\right) \qquad 3.15$$

which gives us

$$V = \frac{e_1}{r}\left(\frac{(a/r)\cos\theta}{1 - [(a/2r)\cos\theta]^2}\right) \qquad 3.16$$

and since $(a/2r) < 1$, Eq. 3.16 simplifies to

$$V = \frac{e_1 a \cos\theta}{r^2} \qquad 3.17$$

The components of the electric field intensity in the radial and transverse directions are then given by

$$\mathbf{E}_r = -\frac{\partial V}{\partial r} = \frac{2e_1 a \cos\theta}{r^3}$$

$$\mathbf{E}_\theta = -\frac{\partial V}{r\partial\theta} = \frac{e_1 a \sin\theta}{r^3} \qquad 3.18$$

In these latter equations, r is the distance from the dipole to the point p, and θ is the angle between the axis of the dipole and the direction r. It is customary to call the quantity $e_1 a$, the product of one of the dipole charges and the distance between the dipole charges, the dipole moment. We will designate it by the symbol μ. The components of the force, which would act on a charge of magnitude e if placed at point p, are accordingly

$$F_r = \frac{2\mu e \cos\theta}{r^3}, \quad F_\theta = \frac{\mu e \sin\theta}{r^3} \qquad 3.19$$

or using meter-kilogram-second (mks) units with $k = 9 \times 10^9$ N m^2/C^2

$$F_r = \frac{2k\mu e \cos\theta}{r^3}, \quad F_\theta = \frac{k\mu e \sin\theta}{r^3} \qquad 3.20$$

Note: In the centimeter-gram-second (cgs) system of units, $k = 1$ dyne (cm^2)/(statcoulomb)^2.

Also note that the electric field intensity due to a dipole varies as the inverse cube of the distance from the dipole, whereas the field of a single charge varies as the inverse square of the distance.

3.6 Inert Cases

Let us turn to a consideration of the inert-gas atoms such as neon or argon. The solids formed by these elements serve as the prototype for the van der Waals crystals, just as crystals of the alkali halides (NaCl, etc.) are the prototype for ionic solids. It is interesting that they crystallize (at low temperatures) in the face-centered cubic system. In these atoms, as in all others, there is a positively charged nucleus surrounded by electrons traveling in orbits. Because of their closed-shell structures, we can consider that over a period of time the negative charges of the electrons are distributed about the nucleus with complete spherical symmetry. The center of gravity of the negative charge on a time-average basis therefore coincides with the center of the positive charge on the nucleus, which means that the inert gas atoms have no average dipole moment. They do, however, have an instantaneous dipole moment because their electrons, in moving around the nuclei, do not have centers of gravity that instantaneously coincide with the nuclei.

3.7 Induced Dipoles

When an atom is placed in an external electrical field, its electrons are, in general, displaced from their normal positions relative to the nucleus. This charge redistribution may be considered equivalent to the formation of a dipole inside the atom. Within limits, the size of the induced dipole is proportional to the applied field, so that we write

$$\mu_I = \alpha \mathbf{E} \qquad 3.21$$

where μ_I is the induced dipole moment, $\mathbf{E}$ is the electrical field intensity, and α is a constant known as the *polarizability*.

When two inert-gas atoms are brought close together, the instantaneous dipole in one atom (due to its electron movements) is able to induce a dipole in the other. This mutual interaction

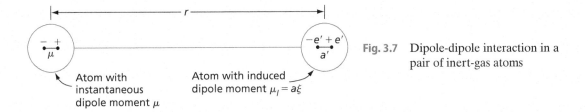

Fig. 3.7 Dipole-dipole interaction in a pair of inert-gas atoms

between the atoms results in a net attractive force between the atoms. Figure 3.7 represents two inert-gas atoms of the same kind (perhaps argon atoms) separated by the distance r. Now let it be assumed that the atom on the left possesses an instantaneous dipole moment μ due to the movement of the electrons around the nucleus. This moment will produce a field $\mathbf{E}$ at the position of the second atom, which, in turn, induces a dipole moment in the latter as given by Eq. 3.21. $\mathbf{E}$ in this equation is the field at the right-hand atom due to the dipole moment in the left-hand atom.

First, consider the force exerted by the left dipole on the right dipole, which may be evaluated as follows. Let us assume, as indicated in Fig. 3.7, that the induced dipole in the right-hand atom is equivalent to the pair of charges $-e'$ and $+e'$ separated by the distance a'. According to this, the induced dipole moment is $e'a'$. Now let $\mathbf{E}$ be the field intensity due to the instantaneous dipole on the left atom at the negative charge $(-e')$ of the induced dipole. The corresponding field at the position of the positive charge $(+e')$ of the induced dipole is $\mathbf{E} + d\mathbf{E}/dr \cdot a'$. The total force on the induced dipole due to the field of the other dipole is

$$f = -e'\mathbf{E} + e'\left(\mathbf{E} + \frac{d\mathbf{E}}{dr}a'\right) = e'a'\frac{d\mathbf{E}}{dr} = \mu_I \frac{d\mathbf{E}}{dr} \qquad 3.22$$

However, substituting for μ_I from Eq. 3.21 gives

$$f = \alpha \, \mathbf{E}\frac{d\mathbf{E}}{dr} \qquad 3.23$$

but in general the field of a dipole is proportional to the inverse cube of the distance, or

$$\mathbf{E} \simeq \frac{\mu}{r^3} \qquad 3.24$$

This leads us to the result

$$f \simeq \alpha \frac{\mu}{r^3} \frac{d}{dr} \frac{\mu}{r^3} \simeq \alpha \frac{\mu^2}{r^7} \qquad 3.25$$

The energy of a pair of inert-gas atoms due to dipole interactions can now be evaluated as follows:

$$\phi \simeq \int_\infty^r \alpha \frac{\mu^2}{r^7} dr \simeq \alpha \frac{\mu^2}{r^6} \qquad 3.26$$

It can be seen, therefore, that the van der Waals energy between a pair of inert-gas atoms due to dipole interactions varies as the square of the dipole moment and the inverse sixth power of their distance of separation; that the force varies as the square of the dipole moment is significant. Over a period of time the average dipole moment for an inert-gas atom must be zero. The square of this quantity does not equal zero, and it is on this basis that inert-gas atoms can interact.

3.8 The Lattice Energy of an Inert-Gas Solid

When the atoms of a rare-gas solid have their equilibrium separation, the van der Waals attraction is countered by a repulsive force. The latter is of the same nature as that which occurs in ionic crystals and is due to the interaction that occurs when closed shells of electrons start to overlap. The cohesive energy of an inert-gas solid may therefore be expressed in the form

$$U = -\frac{A}{r^6} + \frac{B}{r^n}$$ 3.27

where A, B, and n are constants. It was shown, (Ref. 5) over 80 years ago, that if n equals 12, the preceding equation correlates well with the observed properties of rare-gas solids. The first term on the right-hand side represents the total energy for one mole of the crystal caused by the dipole-dipole interactions between all the atoms of the solid. It may be obtained by first computing the energy of a single atom caused by its interactions with its neighbors. This quantity is then summed over all the atoms of the crystal. The computations are lengthy and will not be discussed. The second term in the preceding equation is the molar repulsive energy.

As might be surmised from the second-order nature of the van der Waals interaction, the cohesive energies of the inert-gas solids are quite small, being of the order of $\frac{1}{100}$ of those of the ionic crystals. The rare gases have also very low melting and boiling points, which is to be expected because of their small cohesive energies. Table 3.1 gives these properties for the inert-gas elements, with the exception of helium. Additional property values can be found in Reference 7.

3.9 The Debye Frequency

The zero-point energy of a crystal is its thermal energy when the atoms are vibrating in their lowest energy states. When theoretical and experimental cohesive energies are compared at an assumed temperature of 0 K, this energy, which has previously been neglected, should be included along with other terms. In a crystaline solid there are three vibrational degrees of

Table 3.1 Experimental Cohesive Energies, (Ref. 6) Melting Points, and Boiling Points of Inert Gas Elements.

Element	Cohesive Energy Kcal/mol	KJ/mol	Melting Point °C	Boiling Point °C
Ne	0.450	1.83	−248.6	−246.0
Ar	1.850	7.74	−189.4	−185.8
Kr	2.590	10.84	−157	−152
Xe	3.830	16.03	−112	−108

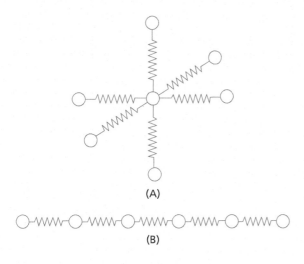

Fig. 3.8 **(A)** Debye model of a simple cubic crystal pictures an atom as a mass joined to its neighbors by springs. **(B)** A one-dimensional crystal model

freedom per atom. This means that an atom is free to vibrate independently in three orthogonal directions. Thus, if one considers an atom to lie at the center of an orthogonal coordinate system, it should be able to vibrate parallel to the x axis without inducing vibrations along the y or z axes. Similarly, it should be able to vibrate independently along either the y or z axes. To each of the three degrees a mode of vibration can be assigned so that there are three modes per atom. A crystal of N atoms is considered as equivalent to $3N$ oscillators of various frequencies v.

The basic reasoning that brought Debye to these conclusions is as follows. First, he assumed that the forces of interaction between a neighboring pair of atoms were roughly equivalent to a linear spring. Pushing the atoms together would have the effect of compressing the spring, and in so doing, a restoring force would be developed that would act to return the atoms to their rest positions. Pulling the atoms apart would produce an equivalent opposite result. On this basis, Debye concluded that the entire lattice might be considered to be a three-dimensional array of masses interconnected by springs. In effect, assuming a simple cubic crystal, each atom would be held in space by a set of three pairs of springs, as indicated in Fig. 3.8A. He next considered how such an array might be able to vibrate. To simplify the presentation, a one-dimensional crystal will be considered, as indicated in Fig. 3.8B, and following Debye, the existence of longitudinal lattice vibrations will be ignored because they are of less significance. The vibration modes of such an array are analogous to the standing waves that can be set up in a string. In a simple string, the number of possible harmonics is theoretically infinite, and there is no lower limit for the wavelengths that might be obtained. According to Debye, this is not true when a series of masses connected by springs are caused to vibrate. Here, as shown in Fig. 3.9, the minimum

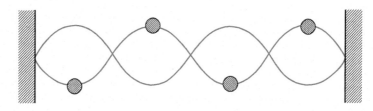

Fig. 3.9 The highest frequency vibration mode for an array of four masses

wavelength or the mode of maximum frequency is obtained when neighboring atoms vibrate against each other. As may be seen in the drawing, the minimum wavelength corresponds to twice the spacing between the atoms, or $\lambda_{min} = 2a$, where a is the interatomic spacing. The (maximum) vibrational frequency associated with this wavelength is

$$v_m = \frac{v}{\lambda} \quad \quad 3.28$$

where v is the velocity of the shortest sound waves. This latter is normally of the order of 5×10^3 m/s. At the same time, the interatomic spacing in metals is about 0.25 nm, so that

$$v_m = \frac{5 \times 10^3}{2(0.25 \times 10^{-9})} = 10^{13} \text{ Hz} \quad \quad 3.29$$

The value, $v_m = 10^{13}$ Hz, is often used in simple calculations to represent the vibration frequency of an atom in a crystal. Since these calculations ordinarily are accurate to only about an order of magnitude (factor of 10), the use of the maximum vibration frequency for the average vibration frequency does not cause serious problems.

3.10 The Zero-Point Energy

In standing waves, the order of the harmonic corresponds to the number of half wavelengths in the standing wave pattern. Observe that in Fig. 3.9, where there are four atoms, there are four half wavelengths, and this system of four atoms will be able to vibrate in only four modes. In the case of a linear array of N_x atoms, there will be N_x half wavelengths when the array vibrates at its maximum frequency. Since this latter frequency corresponds to the N_xth harmonic, the system will have N_x transverse modes of vibration in the vertical plane, which is the assumed plane of vibration in the drawing.

In a three-dimensional crystal of N atoms, each atom inside the crystal can undergo transverse vibrations in three independent directions, as can be deduced by examining Fig. 3.8A; and by reasoning similar to that above, it is possible to show that there are $3N$ independent transverse modes of vibrations.

In a linear crystal, such as that implied in Fig. 3.8B, the density of vibrational modes is the same in any frequency interval dv. However, in a three-dimensional array or crystal the vibrational modes are three-dimensional, and the multiplicity of the standing wave patterns increases with increasing frequency. As a result, in the three-dimensional case the number of modes possessing frequencies in the range v to $v + dv$ is given by

$$f(v)dv = \frac{9N}{v_m^3} v^2 \, dv \quad \quad 3.30$$

where $f(v)$ is a density function, N is the number of atoms in the crystal, v is the vibrational frequency of an oscillator, and v_m is the maximum vibrational frequency.

Figure 3.10 is a schematic plot of the Debye density function $f(v)$ as a function of v. The area under this curve from $v = 0$ to $v = v_m$ equals $3N$, the total number of oscillators. According to the quantum theory, the zero-point energy of a simple oscillator is $hv/2$. The total vibrational energy of the crystal at absolute zero is, accordingly,

$$E_z = \int_0^{v_m} f(v) \frac{hv}{2} dv = \frac{9}{8} Nhv_m \quad \quad 3.31$$

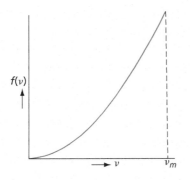

Fig. 3.10 Frequency spectrum of a crystal according to Debye. The maximum lattice frequency is v_m

This quantity should normally be added to experimentally determined cohesive energies (Table 3.1) in comparing them with computed static lattice energies. The correction due to the zero-point energy is about 31 percent or about 0.59 KJ per mol in the case of neon (Ref. 6) so that the lattice energy U_0 should be about 2.47 KJ per mol rather than 1.88 KJ per mol, as shown in Table 3.1. The importance of this correction decreases as the atomic number of the rare-gas element rises, so that for Xe it amounts to about 3 percent.

3.11 Dipole-Quadrupole and Quadrupole-Quadrupole Terms

The van der Waals attractive energy is caused by a synchronization of the motion of the electrons on the various atoms of a solid. As a first approximation it might be considered that this interaction is equivalent to the development of synchronized dipoles on the atoms. The summation of the dipole interactions throughout the crystal then leads to an attractive energy that varies as the inverse sixth power of the distance. Actually, the complex charge distributions that exist in real atoms cannot be accurately represented by picturing them as simple dipoles. Modern quantum mechanical treatments generally use an expression for the van der Waals attractive energy (expressed in terms of a single ion) of the type

$$\phi_{(r)} = -\left(\frac{c_1}{r^6} + \frac{c_2}{r^8} + \frac{c_3}{r^{10}}\right) \qquad 3.32$$

where $c_1, c_2,$ and c_3 are constants. The first term of this expression is the dipole-dipole interaction already considered. The second term in the inverse eighth power of the distance is called the *dipole-quadrupole term* because the interaction between a dipole on one atom and a quadrupole on another will lead to an energy that varies as the inverse eighth power of the distance. A quadrupole is a double dipole consisting of four charges. The last term, varying as the inverse tenth power of the distance, is called the *quadrupole-quadrupole term*. It is, in general, small and amounts to less than 1.3 percent of the total van der Waals attractive energy for all of the inert-gas solids (Refs. 6, 8, and 9). The dipole-quadrupole term, on the other hand, equals approximately 16 percent of the total attractive energy, indicating that, whereas the dipole-dipole term makes up most of the van der Waals attractive energy, the second term is also significant.

3.12 Molecular Crystals

Many molecules form crystals which are held together by van der Waals forces. Among these are N_2, H_2, and CH_4; these are typical covalent molecules in which the atoms share valence electrons to effectively obtain closed shells for each atom in the molecule. The forces of attraction between such molecules are very small and of the order of those in the inert-gas crystals.

The molecules mentioned in the preceding paragraph are nonpolar molecules; they do not have permanent dipole moments. Thus, the attractive force between two hydrogen molecules comes in large measure from the synchronism of electron movements in the two molecules, or from instantaneous dipole-dipole interactions. In addition to these, there are also polar molecules, such as water (H_2O), which possess permanent dipoles. The interaction between a pair of permanent dipoles is, in general, much stronger than that between induced dipoles. This leads to much stronger binding (van der Waals binding) in their respective crystals with correspondingly higher melting and boiling points.

3.13 Refinements to the Born Theory of Ionic Crystals

In the inert-gas and molecular solids considered in the preceding section, van der Waals forces are the primary source of the cohesive energy. These forces exist in other solids, but when the binding due to other causes is strong they may contribute only a small fraction of the total binding energy. This is generally true in ionic crystals, although some types, like the silver halides, may have van der Waals contributions of more than 10 percent. The alkali halides, as may be seen in Table 3.2, have van der Waals energies that amount to only a small percentage of the total energy. This table is of particular interest because it lists the contribution of five terms to the total lattice energy. The first column gives the Madelung energy, or the first term in the simple Born equation. The second is the repulsive energy, caused by the overlapping of closed ion shells. The third and fourth columns are van der Waals terms:

Table 3.2 Contributions to the Cohesive Energies of Certain of the Alkali Halides.*

Alkali Halide Crystal	Madelung	Repulsive	Dipole-Dipole	Dipole-Quadrupole	Zero Point	Total
LiF	285.5	−44.1	3.9	0.6	−3.9	242.0
LiCl	223.5	−26.8	5.8	0.1	−2.4	200.2
LiBr	207.8	−22.5	5.9	0.1	−1.6	189.7
LiI	188.8	−18.3	6.8	0.1	−1.2	176.2
NaCl	204.3	−23.5	5.2	0.1	−1.7	184.4
KCl	183.2	−21.5	7.1	0.1	−1.4	167.5
RbCl	175.8	−19.9	7.9	0.1	−1.2	162.7
CsCl	162.5	−17.7	11.7	0.1	−1.0	155.6

*All values given above are expressed in Kilo calories per mole. From *The Modern Theory of Solids*, by Seitz, F. Copyright 1940, McGraw-Hill Book Company, Inc., New York, p. 88. (Used by permission of the author.)

dipole-dipole and dipole-quadrupole. The fifth column lists zero-point energies: the energy of vibration of the atoms in their lowest energy levels. Finally, the last column is that corresponding to the sum of all five terms, which should equal the internal energy of the crystals at zero degrees absolute and at zero pressure.

3.14 Covalent and Metallic Bonding

In both the ionic and the inert-gas crystals that have been considered, the crystals are formed of atoms or ions with closed-shell configurations of electrons. In these solids, the electrons are considered to be tightly bound to their respective atoms inside the crystal. Because of this fact, ionic and rare-gas solids are easier to interpret in terms of the laws of classical physics. It is only when greater accuracy is needed in computing the physical properties of these crystals that one needs to turn to the more modern quantum mechanical interpretation. Quantum, or wave mechanics, is, however, an essential when it comes to studying the bonding in covalent or metallic crystals. In each of the latter, the bonding is associated with the valence electrons that are not considered to be permanently bound to specific atoms in the solids. In other words, in both of these solids, the valence electrons are shared between atoms. In valence crystals, the sharing of electrons effects a closed-shell configuration for each atom of the solid. The prototype of this form of crystal is the diamond form of carbon. Each carbon atom brings four valence electrons into the crystal. The coordination number of the diamond structure is also four, as shown in Fig. 3.11. If a given carbon atom shares one of its four valence electrons with each of its four neighbors, and the neighbors reciprocate in turn, the carbon atom will, as a result of this sharing of eight electrons, effectively achieve the electron configuration of neon ($1s^2$, $2s^2$, $2p^6$). In this type of crystal, it is often convenient to think of the pairs of electrons which are shared between nearest neighbors as constituting a chemical bond between a pair of atoms. On the other hand, according to the band theory of solids, the electrons are not fixed to specific bonds, but can interchange between bonds. Thus, the valence electrons in a valence crystal can also be thought of as belonging to the crystal as a whole.

The binding associated with these covalent, or *homopolar*, linkages as they are called, is very strong so that the cohesive energy of a solid such as diamond is very large and, in agreement with this fact, these solids are usually very hard and have high melting points.

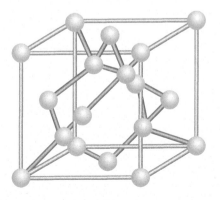

Fig. 3.11 The diamond structure. Each carbon atom is surrounded by four nearest neighbors. *Note*: this structure is the same as that of zincblende (ZnS), Fig. 3.3, except that this lattice contains one kind of atom, instead of two

The covalent bond is also responsible for the cohesion of many well-known molecules, for example, the hydrogen molecule. An idea of how the binding energy develops can be obtained by an elementary consideration of the hydrogen molecule. The lowest atomic energy state is associated with the 1s shell. Two electrons can be accommodated in this state, but only if they have opposite spins. Thus an unexcited helium atom will have both of its electrons in the 1s state but only if the spin vectors of the electrons are opposed. The fact that two electrons can occupy the same quantum state only if the spins are oppositely directed is known as the *Pauli exclusion principle*. Now suppose that two hydrogen atoms are made to approach each other. Then there are two cases to be considered: when the spins of the electrons on the two atoms are parallel, and when the spins are opposed. First consider the latter case. As the atoms come closer and closer together, the electron on either atom begins to find itself in the field of the charge on the nucleus of the other atom. Since the spins of the electrons are opposed, each nucleus is capable of containing both electrons in the 1s ground state. Under these conditions, there is a strong probability that the electrons will spend more time in the neighborhood of one nucleus than the other, and the hydrogen molecule becomes a pair of charged ions—one positive and the other negative. This structure is unstable, especially when the hydrogen atoms are far apart, as may be estimated from the energy required to form a positive and a negative pair of hydrogen ions (−1237 KJ per mol) (Ref. 8). At the normal distance of separation of the atoms in a hydrogen molecule, the ionic structure exists for limited periods of time and contributes about 5 percent of the total binding energy (Ref. 8).

A much more important type of electron interchange occurs when both electrons simultaneously exchange nuclei. The resulting shifting of the electrons back and forth between the nuclei, which occurs at a very rapid rate, is commonly known as a *resonance effect*, and about 80 percent (Ref. 8) of the binding energy of the hydrogen molecule is attributed to it. With the aid of quantum mechanics, the total binding energy of a hydrogen molecule has been computed and, as a result, an insight into the nature of the binding energy associated with this exchange of electrons has been obtained. Quantum mechanics shows that, on the average, the electrons spend more of their time in the region between the two protons than they do on the far sides of the protons. From a very elementary point of view, we may consider that the binding of the hydrogen molecule results from the attraction of the positively charged hydrogen nuclei to the negatively charged electrons, which exists between them.

Attention is called to the interrelation between space and time implied in this discussion concerning the time-average position of the electrons. These ideas are normally expressed in terms of a phase space. This is a coordinate system that includes both the positions of the particles and their momenta (that is, velocities). Thus, for a single particle free to move along a single direction (the *x*-axis), there will be two dimensions in phase space: its position along the axis, and its momentum. For n particles capable of moving in a single direction, there will be $2n$ linear dimensions in the phase space associated with these particles. These are $x_1, x_2, \ldots x_n$, the positions of the particles; and $p_1, p_2, \ldots p_n$, the momenta of the particles. For particles capable of moving in three dimensions, there will be $6n$ degrees of freedom and therefore a corresponding number of dimensions in phase space.

An important theorem in statistical mechanics that bears on the subject of phase space states that the position and momentum average in phase space coincide with the same average over an infinite time. As an example of the application of this theorem, consider the following. The energy of a set of particles is a function of their positions in space and their momenta (velocities). The average energy can be computed by averaging a finite number of position and velocity measurements. The greater the number of these sets of readings that are taken, the closer to the true average will be the result. On the other hand, the positions and velocities are also a function of time.

Therefore their energy can also be considered as a function of time, and the average could be taken from a set of readings made over a very long period of time. This average should agree with that made inside a very short time interval.

In the above, it was assumed that the electrons of the two hydrogen atoms had oppositely directed spins. Now consider the case where two hydrogen atoms with parallel spins are made to approach each other. Here, as the electron on either of the atoms comes within the range of effectiveness of the field on the nucleus of the other atom, it is found that the energy level it would normally occupy is already filled. The situation is similar to that when an electron is brought within the limits of the closed shell of an inert-gas configuration. The normal electronic orbits become badly distorted, or else the second electron moves into a higher energy state, such as $2s$ (Ref. 1). In either case, bringing together two hydrogen atoms with electrons that have parallel spins increases the energy of the system. A stable molecule cannot be formed in this fashion. This is shown in Fig. 3.12, where the uppermost curve represents the hydrogen molecule with parallel spins and both electrons in $1s$ orbits. The lower curve is for opposed electron spins, and it can be observed that this curve has a pronounced minimum, indicating that in this case a stable molecule can be formed. Notice that the curves of Fig. 3.12 represent the total energy of the hydrogen molecule at corresponding values of r. In addition to the ionic and the exchange, or resonance energy, there are other more complicated electrostatic interactions between the two electrons and the two protons. These contribute the remaining 15 percent of the binding energy of the hydrogen molecule.

Valence electrons are also shared between atoms in a typical metallic crystal. Here, however, they are best considered as free electrons and not as electron pairs forming bonds between neighboring atoms. This difference between a metal and a covalent solid is probably a matter of degree, for it is known that the valence electrons in a covalent solid can move from one bond to another and thus are able to move throughout the crystal. It can be shown that the zone or band theory of solids is able to explain adequately the difference between a covalent or homopolar crystal and a metallic one. However, for the present let us assume that in a metal the valence electrons are able to move at will through the lattice, while in a covalent crystal the electrons form directed bonds between neighboring atoms. One result of this difference is that metals tend to crystallize in close-packed lattices (face-centered cubic and close-packed hexagonal structures) in which the directionality of the bonds between atoms is of secondary importance, while covalent crystals form complicated structures so that the bonds between neighboring atoms will give each atom the effect of having a closed-shell configuration of electrons. Thus, carbon, with four valence electrons per atom, crystallizes in the diamond

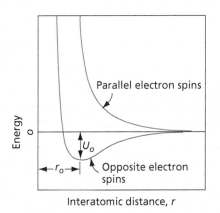

Fig. 3.12 Interaction energy of two hydrogen atoms

lattice with four nearest neighbors, so that each carbon atom has a total of eight shared electrons. In the same manner, arsenic, antimony, and bismuth, which have five valence electrons per atom, need to share electrons with three neighbors in order to attain the eight electrons needed for a closed-shell configuration. The latter substances therefore crystallize with three nearest neighbors. In general, covalent crystals follow what is known as the $(8-N)$ rule, where N is the number of valence electrons, and the factor $(8-N)$ gives the number of nearest neighbors in the structure.

On the assumption that inside a metal the valence electrons are free, we arrive at the following elementary idea of a metal. A metal consists of an ordered array of positively charged ions between which the valence electrons move in all directions with high velocities. Over a period of time, this movement of electrons is equivalent to a more or less uniform distribution of negative electricity which might be thought of as an electron gas that holds the assembly together. In the absence of this gas, the positively charged nuclei would repel each other and the assembly would disintegrate. On the other hand, the electron gas itself could not exist without the presence of the array of positively charged nuclei. This is because the electrons would also repel each other. The cooperative interaction between the electron gas and the positively charged nuclei forms a structure that is stable. The binding forces that hold a metallic crystal together can be assumed to come from the attraction of the positively charged ions for the cloud of negative charge that lies between them. It is also interesting to note that, as the distance between nuclei is made smaller (as the volume of the metal is decreased), the velocities of the free electrons increase with a corresponding rise in their kinetic energy. This leads to a repulsive energy term which becomes large when a metal is compressed. The simplest metals to understand are such alkali metals as sodium and potassium. In them there is only a single valence electron and the positive ions are well separated in the solid so that there is little overlapping of ion shells, and the repulsion caused by the overlapping of shells is therefore small. In these metals, the repulsive energy is chiefly due to the electronic kinetic energy term. In other metallic elements, the theories of the cohesive energies are much more involved. They will not be discussed here.

Additional information concerning bonding in crystals, as well as about an important related area known as the *electron theory of metals*, may be obtained in the books by Hummel (Ref. 3) and Kittel (Ref. 2).

Problems

3.1 The general equation for the force between electrical charges is

$$f = k\frac{e_1 e_2}{(r_{12})^2}$$

where k is a constant with units force $\times$ distance^2 $\div$ charge^2. In the electrostatic or cgs system of units, $k = 1$ dyne cm^2/(statcoulombs)^2. Prove that $k = 9 \times 10^9$ Nm^{2}C in the international (mks) system of units.

3.2 (a) Determine the distance of separation between a positive ion and a negative ion, each carrying a charge equal to that of an electron, if their mutual force of attraction equals -3×10^{-9} N.

(b) What is the coulomb potential energy of this ion pair? Give your answer in joules, calories, electron volts, and joules per mol of ion pairs. See Appendix D.

3.3 (a) Make the actual calculation for the Madelung or attractive energy term of the Born equation for the NaCl lattice using the cgs system. Give your answer in kcal per mol and compare it with the value in Table 3.2.

(b) Also give your answer in joules per mol.

3.4 Using cgs units, compute the repulsive energy term for the NaCl lattice. Assume the Born exponent is 8.00. B can be determined by taking the derivative of the Born equation with respect to r and assuming that at r_0 the forces on the ions are zero. (Coulomb force equals 1 repulsive force.)

3.5 (a) Using the mks system of units, compute the force on an electron, due to a dipole, if the electron is situated at point p in Fig. 3.6, $r = 0.4$ nm, $a = 10^{-3}$ nm and $\theta = 0°$. Assume that the dipole charges e_1 and $-e_2$ are the same as that on an electron.

(b) In what direction does this force act?

(c) What is the magnitude and direction of the force if $\theta = 90°$?

3.6 Determine the magnitude and direction of the force between a dipole and an electron if $r = 0.35$ nm, $\theta = 45°, a = 1.5 \times 10^{-3}$ nm and the dipole charges are the same as that on an electron.

3.7 The zero-point energy of the solid neon crystal is reported to be 590 J/mol. On the basis of this information estimate the maximum lattice vibrational frequency, v_m, of the neon lattice. Use the mks system of units in solving this problem.

3.8 Empirical specific heat data imply that the Debye temperature of pure iron is close to 425 K. The Debye temperature also has been proposed as the temperature at which the energy of the highest vibrational mode of a lattice, hv_m, equals the thermal energy or kT, i.e.,

$$hv = kT$$

On this basis compute a value for iron of v_m using mks units.

3.9 Do you expect a correlation between bonding energy and melting points of metals? Justify your answer.

3.10 The lattice energy of an ionic solid, U, is the amount of energy required to separate a mole of the solid into an ionized gas. As such, it is the opposite of crystallization energy of the solid from ionic gaseous species. Using the following information, show that U for rock salt (NaCl) is 788 kJ/mol. (*Hint*: Construct a Born-Haber cycle.)

Na (s) + $\frac{1}{2}$Cl$_2$ (g) → NaCl (s)	−411 kJ/mol = Enthalpy of formation
Na (s) → Na (g)	108 kJ/mol = Heat of sublimation
Na (g) → Na$^+$ (g) + e	496 kJ/mol = Ionization energy
Cl$_2$ (g) → 2Cl (g)	244 kJ/mol = Bond dissociation energy
Cl (g) + e → Cl$^-$ (g)	−349 kJ/mol = Electron affinity of Cl

References

1. Pearson, W. B., *Crystal Chemistry and Physics of Metals and Alloys*, Wiley-Interscience, New York, 1977.
2. Kittel, C., *Introduction to Solid State Physics*, fifth edition, John Wiley and Sons, New York, 1976.
3. Hummel, R. E., *Electronic Properties of Materials*, Springer-Verlag New York, Inc., New York, 1985.
4. Seitz, F., *Modern Theory of Solids*, McGraw-Hill Book Co., Inc., New York, 1940, p. 80.
5. Lennard-Jones, J. E., *Physica*, 4 941 (1937).
6. Dobbs, E. R., and Jones, G. O., *Reports on Prog. in Phys.*, 20 516 (1957).
7. R. K. Crawford in *Rare Gas Solids*, edited by M. L. Klein and J. A Venables, Academic Press, 1976, vol. 2, chapter 11, pp. 663–728.
8. Pauling, L., *The Nature of the Chemical Bond*, p. 22, Cornell University Press, Ithaca, New York, 1940.
9. Doran, M. B., and Zucker, I. J., *Journal of Physics C: Solid State Physics*, 4 307 (1971).

Chapter 4

Introduction to Dislocations

Learning Objectives

Upon completion of Chapter 4, you will be able to:

1. Develop an understanding of how dislocations contribute to deformation both qualitatively and quantitatively
2. Compare and contrast edge dislocations, screw dislocations, partial dislocations and mixed dislocations
3. Understand vectorial representation of a dislocation, its Burgers vector, and Burgers circuit around a dislocation
4. Explain how kinks and jogs may form upon intersection of dislocations
5. Describe movement of dislocations, their directions and slip planes as a result of an applied force
6. Calculate the force of a dislocation, its stress field, and strain energy for both screw and edge dislocations
7. Delineate the reason (or reasons) for the difference between theoretical and observed yielding of metals

4.1 The Discrepancy Between the Theoretical and Observed Yield Stresses of Crystals

The stress-strain curve of a typical magnesium single crystal, oriented with the basal plane inclined at 45° to the stress axis and strained in tension, is shown in Fig. 4.1. At the low tensile stress of 0.7 MPa, the crystal yields plastically and then easily stretches out to a narrow ribbon which may be four or five times longer than the original crystal. If one examines the surface of the deformed crystal, markings can be seen which run more or less continuously around the

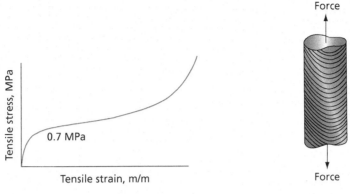

Fig. 4.1 Tensile stress-strain curve for a magnesium single crystal

Fig. 4.2 Slip lines on magnesium crystal

specimen in the form of ellipses. (See Fig. 4.2.) These markings, if viewed at very high magnifications, are recognized as the visible manifestations of a series of fine steps that have formed on the surface. The nature of these steps is shown schematically in Fig. 4.3. Evidently, as a result of the applied force, the crystal has been sheared on a number of parallel planes. Crystallographic analyses of the markings, furthermore, show that these are basal (0002) planes and, therefore, the closest packed plane of the crystal. (Earlier in Chapter 1 the basal plane was given the Miller indices (0001). The notation (0002) is also used, primarily to call attention to the fact that the spacing between basal planes equals one-half the height of the hcp unit cell.) When this type of deformation occurs, the crystal is said to have undergone "slip," the visible markings on the surface are called *slip lines*, or *slip traces*, and the crystallographic plane on which the shear has occurred is called the *slip plane*.

The shear stress at which plastic flow begins in a single crystal is amazingly small when compared to the theoretical shear strength of a perfect crystal (computed in terms of cohesive forces between atoms). An estimate of this strength can be obtained in the following manner.

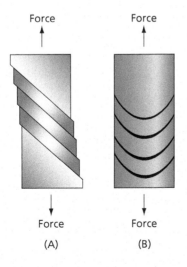

Fig. 4.3 **(A)** Magnified schematic view of slip lines (side view). **(B)** Magnified schematic view of slip lines (front view)

Figure 4.4A shows two adjacent planes of a hypothetical crystal. A shearing stress, acting as indicated by the vectors marked τ, tends to move the atoms of the upper plane to the left. Each atom of the upper plane rises to a maximum position (Fig. 4.4B) as it slides over its neighbor in the plane below. This maximum position represents a saddle point, for continued motion to the left will now be promoted by the forces which pull the atom into the next well. A shear of one atomic distance requires that the atoms of the upper plane in Fig. 4.4A be brought to a position equivalent to that in Fig. 4.4B, after which they move on their own accord into the next equilibrium position, Fig. 4.4C. To reach the saddle point, a horizontal movement of each atom is required equal to an atomic radius. This movement is shown in Fig. 4.4B. Since the separation

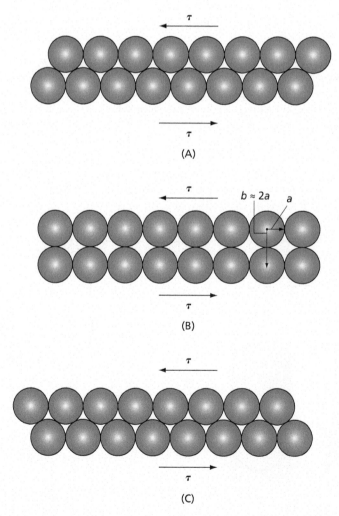

Fig. 4.4 **(A)** Initial position of the atoms on a slip plane. **(B)** The saddle point for the shear of one plane of atoms over another. **(C)** Final position of the atoms after shear by one atomic distance

of the two planes is of the order of two atomic radii, the shear strain at the saddle point is approximately equal to one half. That is,

$$\gamma \simeq \frac{a}{2a} \simeq \frac{1}{2} \qquad 4.1$$

where γ = shear strain. In a perfectly elastic crystal, the ratio of shear stress to shear strain is equal to the shear modulus:

$$\frac{\tau}{\gamma} = \mu \qquad 4.2$$

where γ is shear strain, τ is shear stress, and μ is shear modulus. Substituting the value $\frac{1}{2}$ for γ the shear strain, and the value 17.2 GPa for μ, which is of the order of magnitude of the shear modulus for magnesium, we obtain for τ the stress at the saddle point,

$$\tau = \frac{17{,}200}{2} \simeq 9 \times 10^3 \text{ MPa} \qquad 4.3$$

The ratio of the theoretical stress to start the shear of the crystal to that observed in a real crystal is, therefore, approximately

$$\frac{10^4}{1} = 10{,}000$$

In other words, the crystal deforms plastically at stresses $\frac{1}{10{,}000}$ of its theoretical strength. Similarly, with other metals, real crystals deform at small fractions of their theoretical strengths ($\frac{1}{1000}$ to $\frac{1}{100{,}000}$).

4.2 Dislocations

The discrepancy between the computed and real yield stresses is because real crystals contain defects. Experimental proof for the existence of these defects can be obtained with the aid of the electron microscope. Thus, suppose that a crystal has been deformed so as to form visible slip lines, as indicated in Fig. 4.3. Let us now assume that it is possible to cut from the deformed crystal a transmission electron microscope foil containing a portion of a slip plane.

If the transmission foil has been prepared properly and contains a section of a slip plane, when it is examined in the microscope one may obtain a photograph of the type shown schematically in Fig. 4.5A. In this schematic representation a set of solid dark lines may be seen that start and end at the two dashed lines *a-a* and *b-b*. The latter lines have been drawn on the figure to indicate the positions where the slip plane intersects the foil surfaces. It should be noted that the drawing in Fig. 4.5A is a two-dimensional projection of a three-dimensional specimen. So that the geometrical relations involved in this figure may be better understood, a three-dimensional sketch of the specimen is shown in Fig. 4.5B. This diagram demonstrates that the dark lines in the photograph run across the slip plane from the top to the bottom surfaces of the foil. The fact that these lines are visible in an electron microscope photograph implies that they represent defects in the crystal structure, as was explained previously in Chapter 2.

We may conclude from the preceding discussion that in a crystal which has undergone slip, lattice defects tend to accumulate along the slip planes. These defects are called *dislocations*. The presence of dislocations can also be made evident in another fashion. The points where they intersect a specimen surface can often be made visible by etching the surface with a suitable etching

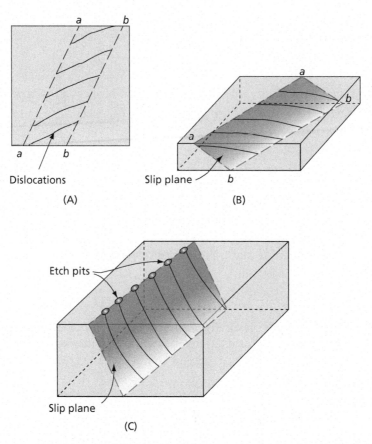

Fig. 4.5 **(A)** Schematic representation of an electron microscope photograph showing a section of a slip plane. **(B)** A three-dimensional view of the same slip plane section. **(C)** Termination of dislocations can also be revealed by etch pits

solution. The fact that the dislocations are defects in the crystal structure tends to make the places where they intersect the surface preferred positions for the attack of the etching solution. As a result, etch pits may form. This is indicated schematically in Fig. 4.5C. Photographs showing etch pits associated with dislocations may be seen in Figs. 5.3 and 6.3.

An electron microscope photographs of piled-up dislocations near a grain boundary and within a grain during the early stages of creep deformation are shown in Fig. 4.6. The pictures are from a modified AISI 316L stainless steel (containing 18 wt% Cr and 12 wt% Ni, with additional small amounts of V and N). The creep deformation was conducted under a constant load of 150 MPa and 650 °C temperature. As described in the reference for the figure, some of the dislocations are multipoles and are held together by the interaction between strain fields of dislocations with Burgers vectors, which are of equal magnitude but opposite in sign.

The fact that dislocations are visible in a transmission electron microscope, and that they may also be revealed by etching a specimen surface, agrees with the assumption that they represent disturbances in the crystal structure. The best evidence now indicates that they are boundaries on the slip planes where a shearing operation has ended. Let us look into the nature of these small shears. Figure 4.7A represents a simple cubic crystal that is assumed to be subjected to shearing

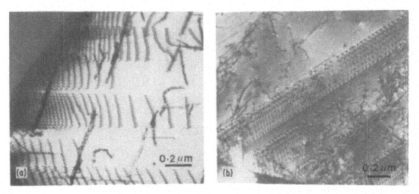

Fig. 4.6 Dislocation pile-up in AISI 316L stainless steel after creep deformation at 150 MPa and 650 °C: **(a)** Closely spaced dislocations in the vicinity of a grain boundary; **(b)** tangled dislocations within a deformed grain. (Printed by permission from Howell, P.R., Nilsson, J.O., Horsewell, A. et al., *J Mater Sci* **16**, 2860–2866 (1981) https://doi.org/10.1007/BF00552971

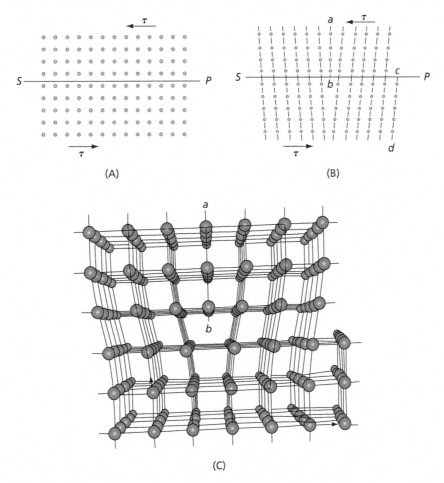

Fig. 4.7 An edge dislocation. **(A)** A perfect crystal. **(B)** When the crystal is sheared one atomic distance over part of the distance *S–P*, an edge dislocation is formed. **(C)** Three dimensional view of slip

stresses, τ, on its upper and lower surfaces, as indicated in the diagram. The line *SP* represents a possible slip plane in the crystal. Suppose that as a result of the applied shear stress, the right-hand half of the crystal is displaced along *SP* so that the part above the slip plane is moved to the left with respect to the part below the slip plane. The amount of this shear is assumed to equal one interatomic spacing in a direction parallel to the slip plane. The result of such a shear is shown in Fig. 4.7B and C. As may be seen in the figure, this will leave an extra vertical half-plane *cd* below the slip plane at the right and outside the crystal. It will also form an extra vertical half-plane *ab* above the slip plane and in the center of the crystal. All other vertical planes are realigned so that they run continuously through the crystal.

Now let us consider the extra half-plane *ab* that lies inside the crystal. An examination of Fig. 4.7B clearly shows that the crystal is badly distorted where this half-plane terminates at the slip plane. It can also be deduced that this distortion decreases in intensity as one moves away from the edge of this half-plane. This is because at large distances from this lower edge of the extra plane, the atoms tend to be arranged as they would be in a perfect crystal. The distortion in the crystal is thus centered around the edge of the extra plane. This boundary of the additional plane is called an *edge dislocation*. It represents one of the two basic orientations that a dislocation may take. The other is called a *screw dislocation* and will be described shortly.

Figure 4.8 represents a three-dimensional sketch of the edge dislocation of Fig. 4.7. The figure clearly shows that the dislocation has the dimensions of a line, in agreement with our discussion of Fig. 4.8. Another important fact shown in Fig. 4.8 is that the dislocation line marks the boundary between the sheared and unsheared parts of the slip plane. This is a basic characteristic of a dislocation. In fact, a dislocation may be defined as a line that forms a boundary on a slip plane between a region that has slipped and one that has not.

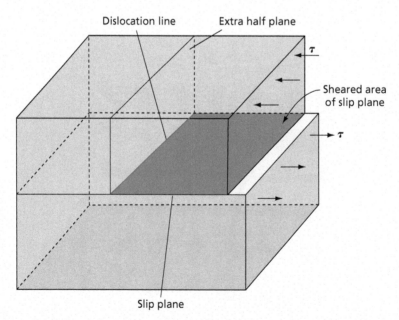

Fig. 4.8 This three-dimensional view of a crystal containing an edge dislocation shows that the dislocation forms the boundary on the slip plane between a region that has been sheared and a region that has not been sheared

Figure 4.9 illustrates how the above dislocation moves through the crystal under an applied shear stress which is indicated by the vectors τ. As a result of the applied stress, atom c may move to the position marked c' in Fig. 4.9B. If it does, the dislocation moves one atomic distance to the left. The plane x, at the top of the figure, now runs continuously from top to bottom of the crystal, while plane y ends abruptly at the slip plane. Continued application of the stress will cause the dislocation to move by repeated steps along the slip plane of the crystal. The final result is that the crystal is sheared across the slip plane by one atomic distance, as is shown in Fig. 4.9C.

Each step in the motion of the dislocation, as can be seen in Figs. 4.9A and B, requires only a slight rearrangement of the atoms in the neighborhood of the edge of the extra plane. As a result, a very small force will move a dislocation. Theoretical calculations show that this force is of the correct order of magnitude to account for the low-yield stresses of crystals.

The existence of dislocations was postulated at least a half of a century before experimental techniques were available to make them visible. In 1934, Orowan (Ref. 1) Polanyi (Ref. 2) and Taylor (Ref. 3) presented papers which are said to have laid the foundation for the modern theory of slip due to dislocations (Ref. 4). This early work in the field of dislocations in metal crystals had as its basis an effort to explain the large discrepancy between the theoretical and observed shear strengths of metal crystals. It was felt that the observed low yield strengths of real crystals could best be explained on the basis that the crystal contained defects in the form of dislocations. An excellent review of the early concepts of dislocations can be found in the proceedings of the commemorative fiftieth anniversary meeting held in London in December 1984 (Ref. 5).

The movement of a single dislocation completely through a crystal produces a step on the surface, the depth of which is one atomic distance. Since an atomic distance in metal crystals is less than a nanometer, such a step will certainly not be visible to the naked eye. Many hundreds or thousands of dislocations must move across a slip plane in order to produce a visible slip line. A mechanism will be given presently to show how it is possible to produce this number of dislocations on a single slip plane inside a crystal. First, however, it is necessary to define a screw dislocation, which is shown schematically in Fig. 4.10A, where each small cube can be considered to represent an atom. Figure 4.10B represents the same crystal with the position of the dislocation line marked by the line DC. The plane $ABCD$ is the slip plane. The upper front portion of the crystal has been sheared by one atomic distance to the left relative to the lower front portion. The designation "screw" for this lattice defect is derived from the fact that the lattice planes of the crystal spiral the dislocation line DC. This statement can be proved by starting at point x in Fig. 4.10A and then proceeding upward and around the crystal in the direction of the arrows. One circuit of the crystal ends at point y; continued circuits will finally end at point z.

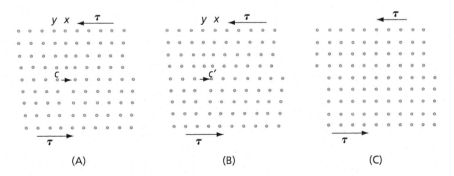

Fig. 4.9 Three stages in the movement of an edge dislocation through a crystal

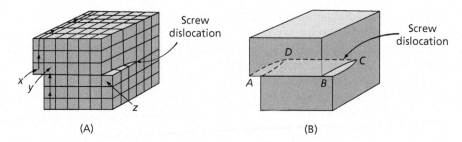

Fig. 4.10 Two representations of a screw dislocation. Notice that the planes in this dislocation spiral around the dislocation like a left-hand screw

Figure 4.10B plainly shows that a dislocation in a screw orientation also represents the boundary between a slipped and an unslipped area. Here the dislocation, centered along line DC, separates the slipped area $ABCD$ from the remainder of the slip plane in back of the dislocation.

A model of a screw dislocation can be readily constructed using a stack of filing cards, or small sheets of paper, and a roll of transparent tape. First, cut halfway through the stack of cards, as indicated in Fig. 4.11. Then tape (along the cut) the left-hand half of the top card to the right-hand half of the card just below it. This is indicated in the right-hand part of the illustration. Now repeat the process through the entire stack. The result should be a continuous spiral plane that goes entirely through the pack. This is a left-hand screw dislocation, as can be checked by placing the thumb of the left hand in the direction of the screw axis and noting that the fingers of the left hand indicate that the direction of advance of the spiral plane is toward the thumb.

The edge dislocation shown in Fig. 4.7B has an incomplete plane which lies above the slip plane. It is also possible to have the incomplete plane below the slip plane. The two cases are differentiated by calling the former a *positive edge dislocation*, and the latter a *negative edge dislocation*. It should be noted that this differentiation between the two dislocations is purely arbitrary, for a simple rotation of the crystal of 180° will convert a positive dislocation into a negative one, and vice versa. Symbols representing these two forms are ⊥ and ⊤, respectively, where the horizontal line represents the slip plane and the vertical line the incomplete plane. There are also two forms of screw dislocations. The screw dislocation shown in Fig. 4.10 has lattice planes that spiral the line DC like a left-hand screw. An equally probable screw dislocation is one in which the lattice spirals in a right-hand fashion around the dislocation line. Both forms of the edge and the screw dislocations, respectively, are shown in Fig. 4.12. The figure also illustrates how the

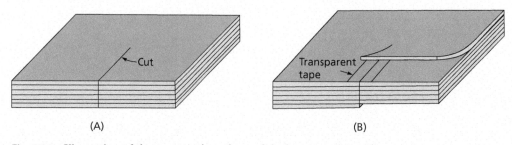

Fig. 4.11 Illustration of the construction of a model of a screw dislocation

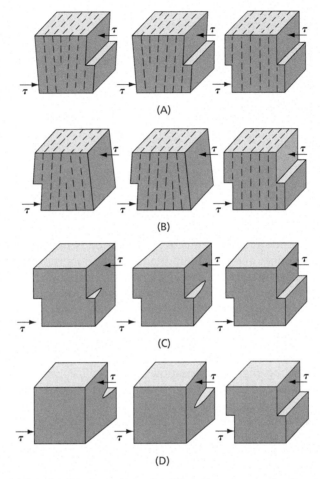

Fig. 4.12 The ways that the four basic orientations of a dislocation move under the same applied stress: **(A)** Positive edge, **(B)** Negative edge, **(C)** Left-hand screw, and **(D)** Right-hand screw

four types move under the same applied shear stress (indicated by the vectors τ). As previously mentioned, Fig. 4.12 shows that a positive edge dislocation moves to the left when the upper half of the lattice is sheared to the left. On the other hand, a negative edge dislocation moves to the right, but produces the identical shear of the crystal. The figure also demonstrates that the right-hand screw moves forward and the left-hand screw moves to the rear, again producing the same shear of the lattice.

In the preceding examples, the dislocation lines have been assumed to run as straight lines through the crystal. As explained at the end of Section 4.2, it is a consequence of the basic nature of dislocations that they cannot end inside a crystal. Thus, the extra plane of an edge dislocation may extend only part way through the crystal, as is shown in Fig. 4.13, and its rear edge b then forms a second edge dislocation. The two dislocation segments a and b thus form a continuous path through the crystal from front to top surfaces.

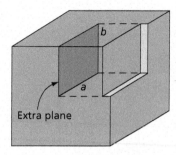

Fig. 4.13 Dislocations can vary in direction. This shaded extra plane forms two dislocations with edge components *a* and *b*

It is also possible for all four edges of an incomplete plane to lie inside a crystal, forming a four-sided closed edge dislocation at the boundaries of the plane. Furthermore, a dislocation that is an edge in one orientation can change to a screw in another orientation, as is illustrated in Fig. 4.14. Figure 4.15 shows the same dislocation viewed from above. Open circles represent atoms lying in the plane just above the slip plane, while dots represent the atoms just below the slip plane. Notice that the lattice is sheared by an atomic distance in the region at the lower right-hand quarter of the figure bounded by the two dislocation segments. Finally, a dislocation does not need to be either pure screw or pure edge, but may have orientations intermediate to both. This fact signifies that dislocation lines do not have to be straight, but can be curved. An example is shown in Fig. 4.16. The drawing, like Fig. 4.15, shows a change in orientation from edge to screw, but here the change is not abrupt.

Consider the closed rectangular dislocation of Fig. 4.17, consisting of the four elementary types of dislocations shown in Fig. 4.12. Sides *a* and *c* are positive and negative edge dislocations, respectively, while *b* and *d* are right-and left-hand screws, respectively. Figure 4.12 shows that, under the indicated shear-stress sense, dislocations *a* and *c* move to the left and right, respectively, while *b* and *d* move forward and to the rear, respectively. The dislocation loop thus opens, or becomes larger, under the given stress. (It would close, however, if the sense of the stress were reversed.) From what has been stated previously, it is evident that the dislocation loop *abcd* does

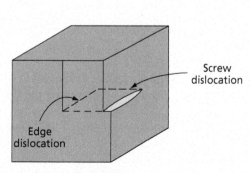

Fig. 4.14 A two-component dislocation composed of an edge and a screw component

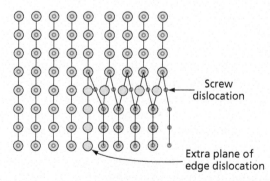

Fig. 4.15 Atomic configuration corresponding to the dislocation of Fig. 4.14 viewed from above. Open-circle atoms are above the slip plane, dot atoms are below the slip plane

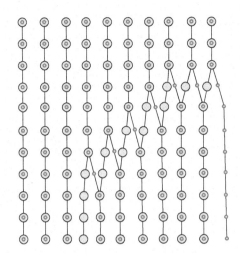

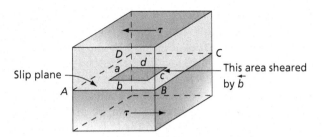

Fig. 4.17 A closed dislocation loop consisting of (a) positive edge, (b) right-hand screw, (c) negative edge, and (d) left-hand screw

Fig. 4.16 A dislocation that changes its orientation from a screw to an edge as viewed from above looking down on its slip plane

not need to be rectangular in order to open under the given shearing stress. A closed curve, such as a circle, would also expand in a similar manner and shear the crystal in the same way.

It has been mentioned that a dislocation cannot end inside a crystal. This is because a dislocation represents the boundary between a slipped area and an unslipped area. If the slipped area on the slip plane does not touch the specimen surface, as in Fig. 4.18, then its boundary is continuous and the dislocation has to be a closed loop. Only when the slipped area extends to the specimen surface, as in Fig. 4.13, is it possible for a single dislocation to have an end point.

4.3 The Burgers Vector

The area inside the rectangle *abcd* of Fig. 4.17, or inside the closed loop of a more general curved dislocation loop, such as that in Fig. 4.18, is sheared by one atomic distance; that is, inside this region the lattice lying above the slip plane (*ABCD*) has slipped one atomic distance to the left relative to the lattice below the slip plane. The direction of this shear is indicated by the vector $\vec{b}$, the length of which is one atomic distance. Outside of the dislocation loop shown in Fig. 4.17, the crystal is not sheared. The dislocation is, therefore, a discontinuity at which the lattice shifts from the unsheared to the sheared state. Although the dislocation varies in orientation in the slip plane *ABCD*, the variation in shear across the dislocation is everywhere the same, and the slip vector $\vec{b}$ is therefore a characteristic property of the dislocation. By definition, this vector is called the *Burgers vector of the dislocation*.

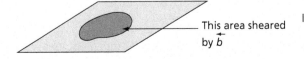

Fig. 4.18 A curved dislocation loop lying in a slip plane

The Burgers vector of a dislocation is an important property of a dislocation because, if the Burgers vector and the orientation of the dislocation line are known, the dislocation is completely described. Figure 4.19 shows a method of determining the Burgers vector applied to a positive edge dislocation. It is first necessary to choose arbitrarily a positive direction for the dislocation. In the present case, let us assume it to be the direction out of the paper. In Fig. 4.19A a counterclockwise circuit of atom-to-atom steps in a perfect crystal closes, but when the same step-by-step circuit is made around a dislocation in an imperfect crystal (Fig. 4.19B), the end point of the circuit fails to coincide with the starting point. The vector b connecting the end point with the starting point is the Burgers vector of the dislocation. This procedure can be used to find the Burgers vectors of any dislocation if the following rules are observed:

1. The circuit is traversed in the same manner as a rotating right-hand screw advancing in the positive direction of the dislocation.
2. The circuit must close in a perfect crystal and must go completely around the dislocation in the real crystal.
3. The vector that closes the circuit in the imperfect crystal (by connecting the end point to the starting point) is the Burgers vector.

The above convention involving a right-hand (RH) circuit around the dislocation line yields a Burgers vector pointing from the finish to the start (FS) of the circuit, and because the closure failure is measured in an imperfect crystal, it is called a *local Burgers vector* or more completely a *RHFS local Burgers vector*. Alternatively, suppose the Burgers circuit is first made to close in Fig. 4.19B, which shows the crystal containing a dislocation. In this case the finishing and starting points will coincide. Then, if a similar step-by-step circuit is made in the perfect crystal of Fig. 4.19A, the starting and finishing points will no longer coincide in this figure. The resulting closure failure now occurs in a crystal where there is no distortion, and therefore, it is known as a true Burgers vector. Since it is measured in a lattice distorted by the strains of the dislocation, the local Burgers vector differs in general from the true Burgers vector. However, if the size of the Burgers circuit around the dislocation is increased, so that the path of the Burgers circuit lies in an almost perfect crystal, the two forms of Burgers vectors will approach each other.

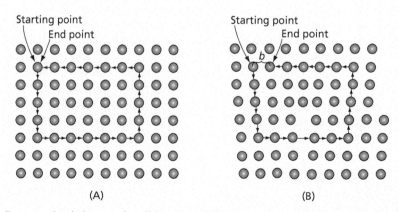

Fig. 4.19 The Burgers circuit for an edge dislocation: **(A)** Perfect crystal and **(B)** crystal with dislocation

It should be understood that a number of arbitrary assumptions are involved in making a Burgers circuit. Thus, one must arbitrarily choose the positive direction of the dislocation line, make a choice on whether the circuit is to be right-handed or left-handed, and decide whether the sense of the vector is determined from the start to the finish of the circuit or vice versa. Unfortunately, in the past not all authors have used the same convention in choosing their Burgers circuits. For more detailed discussion, see Reference 6.

Figure 4.20 shows a Burgers circuit around a left-hand screw dislocation. In Fig. 4.20A, the circuit is indicated for the perfect crystal. Figure 4.20B shows the same circuit transferred to a crystal containing a screw dislocation.

It is now possible to summarize certain characteristics of both edge and screw dislocations.

1. **Edge dislocations:**
   (a) An edge dislocation lies perpendicular to its Burgers vector.
   (b) An edge dislocation moves (in its slip plane) in the direction of the Burgers vector (slip direction). Under a shear-stress sense ⇄ a positive dislocation ⊥ moves to the right, a negative one ⊤ to the left.
2. **Screw dislocations:**
   (a) A screw dislocation lies parallel to its Burgers vector.
   (b) A screw dislocation moves (in the slip plane) in a direction perpendicular to the Burgers vector (slip direction).

A useful relationship to remember is that the slip plane is the plane containing both the Burgers vector and the dislocation. The slip plane of an edge dislocation is thus uniquely defined because the Burgers vector and the dislocation are perpendicular. On the other hand, the slip plane of a screw dislocation can be any plane containing the dislocation because the Burgers vector and dislocation have the same direction. Edge dislocations are thus confined to move or glide in a unique plane, but screw dislocations can glide in any direction as long as they move parallel to their original orientation.

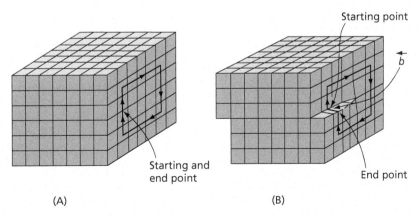

Fig. 4.20 The Burgers circuit for a dislocation in a screw orientation. **(A)** Perfect crystal and **(B)** crystal with dislocation

4.4 Vector Notation for Dislocations

Up to this point we have been interested in dislocations only in a general sense. In an actual crystal, because of the more complex spatial arrangement of atoms, the dislocations are complicated and often difficult to visualize. It is sometimes convenient to neglect all considerations of the geometrical appearance of a complicated dislocation and define the dislocation by its Burgers vector. The vector notation for Burgers vectors is especially convenient here. It has already been pointed out that, in any crystal form, the distance between atoms in a close-packed direction corresponds to the smallest shear distance that will preserve the crystal structure during a slip movement. Dislocations with Burgers vectors equal to this shear are energetically the most favored in a given crystal structure. With regard to the vector notation, the direction of a Burgers vector can be represented by the Miller indices of its direction, and the length of the vector can be expressed by a suitable numerical factor placed in front of the Miller indices. Several explanatory examples will be considered.

Figure 4.21 represents the unit cell for cubic structures. In a simple cubic lattice, the distance between atoms in a close-packed direction equals the length of one edge of a unit cell. Now consider the symbol [100]. This quantity may be taken to represent a vector, such as is shown in Fig. 4.21, with a unit component in the x direction, and is, accordingly, the distance between atoms in the x direction. A dislocation with a Burgers vector parallel to the x direction in a simple cubic lattice is therefore represented by [100]. Now consider a face-centered cubic lattice. Here the close-packed direction is a face diagonal, and the distance between atoms in this direction is equal to one-half the length of the face diagonal. In Fig. 4.21, the indicated face diagonal ob has indices [101]. This symbol also corresponds to a vector with unit components in the x and z directions and a zero component in the y direction. As a result, a dislocation in a face-centered cubic lattice having a Burgers vector lying in the [101] direction should be written $\frac{1}{2}$[101]. In the body-centered cubic lattice, the close-packed direction is a cube diagonal, or a direction of the form $\langle 111 \rangle$. The distance between atoms in these directions is one-half the length of the diagonal, so that a dislocation having a Burgers vector parallel to [111], for example, is written $\frac{1}{2}$[111].

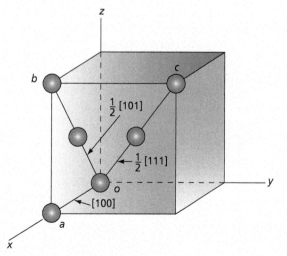

Fig. 4.21 The spacing between atoms in the close-packed directions of the different cubic systems: face-centered cubic, body-centered cubic, and simple cubic

4.5 Dislocations in the Face-Centered Cubic Lattice

The primary slip plane in the face-centered cubic lattice is the octahedral plane {111}. Figure 4.22 shows a plane of this type looking down on the extra plane of an edge dislocation. The purple circles represent the close-packed (111) plane at which the extra plane of the edge dislocation ends, while the blue circles are the atoms in the next following close-packed (111) plane. Notice that in the latter plane a zigzag row of atoms is missing. This corresponds to the missing plane of the edge dislocation. Now consider the plane of atoms (lying directly over the zigzag row of blue atoms) immediately to the left of the missing plane of atoms. The movement of the atoms of this plane through the horizontal distance b displaces the dislocation one unit to the left. The vector b thus represents the Burgers vector of the dislocation, which is designated as $\frac{1}{2}[1\bar{1}0]$, according to the method outlined in the preceding section. Similar movements in succession of the planes of atoms that find themselves to the left of the dislocation will cause the dislocation to move across the entire crystal. As is to be expected, this dislocation movement shears the upper half of the crystal (above the plane of the paper) one unit b to the right relative to the bottom half (below the plane of the paper).

A dislocation of the type shown in Fig. 4.22 does not normally move in the simple manner discussed in the preceding paragraph. As can be deduced with the aid of the ping-pong-ball model of the atom arrangement shown in Fig. 4.22, the movement of a zigzag plane of atoms, such as aa, through the horizontal distance b would involve a very large lattice strain, because each blue atom at the slip plane would be forced to climb over the dark atom below it and to its right. What actually is believed to happen is that the indicated plane of atoms makes the move indicated by the vectors marked c in Fig. 4.23. This movement can occur with a much smaller strain of the lattice. A second movement of the same type, indicated by the vectors marked d, brings the atoms to the same final positions as the single displacement b of Fig. 4.22.

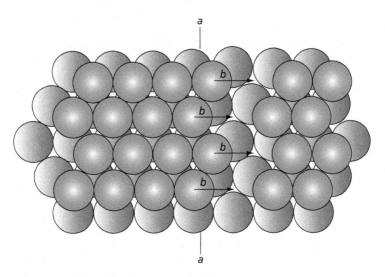

Fig. 4.22 A total dislocation (edge orientation) in a face-centered cubic lattice as viewed when looking down on the slip plane

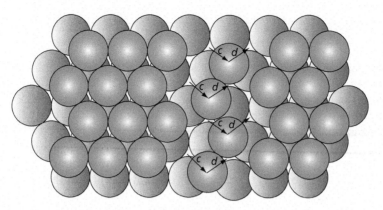

Fig. 4.23 Partial dislocation in a face-centered cubic lattice

The atom arrangement of Fig. 4.23 is particularly significant because it shows how a single-unit dislocation can break down into a pair of partial dislocations. Thus, the isolated single zigzag row of atoms has an incomplete dislocation on each side of it. The Burgers vectors of these dislocations are the vectors c and d shown in the figure. The vector notation for these Burgers vectors can be deduced with the aid of Fig. 4.24. This figure shows the (111) surface of a face-centered cubic crystal in relation to the unit cell of the structure. The positions of the atoms in this surface are indicated by

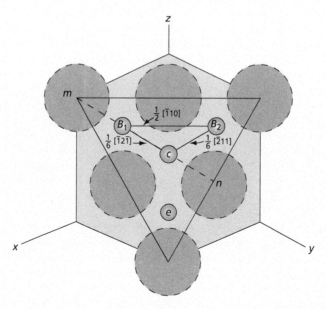

Fig. 4.24 The orientation relationship between the Burgers vectors of a total dislocation and its partial dislocations

circles drawn with dashed lines. The places that would be occupied by the atoms in the next plane above the given one are designated by small circles with the letter B inside them. In the center of Fig. 4.24 is another small circle marked C. The Burgers vector of the total dislocation equals the distance $B_1 B_2$, while the Burgers vectors of the two partial dislocations c and d of Fig. 4.23 are the same as the distances $B_1 C$ and CB_2. The Burgers vector of the total dislocation is $\frac{1}{2}[110]$, which follows from the previous discussion. The line $B_1 C$ lies in the $[\bar{1}2\bar{1}]$ direction. The symbol $[\bar{1}2\bar{1}]$ represents a vector with unit negative components in the x and z directions and a component of 2 in the y direction. This is a vector the length of which is twice the distance mn in Fig. 4.24. Since $B_1 C$ is just one-third of line mn, the Burgers vector for this dislocation is $\frac{1}{6}[\bar{1}2\bar{1}]$. In the same manner, it can be shown that the vector CB_2 can be represented by $\frac{1}{6}[\bar{2}11]$. The total face-centered cubic dislocation $\frac{1}{2}[\bar{1}10]$ is thus able to dissociate into two partial dislocations according to the relation

$$\frac{1}{2}[\bar{1}10] = \frac{1}{6}[\bar{1}2\bar{1}] + \frac{1}{6}[\bar{2}11] \qquad 4.4$$

When a total dislocation breaks down into a pair of partials, the strain energy of the lattice is decreased. This results because the energy of a dislocation is proportional to the square of its Burgers vector (see Eqs. 4.19 and 4.20) and because the square of the Burgers vector of the total dislocation is more than twice as large as the square of the Burgers vector of a partial dislocation. Because the partial dislocations of Fig. 4.23 represent approximately equal lattice strains, a repulsive force exists between them that forces the partials apart. Such a separation will add additional planes of atoms to the single zigzag plane of Fig. 4.23, as shown in Fig. 4.25. A total dislocation that has dissociated into a pair of separated partials like those in the latter figure is known as an *extended dislocation*. An important fact to notice about Fig. 4.25 is that the blue atoms that lie between the two partial dislocations have stacking positions which differ from those of the atoms on the far side of either of the partial dislocations. Thus, if we assume that the purple atoms occupy A positions in a stacking sequence and the blue atoms at either end of the figure, B positions, then the blue atoms between the two partial dislocations lie on C positions. In this region, the $ABCABCABC$... stacking sequence of the face-centered cubic lattice suffers a discontinuity and becomes $ABCA \updownarrow CABCA$.... The arrows indicate the discontinuity. Discontinuities in the stacking

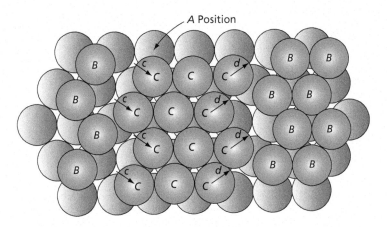

Fig. 4.25 An extended dislocation

order of the {111}, or close-packed planes, are called *stacking faults*. In the present example, the stacking fault occurs on the slip plane (between the purple and blue atoms) and is bounded at its ends by what are known as *Shockley partial dislocations*. In a face-centered cubic lattice, stacking faults may arise in a number of other ways. In all cases, if a stacking fault terminates inside a crystal, its boundaries will form a partial dislocation. The partial dislocations of stacking faults, in general, may be either of the Shockley type, with the Burgers vector of the dislocation lying in the plane of the fault, or of the Frank type, with the Burgers vector normal to the stacking fault. The partial dislocations presently being considered, however, are only those associated with slip and are of the Shockley form. Figure 4.26 shows examples of stacking faults bordered by dislocations in Cu–15.6 at % Al (Ref. 7). The defect marked by A in the figure shows stacking faults bordered by two dislocations in the lower-left- and upper-right-hand sides. The dislocations, which are not straight, are $[10\bar{1}]$ and $[5\bar{2}3]$, respectively (Ref. 7). The defect marked by B shows a narrow region of stacking fault surrounded by Shockley partial dislocations.

Since the atoms on either side of a stacking fault are not at the positions they would normally occupy in a perfect lattice, a stacking fault possesses a surface energy which, in general, is small

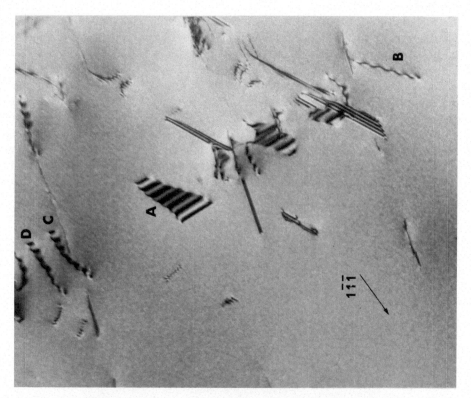

Fig. 4.26 An electron micrograph of a thin foil of lightly deformed Cu + 15.6 at. % Al alloy. The beam direction is close to [101] and *g* is indicated. X 44,000. (From A. K. Head, P. Humble, L. M. Clarebrough, A. J. Morton and C. T. Forwood, *Computed Electron Micrographs and Defect Identification*, American Elsevier Publishing Company, Inc., New York, 1973. Used by permission of the authors.)

compared with that of an ordinary grain boundary, but nevertheless finite. This stacking-fault energy plays an important part in determining the size of an extended dislocation. The larger the separation between the partial dislocations, the smaller is the repulsive force between them. On the other hand, the total surface energy associated with the stacking fault increases with the distance between partial dislocations. The separation between the two partials thus represents an equilibrium between the repulsive energy of the dislocations and the surface energy of the fault. Seeger and Schoeck (Refs. 8 and 9) have shown that the separation of the pair of partial dislocations in an extended dislocation depends on a dimensionless parameter $\gamma_f c/Gb^2$, where γ_f is the specific surface energy of the stacking fault, G is the shear modulus in the slip plane, c is the separation between adjoining slip planes, and b is the magnitude of the Burgers vector. In certain face-centered cubic metals typified by aluminum, this parameter is larger than 10^{-2} and the separation between dislocations is only of the order of a single atomic distance. These metals are said to have high stacking-fault energies. When the parameter is less than 10^{-2}, a metal is said to have a low stacking-fault energy. A typical example of a metal with a low stacking-fault energy is copper. The computed (Ref. 9) separation of the partial dislocations in copper are of the order of twelve interatomic spacings if the extended dislocation is in the edge orientation, and about five interatomic spacings if it is in the screw orientation.

A word should be said about the movement of an extended dislocation through a crystal. The actual movement may be quite complicated for several reasons. First, if the moving dislocation meets obstacles, such as other dislocations, or even second-phase particles, the width of the stacking fault should vary. Second, thermal vibrations may also cause the width of the stacking fault to vary locally along the dislocation, the variation being a function of time. On the assumption that these and other complicating effects can be neglected, an extended dislocation can be pictured as a pair of partial dislocations, separated by a finite distance, which move in consort through the crystal. The first partial dislocation, as it moves, changes the stacking order, while the second restores the order to its proper sequence. After both partials have passed a given point in the lattice, the crystal will have been sheared (at the slip plane) by an amount equal to the Burgers vector b of the total dislocation.

4.6 Intrinsic and Extrinsic Stacking Faults in Face-Centered Cubic Metals

The movement of a Shockley partial across the slip plane of a fcc metal has been shown to produce a stacking sequence $ABCA \updownarrow CABCA \ldots$. In this case the normal stacking sequence can be observed to exist right up to the plane of the fault. A fault of this type, following Frank, is called an *intrinsic stacking fault*. An intrinsic stacking fault may also be developed in a fcc crystal by removing part of a close-packed plane, as shown in Fig. 4.27A. Such a method of developing an intrinsic stacking fault is physically quite possible and can occur by the condensation of vacancies on an octahedral plane. While the fault produced in this manner is the same as that resulting from the slip of a Shockley partial, the partial dislocation surrounding the fault is not the same. In this case it is normal to the {111} slip plane and therefore of the Frank type. Its Burgers vector is equal to one-third of a total dislocation and therefore may be written $\frac{1}{3}\langle 111 \rangle$.

The addition of a portion of an octahedral plane produces a different type of stacking sequence which is $ABCA \updownarrow C \updownarrow BCABC \ldots$. In this fault (Fig. 4.27B) a plane has been inserted that is not correctly stacked with respect to the planes on either side of the fault. This type of fault

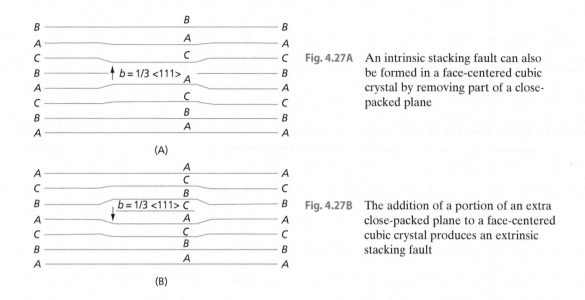

Fig. 4.27A An intrinsic stacking fault can also be formed in a face-centered cubic crystal by removing part of a close-packed plane

Fig. 4.27B The addition of a portion of an extra close-packed plane to a face-centered cubic crystal produces an extrinsic stacking fault

is called an *extrinsic* or double *stacking fault*. The Burgers vector for the extrinsic fault, shown in Fig. 4.27B, is also $\frac{1}{3}\langle 111 \rangle$. An extrinsic stacking fault could be formed by the precipitation of interstitial atoms on an octahedral plane. This is believed to be much less probable than the condensation of vacancies to produce an intrinsic fault. It is also possible to form extrinsic stacking faults by the slip of Shockley partials. However, to do this one normally has to assume that this type of slip occurs on two neighboring planes (Ref. 6).

4.7 Extended Dislocations in Hexagonal Metals

Because the basal plane of a hexagonal metal has exactly the same close-packed arrangement of atoms as the octahedral {111} plane of face-centered cubic metals, extended dislocations also occur in these metals. In the hexagonal system, the dissociation of a total dislocation into a pair of partials on the basal plane is expressed in the following fashion:

$$\tfrac{1}{3}[\bar{1}2\bar{1}0] = \tfrac{1}{3}[01\bar{1}0] + \tfrac{1}{3}[\bar{1}100] \qquad 4.5$$

This equation represents exactly the same addition of Burgers vectors as given in the previous section for the face-centered cubic system, namely,

$$\tfrac{1}{2}[\bar{1}10] = \tfrac{1}{6}[\bar{1}2\bar{1}] + \tfrac{1}{6}[\bar{2}11] \qquad 4.6$$

The only difference between the two equations lies in the present use of the four-digit Miller indices for hexagonal metals. The stacking fault associated with extended dislocations in the hexagonal metals is similar to that of the face-centered metals. The movement of the first partial dislocation through the crystal changes the stacking sequence $ABABABABAB\ldots$ to $ABACBCBCBC\ldots$ on the assumption that the fourth plane slips relative to the third. Notice that the sequence $CBCBCB\ldots$

is a perfectly good hexagonal sequence with alternate planes lying above one another. The stacking fault occurs between the third and fourth planes: between $A \updownarrow C$.... As in the face-centered cubic example, the movement of the second partial dislocation restores the crystal to the proper stacking sequence $ABABAB$....

4.8 Climb of Edge Dislocations

The slip plane of a dislocation is defined as the plane that contains both the dislocation and its Burgers vector. Since the Burgers vector is parallel to a screw dislocation, any plane containing the dislocation is a possible slip plane. (See Fig. 4.28A.) On the other hand, the Burgers vector of an edge dislocation is perpendicular to the dislocation, and there is only one possible slip plane. (See Fig. 4.28B.) A screw dislocation may move by slip or glide in any direction perpendicular to itself, but an edge dislocation can only glide in its single slip plane. There is, however, another method, fundamentally different from slip, by which an edge dislocation can move. This process is called *climb* and involves motion in a direction perpendicular to the slip plane.

Figure 4.29A represents a view of an edge dislocation with the extra plane perpendicular to the plane of the paper and designated by filled circles. In this diagram, a vacancy or vacant lattice site has moved up to a position just to the right of atom a, one of the atoms forming the edge or boundary of the extra plane. If atom a jumps into the vacancy, the edge of the dislocation loses one atom, as is shown in Fig. 4.29B, where atom c, designated with a crossed circle, represents the next atom of the edge (lying just below the plane of the paper). If atom c and all others that formed the original edge of the extra planes move off through interaction with vacancies, the edge dislocation will climb one atomic distance in a direction perpendicular to the slip plane. This situation is shown in Fig. 4.29C. Climb, as illustrated in the above example, is designated as positive climb and results in a decrease in size of the extra plane. Negative climb corresponds to the opposite of the above in that the extra plane grows in size instead of shrinking. A mechanism for negative climb is illustrated in Fig. 4.30A and Fig. 4.30B.

In this case, let us suppose that atom a of Fig. 4.30A moves to the left and joins the extra plane, leaving a vacancy to its right, as is shown in Fig. 4.30B. This vacancy then moves off into the crystal. Notice that this is again an atom by atom procedure and not a cooperative movement

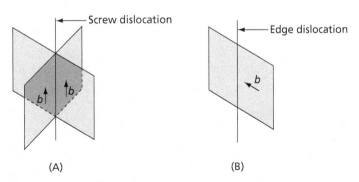

Fig. 4.28 **(A)** Any plane containing the dislocation is a slip plane for a screw dislocation. **(B)** There is only one slip plane for an edge dislocation. It contains both the Burgers vector and the dislocation

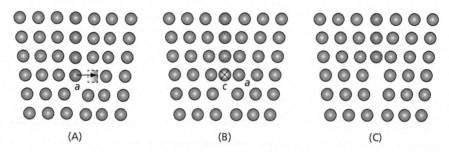

Fig. 4.29 Positive climb of an edge dislocation—an edge atom moves into a vacancy

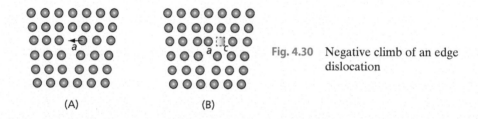

Fig. 4.30 Negative climb of an edge dislocation

of the entire row of atoms lying behind atom *a*. A cooperative movement of all atoms in the row behind *a* corresponds to slip and not to climb.

Because we are removing material from inside the crystal as the extra plane itself grows smaller, the effect of positive climb on the crystal is to cause it to shrink in a direction parallel to the slip plane (perpendicular to the extra plane). Positive climb is therefore associated with a compressive strain and will be promoted by a compressive stress component perpendicular to the extra plane. Similarly, a tensile stress applied perpendicular to the extra plane of an edge dislocation promotes the growth of the plane and thus negative climb. A fundamental difference therefore exists between the nature of the stress that produces slip and that which produces climb. Slip occurs as the result of shear stress; climb as the result of a normal stress (tensile or compressive).

Both positive and negative climb require that vacancies move through the lattice, toward the dislocation in the first case and away from it in the second case. If the concentration of vacancies and their jump rate is very low, then it is not expected that edge dislocations will climb. As we shall see, vacancies in most metals are practically immobile at low temperatures (one jump in eleven days in copper at room temperature), but at high temperatures they move with great rapidity, and their equilibrium number increases exponentially by many powers. Climb, therefore, is a phenomenon that becomes increasingly important as the temperature rises. Slip, on the other hand, is only slightly influenced by temperature.

4.9 Dislocation Intersections

The dislocations in a metal constitute a three-dimensional network of linear faults. On any given slip plane, there will be a certain number of dislocations that lie in this plane and are capable of producing slip along it. At the same time, there will be many other dislocations that intersect it at various angles. Consequently, when a dislocation moves it must pass through those dislocations that intersect its slip plane. The cutting of dislocations by other dislocations is an important subject because, in general, it requires work to make the intersections. The relative ease or difficulty with which slip occurs is thus determined in part by the intersection of dislocations.

For a simple example of the result of a dislocation intersection, see Fig. 4.31. In the drawing it is assumed that a dislocation has moved across the slip plane *ABCD*, thereby shearing the top half of the rectangular crystal relative to the bottom half by the length of its Burgers vector *b*. A second (vertical) dislocation, having a loop that intersects the slip plane at two points, is shown in Fig. 4.31. It is assumed, for the sake of convenience, that this loop is in the edge orientation where it intersects the slip plane. The indicated displacement of the crystal, shown in Fig. 4.31, also shears the top half of the vertical-dislocation loop relative to its bottom half by the amount of the Burgers vector *b*. This shearing action cannot break the loop into two separated half-loops because, according to rule, a dislocation cannot end inside a crystal. The only alternative is that the displacement lengthens the vertical-dislocation loop by an amount equal to the two horizontal steps shown in Fig. 4.31. This result is characteristic of the intersection of dislocations, for whenever a dislocation cuts another dislocation, both dislocations acquire steps of a size equal to the other's Burgers vector.

Let us now consider some of the simpler types of steps, formed by dislocation intersections. The two basic cases are, first, where the step lies in the slip plane of a dislocation and, second, where the step is normal to the slip plane of a dislocation. The first type is called a *kink*, while the second is called a *jog*. The first is treated in Fig. 4.32. Here (A) represents a dislocation in the edge orientation, and (B) a dislocation in the screw orientation, both of which have received kinks (*on*) as a result of intersections with other dislocations. The kink in the edge dislocation has a screw orientation (Burgers vector parallel to line *on*), while the step in the screw dislocation has an edge orientation (Burgers vector normal to line *on*). Both of these steps can easily be eliminated, for example, by moving line *mn* over to the position of the dashed line. This movement in both cases can occur by simple slip. Since the elimination of a step lowers the energy of the crystal by the

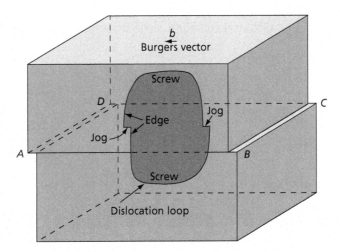

Fig. 4.31 In the figure a dislocation is assumed to have moved across the horizontal plane *ABCD* and, in cutting through the vertical-dislocation loop, it forms a pair of jogs in the latter

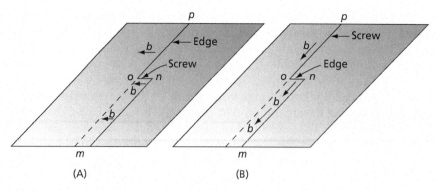

Fig. 4.32 Dislocations with kinks that lie in the slip plane of the dislocations

amount of the strain energy associated with a step, it can be assumed that steps of this type may tend to disappear.

An edge and a screw dislocation, with steps normal to the primary slip plane, are shown in Figs. 4.33A and 4.33B. This type of discontinuity is called a *jog*. It should be noted that the jog of Fig. 4.33B is also capable of elimination if the dislocation is able to move in a plane normal to the indicated slip plane, that is, in a vertical plane. This follows from the fact that if this stepped dislocation is viewed as lying in a vertical plane, then we have the same step arrangement as that shown in Fig. 4.32B. However, let us assume that both dislocations of Fig. 4.33 are not capable of gliding in a vertical plane. What then is the effect of the steps on the motion of the dislocations in their horizontal-slip surfaces? In this respect, it is evident that the edge dislocation with a jog, shown in Fig. 4.33A, is free to move on the stepped surface shown in the figure, for all three segments of the dislocation, *mn, no*, and *op*, are in a simple edge orientation with their respective Burgers vectors lying in the crystal planes that contain the dislocation segments. The only difference between the motion of this dislocation and an ordinary edge dislocation is that instead of gliding along a single plane it moves over a stepped surface.

The screw dislocation with a jog, shown in Fig. 4.33B, represents quite a different case. Here the jog is an edge dislocation with an incomplete plane lying in the stepped surface. For the sake

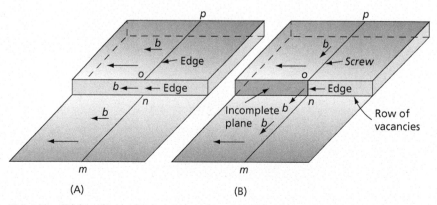

Fig. 4.33 Dislocations with jogs normal to their slip planes

of argument, let us assume that its extra plane lies to the left of line *no* and thus corresponds to the cross-hatched area in Fig. 4.33B. Now there are two basic ways that an incomplete plane that is only one atomic spacing high can be formed. In the first case, it can be assumed that the shaded area corresponds to a row of interstitial atoms which ends at *no*. Alternatively, the shaded step may be considered a part of a normal continuous lattice plane, and then the stepped surface to the right of *no* would have to be a row of vacancies. The latter case is the one indicated in the drawing in Fig. 4.33B. Which of these two alternatives eventuates depends on the relative orientation of the Burgers vectors of the two dislocations, the intersection of which caused the jog. In this respect, it should be mentioned that these jogs are formed when one screw dislocation intersects another screw dislocation. The other steps illustrated in Figs. 4.32 and 4.33A are formed by various intersections involving edge dislocations with other edges, or edge dislocations with screws. Read (Ref. 10) discusses these different possibilities in detail and shows in particular how the intersection of two screw dislocations can lead, at least in principle, to the production of either a row of vacancies or a row of interstitial atoms. Relative to these rows of vacancies, or interstitial atoms, it should be pointed out that after one screw dislocation has intersected another, and if it moves on away from the point of intersection, the row of point defects (vacancies or interstitial atoms) stretches in a line from the moving dislocation back to the stationary one. This, of course, is on the assumption that thermal energy does not cause the point defects to diffuse off into the lattice. In terms of Fig. 4.33B, this signifies that if the indicated dislocation is moving to the left, the row of vacancies to its right extends back to another screw dislocation, whose intersection by the moving dislocation caused the row of vacancies to be formed.

Although the edge dislocation with a jog normal to its slip plane is capable of moving by simple glide along the stepped surface of Fig. 4.33A, this is not the case of the stepped screw dislocation of Fig. 4.33B. Here the jog (line *no*), which is in an edge orientation, is not capable of gliding along the vertical surface shown in the figure because its Burgers vector is not in the surface of the step but is normal to it. The only way that the jog can move across the surface of the step is for it to move by dislocation climb. Thus, in Fig. 4.33B, if the jog is to move to the left with the rest of the dislocation, additional vacancies will have to be added to the row of vacancies (to the right of line *no*). Alternatively, if the extra plane had been formed by a row of interstitial atoms to the left of the jog, the movement of the jog (in this case to the right) requires the creation of additional interstitial atoms. In general, when dislocations pass each other, they leave various defects or "debris" behind (Ref. 11).

4.10 The Stress Field of a Screw Dislocation

The elastic strain of a screw dislocation is shown in Fig. 4.34. In this form of dislocation, the lattice spirals around the center of the dislocation, with the result that a state of shear strain is set up in the lattice. This strain is symmetrical about the center of the dislocation and its magnitude varies inversely as the distance from the dislocation center. Vector diagrams are drawn on the front and top of the cylindrical crystal to indicate the shear sense.

That the strain decreases as one moves away from the dislocation line is deduced as follows. Consider the circular Burgers circuit shown in Fig. 4.34. Such a path results in an advance (parallel to the dislocation line) equal to the Burgers vector b. The strain in the lattice, however, is the advance divided by the distance around the dislocation. Thus,

$$\gamma = \frac{b}{2\pi r} \qquad 4.7$$

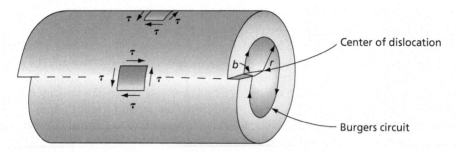

Fig. 4.34 Shear strain associated with a screw dislocation

where *r* is the radius of the Burgers circuit. This strain is accompanied by a corresponding state of stress in the crystal. A great deal of useful information about the nature of the stress fields produced by dislocations has been obtained by assuming the crystals to be homogeneous isotropic bodies. If this is done, the elastic stress field surrounding a screw dislocation is written:

$$\tau = \mu\gamma = \frac{\mu b}{2\pi r} \qquad 4.8$$

where μ is the shear modulus of the material of the crystal. This equation gives a reasonable approximation of the stress at distances greater than several atomic distances from the center of the dislocation. As the center of the dislocation is approached, however, representation of the crystal as a homogeneous and isotropic medium becomes less and less realistic. At positions close to the dislocation, atoms are displaced long distances from their normal lattice positions. Under these conditions, the stress is no longer directly proportional to strain, and it becomes necessary to think in terms of forces between individual atoms. The analysis of the stresses close to the center of the dislocation is extremely difficult, and no completely satisfactory theory has yet been developed. In this respect, it should be mentioned that the infinite stress predicted by Eq. 4.8 at zero radius has no meaning. While the exact stress at the center of the dislocation is not known, it cannot be infinite.

4.11 The Stress Field of an Edge Dislocation

The stress field surrounding a dislocation in an edge orientation is more complex than that of a dislocation in a screw orientation. It will now be assumed that an edge dislocation lies in an infinitely large and elastically isotropic material and that the dislocation line coincides with the *z* axis of a Cartesian coordinate system. Under these conditions the stress can be considered to be independent of position along the *z* direction. In other words, the stress should be simply a function of its position in the *x*, *y* plane, the plane normal to the dislocation. In this regard, consider Fig. 4.35A, where an edge dislocation is shown lying at the origin of a two-dimensional *x*, *y* coordinate system. With the aid of elasticity theory it may be shown that the stress at some point, with coordinates *x* and *y*, has the following components:

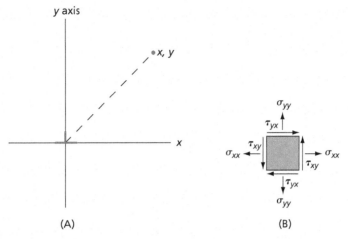

Fig. 4.35 **(A)** An edge dislocation aligned along the z-axis. **(B)** The stress components at point x, y

$$\sigma_{xx} = \frac{-\mu b}{2\pi(1-\nu)} \frac{y(3x^2 + y^2)}{(x^2 + y^2)^2}$$

$$\sigma_{yy} = \frac{\mu b}{2\pi(1-\nu)} \frac{y(x^2 - y^2)}{(x^2 + y^2)^2} \qquad 4.9$$

$$\tau_{xy} = \frac{\mu b}{2\pi(1-\nu)} \frac{x(x^2 - y^2)}{(x^2 + y^2)^2}$$

where σ_{xx} and σ_{yy} are tensile stress components in the x and y directions, respectively, and τ_{xy} is the shear stress as shown in Fig. 4.35B.

In the general case, that is, for an arbitrary position with coordinates x and y, the stress will contain both normal and shear stress components and σ_{xx} may not equal σ_{yy}. However, for points along the x-axis the normal stress components both vanish and the state of stress is pure shear. Also note that the sense of the shear stress along the x axis, which corresponds to the slip plane, is reversed if one moves from the right of the dislocation to its left, as may be seen in Fig. 4.36. Also, as may be seen in Fig. 4.36, above and below the dislocation, that is, along the y axis, there is no shear stress component. Here the stress is a biaxial normal stress with $\sigma_{xx} = \sigma_{yy}$. Above the dislocation the lattice is under a compressive stress, while below it the lattice is in a state of tensile stress. Note that the magnitude of the stress depends only on the distance from the dislocation, r, and varies as 1/r.

A significant fact is that the stress field around a dislocation can be described in a somewhat simpler fashion if one uses polar coordinates as defined in Fig. 4.37. In this case we have

$$\sigma_{rr} = \sigma_{\theta\theta} = \frac{-\mu b}{2\pi(1-\nu)} \cdot \left(\frac{\sin\theta}{r}\right)$$

$$\tau_{r\theta} = \frac{\mu b}{2\pi(1-\nu)} \cdot \left(\frac{\cos\theta}{r}\right) \qquad 4.10$$

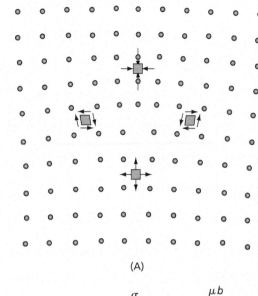
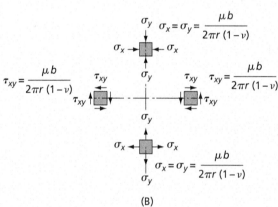

Fig. 4.36 Stress and strain associated with an edge dislocation. (In the above equations, μ is the shear modulus in Pa, b the Burgers vector of the dislocation, ν is Poisson's ratio, and r the distance from the center of the dislocation.)

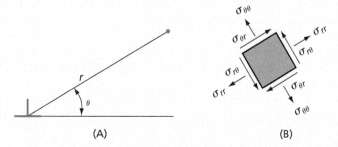

Fig. 4.37 **(A)** An edge dislocation in polar coordinates, **(B)** The corresponding stresses

4.12 The Force on a Dislocation

The concept of the virtual force that a stress applied to a crystal exerts on a dislocation is important. The force exerted on a straight screw dislocation by a shear stress will be examined first. Figure 4.38 shows a right-hand screw dislocation, which is assumed to lie far enough away from the crystal ends that surface end effects may be ignored. The length of this dislocation, L, equals the crystal width. Now imagine that the dislocation moves along the slip plane through a distance Δy. This causes a section of the top half of the crystal of width L and length Δy to be displaced to the left by a Burgers vector, b, relative to the bottom half. The external work W done by the applied stress in this movement of the dislocation is equivalent to that of a force $\tau L \Delta x$ moving through a distance b or

$$W = \tau L \Delta x b \qquad 4.11$$

where τ is the applied shear stress, L the crystal width, and Δx the distance the dislocation moves. The internal work performed, as the dislocation moves, can be expressed as $fL\Delta x$, where f is the virtual force per unit length on the dislocation, L the dislocation length, and Δx the distance through which the total force on the dislocation, fL, moves. Equating the internal work to the external gives

$$fL\Delta x = \tau L \Delta x b$$

or

$$f = \tau b \qquad 4.12$$

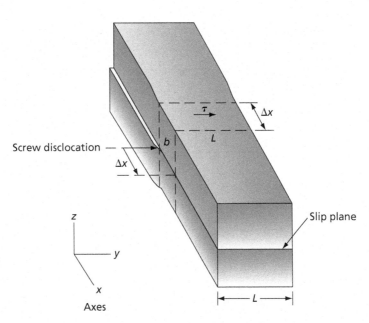

Fig. 4.38 A right-hand screw dislocation in a long crystal

The force per unit length on the dislocation, f, lies in the slip plane and is normal to the dislocation line, that is, it is toward the front of the crystal.

Now consider Fig. 4.39, in which a positive edge dislocation is assumed to move through a distance Δx under an applied shear stress τ. The dislocation line length and crystal width are again assumed to equal L. By an argument similar to that just used for the screw dislocation, it may be easily shown that the force per unit length on the edge component of a dislocation line is also

$$f = \tau b \qquad 4.13$$

This force also lies in the slip plane and is normal to the dislocation line. Note that since the screw and edge components of the same dislocation are normal to each other, the forces on these components must also be normal to each other. It can also be shown that if a part of a dislocation has a mixed—part edge and part screw—character, the corresponding force per unit length, f, on the segment is also normal to it. Thus, a dislocation loop, such as that shown in Fig. 4.18, will be subjected to a force normal to the dislocation $f = \tau b$ everywhere around the loop.

There remains to be considered the climb force on an edge dislocation. Here the stress applied to the crystal is a normal (tensile or compressive) stress. Thus, consider Fig. 4.40, where a tensile stress σ is shown applied to a crystal containing a positive edge dislocation. The given tensile stress will act to make the extra plane increase in size or undergo negative climb. As a result, the dislocation line moves downward. In this case it may be readily shown that the force per unit length on the dislocation is given by

$$f = -\sigma b \qquad 4.14$$

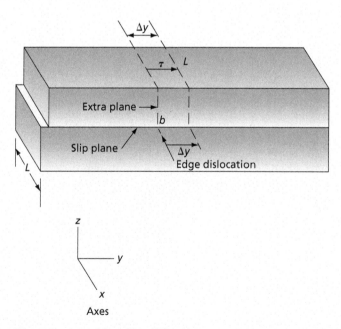

Fig. 4.39 A positive edge dislocation

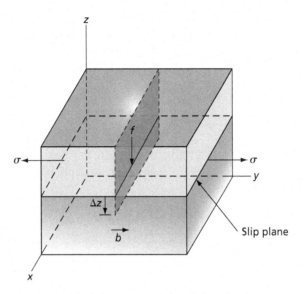

Fig. 4.40 Climb force on an edge dislocation

where f is the force on the dislocation line, σ the tensile stress, and b the Burgers vector of the dislocation. The climb force on a dislocation is not only normal to the dislocation line but it is also normal to the slip plane of the dislocation. This is in contrast to the direction of f in slip, which is also always normal to the dislocation line but lies in the slip plane. If the applied stress is purely compressive, the climb force will point toward the positive z direction and the dislocation should climb upward.

In the general case, however, a dislocation may have a mixed Burgers vector, so that it is neither pure screw nor pure edge, and the dislocation line may lie in any direction. Thus, it is common practice to define the orientation of the dislocation line, at a given point along it, by the unit vector tangent to the dislocation line at the point. This vector is designated ζ and has the components ζ_x, ζ_y, and ζ_z. Furthermore, the Burgers vector can be expressed in the form

$$b = b_x i + b_y j + b_z k \qquad 4.15$$

where b_x, b_y, and b_z are three components of the Burgers vector and i, j, and k are unit vectors in the x, y, and z directions, respectively.

When a general stress Σ, which may have both normal and shear components in all three directions, is applied to a crystal containing a dislocation, the force on the dislocation is given by the cross-product of Σ and ζ, which can be written as

$$f = \Sigma X \zeta = \begin{vmatrix} i & j & k \\ \Sigma_x & \Sigma_y & \Sigma_z \\ \zeta_x & \zeta_y & \zeta_z \end{vmatrix} \qquad 4.16$$

where f is the force per unit length on the dislocation, and Σ_x, Σ_y, Σ_z are given by the following summations:

$$\Sigma_x = \sigma_{xx}b_x + \tau_{xy}b_y + \tau_{xz}b_z$$
$$\Sigma_y = \tau_{yx}b_x + \sigma_{yy}b_y + \tau_{yz}b_z \qquad 4.17$$
$$\Sigma_z = \tau_{zx}b_x + \tau_{zy}b_y + \sigma_{xx}b_z$$

Because of the cross-product between Σ and ζ in Eq. 4.16, the force on the dislocation is normal to the dislocation. The relation between the dislocation, its Burgers vector, and the stress, given above, can be written

$$F = (b \cdot G)X\zeta \qquad 4.18$$

This is known as the Peach–Kohler equation. More detailed discussion on this subject can be found elsewhere (Refs. 6 and 12).

4.13 The Strain Energy of a Screw Dislocation

An important property of a dislocation is its strain energy, normally expressed as its energy per unit length. This parameter can be readily obtained for a screw dislocation located at the center of an infinitely long, large cylindrical crystal, see Fig. 4.34. According to linear elasticity theory, the strain energy density in the stress field of a screw dislocation is $\tau^2/2\mu$, and by Eq. 4.8 the stress, τ, is $\mu b/2\pi r$. This suggests that the strain energy per unit length of this screw dislocation might be estimated with the following integration.

$$w_s = \int_{r_0}^{r'} \left(\frac{\mu b}{2\pi r}\right)^2 \left(\frac{1}{2\mu}\right) 2\pi r \, dr = \frac{\mu b^2}{4\pi} \ln \frac{r'}{r_0} \qquad 4.19$$

where w_s is the energy per unit length of the screw dislocation, μ the shear modulus, b the Burgers vector, r_0 an inner radius that excludes the dislocation core, and r' an outer limiting radius for the integration. A unit thickness is assumed for the volume over which the integration is made.

The choices of r_0 and r' involve the consideration of a number of factors. First, with regard to r_0, Eq. 4.19 clearly indicates that as $r_0 \to 0$, $w \to \infty$, implying that a lower limit for r_0 is necessary. This conclusion is also strengthened by the fact that near the dislocation center, the atomistic character of a crystal becomes more and more significant and the assumption that the crystal is a simple elastic continuum becomes less and less tenable. Further, while the strain certainly must be very severe, in this region, it can never be infinite as predicted by Eq. 4.7 for $r_0 = 0$. For these reasons it is normally assumed that linear elasticity does not hold below $r_0 \sim b$, where b is the Burgers vector. However, in order to include the energy in the highly strained region of the dislocation core it has been suggested (Ref. 6) that one take $r_0 = b/\alpha$, where α is a constant. The value of α has been variously suggested to be between 2 and 4. For the present let $\alpha = 4$.

Equation 4.19 also indicates that $w \to \infty$ as $r' \to \infty$. Fortunately, a crystal normally possesses a finite dislocation density. Even a well-annealed crystal may contain about 10^8 cm/cm^3 (10^{12} m/m^3) of dislocations. Under these conditions the stress field of a dislocation may be assumed to be neutralized by those of its neighbors at a distance r' equal to one-half the average spacing between dislocations. This assumption is reasonable if the numbers of the dislocations with opposite signs are approximately equal.

Assuming that one has a soft iron crystal with an array of infinitely long straight screw dislocations, with equal numbers of opposite signs, and whose density $\rho = 10^8$ cm/cm^3, the average distance between dislocations would be approximately 10^{-6} m so that $r' = 5 \times 10^{-7}$ m. The shear modulus of iron is about 8.6×10^{10} Pa and its Burgers vector equals 2.48×10^{-10} m. With the aid of these data Eq. 4.19 gives $w_s = 3.79 \times 10^{-9}$ J/m (3.79×10^{-11} J/cm, 3.79×10^{-4} ergs/cm). If one multiplies w_s by the dislocation density, a rough estimate of the stored strain energy in a unit volume may also be computed. Thus for $\rho = 10^8$ cm/cm^3 the stored energy in a cm^3 should be about 3.79×10^4 erg/cm^3 (3.79×10^{-3} J/cm).

4.14 The Strain Energy of an Edge Dislocation

An equation for the strain energy per unit length may also be derived for an infinitely long edge dislocation using an approach similar to that used to obtain Eq. 4.19. The result is

$$w_e = \frac{\mu b^2}{4\pi(1-\nu)} \ln \frac{4r'}{b} \qquad 4.20$$

where w_e is the strain energy per unit length of an edge dislocation, μ the shear modulus, ν Poisson's ratio, b the Burgers vector, and r' the outer radius of the volume over which the integration is carried out. As in the case of the screw dislocation, the inner limit r_0 is taken as $b/4$.

Note that the strain energy for the edge dislocation differs from that of the screw dislocation by a factor of $1/(1-\nu)$. Since ν for most metals is near 1/3, the strain energy for the edge dislocation is thus about 50 percent larger than that for the screw dislocation.

Problems

4.1 Prove that 1,000 psi = 6.9 MPa using conversion factors given in Appendix D.

4.2 A single crystal of copper yields under a shear stress of about 0.62 MPa. The shear modulus of copper is approximately 7.9×10^6 psi. With these data compute an approximate value for the ratio of the theoretical to the experimental shear stresses in copper.

4.3 Make a model of a left-hand screw dislocation following the technique shown in Fig. 4.11.

4.4 (a) Using an RHSF Burgers circuit, illustrate how to determine the true Burgers vector of an edge dislocation.

(b) Is the sense of the Burgers vector direction in part (a) of this problem the same as that of the RHFS local Burgers vector in Fig. 4.19.

4.5 (a) How many equivalent {111} ⟨1$\bar{1}$0⟩ slip systems are there in the fcc lattice?

(b) Identify each system by writing out its slip plane and slip direction indices.

4.6

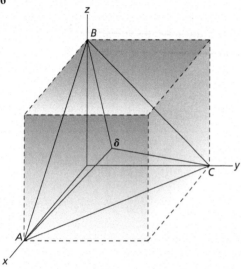

Assume that the triangle in the drawing lies on the (111) plane of a face-centered cubic crystal, and that its edges are equal in magnitude to the Burgers vectors of the three total dislocations that can glide in this plane. Then, if δ lies at the centroid of this triangle, lines Aδ, Cδ, and Bδ, accordingly, correspond to the three possible partial dislocations of this plane.

(a) Identify each line (AB, Aδ, etc.) with its proper Burgers vector expressed in the vector notation.

(b) Demonstrate by vector addition that

$$B\delta + \delta C = BC$$

4.7

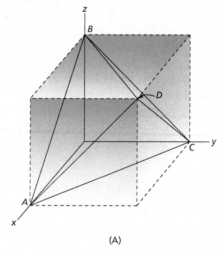

(A)

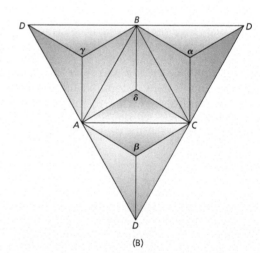

(B)

Figure A represents the Thompson tetrahedron as seen in three dimensions. In Fig. B, the sides of the tetrahedron have been hypothetically folded out so that all four surfaces can be easily viewed. The symbols α, β, γ, and δ represent the centroid of each face, respectively. As in Problem 4.6, lines such as BD represent total dislocations, that is, $\frac{1}{2}[110]$, and lines such as $B\gamma$ represent partial dislocations, that is, $\frac{1}{6}[21\bar{1}]$.

(a) Write the Miller indices symbols for CD and DC.

(b) Do the same for BD and DB.

(c) Show that $BD + DC = BC$.

4.8 List all the primary or total Burgers vectors available for slip in a fcc crystal. Use the vector or Miller indices notation to describe each Burgers vector.

4.9 An important slip plane in the hexagonal metals, titanium and zirconium, is the $\{11\bar{2}2\}$ plane on which dislocations with a $\frac{1}{3}\langle11\bar{2}\bar{3}\rangle$ Burgers vector may move. Prove that the $\frac{1}{3}\langle11\bar{2}\bar{3}\rangle$ Burgers vector can be considered as the sum of a basal slip Burgers vector and a unit Burgers vector in the c-axis direction.

4.10 (a) If a vacancy disc forms on a basal plane of a hexagonal close-packed metal, what stacking sequence of basal planes would result across the disc?

(b) Why would the strain energy of the resulting stacking fault be very high?

(c) Explain how a simple shear along the basal plane, equal to that of a Shockley partial, could eliminate this high-energy stacking fault and replace it with one of lower energy.

(d) What then would be the stacking arrangement of the basal planes at the fault?

(e) Is the result in (d) unique or are there several basic possibilities? Explain.

4.11 (a) Write a simple computer program that gives the shear stress of a screw dislocation as a function of the perpendicular distance from the dislocation (see Eq. 4.8). Assuming the shear modulus of iron is 86 GPa and the Burgers vector is 0.248 nm, use the program to obtain the shear stress at the following values of r: 50, 100, 150, and 200 nm, respectively. Plot the resulting τ versus r data, and with the aid of this curve, determine the distance from the dislocation where τ is 4000 psi, the shear stress at which an iron crystal will begin to undergo slip.

(b) To how many Burgers vectors does this distance correspond?

4.12 Equations 4.9 are the stress field equations of an edge dislocation, in Cartesian coordinates. Write a computer program based on these equations that gives simultaneous values of σ_{xx}, σ_{yy}, τ_{xy} for an edge dislocation in an iron crystal assuming $\mu = 86$ GPa, $b = 0.248$ nm, $\nu = 0.3$, and $r = 40b$, and letting $x = r \cos \theta$ and $y = r \sin \theta$. Simplify the equations so that they can be expressed as a function of only the angle θ. Now develop a figure in which curves are plotted for all three of the stress components over the range of angles from 0 to 2π radians.

4.13 Now solve Problem 4.11 using the stress field equations of an edge dislocation expressed in polar coordinates.

4.14 (a) Consider that two infinitely long parallel positive edge dislocations are viewed on a simple two-dimensional x, y diagram such as Fig. 4.36. One dislocation is located at $x = 0$, $y = 0$. Since this is a positive edge, its slip plane is horizontal and contains the x axis. The other dislocation also lies in a horizontal slip plane but this plane is separated from that of the first dislocation by a vertical distance (y) of $10b$, where b is the Burgers vector for both dislocations. Now assume the first dislocation is fixed in place and the second can move parallel to the x axis. With the aid of a computer, plot F_x, the x component of the force (per unit length) between the dislocations, as a function of x from $x = -240$ nm to $x = +240$ nm. Let $\mu = 86$ GPa, $b = 0.248$ nm, and $\nu = 0.3$.

(b) Discuss the significance of the variation of F_x with distance as the mobile dislocation is moved from $x = -\infty$ to $+\infty$.

4.15 The strain energy of a dislocation normally varies as the square of its Burgers vector. One may see this by examining Eqs. 4.19 and 4.20. This relationship between the dislocation strain energy and the Burgers vector is known as *Frank's rule*. Thus, if $b = a[hkl]$, where a is a numerical factor, then

$$\text{Energy/cm} \sim a^2 \{h^2 + k^2 + l^2\}.$$

Show that in an f.c.c. crystal the dissociation of a total dislocation into its two partial dislocations is energetically feasible. See Eq. 4.4.

4.16 The *c/a* ratio of the hcp zinc crystal is 1.886. Determine the ratio of the strain energy in zinc of a dislocation with a $\frac{1}{3}\langle 11\bar{2}3\rangle$ Burgers vector to that of a basal slip dislocation.

4.17 (a) Consider Eq. 4.19, which gives the strain energy per unit length of a screw dislocation. Assume that one has a very large square array of long, straight, parallel screw dislocations of alternating signs so that the effective outer radius r' of the strain field of a dislocation may be taken as $1/2\sqrt{\rho}$. With the aid of a computer determine the strain energy per unit length of a screw dislocation as a function of the dislocation density, ρ, between $\rho = 10^{11}$ and $\rho = 10^{18}$ m/m^3.

(b) Plot the line energy against the dislocation density. Assume that $\mu = 86$ GPa and $b = 0.248$ nm.

(c) Now plot the energy per unit volume as a function of ρ, assuming that the former can be equated to ρw, where w is the energy per unit length of the screw dislocations.

References

1. Orowan, E., *Z. Phys.*, **89** 634 (1934).
2. Polanyi, *Z. Phys.*, **89** 660 (1934).
3. Taylor, G. I., *Proc. Roy. Soc.*, **A145** 362 (1934).
4. Nabarro, F. R. N., *Theory of Crystal Dislocations*, p. 5, Oxford University Press, London, 1967.
5. *Dislocations and Properties of Real Materials*, published by the Institute of Metals, 1985.
6. Hirth, J. P. and Lothe, J., *Theory of Dislocations*, 2nd ed., John Wiley and Sons, New York, 1982.
7. *Computed Electron Micrographs and Defect Identification*, A. K. Head et al., North-Holland Publishing Company, 1973, p. 176.
8. Seeger, A., and Schoeck, G., *Acta Met.* **1** 519 (1953).
9. Seeger, A., *Dislocations and Mechanical Properties of Crystals*, John Wiley and Sons, Inc., New York, 1957.
10. Read, W. T., Jr., *Dislocations in Crystals*, McGraw-Hill Book Company, Inc., New York, 1953.
11. *Dislocations and Properties of Real Materials*, published by The Institute of Metals, London, 1985, p. 145.
12. Weertman, J., and J. R. Weertman, *Elementary Dislocation Theory*, p. 55, Macmillan, New York, 1964.

Chapter 5

Dislocations and Plastic Deformation

> **Learning Objectives**
>
> Upon completion of Chapter 5, you will be able to:
> 1. Develop an understanding of dislocation generation via Frank-Read mechanisms.
> 2. Explain slip planes, slip directions and slip systems
> 3. Differentiate between cross slip, double cross-slip and slip bands
> 4. Understand development of cross-slip by dissociation of extended dislocations
> 5. Calculate resolved shear stress of a tensile force on a given slip plane and slip direction
> 6. Differentiate engineering stress-strain curve from true-stress and true strain curve
> 7. Explain mechanism of work (or strain) hardening
> 8. Understand the relations between dislocation density and applied stress

The dislocation topics in the preceding chapter dealt primarily with (1) the structure or geometry of dislocations and (2) their stress and strain fields. This chapter is concerned with the relation of dislocations to plastic deformation. The first topic to be considered involves a mechanism proposed some years ago to explain why the amount of shear strain that is obtained when a soft (annealed) crystal specimen is plastically deformed is normally many times greater than could be attained from slip due to the glide of the dislocations that existed in the crystal before the start of the deformation. This clearly indicates that dislocations must be created during plastic deformation. While there are a number of ways that dislocations may be created as a result of plastic deformation, the Frank-Read source is one of considerable historical significance.

5.1 The Frank-Read Source

Suppose that a positive edge dislocation xy, lying in plane $ABCD$ of Fig. 5.1, is connected to two other edge dislocations running vertically to the upper surface of the crystal. The two vertical edge dislocations cannot move under the applied shear stress τ. This is because, in general, the movement of a dislocation results in the displacement of a part of a crystal relative to an adjacent part. When an externally applied stress causes such a movement, the externally applied stress moves and performs work. Movements of the vertical dislocation segments will cause shear displacements between front and back parts of the crystal. The applied shear stress τ on the top and bottom of the crystal cannot cause this type of displacement. The vertical segments would move, however, under a shear stress in the horizontal direction applied to the front and back of the crystal. Since the dislocation segment xy is a positive edge, the stress τ will tend to move it to the left, causing the line to form an arc with ends at the fixed endpoints x and y. This arc is indicated in Fig. 5.2 by the symbol a. Further application of the stress causes the curved dislocation to expand to the successive positions b and c. At c the loop intersects itself at point m, but since one intersecting segment is a left-hand screw and the other a right-hand screw, the segments cancel each other at the point of intersection. (A cancellation always occurs when opposite forms of dislocations lying in the same plane intersect. This fact can be easily demonstrated for the case of positive and negative edge dislocations since their intersections form a complete lattice plane from two incomplete planes.) The cancellation of the dislocation segments at the point of contact m breaks the dislocation into two segments marked d, one of which is circular and expands to the surface of the crystal, producing a shear of one atomic distance. The other component remains as a regenerated positive edge, lying between points x and y where it is in a position to repeat the cycle. In this manner many dislocation loops can be generated on the same slip plane, and a shear can be produced that is large enough to account for the large size of observed slip lines. A dislocation generator of this type is called a *Frank-Read source*.

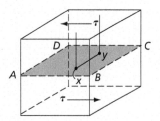

Fig. 5.1 Frank-Read source. The dislocation segment xy may move in plane $ABCD$ under the applied stress. Its ends, x and y, however, are fixed

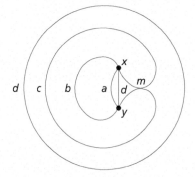

Fig. 5.2 Various stages in the generation of a dislocation loop at a Frank-Read source

5.2 Nucleation of Dislocations

Experimental evidence shows that Frank-Read sources actually exist in crystals (Ref. 1). How important these dislocation generators are in the plastic deformation of metals is not known, but other evidence shows that dislocations can also be formed without the aid of Frank-Read or similar sources (Ref. 2). If dislocations are not formed by dislocation generators, then they must be created by a nucleation process. As with all nucleation phenomena, dislocations can be created in two ways: homogeneously or heterogeneously. In the case of dislocations, *homogeneous nucleation* means that they are formed in a perfect lattice by the action of a simple stress, no agency other than stress being required. *Heterogeneous nucleation*, on the other hand, signifies that dislocations are formed with the help of defects present in the crystal, perhaps impurity particles. The defects make the formation of dislocations easier by lowering the applied stress required to form dislocations. It is universally agreed that homogeneous nucleation of dislocations requires extremely high stresses, stresses that theoretically are of the order of $\frac{1}{10}$ to $\frac{1}{20}$ of the shear modulus of a crystal (Ref. 3). Since the shear modulus of a metal is usually about 7 to 70 GPa, the stress to form dislocations should be of the order of 0.7 GPa. However, the actual shear stress at which metal crystals start to deform by slip is usually around 70 MPa. This evidence certainly favors the opinion that if dislocations are not formed by Frank-Read sources, then they must be nucleated heterogeneously.

It should be noted that dislocations can also nucleate heterogeneously at the crystal surface and at critical stresses much less than that required for homogenous nucleation within the crystal. Moreover, the critical stress is further reduced when the crystal surface is not perfectly flat and contains heterogeneities, such as atomic scale surface steps. For example, the recent atomic scale calculations by Guanshui Xu and co-workers (Refs. 4 and 5) indicated that, compared to nucleation of screw dislocation at a perfectly flat surface, the required critical stress is reduced by about an order of magnitude at the presence of surface steps. Stress concentration at the heterogeneities and thermal fluctuations are thought to play a critical role in the nucleation of such dislocations.

Metal crystals are not particularly suited to an investigation of nucleation phenomena. When prepared by solidification or other means, they usually possess a relatively high density of grown-in dislocations in the form of a more-or-less random network that extends throughout the specimen. Our present interest does not lie in these networks, but in the creation of new, independent dislocation loops. A high concentration of grown-in dislocations, however, complicates the observation of nucleation phenomena. Also, in metals, slip occurs readily so that once the yield point is reached, many dislocations usually form at the same time. These experimental difficulties are almost eliminated if dislocation nucleation is studied in crystals of lithium fluoride (Ref. 6) which is an ionic salt that crystallizes in the simple cubic (rock salt) lattice. This material may be prepared in single-crystal form with a high degree of perfection so that it has a low density of grown-in dislocations (5×10^4 cm of dislocations per cubic cm). In addition, crystals of LiF are rigid enough at room temperature to be handled without distortion, and they are only slightly plastic at this temperature. Thus, with a small stress (5 to 7 MPa) applied for a short period of time, dislocations can be formed in them in controlled small numbers.

One of the best and simplest ways of observing dislocations in crystals is through the use of an etching reagent, which forms an etch pit on the surface of a crystal at each point where a dislocation intersects the surface. This method is not without its difficulties for there is often no way of knowing whether the etch reveals all dislocations, or whether some of the pits are due to other defects. In lithium fluoride, the etch-pit method seems to be highly reliable (Ref. 6). Several etching solutions have been developed (Ref. 7) for use in LiF, one of which is capable of distinguishing between grown-in dislocations and newly formed dislocations. The action of this solution can be seen in Fig. 5.3. The large square pits that run in two horizontal rows are associated with newly formed dislocations. The horizontal rows define the intercept of the slip plane of these dislocations with the surface. In addition to the large pits, there are two intersecting curved rows of closely spaced smaller pits. The latter outline what is known as a *low-angle grain boundary*. (See Chapter 6.) The boundaries actually consist of a number of closely spaced dislocations. It should be noted that the given etching solution forms large pits at new dislocations and small pits at network dislocations. The reason for this ability of the etch to distinguish between the two types of dislocations is not understood, but it might be related to the fact that impurity atoms tend to collect around dislocations. This segregation of impurity atoms cannot normally occur in a reasonable length of time at low temperatures because the atoms of the solid do not diffuse or move fast enough at low temperatures to let them collect around dislocations. However, at higher temperatures the movement of impurity atoms to dislocations can occur quite rapidly. Thus, dislocations formed at high temperatures are more likely to have impurity atoms segregated around them than dislocations created in the lattice at room temperature. This ability of an etching solution to distinguish between the grown-in and newly formed dislocations is one of the distinct advantages offered by the use of LiF in studying nucleation phenomena.

Another interesting facet to the use of etch pits-in observing dislocations in LiF is that with the proper technique one can follow the movement of a dislocation under the action of an applied

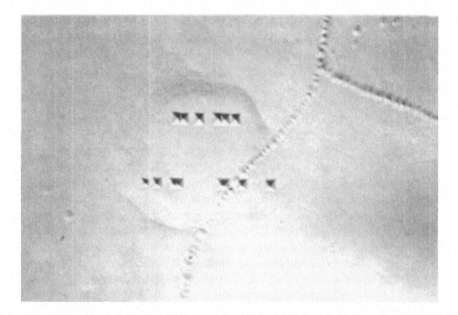

Fig. 5.3 The large square etched pits in horizontal rows correspond to dislocations formed in LiF at room temperature, while the smaller, closely spaced pits lying in curved rows were grown into the crystal when it was manufactured (Gilman, J. J., and Johnson, W. G., *Dislocations and Mechanical Properties of Crystals*, p. 116, John Wiley and Sons, Inc., New York, 1957. Used by permission of the author.)

stress. This usually requires several repetitions of the etching process. Thus, the surface of a specimen is first etched to reveal the positions of the dislocations at a given time. The pits that form are usually observed on {001} surfaces: LiF crystals are easily split along {001} planes. On this type of surface, the pits form as four-sided pyramids with a sharp point at their lower extremity. If a stress is applied to the specimen now, the dislocations will move away from their pits. A second etch will both reveal the new positions of the dislocations and enlarge the old pits representing the original positions. The two sets of pits have a distinct difference in appearance, however. The pits actually connected with the dislocation always have pointed extremities, while those from which the dislocations have moved have flat bottoms. See Fig. 5.4.

Johnston and Gilman (Ref. 7) took advantage of their technique for following the movement of a dislocation in LiF, and were able to measure the velocity of a dislocation moving under a fixed applied stress. They simply divided the distance that a dislocation moved by the time that the stress pulse was applied to the crystal containing the dislocation. This subject will be discussed further in Section 5.21.

The work on lithium fluoride has shown that, in this particular material, the grown-in dislocations are usually firmly anchored in place and do not take part in the plastic deformation processes. The immobility of network dislocations can be credited to the presence of atmospheres of impurity atoms that segregate around each dislocation. This subject is considered in more detail in Chapter 9.

It has also been demonstrated (Ref. 6) experimentally that very large stresses can be applied to LiF crystals without homogeneously nucleating dislocations. For example, when a small, carefully cleaned sphere (made of glass) is pressed on a dislocation-free region of the surface of a LiF crystal, it is possible to attain shear stresses estimated to be as large as 760 MPa, and still not create dislocations. This stress, it is to be noted, is more than 100 times larger than the stress normally required to cause yielding in this material. It is further believed that even this high stress, which was limited by experimental difficulties caused by breaking the glass ball, does not define the stress required to homogeneously nucleate dislocations in LiF. At any rate, it is clear that nucleation of dislocations by an unaided stress is very difficult. Because the yield stress, or the stress at which dislocations normally start to move, is much lower than that required to homogeneously nucleate dislocations, it is clear that the majority of dislocations must be nucleated heterogeneously. Gilman concludes (Ref. 6) that in LiF small foreign heterogeneities cause most of the dislocation nucleation. The most important of these are probably small impurity particles. Experimental evidence (Ref. 6) for the formation of dislocations at inclusions in LiF crystals has actually been attained.

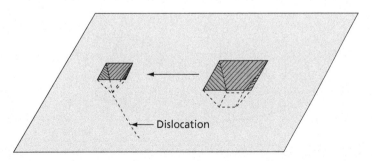

Fig. 5.4 Dislocation movement in LiF as revealed by repeated etching (Reprinted with permission from J. J. Gilman and W. G. Johnson, *Journal of Applied Physics*, Vol. 30, Issue 2, Page 129, Copyright 1959, American Institute of Physics)

5.3 Bend Gliding

It has been known for a good many years that crystals can be plastically bent and that this form of plastic deformation occurs as a result of slip. The bending of crystals can be explained in terms of Frank-Read or other sources.

Let equal couples (of magnitude M) be applied to the ends of the crystal shown in Fig. 5.5. The effect of these couples is to produce a uniform bending moment (M) throughout the length of the crystal. Until the yield point of the crystal is exceeded, the deformation will be elastic. The stress distribution across any cross-section, such as aa, is given by the equation

$$\sigma_x = \frac{My}{I} \qquad 5.1$$

where y is the vertical distance measured from the neutral axis of the crystal (dotted horizontal center line), M is the bending moment, and I the moment of inertia of the cross-section of the crystal ($\pi r^4/4$ for crystals of circular cross-section). Figure 5.5B shows the stress distribution, which varies uniformly from a maximum compressive stress at the upper surface, to zero at the neutral axis, and then to a maximum tensile stress at the lower surface. Figure 5.6 represents the same crystal in which the line mn is assumed to represent the trace of a slip plane. The curvature of the crystal is not shown in order to simplify the figure. For convenience, it is further assumed that the slip plane is perpendicular to the plane of the paper and that the line mn is also the slip direction. The horizontal vectors associated with line mn represent the same stress distribution as that of Fig. 5.5B. Figure 5.6 also shows, along line op, the shear-stress component (parallel to the slip plane) of the stress distribution. Notice that the sense of the shear stress changes its sign as it crosses the neutral axis. Furthermore, the shear stress is zero at the neutral axis and a maximum at the extreme ends of the slip plane. Because of the shear-stress distribution, the first dislocation loops will form at Frank-Read or other sources close to either the upper or lower surface. The manner in which these dislocation loops move, however, depends on whether the dislocation lies above or below the neutral axis of the specimen. In both cases, the positive edge components of all dislocation loops move toward the surface, while the negative edge components move toward the specimen's neutral axis. (See Fig. 5.7). The

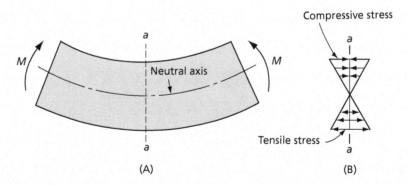

Fig. 5.5 **(A)** The elastic deformation of a crystal subject to two equal moments (M) applied at its ends. **(B)** The normal stress distribution on a cross-section such as aa

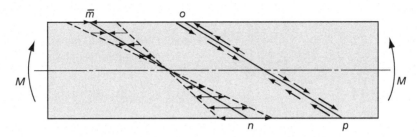

Fig. 5.6 The stress distribution on slip planes corresponding to the elastic deformation shown in Fig. 5.5

negative edge dislocations move toward a region of decreasing shear stress and must eventually stop. The positive edge dislocations, on the other hand, are in a region of high stress, as are the right- and left-hand screw components of each dislocation loop, which move under the applied stress in a direction either into or out of the paper. All three of these last components (positive edge and right- and left-hand screws) can be assumed to move to the surface and leave the crystal. For example, under the applied bending-stress distribution, closed dislocation loops originating at Frank-Read sources eventually become negative-edge dislocations that move toward the neutral axis of the crystal. As the crystal is bent further and further, the negative-edge dislocations will be driven further and further along the slip planes toward the center of the crystal. Eventually an orderly sequence of dislocations will be found on each active slip plane, with the dislocations having a more or less uniform separation, the minimum spacing of which is dependent on the fact that dislocations of the same type and same sign mutually repel if they lie on the same slip plane. Figure 5.8 illustrates the general nature of the dislocation distribution. The narrow section surrounding the neutral axis is presented free of dislocations, in agreement with the fact that this region, under moderate bending stresses, will not be stressed above the elastic limit, and the deformation will be elastic and not plastic.

Now each negative-edge dislocation of a sequence lying on a given slip plane represents an extra plane that ends at the slip plane. Each of these extra planes lies on the left of its slip plane. In order to accommodate these extra planes (all on one side of a given slip plane), the slip planes must assume a curvature that is convex downward and to the left, and the crystal as a whole assumes a convex curvature downward.

If the couples applied to the crystal in Figs. 5.5 to 5.8 were reversed, an excess of positive dislocations would develop along the slip planes. The slip planes and the crystal would then assume a curvature the reverse of that described above.

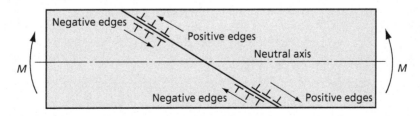

Fig. 5.7 The effect of the stress distribution on the movement of dislocations. Positive-edge components move toward the surface; negative edges toward the neutral axis

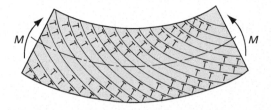

Fig. 5.8 Distribution of the excess edge dislocations in a plastically bent crystal

In the preceding discussion, the crystal was assumed to be of macroscopic dimensions and the bending deformation was assumed to be uniformly distributed over the length of the crystal. The phenomenon that has been described is not, however, restricted to large crystals. Bending of crystal planes through accumulation of an excess number of edge dislocations of the same sign may occur in quite small crystals, or even in extremely small areas of crystals. A number of related phenomena that have been observed in metal crystals can be explained in terms of localized lattice rotations. In each case, there is an accumulation of edge dislocations of the same sign upon slip planes of the crystal. Among these are kinks, bend planes, and deformation bands. A detailed description of the latter is beyond the scope of the present text, but the fact that they are frequently observed emphasizes the fact that plastic bending of metal crystals is an important deformation mechanism.

5.4 Rotational Slip

Thus far we have seen that dislocations are capable of producing simple shear (as in Fig. 4.12) and bending (as in Fig. 5.8). A third type of deformation that can be developed by dislocations is shown in Fig. 5.9. In this sketch, the crystal is assumed to be a cylinder with an active slip plane

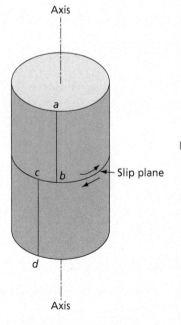

Fig. 5.9 A single crystal can be rotated about an axis normal to a slip plane that contains several slip directions

perpendicular to its axis. As may be seen in the drawing, the top half of the crystal has been rotated clockwise relative to the bottom half. This is indicated by the horizontal displacement of the line *abcd* between the points *b* and *c*. This type of deformation therefore corresponds to a rotation of a crystal on its slip plane about an axis normal to the slip plane. Torsional deformation such as this can be explained in terms of screw dislocations lying on the slip plane. However, unlike the case of bending, more than one set of dislocations is required. This, in turn, signifies that the slip plane must contain more than one possible slip direction. The basal plane in a hexagonal close-packed metal and the {111} planes in face-centered cubic metals, with their three slip directions, are almost ideal for producing this type of deformation.

The need for more than one set of screw dislocations in order to explain rotational slip can be seen with the aid of Figs. 5.10 and 5.11. These diagrams are drawn in a manner similar to Figs. 4.15 and 4.16 and therefore correspond to a view looking down from above on the slip plane in a simple cubic lattice. As before, open circles represent atoms just above the slip plane, while dots correspond to atoms just below it. Figure 5.10 corresponds to a single array of (horizontal) parallel screw dislocations. As may be seen in the illustration, such a dislocation arrangement shears the material

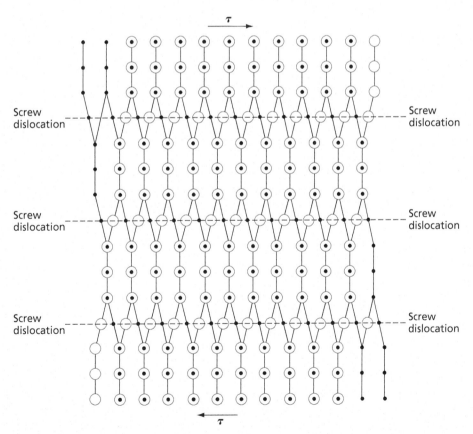

Fig. 5.10 An array of parallel screw dislocations. Open circles represent atoms just above the slip plane, while dots correspond to atoms just below it

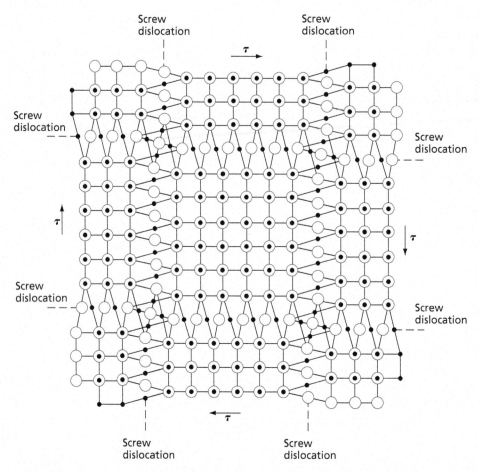

Fig. 5.11 A double array of screw dislocations. This array does not have a long-range strain field; open circles show atoms above the slip plane, while dots represent those below the plane

above the slip plane relative to that below it only in the horizontal direction. For a true rotation one needs a similar component of shear at 90° to this direction. A simple double array of screw dislocations is shown in Fig. 5.11. This gives the desired rotational deformation. Further, it should be noted that the two arrays of screw dislocations at 90° to each other have strain fields that tend to compensate each other, either above or below the slip plane. As a result, the strain field of the double array tends to be very small at any reasonable distance from the slip plane; and the array, therefore, corresponds to one of low strain energy. This is not true, however, of a single array, as shown in Fig. 5.10. Here the strain fields of the individual parallel dislocations are additive and the array is not one of low energy.

While rotational slip has not been extensively studied, it does represent a basic way in which a crystal can be deformed. The amount of slip deformation one can obtain by this mechanism can be very large. In fact, it is possible to twist a 1 cm diameter zinc crystal about its basal plane pole through as many as 10 or more revolutions per inch. This deformation, of course, occurs on many slip planes distributed over the gage section and not on one plane, as shown in Fig. 5.9.

5.5 Slip Planes and Slip Directions

It is an experimental fact that in metal crystals slip, or glide, occurs preferentially on planes of high-atomic density. It is a general rule that the separation between parallel lattice planes varies directly as the degree of packing in the planes. Therefore, crystals are sheared most easily on planes of wide separation. This statement does not mean that slip cannot occur in a given crystal on planes other than the most closely packed plane. Rather, it means that dislocations move more easily along planes of wide spacing where the lattice distortion due to the movement of the dislocation is small.

Not only does slip tend to take place on preferred crystallographic planes, but the direction of shear associated with slip is also crystallographic. The slip direction of a crystal (shear direction) has been found to be almost exclusively a close-packed direction, a lattice direction with atoms arranged in a straight line, one touching the next. This tendency for slip to occur along close-packed directions is much stronger than the tendency for slip to occur on the most closely packed plane. For practical purposes, it can usually be assumed that slip occurs in a close-packed direction.

The fact that the experimentally determined slip direction coincides with the close-packed directions of a crystal can be explained in terms of dislocations. When a dislocation moves through a crystal, the crystal is sheared by an amount equal to the Burgers vector of the dislocation. After the dislocation has passed, the crystal must be unchanged in the geometry of the atoms; that is, the symmetry of the crystal must be retained. The smallest shear that can fulfill this condition equals the distance between atoms in a close-packed direction.

In order to explain this point more clearly, let us consider a hard-ball model of a simple cubic crystal structure. Line mn of Fig. 5.12A is a close-packed direction. In Fig. 5.12B the upper half of the lattice has been sheared to the right by a, the distance between atoms in the direction mn. The shear, of course, has not changed the crystal structure. Consider now an arbitrarily chosen nonclose-packed direction such as qr in Fig. 5.12A. Figure 5.12C shows that a shear of c (the distance between atom centers in this direction) also preserves the lattice. However, c is larger than a ($c = 1.414a$). Furthermore, c and a equal the respective sizes of the Burgers vectors of the dislocation capable of producing the two shears. The dislocation corresponding to the shear in the close-packed direction thus has the smallest Burgers vector. However, the lattice distortion and strain energy associated with a dislocation are functions of the size of the Burgers vector, and it has been shown by Frank that the strain energy varies directly as the square of the Burgers vector.

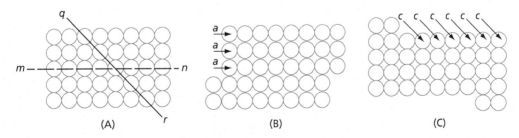

Fig. 5.12 Two ways in which a simple cubic lattice can be sheared while still maintaining the lattice symmetry: **(A)** Crystal before shearing, **(B)** shear in a close-packed direction, and **(C)** shear in a non-close-packed direction

In the present case, the strain energy of a dislocation of Burgers vector c is twice that of a dislocation of Burgers vector a (that is, $c^2 = (1.414)^2 a^2$). Thus, a dislocation with a Burgers vector equal to the spacing of atoms in a close-packed direction would be unique. It possesses the smallest strain energy of all dislocations whose movement through the crystal does not disturb the crystal structure. The fact that it possesses the least strain energy should make this form of dislocation much more probable than forms of higher strain energy. It should also account for the experimentally observed fact that the slip direction in crystals is almost always a close-packed direction.

5.6 Slip Systems

The combination of a slip plane and one of its close-packed directions defines a possible *slip mode* or *slip system*. If the plane of the paper in Fig. 5.13 is considered to define a slip plane, then there will be three slip systems associated with the indicated close-packed plane, one mode corresponding to each of the three slip directions. All of the modes of a given slip plane are crystallographically equivalent. Further, all slip systems in planes of the same form [(111), (1$\bar{1}$1), ($\bar{1}$11), and (11$\bar{1}$)] are also equivalent. However, the ease with which slip can be produced on slip systems belonging to planes of different forms [(111) and (110)] will, in general, be greatly different.

5.7 Critical Resolved Shear Stress

It is a well-known fact that polycrystalline metal specimens possess a yield stress that must be exceeded in order to produce plastic deformation. It is also true that metal single crystals need to be stressed above a similar yield point before plastic deformation by slip becomes macroscopically measurable. Since slip is caused by shear stresses, the yield stress for crystals is best expressed in terms of a shear stress resolved on the slip plane and in the slip direction. This stress is called the *critical resolved shear stress*. It is the stress that will cause sufficiently large numbers of dislocations to move so that a measurable strain can be observed. Most crystal specimens are not tested directly in shear, but in tension. There are good reasons for this. The most important reason is that it is almost impossible to test a crystal in direct shear without introducing bending moments where the specimen is gripped. The effect of these bending moments is to produce shear stress components on slip

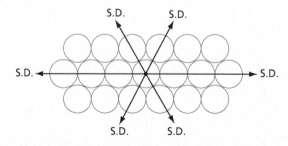

Fig. 5.13 The three slip directions (S.D.) in a plane of closest packing. Notice that this type of plane occurs in both the hexagonal close-packed and the face-centered cubic lattices

planes other than the one on which it is desired to study slip. If slip on these planes is not measurably more difficult than on the plane to be tested, one obtains a condition where slip occurs on several slip planes over those parts of the specimen near the grips. The effect of this deformation may be to cause bending of the specimen near the grips, and one is thus left with a deformation that is far from homogeneous. There are also problems associated with the use of single crystal tensile specimens, but these are less serious and, by proper design of the grips, they may be largely eliminated.

An equation will now be derived that relates the applied tensile stress to the shear stress resolved on the slip plane and in the slip direction.

Let the inclined plane at the top of the cylindrical crystal in Fig. 5.14 correspond to the slip plane of the crystal. The normal to the slip plane and the slip direction are indicated by the lines p and d, respectively. The angle between the slip plane normal and the stress axis is represented by θ, and that between the slip direction and the stress axis by ϕ. The axial tensile force applied to the crystal is designated by f_n.

The cross-section area of the specimen perpendicular to the applied tensile force is to the area of the slip plane as the cosine of the angle between the two planes. This angle is the same as the angle between the normals to the two planes in question and is the angle θ in the figure. Thus,

$$\frac{A_n}{A_{sp}} = \cos\theta \qquad 5.2$$

or

$$A_{sp} = \frac{A_n}{\cos\theta}$$

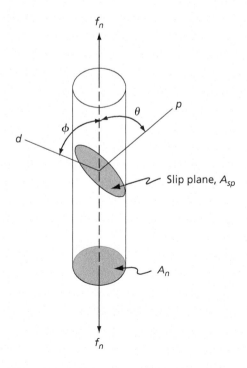

Fig. 5.14 A figure for the determination of the critical resolved shear stress equation

where A_n is the cross-section area perpendicular to the specimen axis, and A_{sp} the area of the slip plane. The stress on the slip plane equals the applied force divided by the area of the slip plane:

$$\sigma_A = \frac{f_n}{A_{sp}} = \frac{f_n}{A_n} \cos \theta$$

where σ_A is the stress on the slip plane in the direction of the original force f_n. This is not, however, the shear stress that acts in the slip direction, but the total stress acting on the slip plane. The component of this stress parallel to the slip direction is the desired shear stress and may be obtained by multiplying σ_A by $\cos \phi$, where ϕ is the angle between σ_A and τ the resolved shear stress. As a result of the above, we can now write

$$\tau = \sigma_A \cos \phi = \frac{f_n}{A_n} \cos \theta \cos \phi$$

where τ is the shear stress resolved on the slip plane and in the slip direction. Finally, since f_n/A_n is the applied tensile force divided by the area normal to this force, this term may be replaced by σ the normal tensile stress:

$$\tau = \sigma \cos \theta \cos \phi \qquad 5.3$$

Several important conclusions may be drawn from the Eq. 5.3. If the tensile axis is perpendicular to the slip plane, the angle ϕ is 90° and the shear stress is zero. Similarly, if the stress axis lies in the slip plane, the angle θ is 90° and the shear stress is again zero. Thus, it is not possible to produce slip on a given plane when the plane is either parallel or perpendicular to the axis of tensile stress. The maximum shear stress that can be developed equals $0.5\ \sigma$ and occurs when both θ and ϕ equal 45°. For all other combinations of these two angles, the resolved shear stress is smaller than one-half the tensile stress.

In regard to Eq. 5.3, it is important to note that the resolved shear stress τ due to a tensile stress σ depends on the cosines of the angles between two pairs of directions; that is, the cosine of the angle θ between the tensile stress axis and the slip plane pole and the cosine of the angle ϕ between the tensile stress axis and the slip direction. Assuming that one has a cubic crystal and the direction indices for these three directions are known, one can can easily compute these cosines with the aid of the following equation:

$$\cos \phi = \frac{h_1 \cdot h_2 + k_1 \cdot k_2 + l_1 \cdot l_2}{\sqrt{h_1^2 + k_1^2 + l_1^2} \cdot \sqrt{h_2^2 + k_2^2 + l_2^2}} \qquad 5.4$$

where h_1, k_1, and l_1 are the direction indices of one of the directions and h_2, k_2, and l_2 are the direction indices of the other direction.

Thus, to illustrate the use of Eq. 5.4, suppose that one wishes to find the cosine of the angle between the [121] and [30$\bar{1}$] directions. In this case we have

$$\cos \phi = \frac{1 \cdot 3 + 2 \cdot 0 - 1 \cdot 1}{\sqrt{1^2 + 2^2 + 1^2} \cdot \sqrt{3^2 + 0 + 1^2}} = \frac{2}{\sqrt{6} \cdot \sqrt{10}}$$

or $\cos \phi = 0.258$ and $\phi = 75.04°$.

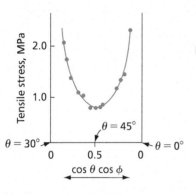

Fig. 5.15 The tensile yield point for magnesium single crystals of different orientations. Abscissae are values of the function cos θ cos φ. Smooth curve is for an assumed constant critical resolved shear stress of 63 psi (Burke, E. C., and Hibbard, W. R., Jr., *Trans. AIME*, **194**, 295 [1952].)

It has been experimentally verified that the critical resolved shear stress for a given crystallographic plane is independent of the orientation of the crystal for some metals. Thus, if a number of crystals, differing only in the orientation of the slip plane to the axis of tensile stress, are pulled in tension, and the shear stress at which they yield is computed with the above equation, it will be found that the yield stress is a constant. Figure 5.15 shows the critical resolved shear-stress data of Burke and Hibbard for magnesium single crystals of 99.99 percent purity. The ordinate of this curve is the tensile stress at which yielding was observed, while the abscissae give corresponding values of the function cos θ cos φ. A smooth curve is plotted through the data corresponding to a constant yield stress (shear stress) of 0.43 MPa. The experimental points fall on this curve with remarkable accuracy. Some work (Refs. 8–10) on bcc metals has indicated that, in these metals, the critical resolved shear stress may be a function of orientation as well as of the type of stress. In other words, the yield stress for this type of crystal can be different depending upon whether the applied stress is tensile or compressive.

In some metals the critical resolved shear stress for slip on a given type of plane is remarkably constant for crystals of the same composition and previous treatment. However, the critical resolved shear stress is sensitive to changes in composition and handling. In general, the purer the metal the lower the yield stress, as may be seen quite clearly from the curves of Fig. 5.16 for silver and copper single crystals. The silver data in particular show that changing the composition from a purity of 99.999 to 99.93 percent raises the critical resolved shear stress by a factor of more than three.

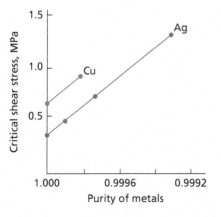

Fig. 5.16 Variation of the critical resolved shear stress with purity of the metal (After Rosi, F. D., *Trans. AIME*, **200**, 1009 [1954].)

The critical resolved shear stress is a function of temperature. In the case of face-centered cubic crystals, this temperature dependence may be small. Metal crystals belonging to other crystal forms (body-centered cubic, hexagonal, and rhombohedral) show a larger temperature effect. The yield stress in these crystals increases as the temperature is lowered, with the rate of increase generally becoming greater as the temperature drops.

5.8 Slip on Equivalent Slip Systems

It has been empirically determined that when a crystal possesses several crystallographically equivalent slip systems, slip will start first on the system having the highest resolved shear stress. It has also been found that if several equivalent systems are equally stressed, slip will usually commence simultaneously on all of these systems.

5.9 The Dislocation Density

Even in a deformed single crystal, only a small fraction of the dislocations formed during deformation come to the surface and are lost. This means that with continued straining, the number of dislocations in a metal increases. As will be shown later, this increase in the number of retained dislocations results in a strengthening of the metal. In other words, the increase in hardness or strength of a metal with deformation is closely associated with an increase in the concentration of the dislocations. The parameter commonly used to express this quantity is ρ, the dislocation density, which is defined as the total length of all the dislocation lines in a unit volume. Its dimensions are thus cm/cm^3 or cm^{-2}.

The dislocation density is often determined by estimating the length of the dislocation line visible in a transmission electron microscope photograph of a metal foil specimen of known thickness after making corrections for the magnification of the microscope. An alternate procedure is to measure the number of dislocation etch pits visible on a suitably etched specimen surface. In this case the dislocation density is given as the number of pits per cm^2, that is, $\rho = N/\text{cm}^2$, where N is the number of pits.

5.10 Slip Systems in Different Crystal Forms

Face-Centered Cubic Metals The close-packed directions are the $\langle 110 \rangle$ directions in the face-centered cubic structure. These are directions that run diagonally across the faces of the unit cell. Figure 5.13 shows a segment of a plane of closest packing. There are four of these planes in the face-centered cubic lattice, called *octahedral planes*, with indices (111), (1$\bar{1}$1), (11$\bar{1}$), and ($\bar{1}$11). Each octahedral plane contains three close-packed directions, as can be seen in Fig. 5.13, and, therefore, the total number of octahedral slip systems is 4 × 3 = 12. The number of octahedral slip systems can also be computed in a different manner. There are 6$\langle 110 \rangle$ directions, and since each close-packed direction lies in two octahedral planes, the number of slip systems is therefore 12.

The only important slip systems in the face-centered cubic structure are those associated with slip on the octahedral plane. There are several reasons for this. First, slip can occur much more easily on a plane of closest packing than on planes of lower atomic density; that is, the critical resolved shear stress for octahedral slip is lower than for other forms. Second, there are twelve different ways that octahedral slip can occur, and the twelve slip systems are well distributed in space. It is, therefore, almost impossible to strain a face-centered cubic crystal and not have at least one {111} plane in a favorable position for slip.

Table 5.1 lists the critical resolved shear stress, measured at room temperature, for several important face-centered cubic metals. Table 5.1 clearly shows that the critical resolved shear stresses of face-centered cubic metals in the nearly pure state are very small.

In general, plastically deformed face-centered cubic crystals slip on more than one octahedral plane because of the large number of equivalent slip systems. In fact, even in a simple tensile test it is very difficult to produce strains of over a few percent without inducing glide simultaneously on several planes. However, when slip occurs at the same time on several intersecting slip planes, the stress required to produce additional deformation rises rapidly. In other words, the crystal strain hardens. Figure 5.17 shows typical tensile stress-strain curves for a pair of face-centered cubic crystals. Curve a corresponds to a crystal whose original orientation lies close to $\langle 100 \rangle$. In this crystal several slip systems have nearly equal resolved shear stresses. As a consequence, plastic deformation occurs by slip on several slip planes and the curve has a steep slope from the beginning of the deformation. On the other hand, curve b corresponds to a crystal whose orientation falls in the center of the stereographic triangle and is representative of crystals in which one slip plane is more highly stressed than all the others at the start of deformation. The region marked 1 of this curve corresponds to slip on this plane only; the other slip planes are inactive. The small slope of the curve in stage 1 shows that the strain hardening is minor when slip occurs on a single crystallographic plane. Stage 2 of curve b, which appears after strains of several percent, has a much steeper slope, and the crystal hardens rapidly with increasing strain. In this region, slip on a single plane ceases, and multiple glide on intersecting slip planes begins. Here the dislocation density grows rapidly with increasing strain. Finally, stage 3 represents a region where the rate of strain hardening progressively decreases. In this region, the rate of increase of the dislocation density becomes smaller with increasing strain.

Table 5.1 Critical Resolved Shear Stresses for Face-Centered Cubic Metals

Metal	Purity	Slip System	Critical Resolved Shear Stress MPa
Cu*	99.999	{111} $\langle 110 \rangle$	0.63
Ag[†]	99.999	{111} $\langle 110 \rangle$	0.37
Au[‡]	99.99	{111} $\langle 110 \rangle$	0.91
Al[§]	99.996	{111} $\langle 110 \rangle$	1.02

*Rosi, F. D., *Trans. AIME*, **200**, 1009 (1954).

[†]daC. Andrade, E. N., and Henderson, C., *Trans. Roy. Soc. (London)*, 244, 177 (1951).

[‡]Sachs, G., and Weerts, J., *Zeitschrift für Physik*, **62**, 473 (1930).

[§]Rosi, F. D., and Mathewson, C. W., *Trans. AIME*, **188**, 1159 (1950).

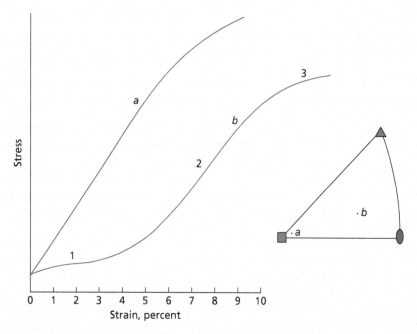

Fig. 5.17 Typical face-centered cubic single crystal stress-strain curves. Curve *a* corresponds to deformation by multiple glide from start of deformation; curve *b* corresponds to multiple glide after a period of single slip (easy glide). Crystal orientations are shown in the stereographic triangle

Region 1 of curve *b*, where slip occurs on a single plane, is called the region of *easy glide*. The extent of the region of easy glide depends on several factors, among which are size of specimen and purity of the metal. When the cross-section diameter of a crystal specimen is large, or the metal very pure, the region of easy glide tends to disappear. In any case, the region of easy glide, or single slip, rarely exceeds strains of several percent in face-centered cubic crystals, and, for all practical purposes, it can be assumed that these metals deform by multiple glide on a number of octahedral systems. This deformation is especially true in the case of polycrystalline face-centered cubic metals.

The plastic properties of pure face-centered cubic metals are as follows. Low critical resolved shear stresses for slip on octahedral planes signifies that plastic deformation of these metals starts at low stress levels. Multiple slip on intersecting slip planes, however, causes rapid strain hardening as deformation proceeds.

Hexagonal Metals

Since the basal plane of the close-packed hexagonal crystal and the octahedral {111} plane of the face-centered cubic lattice have identical arrangements of atoms, it would be expected that slip on the basal plane of hexagonal metals would occur as easily as slip on the octahedral planes of face-centered cubic metals. In the case of the three hexagonal metals, zinc, cadmium, and magnesium, this is actually the case. Table 5.2 lists the critical resolved shear stress for basal slip of these metals as measured at room temperature. The hexagonal Miller indices of the basal plane are (0001), and the close-packed, or slip, directions are $\langle 11\bar{2}0 \rangle$.

Table 5.2 definitely confirms that plastic deformation by basal slip in these three hexagonal metals begins at stresses of the same order of magnitude as those required to start slip in face-centered cubic metals.

Table 5.2 Critical Resolved Shear Stress for Basal Slip

Metal	Purity	Slip Plane	Slip Direction	Critical Resolved Shear Stress MPa
Zinc*	99.999	(0001)	⟨11$\bar{2}$0⟩	0.18
Cadmium†	99.996	(0001)	⟨11$\bar{2}$0⟩	0.57
Magnesium‡	99.95	(0001)	⟨11$\bar{2}$0⟩	0.43

* Jillson, D. C., *Trans. AIME*, 188, 1129 (1950).
† Boas, W., and Schmid, E., *Zeits. für Physik*, 54, 16 (1929).
‡ Burke, E. C., and Hibbard, W. R., Jr., *Trans. AIME*, 194, 295 (1952).

Two other hexagonal metals of interest are titanium and beryllium, in which the room-temperature critical resolved shear stress for basal slip is very high [approximately 110 MPa in the case of titanium, (Ref. 11) and 39 MPa for beryllium (Ref. 12)]. Furthermore, it has been established that in titanium the critical resolved shear stress for slip on {10$\bar{1}$0} prism planes in ⟨11$\bar{2}$0⟩ close-packed directions is about 49 MPa (Ref. 11). The latter plane is therefore the preferred slip plane in titanium. The basal slip of zirconium, still another hexagonal metal, has thus far not been observed. It appears that slip occurs primarily on {10$\bar{1}$0} ⟨11$\bar{2}$0⟩ slip systems. The critical resolved shear stress for this form of slip is about 6.2 MPa (Ref. 13). The question is how the differences in the slip behavior of magnesium, zinc, and cadmium on the one hand, and beryllium, titanium, and zirconium on the other, can be explained. A complete solution to this problem is not at hand, but the following is undoubtedly related to this effect.

Figure 1.16 of Chapter 1 shows the unit cell of the hexagonal lattice. In this figure, distance *a* equals the distance between atoms in the basal plane, while *c* is the vertical distance between atoms in every other basal plane. The ratio *c/a* is, therefore, a dimensionless measure of the spacing between basal planes. If the atoms of hexagonal metals were truly spherical in shape, the ratio *c/a* would be the same in all cases (1.632). Table 5.3 shows, however, that this value is not the same, but that it varies from 1.886 in the case of cadmium to 1.586 in the case of beryllium. Only magnesium has an atom that approaches a true spherical shape, *c/a* = 1.624. Cadmium and zinc have a basal-plane separation greater than that of packed spheres, while beryllium, titanium, and zirconium have a smaller one. It is significant that the hexagonal metals with small separations between basal planes are those with very high critical resolved shear stresses for basal slip.

The hexagonal metals zinc and cadmium have been observed to slip on a unique slip system when deformed in such a manner as to make the resolved shear stress on the basal plane very small. This deformation can be accomplished, for example, by placing the tensile stress axis nearly parallel to

Table 5.3 The *c/a* Ratio for Hexagonal Metals

Metal	c/a
Cd	1.886
Zn	1.856
Mg	1.624
Zr	1.590
Ti	1.588
Be	1.586

the basal plane. The observed slip plane for this type of deformation is {11$\bar{2}$2}, while the slip direction is ⟨11$\bar{2}$3⟩. Figure 5.18 shows the position of one of these slip planes and one of these slip directions in a hexagonal unit cell. The significant characteristic of this mode or deformation is that the ⟨11$\bar{2}$3⟩ direction is not the closest packed direction in the hexagonal crystal structure. Prior to the observation of {11$\bar{2}$2} ⟨11$\bar{2}$3⟩ slip, the slip direction in all metals had been almost universally observed as the direction of closest packing. This second-order pyramidal glide was first observed by Bell and Cahn (Ref. 14) on macroscopic zinc crystals, but it has also been verified in both zinc and cadmium by Price (Refs. 15 and 16), who used the transmission electron microscope. The work of Price has not only confirmed the existence of this kind of slip, but has actually shown the dislocations responsible for it. For photographs of these dislocations, one is referred to the original publications (Refs. 15 and 16).

Easy Glide in Hexagonal Metals The metals zinc, cadmium, and magnesium are unique in that they possess both a low critical resolved shear stress and a single primary slip plane, the basal plane. Provided that this slip plane is suitably oriented with respect to the stress axis, strains of very large magnitude can be developed by basal slip. Simultaneous slip on intersecting primary slip planes is not a problem in these metals with a single slip plane, and the rate of strain hardening is therefore much smaller than in face-centered cubic crystals. The region of easy glide in a tensile stress-strain curve, instead of extending only several percent, may exceed values of 100 percent. In fact, it is quite possible to stretch a magnesium crystal ribbonlike four to six times its original length.

The very large plasticity of single crystals of these three hexagonal metals does not carry over to their polycrystalline form. Polycrystalline magnesium, zinc, or cadmium have low ductilities. In the previous paragraph it was pointed out that the large ductility of the single crystals is due to the fact that slip occurs on a single crystallographic plane. However, in polycrystalline material, plastic deformation is much more complicated than it is in single crystals. Each crystal must undergo a deformation that allows it to conform to the changes in the shape of its neighbors. Crystals with only a single slip plane do not have enough plastic degrees of freedom for extensive deformation under the conditions that occur in polycrystalline metals.

Body-Centered Cubic Crystals The body-centered cubic crystal is characterized by four close-packed directions, the ⟨111⟩ directions, and by the lack of a truly close-packed plane such as the octahedral plane of the face-centered cubic lattice, or the basal plane of the hexagonal lattice. Figure 5.19

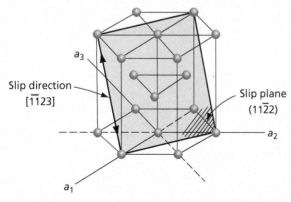

Fig. 5.18 {1122} ⟨11$\bar{2}$3⟩ slip in hexagonal metals

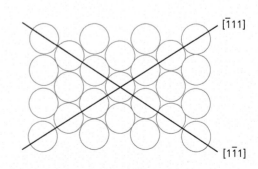

Fig. 5.19 The (110) plane of the body-centered cubic lattice

shows a hard-ball model of the body-centered cubic (110) plane, the most closely packed plane in this lattice. Two close-packed directions lie in this plane, the $[\bar{1}11]$ and the $[1\bar{1}1]$.

The slip phenomena observed on body-centered cubic crystals corresponds closely to that expected in crystals with close-packed directions and no truly close-packed plane. The slip direction in the body-centered cubic crystals is the close-packed direction, $\langle 111 \rangle$; the slip plane, however, is not well defined. Body-centered cubic slip lines are wavy and irregular, often making the identification of a slip plane extremely difficult. The {110}, {112}, and {123} planes have all been identified as slip planes in body-centered cubic crystals, but work on iron single crystals indicates that any plane that contains a close-packed $\langle 111 \rangle$ direction can act as a slip plane. In further agreement with the lack of a close-packed plane is the high critical resolved shear stress for slip in body-centered cubic metals. In iron it is approximately 28 MPa at room temperature.

5.11 Cross-Slip

Cross-slip is a phenomenon that can occur in crystals when there are two or more slip planes with a common slip direction. As an example, take the hexagonal metal magnesium in which, at low temperatures, slip can occur either on the basal plane or on $\{10\bar{1}0\}$ prism planes. These two types of planes have common slip directions—the close-packed $\langle 11\bar{2}0 \rangle$ directions. The relative orientations of the basal plane and one prism plane are shown in Fig. 5.20A and 5.20B, where each diagram is supposed to represent a crystal in the same basic orientation (basal plane parallel to the top and bottom surfaces). In the first sketch, the crystal is sheared on the basal plane, while in the second it is sheared on the prism plane. The third illustration, Fig. 5.20C, shows

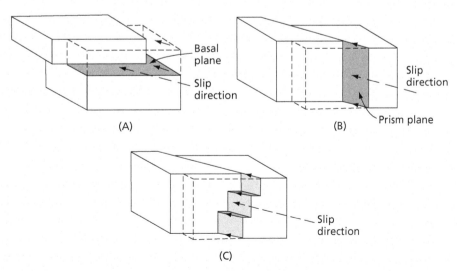

Fig. 5.20 Schematic representation of cross-slip in a hexagonal metal: **(A)** Slip on basal plane, **(B)** slip on prism plane, and **(C)** cross-slip on basal and prism planes

the nature of cross-slip. Here it is observed that the actual slip surface is not a single plane, but is made up of segments, part of which lie in the basal plane and part in the prism plane. The resulting profile of the slip surface has the appearance of a staircase. A simple analogy for cross-slip is furnished by a drawer in a piece of furniture. The sliding of the sides and bottom of the drawer relative to the frame of the piece is basically similar to the shearing motion that results from cross-slip.

A photograph of a magnesium crystal showing cross-slip is given in Fig. 5.21, where the plane of the photograph is equivalent to the forward faces of the schematic crystals shown in Fig. 5.20. Notice that while the slip in this specimen has occurred primarily along a prism plane, cross-slip segments on the basal plane are clearly evident.

During cross-slip the dislocations producing the deformation must, of necessity, shift from one slip plane to the other. In the example given above, the dislocations move from prism plane to basal plane and back to prism plane. The actual shift of the dislocation from one plane to another can only occur for a dislocation in the screw orientation. Edge dislocations, as was pointed out earlier, have their Burgers vector normal to their dislocation lines. Since the active slip plane must contain both the Burgers vector and the dislocation line, edge dislocations are confined to move in a single slip plane. Screw dislocations, with their Burgers vectors parallel to the dislocation lines, are capable of moving in any plane that passes through the dislocation line. The manner in which a screw dislocation can produce a step in a slip plane is shown schematically in Fig. 5.22.

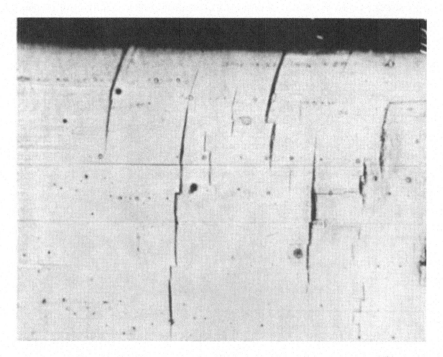

Fig. 5.21 Cross-slip in magnesium. The vertical slip plane traces correspond to the $\{10\bar{1}0\}$ prism plane, whereas the horizontal slip plane traces correspond to the basal plane (0002). 290× (Reed-Hill, R. E., and Robertson, W. D., *Trans. AIME*, **209** 496 [1957].)

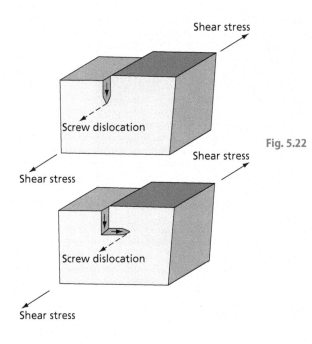

Fig. 5.22 Motion of a screw dislocation during cross-slip. In the upper figure the dislocation is moving in a vertical plane, while in the lower figure it has shifted its slip plane so that it moves horizontally

5.12 Slip Bands

A slip band is a group of closely spaced slip lines that appears, at low magnification, to be a single large slip line. In many metals slip bands tend to be wavy and irregular in appearance; this is evidence that the dislocations that produce the bands in these metals are not so confined as to move in a single plane. The shifting of the dislocations from one slip plane to another is usually the result of the cross-slip of screw dislocations.

The resolution of the individual slip lines in a slip band is normally a task requiring the use of an electron microscope. However, when a slip band is observed on a surface that is nearly parallel to the active slip plane, it is sometimes possible to partially resolve the components of a slip band with a light microscope. Figure 5.23 provides good examples for slip bands during in-situ three-point bending of an age-hardened duplex stainless steel which contained both body-centered cubic (BCC) ferritic grains as well as face-centered cubic (FCC) austenitic grains. The images were taken using high resolution electron backscatter diffraction (HR-EBSD). The sample, with $1.5 \times 12 \times 0.8$ mm in dimensions, was bent consecutively at two bending displacements of (a) 1.2 mm and (b) 1.5 mm. The increase in the density, waviness and height of the slip bands can easily be seen in these figures.

5.13 Double Cross-Slip

Another very important aspect of the work done on LiF crystals (Ref. 17) is that it has shown that moving dislocations can multiply. The mechanism that seems most adequately to explain this dislocation multiplication is *double cross-slip*. As mentioned in Sec. 5.2, dislocations in LiF appear to nucleate at impurity particles. The slip process that develops from these nucleated dislocations first forms narrow slip planes that grow into slip bands of finite width with continued straining. These slip bands, in turn, can expand so as to cover effectively the complete crystal. Several rather

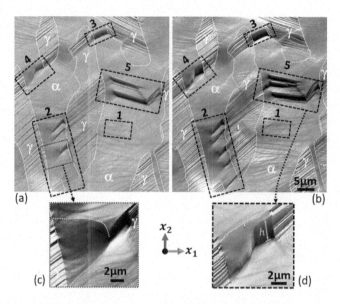

Fig. 5.23 High resolution electron backscatter diffraction (HR-EBSD) during in-situ three-point bending of an age-hardened duplex stainless-steel containing ferrite (α) and austenite (γ) grains; **(a)** shows slip planes after bending displacement of 1.2 mm, and **(b)** after 1.5 mm displacement. Comparing different boxes and enlarged views show increased in wavy slips in α, increased number of slip bands and increased step heights "h". (Printed by permission from Abdalrhaman Koko, Elsiddig Elmukashfi, Thorsten H. Becker, Phani S. Karamched, Angus J. Wilkinson, T. James Marrow, *Acta Materialia* 239 (2022) 118284. https://doi.org/10.1016/j.actamat.2022.118284)

narrow bands, as well as some that have grown to a moderate width, may be seen in the photograph of Fig. 5.24. Note that the presence of the slip bands is revealed by the rows or etch pits.

The development of the double cross-slip mechanism, which was originally proposed by Koehler (Ref. 18) and later by Orowan (Ref. 19), is shown in Fig. 5.25. In sketch (A), a dislocation loop is assumed to be expanding on the primary slip plane. In part (B), a segment of the dislocation loop that is in the

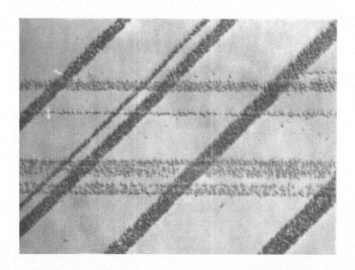

Fig. 5.24 Slip bands in LiF. Bands formed at −196 °C and 0.36 percent strain (Reprinted with permission from J. J. Gilman and W. G. Johnson, *Journal of Applied Physics*, Vol. 30, Issue 2, Page 129, Copyright 1959, American Institute of Physics)

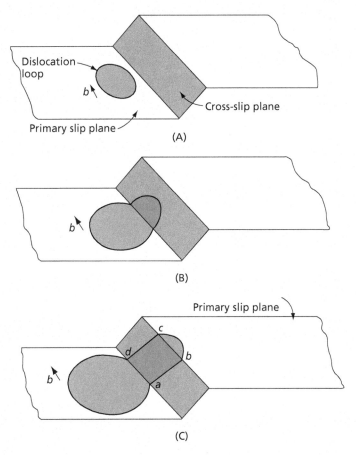

Fig. 5.25 Double cross-slip

screw orientation is shown cross-slipping onto the cross-slip plane. Finally, in (C), the dislocation moves back into a plane parallel to its original slip plane. Attention is now called to the similarity of the dislocation configuration, starting at point b and ending at point c, to the Frank-Read source configuration shown in Fig. 5.1. This is, in fact, a classical dislocation generator, and dislocations can be created with it on the new slip plane. It is also true that this dislocation configuration can also act as a Frank-Read source in the slip plane of the original loop. Proof of this last statement is left as an exercise.

An important feature of the double cross-slip mechanism is that Frank-Read sources associated with this mechanism involve freshly created dislocations; that is, dislocations which, in general, will not have had time enough to become pinned by impurity atoms. The operation of a Frank-Read source with this type of dislocation is much more probable than that involving grown-in dislocations.

5.14 Extended Dislocations and Cross-Slip

While a total $\frac{1}{2}\langle 110 \rangle$ dislocation in a fcc metal can readily cross-slip between a pair of octahedral planes, this is not true of an extended dislocation. The reason for this is shown in the diagrams of Fig. 5.26. In part (A) of this illustration, a total $\frac{1}{2}\langle \bar{1}10 \rangle$ dislocation is shown cross-slipping from the primary slip plane (111) to the cross-slip plane (11$\bar{1}$). Note that the Burgers vector of

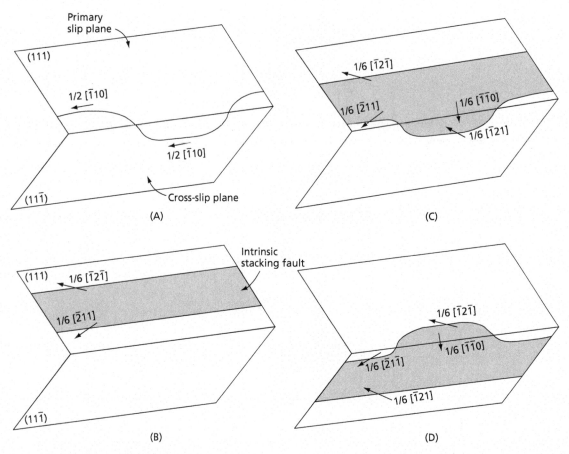

Fig. 5.26 The cross-slip of an extended dislocation

this dislocation lies along a direction common to both slip planes. In part (B) of this diagram, a corresponding extended dislocation is shown moving downward on the (111) primary slip plane. Assume, as shown in Fig. 5.26C, that the leading partial dislocation of this extended dislocation moves off on the cross-slip plane. Since the Burgers vector of this partial dislocation is $\frac{1}{6}[\bar{2}11]$ and this vector has a direction that does not lie in the $(11\bar{1})$ cross-slip plane, a dislocation reaction has to occur that results in the creation of two new dislocations. This reaction is

$$\tfrac{1}{6}[\bar{2}11] \to \tfrac{1}{6}[\bar{1}21] + \tfrac{1}{6}[\bar{1}\bar{1}0] \qquad 5.5$$

One of these dislocations is a Shockley partial that is free to move on the cross-slip plane and has a Burgers vector $\frac{1}{6}[\bar{1}21]$. The other is immobile (sessile) and remains behind along the line of intersection between the primary and cross-slip planes. This latter dislocation is called a *stair-rod* dislocation, and its Burgers vector is $\frac{1}{6}[\bar{1}\bar{1}0]$. Stair-rod dislocations, of which the present example represents only one of a number of different types, are always found when a stacking fault runs from one slip plane over into another, as shown in Fig. 5.26C. Note that the $\frac{1}{6}[1\bar{1}0]$ of the stair-rod dislocation in

Fig. 5.26C is perpendicular to the line of intersection of the two slip planes and that it also does not lie in either slip plane.

In the cross-slip of an extended dislocation, the stair-rod dislocation is removed when the trailing partial dislocation arrives at the line of intersection of the two slip planes and then moves off on the cross-slip plane, as shown by the following equation:

$$\tfrac{1}{6}[\bar{1}2\bar{1}] + \tfrac{1}{6}[\bar{1}\bar{1}0] = \tfrac{1}{6}[\bar{2}\bar{1}] \qquad 5.6$$

This reaction is illustrated in Fig. 5.26C.

Since additional strain energy is required to create the stair-rod dislocation associated with the cross-slip of an extended dislocation, it should be much easier for a total dislocation than for an extended dislocation to cross-slip.

5.15 Crystal Structure Rotation during Tensile and Compressive Deformation

When a single crystal is deformed in either tension or compression, the crystal lattice usually suffers a rotation, as indicated schematically in Fig. 5.27. In tension, this tends to align the slip plane and the active slip direction parallel to the tensile stress axis. It is often possible to identify the active slip plane and slip direction as a result of this rotation. Thus, suppose that, before it is deformed, an fcc crystal has the orientation marked a_1 shown in the standard projection of Fig. 5.28. If it is now strained by a small amount and its orientation is redetermined by a Laue back-reflection photograph, the stress axis should plot (in the standard stereographic projection) at a new position such as a_2. A second similar deformation should produce a third stress axis

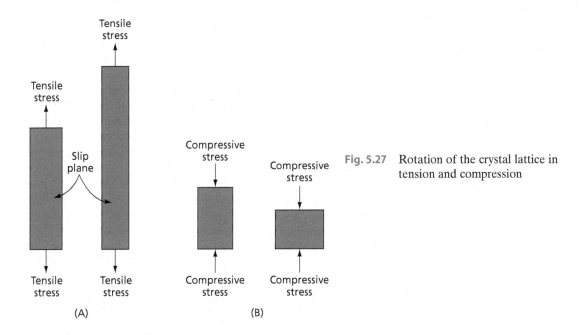

Fig. 5.27 Rotation of the crystal lattice in tension and compression

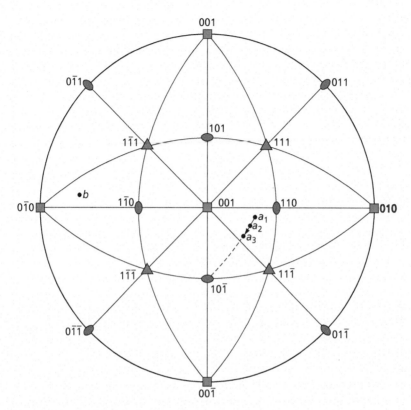

Fig. 5.28 In tension the lattice rotation is equivalent to a rotation of the stress axis (*a*) toward the slip direction. This stereographic projection shows this rotation in a face-centered cubic crystal

orientation at a_3. These three stress axis positions will normally fall along a great circle and, if this great circle is projected ahead, it should pass through the active slip direction. In the present case, as may be seen in Fig. 5.28, the slip direction is [10$\bar{1}$]. As shown in Fig. 5.29, this slip direction lies in both the (111) and the (1$\bar{1}$1) planes. Since the (111) pole makes an angle of nearly 45° with the stress axis, while the (1$\bar{1}$1) pole makes a corresponding angle closer to 90°, one can deduce that the resolved shear stress should be larger on the former. This means that the active slip plane should be (111). Normally it should also be possible to verify this fact by measuring the orientation of the slip line traces on the specimen surface.

Note that in the example of Fig. 5.28 the active slip direction is the closest ⟨110⟩ direction to the stress axis that can be reached by crossing a boundary of the stereographic triangle that contains the stress axis. This fact can be considered to define a general rule that is applicable to the tensile deformation of fcc crystals. As an added example, consider the possible stress axis orientation, marked *b*, lying in the stereographic triangle at the left of the figure and near the (0$\bar{1}$0) pole. Following the above rule, the slip direction should be [0$\bar{1}$1]. The corresponding slip plane pole should be (1$\bar{1}$1). Once the slip direction has been obtained, the pole of the slip plane can easily be determined in the case of fcc crystals. It should lie on the other side of the stress axis from the slip direction and all three directions should lie roughly on a common great circle. Finally, the indicated combinations of slip directions and

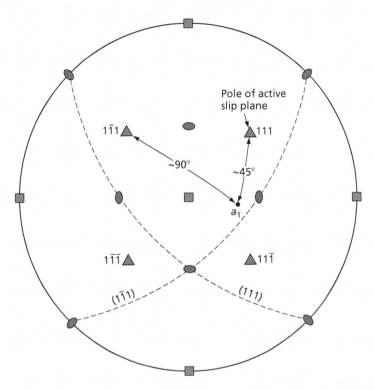

Fig. 5.29 The original stress axis orientation in Fig. 5.28 lies about 45° from the pole of the (111) plane and about 90° from the pole of (1̄11). These are the two slip planes that contain the active slip direction

slip planes yield the active slip systems because in each case they represent the slip system that has the highest resolved shear stress acting on it in a given stereographic triangle.

Now let us consider the case of a crystal deformed in compression. As shown in Fig. 5.27B, the slip plane rotates so that it tends to become perpendicular to the stress axis. In this case, if the orientation of the stress axis is plotted in a standard stereographic projection, the stress axis will be found to follow a path that passes through the position of the slip plane normal. This is illustrated in Fig. 5.30, using once again as an example a single crystal of a face-centered cubic metal.

5.16 The Notation for the Slip Systems in the Deformation of fcc Crystals

The preceding section has shown that, for an orientation in the center of a stereographic triangle, one slip system is favored because the resolved shear stress is highest on this system. This slip system is called the *primary slip system*. In the example of Fig. 5.28, the primary slip system for the original stress axis orientation is (111) [10̄1]. If the dislocations of this system were to cross-slip, they would have to move onto the other slip plane that contains the [10̄1] slip direction. As may be seen in Fig. 5.29, this is the (1̄11) plane. The cross-slip system for the stress axis orientation a_1 is, accordingly, (1̄11) [10̄1].

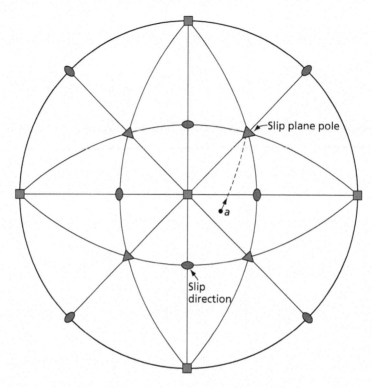

Fig. 5.30 In compression, the stress axis (*a*) rotates toward the pole of the active slip plane

Another important slip system is called the *conjugate slip system*. This is the system that becomes the preferred slip system once the rotation of the crystal structure, relative to its tensile stress axis, results in moving the orientation of the stress axis out of its original stereographic triangle into the one adjoining it. This movement of the stress axis on the stereographic projection is shown in Fig. 5.31, which is an enlarged view of the standard projection showing only the two stereographic triangles of present interest.

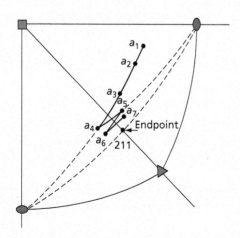

Fig. 5.31 When the stress axis leaves its original stereographic triangle, a second or conjugate slip system becomes more highly stressed than the primary slip system. After this occurs, deformation occurs alternately on both systems

As indicated above, the stress axis position on the stereographic projection tends to move toward the active slip direction through the positions a_1, a_2, and a_3. Continued motion of this type carries it to a position such as a_4, where it lies in another stereographic triangle where the resolved shear stress is greater on the $(1\bar{1}1)$ [110] slip system. As a result, slip will be instituted on this slip system, called the conjugate slip system, and the stress axis will now move along a path toward the [110] slip direction. Theoretically, while this shift in the slip system should occur once the stress axis crosses the boundary separating the two stereographic triangles, there is always some degree of overshoot before the conjugate slip system takes over. The degree of this overshoot depends on a number of factors that will not be discussed here. The diagram shows clearly that slip on the second (or conjugate) system will bring the stress axis back into the original stereographic triangle where the primary slip system is again favored. It is therefore to be expected that the primary slip system will eventually predominate at a point such as a_5 and the stress axis will again move toward the slip direction of the primary system. This oscillation of the stress axis will be repeated a number of times, with the result (shown in the figure) that the axis eventually reaches the [211] direction, a direction that lies on the same great circle as the conjugate and primary slip directions and midway between them. This is a stable end orientation for the crystal, and once it has obtained this orientation, further deformation will not change the orientation of the crystal relative to the tensile stress axis.

In the above discussion, the primary, conjugate, and cross-slip slip systems have been identified for a given starting orientation of the axis of a crystal. These have involved three of the possible slip planes in the face-centered cubic structure. The fourth plane, which is $(11\bar{1})$ in the example in Fig. 5.28, is called the *critical plane*.

5.17 Work Hardening

The drawing in Fig. 5.32 is assumed to represent a typical engineering stress-strain curve of a polycrystalline pure metal. Note that after the metal begins to deform plastically above the proportional limit at point a, the stress still continues to rise. This rise in the flow-stress reflects an increase in the strength of the metal caused by the deformation. That this strength increase is real can be shown by unloading the specimen from a point such as b on the curve and then by reloading the specimen. Provided that this experiment does not occur at a temperature where recovery rates are rapid, the stress on the specimen will have to be returned to the value σ_b that it had prior to the unloading before macroscopic plastic flow will again occur. Deforming the specimen to a plastic strain ϵ_b has, accordingly, raised the stress at which the metal will flow from σ_a to σ_b. As will be shown shortly, the stress at which a metal will flow, the *flow stress*, is intimately connected with changes in the dislocation structure in the metal resulting from deformation. However, it is necessary that we first consider a different way of representing the stress-strain data. The engineering stress-strain curve expresses both stress and strain in terms of the original specimen dimensions, a very useful procedure when one is interested in determining strength and ductility data for purposes of engineering design. On the other hand, this type of representation is not as convenient for showing the nature of the work hardening process in metals. A better set of parameters are true stress and true strain. True stress (σ_t) is merely the load divided by the instantaneous cross-section area, or

$$\sigma_t = \frac{P}{A} \qquad 5.7$$

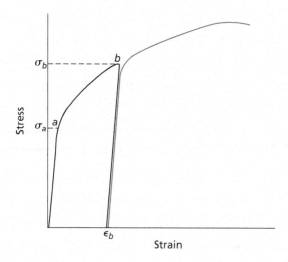

Fig. 5.32 Normally when a metal is deformed to a strain such as ϵ_b and then it is unloaded, it will not begin to deform until the stress is raised back to σ_b. The strain ϵ_b raises the flow stresses from σ_a to σ_b.

where P is the load and A is the cross-section area. If one assumes that during plastic flow the volume remains effectively constant, then we have $A_0 l_0 = Al$, where A_0 and l_0 are the original cross-section area and gage length, respectively, and A and l are the corresponding values of these quantities at any later time. With this assumption, it follows that

$$\sigma_t = \frac{P}{A} = \frac{Pl}{A_0 l_0} = \frac{P}{A_0} \frac{(l_0 + \Delta l)}{l_0} = \sigma(1 + \epsilon) \qquad 5.8$$

where Δl is the increase in length of the specimen, σ is the engineering stress, and ϵ is the engineering strain. This equation simply states that the true stress is equal to the engineering stress, times one, plus the engineering strain. True strain is defined by the relationship

$$\epsilon_t = \int_{l_0}^{l} \frac{dl}{l} = \ln \frac{l}{l_0} = \ln(1 + \epsilon) \qquad 5.9$$

which states that the true strain is equal to the natural logarithm of one plus the engineering strain.

The preceding equations for the true stress and true strain are valid as long as the deformation of the gage section is essentially uniform. In this regard, it is generally assumed that the gage section deforms uniformly to approximately the point of maximum stress (point m in Fig. 5.33) on an engineering stress-strain diagram. Beyond this point the specimen begins to neck and the strain is restricted to the necked region. It is also possible to follow the relation between the true stress and true strain beyond the point of maximum load by considering the stress-strain behavior of the metal only in the necked region. One has, however, to correct for the triaxiality of the stress due to the effective notch produced by the neck, and the strain should be measured in terms of the specimen diameter at the neck, rather than in terms of the specimen gage length. If these factors are considered, one obtains a stress-strain curve of the form of the upper curve in Fig. 5.33. Note that the true stress continues to rise with increasing stress until the point of fracture. There seems to be good evidence for assuming that between the point of maximum load and the point where fracture occurs, the true-stress true-strain curve is approximately a straight line (Ref. 20). The curve of Fig. 5.33 shows that in a simple tensile test, the true strength of the metal normally increases with increasing strain until it fractures.

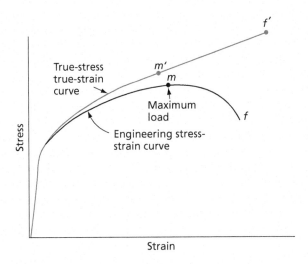

Fig. 5.33 A comparison between an engineering stress-strain curve and the corresponding true-stress and true-strain curve.

5.18 Considère's Criterion

A condition for the onset of necking was originally proposed by Considère (Ref. 21) based on the assumption that necking begins at the point of maximum load. At the maximum load

$$dP = A d\sigma_t + \sigma_t dA = 0 \qquad 5.10$$

where P is the load, A is the cross-section area, and σ_t is the true stress. In effect, this relationship states that for a given strain increment, the resulting decrease in specimen area reduces the load-carrying ability of the cross-section by the same amount that the load-carrying ability of the specimen is increased by the rise in its strength due to strain hardening. The above relationship may be rearranged to read

$$d\sigma_t = -\sigma_t \frac{dA}{A} \qquad 5.11$$

In plastic deformation, a reasonable assumption is that the volume remains constant, so that

$$dV = d(Al) = l dA + A dl = 0 \qquad 5.12$$

or

$$\frac{dA}{A} = -\frac{dl}{l} \qquad 5.13$$

where l is the specimen gage length and dl/l is ϵ_t, the true strain. Therefore,

$$\frac{d\sigma_t}{d\epsilon_t} = \sigma_t \qquad 5.14$$

This is Considère's criterion for necking. When the slope of the true-stress true-strain curve, $d\sigma_t/d\epsilon_t$, is equal to the true stress σ_t, necking should begin. As will be shown presently, this relationship can help to rationalize a basic difference in the observed stress-strain behavior of face-centered cubic and body-centered cubic metals.

5.19 The Relation Between Dislocation Density and the Stress

With the development of the transmission electron microscope technique, it has been possible to make direct studies of the dislocation structure in deformed metals. These investigations have indicated that for a very wide range of metals there exists a rather simple relationship between the dislocation density and the flow-stress of a metal. Thus, let us assume that Fig. 5.34 represents the general shape of the stress-strain curve of a metal and that a series of specimens are deformed to different strains, as indicated by the marked points along the curve. Furthermore, let us assume that on reaching the specified strains, they are unloaded, sectioned for observation in the electron microscope, and that dislocation density measurements are made on the foils. Figure 5.35 shows the actual experimental results obtained using a set of titanium specimens. This data corresponds to specimens of three different grain sizes. Note that all of the data plots on the same straight line. Data such as this supports the assumption that the stress varies directly as the square root of the dislocation density, or

$$\sigma = \sigma_0 + k\rho^{1/2} \qquad 5.15$$

where ρ is the measured dislocation density in centimeters of dislocation per unit volume, k is a constant, and σ_0 is the stress obtained when $\rho^{1/2}$ is extrapolated to zero. This result is good evidence that the work hardening in metals is directly associated with the build-up of the dislocation density in the metal. While the above relationship corresponds to data from polycrystalline specimens, the relationship has also been observed in single-crystal specimens. In this case, it is more proper to express the relationship in terms of the resolved stress on the active slip plane τ. This gives us

$$\tau = \tau_0 + k\rho^{1/2} \qquad 5.16$$

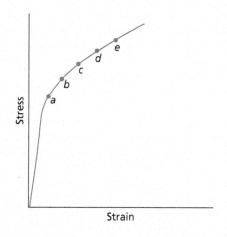

Fig. 5.34 To determine the variation of the dislocation density with strain during a tensile test, a set of tensile specimens are strained to a number of different positions along the stress-strain curve, such as points *a* to *e* in this diagram. These specimens are then sectioned to obtain transmission electron microscope foils

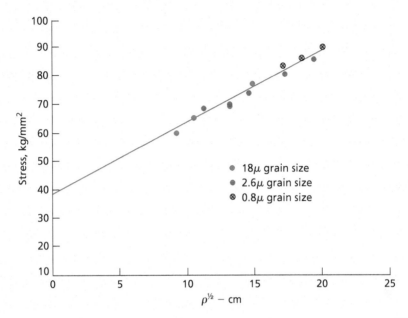

Fig. 5.35 The variation of the flow-stress σ with the square root of the dislocation density $\rho^{1/2}$ for titanium specimens deformed at room temperature, and at a strain rate of 10^{-4} sec^{-1} (After Jones, R. L., and Conrad, H., *TMS-AIME*, **245** 779 [1969].)

where τ_0 is the extrapolated shear stress corresponding to a zero dislocation density. Actually, if the dislocation density were zero, then the metal could not be deformed. As a consequence, σ_0 or τ_0 are best considered as convenient constants rather than as simple physical properties.

5.20 Taylor's Relation

In 1934, Taylor (Ref. 22) proposed a theoretical relationship that is basically equivalent to the experimentally observed functional relationship between the flow stress and the dislocation density. In the model that he used, it was assumed that all the dislocations moved on parallel slip planes and the dislocations were parallel to each other. This model has since been elaborated by Seeger (Ref. 23) and his collaborators. In brief, this approach assumes that if the dislocation density is expressed in numbers of dislocations intersecting a unit area, then the average distance between dislocations is proportional to $\rho^{-1/2}$. As shown earlier in Sec. 4.10, the stress field of a dislocation varies as $1/r$, or in general we may write

$$\tau \approx \frac{\mu b}{r} \qquad 5.17$$

where μ is the shear modulus, b is the Burgers vector, and r is the distance from the dislocation. Now consider two edge dislocations on parallel slip planes. If they are of the same sign, they will exert a repulsive force on each other. If they are of opposite sign, the force will be attractive. In either case, this interaction must be overcome in order to allow the dislocations to continue to glide on their respective slip planes. Since, as shown above, the average distance between dislocations is proportional to $\rho^{-1/2}$, we have

$$\tau = \alpha\mu b\rho^{1/2} \qquad 5.18$$

or

$$\tau = k\rho^{1/2} \qquad 5.19$$

where k is a constant of proportionality equal to $\alpha\mu b$.

5.21 The Orowan Equation

A relationship between the velocity of the dislocations in a test specimen and the applied strain rate will now be derived. This expression is known as the Orowan equation.

As shown in Figs. 5.36A and 5.36B, when an edge dislocation moves completely across its slip plane, the upper half of the crystal is sheared relative to the lower half by an amount equal to one Burgers vector. It can also be deduced and rigorously proved (Ref. 24) (Fig. 5.36C) that if the dislocation moves only through a distance Δx, then the top surface of the crystal will be sheared by an amount equal to $b(\Delta x/x)$, where x is the total distance across the slip plane. In other words, the displacement of the upper surface will be in proportion to the fraction of the slip plane surface that the dislocation has crossed, or to $b(\Delta A/A)$, where A is the area of the slip plane and ΔA is the fraction of it passed over by the dislocation. Since the shear strain γ given to the crystal equals the displacement $b(\Delta A/A)$ divided by the height z of the crystal, we have

$$\Delta\gamma = \frac{b\Delta A}{Az} = \frac{b\Delta A}{V} \qquad 5.20$$

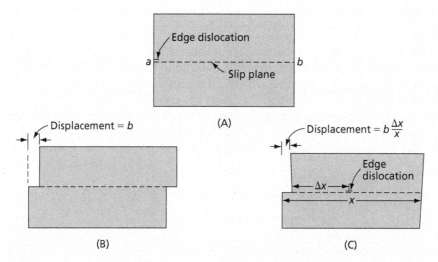

Fig. 5.36 The displacement of the two halves of a crystal is in proportion to the distance that the dislocation moves on its slip plane

since Az is the volume of the crystal. For the case where n edge dislocations of length l move through an average distance $\Delta \bar{x}$, this relation becomes

$$\Delta \gamma = \frac{bnl\Delta \bar{x}}{V} = \rho b \Delta \bar{x} \qquad 5.21$$

where ρ, the dislocation density, is equal to nl/V. If, in a time interval Δt, the dislocations move through the average distance $\Delta \bar{x}$, we have

$$\frac{\Delta \gamma}{\Delta t} = \dot{\gamma} = \rho b \bar{v} \qquad 5.22$$

where $\dot{\gamma}$ is the shear strain rate and $\bar{v}$ is the average dislocation velocity.

This expression, derived for the specific case of parallel edge dislocations, is a general relationship, and it is customary to consider that ρ represents the density of all the mobile dislocations in a metal whose average velocity is assumed to be $\bar{v}$. Furthermore, if $\dot{\epsilon}$ is the tensile strain rate in a polycrystalline metal, a reasonable assumption is that

$$\dot{\epsilon} = \frac{1}{2}\gamma = \frac{1}{2}\rho b \bar{v} \qquad 5.23$$

where the factor 1/2 is an approximate Schmid orientation factor.

Problems

5.1 If the shear vectors, τ, in Fig. 5.1, were moved from the top and bottom faces of the crystal and applied to the front and rear surfaces, with the forward vector pointing up and the rear one down, could any of the Frank-Read dislocation segments move as a result of this shearing stress? Explain.

5.2 (a) Again, with reference to Fig. 5.1, describe what might be expected to happen to the dislocation configuration of this crystal if a horizontal tensile stress were to be applied to the right and left faces of the crystal.

(b) What would be the effect on the dislocation configuration of Fig. 5.1 if the tensile stress in part (a) of this problem were to be changed to a compressive stress?

5.3 There are three slip systems on an fcc octahedral plane. Assume a 2 MPa tensile stress is applied along the [100] direction of a gold crystal, whose critical resolved shear stress is 0.91 MPa. Demonstrate quantitatively that measurable slip will not occur on any of the three slip systems in the (111) plane as a result of this applied stress.

5.4 On a 100 standard projection of a cubic crystal (see Fig. 1.31), plot the (111) pole as well as the great circle corresponding to this plane. Mark the three $\langle 1\bar{1}0 \rangle$ slip directions on the (111) great circle. Then plot the position of the [310] direction on the standard projection. If a tensile stress is applied along the [310] direction, what would be the magnitude of the Schmid factor (i.e., $\cos \theta \cos \phi$) for the (111)[$10\bar{1}$] slip system with this stress axis orientation?

5.5 Deformation twins also form along the {111} planes of fcc crystals as a result of shear stresses. The twinning shear directions are ⟨112⟩.

(a) Prove, using Eq. 5.4, that $[1\bar{2}1]$ and $[11\bar{2}]$ are directions that lie in the (111) plane.

(b) Determine the Schmid factors for the (111) $[\bar{2}11]$, (111) $[1\bar{2}1]$, and (111) $[11\bar{2}]$ twinning systems, if a tensile stress is applied along the [711] direction.

5.6 With the aid of Eq. 5.4, prove that the $(\bar{4}22)$ plane of a cubic crystal belongs to the zone whose axis is [111].

5.7 (a) Determine the angle between the [123] and [321] directions in a cubic crystal. Check Appendix A to see if your answer is correct. **(b)** Find a combination of two ⟨321⟩ directions that make an angle of 85.90° with each other.

5.8 A 10 mm diameter cylindrical zinc single crystal has a longitudinal axis that makes an angle of 85° with the pole of the basal plane and a 7° angle with the closest ⟨11$\bar{2}$0⟩ slip direction in the basal plane. If the critical resolved shear stress of zinc is 0.20 MPa, at what axial load should the crystal begin to deform by basal slip in:

(a) Newtons.

(b) Kilograms force.

5.9 With regard to slip in hcp crystals, answer the following:

(a) Would it be possible for rotational slip to occur with the pole of the {11$\bar{2}$2} plane as an axis of rotation?

(b) Would bend gliding be possible with (11$\bar{2}$2) $[\bar{1}\bar{1}23]$ slip? Explain.

5.10 (a) Would it be theoretically possible to deform a zinc crystal in compression if the stress axis is along its [0001] axis using only the three (0001) [11$\bar{2}$0] slip systems? Explain.

(b) If the (11$\bar{2}$2) $[\bar{1}\bar{1}23]$ slip system were to operate could the crystal be deformed in the direction of its basal plane pole?

5.11 The total line length of the dislocations visible in a 4 cm by 4 cm TEM photograph of a metal foil, taken at a magnification of 20,000× is measured as 400 cm. The foil specimen imaged by this picture had a thickness of 300 nm. Determine the dislocation density in the specimen.

5.12 Identify the dislocation, in terms of its Burgers vector (using vector notation), that can cross slip between the (111) and (1$\bar{1}$1) planes of an fcc crystal.

5.13 In some hcp metals, a dislocation with a $\frac{1}{3}\langle 11\bar{2}3\rangle$ Burgers vector has been observed to cross slip between the (0001), {10$\bar{1}$0}, and {10$\bar{1}$1} planes. Identify the specific planes on which a dislocation with a $\frac{1}{3}[\bar{1}\bar{1}23]$ Burgers vector may move.

5.14

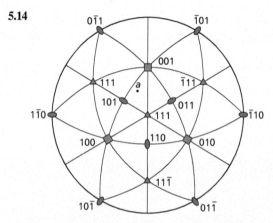

(a) The above diagram shows a 111 standard projection of an fcc crystal in which the standard stereographic triangles are outlined. Assuming that the point, a, in the figure represents the orientation of the tensile stress axis, indicate on a copy of this diagram the path that the crystal axis should follow during tensile deformation of the crystal.

(b) Give the Miller indices of the final stress axis orientation.

5.15 Now consider that the point, a, represents the stress axis during compressive deformation. Show the path on the stereographic projection that this axis will follow and identify its end orientation.

5.16 For the case of tensile deformation considered in Problem 5.14, determine the indices of the primary, conjugate, and cross-slip slip systems as well as that of the critical plane.

5.17 Johnston and Gilman have reported that in a grown LiF crystal that has been subjected to a constant stress of 1100 gm/mm² (10.8 MPa), the dislocation velocity at 249.1 K was 6×10^{-3} cm/s (6×10^{-5} m/s) and at 227.3 K the velocity was 10^{-6} cm/s (10^{-8} m/s). They also observed that their data suggested an Arrhenius relationship between the dislocation velocity and the absolute temperature so that one might write $v = A \exp(-Q/RT)$, where v is the dislocation velocity, A is a constant of proportionality, Q an effective activation energy in J/mol, R the international gas constant (8.314 J/mol·K) and T is in K. Use the velocity vs. temperature data of Johnston and Gilman, given above, to determine Q and A for their LiF crystal, stressed at 1100 gm/mm². Note that Q may be obtained using the relation

$$\ln v_1 - \ln v_2 = Q/R(1/T_2 - 1/T_1)$$

Once Q has been determined, A may be obtained by substitution back into the Arrhenius equation.

5.18 (a) Johnston and Gilman also observed that, at a constant temperature, the dislocation velocity obeyed a power law. Assuming that the dislocation velocity exponent, m, is 16.5 and that the stress, D, for a velocity of 1 cm/s is 5.30 MPa, determine the stress in MPa needed to obtain a velocity that is 5 times greater.

(b) Also give your answer in psi.

5.19 A typical cross-head speed in a tensile testing machine is 0.2 in./min.

(a) What is the nominal engineering strain rate imposed by this cross-head speed on a typical engineering tensile specimen with a 2 inch gage length?

(b) Estimate the dislocation velocity that would be obtained at this strain rate in an iron specimen with a dislocation density of 10^{10} cm/cm³. Assume that the Burgers vector of iron is 0.248 nm.

(c) If in a very slow tensile test a strain-rate of $10^{-7} \, \text{s}^{-1}$ is used, what dislocation velocity would be expected in the above iron specimen?

5.20 Necking in a tensile specimen begins at an engineering strain of 0.20. The corresponding engineering stress at this point is 1000 MPa. Determine the work hardening rate at the beginning of necking.

5.21 A tensile test was made on a specimen that had a cylindrical gage section with a diameter of 10 mm and a length of 40 mm. After fracture the total length of the gage section was found to be 50 mm, the reduction in area 90 percent, and the load at fracture 1000 N. Compute:

(a) The specimen elongation.

(b) The engineering fracture stress.

(c) The true fracture stress, ignoring the correction for triaxiality at the neck.

(d) The true strain at the neck.

5.22 The Hollomon equation

$$\sigma_t = k\epsilon_t^m$$

where k and m are constants, is capable of roughly approximating the shape of some stress-strain curves.

(a) Assume $k = 750$ MPa and $m = 0.6$. If the true stress at the point of maximum load is 552 MPa, what is the true strain at the maximum load?

(b) Compare m with ϵ_t at the maximum load.

(c) Prove that in general $m = \epsilon_t$ at the maximum load.

5.23 The slope, m, of the curve drawn through the data points in Fig. 5.35 is approximately equal to $2.55 \times 10^{-4} \, \dfrac{\text{kg}}{\text{mm}^2} \cdot \text{cm}$. Compute the increase in the dislocation density that would correspond to an increase in flow stress from 588 to 784 MPa (use the titanium data of Jones and Conrad).

References

1. Dash, W. C., *Dislocations and Mechanical Properties of Crystals*, p. 57, John Wiley and Sons, Inc., 1957.
2. Gilman, J. J., *Jour. Appl. Phys.*, **30** 1584 (1959).
3. Kelly, A., Tyson, W. R., and Cottrell, A. H. *Can. Jour. Phys.*, **45** No. 2, Part 3, p. 883 (1967).

4. Li, C., and Xu, G., *Philosophical Magazine*, **86**, No. 20, p. 2957 (2006).
5. Xu, G., and Zhang, C., *Journal of the Mechanics and Physics of Solids*, **51**, p. 1371 (2003).
6. Gilman, J. J., *Jour. Appl. Phys.*, **30** 1584 (1959).
7. Gilman, J. J., and Johnston, W. G., *Dislocations and Mechanical Properties of Crystals*, p. 116, John Wiley and Sons, Inc., New York, 1957.
8. Stein, D. F., *Canadian J. Phys.*, **45**, No. 2, Part 3, 1063 (1967).
9. Sherwood, P. J., Guiu, F., Kim, H. C., and Pratt, P. L., *Ibid*, p. 1075.
10. Hull, D., Byron, J. F., and Noble, F. W., *Ibid*, p. 1091.
11. Anderson, E. A., Jillson, D. C., and Dunbar, S. R., *Trans. AIME*, **197** 1191 (1953).
12. Tuer, G. R., and Kaufmann, A. R., *The Metal Beryllium*, ASM Publication, Novelty, Ohio (1955) p. 372.
13. Rapperport, E. J., and Hartley, C. S., *Trans. Metallurgical Society, AIME*, **218** 869 (1960).
14. Bell, R. L., and Cahn, R. W., *Proc. Roy. Soc.*, **A239** 494 (1957).
15. Price, P. B., *Phil. Mag.*, **5** 873 (1960).
16. Price, P. B., *Jour. Appl. Phys.*, **32** 1750 (1961).
17. Johnston, W. G., and Gilman, J. J., *Jour. Appl. Phys.*, **31** 632 (1960).
18. Koehler, J. S., *Phys. Rev.*, **86** 52 (1952).
19. Orowan, E., *Dislocations in Metals*, p. 103, American Institute of Mining, Metallurgical and Petroleum Engineers, New York, 1954.
20. Glen, J., *J. of the Iron and Steel Institute*, **186** 21 (1957).
21. Considère, *Ann. ponts et chaussees*, 9, ser 6 p. 574 (1885).
22. Taylor, G. I., *Proc. Roy. Soc.*, **A145** 362, 388 (1934).
23. Seeger, A., *Dislocations and Mechanical Properties of Crystals*, p. 243, John Wiley and Sons, New York, 1957.
24. Cottrell, A. H., *Dislocations and Plastic Flow in Crystals*, p. 45, Oxford Press, London, 1953.

Chapter 6

Elements of Grain Boundaries

Learning Objectives

Upon completion of Chapter 6, you will be able to:
1. Summarize the key features of polycrystalline metals and grain boundaries
2. Derive the equations to describe the dislocation model of low angle grain boundaries
3. Explain how grain boundaries can exist across five degrees of freedom
4. Understand the origin and calculation of stress field of grain boundary
5. Express grain boundary energy as a function of mismatch angle
6. Discuss the influence of surface tension on tri-grain junction angles in single phase and multiphase materials
7. Explain mechanism of recovery, formation of cell boundaries and low energy dislocations
8. Discuss dynamic recovery and the influence of stacking fault energies
9. Relate the effect of grain size to mechanical properties and applicability of Hall-Petch relations at different grain length scales from micro- to nanocrystalline

6.1 Grain Boundaries

In the previous chapters we have been concerned primarily with a study of the properties of very pure metals in the form of single crystals. Single crystal studies are important because they lead to quicker understanding of many basic phenomena. However, single crystals are essentially a laboratory tool and are rarely found in commercial metal objects which, as a rule, consist of many thousands of microscopic metallic crystals. Figure 6.1 shows the crystalline structure of a typical *polycrystalline* (many crystal) specimen magnified 350 times. The average diameter of the crystals is approximately 0.05 mm, and each crystal is seen to be separated from its neighbors by the grain boundaries. Note that grain boundaries in a highly polished polycrystalline specimen of a pure metal that has not been etched will appear entirely "white" under the microscope; that is, it will not show grain boundaries. The width of a grain boundary is, therefore, very small.

Fig. 6.1 A polycrystalline brass specimen photographed with polarized light microscope. In this photograph, individual Crystals(grains) can be distinguished by a difference in color shading, bounded by thin lines representing grain boundaries. Note that most of the triple junctions form 120° angles. The straight parallel sets of lines within grains are twin boundaries (see Chapter 17). Magnification: x135 at 35 mm size. (G. Muller, Struers GmbH/Science Source)

Grain boundaries, which disrupt long-range crystalline order, play an important part in determining the properties of a metal. For example, at low temperatures the grain boundaries are, as a rule, quite strong and do not weaken metals. In fact, heavily strained pure metals, and most alloys, fail at low temperatures by cracks that pass through the crystals and not the boundaries. Fractures of this type are called *transgranular*. However, at high temperatures and slow strain rates, the grain boundaries lose their strength more rapidly than do the crystals, with the result that fractures no longer traverse the crystals but run along the grain boundaries. Fractures of the latter type are called *intergranular fractures*. A number of other important grain-boundary aspects will be covered in the following paragraphs.

6.2 Dislocation Model of a Small-Angle Grain Boundary

In 1940, both Bragg and Burgers introduced the idea that boundaries between crystals of the same structure might be considered as arrays of dislocations. If two grains differ only slightly in their relative orientation, it is possible to draw a dislocation model of the boundary without difficulty. Figure 6.2A shows an example of such an elementary boundary in a simple cubic lattice in which the crystal on the right is rotated with respect to that on the left about the [100] direction (the direction perpendicular to the plane of the paper). The line between points *a* and *b* corresponds to the boundary. The lattice on both sides of the boundary is seen to be inclined downward toward the boundary, with the result

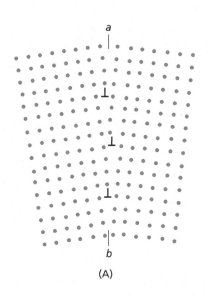
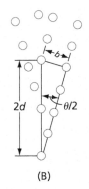

Fig. 6.2 **(A)** Dislocation model of a small-angle grain boundary. **(B)** The geometrical relationship between θ, the angle of tilt, and d, the spacing between the dislocations

that certain nearly vertical lattice planes of both crystals terminate at the boundary as positive-edge dislocations. The greater the angular rotation of one crystal relative to the other, the greater is the inclination of the planes that terminate as dislocations at the boundary, and the closer is the spacing of the dislocations in the vertical boundary. The dislocation spacing in the boundary thus determines the angular inclination between the lattices. It is easy to derive a simple expression for this relationship. Thus, referring to Fig. 6.2B, it can be seen that

$$\sin \theta/2 = b/2d \qquad 6.1$$

where b is the Burgers vector of a dislocation in the boundary and d is the spacing between the dislocations. If the angle of rotation of the crystal structure across the boundary is assumed to be small, then $\sin \theta/2$ may be replaced by $\theta/2$, and the equation relating the angle of tilt across a boundary composed of simple edge dislocations becomes

$$\theta = b/d \qquad 6.2$$

As was shown in Sec. 5.2, experimental evidence for small-angle boundaries has been obtained through the use of suitable etching reagents that locally attack the surface of metals at positions where dislocations cut the surface. After etching with one of these reagents, small-angle boundaries, consisting of rows of edge dislocations, appear as arrays of well-defined etch pits. According to the discussion in the preceding paragraph, the separation of the etch pits (dislocations) should determine the angular rotation of the lattice on one side of the boundary relative to that on the other. The fact that X-ray diffraction determinations of the same lattice rotation are in agreement with the value predicted by dislocation spacings serves as an excellent check on the validity of the dislocation model for small angle boundaries. Figure 6.3 shows a photograph of low-angle boundaries in a magnesium crystal. Low-angle boundaries may also be observed by transmission electron microscopy; an example is shown in Fig. 6.4.

The grain boundary represented in Fig. 6.2 is a very special boundary in which the two lattices are considered to be rotated only slightly with respect to each other about a common lattice

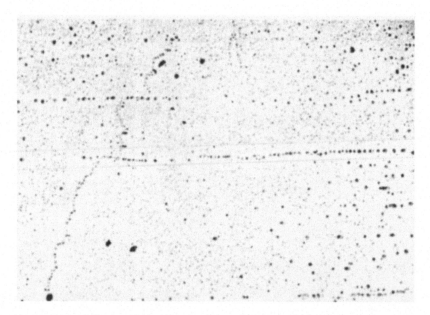

Fig. 6.3 Low-angle boundaries in a magnesium specimen. The rows of etch pits correspond to positions where dislocations intersect the surface

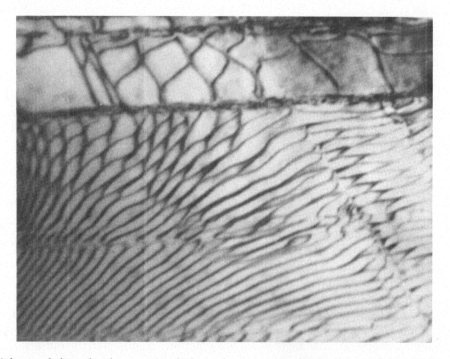

Fig. 6.4 A low-angle boundary in a copper–13.2 atomic percent aluminum specimen deformed 0.7 percent in tension as observed in the transmission electron microscope. This boundary has both a tilt and a twist character. Magnification: 32,000×. (Photograph courtesy of J. Kastenbach and E. J. Jenkins)

direction (a cube edge [100]). This, of course, does not represent the general case of a grain boundary. More complicated grain boundaries with large angles of misfit between grains must involve very complicated arrays of dislocations, and no simple picture of their structure has yet been worked out.

Small-angle boundaries commonly have less than 10–15° of misorientation and can be described by a dislocation model such as the one shown in Fig. 6.2. When the angle becomes large, the principal difficulty in postulating a suitable dislocation model is that as the angle of mismatch between grains becomes larger, the dislocations must come closer together and, in so doing, they tend to lose their identity.

6.3 The Five Degrees of Freedom of a Grain Boundary

The boundary shown in Fig. 6.2 is special in more than one sense. Not only is the angle of misorientation across the boundary small, but also the boundary has only a single degree of freedom. Actually, there are five degrees of freedom of a grain boundary, as illustrated in Fig. 6.5.

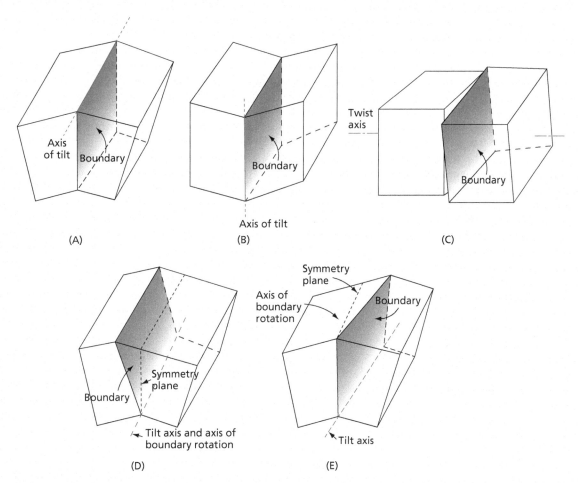

Fig. 6.5 The five degrees of freedom of a grain boundary

The boundary of Fig. 6.2 is shown again in part A of this illustration. Note that it is symmetrically positioned between the two crystals that are tilted with respect to each other about a horizontal axis that runs out of the plane of the paper. This is called a simple symmetrical *tilt boundary*. In part B, a symmetrical tilt boundary with a vertical tilt axis is shown. Part C shows a basically different way of orienting two crystals relative to each other. In this case they are rotated about an axis normal to the boundary, instead of about an axis lying in the boundary, as in the tilt boundaries in A and B. This is called a *twist boundary*. Such a boundary will normally contain at least two different arrays of screw dislocations, in agreement with our earlier discussion on rotational slip in Sec. 5.4.

The examples just shown represent several basically different ways that two crystals can be oriented with respect to each other. In each case, the boundary was assumed to be symmetrically positioned with respect to the two crystals. In addition, the boundary itself has two degrees of freedom, as indicated in Figs. 6.5D and 6.5E. These show that the boundary does not have to be in a symmetry position between the two crystals, but can be rotated about either of two axes at 90° to each other. Thus, as shown in Fig. 6.5, there are three ways that we can tilt or twist one crystal relative to another, and two ways that we can align the boundary between the crystals. An average grain boundary in a polycrystalline metal will normally involve, in varying degree, all five of these degrees of freedom. It is obvious that the general grain boundary is complex.

6.4 The Stress Field of a Grain Boundary

It is well known that, except at distances very close to the boundaries, the interiors of grains or subgrains are free of long-range stresses due to their boundaries. This is equivalent to saying that the boundaries do not possess long-range stress fields. This fact will now be demonstrated for the simple tilt boundary in Fig. 6.2, which is shown again in Fig. 6.6 in a more schematic form. Note that in this latter diagram, a slip plane is shown that passes through one of the boundary dislocations.

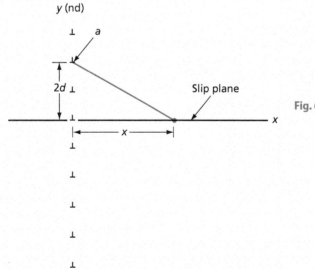

Fig. 6.6 A diagram defining the parameters used in computing the stress due to a simple tilt boundary. The y scale is in units of d, the distance between adjacent boundary dislocations, and n in the number

Now consider the shear stress on this slip plane at a distance, x, to the right of the boundary. The contribution to the shear stress at this point due to a specific boundary dislocation, such as that at point a, may be obtained using the shear-stress equation of Eqs. 4.9, or

$$\tau_{xy} = \frac{\mu b}{2\pi(1-\nu)} \cdot \frac{x(x^2 - y^2)}{(x^2 + y^2)^2} \qquad 6.3$$

where τ_{xy} is the shear stress on the slip plane, μ the shear modulus, and ν Poisson's ratio. For the geometry in Fig. 6.6, it is convenient to assume that $y = -nd$, where d is the distance between adjacent tilt boundary dislocations and n is an ordinal number defining the positions of the dislocations in the boundary. Thus, for the dislocation at point a in Fig. 6.6, $n = 2$ and Eq. 6.3 becomes

$$\tau_{xy} = \frac{\mu b}{2\pi(1-\nu)} \cdot \frac{x(x^2 - 4d^2)}{(x^2 + 4d^2)^2} \qquad 6.4$$

The total shear stress at a distance x from the tilt boundary is now obtained by summing the stress contributions from all the dislocations in the boundary. For this purpose we will assume the boundary to have an infinite extent. This leads to the following equation:

$$\tau_{xy} = \frac{\mu b}{2\pi(1-\nu)} \sum_{n=-\infty}^{n=+\infty} \frac{x[x^2 - (-nd)^2]}{[x^2 + (-nd)^2]^2} \qquad 6.5$$

The solution to this equation may be found in standard texts on dislocation theory (Ref. 1) and is

$$\tau_{xy} = \frac{\mu b}{2(1-\nu)} \frac{\pi x}{d^2(\sinh^2(\pi x/d))} \qquad 6.6$$

where τ_{xy} is the shear stress on the slip plane, μ the shear modulus, x the distance on the slip plane from the boundary, ν Poisson's ratio, and d the distance between edge dislocations in the boundary. A plot of τ_{xy} as a function of x is shown in Fig. 6.7A in which the metal is assumed to be iron, with $\mu = 86$ GPa, $\nu = 0.3$, and $b = 0.248$ nm. It is also assumed that the spacing between tilt boundary dislocations, d, is $22b$. The curve in question is at the lower left part of the diagram. Note that the shear stress due to the boundary falls very rapidly with increasing x. In order to show how rapid this decrease actually is, a second curve is drawn that shows the magnitude of the shear stress due to a single edge dislocation located at the intersection of the boundary with the slip plane. This curve lies above and to the right of that due to the boundary. Note, however, that the boundary stress approaches that of the single dislocation as x becomes very small. Finally, to further demonstrate the rapid rate of decrease of the boundary stress with x, Fig. 6.7B shows a second plot in which the scale of the stress axis has been expanded by a factor of 10. On both diagrams a horizontal line has also been drawn to show the approximate size of the critical resolved shear stress (CRSS) of iron. Note that the boundary stress equals the CRSS at a distance of only about $25b$. Further note that the boundary stress is negligible when x is greater than approximately $50b$.

6.5 Grain-Boundary Energy

As demonstrated in Sections 4.13 and 4.14, a strain energy is associated with a dislocation when it is in either its screw or edge orientation because the atoms of the crystal around a dislocation are displaced from their normal equilibrium positions. It is also generally true that any dislocation,

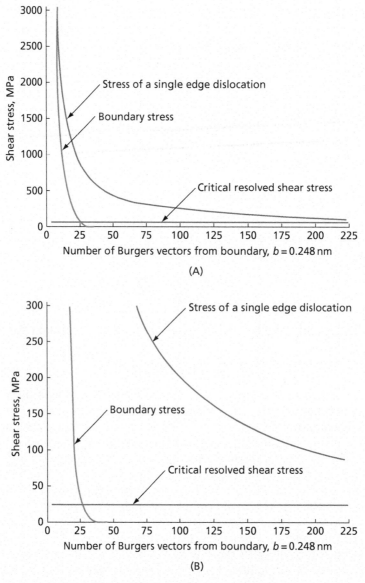

Fig. 6.7 **(A)** The stress, τ_{xy}, due to a tilt boundary and due to a single edge dislocation as functions of the distance measured in Burgers vectors. **(B)** Same as in (A) but at an expanded stress $d = 22b$

no matter what its orientation, possesses a strain energy. Since a grain boundary can consist of an array of dislocations, grain boundaries should also have strain energies. Because of the inherent two-dimensional nature of a grain boundary, its energy is normally expressed in terms of an energy per unit area (J/m²).

In order to demonstrate the nature of this grain-boundary energy we shall now consider the simple tilt boundary of Figs. 6.2 and 6.6. Equation 6.6 can be taken as the starting point for the computation of the energy of a tilt boundary. Following the derivation in Hirth and Lothe (Ref. 1), it will first be assumed that there are two infinitely long parallel tilt boundaries, one composed of positive edge dislocations and the other of negative edge dislocations, which are arranged such that for each positive edge dislocation in one boundary there is a negative edge dislocation on its slip plane in the other boundary. If the distance between these tilt boundaries is very large, the specific energy of formation of the pair of boundaries should be just twice that needed for the formation of one of the boundaries.

Now suppose that there is a positive and negative pair of edge dislocations on the same slip plane, and that the pair is brought together until their separation is $r_0 = b/\alpha$, where α is the factor that accounts for the strain energy at the dislocation cores. At this distance the core energies of the two dislocations should cancel each other. Consequently, on separating the dislocations from this position, the work of separation should equal the interaction energy of the pair. Next consider that the positive dislocation lies in an infinitely long tilt boundary and that the negative edge dislocation is a single dislocation. The force (per unit length) attracting the negative dislocation toward the positive boundary is $\tau_{xy} b$. Therefore, it may be deduced that the energy per unit length of a dislocation in either boundary equals one half the total interaction energy obtained by the following equation:

$$w_{bd} = \frac{1}{2} \int_{r_0}^{\infty} \tau_{xy} b\, dx \qquad 6.7$$

where τ_{xy} is obtained from Eq. 6.6. Now if we let $\eta = \pi x/d$ and $\eta_0 = \pi b/\alpha d$, Eq. 6.7 becomes:

$$w_{bd} = \frac{\mu b^2}{4\pi(1-\nu)d} \int_{\eta_0}^{\infty} \frac{\eta\, d\eta}{\sinh^2 \eta} \qquad 6.8$$

Multiplying the solution to this equation by $1/d$, the number of dislocations per unit area, yields γ_b, the energy per unit area of the boundary. Thus,

$$\gamma_b = w_{bd}/d = \frac{\mu b^2}{4\pi(1-\nu)} [\eta_0 \coth \eta_0 - \ln(2 \sinh \eta_0)] \qquad 6.9$$

If the use of Eq. 6.9 is limited to tilt boundaries with a tilt angle smaller than a few degrees, then by Eq. 6.2, we may take $\theta = b/d$ where b is the Burgers vector and d the separation between a pair of adjacent boundary dislocations. In this case, η_0 will be small and Eq. 6.9 may be written:

$$\gamma_b = \frac{\mu b}{4\pi(1-\nu)} \theta(\ln \alpha/2\pi - \ln \theta + 1) \qquad 6.10$$

where γ_b is the energy per unit area of the boundary, μ the shear modulus, b the Burgers vector, θ the tilt angle of the boundary, α a factor accounting for the dislocation core energy, and ν Poisson's ratio. This is basically the Shockley-Read equation used to obtain the historically significant plot shown in Fig. 6.8. The solid line in this figure represents the Shockley-Read equation, and the data points are experimental values. The good correlation between the data and the theoretical curve in Fig. 6.8, at higher values of θ, is now believed to be fortuitous. Theoretically, Eq. 6.10 should only be accurate to within the experimental error of the data for angles less than a few degrees. However, it is interesting to compare a plot of the small-angle solution, Eq. 6.10, with that of the large-angle solution, Eq. 6.9. This is done in Fig. 6.9, where it is again assumed that the metal is iron, with $\mu = 86$ GPa, $b = 0.248$ nm, $\nu = 0.3$, and $\alpha = 4$. Note that, under these conditions, the two curves do not appear to deviate significantly for angles less than 0.15 radians (8.6°).

A significant factor in the γ_b against θ equations is the size of the lower integration limit, r_0, in Eq. 6.7. Since $r_0 = b/\alpha$, where α is selected to account for the core energy, α actually

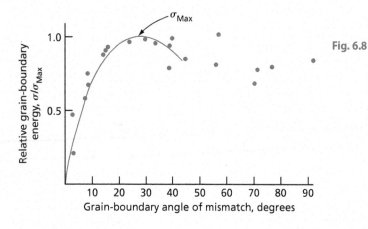

Fig. 6.8 Relative grain-boundary energy as a function of the angle of mismatch between the crystals bordering the boundary. Solid-line theoretical curve; dots experimental data of Dunn for silicon iron. (After *Dislocations* in *Crystals*, by Read, W. T., Jr. Copyright 1953. McGraw-Hill Book Co., Inc., New York)

determines r_0. That α is an important factor is clearly shown in Fig. 6.10, where three plots of γ_b against θ, corresponding to α values of 1, 2, and 4, respectively, made using the large-angle equation (Eq. 6.9), are given. These plots also involve the same values of the other parameters used in Fig. 6.9. In this figure the curve with the highest surface energy for a given angle of tilt corresponds to $\alpha = 4$ and that with the lowest to $\alpha = 1$.

6.6 Low-Energy Dislocation Structures, LEDS

It is evident from the preceding section that a tilt boundary has an inherent surface energy, as is true for grain boundaries in general. Because of the simplicity of the tilt boundary, one can learn a great deal about the basic characteristics of grain boundaries from a study of this type of boundary. It has already been demonstrated that a tilt boundary does not possess a long-range

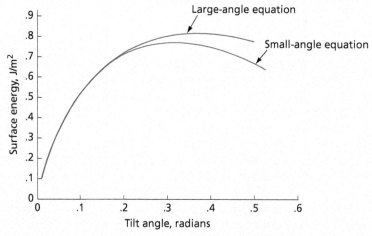

Fig. 6.9 The surface energy of a tilt boundary, γ_b, as a function of its angle of tilt, θ, as obtained with the small-angle and large-angle equations

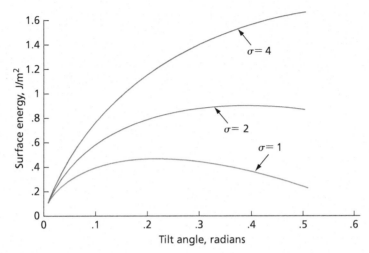

Fig. 6.10 The effect of α on the γ_b against θ curves

shear stress field. This is true in spite of the fact that it is composed entirely of edge dislocations of the same sign and Burgers vector. The reason for this is that the sign of the contribution to τ_{xy} at point x of a given boundary dislocation in Fig 6.6 depends on the latter's position in the tilt boundary. This is demonstrated in Fig. 6.11, where the magnitudes of the contributions to τ_{xy} of the dislocations in the boundary are plotted against n, the ordinal number of the dislocations, over the interval from $n = -50$ and $n = +50$. Note that large positive contributions come from dislocations with small n values, however; as n increases either positively or negatively, the contributions become negative. The result is that the net shear stress τ_{xy} at point x becomes very small when x lies at any reasonable distance from the tilt boundary.

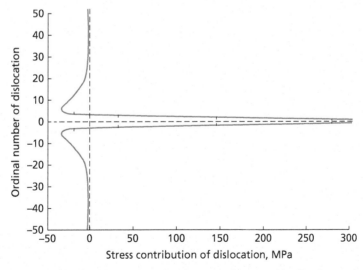

Fig. 6.11 The magnitude of the shear-stress contribution, at point x on the slip plane, of a dislocation in the tilt boundary as a function of its location in the boundary

There remains to be considered the two normal stress components of the boundary stress at point x, namely σ_{xx} and σ_{yy}. Because of the anti-symmetrical nature of the equations for σ_{xx} and σ_{yy} (see Eqs. 4.9), the net normal stress contributions from all the dislocations in a tilt boundary sum up to zero for both of the boundary stresses. In brief, both σ_{xx} and σ_{yy}, due to the edge dislocation lying on the slip plane, are automatically zero at point x because $n = 0$ and thus, $y = nd = 0$. At the same time, for all other dislocations, the contribution from a dislocation with an ordinal number $+n$ is cancelled by that from the dislocation with the ordinal number $-n$. As a result we may conclude that there is no effective long-range stress due to a tilt boundary.

Now consider the effect of a random dislocation on the magnitude of the strain energy when the dislocation is incorporated into a tilt boundary. The energy per unit length of a tilt boundary dislocation, w_{bd}, can be determined by simply multiplying Eq. 6.9 by d to remove the multiplying factor, $1/d$, which gives the number of dislocations per meter in the tilt boundary. One may then compare w_{bd} with w_e, the energy per unit length of a random dislocation, to obtain the energy of a boundary dislocation relative to that of a dislocation that is well removed from other dislocation. By Sec. 4.14, w_e, the energy per unit length of such an edge dislocation in an iron crystal, is 5.41 J/m², (if it is assumed that it contains equal numbers of positive and negative edges at a dislocation density $\rho = 10^{12}$ m/m³). Dividing w_e into w_{bd}, using the same iron parameters gives the desired ratio. Since w_{bd} depends on d, the distance between dislocations in the tilt boundary, it is best to consider the variation of w_{bd}/w_e with d. (See Fig. 6.12.) Note that even for a spacing, d, between boundary dislocations of 500 Burgers vectors, there is about a 25 percent decrease in the energy of the boundary dislocation over that of a random dislocation. This spacing corresponds to a tilt angle of only 6.9 min. An important feature of Fig. 6.12 is that it shows that as d decreases, the energy per dislocation also decreases, indicating that there is a driving force that attracts the edge dislocations to a tilt boundary.

The tilt boundary considered here is only one of a very large number of dislocation arrays that possess the property of having a low strain energy. Kuhlmann-Wilsdorf (Refs. 2 and 3) has proposed that these be known as low-energy dislocation structures, or LEDS. A very significant feature of

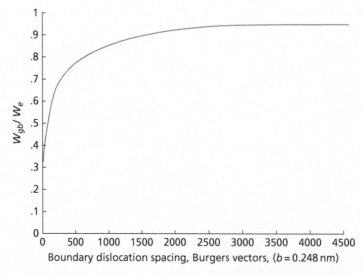

Fig. 6.12 The variation of W_{gb}/W_e with the spacing between the dislocations in a tilt boundary

these LEDS is that grain and subgrain boundaries normally fall into the LEDS category. One of the earliest forms of LEDS was proposed by G. I. Taylor in 1934, the year that dislocations were first used in the field of metallurgy. This consists of an equilibrium array of alternating rows of positive and negative edge dislocations arranged so that the four nearest neighbors of a given dislocation have the sign opposite to that of the given dislocation. This low-energy dislocation structure is illustrated in Fig. 6.13. Several other LEDS are shown in Fig. 6.14. These include a simple dipolar mat in Fig. 6.14A that may be formed by the locking of dislocations of opposite sign moving on adjacent slip planes because of the screening interaction of the stress fields of dislocation of opposite sign. Figure 6.14B shows another simple form of a LEDS. This is a dipolar wall formed by kinking on one of the crystallographic slip planes of a crystal. These kinks are a secondary mechanism of plastic deformation in many crystals, particularly when there are only a limited number of slip systems. Note that in the kink band, the edge dislocations possess a different orientation than in Fig. 6.14A. The result is a dipolar wall composed of two parallel tilt boundaries containing dislocations of opposite sign.

The LEDS shown previously are normally observed when slip occurs on a single slip system. When the plastic deformation involves a number of different slip planes and Burgers vectors, the LEDS can become much more complicated and appear as dislocation tangles in which the geometry of the dislocation arrangement is difficult to perceive. The significant factor is that, due to the mutual screening action of the stress fields of the various dislocations, there is a signicant decrease in the total strain energy associated with the dislocations. During plastic deformation there is, thus, a tendency to form cells with a low internal dislocation density and boundaries between the cells composed of dislocation tangles. A typical example of this type of microstructure is shown in the electron micrograph of Fig. 6.15. The driving force for the formation of this structure is the strain energy decrease associated with the formation of the tangles. A very significant factor is that,

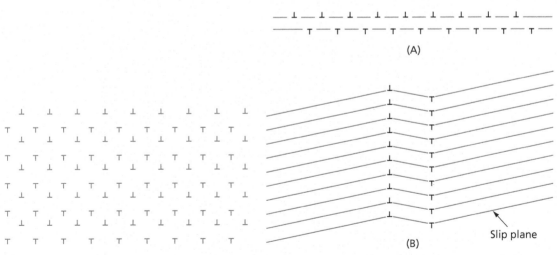

Fig. 6.13 The LEDS known as the Taylor lattice

Fig. 6.14 **(A)** A dipolar mat that can form as a result of the interaction between dislocations of opposite sign moving on a pair of adjacent and parallel slip planes. **(B)** When a kink band forms in a crystal, a dipolar array of edge dislocations of a different type is created

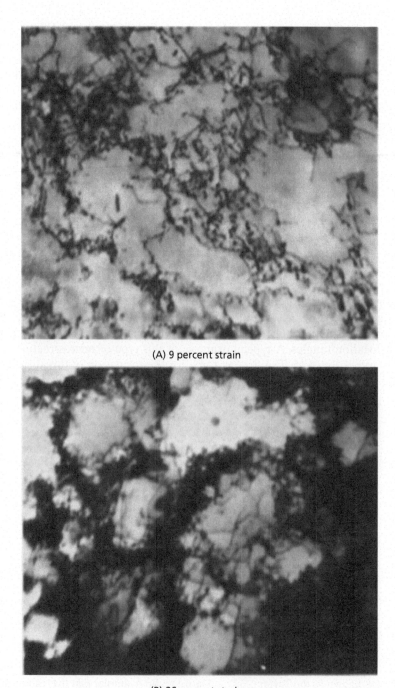

(A) 9 percent strain

(B) 26 percent strain

Fig. 6.15 These transmission electron micrographs illustrate dynamic recovery in nickel deformed at 77 K. Note that even at this low temperature there is a definite tendency to form a cell structure that increases at high strains. (This material was published in Scripta Metallurgica, Vol. 4, Issue 10, W. P. Longo and R.E. Reed-Hill, Work softening in polycrystalline metals, 765–770. Copyright Elsevier Science Ltd. (1970). https://doi.org/10.1016/0036-9748(70)90057-8)

with increasing plastic deformation and accompanying increases in dislocation density, the cell size decreases while the number of cells increases. There exists a significant empirical relationship (Ref. 2) between the cell size and the dislocation density, which is

$$\vartheta = \kappa/\sqrt{\rho} \qquad 6.11$$

where ϑ is the average cell diameter, κ a constant, and ρ the dislocation density.

It should be pointed out that the formation of LEDS can occur as a result of the solidification of a metal from its melt and can be influenced by the process of annealing or heating of a cold worked metal. This latter subject is considered in Chapter 8 (Annealing).

6.7 Dynamic Recovery

As may be seen in Figs. 8.13 and 8.14 in Chapter 8 on annealing, the basic effect of high temperature recovery is the movement of the dislocations resulting from plastic deformation into subgrain or cell boundaries. In many cases, this process can actually start during plastic deformation. When this happens, the metal is said to undergo *dynamic recovery*. The tendency for dislocations to form LEDS with a cell structure is quite strong in many pure metals and exists even to very low temperatures, as can be seen in Fig. 6.15. This illustration shows two transmission electron microscope foils taken from nickel specimens deformed at the temperature of liquid nitrogen (77 K). Note that for small strains (9 percent) the dislocation arrangement tends to be roughly uniform. However, when the strain becomes large (26 percent) a definite trend toward a cell structure becomes clearly evident. The size of the cells in this structure is small.

At more elevated temperatures, the effects of dynamic recovery naturally become stronger because the mobility of the dislocations increases with increasing temperature. As a result, the cells tend to form at smaller strains, the cell walls (LEDS) become thinner and much more sharply defined, and the cell size becomes larger. Dynamic recovery is therefore often a strong factor in the deformation of metals under hot working conditions.

Dynamic recovery has a strong effect on the shape of the stress-strain curve. This is because the movement of dislocations from their slip planes into LEDS lowers the average strain energy associated with the dislocations. The effect of this is to make it less difficult to nucleate the additional dislocations that are needed to further strain the material. Dynamic recovery thus tends to lower the effective rate of work hardening.

The role of dynamic recovery is not the same in all metals. Dynamic recovery occurs most strongly in metals of high stacking-fault energies and is not readily observed in metals of very low stacking-fault energy. These latter are normally alloys such as brass (copper plus zinc). The dislocation structure resulting from cold work in these latter materials often shows the dislocations still aligned along their slip planes. An example is shown in Fig. 6.16. In this case the alloy is one of nickel and aluminum.

The correspondence between the ability of a metal to undergo dynamic recovery and the magnitude of its stacking-fault energy strongly suggests that the primary mechanism involved in dynamic recovery is thermally activated cross-slip. This mechanism was considered in Chapter 5. At present, it is important to note the basic underlying difference between dynamic recovery and static recovery, such as occurs in annealing after cold work. In static recovery the movement of the dislocations into the cell walls occurs as a result of the interaction stresses between the dislocations themselves. In

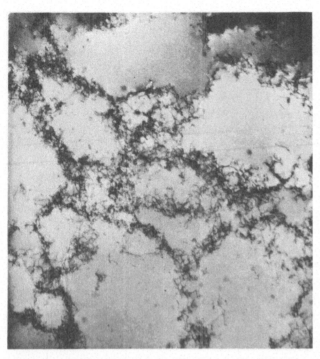

(A)

(B)

Fig. 6.16 Alloying normally reduces the stacking-fault energy of a metal. This can have a pronounced effect on the dislocation structure, as can be seen in these two electron micrographs. **(A)** Pure nickel strained 3.1 percent at 293 K. Magnification: 25,000X. **(B)** Nickel −5.5 wt. percent aluminum alloy strained 2.7 percent at 293 K. Magnification: 37,500X. (Photographs courtesy of J. O. Stiegler, Oak Ridge National Laboratories, Oak Ridge, Tenn.)

dynamic recovery, the applied stress causing the deformation is added to the stresses acting between the dislocations. As a result, dynamic recovery effects may be observed at very low temperatures, and at these temperatures the applied stresses can be very large.

6.8 Surface Tension of the Grain Boundary

The surface energy of a grain boundary, γ_G, has the units of ergs/cm² or J/m², where 1 erg/cm² = 10^{-3} J/m². That is,

$$\gamma_G = \frac{J}{m^2} \qquad 6.12$$

which is equivalent to

$$\gamma_G = \frac{N \cdot m}{m^2} = \frac{N}{m} \qquad 6.13$$

but the units N/m are those of a surface tension. It is often reasonable to assume that solid grain boundaries possess a surface tension equivalent to that of liquid surfaces. The surface tension of crystal boundaries is an important metallurgical phenomenon. The experimental data of Fig. 6.8 show that the grain-boundary surface tension is an increasing function of the angle of mismatch between grains to an angle of approximately 20°, and then it is essentially constant for all larger angles.

Absolute or numerical values of the surface tension of metal surfaces are difficult to determine. However, measurements have been made on copper, silver, and gold which show that their surface tensions of free surfaces (external surfaces) are of the order of 1.2 to 1.8 J/m. External metal surfaces, therefore, possess relatively large surface tensions; they are approximately 20 times larger than that of liquid water. It has been determined further that the surface tensions of large-angle grain boundaries equal approximately one-third of the free surface values, and are therefore of the order of 0.3 to 0.5 J/m.

While absolute values of the surface tension of grain boundaries are difficult to measure, relative values of boundary-surface energy may be estimated with the aid of a simple relationship. Let the three lines of Fig. 6.17 represent grain boundaries that lie perpendicular to the plane of the paper and meet in a line the projection of which is o. The three vectors γ_a, γ_b, and γ_c originating at

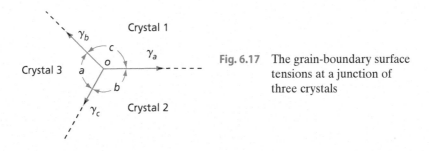

Fig. 6.17 The grain-boundary surface tensions at a junction of three crystals

point *o* represent, by their directions and magnitudes, the surface tensions of the three boundaries. If these three force vectors are in static equilibrium, then the following relationship must be true:

$$\frac{\gamma_a}{\sin a} = \frac{\gamma_b}{\sin b} = \frac{\gamma_c}{\sin c} \qquad 6.14$$

where *a*, *b*, and *c* are the dihedral angles between boundaries.

Since crystal boundaries are regions of misfit or disorder between crystals, it is to be expected that atom movements across and along boundaries should occur quite easily. The boundary is caused to move by the simple process whereby atoms leave one crystal and join another crystal on the other side of a boundary. Crystal boundaries in solid metals can move and should not be thought of as fixed in space. The speed with which crystal boundaries move depends on a number of factors. The first of these is the temperature, as the energy that makes it possible for an atom to move from one equilibrium grain-boundary position to another comes from thermal vibrations. We have, therefore, an analogous situation to that of atom movements inside crystals where the atoms diffuse into lattice vacancies as a result of thermally activated jumps. The rate of movement must increase rapidly with rising temperature. However, other things being equal, it is to be expected that the same number of atoms would cross a boundary in one direction as in the other, and that the boundary would be relatively fixed in space. Actually, grain-boundary movements occur only if the energy of the metal as a whole is lowered by a greater movement of atoms in one direction than in the other. One way that the energy of a specimen can be lowered by the motion of a grain boundary occurs when it moves into a deformed crystal, leaving behind a strain-free crystal (This is covered in Sec. 8.9.) Another driving force for movement lies in the energy of the boundaries themselves. A metal can approach a more stable state by reducing its grain-boundary area. There are two ways in which this may be achieved. First, boundaries may move so as to straighten out sharply curved regions; and, second, they may move in such a way that some crystals are caused to disappear, while others grow in size. The latter phenomenon, which results in a decrease in the total number of grains, is called *grain growth*. (See Sec. 8.20.)

One of the consequences of grain-boundary movement is that if a metal is heated at a sufficiently high temperature for a long enough time, the equilibrium relationship between the surface tensions and the dihedral angles can actually be observed. (See Eq. 6.14.) In pure metals with randomly oriented crystals, low-angle boundaries occur infrequently, and it may be assumed that the grain-boundary energy is constant for all boundaries. (See Fig. 6.8.) If the surface energies (surface tension) are equal in each of three boundaries that meet at a common line, and if the boundaries have attained an equilibrium configuration, the three dihedral angles must be equal. An examination of the junctions where three boundaries meet in the photograph, Fig. 6.1, shows a large number of 120° dihedral angles. This fact is surprising when it is considered that many of the grain boundaries are not normal to the plane of the photograph. In fact, if it were possible to cut the surface so that all the boundaries were perpendicular to the surface, a good agreement between the experimental and predicted angles would be observed. It may be concluded that in a well-annealed pure metal, that is, one that has been heated for a long time at a high temperature, the grain-boundary intersections form angles very close to 120°.

6.9 Boundaries between Crystals of Different Phases

A *phase* is defined as a homogeneous body of matter that is physically distinct. The three states of matter (liquid, solid, and gas) all correspond to separate phases. Thus, a pure metal, for example, copper, can exist in either the solid, liquid, or gaseous phase, each stable in a different temperature range. In this respect, it would appear that there is no difference between the concepts of phase and state. However, a number of metals are allotropic (polymorphic); that is, they are able to exist in different crystal structures, each stable in a different temperature range. Each crystal structure in these metals corresponds to a separate phase, and allotropic metals, therefore, possess more than three possible phases. When pure metals are combined to form alloys, additional crystal structures can result in certain composition and temperature ranges. Each of these crystal structures in itself constitutes a separate phase. Finally, it should be pointed out that a solid solution (a crystal containing two or more types of atoms in the same lattice) also satisfies the definition of a phase. A good example of a solid solution occurs when copper and nickel are melted together and then frozen slowly into the solid state. The crystals that form contain both nickel and copper atoms in the same ratio as the original liquid mixture; both types of atoms occupy the lattice sites of the crystal indiscriminately.

For the present, the concept of phases has been introduced in order to explain additional surface-tension phenomena. The study of the phases of alloy systems is a topic that will be considered in Chapters 10 and 11.

The grain boundaries of alloys containing crystals of only a single phase should behave in a manner analogous to those of a pure metal. The crystals of a well-annealed solid-solution alloy, like the copper-nickel alloy mentioned previously, appear under the microscope like those of a pure metal. At all points where three grains meet there is a mean dihedral angle of 120°.

In alloys of two phases, two types of boundaries are possible: boundaries separating crystals of the same phase, and boundaries separating crystals of the two phases. Three grains of a single phase can still intersect on a line, but there is the additional possibility of two grains of one phase meeting a grain of the other phase at a common intersection. A junction of this type is shown in Fig. 6.18. If the surface tensions in the boundaries are in static equilibrium, then

$$\gamma_{11} = 2\gamma_{12} \cos \frac{\theta}{2} \qquad 6.15$$

where γ_{11} is the surface tension in the single-phase boundary, γ_{12} the surface tension in the two-phase boundary, and θ the dihedral angle between the two boundaries that separate phase 2 from phase 1.

Let us solve the above expression for the ratio of the surface tension of the two-phase boundary to that of the single-phase boundary. This yields

$$\frac{\gamma_{12}}{\gamma_{11}} = \frac{1}{2 \cos \frac{\theta}{2}} \qquad 6.16$$

The ratio γ_{12}/γ_{11} is plotted as a function of the dihedral angle θ in Fig. 6.19. Notice that as the surface tension of the boundary between two phases approaches half of that of the single phase, the

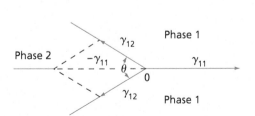

Fig. 6.18 The grain-boundary surface tensions at a junction between two crystals of the same phase with a crystal of a different phase

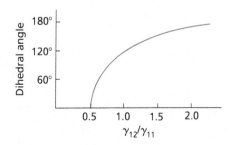

Fig. 6.19 The dependence of the two-phase dihedral angle on the ratio of the two-phase surface tension to the single-phase surface tension

dihedral angle falls rapidly to zero. The significance of small dihedral angles is readily apparent in Fig. 6.20, where the shape of the intersection is shown for angles of 10° and 1°. It is apparent that as the angle approaches zero, the second phase moves to form a thin film between the crystals of the first phase. Further, if the surface tension of the two-phase boundary (γ_{12}) falls to a value less than $\frac{1}{2}$ of the single-phase surface tension γ_{11}, static equilibrium of the three forces is not possible and point *o* moves to the left, which is equivalent to saying that when the dihedral angle becomes zero, the second phase penetrates the single-phase boundaries and isolates the crystals of the first phase. Furthermore, the extent of penetration also depends on the grain-boundary energy. For example, it has been shown (Ref. 4) that for sn-rich particles immersed in molten pb-rich liquid, about two-thirds of the boundaries were wetted with liquid. The penetration may occur even if the second phase is present in almost negligible quantities. A good example occurs in the case of bismuth in copper. The surface tension of a bismuth-copper interface is so low that the dihedral angle is zero and a minute quantity of bismuth is capable of forming a thin film between the copper crystals. Whereas copper is ordinarily a metal of high ductility and capable of extensive plastic deformation, bismuth is not. In fact, bismuth is a very brittle metal, and when it forms a continuous film around copper crystals, the copper loses its ductility, even though the total amount of the bismuth impurity is less than 0.05 percent. This loss in ductility is observed at all temperatures at which copper is worked.

In a number of important metallurgical cases, second-phase impurities remain in the liquid state until a temperature is reached well below the freezing point of the major phase. The amount of harm that these impurities (in small percentages) can do to the plastic properties of metals is a function of the surface tension between liquid and solid. If this interfacial energy is high, the liquid tends to form discrete globules with little effect on the hot working properties of metals. On the other hand, low interfacial energies lead to liquid grain-boundary films. These, of course, are very harmful to the plastic properties of metals.

Let us take the case of iron containing small quantities of sulphur as an impurity. Iron sulfide is liquid at temperatures well below the freezing point of iron. This range includes the temperature range normally used for hot rolling iron and steel products. Unfortunately, the surface energy of the iron sulfide to iron boundary is very close to one-half that of the boundary between iron crystals, and the liquid sulfide forms a grain-boundary film that almost completely separates the iron crystals. Since a liquid possesses no real strength, iron or steel in this condition is brittle and cannot be hot-worked without disintegrating. In such a case as this, when a metal becomes brittle at high temperatures, the metal is said to be "hot short," that is, its properties are deficient at hot-working temperatures.

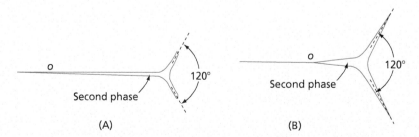

Fig. 6.20 When the dihedral angle is small, the second phase (even if present in small quantities) tends to separate crystals of the first phase. **(A)** dihedral angle 1°, **(B)** dihedral angle 10°

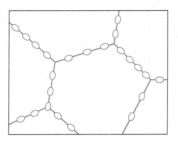

Fig. 6.21 When the dihedral angle is large (above 60°), the second phase (when present in small quantities) tends to form small discrete particles, usually in the boundaries of the first phase

With regard to the problem caused by sulfur in promoting hot shortness, it should be pointed out that this problem has for many years been alleviated by incorporating in the steel a small amount of manganese as an alloying element. Manganese has a very strong ability to combine with the sulfur in a steel to form globules, which are solid at the rolling temperatures of steel. The MnS particles have a much less deleterious effect on the hot rolling properties.

It has already been explained that a second phase in small quantities forms thin intergranular films when the dihedral angle is zero. For dihedral angles greater than zero but less than 60°, the equilibrium configuration is one where small quantities of the second phase run as a continuous network along the grain edges of the first phase; that is, along the lines where three grains of the first phase would meet in the absence of the second phase. According to Fig. 6.19, a dihedral angle of 60° corresponds to a ratio of interphase to single-phase surface tensions of 0.582. Many liquid-solid interfaces possess surface tensions less than 0.600 of the solid-boundary surface tension. Therefore, when liquids and solids coexist, there is a strong possibility that the liquid will form a continuous network throughout the metal. Since diffusion in liquids is more rapid than in solids, this network forms a convenient path for rapid diffusion of elements (both good and bad) into the center of the metal.

When the dihedral angle between the interphase boundaries is greater than 60°, the second phase no longer forms a continuous network unless it is present as the major phase. The second phase, if present in small quantities, now forms discrete globular particles, usually along the boundaries of the first-phase crystals. (See Fig. 6.21.)

Below approximately 1273 K, iron sulfide is no longer a liquid, but a solid. Solid iron sulfide forms an equilibrium angle with solid iron greater than 60° and, therefore, tends to crystallize as discrete particles of FeS instead of a continuous grain-boundary film. Small discrete particles of a second phase, even though they may be brittle in themselves, are far less damaging to the properties of a metal than are continuous brittle grain-boundary films. Iron containing sufficient FeS to embrittle it at hot-working temperatures will retain much of its ductility at room temperature.

6.10 The Grain Size

The size of the average crystal is a very important structural parameter in a polycrystalline aggregate of a pure metal or single-phase metal. Unfortunately, this is a difficult variable to define precisely. In many cases, the three-dimensional shape of a grain is very complex, and even in those cases where the grains appear to be of nearly equal size in a two-dimensional micrograph, the grains can vary through a very wide range of sizes. Hull (Ref. 5) has actually demonstrated this by allowing mercury to penetrate along the grain boundaries of a high zinc brass (beta brass) which embrittled the boundaries and allowed the individual grains to be separated from each other. Fig. 6.22 shows a set of his photographs corresponding to three of some 15 size classifications into which he grouped

Fig. 6.22 Grains removed from a beta brass specimen that appeared to be of a nearly uniform grain size on a metallographic section. Three of some 15 size classifications are shown in these photographs. They represent the smallest, the largest, and a size from the middle of the range. (The original photographs are from F. Hull, Westinghouse Electric Company Research Laboratories, Pittsburgh. Copies were furnished courtesy of K. R. Craig)

the grains. The upper picture shows the smallest grains and the bottom the largest. The middle photograph represents an average size. Most metallographic structures are observed on planar sections, and linear measurements made on such surfaces are not normally capable of being accurately related to the grain diameter, which is a property of a complex, three-dimensional quantity.

In spite of the problems indicated above, some sort of measurement to define the size of the structural unit in a polycrystalline material is needed, and the concept of an average grain size is widely used in the literature.

Perhaps the most useful parameter for indicating the relative size of the grains in a microstructure is $\bar{l}$ determined by the linear intercept method. This quantity is called the mean grain intercept and is the average distance between grain boundaries along a line laid down on a photomicrograph. In making this measurement, one can lay a straight edge of perhaps 10 cm in length down on the photograph and then count the number of boundaries that the edge of the instrument crosses. This measurement should then be repeated several times, placing the straight edge down on the photograph in a random fashion. The total number of intersections, when divided by the total line length over which the linear measurement was made, and multiplied by the magnification in the photograph, yields the quantity $\overline{N_l}$, the average number of grain boundaries intercepted per centimeter. The reciprocal of this quantity is $\bar{l}$, which is often used as a parameter for indicating the approximate grain size. Thus we have

$$\bar{l} = \frac{1}{\overline{N_l}} \qquad 6.17$$

Even though there is actually no geometrical relationship between $\bar{l}$ and the actual average grain diameter d, this quantity is widely used to represent the grain diameter. In this regard, it might be mentioned that such a relationship would exist if all grains had the same shape and size. On the other hand, quantitative metallography (Refs. 6 and 7) has shown that the reciprocal of $\bar{l}$, that is $\overline{N_l}$, is directly related to the amount of grain-boundary surface area in a unit volume. The relationship in question is

$$S_v = 2\overline{N_l} \qquad 6.18$$

where S_v is the surface area of the grain boundaries per unit volume. It is therefore well to recognize that when one determines $\bar{l}$, what is actually determined is the reciprocal of the grain-boundary surface area in a unit volume.

6.11 The Effect of Grain Boundaries on Mechanical Properties: Hall-Petch Relation

Polycrystalline metals almost always show a strong effect of grain size on hardness and strength, except possibly at very elevated temperatures. The smaller the grain size the greater the hardness or flow-stress, where the flow-stress is the stress in a tension test corresponding to some fixed value of strain. Figures 6.23 and 6.24 support this statement. In the first of these two illustrations, the hardness of titanium is plotted as a function of the reciprocal of the square root of the grain size. The hardness measurements corresponding to these data were obtained using a Vickers 138° diamond-pyramid indentor. According to the illustration, it is possible to write an empirical relationship of the form

$$H = H_o + k_H d^{-1/2} \qquad 6.19$$

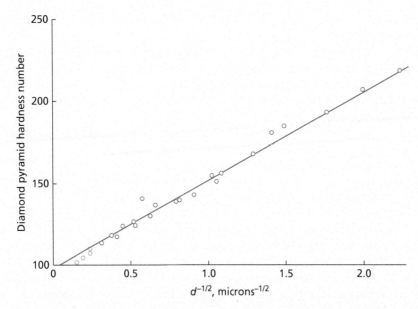

Fig. 6.23 The hardness of titanium as a function of the reciprocal of the square root of the grain size. (From the data of H. Hu and R. S. Cline, *TMS-AIME*, **242** 1013 [1968]. This data has been previously presented in this form by R. W. Armstrong and P. C. Jindal, *TMS-AIME*, **242** 2513 [1968].)

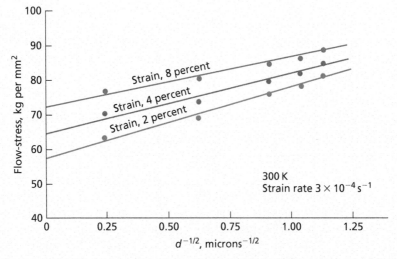

Fig. 6.24 The flow-stress of titanium as a function of the reciprocal of the square root of the grain size. (After Jones, R. L. and Conrad, H., *TMS-AIME*, **245** 779 [1969].)

where H is the hardness, d is the average grain diameter, k_H is the slope of the straight line drawn through the data, and H_o is the intercept of the line with the ordinate axis and corresponds to the hardness expected at a hypothetical infinite grain size. It is not necessarily the hardness of a single crystal because the mechanical properties of a crystal are in general nonisotropic, so that the hardness of a crystal may vary with the orientation of the crystal.

In Fig. 6.24, tensile test data is plotted for titanium specimens tested at room temperature. Here the flow-stress corresponding to three different strains (2 percent, 4 percent, and 8 percent) are plotted against $d^{-1/2}$, and straight lines are drawn through the data corresponding to linear relationships of the form

$$\sigma = \sigma_o + kd^{-1/2} \qquad 6.20$$

where σ is the flow-stress and σ_o and k are constants equivalent to H_o and k_H in the previous equation.

A linear relationship between the flow-stress and square root of the dislocation density was originally proposed by both Hall (Ref. 8) and Petch (Ref. 9) and, as a consequence, it is now generally known as the Hall-Petch equation. Such a relationship can be rationalized by dislocation theory assuming that grain boundaries act as obstacles to slip dislocations, causing dislocations to pile up on their slip planes behind the boundaries. The number of dislocations in these pile-ups is assumed to increase with increasing grain size and magnitude of the applied stress. Furthermore, these pile-ups should produce a stress concentration in the grain next to that containing a pile-up that varies with the number of dislocations in the pile-up and the magnitude of the applied stress. Thus, in coarse-grained materials, the stress multiplication in the next grain should be much greater than that in fine-grained materials. This means that in the fine-grained materials a much larger applied stress is needed to cause slip to pass through the boundary than is the case with coarse-grained materials.

Although the Hall-Petch relation has been widely accepted, it has by no means been completely verified. While in many cases grain size data can be plotted to yield an apparent linear relationship between σ and $d^{-1/2}$, as in Fig. 6.24, Baldwin (Ref. 10) has shown that the general scatter normally observed with this type of data can, in many cases, give equally good linear plots when σ is plotted against d^{-1} or $d^{-1/3}$.

Overriding the question of whether a variation of σ or H with $d^{-1/2}$ has true physical significance is the fact that plots such as those in Figs. 6.23 and 6.24 clearly demonstrate the dependence of the hardness and flow-stress on the grain size. Note, for example, that the tensile flow-stress of titanium at 2 percent strain increases from about 448 MPa at a grain size of 17 microns to about 565 MPa at a grain size of 0.8 microns (10^{-4} cm). Nevertheless, as will be discussed in the following section, there are now clear indications for deviations from the classical Hall-Petch relation as grain sizes approach nanometer dimensions.

6.12 Grain Size Effects in Nanocrystalline Materials

Nanocrystalline is a terminology that has been used in recent years to describe materials whose grain size does not exceed 10 nm. This upper limit of the grain size is chosen arbitrarily for convenience rather than being based on a physical significance or a characteristic length scale. The central question in understanding the influence of the grain size on the mechanical properties of nanocrystalline materials is whether the deformation mechanisms are similar to those of coarse-grained materials, and if so, whether the Hall-Petch relation as defined in Eq. 6.20 holds true for nanograins. In other words, the question is whether or not the relation as derived for coarse-grained materials can be extrapolated to determine mechanical properties in the nanocrystalline range.

As discussed in the previous section, the Hall-Petch relation can be rationalized based on the assumption that grain boundaries act as obstacles to slip dislocations, causing their pile-up on their slip planes behind boundaries. The number of the piled-up dislocations is also assumed to be large and to increase with increasing grain size and magnitude of the applied stress. As such, much larger applied stress is needed in fine-grained materials to force slip through the grain boundary and initiate slip in the neighboring grain. From a geometrical point of view, it is obvious that these assumptions cannot hold true when the grains become so small that they cannot accommodate many dislocations. Furthermore, grain boundaries in nanocrystalline materials occupy a larger fraction of the volume than that in coarse-grained materials. As such, grain-boundary contributions to the deformation can cause deviations from the classical Hall-Petch relation.

Sanders et al. (Ref. 11) have investigated tensile behavior of high-purity Cu and Pd nanocrystalline metals with grain sizes in the range of 10 to 110 nm. The samples were produced by inert gas condensation followed by warm compaction. The hardness of the samples was found to follow the coarse-grained Hall-Petch relation down to grain size around 16 nm. Below this size, the hardness values were found to deviate from the relation and level off, as shown in Fig. 6.25. The authors indicate that the drop in the hardness could be indicative of a change in the deformation mechanism as grains fall into this nanometer range or due to the change in the processing conditions of the samples with finer grains. The tensile yield stress of samples with 110 nm grain size was also found by Sanders et al. to be close to the value predicted from the extrapolation of the Hall-Petch

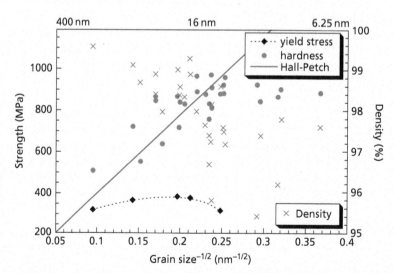

Fig. 6.25 Plot of yield strength and Vickers microhardness (divided by 3) as a function of the inverse square root of the grain size. Also shown is an extrapolation to small grain sizes of the coarse-grained Hall-Petch relation for Cu and the change in the bulk density. (This material was published in *Acta Materialia*, Vol. 45, Issue 10, P. G. Sanders, J. A. Eastman and J. R. Weertman, Elastic and tensile behavior of nanocrystalline copper and palladium, 4019–4025. Copyright Elsevier Science Ltd. (1997). https://doi.org/10.1016/S1359-6454(97)00092-X)

relation for coarse-grained copper, as shown in the figure. However, the yield strength dropped off dramatically as grain size fell below 110 nm. The failure of the finer-grained samples also occurred with little plastic deformation with a fracture surface perpendicular to the stress axis. The authors believe that this observation may indicate that dislocation-assisted deformation mechanisms that operate in coarse-grained materials may not operate effectively in nanocrystalline materials. However, the authors caution that the reduction in the yield strength and brittle failures could be also due to the flaws and defects introduced during processing of the finer-grain materials. Indeed, as shown in Fig. 6.25, the bulk density of the finer grained materials was appreciably lower than that of the coarser-grained ones, indicating the presence of appreciable porosity in the samples. As such, the tensile measurements alone may not have sufficient accuracy to ascertain the applicability of the Hall-Petch relation to nanocrystalline materials.

Figure 6.26 shows yield stress data as a function of grain size for several metallic systems, as compiled by Masumura et al (Ref. 12). The line with slope of one in the figure represents the Hall-Petch relation given by Eq. 6.2. According to these authors, for grain sizes larger than about 1 μm, the data follow the Hall-Petch relation; the yield strength increases as grain size decreases. At grain sizes between 1 μm and 30 nm, the Hall-Petch relation roughly holds, but the increase in the strength with grain size is less than that predicted by the Hall-Petch relation. In contrast, for grain sizes smaller than about 30 nm, the yield strength actually decreases with decreasing grain size.

The deviations from the classical Hall-Petch relation presented in Figs. 6.25 and 6.26 indicate that the relation may not hold true below a certain critical grain size. Several models and arguments have been advanced to explain the deviation of mechanical behavior of nanocrystalline

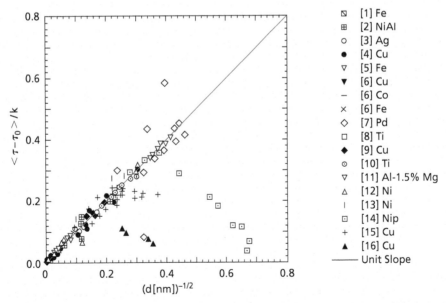

Fig. 6.26 Compilation of yield stress data for several metallic systems. (This article was published in Acta Materialia, R. A. Masumura, P. M. Hazzledine and C. S. Pande, Vol. 46, No. 10, Yield Stress of Fine Grained Materials, pp. 4527–4534. Copyright Elsevier Science Ltd. (1998) http://www.sciencedirect.com/science/journal/13596454)

material below a certain critical grain size from the classical Hall-Petch relation. Among them are the possibility of diffusional creep at room temperature (Ref. 13) and dislocation accommodated boundary sliding (Ref. 14).

As can be surmised from the above discussion, the exact influence of the grain size in nanoscale range on the strength of the material is a matter of dispute. Part of the difficulty in answering this question is that the property measurements for nanocrystalline materials, particularly those done under tensile conditions, do not have sufficient accuracy to ascertain the applicability of the Hall-Petch relation in this region. Another difficulty is that nanocrystalline materials often include porosities, flaws, or impurities. As such, a direct comparison with high-purity coarse-grained counterparts cannot be made. As Andrievski and Glezer conclude (Ref. 15), further studies are required to determine the nature of size effects in nanocrystalline materials.

6.13 Coincidence Site Boundaries

In 1949 Kronberg and Wilson (Ref. 16), as a result of a study of secondary recrystallization in copper, were able to call attention to an important form of grain boundary: the *coincident site boundary*. In their investigation they first very heavily cold rolled some copper plates, which were then annealed at 400 °C (673 K) to produce a fine-grained microstructure with an approximate grain size of 0.03 mm (3 μm). This structure had a very pronounced texture; that is, nearly all the grains of the metal had close to the same orientation in space. Moreover, the grains were aligned so that each had a cube plane (i.e., a {100} plane) almost parallel to the rolling plane of the sheet. In addition, a ⟨010⟩ direction of these grains tended to be closely aligned to the direction in which the sheet had been rolled. Actually, this type of texture has been observed in a number of fcc metals and is known as a *cube texture*. When these sheets with the cube texture were heated to between 800 and 1000 °C (1073 to 1273 K), a new set of larger crystals of a different orientation appeared. This form of grain growth is called *secondary recrystallization*.

Kronberg and Wilson next quantitatively compared the orientations of these new grains to those of the original cube texture and found that their two orientations were rather simply related. They also observed that a secondary recrystallized grain orientation could be obtained from that of an original grain by a rotation of either 22° or 38° about a ⟨111⟩ axis, or, in a few cases, by a 19° rotation about the ⟨010⟩ axis aligned parallel to the rolling direction.

Kronberg and Wilson then demonstrated that these rotations could produce surfaces, separating the new crystal from the old, that contained a number of positions where the atoms in both crystals were in coincidence. These were called *coincidence sites*. As an example, consider Fig. 6.27 based on a drawing in the Kronberg and Wilson paper. This assumes a 22° rotation about a ⟨111⟩ axis in an fcc crystal. Net A corresponds to the atomic arrangement on the (111) plane of the parent crystal where the atom centers are indicated by open circles. Net B, on the other hand, corresponds to the (111) plane in a secondarily recrystallized crystal. Here the atom centers are designated by filled-in circles. The two nets are rotated relative to each other by 22° about the [111] pole, which is assumed to lie at the center of the drawing. The points where atoms on the two nets coincide are shown as larger open circles. Note that the coincidence sites also define a similar but larger hexagonal net than the nets A and B. In other words, the coincidence sites form a net that is a multiple of the primitive net. A boundary of this type, with a large fraction of coincidence sites, is considered to have valuable properties, such as a higher mobility or the ability to support a rapid rate of grain growth.

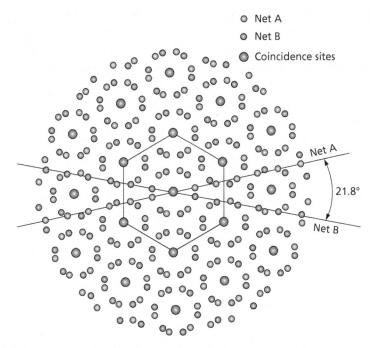

Fig. 6.27 A coincident site boundary obtained by a 22° rotation across a (111) plane (i.e., about a ⟨111⟩ axis). The reciprocal density, Σ, of the coincident sites, shown as large open circles (colored purple), is 7. (After Kronberg and Wilson)

6.14 The Density of Coincidence Sites

By a direct count, Kronberg and Wilson were able to show that the number of coincidence sites in the boundary of Fig. 6.27 was equal to 1/7 of the atoms in either net A or B. What is more, they also pointed out that the remaining atoms of the two nets, those not in coincidence, could be brought into alignment by relatively small atom movements of the order of 1/3 of an atomic distance.

The fraction of atoms in coincidence, at a boundary of this type, is generally known as the *density of coincidence sites*. The reciprocal of the density, however, is more commonly used as a parameter to describe a coincidence site boundary. This is usually designated by the Greek symbol Σ. Thus, for the boundary in Fig. 6.27, Σ = 7.

6.15 The Ranganathan Relations

The theory of coincidence site boundaries received a large advance from a paper published by Ranganathan (Ref. 17) in 1966. In this paper he pointed out that not only could rotational symmetry operations, such as a 90° rotation about a ⟨100⟩ pole or a 120° rotation about a ⟨111⟩ pole, bring a lattice back into complete self-coincidence, but also partial self-coincidence might be attained with specific rotations about other axes. He then went on to formulate simple ground rules that allow one to define these other rotations.

Ranganathan also pointed out that these latter rotations not only were able to produce a coincidence site net at a boundary between two crystals, but they were also capable of yielding the concept of a coincidence site three-dimensional lattice. In this latter case, one needs only to consider that two identical crystals, originally in perfect coincidence, are rotated relative to each other about the same axis. Thus, the coincidence site net in Fig. 6.27 represents a planar section through a coincidence site lattice.

There are four basic factors involved in a coincidence site lattice. The first is the axis $[hkl]$ about which the rotation occurs; the second is the rotation angle, θ, about this axis; the third is the coordinates of a coincidence site in the coincident site net on (hkl), the plane normal to the axis of rotation; and the fourth is Σ, the reciprocal of the density of coincidence sites in the (hkl) net. These four factors are not all independent. Actually, Ranganathan was able to formulate equations giving both θ and Σ in terms of the Miller indices and the coordinates (x, y) of a coincidence site in the plane (hkl). These equations are

$$\theta = (2 \tan^{-1} (y/x))(N^{1/2}) \qquad 6.21$$

and

$$\Sigma = x^2 + y^2 N \qquad 6.22$$

where

$$N = h^2 + k^2 + l^2 \qquad 6.23$$

6.16 Examples Involving Twist Boundaries

The application of the Ranganathan relations will now be demonstrated with the aid of several elementary examples involving twist boundaries. First consider a simple cubic lattice rotated about one of its cube axes, [100]. This case is treated in Fig. 6.28 in which the drawing shows the (100) plane. In this illustration the x and y axes are the [010] and [001] directions, respectively. The unit of measurement along these axes is the length of one edge of the simple cubic unit cell. Rotation of the lattice through an arbitrary angle θ about [100] will cause the x and y axes to assume the rotated positions x' and y'. Since a rotation about [100] is being considered, $h = 1$ while k and l are 0, so that $N = 1$. A coincidence site lattice can be formed if one takes $x = 2$ and $y = 1$. In this case $\theta = 2 \tan^{-1}(1/2) \sqrt{1} = 53.1°$ while $\Sigma = 2^2 + 1^2(1) = 5$. In Fig. 6.28 the atoms corresponding to the x, y coordinates are open circles, the atoms in the rotated x'/y' coordinates are filled-in circles, and coincidence sites are represented by larger open circles. Notice that the coincidence sites also define a simple lattice with a cell whose sides in the (100) plane equal $\sqrt{5}a$, where a is the lattice constant of the primitive lattice. Thus, the cell size of the coincidence site lattice is five times larger than that of the primitive lattice. Note that the ratio of the cell sizes also equals the value of Σ.

Another possibility, involving a rotation about [100], occurs when one chooses $x = 3$ and $y = 1$. This is illustrated in Fig. 6.29, where $\theta = 2 \tan^{-1}(1/3) \sqrt{1} = 36.9°$ and $\Sigma = 3^2 + 1^2 \cdot 1 = 10$. However, in regard to the value of 10 for Σ, it has been demonstrated that in a cubic lattice Σ can only take odd values. That is, if Σ is even, it should be divided by multiples of 2 until an odd number is attained.

6.16 Examples Involving Twist Boundaries

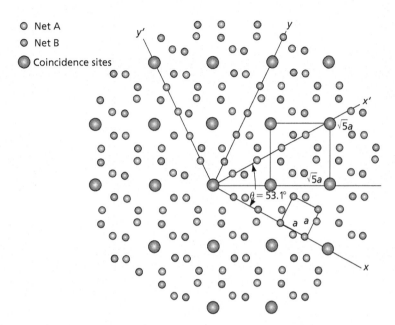

Fig. 6.28 A rotation of 53.1° about a ⟨100⟩ axis of a simple cubic crystal gives this coincident site boundary with $\Sigma = 5$. The coincident sites also form a lattice with a unit cell whose sides are equal to $\sqrt{5}a$ in the boundary. The cell size of the reciprocal lattice is thus five times larger than that of the primitive lattice

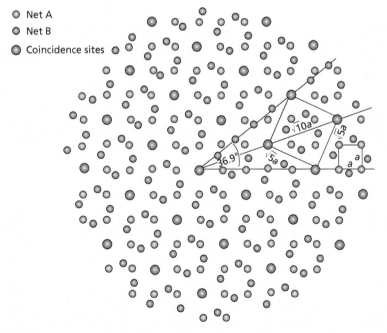

Fig. 6.29 A 36.9° rotation about a ⟨100⟩ axis also produces a coincident site boundary with $\Sigma = 5$ due to the symmetry of the simple cubic lattice

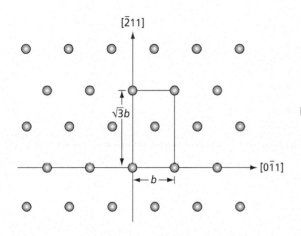

Fig. 6.30 The outline of the tetragonal fcc unit cell on a (111) plane

Thus, the proper value for Σ is 10/2 or 5. This is the same as for the first rotation where $x = 2$ and $y = 1$. Note that the same coincidence site lattice may be obtained by either a rotation of 53.1° or 36.9°. The sum of these angles equals 90°. This result is related to the four-fold symmetry about a ⟨100⟩ cubic axis.

We will now consider another example involving a rotation about a ⟨111⟩ axis of an fcc crystal. This is the Kronberg and Wilson boundary in Fig. 6.27. In this case, one uses a tetragonal unit cell, which has a rectangular cross-section on a {111} plane as indicated in Fig. 6.30. Here the x and y axes are assumed to lie respectively along the [011] and [$\bar{2}$11] directions of the (111) plane. The length of the unit cell along the x axis is the distance between atom centers in a close-packed direction of the crystal and thus is equal to b, the magnitude of an fcc crystal's Burgers vector. In the y or [$\bar{2}$11] direction, the length of the unit cell is larger and equals $\sqrt{3}b$. These x and y units may be used in Eqs. 6.12 and 6.13. The coincidence site lattice deduced graphically by Kronberg and Wilson and shown in Fig. 6.27 can be obtained by letting $x = 9$ and $y = 1$ with $N = 1^2 + 1^2 + 1^2 = 3$. This gives $\Sigma = 9^2 + 1^2 \cdot 3 = 84$. Dividing 84 by 12 to obtain an odd number makes $\Sigma = 7$. The corresponding value of θ is $2 \tan^{-1} (1/9) \sqrt{3} = 21.8°$, which is close to the 22° reported by Kronberg and Wilson. The alternate rotation of 38° observed by these authors to give this same lattice, with $\theta = 7$, corresponds to $x = 5$ and $y = 1$, which gives $\theta = 38.2°$ using the Ranganathan equations.

6.17 Tilt Boundaries

The coincidence site nets in Figs. 6.27 to 6.29 represent boundaries between crystals that are rotated relative to each other across these nets; consequently, they are twist boundaries. The concept of interfaces containing coincidence sites to develop tilt boundaries can also be used. However, the arrangement or pattern that the coincidence sites may take in a tilt boundary may vary considerably. In the twist boundary obtained by a 53.1° rotation about a ⟨100⟩ pole in Fig. 6.28, the coincidence sites form a two-dimensional square array with a cell size area five times larger than the cell area of the primitive lattice; that is, $\Sigma = 5$. A tilt boundary can also be created by rotating two sections of the same crystal through 53.1° and joining them together, as shown in Fig. 6.31. The grain boundary in this case is not in the plane of the paper as it was for the twist boundary. Rather, it is in a horizontal plane normal to the plane of the figure. In Fig. 6.31, the atom positions in the B section are shown as open small circles, those of the A section are shown by filled-in small circles, and the coincident sites are illustrated by

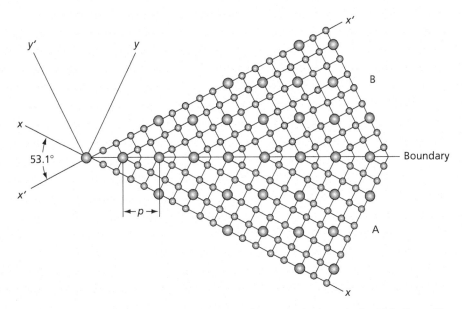

Fig. 6.31 The coincident site boundaries in Figs. 6.27 to 6.29 are twist boundaries. This figure illustrates a coincident site boundary formed by a tilt of 53.1° about a ⟨100⟩ axis of a simple cubic crystal. Σ also equals 5 for this boundary and its structural periodicity, p, is equal to an edge of the coincident site lattice

larger open circles. On the assumption that the B section of the crystal lies above the grain boundary and the A section below, all of the coincidence sites that actually exist will lie in the boundary. However, in order to clarify the relation of these coincidence sites to the coincidence site lattice, the pattern of the coincidence sites is shown in both the A and B sections of the crystal.

Note that the boundary itself is drawn as a series of steps that conform, in both the A and B crystal sections, to the Ranganathan coordinates $x = 2$ and $y = 1$. Thus, the distance between steps and thus, between coincidence sites on the boundary, equals $\sqrt{x^2 + y^2} = \sqrt{5}a$. This distance, p, is called the *structural periodicity*. As may be seen in Fig. 6.31, the coincident sites in the boundary, which is normal to the plane of the figure, have a separation p. The coincidence sites lie in close-packed rows on the plane of the boundary. Thus, each coincidence site along the boundary in Fig. 6.31 is only the outermost of a row of coincidence sites separated by a distance a, where a is the simple cubic lattice constant.

Note that the structural periodicity, p, of the boundary in Fig. 6.31 equals the length of an edge of the unit cell of the coincidence site lattice. As shown earlier, a coincidence site lattice with the same Σ—for example, 5—may be obtained with a twist rotation of 36.9° about a ⟨100⟩ direction. In this case, the Ranganathan coordinates are $x = 3$ and $y = 1$. The corresponding tilt boundary is shown in Fig. 6.32. Notice that its structural periodicity, p, is equal to a diagonal of a coincidence site lattice cell. For a 53.1° rotation, p equaled an edge of this cell. Thus, the 36.9° and 53.1° rotations yield tilt grain boundaries with different structural periodicities.

In the tilt boundaries of Figs. 6.31 and 6.33, the circles actually only represent atom centers. If one considers the atoms as touching spheres, the illustrated boundaries would contain a pair of overlapping atoms just to the right of each coincidence site on these boundaries, as is

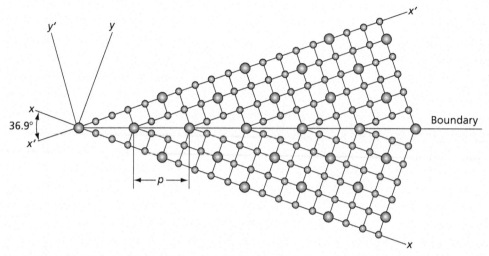

Fig. 6.32 A tilt of 36.9° about ⟨100⟩ of a simple cubic lattice also yields a coincident site boundary, but in this case the structural periodicity, p, is equal to the diagonal of the coincident site unit cell. Σ again is equal to 5

demonstrated in Fig. 6.33 for the 36.9° boundary of Fig. 6.32. As pointed out by Aust (Ref. 18), the problems of this overlap could be relieved by removing an atom from each of these pairs or by a relative translation of the lattices above and below the grain boundary as suggested in Fig. 6.34. This results in an alternation of the ledges between the two crystal halves and the resulting boundary is known as a *relaxed coincidence boundary*. Such a boundary is still considered to have a structural periodicity equivalent to that of the boundary where the atoms of the two halves are shared at the coincidence sites as in Fig. 6.31. It should also have a lower grain boundary energy.

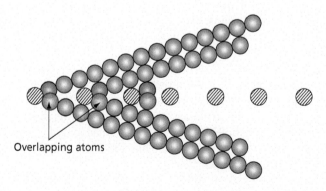

Fig. 6.33 If the atoms of the diagram in Fig. 6.32 are drawn as hard balls instead of points, one finds that there is an overlap of the atoms just to the right of the coincident sites on the boundary

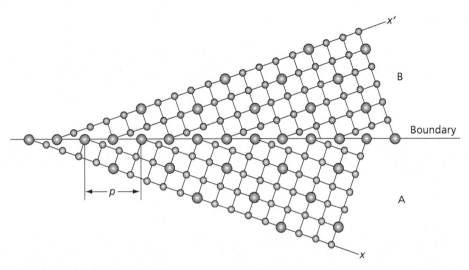

Fig. 6.34 As suggested by Aust (Ref. 18), the overlapping of the atoms at the boundary could be relieved by a relative translation of the lattices above and below the boundary

Problems

6.1 (a) Given a small-angle tilt boundary whose angle of tilt is 0.1°, find the spacing between the dislocations in the boundary if the Burgers vector of the dislocations is 0.33 nm.

(b) On the assumption that the dislocations conform to the conditions involved in Eq. 4.20, that $r' = d/2$, $\mu = 8.6 \times 10^{10}$ MPa and $\nu = 0.3$, determine an approximate value for the surface energy of the tilt boundary. Give your answer in both J/m² and ergs/cm².

6.2 According to quantitative metallography, N_1, the average number of grain-boundary intercepts per unit length of a line laid over a microstructure is directly related to S_v, the surface area per unit volume, by the relation

$$S_v = 2N_1$$

(a) Determine the value of N1 for the microstructure in Fig. 6.1 if the magnification of the photograph is 350×.

(b) Assuming that the grain-boundary energy of zirconium is about 1 J/m², what would be the grain-boundary energy per unit volume in J/m³ of the zirconium in the specimen?

6.3 A very fine-grained metal may have a mean grain intercept of the order of 1 micron or 10^{-6} m. Assuming the grain-boundary energy of the metal is 0.8 J/m², what would be the approximate value of its grain-boundary energy per unit volume? Give your answer in both J/m³ and calories per cm³.

6.4 (a) Consider Fig. 6.8. If the grain-boundary energy of a boundary between two iron crystals is 0.78 J/m², while that between iron and a second phase particle is 0.40 J/m², what angle θ should occur at the junction?

(b) If the surface energy between the iron and the second phase particle were to be 0.35, what would the angle be?

6.5 The following data are taken from Jones, R. L. and Conrad, H., *TMS-AIME*, **245** 779 (1969) and give the flow stress σ, at 4 percent strain, as a function of the grain size of a very high purity titanium metal. Make a plot of σ versus $d^{-1/2}$, and from this determine the Hall-Petch parameters k and σ_0. Express k in $N/m^{3/2}$.

6.6 Plot the data of the preceding problem showing σ as a function of d^{-1}, as well as of $d^{-1/3}$. Do these plots indicate that there is some justification for Baldwin's comments on the Hall-Petch relationship (see Sec. 6.11)?

Grain Size in Micron, μ	Stress, σ in MPa
1.1	321
2.0	279
3.3	255
28.0	193

6.7 (a) With regard to the coincident site twist boundary on a {111} fcc metal plane shown in Fig. 6.25, show, using the Ranganathan relationships (Eqs. 6.21, 6.22, and 6.23), that a choice of $x = 2$ and $y = 1$ will also give this $\Sigma = 7$ boundary.

(b) To what angle of twist does $x = 2$ and $y = 1$ correspond?

(c) Is the fact that $x = 2$ and $y = 1$ are able to produce an equivalent coincident site boundary to that in Fig. 6.25 related to the symmetry of the atomic arrangement on the {111} plane? Explain.

6.8 This problem concerns a coincidence site boundary on a {210} plane of a cubic lattice. Note that the determination does not depend on whether the lattice is simple, face-centered, or body-centered cubic. It holds for all three cases. The basic cell in this plane has $x = a$ and $y = \sqrt{5}a$. A coincidence site lattice can be formed with $x = 2$ and $y = 1$. Assuming a twist boundary, determine:

(a) The twist angle θ for the coincidence structure.

(b) Σ, the reciprocal of the density of coincidence sites.

(c) Check your answer for θ and Σ using two drawings of the atomic structure as revealed on a {210} plane. Note that this involves an array of rectangular cells in which $x = a$ and $y = \sqrt{5}a$, where a is the lattice constant of the crystal. In this operation, note that one drawing of the structure should be made on tracing paper so that it can be placed on and rotated over the other so as to reveal the coincidence sites.

References

1. Hirth, J. P., and Lothe, J., *Theory of Dislocations*, 2nd ed., p. 731, John Wiley and Sons, New York, 1982.
2. Kuhlmann-Wilsdorf, D., *Mat. Sci. and Eng.*, **86** 53 (1987).
3. Hansen, N., and Kuhlmann-Wilsdorf, D., *Mat. Sci. and Eng.*, **81** 141 (1986).
4. Rowenhorst, D. J., and Voorhees, R. W., *Met. and Mat. Trans. A*, 36A 2127 (2005).
5. Hull, F., Westinghouse Electric Corporation, Research and Development Center, Pittsburgh, Pa. These experimental results were demonstrated at the Quantitative Microscopy Symposium, Gainesville, Fla., Feb., 1961.
6. DeHoff, R. T., and Rhines, F. N., *Quantitative Microscopy*, McGraw-Hill Book Company, New York, 1968.
7. Fullman, R. L., *Trans. AIME*, **197** 447–53 (1953).

8. Hall, E. O., *Pro. Phys. Soc. London*, **B64** 747 (1951).
9. Petch, N. J., *J. Iron and Steel Inst.*, **174** 25 (1953).
10. Baldwin, W. M., Jr., *Acta Met.*, **6** 141 (1958).
11. Sanders, P. G., Eastman, J. A., and Weertman, J. R., *Acta Mater.*, **45** 4019 (1997).
12. Masumura, R. A., Hazzledine, P. M., and Pande, C. S., *Acta Mater.* **46** 4527 (1998).
13. Chokshi, A. H., Rosen, A., Karch, J., and Gleiter, H., *Scripta Met.*, 23 1679 (1989).
14. Mohamed, F. A., and Chauhan, M., *Met. and Mat. Trans.* 37A 3555 (2006).
15. Andrievski, R. A., and Glezer, A. M., *Scripta Mat.*, 44 1621 (2001).
16. Kronberg, M. L., and Wilson, F. H., *Trans. AIME*, **185** 501 (1949).
17. Ranganathan, S., *Acta Cryst.*, **21** 197 (1966).
18. Aust, K. T., *Prog. in Mat. Sci.*, p. 27 (1980).

Chapter 7

Vacancies and Thermodynamics

> **Learning Objectives**
>
> Upon completion of Chapter 7, you will be able to:
> 1. Understand thermodynamics as a concept and contrast it with statistical mechanics and with kinetics theory
> 2. Relate entropy, enthalpy and free energy changes for equilibrium and spontaneous reactions
> 3. Understand the origin and driving force for the formation of vacancies in a crystalline material
> 4. Differentiate between mixing entropy and vibrational entropy
> 5. Discuss the influence of temperature on equilibrium number of vacancies of temperature
> 6. Summarize the key features of vacancy motion and the required activation energy for its migration
> 7. Differentiate between vacancies, divacancies and interstitials

7.1 Thermal Behavior of Metals

Many important metallurgical phenomena depend strongly on the temperature at which they occur. An important example of practical importance occurs in the softening of hardened metals by heating. A brass specimen, hardened by hammering, can be softened to its original hardness in a few minutes if exposed to a temperature of 900 K. At 600 K it may take several hours to achieve the same loss of hardness, while at room temperature it might easily take several thousand years. For all practical purposes it can therefore be said that brass will not soften, or anneal, at room temperature.

In recent years, large advances have been made toward theoretical explanations of such highly temperature-dependent phenomena. Each of the three sciences of heat—*thermodynamics, statistical mechanics,* and *kinetic theory*—has contributed to this knowledge and each approaches the subject of heat differently.

Thermodynamics is based upon laws that are postulated from experimental evidence. Since the experiments that lead to the laws of thermodynamics were performed on bodies of matter containing very large numbers of atoms, thermodynamics is not directly concerned with what happens on an atomic scale; it is more concerned with the average properties of large numbers of atoms, and mathematical relationships are developed between such thermodynamical functions as temperature, pressure, volume, entropy, internal energy, and enthalpy, without consideration of atomic mechanisms. This neglect of atomic mechanisms in thermodynamics has both advantages and disadvantages. It makes computations easier and more accurate, but, unfortunately, it tells us nothing about what makes things happen as they do. As a simple example, consider the *equation of state* for an ideal gas

$$PV = nRT \qquad 7.1$$

where P is the pressure, V is the volume, n is the number of moles of gas, R the universal (gas) constant, and T the absolute temperature. In thermodynamics, this equation is derived from experiments (Boyle's law, law of Gay-Lussac, and Avogadro's law). No explanation as to the reasons for the existence of this relationship is given.

Kinetic theory, in contrast to thermodynamics, attempts to derive relationships such as this equation of state, starting with atomic and molecular processes. In college physics textbooks, classical derivations of Eq. 7.1 that use a simple kinetic approach can be found. In these derivations, the gas atoms are assumed to behave as elastic spheres, to move at high speeds with random velocities, and to be separated by distances that are large compared with the size of the atoms. Using these assumptions, it can be shown that the pressure exerted by the gas equals the time rate of change of momentum due to the collision of gas atoms with the walls of the container. Thus, the pressure is merely the average force exerted on the walls by the collision of the gas atoms with the walls. It can also be shown that the average kinetic energy of the atoms is directly proportional to the absolute temperature. Kinetic theory thus gives us an insight into the significance of two important thermodynamical functions: temperature and pressure. Thermodynamics does not give us this same ability to understand gas phenomena.

The third science of heat, *statistical mechanics*, applies the realm of statistics to heat problems. Kinetic theory is concerned with the explanation of heat phenomena in terms of the mechanics of individual atoms. The mechanics of the disordered atomic motions that are ascribed to heat phenomena are attacked from probability considerations. This approach is feasible because most practical problems involve quantities of matter containing large numbers of atoms or molecules. It is thus possible to think in terms of the behavior of the group as a whole, as does a life insurance actuary who predicts the vital statistics of a large population.

In one of the following sections, a physical interpretation will be given for the thermodynamical function called *entropy*. When approached from the point of view of statistical mechanics, this quantity is given a real significance that is not apparent in classical thermodynamics.

It should now be mentioned that thermodynamics and statistical mechanics are only applicable to problems involving equilibrium, and cannot predict the speed of a chemical or metallurgical reaction. This latter is the special province of the kinetic theory.

As a simple example of a system in equilibrium, consider a closed system containing a liquid metal and its equilibrium vapor in which the average number of metal atoms leaving the liquid to join the vapor equals the corresponding number traveling in the opposite direction. The concentration of atoms in the vapor, and therefore the vapor pressure, is a constant with respect to time. Under conditions such as these, thermodynamics and statistical mechanics are able to produce much useful information; for example, how equilibrium-vapor pressure changes with a change in temperature. Suppose, however, that liquid metal is placed inside the bell jar of a vacuum system so that the vapor is swept away as fast as it forms. In this case, there can be no equilibrium because atoms will leave the liquid at a much faster rate than they return to it. Because the liquid-vapor system is no longer in equilibrium, thermodynamics and statistical mechanics can no longer be used. Questions relating to how fast the metal atoms evaporate belong in the realm of the kinetic theory. Kinetic theory is thus most useful when the rates at which atomic changes take place are being studied.

7.2 Internal Energy

The preceding paragraphs have discussed the interrelationships of the three branches of the science of heat. The principal objective of this discussion was to show that physical meanings can be given to thermodynamical functions. Let us now consider a solid crystalline material. An important thermodynamical function that will be needed in the following sections is *internal energy*, which shall be denoted by the symbol U. This quantity represents the total kinetic and potential energy of all the atoms in a material body, or system. In the case of crystals, a large part of this energy is associated with the vibration of the atoms in the lattice. Each atom can be assumed to vibrate about its rest position with three degrees of freedom (x, y, and z directions). According to the Debye theory, the atoms do not vibrate independent of each other, but rather as the result of random elastic waves traveling back and forth through the crystal, and, because they have three degrees of vibrational freedom, the lattice waves can be considered to be three independent sets of waves traveling along the x, y, and z axes, respectively. When the temperature of the crystal is raised, the amplitudes of the elastic waves increase, with a corresponding increase in internal energy. The intensity of the lattice vibrations is therefore a function of temperature.

7.3 Entropy

In thermodynamics, the entropy change, ΔS, may be defined by the following equation:

$$\Delta S = S_B - S_A = \int_A^B \frac{dQ}{T} \bigg]_{rev} \qquad 7.2$$

where S_A is the entropy in state A, S_B is the entropy in state B, T is the absolute temperature, and dQ is the heat added to system. The integration is assumed to be taken over a reversible path between the two equilibrium states A and B.

The entropy is a function of state; that is, it depends only on the state of the system. This signifies that the entropy difference $(S_B - S_A)$ is independent of the way that the system is carried from state A to state B. If the system moves from A to B as a result of an irreversible reaction, the entropy change is still $(S_B - S_A)$. However, the entropy change equals the right-hand side of Eq. 7.2 only if the path is reversible. Thus, to measure the entropy change of a system (in going between states A and B), one needs to integrate the quantity dQ/T over a reversible path. The integral of this same quantity over an irreversible path does not equal the entropy change. In fact, it is shown in all standard thermodynamical textbooks that

$$\Delta S = S_B - S_A > \int_A^B \frac{dQ}{T} \bigg]_{irrev} \qquad 7.3$$

for an irreversible path between A and B.

Differentiating the preceding two equations yields the following:

$$dS = \frac{dQ}{T} \quad \text{(reversible change)} \qquad 7.4$$

$$dS > \frac{dQ}{T} \quad \text{(irreversible change)} \qquad 7.5$$

for an infinitesimal change in state.

7.4 Spontaneous Reactions

Consider the transformation of water from liquid to solid. The equilibrium temperature for this reaction at atmospheric pressure is 0°C, by definition, which is approximately 273 K. At the equilibrium temperature, liquid water and ice can be maintained in the same isolated container for indefinite periods of time as long as heat is neither added nor taken away from the system. Ice and water under these conditions furnish a good example of a system at equilibrium. Now, if heat is slowly applied to the container, some of the ice will melt and become liquid; or if heat is abstracted from the container, more ice will form at the expense of the liquid. In either case, a reversible exchange of heat (heat of fusion) between the liquid-solid system and its surroundings is necessary in order to change the ratio of liquid to solid. This transformation of liquid to solid at 0°C is an example of an equilibrium reaction.

Consider now the case of a thermally isolated container with liquid water that is supercooled below the equilibrium freezing point (0°C). Even though ice may be thermally isolated from its surroundings, freezing can begin spontaneously without loss of heat to the surroundings. The heat of fusion, released by the portion of the water that freezes, will, in this case, raise the temperature of the system back toward the equilibrium freezing temperature. The difference between freezing at the equilibrium freezing point and freezing at temperatures below it is an important one. In one case, freezing occurs spontaneously; in the other, it is spontaneous only if the heat of fusion is removed from the system. It should also be pointed out that the reverse transformation, in which ice melts at a temperature above the equilibrium freezing point, also occurs spontaneously. In this case, if the system is isolated (so that heat cannot be added to the system during the melting of the ice), the heat of fusion will cause the temperature to fall back toward the equilibrium temperature (0°C).

Reactions that occur spontaneously are always irreversible. Liquid water will transform to ice at −5°C (≈268 K), but the reverse reaction is, of course, impossible. Spontaneous reactions occur frequently in metallurgy, sometimes with drastic results, often with quite beneficial results. In either case, it is important to know the conditions that bring about spontaneous reactions, and to have a yardstick for measuring the driving force of this type of reaction. The yardstick that is most valuable for this purpose is called *Gibbs free energy*.

7.5 Gibbs Free Energy

Gibbs free energy is defined by the following equation:

$$G = U + PV - TS \qquad 7.6$$

where G is Gibbs free energy, U is internal energy, P is pressure, V is volume, T is absolute temperature, and S is entropy. The sum, $U + PV$, in Eq. 7.6 is a combination that occurs so often in relations dealing with systems at constant pressure that it has been designated with the special symbol H and is called the *enthalpy*. Thus,

$$H = U + PV \qquad 7.7$$

Accordingly, Eq. 7.6 may be written

$$G = H - TS \qquad 7.8$$

where G is the Gibbs free energy, H the enthalpy, T the temperature in degrees Kelvin, and S the entropy.

Most metallurgical processes of interest such as freezing occur at constant pressure (atmospheric). Furthermore, since we are primarily interested in solids and liquids, the volume changes in metallurgical reactions are usually very small. As an illustrative example consider water in equilibrium with its solid form (ice). Let G_2 be the free energy of a mole of solid water (ice), and G_1 that of a mole of liquid water. When a mole of water changes into ice, the free energy change is

$$\Delta G = G_2 - G_1 = (H_2 - TS_2) - (H_1 - TS_1) \qquad 7.9$$

where H_2 and H_1 are the enthalpies of the solid and liquid, respectively, S_2 and S_1 their respective entropies, and T the temperature (which remains constant during the reaction). Equation 7.9 may also be written

$$\Delta G = \Delta H - T\Delta S \qquad 7.10$$

Water freezing to ice at the equilibrium freezing point, T_e, is a reversible reaction, and under these conditions we have seen that the entropy change is given by

$$\Delta S = \int_A^B \frac{dQ}{T_e} \qquad 7.11$$

which, in this case, reduces to

$$\Delta S = \frac{\Delta Q}{T_e} \qquad 7.12$$

where ΔQ is the latent heat of freezing for water. Also, by the first law of thermodynamics, we have

$$dU = dW + dQ \qquad 7.13$$

where dU is the change in internal energy, dW is the work done on system, and dQ is the heat added to system.

The first law may also be written in terms of the enthalpy instead of the internal energy. To do this first take the derivative of H, which gives

$$dH = dU + PdV + VdP \qquad 7.14$$

where at constant pressure VdP is by definition zero and PdV is equivalent to dW in Eq. 7.13. However, in the freezing of water, the only external work that is done is against the pressure of the atmosphere due to the expansion when water changes from liquid to solid. This can be neglected because of its small size, so setting dW equal to zero, we obtain

$$dH = dU = \Delta Q \qquad 7.15$$

Substituting ΔS and ΔQ in Eq. 7.8 for the free energy gives

$$dG = \Delta Q - T_e \frac{\Delta Q}{T_e} = \Delta Q - \Delta Q = 0 \qquad 7.16$$

The free-energy change in this reversible reaction (the freezing of water at 0°C) is zero. It may also be shown, with the aid of thermodynamics, that the Gibbs free-energy change is zero for any reversible reaction that takes place at constant temperature and pressure. If liquid water is now cooled to a temperature well below 273 K and allowed to freeze isothermally, the transformation will be made under irreversible conditions. But, in an irreversible reaction,

$$\Delta S > \frac{\Delta Q}{T} \qquad 7.17$$

or

$$T\Delta S > \Delta Q$$

The free-energy equation tells us that for this reaction

$$\Delta G = \Delta H - T\Delta S \qquad 7.18$$

where ΔH again equals ΔQ as in the previous example. Therefore,

$$\Delta G = \Delta Q - T\Delta S \qquad 7.19$$

If $T\Delta S$ is greater than ΔQ, however, ΔG must be negative. This fact is important. The free-energy change for this spontaneous reaction is negative, which means that the system reacts so as to lower its free energy. This result is true not only in the above simple system, but also for all spontaneous reactions. A spontaneous reaction occurs when a system can lower its free energy.

While the free energy tells us whether or not a spontaneous reaction is possible, it cannot predict the speed of the reaction. An excellent example of this fact can be seen in the case of the two phases of carbon—diamond and graphite. Graphite is the phase with the lower free energy, and diamond should, therefore, transform spontaneously into graphite. The rate is so slow, however, that there is no need to consider it. Rate problems such as these are treated in the kinetic theory of matter and are not in the realm of thermodynamics.

7.6 Statistical Mechanical Definition of Entropy

The significance of the entropy will now be considered. Let us take a two-chamber box, each chamber filled with a different monatomic ideal gas. Gas A is in chamber I, and gas B in chamber II. Let the partition between the chambers be removed and the gases will mix by diffusion. Such mixing occurs at constant temperature and constant pressure if the gases have the same original temperature and pressure. No work is done and no heat is transferred to or from the gases; therefore, the internal energy of the gaseous system does not change. This fact is in agreement with the law of conservation of energy (first law of thermodynamics), which is expressed in the following equation:

$$dH = dQ + dW \qquad 7.20$$

$$dH = 0 \qquad 7.21$$

where dQ is the heat absorbed by the system (gases), dH is the change of the enthalpy of the system (gases), dW is the work done on gases by the surroundings. A fundamental change in the system occurs as a result of the diffusion. That this is true can be judged from the fact that considerable effort is required to separate this mixture of gases into its components again. Like the freezing of water at temperatures below 273 K, the mixing of gases is a spontaneous, or irreversible reaction, and, as in all irreversible reactions, the free energy must decrease. The free-energy equation states that

$$dG = dH - TdS \qquad 7.22$$

However, it was shown above that dH is zero and, therefore

$$dG = -TdS \qquad 7.23$$

A decrease in free energy can only mean that dS must be positive. In other words, the entropy of the system has increased by the mixing of the gases. The entropy increase involved in this reaction is known as *an entropy of mixing*. It is only one of a number of forms of entropy. All forms, however, have one thing in common. When the entropy of a system increases, the system becomes more disordered.

In the above example, a disorder in the spatial distribution of two kinds of gas atoms has been considered, but entropy can also be associated with disorder in the motion of atoms, with respect to the directions and the magnitudes of the velocities of the atoms.

As a hypothetical example, let us consider that it is possible to introduce gas molecules into a chamber in such a manner that all of them begin to travel in the same direction and with the same speed back and forth between two opposite walls of the box. Collisions of the molecules with each other and with the walls quickly bring about a random distribution in the directions and magnitudes of the molecular velocities. An increase in entropy necessarily accompanies this irreversible change from uniformly ordered motion to random motion. The question now arises: why do two unmixed gases seek a random distribution? The answer lies in the fact that, on removal of the partition, each gas atom is free to move through both compartments and has equal probability of being found in either. Once the barrier has been removed, the probability is extremely small that all of the A atoms will be found in compartment A and, at the same time, all of the B atoms in compartment B. Furthermore, the probability of the atoms maintaining such segregation is even more remote. On the other hand, the chance of finding a random distribution throughout the box is almost a certainty.

Since a shift from a state of low probability (two unmixed gases in contact) to a state of high probability (random mixture) accompanies an entropy increase, it would appear that there is a close relationship between entropy and probability. This relationship does exist and was first expressed mathematically by Boltzmann, who introduced the following equation:

$$S = k \ln P \qquad 7.24$$

7.6 Statistical Mechanical Definition of Entropy

where S is the entropy of a system in a given state, P is the probability of the state, and k is Boltzmann's constant (1.38×10^{-23} joules/deg (K)).

The change in entropy (mixing entropy) resulting from the mixing of gas A and gas B may be expressed in terms of the Boltzmann equation:

$$\Delta S = S_2 - S_1 = k \ln P_2 - k \ln P_1 \qquad 7.25$$

$$\Delta S = k \ln \frac{P_2}{P_1} \qquad 7.26$$

where S_1 is the entropy of unmixed gases, S_2 is the entropy of mixed gases, P_1 is the probability of unmixed state, and P_2 is the probability of mixed state.

The probability of finding the atoms in the unmixed, or segregated, state is computed as follows:

Let V_A = the volume originally occupied by the atoms of gas A
V_B = the volume originally occupied by the atoms of gas B
V = the total volume of the box

If one A atom is introduced into the undivided box, the probability of finding it in V_A is V_A/V. If a second A atom is now added to the box, the chance of finding both at the same time in V_A is $(V_A/V) \times (V_A/V)$. This problem is similar to that of producing two heads on the toss of a pair of coins, where the chance of a head on either coin is 1/2, but for the pair it is $1/2 \times 1/2 = 1/4$. A third A atom reduces the probability of finding all A atoms in V_A to $(V_A/V)^3$, and if n_A is the total number of atoms of gas A, the probability of finding all n_A in V_A is $(V_A/V)^{n_A}$. Now if an atom of B is added to the box, the chance of finding it in V_B is (V_B/V), and the chance of finding it in V_B, and at the same time finding all the A atoms in V_A, is $(V_A/V)^{n_A} \times (V_B/V)$. Finally, the probability of finding all atoms of gas A in V_A, while all atoms of gas B are in V_B is

$$P_1 = \left(\frac{V_A}{V}\right)^{n_A} \cdot \left(\frac{V_B}{V}\right)^{n_B} \qquad 7.27$$

where n_A is the number of A atoms, and n_B is the number of B atoms.

It is now necessary to consider the probability of the homogeneous mixture. Thought must first be given to the meaning of the term *homogeneous mixture*. A very accurate experimental analysis might possibly be able to detect a variation in composition of the mixture of the order of one part in 10^{10}, but be unable to detect a smaller variation. Therefore, an experimentally homogeneous mixture is not one with a perfectly constant ratio of A to B atoms, but one with a ratio that does not vary sufficiently from the mean value to be detectable. Such a mixture is extremely probable when the number of atoms is large, as in most real systems, where the number usually exceeds 10^{20} (approximately 10^{-3} moles). The importance of large numbers in statistics can easily be seen in a somewhat analogous case: that of flipping coins and counting the number that come "heads up." If 10 coins are tossed together, it may be shown that the chance of getting 5 heads is 0.246, while that of obtaining any of the numbers in the group of 4, 5, or 6 heads is 0.666. Thus, even with 10 coins, it is apparent that a number close to the mean distribution is quite probable. If the number of tossed coins is increased to 100, the probability of finding somewhere between 40 and 60 heads is 0.95. Increasing the number of coins by one factor of 10 clearly increases the probability of finding a distribution near the mean. Finally, let us increase the number of coins to approximately that found in a real system of atoms (10^{20}). In this case, the mean distribution is 5×10^{19} heads. The probability of finding a number of heads within 3.5 parts in 10^{10} of this number is 0.999999999997. In other words, statistics tells us that the chance of finding between

$$50{,}000{,}000{,}035{,}000{,}000{,}000$$

and

$$49{,}999{,}999{,}965{,}000{,}000{,}000$$

heads is within 3 parts in 10^{12} of unity. It also tells us that the chance of finding a distribution containing a number of heads that lies outside the above range is just about nil.

Consider again the box containing the mixture of A and B gas atoms. The chance of finding a head on the flip of a coin is exactly analogous to the chance of finding an A atom in one half of the box. It is also analogous to the chance of finding a B atom in the same half. It can be concluded, therefore, that both kinds of atoms, when present in very large numbers, will seek a mean distribution in which the atoms are uniformly distributed in the box. Deviations from this mean value will be very small and the probability of an experimentally homogeneous mixture extremely high. It can be assumed, for the purposes of calculation, that this probability is equal to one.

Returning to the mixing-entropy equation given in Eq. 7.26, and based on the above considerations, P_2 can be taken as unity. Thus,

$$\Delta S = k \ln \frac{1}{P_1} = -k \ln P_1$$

Substituting for P_1 from Eq. 7.27 results in

$$\Delta S = -k \ln \left(\frac{V_A}{V}\right)^{n_A} \cdot \left(\frac{V_B}{V}\right)^{n_B}$$

$$= -k \ln \left(\frac{V_A}{V}\right)^{n_A} - k \ln \left(\frac{V_B}{V}\right)^{n_B}$$

$$= -k n_A \ln \left(\frac{V_A}{V}\right) - k n_B \ln \left(\frac{V_B}{V}\right)$$

But since we have assumed ideal gases at the same temperature and pressure, the volumes occupied by the gases must be proportional to the numbers of atoms in the gases. Thus, we have

$$\frac{n_A}{n} = \frac{V_A}{V}$$

and

$$\frac{n_B}{n} = \frac{V_B}{V}$$

where n is the total number of atoms of both kinds. However, the ratios n_A/n and n_B/n are the mean chemical fractions of atoms A and B in the box, and thus we may set

$$\frac{n_A}{n} = \frac{V_A}{V} = X_A \qquad \qquad 7.28$$

and

$$\frac{n_B}{n} = \frac{V_B}{V} = (1 - X_A) = X_B \qquad \qquad 7.29$$

where X_A is the mole fraction of A, and $X_B = (1 - X_A)$ is the mole fraction of B. The entropy of mixing can now be expressed in terms of concentrations as follows:

$$\Delta S = -kn\left(\frac{n_A}{n}\right) \ln X_A - kn\left(\frac{n_B}{n}\right) \ln (1 - X_A)$$
$$= -kn\, X_A \ln X_A - kn(1 - X_A) \ln (1 - X_A) \qquad 7.30$$

If it is now assumed that we have one mole of gas, then the number of atoms n becomes equal to N where N is Avogadro's number. Also, k (Boltzmann's constant) is the gas constant for one atom; that is,

$$k = \frac{R}{N} \qquad 7.31$$

where R is the universal gas constant (8.31 joules per mol) and N is Avogadro's number. As a result, we have

$$kn = kN = R$$

We therefore write the mixing-entropy equation in its final form

$$\Delta S = -R[X_A \ln X_A + (1 - X_A) \ln (1 - X_A)]$$
$$= -R[X_A \ln X_A + X_B \ln X_B] \qquad 7.32$$

7.7 Vacancies

Metal crystals are never perfect. It is now well understood that they may contain many defects. These are classified according to a few important types. One of the most important is called a *vacancy*. The existence of vacancies in crystals was originally postulated to explain solid-state diffusion in crystals. Because of the mobility of gas atoms and molecules, gaseous diffusion is easy to comprehend, but the movement of atoms inside crystals is more difficult to grasp. Still, it is a well known fact that two metals, for example, nickel and copper, will diffuse into each other if placed in intimate contact and heated to a high temperature. Several mechanisms have been proposed to explain these diffusion phenomena, but the most generally acceptable has been the vacancy mechanism.

Diffusion in crystals is explained, in terms of vacancies, by assuming that the vacancies move through the lattice, thereby producing random shifts of the atoms from one lattice position to another. The basic principle of vacancy diffusion is illustrated in Fig. 7.1, where three successive steps in the movement of a vacancy from position I to II are shown. In each case, it can be seen that the vacancy moves as a result of an atom jumping into a hole from a lattice position bordering the hole. In order to make the jump, the atom must overcome the net attractive force of its neighbors on the side opposite the hole. Work is therefore required to make the jump into the hole, or, as it may also be stated, an energy barrier must be overcome. Energy sufficient to overcome the barrier is furnished by the thermal or heat vibrations of the crystal lattice. The higher the temperature, the more intense the thermal vibrations, and the more frequently are the energy barriers overcome. Vacancy motion at high temperatures is very rapid and, as a consequence, the

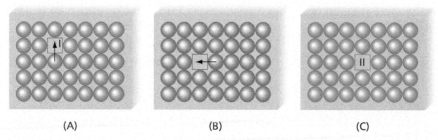

Fig. 7.1 Three steps in the motion of a vacancy through a crystal from position I to position II

rate of diffusion increases rapidly with increasing temperature. An equation will now be derived that gives the equilibrium concentration of vacancies in a crystal as a function of temperature.

Let us assume that in a crystal containing n_0 atoms there are n_v vacant lattice sites. The total number of lattice sites is, accordingly, $n_0 + n_v$, or the sum of the occupied and unoccupied positions. Suppose that vacancies are created by movements of atoms from positions inside the crystal to positions on the surface of the crystal, in the manner shown in Fig. 7.2. When a vacancy has been formed in this manner, *a Schottky defect* is said to have been formed. Let the work required to form a Schottky defect be represented by the symbol w. A crystal containing n_v vacancies will therefore have an internal energy greater than that of a crystal without vacancies by an amount $n_v w$.

The free energy of a crystal containing vacancies will be different from that of a crystal free of vacancies. This free-energy increment may be written as follows:

$$G_v = H_v - TS_v \qquad 7.33$$

where G_v is the free energy change due to vacancies, H_v is the enthalpy increase due to the vacancies, and S_v is the entropy change due to the vacancies.
But, according to the above,

$$H_v = n_v w$$

and thus,

$$G_v = n_v w - TS_v \qquad 7.34$$

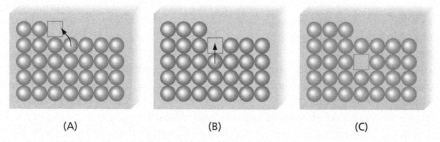

Fig. 7.2 The creation of a vacancy by moving an atom to the surface

7.7 Vacancies

Now the entropy of the crystal is increased in the presence of vacancies for two reasons. First, the atoms adjacent to each hole are less restrained than those completely surrounded by other atoms and can, therefore, vibrate in a more irregular or random fashion than the atoms removed from the hole. Each vacancy contributes a small amount to the total entropy of the crystal.

Let us designate the vibrational entropy associated with one vacancy by the symbol s. The total increase in entropy arising from this source is $n_v s$, where n_v is the total number of vacancies. While a consideration of this vibrational entropy is important in a thorough theoretical treatment of vacancies, it will be omitted in our present calculations because its effect on the number of vacancies present in the crystal is of secondary importance.

The other entropy form arising in the presence of vacancies is an *entropy of mixing*. The entropy of mixing has already been derived for the mixing of two ideal gases, and is expressed by the equation

$$S_m = \Delta S = -nk[X_A \ln X_A + (1 - X_A) \ln (1 - X_A)] \qquad 7.35$$

where S_m is the mixing entropy, n is the total number of atoms $(n_A + n_B)$, k is the Boltzmann's constant, X_A is the concentration of atom $A = n_A/n$, and $(1 - X_A)$ is the concentration of atom $B = n_B/n$. The above equation applies directly to the present problem if we consider mixing of lattice points, instead of atoms, in which there are two types of lattice points: one occupied by atoms, the other unoccupied. If there are n_o occupied sites and n_v unoccupied, the unmixed state will correspond to one in which a lattice of $n_o + n_v$ sites has all positions on one side filled and all on the other empty. This is equivalent to a box problem, shown schematically in Fig. 7.3A, in which there are two compartments, one filled with occupied lattice positions and the other filled with empty positions. The corresponding mixed state in the box is shown in Fig. 7.3B.

The present problem consists of mixing n_o objects of one kind with n_v of another kind, with a total number $(n_o + n_v)$ to be mixed. Therefore, we may make the following substitutions in the mixing-entropy equation:

$$n = n_o + n_v$$

$$X_A = X_v = \frac{n_v}{n_o + n_v}$$

$$(1 - X_A) = X_o = \frac{n_o}{n_o + n_v}$$

where X_v is the concentration of vacancies, and X_o is the concentration of occupied lattice positions.

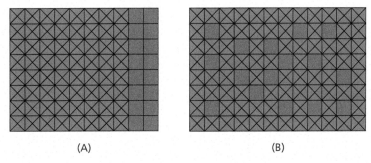

Fig. 7.3 The box analogy of a crystal. **(A)** Vacancies and atoms in the segregated state. Atoms to the left, vacancies to the right. **(B)** The mixed state

If the above quantities are set into the mixing-entropy equation, we have

$$S_m = -(n_o + n_v) k \left[\frac{n_v}{n_o + n_v} \ln \frac{n_v}{n_o + n_v} + \frac{n_o}{n_o + n_v} \ln \frac{n_o}{n_o + n_v} \right]$$

which becomes after simplification

$$S_m = k[(n_o + n_v) \ln (n_o + n_v) - n_v \ln n_v - n_o \ln n_o] \qquad 7.36$$

The free-energy equation for vacancies may now be written

$$\begin{aligned} G_v &= n_v w - S_m T \\ &= n_v w - kT[(n_o + n_v) \ln (n_o + n_v) - n_v \ln n_v - n_o \ln n_o] \end{aligned} \qquad 7.37$$

This free energy must be a minimum if the crystal is in equilibrium; that is, the number of vacancies (n_v) in the crystal will seek the value that makes G_v a minimum at any given temperature. As a result, the derivative of G_v with respect to n_v must equal zero with the temperature being held constant. Thus,

$$\frac{dG_v}{dn_v} = w - kT \left[(n_o + n_v) \frac{1}{(n_o + n_v)} + \ln (n_o + n_v) - n_v \frac{1}{n_v} - \ln n_v - 0 \right]$$

$$0 = w + kT \left[\ln \frac{n_v}{n_o + n_v} \right]$$

which, when expressed in exponential form, becomes

$$\frac{n_v}{n_o + n_v} = e^{-w/kT} \qquad 7.38$$

In general, it has been found that the number of vacancies in a metal crystal is very small when compared with the number of atoms (the number of occupied sites n_o). Therefore Eq. 7.38 can be written in terms of n_v/n_o, the ratio of vacancies to atoms in the crystal.

$$\frac{n_v}{n_o} = e^{-w/kT} \qquad 7.39$$

where w is the work to form one vacancy in J, k is the Boltzmann's constant in J/K, T is the absolute temperature in K, n_v is the number of vacancies, and n_o is the number of atoms. If both the numerator and the denominator of the exponent of Eq. 7.39 are now multiplied by N, Avogadro's number (6.03×10^{23}), the equation will be unaltered with respect to the functional relationship between the concentration of vacancies and the temperature. However, the quantities in the exponent will then correspond to standard thermodynamical notation. Therefore, let

$$H_f = Nw$$
$$R = kN$$
$$\frac{n_v}{n_o} = e^{-Nw/NkT} = e^{-H_f/RT}$$

where H_f is the heat of activation; that is, the work required to form one mole of vacancies, in joules per mole, N is the Avogadro's number, k is the Boltzmann's constant, and R is the gas constant = 8.31 joules per mole −K. Therefore,

$$\frac{n_v}{n_o} = e^{-H_f/RT} \qquad 7.40$$

The experimental value for the activation enthalpy for the formation of vacancies in copper is approximately 83,700 joules per mole of vacancies. This value may be substituted into Eq. 7.40 in an attempt to determine the effect of temperature on the number of vacancies. Remembering that $R \approx 8.37$ joules per mole −K,

$$\frac{n_v}{n_o} = e^{-H_f/RT} = e^{-83,700/8.37T} = e^{-10,000/T}$$

At absolute zero the equilibrium number of vacancies should be zero, for in this case

$$\frac{n_v}{n_o} = e^{-10,000/0} = e^{-\infty} = 0$$

At 300 K, approximately room temperature

$$\frac{n_v}{n_o} = e^{-10,000/300} = e^{-33} = 4.45 \times 10^{-15}$$

However, at 1350 K, six degrees below melting point,

$$\frac{n_v}{n_o} = e^{-10,000/1350} = e^{-7.40} = 6.1 \times 10^{-4} \simeq 10^{-3}$$

Therefore, just below the melting point there is approximately one vacancy for every 1000 atoms. While at first glance this appears to be a very small number, the mean distance between vacancies is only about 10 atoms. On the other hand, the room-temperature equilibrium concentration of vacancies (4.45×10^{-15}) corresponds to a mean separation between vacancies of the order of 100,000 atoms. These figures show clearly the strong effect of temperature on the number of vacancies.

Two questions need to be considered: first, why should there be an equilibrium number of vacancies at a given temperature, and, second, why does the equilibrium number change with temperature? Figure 7.4 answers the first of these questions in terms of curves of the functions G_v, $n_v w$, and $-TS_m$ as functions of the number of vacancies n_v. In this figure, the temperature is assumed close to the melting point. Whereas the work to form vacancies ($n_v w$) increases linearly with the number of vacancies, at low concentrations the entropy component ($-TS$) increases very rapidly with n_v, but less and less rapidly as n_v grows larger. At the value marked n_1 in the figure, the two quantities n_{vw} and $-TS$ become equal, and, at this point, the free energy G_v, the sum of ($n_v w - TS$), equals zero. For all values of n_v greater than n_1, the free energy is positive, and for all values of n_v smaller than n_1, the free energy is negative. Furthermore, in the region of negative free energy, a minimum occurs corresponding to the concentration marked n_e in the figure. This is the equilibrium concentration of vacancies.

Figure 7.4 shows a second set of dashed line curves for a lower-temperature curves ($T_B < T_A$). The lowering of the temperature does not change the $n_v w$ curve, but makes all ordinates of the $-TS$ curve smaller in the ratio T_B/T_A. As a result, both n_1 and n_e of the free-energy curve (G_v) are shifted to the left. Thus, with decreasing temperature, the equilibrium number of vacancies becomes smaller because the entropy component ($-TS$) decreases.

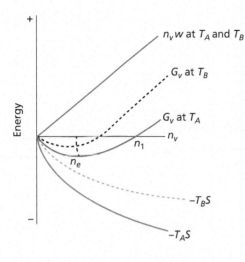

Fig. 7.4 Free energy as a function of the number of vacancies, n_v, in a crystal at two temperatures, $T_B < T_A$

It should be noted that Eq. 7.40 was derived assuming that the vibrational entropy contribution is negligible. When this entropy is included, a similar derivation would show the equilibrium vacancy concentration, X_c, as

$$X_c = \frac{n_v}{n_o} = e^{S_v/k} \cdot e^{-H_v/kT} \qquad 7.41$$

where H_v and S_v are the enthalpy and entropy of formation of a single vacancy, respectively. The above equation, called the Arrhenius law, when plotted as $\ln X_c$ versus $1/T$ would give a straight line, see Fig. 7.5. The slope of the line is $-H_v/k$, whereas its intercept with the Y axis would give S_v/k. Such a plot can be used to determine the enthalpy and entropy of vacancy formation.

The most direct method of determination of the equilibrium vacancy concentration involves measurement of the actual linear thermal expansion coefficient ($\Delta L/L$) and the lattice parameter expansion coefficient ($\Delta a/a$) by X-ray techniques at high temperatures (Ref. 1). The measured thermal expansion contains two components: one due to the increase of the lattice constant with temperature and the other due to the increase in the number of lattice sites as vacancies are produced. As such, the difference between the two coefficients can be used to determine the vacancy concentration. Other techniques, such as quenching of the sample to low

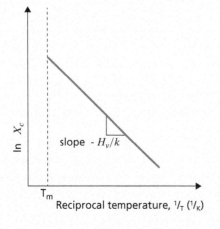

Fig. 7.5 Temperature dependence of vacancy concentration

Table 7.1 Enthalpy and Entropy of Formation of Single Vacancy in Selected Metals.

Element	H_v (eV)	S_v (units of K)
Aluminum	0.60–0.77	0.7–1.76
Cadmium	0.39–0.47	1.5–2.8
Cobalt	1.34	—
Copper	1.04–1.31	1.5–2.8
Gold	0.89–1.0	1.1
Iron, bcc	1.4–1.6	—
Molybdenum	3.0–3.24	—
Niobium	2.6 ≥ 2.7	—
Nickel	1.45–1.74	—
Platinum	1.15–1.6	1.3–4.5
Silver	1.09–1.19	—
Tungsten	3.1–4.0	2.3

Source: Wollenberger (Ref. 5)

temperatures such that vacancies are immobile and positron-annihilation spectroscopy have also been used to measure vacancy concentration as a function of temperature (Refs. 2 through 5). Table 7.1 gives the enthalpy and entropy values for selected metals. The enthalpy varies from about 0.4 eV for low melting temperature metal such as cadmium to about 4 for tungsten. The trend shows the influence of the bonding strength on the energy needed to form a vacancy and on the melting temperature.

7.8 Vacancy Motion

We have determined that the equilibrium ratio of vacancies to atoms at a given temperature T is given by Eq. 7.40 or 7.41.

Nothing, however, has been said about the time required to attain an equilibrium number of vacancies. This may be very long at low temperatures or very rapid at high temperatures. Since the movement of vacancies is caused by successive jumps of atoms into vacancies, a study of the fundamental law governing the jumps is important. It was mentioned earlier that an energy barrier, shown schematically in Fig. 7.6, must be overcome when a jump is made, and that the required energy is supplied by the thermal, or heat, vibrations of the crystal.

If q_o is the height of the energy barrier that an atom, such as atom a in Fig. 7.6, must overcome in order to jump into a vacancy, then the jump can only occur if atom a possesses vibrational energy greater than q_o. Conversely, if the vibrational energy is lower than q_o, the jump cannot occur. The chance that a given atom possesses an energy greater than q_o has been found to be proportional to the function $e^{-q_o/kT}$, or

$$p = \text{const } e^{-q_o/kT} \qquad 7.42$$

where p is the probability that an atom possesses an energy equal to or greater than a given energy q_o, k is Boltzmann's constant, and T is the absolute temperature. Equation 7.42 was originally derived for the energy distribution of atoms in a perfect gas (Maxwell-Boltzmann distribution).

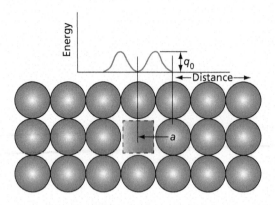

Fig. 7.6 The energy barrier that an atom must overcome in order to jump into a vacancy

However, this same function has also been found to predict quite accurately the vibrational energy distribution of the atoms in a crystalline solid.

Since the above function is the probability that a given atom has an energy greater than that required for a jump, the probability of jumping must be proportional to this function. Therefore, we can write the following equation:

$$r_v = Ae^{-q_o/kT} \qquad 7.43$$

where r_v is the number of atom jumps per second into a vacancy, A is a constant, q_o is the activation energy per atom (height of the energy barrier), and k and T have the usual significance. If the numerator and denominator of the exponent in Eq. 7.43 are both multiplied by N, Avogadro's number 6.02×10^{23}, we have

$$r_v = Ae^{-H_m/RT} \qquad 7.44$$

where H_m is the activation enthalpy for the movement of vacancies in joules per mole, and R the gas constant (8.31 J per mole K).

The constant A in Eq. 7.44 depends on a number of factors. Among these is the number of atoms bordering the hole. The larger the number of atoms able to jump, the greater the frequency of jumping. A second factor is the vibration frequency of the atoms. The higher the rate of vibration, the more times per second an atom approaches the hole, and the greater is its chance of making a jump.

Let us consider Eq. 7.44 with respect to an actual metal. The value of A for copper is approximately 10^{15}, and the activation enthalpy H_m = 121 KJ per mole. If these values are substituted into the jump-rate equation, we have at 1350 K (just below melting point of copper)

$$r_v \simeq 2 \times 10^{10} \text{ jumps/s}$$

at 300 K (room temperature)

$$r_v \simeq 10^{-6} \text{ jumps/s}$$

The tremendous difference in the rate with which vacancies move near the melting point, compared with their rate at room temperature, is quite apparent. In one second at 1350 K the vacancy moves approximately 30 billion times, while at room temperature a time interval of about 10^6 sec, or 11 days, occurs between jumps.

While the jump rate of vacancies is of some importance, we are really more interested in how many jumps the average atom makes per second when the crystal contains an equilibrium number of vacancies. This quantity equals the fractional ratio of vacancies to atoms n_v/n_o times the number of jumps per second into one vacancy, or

$$r_a = \frac{n_v}{n_o} A e^{-H_m/RT}$$

where r_a is the number of jumps per second made by an atom, n_v is the number of vacancies, n_o the number of atoms. However, we have seen that

$$\frac{n_v}{n_o} = e^{-H_f/RT} \qquad 7.45$$

where H_f is the activation enthalpy for the formation of vacancies. Therefore,

$$r_a = A e^{-H_m/RT} \times e^{-H_f/RT} = A e^{-(H_m + H_f)/RT} \qquad 7.46$$

The rate at which an atom jumps, or moves from place to place in a crystal, thus depends on two energies: H_f, the work to form a mole of vacancies, and H_m, the energy barrier that must be overcome in order to move a mole of atoms into vacancies. Since the two energies are additive, the atomic jump rate is extremely sensitive to temperature. This last statement can be made in a different form. It has previously been shown that in copper the ratio of vacancies to atoms is about one to a thousand at 1350 K, while at 300 K, the ratio is one to approximately 5×10^{15}, a decrease by a factor of about 10^{12}. During the same temperature interval, the number of jumps per second into a vacancy decreases by a factor of approximately 10^{16}. Between the melting point and room temperature, the average rate of atomic movement thus decreases by a factor of about 10^{28}. From a more simple point of view, at 1350 K the vacancies in copper are approximately 10 atoms apart, and atoms jump into these vacancies at the rate of about 30 billion jumps per second; at 300 K the vacancies are 100,000 atoms apart, and atoms jump into them at the rate of one jump in 11 days.

The above discussion leads to one firm conclusion: those physical properties of copper that can be changed through diffusion of copper atoms can be considered unchangeable at room temperature.

Similar conclusions can be drawn for other metals, but the degree of change will depend upon the metal in question. For example, vacancies in lead at room temperature may be shown, by calculations similar to the above, to jump at a rate of approximately 22 jumps per second, with a mean spacing between vacancies of about 100 atoms. Appreciable atomic diffusion therefore occurs in lead at room temperature.

7.9 Interstitial Atoms and Divacancies

Next to a vacancy, the most important point defects in a metal crystal are an *interstitial atom* and a *divacancy*. Let us briefly consider the first of these.

An interstitial atom is one that occupies a place in a crystal that normally would be unoccupied. Such places are the holes or interstitial positions that occur between the atoms lying on the normal lattice sites. Such a hole occurs even in the close-packed face-centered cubic lattice at the center of the unit cell, as indicated in Fig. 7.7. An equivalent hole exists between any pair of atoms lying at the corners of the unit cell. The hole between the two corner atoms along the right-hand front vertical edge of the cell is indicated in Fig. 7.7. This interstitial site can also be seen in Fig. 1.2.

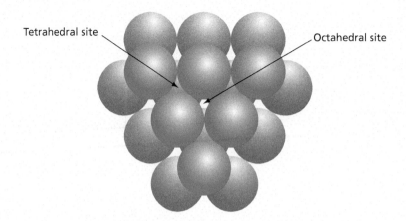

Fig. 7.7 Interstitial sites in a face-centered cubic crystal. Tetrahedral site is surrounded by four atoms, whereas octahedral site is surrounded by six atoms.

It is necessary to differentiate between two basically different types of atoms that can occupy these interstitial sites. The first of these is composed of atoms whose sizes are small, such as carbon, nitrogen, hydrogen, and oxygen. Atoms of this type can sometimes occupy a small fraction of the interstitial sites in a metal. When this occurs, it is said to be an interstitial solid solution. (This subject is considered further in Chapter 9.) The other type of interstitial atom is our present concern: this consists of an atom that would normally occupy a regular lattice site. In the case of a pure metal like copper, such an interstitial atom would be a copper atom. The hard-ball model of the {100} plane of an fcc crystal in Fig. 7.8 clearly shows that the interstitial sites are too small to hold such a large atom without badly distorting the lattice. As a consequence, while the activation energy to form vacancies in copper is a little less than 1 eV, that for the formation of interstitital copper atoms is probably about 4 eV. This difference in activation energy means that interstitial atoms produced by thermal vibrations should be very rare in a metal such as copper. This can easily be shown by a computation using the same form of equation as that developed earlier in this chapter to compute the equilibrium ratio of vacancies to atoms.

While interstitial defects of the type under consideration do not normally exist in sufficient concentrations in a metal to be significant, they can be readily produced in metals as a result of radiation damage. Collisions between fast neutrons and a metal can result in knocking atoms out of their normal

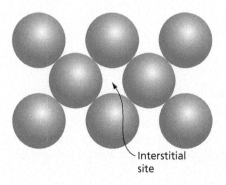

Fig. 7.8 The size of the interstitial site is much smaller than the size of the solvent atoms

lattice sites. The removal of an atom from its lattice position, of course, produces both an interstitial atom and a vacancy. It has been estimated (Ref. 6) that each fast neutron can result in the creation of about 100–200 interstitials and vacancies.

While it is very difficult to create an interstitial defect of the type presently being considered, they are very mobile once they are produced. The energy barrier for their motion is small and, in the case of copper, it has been estimated (Ref. 7) to be of the order of 0.1 eV. The reason for this high mobility is shown in Fig. 7.9. In this figure it is assumed that one is observing the {100} plane of an fcc crystal. In Fig. 7.9A the interstitial atom is shown as a shaded circle and is designated by the symbol a. Suppose that this atom were to move to the right. This motion would push atom b into an interstitial position, as shown in Fig. 7.9B. In this process atom a would return to a normal

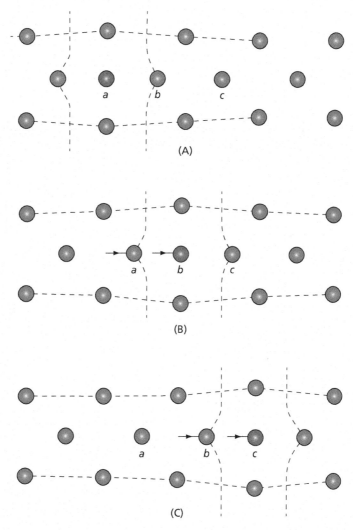

Fig. 7.9 The configuration when an interstitial atom can move readily through the crystal. The atoms surrounding the interstitials are squeezed apart

lattice site. Further motion of b in the same direction would make an interstitial atom out of c. By this type of motion, the configuration associated with an interstitial atom can readily move through the lattice. Any given interstitial atom only moves a small distance, and because the distortion of the lattice is severe around the interstitial atom, this type of motion requires little energy. The effect of the lattice distortion around the defect in making the motion easier is exactly equivalent to that observed in the motion of dislocations.

If a pair of vacancies combine to make a single point defect, a *divacancy* is said to be created. It is difficult to estimate or measure the binding energy of this type of defect. One calculation, (Ref. 8) however, places it at about 0.3–0.4 cV in the case of copper. In a metal where the vacancies and divacancies are in equilibrium, one can approximately compute (Ref. 6) the ratio of divacancies to vacancies using the equation

$$\frac{n_{dv}}{n_v} = 1.2 z e^{-q_b/kT} \qquad 7.47$$

where n_{dv} is the concentration of divacancies, z is the coordination number, and q_b is the binding energy of a divacancy. The recent first principles calculations by P. Zhang et al. (Ref. 9) indicate that the formation energies for mono- and divancies correlate well with the melting temperatures for bcc and hcp metals, but not for fcc metals.

Problems

7.1 Gold has a melting point of 1,063°C and its latent heat of fusion is 12,700 J/mol. Determine the entropy change due to the freezing of one mole of gold and indicate whether it is positive or negative.

7.2 When a mole of gold freezes at its freezing point is there a change in its internal energy? Explain.

7.3 The *Handbook of Chemistry and Physics*, 57th edition (1976–1977) lists on pages D-61 to D-63 a section on the thermodynamic properties of the elements. These include an empirical equation for calculating the enthalpy of the elements as a function of the temperature at constant pressure. The equation has the form

$$H_T - H_{298} = aT + (1/2)(b \times 10^{-3})T^2 + (1/3)(c \times 10^{-6})T^3 - A$$

where H_T is the enthalpy at a temperature T; H_{298} is the enthalpy at 298 K, the reference temperature; a, b, c and A are empirical constants obtained from specific heat at constant pressure; C_p data; and T is the absolute temperature in degrees K. In the case of solid gold and when enthalpies are expressed in J/gmol, these constants are $a = 25.69$, $b = -0.732$, $c = 3.85$, and $A = 7661$. For liquid gold the pertinent constants are $a = 29.29$ and $A = -2640$. The constants b and c may be assumed to equal zero. The melting point of gold is 1336 K. Write a computer program designed to give (1) ΔH the difference, at a given temperature, between $H_T - H_{298}$, for the liquid and for the solid and (2) the corresponding entropy difference $\Delta S = \Delta H/T$, in both cases, at 20 degree intervals between 1036 and 1336 K.

7.4 The *Handbook of Chemistry and Physics* also gives the empirical equation for computing the entropy (referenced to 298 K). It is

$$S_T = a \ln T + (b \times 10^{-3})T + \tfrac{1}{2}(c \times 10^{-6})T^2 - B$$

where S_T is the entropy at temperature T; a, b, and c are constants with the same values as in the enthalpy equation of Prob. 7.3; B is a constant; and T is the absolute temperature.

(a) Taking the values of a, b, and c for the solid and the liquid gold given in Prob. 7.3 and $B = 98.95$ and 112.9 J/K-gmol for solid and liquid gold, respectively, write a computer program giving the entropy difference between the liquid and solid at 20 deg. intervals between 1036 and 1336 K.

(b) For the ΔS values obtained in this and the preceding problem, plot the values on the same graph as functions of the temperature and rationalize your results. Note that at 1336 K both problems give the same values for ΔS.

7.5 Consider an insulated chamber with two equally sized compartments that are separated from each other by a removable partition. Initially one of the compartments is assumed to be evacuated completely while the other is filled with a mole of an ideal gas under standard atmospheric conditions. Now consider that the partition is removed so that the gas can expand to fill the two chambers.

(a) Will there be a change in the temperature of the gas? Explain.

(b) Compute the value of the entropy change.

7.6 Consider that the two compartments of the insulated box of Prob. 7.5 initially contain on one side a mole of ideal gas A and on the other side a mole of ideal gas B, with both gases at a pressure of 0.1013 MPa and a temperature of 298 K. Now assume that the partition is removed.

(a) How large will the enthalpy change be?

(b) Compute the entropy change.

(c) Compute the Gibbs free-energy change. Is the magnitude of this change significant? Explain.

7.7 With the aid of a suitable computer program compute the entropy of mixing (see Eq. 7.32) as a function of the concentration X from $X = 0$ to 1 and plot S_m in J/K-gmol versus X within these limits.

7.8 Compute the equilibrium concentration of vacancies in pure copper at 700°C.

7.9 Bradshaw, F. J., and Pearson, S., *Phil. Mag.*, **2** 379 (1957) reported that the activation enthalpy, H_f, for the formation of a mole of vacancies in gold is 0.95 eV/mol.

(a) First convert their activation enthalpy in eV/mol to J/mol and then compute the equilibrium concentration of vacancies in gold at 1000 K.

(b) What is the number of vacancies in 1 mol of the gold at 1000 K?

7.10 The following table lists six metallic elements, their enthalpies for the formation of a vacancy, and their melting points. **(a)** Determine the vacancy concentration of each element at its melting point. To make these calculations it is convenient to write a short computer program into which one may input H_f and T_m as listed in the table.

Element	H_f, eV	Melting Point,°C
Al	0.76	660
Ag	0.92	961
Cu	0.90	1083
Au	0.95	1063
Ni	1.4	1453
Pt	1.4	1769

(b) There is some evidence that the vacancy concentration at the melting point may become so large and the jump rate so high that it is no longer possible for a macroscopic crystal to exist. Do the results of this problem support this view? Explain.

7.11 (a) It has been estimated that the enthalpy for the formation of self-interstitial atoms in copper is about 385,000 J/mole. Compute the equilibrium concentration of these interstitial atoms in copper at 1000 K.

(b) The activation enthalpy for the movement of the self-interstitial atoms in copper is believed to be about 9,640 J/mol. Estimate the jump frequency of these interstitials at 1000 K.

References

1. Simmons, R. O., and Balluffi, R. W., *Phys. Rev.* 129 No. 4, pp. 1533–1544 (1963).
2. Hehnkamp, Th., et al., *Phys. Rev. B*, 45 No. 5, pp. 1998–2003 (1992).
3. Matter, H., Winter, J., and Triftshauser, W., *Appl. Phys.* 20, pp. 135–140 (1979).
4. McLellan, R. B., and Angel, Y. C., *Acta Metall. Mater*, 45 No. 10, pp. 3721–3725 (1995).
5. Wollenberger, H., "Point Defects," in *Physical Metallurgy*, 4th ed., R. W. Cahn and P. Haasen (eds.), North-Holland, Amsterdam (1996), pp. 1621–1721.
6. Cottrell, A. H., *Vacancies and Other Point Defects in Metals and Alloys*, p. 1. The Institute of Metals, London, 1958.
7. Huntington, H. B., *Phys. Rev.*, **91**, 1092 (1953).
8. Seeger, A., and Bross, H., *Z. Physik*, **145** 161 (1956).
9. Zhang, P. et al., *J. Nuclear Materials*, Vol. **538** (2020) 152253.

Chapter 8

Annealing

> **Learning Objectives**
>
> Upon completion of Chapter 8, you will be able to:
> 1. Understand the influence of cold working on stored energy and its release during annealing
> 2. Distinguish between recovery, recrystallization, and grain growth
> 3. Differentiate between activation energy for recrystallization and motion of vacancies
> 4. Explain polygonization phenomena and dislocation movements during annealing
> 5. Relate the effects of time, temperature, and deformation amount on recrystallization rate
> 6. Understand the nucleation and growth reactions during recrystallization
> 7. Explain why recrystallization temperature is not a property of a metal
> 8. Describe grain boundary movement rate and the effects of impurities, inclusions and pores on grain growth
> 9. Discuss the interactions between grain boundary migration and inclusions

8.1 Stored Energy of Cold Work

When a metal is plastically deformed at temperatures that are low relative to its melting point, it is said to be *cold worked*. The temperature defining the upper limit of the cold working range cannot be expressed exactly, for it varies with composition as well as the rate and the amount of deformation. A rough rule-of-thumb is to assume that plastic deformation corresponds to cold working if it is carried out at temperatures lower than one-half of the melting point measured on an absolute scale.

Most of the energy expended in cold work appears in the form of heat, but a finite fraction is stored in the metal as strain energy associated with various lattice defects created by the deformation. The amount of energy retained depends on the deformation process and a number of other variables, for example, composition of the metal as well as the rate and temperature of deformation. A number of investigators have indicated that the fraction of the energy that remains in the metal varies from a low percentage to somewhat over 10 percent. Figure 8.1 shows the relationship between the stored energy and the amount of deformation in a specific metal (polycrystalline 99.999 percent pure copper) for a specific type of deformation (tensile strain). The data, from the work of Gordon, (Ref. 1) show that the stored energy increases with increasing deformation, but at a decreasing rate, so that the fraction of the total energy stored decreases with increasing deformation. The latter effect is shown by a second curve plotted in Fig. 8.1.

The maximum value of the stored energy in Fig. 8.1 is only 6 cal (25 J) per mole, which represents the strain energy left in a very pure metal after a moderate deformation (30 percent extension) at room temperature. The amount of stored energy can be greatly increased by increasing the severity of the deformation, lowering the deformation temperature, and by changing the pure metal to an alloy. Thus, metal chips formed by drilling an alloy (82.6 percent Au–17.4 percent Ag) at the temperature of liquid nitrogen are reported (Ref. 2) to have a stored-energy content of 200 cal (837 J) per mole.

Let us consider the nature of the stored energy of plastic deformation. Cold working is known to increase greatly the number of dislocations in a metal. A soft annealed metal can have dislocation densities of the order of 10^{10} to 10^{12} m^{-2}, and heavily cold-worked metals can have approximately 10^{16}. Accordingly, cold working is able to increase the number of dislocations in a metal by a factor as large as 10,000 to 1,000,000. Since each dislocation represents a crystal defect with an associated lattice strain, increasing the dislocation density increases the strain energy of the metal.

The creation of point defects during plastic deformation is also recognized as a source of retained energy in cold-worked metals. One mechanism for the creation of point defects in a crystal has already been described. In Sec. 4.9 it was mentioned that a screw dislocation that cuts another screw dislocation may be capable of generating a close-packed row of either vacancies or interstitials as it glides, with the type of point defect produced depending on the relative Burgers vectors of the intersecting dislocations. Since the strain energy associated with a vacancy is much smaller than that associated with an interstitial atom, it can be assumed that vacancies will be formed in greater numbers than interstitial atoms during plastic deformation.

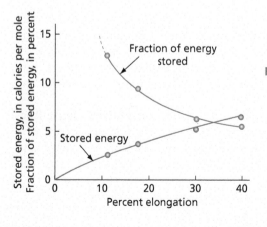

Fig. 8.1 Stored energy of cold work and fraction of the total work of deformation remaining as stored energy for high purity copper plotted as functions of tensile elongation. (Data of Gordon, P., *Trans. AIME*, **203** 1043 (1955).)

8.2 The Relationship of Free Energy to Strain Energy

The free energy of deformed metal is greater than that of an annealed metal by an amount approximately equal to the stored strain energy. While plastic deformation certainly increases the entropy of a metal, the effect is small compared to the increase in internal energy (the retained strain energy). The term $-T\Delta S$ in the free-energy equation may, therefore, be neglected and the free-energy increase equated directly to the stored energy. Therefore,

$$\Delta G = \Delta H - T\Delta S$$

becomes

$$\Delta G \approx \Delta H \qquad 8.1$$

where ΔG is the free energy associated with the cold work, ΔH is the enthalpy, or stored strain energy, S is the entropy increase due to the cold work, and T is the absolute temperature.

Since the free energy of cold-worked metals is greater than that of annealed metals, they may soften spontaneously. A metal does not usually return to the annealed condition by a single simple reaction because of the complexity of the cold-worked state. A number of different reactions occur, the total effect of which is the regaining of a condition equivalent to that possessed by the metal before it was cold worked. Many of these reactions involve some form of atom, or vacancy, movement and are, therefore, extremely temperature sensitive. The reaction rates of these reactions may usually be expressed as simple exponential laws similar to that previously written for vacancy movement. Heating a deformed metal, therefore, greatly speeds up its return to the softened state.

8.3 The Release of Stored Energy

Valuable information about the nature of the reactions that occur as a cold-worked metal returns to its original state may be obtained through a study of the release of its stored energy. There are several basically different methods of accomplishing this. Two of the more important will now be briefly indicated. In the first, the *anisothermal anneal* method, the cold-worked metal is heated continuously from a lower to a higher temperature and the energy release is determined as a function of temperature. One form of the anisothermal anneal measures the difference in the power required to heat two similar specimens at the same rate. One specimen of the two is cold worked before the heating cycle, while the other serves as a standard and is not deformed. During the heating cycle, the cold-worked specimen undergoes reactions that release heat and lower the power required to heat it in comparison with that required to heat the standard specimen. Measurements of the difference in power give direct evidence of the rate at which heat is released in the cold-worked specimen. Figure 8.2 shows a typical anisothermal anneal curve for a commercially pure copper (99.97 percent copper). (Ref. 3) It is noteworthy that some heat is released at temperatures only slightly above room temperature. The significance of this, and of the pronounced maximum that appears in the curve, will be discussed presently.

The other method of studying energy release involves *isothermal annealing*. Here the energy is measured while the specimen is maintained at a constant temperature. Figure 8.3 is

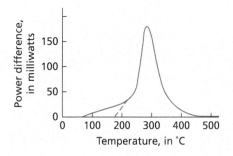

Fig. 8.2 Anisothermal anneal curve. Electrolytic copper. (From Clarebrough, H. M., Hargreaves, M. E., and West, G. W., *Proc. Roy. Soc.*, London, **232A**, 252 (1955).)

representative of the type of curves obtained in an isothermal anneal. The data for this particular curve were obtained with the aid of a microcalorimeter with a sensitivity capable of measuring a heat flow as low as 13 mJ per hr.

Both the anisothermal anneal and the isothermal anneal curves of Figs. 8.2 and 8.3 show maxima corresponding to large energy releases. Metallographic specimens prepared from samples annealed by either method show that an interesting phenomenon occurs in the region of maximum energy release. These large energy releases appear simultaneously with the growth of an entirely new set of essentially strain-free grains, which grow at the expense of the original badly deformed grains. The process by which this occurs is called *recrystallization* and may be understood as a realignment of the atoms into crystals with a lower free energy.

While the major energy release of the curves of Figs. 8.2 and 8.3 correspond to recrystallization, both curves show that energy is released before recrystallization. In this regard, dashed lines have been drawn schematically on both curves in order to delineate the recrystallization portion of the energy release. The area under each solid curve that lies to the left and above the dashed lines represents an energy release not associated with recrystallization. In the anisothermal anneal curve, this freeing of strain energy starts at temperatures well below those at which recrystallization starts. Similarly, in the isothermal anneal curve it begins at the start of the annealing cycle and is nearly completed before recrystallization starts. The part of the annealing cycle that occurs before recrystallization is called *recovery*. It should be recognized, however, that the reactions that occur during the recovery stages are able to continue during the progress of recrystallization; not in those regions that have already recrystallized, but in those that have not yet been converted into new grains. Recovery reactions will be described in the next section. First, however, it is necessary to define the third stage of annealing—*grain growth*. Grain growth occurs when annealing is continued after recrystallization has been completed. In grain growth, certain of the recrystallized grains continue to grow in size, but only at the expense of other crystals which must, accordingly, disappear.

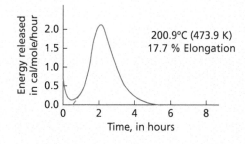

Fig. 8.3 Isothermal anneal curve. High purity copper. (From data of Gordon, P., *Trans. AIME*, **203** 1043 (1955).)

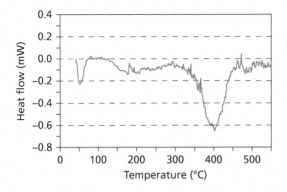

Fig. 8.4 Heat flow versus temperature for 80 percent cold-rolled ultra-high-purity iron. (Reprinted from Scholz, F., Driver, J.H., and Woldt, E., "The stored energy of cold rolled ultra high purity iron," *Scripta Mat.*, **40**, 949 (1999), with permission from Elsevier. http://www.sciencedirect.com/science/journal/00369748)

The release of stored energy in cold-rolled high-purity iron, measured by Scholz et al. (Ref. 4) using a differential scanning calorimeter (DSC), is shown in Fig. 8.4. The DSC measurements were conducted with a heating rate of 20 K/min from 40 to 650°C. In this figure, where the heat generation by the metal shows as a negative deviation in the heat flow curve, after the initial transient peak below 100°C, a broad exothermic peak extending from about 100 to 280°C can be seen. This broad exothermic peak is followed by a larger one extending from about 300 to 480°C. The energy released in these two peaks has been calculated by the authors as 3.9 and 15.1 J/mol, corresponding to 19 J/mol for the total stored energy. The changes in the microstructure of the cold-rolled iron during heating, revealed by interrupted quenching, are shown in Figs. 8.5A–F. As can be seen, the deformed structure is retained up to 300°C, but new grains form as the metal is heated further. The new-formed grains can be clearly seen as isolated spots in Fig. 8.5C, while other regions still show the deformed microstructure. At 460°C (Fig. 8.5E), most of the deformed regions have been replaced by new grains which grow considerably larger at the higher temperature, as seen in Fig. 8.5F.

The three stages of annealing—recovery, recrystallization, and grain growth—have now been defined. Some of the important aspects of each will now be considered.

8.4 Recovery

When a metal is cold worked, changes occur in almost all of its physical and mechanical properties. Working increases strength, hardness, and electrical resistance, and it decreases ductility. Furthermore, when plastically deformed metal is studied using X-ray diffraction techniques, the X-ray reflections become characteristic of the cold-worked state. Laue patterns of deformed single crystals show pronounced asterism corresponding to lattice curvatures. Similarly, Debye-Scherrer photographs of deformed polycrystalline metal exhibit diffraction lines that are not sharp, but broadened, in agreement with the complicated nature of the residual stresses and deformations that remain in a polycrystalline metal following cold working.

In the recovery stage of annealing, the physical and mechanical properties that suffered changes as a result of cold working tend to recover their original values. For many years, the fact that hardness and other properties could be altered without an apparent change in the microstructure, as signified by recrystallization, was regarded as a mystery. That the various physical and mechanical properties do not recover their values at the same rate is indicative of the complicated

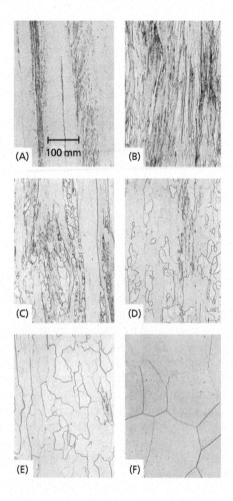

Fig. 8.5 Optical view of microstructure of deformed iron at different annealing temperatures: **(A)** as cold rolled, **(B)** annealed at 300°C, **(C)** annealed at 370°C, **(D)** annealed at 410°C, **(E)** annealed at 460°C, and **(F)** annealed at 650°C. (Reprinted from Scholz, F., Driver, J. H., and Woldt, E., "The stored energy of cold rolled ultra high purity iron", *Scripta Mat.*, **40**, 949 (1999), with permission from Elsevier. http://www.sciencedirect.com/science/journal/00369748)

nature of the recovery process. Figure 8.6 shows another anisothermal anneal curve corresponding to the energy released on heating cold-worked polycrystalline nickel. (Ref. 3) The peak at point c defines the region of recrystallization. The fraction of the energy released during recovery in this metal is much larger than that in the example of Fig. 8.2. Plotted on the same diagram are curves indicating the change in electrical resistivity and hardness as a function of the annealing temperature. Notice that the resistivity is almost completely recovered before the state of recrystallization. On the other hand, the major change in the hardness occurs simultaneously with recrystallization of the matrix.

8.5 Recovery in Single Crystals

The complexity of the cold-worked state is directly related to the complexity of the deformation that produces it. Thus, the lattice distortions are simpler in a single crystal deformed by easy glide than in a single crystal deformed by multiple glide (simultaneous slip on several systems), and lattice distortions may be still more severe in a polycrystalline metal.

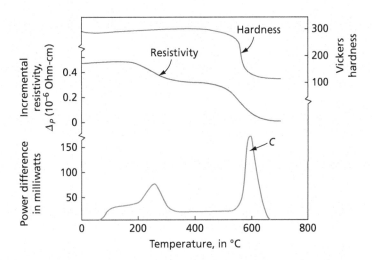

Fig. 8.6 Anisothermal-anneal curve for cold-worked nickel. At the top of the figure curves are also drawn to show the effect of annealing temperature on the hardness and incremental resistivity of the metal. (From Clarebrough, H. M., Hargreaves, M. E., and West, G. W., *Proc. Roy. Soc.*, London, **232A**, 252 (1955).)

If a single crystal is deformed by easy glide (slip on a single plane) in a manner that does not bend the lattice, it is quite possible to completely recover its hardness without recrystallization of the specimen. In fact, it is generally impossible to recrystallize a crystal deformed only by easy glide, even if it is heated to temperatures as high as the melting point. Figure 8.7 shows schematically a stress-strain curve for a zinc single crystal strained in tension at room temperature where it deforms by basal slip.

Let us suppose that the crystal was originally loaded to point *a* and then the load was removed. If the load is reapplied after only a short rest period (about half a minute), it will not

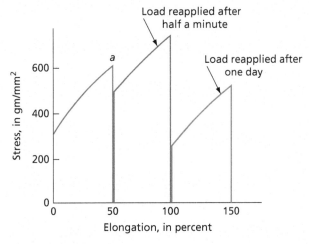

Fig. 8.7 Recovery of the yield strength of a zinc single crystal at room temperature. (After Schmid, E., and Boas, W., *Kristallplastizität*, Julius Springer, Berlin, 1935.)

yield plastically until the stress almost reaches the value attained just before the load was removed on the first cycle. There is, however, a definite decrease in the stress at which the crystal begins to flow the second time. This *flow-stress* would only equal that attained at the end of the previous loading cycle if it could be unloaded and reloaded without loss of time. Recovery of the yield point thus begins very rapidly. The yield point can be completely recovered in a zinc crystal at room temperature in a period of a day. This is shown by the third loading cycle in Fig. 8.7. These stress-strain diagrams indicate a well known fact: the rate at which a property recovers isothermally is a decreasing function of the time. Figure 8.8 illustrates this effect graphically by plotting the yield point of deformed zinc crystals as a function of the time for several different recovery temperatures. In this case the crystals were plastically deformed by easy glide at 223 K and isothermally annealed at the indicated temperatures. Notice that the rate of recovery is much faster at 283 K than it is at 253 K. This confirms our previous observations of the effect in temperature on the rate of recovery. In fact, the recovery in zinc crystals deformed by simple glide can be expressed in terms of a simple activation, or Arrhenius-type law, of the form

$$\frac{1}{\tau} = A e^{-Q/RT} \qquad 8.2$$

where τ is the time required to recover a given fraction of the total yield point recovery; Q the activation energy; R the universal gas constant; T the absolute temperature; and A a constant.

Let us now assume that the reaction has proceeded to the same extent at two different temperatures. Then we have

$$A e^{-Q/RT_1} \tau_1 = A e^{-Q/RT_2} \tau_2$$

which is equivalent to

$$\frac{\tau_1}{\tau_2} = \frac{e^{-Q/RT_2}}{e^{-Q/RT_1}} = e^{-\frac{Q}{R}\left(\frac{1}{T_2} - \frac{1}{T_1}\right)} \qquad 8.3$$

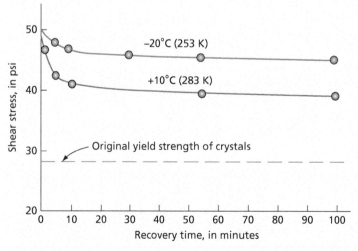

Fig. 8.8 Recovery of the yield strength of zinc single crystals at two different temperatures. (From the data of Drouard, R., Washburn, J., and Parker, E. R., *Trans. AIME*, **197** 1226 [1953].)

According to Drouard et al., (Ref. 5) the activation energy Q for recovery of the yield point in zinc is 83,140 J per mole. Thus, if a strained zinc crystal recovers one-fourth of its original yield point in 5 min at 0°C (273 K), we would expect that the same amount of recovery at 27°C (300 K) would take

$$\tau_1 = 5\,e^{-\frac{83,140}{8.314}\left(\frac{1}{273}-\frac{1}{300}\right)} = 0.185 \text{ min}$$

On the other hand, at $-50°C$ (223 K)

$$\tau_1 = 25,000 \text{ min or 17 days}$$

8.6 Polygonization

Recovery associated with a simple form of plastic deformation has been considered in the preceding section. Recovery in this case is probably a matter of annihilating excess dislocations. Such annihilation can occur by the coming together of dislocation segments of opposite sign (that is, negative edges with positive edges and left-hand screws with right-hand screws). In this process it is probable that both slip and climb mechanisms are involved.

Another recovery process is called *polygonization*. In its simplest form it is associated with crystals that have been plastically bent. Because the X-ray beam is reflected from curved planes, bent crystals give Laue photographs with elongated, or asterated, spots. (See Chapter 2.) Many workers have shown that the Laue spots of bent crystals assume a fine structure after a recovery anneal (an anneal that does not recrystallize the specimen). This is shown schematically in Fig. 8.9, where the left-hand figure represents the Laue pattern of a bent crystal before annealing, and the right-hand figure the pattern after annealing. Each of the elongated, or asterated spots of the deformed crystal is replaced by a set of tiny sharp reflections in the annealed crystal.

In a Laue photograph, each spot corresponds to the reflection from a specific lattice plane. When a single crystal is exposed to the X-ray beam of a Laue camera, a finite number of spots is obtained with a pattern on the film characteristic of the crystal and its orientation. If a Laue beam straddles the boundary between two crystals, a double pattern of spots is observed, each characteristic of the orientation of its respective crystal. Furthermore, if the two crystals have nearly identical orientations, as is the case when the boundary is a low-angle boundary or low-energy

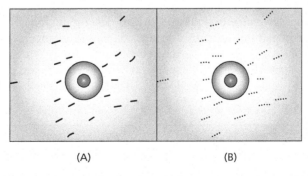

Fig. 8.9 Schematic Laue patterns showing how polygonization breaks up asterated X-ray reflections into a series of discrete spots. The diagram on the left corresponds to reflections from a bent single crystal; that on the right corresponds to the same crystal after an anneal that has polygonized the crystal

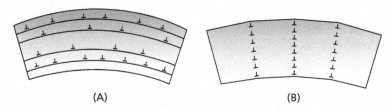

Fig. 8.10 Realignment of edge dislocations during polygonization **(A)** The excess dislocations that remain on active slip planes after a crystal is bent. **(B)** The rearrangement of the dislocations after polygonization

dislocation structure (i.e., LEDS; see Sec. 6.6), the two patterns will almost coincide, and the photograph will show a set of closely spaced double spots. Finally, if the X-ray beam falls on a number of very small crystalline areas, each separated from its neighbors by LEDS, a pattern such as that in Fig. 8.9B can be expected. Evidently, when a bent crystal is annealed, the curved crystal breaks up into a number of closely related small perfect crystal segments. This process has been given the name *polygonization*.[6]

The polygonization phenomenon can also be explained with drawings such as those of Fig. 8.10. The left-hand figure represents a portion of a plastically bent crystal. For simplicity, the active slip plane has been assumed parallel to the top and the bottom surface of the crystal. A plastically bent crystal must contain an excess of positive edge dislocations that lie along active slip planes in the manner suggested by the figure. The dislocation configuration of Fig. 8.10A is one of high strain energy. A different arrangement of the same dislocations that constitutes a simple form of LEDS is shown in Fig. 8.10B. Here the excess edge dislocations are found in arrays that run in a direction normal to the slip planes. Configurations of this nature constitute low-angle grain boundaries. (See Chapter 6.) When edge dislocations of the same sign accumulate on the same slip plane, their strain fields are additive.

In addition to lowering the strain energy, the regrouping of edge dislocations into low-angle boundaries has a second important effect. This is the removal of general lattice curvature. As a result of polygonization, crystal segments lying between a pair of low-angle boundaries approach the state of strain-free crystals with flat uncurved planes. However, each crystallite possesses an orientation slightly different from its neighbors because of the low-angle boundaries that separate them from each other. When an X-ray beam strikes the surface of a polygonized crystal it falls on a number of small, relatively perfect crystals of slightly different orientation. The result is a Laue pattern of the type shown in 8.9B.

It is customary to call low-angle boundaries, such as develop in polygonization, *subboundaries*, and the crystals that they separate, *subgrains*. The size, shape, and arrangement of the subgrains constitute the substructure of a metal. The difference between the concepts of grains and subgrains is an important one: subgrains lie inside grains. Originally a subgrain structure was called a *mosaic structure*.

8.7 Dislocation Movements in Polygonization

An edge dislocation is capable of moving either by slip on its slip plane or by climb in a direction perpendicular to its slip plane. Both are required in polygonization, as shown schematically in Fig. 8.11, where the indicated vertical movement of each dislocation represents climb, and the horizontal movement slip. The driving force for these movements comes from the strain energy of the dislocations, which decreases as a result of polygonization. From an equivalent viewpoint,

8.7 Dislocation Movements in Polygonization

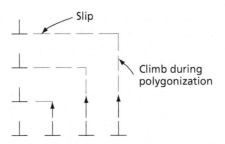

Fig. 8.11 Both climb and slip are involved in the rearrangement of edge dislocations

we may say that the strain field of dislocations grouped on slip planes produces an effective force that makes them move into subboundaries or LEDS. This force exists at all temperatures, but at low temperatures edge dislocations cannot climb. However, since dislocation climb depends on the movement of vacancies (an activated process), the rate of polygonization increases rapidly with temperature. Increasing temperature also aids the polygonization process in another manner, for the movement of dislocations by slip also becomes easier at high temperatures. This fact is observable in the fall of the critical resolved shear stress for slip with rising temperature.

The photographs of Fig. 8.12 are of special interest because they show the polygonization process in an actual metal (a silicon-iron alloy with 3.25 percent Si). The four photographs show the surface of crystals plastically bent to a fixed radius of curvature, and then annealed. Each specimen was given a 1-hr anneal at a different temperature in order to bring out the various stages of

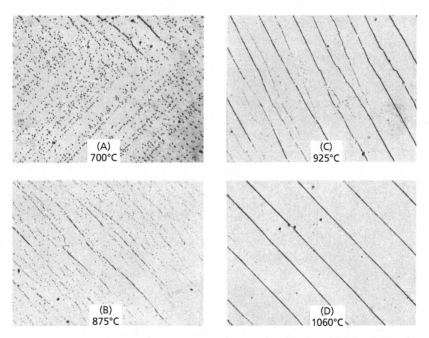

Fig. 8.12 Polygonization in bent and annealed iron-silicon single crystals. All specimens annealed for one hour at the indicated temperatures. Note that polygonization is more complete the higher the temperature of the anneal. All pictures were taken at 750×. (Reprinted from Hibbard, W. R. Jr, and Dunn, C. G., "A study of <112> edge dislocations in bent silicone-iron single crystals," *Acta Met.* **4** 306 (1956), Fig. 11, 14, 16 and 19 with permission of Elsevier. https://doi.org/10.1016/0001-6160(56)90068-2.)

polygonization. The higher the temperature the more complete is the polygonization process. The plane of the photograph is perpendicular to the axis of bending and corresponds to the front face of the crystal, shown diagrammatically in Fig. 8.13. The surface in question is perpendicular to the slip plane (01$\bar{1}$) of this body-centered cubic metal, and also to the plane along which the subboundaries form (111). The orientations of both planes are shown in the figure and it is clear that they make 45° angles with the horizontal: the slip plane having a positive slope and the subboundary a negative slope. Figure 8.12A shows the effect of a 1-hr anneal at 973 K. Each black dot of the illustration is a pit developed by etching the specimen in a suitable etching reagent, and shows the intersection of a dislocation with the specimen surface. Notice that the dislocations are, for the most part, associated with the slip planes, although several well-defined subboundaries can be seen near the top of the photograph. Many small groups containing three or four dislocations aligned perpendicular to the slip planes can also be seen in all areas of the picture. Figure 8.12B shows a more advanced stage in the polygonization process corresponding to an anneal at a higher temperature (1048 K). In this photograph, all the dislocations lie in the subboundaries, or polygon walls.

When all of the dislocations have dissociated themselves from the slip planes and have aligned themselves in low-angle boundaries, the polygonization process is not complete. The next stage is a coalescence of the low-angle boundaries where two or more subboundaries combine to form a single boundary. The angle of rotation of the subgrains across the boundary or LEDS must, of course, grow in this process. Well-defined subboundaries formed by coalescence can be seen in Fig. 8.12C, where the surface corresponds to a crystal annealed for an hour at 1198 K. Notice that, although this photograph and that of Fig. 8.12B were taken at the same magnification, the number of subboundaries is now much less. On the other hand, the density of dislocations in the boundaries is much higher, so that it is not generally possible to see individual dislocation etch pits.

The coalescence of subboundaries results from the fact that the strain energy associated with a combined boundary is less than that associated with two separated boundaries. The movement of subboundaries that occurs in coalescence is not difficult to understand, for the boundaries are arrays of edge dislocations which, in turn, are fully capable of movement by either climb or glide at high-annealing temperatures. In Fig. 8.12C, several junctions of subboundary pairs can be seen in the upper central portion of the photograph. Coalescence is believed to occur by the movement of these y-junctions. In the present example, movement of the junctions toward the bottom of the photograph will combine the pairs of branches into single polygon walls.

As the polygon walls become more widely separated, the rate of coalescence becomes a decreasing function of time and temperature so that the polygonization process approaches a more or less stable state with widely spaced, approximately parallel (in a single crystal deformed by simple bending) subboundaries. This state is shown in Fig. 8.12D and corresponds to a 1-hr anneal of a bent silicon-iron crystal at 1333 K.

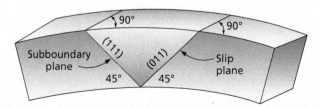

Fig. 8.13 Orientation of the iron-silicon crystal shown in the photographs of Fig. 8.12

The photographs in Fig. 8.12 have shown the polygonization process in a single crystal deformed by simple bending. In polycrystalline metals deformed by complex methods, polygonization may still occur. The process is complicated by the fact that slip occurs on a number of intersecting slip planes, and lattice curvatures are complex and vary with position in the crystal. The effect of such complex deformation on the polygonization process is shown in Fig. 8.14, which reveals a substructure resulting from deforming a single crystal in a small rolling mill after being annealed at a very high temperature (1373 K). The boundaries in this photograph are subboundaries and are not grain boundaries. No recrystallization is involved in this highly polygonized single crystal. Substructures such as this can be expected in cold-worked polycrystalline metals that have been annealed at temperatures high enough to cause polygonization, but not high enough, or long enough, to cause recrystallization.

A typical substructure, as revealed by the transmission electron microscope, is shown in Fig. 8.15. This iron plus 3 percent silicon single crystal was originally cold rolled 60 percent in the (001) [110] orientation and then annealed at elevated temperatures. Rolling tends to form high-density dislocations with no well defined cell boundaries, as shown in Fig. 8.15A. When the sample is annealed at 400°C for 1280 minutes, the dislocation density is reduced and random arrays of dislocations are visible, as seen in Fig. 8.15B. Subgrains and well defined polygonization become evident, Fig. 8.15C, when the sample is annealed at a higher temperature of 600°C for 1280 minutes. Annealing at 800°C for 5 minutes, on the other hand, causes growth of the subgrains with concomitant increase of the average subgrain diameter, as shown in Fig. 8.15D. An important feature of this photograph is that the dislocations resulting from this large strain

Fig. 8.14 Complex polygonized structure in a silicon-iron single crystal that was formed 8 percent by cold rolling before it was annealed 1 hr at 1373 K. (From Hibbard, W. R. Jr, and Dunn, ASM Seminar, *Creep and Recovery*, 1957, p. 52. Reprinted with permission of ASM International®. All rights reserved. www.asminternational.org)

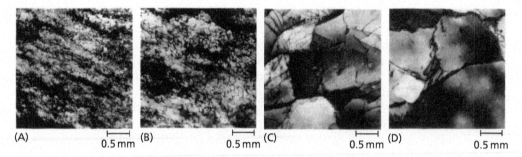

Fig. 8.15 Effect of annealing time and temperature on the microstructure of iron–3% Si single crystal cold rolled in the (00!) [100] orientation. (From Metallurgy and Microstructures, *ASM Handbook*, published by ASM International, 2004, page 209. Reprinted with permission of ASM International®. All rights reserved. www.asminternational.org)

(60 percent) have almost completely entered the cell walls. As a result, most subgrains have interiors that are basically free of dislocations. Where groups of dislocations are visible, it is probable that one is looking through a part of a cell wall.

8.8 Recovery Processes at High and Low Temperatures

Polygonization is too complicated a process to be expressed in terms of a simple rate equation, such as that used to describe the recovery process after easy glide. Because polygonization involves dislocation climb, relatively high temperatures are required for rapid polygonization. In deformed polycrystalline metals, high-temperature recovery is considered to be essentially a matter of polygonization and annihilation of dislocations.

At lower temperatures other processes such as occur in dynamic recovery (see Sec. 6.7) are of greater importance. At these temperatures, current theories picture the recovery process as primarily a matter of reducing the number of point defects to their equilibrium value. The most important point defect is a vacancy which may have a finite mobility even at relatively low temperatures.

8.9 Recrystallization

Recovery and recrystallization are two basically different phenomena. In an isothermal anneal, the rate at which a recovery process occurs always decreases with time; that is, it starts rapidly and proceeds at a slower and slower rate as the driving force for the reaction is expended. On the other hand, the kinetics of recrystallization are quite different, for it occurs much like a nucleation and growth process. In agreement with other processes of this type, recrystallization during an isothermal anneal begins very slowly, and builds up to a maximum reaction rate, after which it finishes slowly. This difference between the isothermal behavior of recovery and recrystallization is clearly evident in Fig. 8.4, where recovery processes start at the beginning of the annealing cycle and account for the initial energy release, while recrystallization starts later and accounts for the second (larger) energy release.

8.10 The Effect of Time and Temperature on Recrystallization

One way to study the recrystallization process is to plot isothermal recrystallization curves of the type shown in Fig. 8.16. Each curve represents the data for given temperature and shows the amount of recrystallization as a function of time. Data for each curve of this type are obtained by holding a number of identical cold-worked specimens at a constant temperature for different lengths of time. After removal from the furnace and cooling to room temperature, each specimen is examined metallographically to determine the extent of recrystallization. This quantity (in percent) is then plotted against the logarithm of the time. The effect of increasing the annealing temperature is shown clearly in Fig. 8.16. The higher the temperature, the shorter the time needed to finish the recrystallization. The S-shaped curves of Fig. 8.16 are similar to those of nucleation and growth processes. A similar curve will be shown for the nucleation and growth reaction involved in the precipitation of a second phase from a supersaturated solid solution. (See Figure 16.4 in Chapter 16.)

A horizontal line drawn through the curves of Fig. 8.16 corresponds to a constant fraction of recrystallization. Let us arbitrarily draw such a line corresponding to 50 percent recrystallization. The intersection of this line with each of the isothermal recrystallization curves gives the time at a given temperature required to recrystallize half of the structure. Let us designate this time interval by the symbol τ. Figure 8.17 shows τ plotted as a function of the reciprocal of the absolute temperature. The curve of Fig. 8.17 is a straight line, which shows that the recrystallization data for this metal (pure copper) can be expressed as an empirical equation of the form

$$\frac{1}{T} = K \log_{10} \frac{1}{\tau} + C \qquad 8.4$$

where K (the slope of the curve) and C (the intercept of the curve with the ordinate axis) are both constants. Equation 8.4 can also be expressed in the form

$$\frac{1}{\tau} = A e^{-Q_r/RT} \qquad 8.5$$

where $1/\tau$ is the rate at which 50 percent of the structure is recrystallized, R is the gas constant (8.37 J/mol-K), and Q_r is called the activation energy for recrystallization.

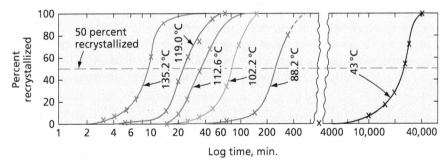

Fig. 8.16 Isothermal transformation (recrystallization) curves for pure copper (99.999 percent Cu) cold-rolled 98 percent. (From Decker, B. F., and Harker, D., *Trans. AIME*, **188** 887 [1950].)

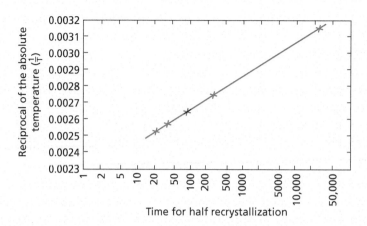

Fig. 8.17 Reciprocal of absolute temperature (K) vs. time for half recrystallization of pure copper. (From Decker, B. F., and Harker, D., *Trans. AIME*, **188** [1950].)

The quantity τ in Eq. 8.5 is not restricted to representing the time to recrystallize half of the matrix. It may represent the time to complete any fraction of the recrystallization process, such as the time to start the formation of new grains (several percent), or the time to complete the process (100 percent).

It is necessary to point out the difference between the activation energy Q_r for recrystallization and the previously discussed activation enthalpy for the motion of vacancies. The latter quantity can be directly related to a simple physical property: the height of the energy barrier that an atom must cross to jump into a vacancy. In the present case, the physical significance of Q_r is not completely understood. There is good reason to believe that more than one process is involved in recrystallization, so that Q_r cannot be related to a single simple process. It is best, therefore, to consider the recrystallization activation energy as an empirical constant. Furthermore, modern recrystallization theory indicates that the activation energy generally does not remain a constant throughout the recrystallization process. In most cases, the activation energy changes continuously during recrystallization as the driving force for recrystallization, the stored energy of cold work, is depleted. In addition, although recrystallization tends to follow a pattern like a nucleation and growth phenomenon, the formation of a nucleus—in the classical sense of an atom by atom addition to an embryo until a stable nucleus is formed that then grows into a newly formed recrystallized grain—does not occur. The origin of a recrystallized grain is always a prexisting region that is highly misoriented in relation to the material surrounding it. This high degree of misorientation gives the region from which the new grain originates the needed growth mobility.

Equation 8.5 is exactly equivalent to the empirical equation that holds for the recovery of zinc single crystals after easy glide. Similar empirical rate equations have been found to describe the recrystallization process of a number of metals besides pure copper. Although it is not possible to generalize this equation and state that it accurately applies to recrystallization phenomena in all metals, we can consider that it is roughly descriptive of the relationship between time and temperature in recrystallization.

8.11 Recrystallization Temperature

A frequently used metallurgical term is *recrystallization temperature*. This is the temperature at which a particular metal with a particular amount of cold deformation will completely recrystallize in a finite period of time, usually 1 hr. Of course, in light of Eq. 8.5, the recrystallization temperature has no meaning unless the time allowed for recrystallization is also specified. However, because of the large activation energies encountered, recrystallization actually appears to occur at some definite minimum temperature. Suppose for a given metal $Q_r = 200,000$ J per mole, and that recrystallization is completed in 1 hr at 600 K. Then it may be shown, with the aid of Eq. 8.5, that if the annealing is carried out at a 10 K lower temperature (590 K), complete recrystallization will require slightly over 2 hr. A specimen of this particular metal will only be partly recrystallized at the end of an anneal of 1 hr at 10 K below its recrystallization temperature (600 K). On the other hand, an hour's anneal is more than enough to recrystallize the metal at any temperature above 600 K. In fact, a 10 K rise in temperature to 610 K shortens the recrystallization time to half an hour, and a 20 K rise to approximately 15 min. To the practical engineer, this sensitivity of the recrystallization process to small changes in temperature makes it appear as though the metal has a fixed temperature, below which it will not recrystallize, and for this reason, there is a tendency to regard the recrystallization temperature as a property of the metal and to neglect the time factor in recrystallization.

8.12 The Effect of Strain on Recrystallization

Two rate-of-crystallization curves similar to that of Fig. 8.17 are plotted in Fig. 8.18. These differ only in that they represent the recrystallization of zirconium instead of copper, and the rate at which complete recrystallization is observed instead of half-recrystallization. The data are plotted in the usual sense (log $1/\tau$ vs $\cdot 1/T$), but for convenient reading, the abscissae and ordinates are expressed in degrees Kelvin and hours, respectively. The two curves represent data from specimens cold worked by different amounts. In both cases, the zirconium metal was cold worked by swaging. *Swaging* is a means of mechanical deformation used on cylindrical rods in which

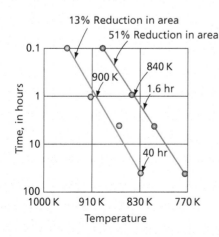

Fig. 8.18 Temperature-time relationships for recrystallization of zirconium (iodide) corresponding to two different amounts of prior cold work. (Treco, R. M., Proc., 1956, *AIME* Regional Conference on Reactive Metals, p. 136.)

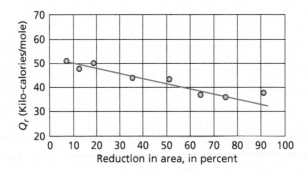

Fig. 8.19 Activation energy for the recrystallization of zirconium (iodide) as a function of the amount of cold work. (Treco, R. M., Proc., 1956, *AIME* Regional Conference on Reactive Metals, p. 136.)

the diameter of the rod is reduced uniformly by a mechanical hammer equipped with rotating dies. The amount of this cold work is measured in terms of the percentage reduction in area of the cylindrical cross-section. The left-hand curve of Fig. 8.18 corresponds to specimens with a cross-section area reduced 13 percent, while the right-hand curve represents specimens that suffered a larger reduction in area (51 percent). The two curves show clearly that recrystallization is promoted by increasing amounts of cold work. When annealed at the same temperature, the metal with the larger amount of cold work recrystallizes much faster than that with the lesser reduction. As an example, at 826 K the times for completion of recrystallization are 1.6 and 40 hr for the larger and the smaller reductions in area, respectively. Similarly, the temperature at which the metal recrystallizes completely within an hour is lower for a greater amount of cold work; 840 K as compared to 900 K.

A close examination of the two curves of Fig. 8.18 shows that these straight lines do not have the same slope. This means that the temperature dependence of recrystallization varies with the amount of cold work, or that the activation energy for recrystallization is a function of the amount of deformation. This fact is further emphasized by the data of Fig. 8.19, which show the variation of the activation energy for this same metal (zirconium) as a function of the percent reduction in area over the range of deformation from less than 10 percent to greater than 90 percent. The fact that the activation energy Q_r varies with the amount of cold work lends additional confirmation to the previous statement about the complex nature of Q_r.

8.13 The Rate of Nucleation and the Rate of Nucleus Growth

The preceding sections have shown that the recrystallization reaction in many metals can be described by a simple activation equation having the form

$$\text{rate} = Ae^{-Q_r/RT} \qquad 8.6$$

Unfortunately, this empirical equation reveals very little about the atomic mechanisms that occur during recrystallization. This is because of the dual character of a nucleation and growth reaction. The rate at which a metal recrystallizes depends on the rate at which nuclei form, and also on the rate at which they grow. These two rates also determine the final grain size of a recrystallized metal. If nuclei form rapidly and grow slowly, many crystals will form before their mutual impingement completes the recrystallization process. In this case, the final grain size is small. On the other hand, it will be large if the rate of nucleation is small compared to the rate of growth. Since the kinetics of recrystallization can often be described in terms of these two rates, a number of investigators have measured these quantities under isothermal conditions in the hope of learning more about the mechanism of recrystallization. This requires the introduction of two parameters: N, the rate of nucleation and G, the rate of growth.

It is customary to define the nucleation frequency, N, as the number of nuclei that form per second in a cubic centimeter of unrecrystallized matrix. This parameter is referred to the unrecrystallized matrix because the recrystallized portion is inactive with regard to further nucleation. The linear rate of growth, G, is defined as the time rate of change of the diameter of a recrystallized grain. In practice, G is measured by annealing for different lengths of time a number of identical specimens at a chosen isothermal temperature. The diameter of the largest grain in each specimen is measured after the specimens are cooled to room temperature and prepared metallographically. The variation of this diameter with isothermal annealing time gives the rate of growth, G. The rate of nucleation can be determined from the same metallographic specimens by counting the number of grains per unit area on the surface of each. These surface-density measurements can then be used to give the number of recrystallized grains per unit volume. Of course, each determination must be corrected for the volume of the matrix that has recrystallized.

Several equations (Ref. 7) have been derived starting with the parameters N and G, which express the amount of recrystallization as a function of time. (See Fig. 8.16.) Because the theories on which these equations are based diverge to some extent, and because space is limited, they will not be discussed further. The concepts of nucleation rate and growth rate are useful, however, in explaining the effects of several other variables on the recrystallization process.

8.14 Formation of Nuclei

In recrystallization, an entirely new set of grains is formed. New crystals appear at points of high-lattice-strain energy, such as slip-line intersections, deformation twin intersections, and in areas close to grain boundaries. In each case, it appears that nucleation occurs at points of strong lattice curvature. In this regard it is interesting to note that bent, or twisted single crystals recrystallize more readily than do similar crystals that have been bent, or twisted, and then unbent, or untwisted.

Because the new grains form in regions of severe localized deformation, the sites where they appear are apparently predetermined. Nuclei of this type are called *preformed nuclei*. (Ref. 8) A number of models have been proposed to show how it is possible to form a small, strain-free volume that can grow out and consume the deformed matrix around it. These models are in general agreement on two points. First, a region of a crystal can become a nucleus and grow only if its size exceeds some minimum value. For example, Detert and Zieb (Ref. 9) have computed that in a deformed metal with a dislocation density of $10^{12}\,\text{cm}^{-2}$, a preformed nucleus has to have a diameter greater than about 15 nm for it to be able to expand. (This general concept of the critical size of a nucleus will be considered in Chapters 15 and 16.)

The other condition for the formation of a nucleus is that it become surrounded, at least in part, by the equivalent of a high-angle grain boundary. This condition is required because the mobility of an arbitrary low-angle grain boundary is normally very low. Beyond these two points, the various models of the nucleation process vary to a considerable degree. It is possible that most of these mechanisms may operate and that the preferred one in a given situation will depend largely on the nature of the deformed specimen being recrystallized. In this regard, a single crystal lacks the sites along grain boundaries and along lines where three grains meet that are available for nucleation in a polycrystalline metal. Both grain boundaries and these triple lines are regions where high-angle boundaries already exist, so that one of the criteria for the formations of a nucleus is effectively satisfied. A typical mechanism applicable to polycrystals is that of Bailey and Hirsch, (Ref. 10) who propose that if a difference in dislocation density exists across a grain boundary in a cold-worked metal, then during annealing a portion of the more perfect grain might migrate into the less perfect grain under the driving force associated with the strain energy difference across the boundary. This would be accomplished by the forward movement of the grain boundary so as to form a bulge, as indicated in Fig. 8.20. This boundary movement should effectively sweep up the dislocations in its path, thereby creating a small, relatively strain-free volume of crystal. If this bulge exceeds the critical nucleus size, both primary conditions for the formation of a nucleus would be satisfied.

It is not possible to consider in detail all of the proposed nucleation models. We will discuss briefly two mechanisms that are probably more applicable to single crystals than to polycrystals. The first is due to Cahn (Ref. 11) and to Beck. (Ref. 12) This mechanism, which predates the transmission electron microscopy studies of recrystallization, simply proposed that, as a result of polygonization, it might be possible to produce a subgrain capable of growing out into the surrounding polygonized matrix. The other mechanism that should be applicable to single crystals involves the concept of subgrain coalescence, or the combination of subgrains to form a strainfree region large enough in size to grow. As postulated in this theory, (Ref. 13) the elimination of a subgrain boundary must result in a relative rotation of the two subgrains that are combined.

For polycrystalline materials, nucleation will again take place preferentially at highly energetic sites, such as grain-boundary triple points, original grain boundaries, and boundaries between deformation bands. The nuclei may form by subgrain growth and/or grain boundary migration, as schematically shown in Fig. 8.21. The nucleus shown in Fig. 8.21A has formed by growth of the subgrains to the right of the original grain boundary, whereas the nucleus in Fig. 8.21B has formed by grain boundary migration to the right and subgrain growth to the left. For both cases, the nucleus is surrounded by high-angle boundaries. For the nucleus shown in Fig. 8.21C, on the other hand, the subgrain growth to the left has taken place without forming a new high-angle boundary.

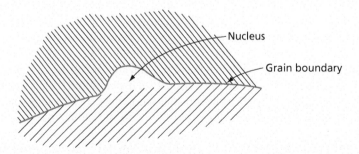

Fig. 8.20 The bulge mechanism for the formation of a nucleus at a grain boundary. (After Bailey, J. E., and Hirsh, P. B., *Proc. Roy Soc.*, **A267**, 11 (1962).)

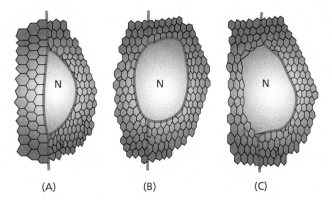

Fig. 8.21 Schematic drawing of grain boundary nuclei, marked by N, between two deformed grains with hexagonal subgrains. (From Ray, B. and Hansen, N., *Met. Trans. A*, **15A** 293 (1984) with kind permission of Springer Science and Business Media.)

8.15 Driving Force for Recrystallization

The driving force for recrystallization comes from the stored energy of cold work. In those cases where polygonization is essentially completed before the start of recrystallization, the stored energy can be assumed to be confined to the dislocations in polygon walls. The elimination of the subboundaries is a basic part of the recrystallization process.

8.16 The Recrystallized Grain Size

Another important factor in recrystallization studies is the recrystallized grain size. This is the crystal size immediately at the end of recrystallization, that is, before grain growth proper has had a chance to occur. Figure 8.22 shows that the recrystallized grain size depends upon the amount of deformation given to the specimens before annealing. The significant part of this curve is that the grain size grows rapidly with decreasing deformation. Too little deformation, however, will make recrystallization impossible in any reasonable length of time. This leads to the concept of the critical amount of cold work, which may be defined as the minimum amount of cold deformation that allows the specimen to recrystallize (within a reasonable time period). In Fig. 8.22 it corresponds to about 3 percent elongation of the polycrystalline brass in a tensile test. The critical deformation, like the recrystallization temperature, is not a property of a metal, since its value varies with the type of deformation (tension, torsion, compression, rolling, etc.). In single crystals of hexagonal metals, when deformation occurs by easy glide, the critical deformation may exceed several hundred percent. Twisting the same crystal to a few percent strain, however, may make it possible to recrystallize the specimen.

The concept of a critical deformation is important because the very large grain size associated with it is usually undesirable in metals that are to be further deformed. This is particularly true for sheet metal that is to be cold-formed into complicated shapes. If the grain size of a metal is very small (less than about 0.05 mm in diameter), plastic deformation occurs without appreciable

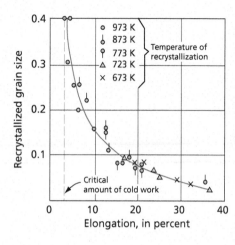

Fig. 8.22 Effect of prior cold work on the recrystallized grain size of alpha-brass. Notice that the grain size at the end of recrystallization does not depend on the temperature of recrystallization. (Smart, J. S., and Smith, A. A., *Trans. AIME*, **152** 103 [1943].)

roughing of the surface (assuming that deformation does not occur by movement of Lüders bands; see section 9.11). On the other hand, if the diameter of the average grain is large, cold working produces a roughened, objectionable surface. Such a phenomenon is frequently identified by the term *orange-peel effect* because of the similarity of the roughened surface to that of the peel of a common orange. The anisotropic nature of plastic strain inside crystals is directly responsible for the orange-peel effect, and the larger the crystals the more evident will be the nonhomogeneous nature of the deformation.

In metal that is cold-rolled (sheets) or drawn cold through dies (wires, rods, and pipe), it is relatively easy to avoid a critical amount of cold work because the metal is more or less uniformly deformed. On the other hand, if only a portion of a metallic object is deformed cold, a region containing a critical amount of cold work must exist between the worked and unworked areas. Annealing in this case can easily lead to a localized, very coarse-grain growth.

On the other hand, it should be noted that the concept of a critical strain can sometimes be most useful, as in the production of single crystals for use in experimental studies of basic crystal properties. This is also true for the production of bi-crystals and other large-grained specimens to be used in experimental investigations of grain boundaries.

The ratio of the rate of nucleation to the rate of growth, N to G, is frequently used in the interpretation of recrystallization data. If it is assumed that both N and G are constant or are average values for an isothermal recrystallization process, then the recrystallized grain size can be deduced from this ratio. If the ratio is high, many nuclei will form before the recrystallization process is completed, and a fine-grain size will result. On the other hand, a low ratio corresponds to a slow rate of nucleation relative to the rate of growth, and to a coarse crystal size in a recrystallized specimen. In Fig. 8.23, the rate of nucleation, N, the rate of growth, G, and the ratio of N to G are all plotted as a function of strain for a specific metal (aluminum). These curves show that, as the deformation before annealing is reduced to smaller and smaller values, the rate of nucleation falls much faster than the rate of growth. As a consequence, the ratio N to G decreases in magnitude with decreasing strain and, for the data of Fig. 8.23, is effectively zero at several percent elongation.

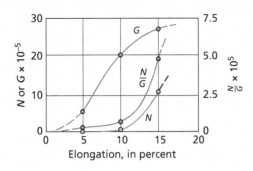

Fig. 8.23 Variation of the rate of nucleation (N), the rate of growth (G), and their ratio (N/G) as a function of deformation before annealing. (Data for aluminum annealed at 350°C.) (From Anderson, W. A., and Mehl, R. F., *Trans. AIME*, **161** 140 [1945].)

Thus it can be concluded that a critical amount of cold work corresponds to an amount just capable of forming the nuclei needed for recrystallization. This agrees well with the fact that nuclei form at points of high-strain energy in the lattice. The number of these points certainly should increase with the severity of the deformation and at low strains the number should almost vanish.

Another very important factor concerning the recrystallized grain size (at the end of recrystallization and before the beginning of grain growth) is apparent in Fig. 8.22: the recrystallized grain size in the case of the brass specimens used for the data of Fig. 8.22 is independent of the temperature of recrystallization. Notice that these data correspond to five different annealing temperatures and that all temperatures give values that fall on a single curve. This relationship also holds, within limits, for many other metals and to a first approximation we can assume that the grain size of a metal at the end of recrystallization is independent of the recrystallization temperature.

8.17 Other Variables in Recrystallization

It has been shown that the rate of recrystallization is dependent upon two variables: (1) temperature of annealing and (2) amount of deformation.

Similarly, it has been shown that, for many metals, the recrystallized grain size is independent of the annealing temperature, but sensitive to the amount of strain. The recrystallization process is also dependent upon several other variables. Two of the more important of these are (1) purity, or composition of the metal; and (2) the initial grain size (before deformation). These factors will be considered briefly.

8.18 Purity of the Metal

It is a well known fact that extremely pure metals have very rapid rates of recrystallization. This is apparent in the sharp dependence of the recrystallization temperature on the presence of solute. As little as 0.01 percent of a foreign atom in solid solution can raise the recrystallization temperature by several hundred degrees. Conversely, a spectroscopically pure metal recrystallizes in a fixed interval at much lower temperatures than a metal of commercial purity.

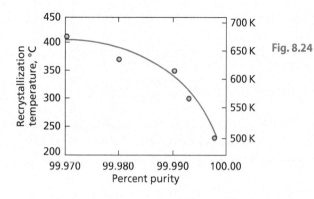

Fig. 8.24 Effect of impurities on the recrystallization temperature (30 minutes of annealing) of aluminum cold-rolled 80 percent. (From Perryman, E. C. W., ASM Seminar, *Creep and Recovery*, 1957, p. 111. Reprinted with permission of ASM International®. All rights reserved. www.asminternational.org)

The effect of solute atoms on the rate of recrystallization is most apparent at very small concentrations. This is shown clearly in Fig. 8.24 for aluminum of various degrees of purity. It has also been observed that the increase in the recrystallization temperature caused by the presence of foreign atoms depends markedly upon the nature of the solute atoms. Table 8.1 shows the increase in the recrystallization temperature corresponding to the addition of the same amount (0.01 atomic percent) of several different elements in pure copper.

The fact that very small numbers of solute atoms have such a pronounced effect on recrystallization rates is believed to indicate that the solute atoms interact with grain boundaries. The proposed interaction is similar to that between dislocations and solute atoms. When a foreign atom migrates to a grain boundary, both its elastic field, as well as that of the boundary, are lowered. In recrystallization, grain-boundary motion occurs as the nuclei form and grow. The presence of foreign atoms in atmospheres associated with these boundaries strongly retards their motion, and therefore lowers the recrystallization rates.

Table 8.1 Increase in the Recrystallization Temperature of Pure Copper by the Addition of 0.01 Atomic Percent of the Indicated Element.*

Added Element	Increase in Recrystallization Temperature K
Ni	0
Co	15
Fe	15
Ag	80
Sn	180
Te	240

*Data of Smart, J. S., and Smith, A. A., *Trans. AIME*, **147** 48 (1942); **166** 144 (1946).

8.19 Initial Grain Size

When a polycrystalline metal is cold worked, the grain boundaries act to interrupt the slip processes that occur in the crystals. As a consequence, the lattice adjacent to the grain boundaries is, on the average, much more distorted than in the center of the grains. Decreasing the grain size increases the grain-boundary area and, as a consequence, the volume and uniformity of distorted metal (that adjacent to the boundaries). This effect increases the number of possible sites of nucleation and, therefore, the smaller the grains of the metal before cold work, the greater will be the rate of nucleation and the smaller the recrystallized grain size for a given degree of deformation.

8.20 Grain Growth

It is now generally recognized that in a completely recrystallized metal, the driving force for grain growth lies in the surface energy of the grain boundaries. As the grains grow in size and their numbers decrease, the grain boundary area diminishes and the total surface energy is lowered accordingly. The growth of cells in a foam of soap also occurs as a result of a decrease in surface energy: the surface energy of the soap film. Because a number of complicating factors that influence the growth of metal crystals do not apply in the case of soap films, the growth of soap bubbles may be taken as a rather ideal case of cellular growth. For this reason, the growth of soap cells will be considered before the more complicated case of metallic grain growth.

First, consider a single spherical soap bubble. The gas enclosed by the soap film is always at a greater pressure than that on the outside of the bubble because of the surface tensions in the soap film. In an elementary physics course it is shown that this pressure difference between the inside and outside of the soap bubble can be expressed by the simple equation

$$\Delta p = \frac{4\gamma}{R} = \frac{8\gamma}{D} \qquad \text{8.7}$$

where γ is the surface tension of one surface of the film (soap films have two surfaces), R is the radius of the soap bubble, and D is its diameter. This equation shows clearly that the smaller the bubble, the greater the excess pressure inside the bubble.

Because of the pressure difference that exists across a curved soap film, gaseous diffusion occurs; the net flow occurring through the film from the high- to the low-pressure side. In other words, the atoms diffuse from the inside to the outside of the bubble, resulting in a decrease in size of the bubble and a movement of its walls inward toward their center of curvature.

The above discussion can now be carried over to the more general example, a soap froth. In the froth, the cells contain curved walls, and the curvature varies from cell to cell and within the film surrounding any one cell, depending on the relative size and shape of the neighboring cells. (These factors will be discussed in more detail presently.) In all cases, however, a pressure difference exists across each curved wall, with the greater pressure on the concave side. Gas diffusion resulting from this pressure difference, in turn, causes the walls to move but always in a direction toward their center of curvature.

Let us find out how the movement of cell walls causes the cells in the soap froth to grow in size. For the sake of simplicity, consider a network of two-dimensional cells, those with walls perpendicular

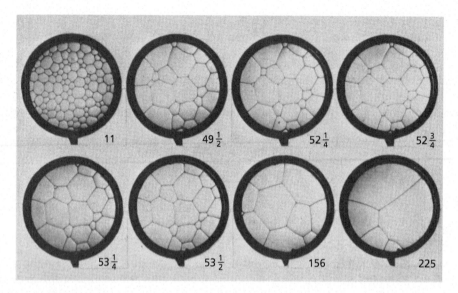

Fig. 8.25 Growth of soap cells in a flat container. (From Smith, C. S., ASM Seminar, *Metal Interfaces*, 1952, p. 65. Reprinted with permission of ASM International®. All rights reserved. www.asminternational.org)

to the plane of view. A froth of this type can be formed between two closely spaced parallel plates of glass. This simplification greatly reduces geometrical complexity, and still permits the more important principles of cellular growth to be observed.

Figure 8.25 presents a sequence of pictures from the work of C. S. Smith, showing the growth of the cells in a two-dimensional soap froth formed in a small flat glass cell. The figure at the lower right-hand corner of each photograph represents the number of minutes from the time that agitation of the cell to form the froth was ended. It also represents the time during which cell growth has taken place. In several of the photographs, small, three-sided cells can be observed. In the third photograph from the left in the upper row one appears at approximately 9 o'clock, and another at about 10 o'clock in the first photograph from the left in the lower row. An enlarged sketch of one of these cells is shown in Fig. 8.26. Notice that in order to maintain the equilibrium angle of 120° required when three surfaces with equal surface tensions meet at a common junction, the cell walls of the three-sided cell have been forced to assume a rather pronounced curvature. Since this curvature is concave toward the center of the cell, the wall can be expected to migrate, thus decreasing the volume of the cell and causing it to

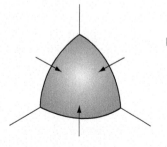

Fig. 8.26 Sketch of a three-sided soap cell. Notice the pronounced curvature of the boundaries, which is concave toward the center of the cell. Arrows show direction of cell boundary movement

disappear completely. That this actually occurs can be seen by studying the photographs just to the right of the two photographs mentioned previously. In each case the triangular grains are no longer visible.

Further study of the photographs of Fig. 8.25 reveals that cells with less than six sides have walls that are primarily concave toward their centers. Those cells with more than six sides have walls convex toward their centers; this effect is more pronounced the larger the number of sides above six. This confirms the fact that the only two-dimensional geometrical figure formed by straight lines that can have an average internal angle of 120° is the hexagon. In a two-dimensional structure, such as that shown in the photographs of Fig. 8.25, all cells with less than six sides are basically unstable and tend to shrink in size, while those with more than six sides tend to grow in size. Another interesting fact is that there is a definite correspondence between the size of the cells and the number of sides that they contain. The smaller cells usually have the fewest number of sides. It is no wonder that three-sided cells disappear so rapidly, for both their small size and minimum number of sides require that their walls have very large curvatures, with accompanying high-pressure differentials, diffusion rates, and rates of wall migration. The photographs show that, in general, four- and five-sided cells do not disappear as a unit, but first change to three-sided cells which rapidly disappear.

Another important aspect of cellular growth can be seen in the photographs of Fig. 8.25. Over a period of time the number of sides that any given grain possesses continually changes. The number of sides may increase or decrease, as can be seen by considering the mechanism of Burke and Turnbull (Ref. 14) illustrated in Fig. 8.27. Because of the curvature of the boundaries that separate cells B and D from A and C, respectively, the boundaries migrate, thus eliminating the boundary between cells B and D, and then they create a new boundary between A and C. These steps are indicated in Figs. 8.27B and 8.27C, respectively. As a consequence of this process, cells B and D each lose a side, while cells A and C each gain a side. Another method by which the number of sides of a cell can be changed has already been described. Each time a three-sided cell disappears, each of its neighboring cells loses one side.

8.21 Geometrical Coalescence

Nielsen (Ref. 15) has proposed that the geometrical coalescence of grains is an important phenomenon in recovery, recrystallization, and grain growth. Subgrain coalescence has already been discussed in relation to the formation of nuclei during recrystallization. The mechanism that Nielsen favors does not require that the subgrains or grains rotate through an appreciable angle relative

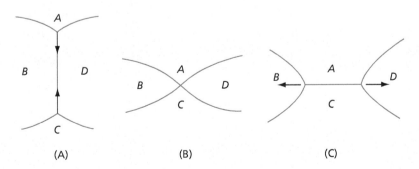

Fig. 8.27 A mechanism that changes the number of sides of a grain during grain growth

to each other. Geometrical coalescence can be simply described as an encounter of two grains whose relative orientations are such that the boundary formed between the two grains is one of much lower surface energy, γ_G, than that of the average boundary. In a polycrystalline metal, such a boundary would be equivalent to a subgrain boundary. The effect of producing this type of boundary on the microstructure is indicated schematically in Fig. 8.28. First, let us imagine that grains A and B encounter each other during a process of grain growth in a metal where it is assumed that the boundaries have the two-dimensional character of the soap froth in a flat cell. Before the encounter, the two grains are separated, as shown in Fig. 8.28B. If the boundary that is produced when they meet is a typical high-angle boundary, the grain boundary surface energy γ_G will be effectively the same as that of the other boundaries, and a boundary configuration such as that in Fig. 8.28C is expected. On the other hand, if a very low-energy boundary is formed, then, because of the low value of γ_G in the boundary ab, the effective surface tension forces along ab will be very small and the grain boundary configuration (shown in Fig. 8.28D) should result. From this illustration it can be seen that a geometrical coalescence should result in the sudden development of a much larger grain. Note that in this two-dimensional example, this large grain has nine sides. Grains formed by geometrical coalescence should thus have the possibility of continued rapid growth.

Geometrical coalescence, if it occurs to any extent, should have a strong effect on grain-growth kinetics. In a metal containing a more or less random set of crystal orientations, geometrical coalescence should probably occur infrequently. On the other hand, geometrical coalescence may be an important phenomenon in a highly textured metal: one that has a strong preferred orientation. In this case the chances of two grains of nearly identical orientation encountering each other during grain growth is certainly much greater. It is also much greater in the case of

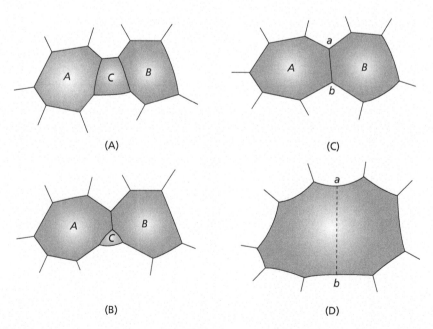

Fig. 8.28 Geometrical coalescence. Two grains, A and B, encounter as a result of the disappearance of grain C. If grains A and B have nearly identical orientation, then boundary ab becomes the equivalent of a subboundary and grains A and B may be considered the equivalent of a single grain

subgrain growth during recrystallization. It should be noted that, in addition to grain boundary curvatune grain growth and geometrical coalescence, grain rotation may also contribute to the coalescence of neighbouring grains. (Ref. 16)

8.22 Three-Dimensional Changes in Grain Geometry

In a metal, the grains are not two-dimensional in character, but three-dimensional. The five basic mechanisms by which the geometrical properties of the three-dimensional grains can change have been outlined by Rhines. (Ref. 17) A brief summary of these mechanisms due to DeHoff is shown in Fig. 8.29. The analog of the three-sided two-dimensional grain of Fig. 8.26 is shown in Fig. 8.29A. It is a four-sided or tetrahedral grain. Its disappearance results in the loss of four grain boundaries. Figure 8.29B is the corresponding three-dimensional analog of the mechanism illustrated in Fig. 8.27. Note that the grain boundary BD in Fig. 8.27 becomes a three-grain juncture or triple line in the three-dimensional example. If the upper and lower grains should come together, this line is removed and replaced by a horizontal grain boundary lying between the upper and lower grains. The result is a gain of one grain boundary. The third case, in Fig. 8.29C, is simply the inverse of that which has just been discussed.

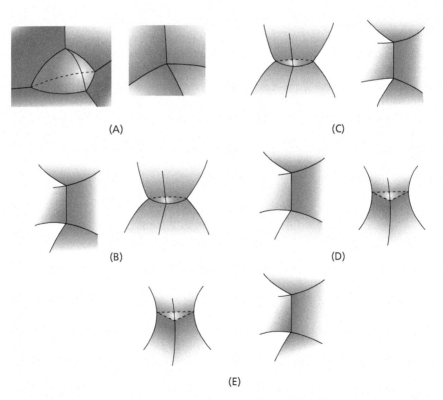

Fig. 8.29 The five basic three-dimensional geometrical processes in grain growth. (After Rhines, F. N., and DeHoff, R. T.)

An interesting mechanism is shown in 8.29D. This is the three-dimensional example of the geometrical coalescence of grains discussed in the preceding section. In this case, the upper and lower grains are again assumed to be able to approach each other as in the case of Fig. 8.29B, but in this case the boundary formed between the two grains is a low-energy boundary. The result is an effective coalescence of the upper and lower grains. Finally, Fig. 8.29E shows the inverse of geometrical coalescence. Here one grain necks and separates into two grains.

8.23 The Grain Growth Law

Another interesting aspect of the grain growth pictures in Fig. 8.25 is that, while the number of cells keeps decreasing as time goes on, the cellular arrays are geometrically similar at all times. This would be even more evident if a larger sample of soap froth were available for our study. At any rate, it is apparent that at any given instant the cells vary in size about a mean and that this mean size grows with time. The mean cell diameter serves as a convenient measure of the cell size of an aggregate. Therefore, when one refers to the cell size of a froth, it is the diameter of the average cell that is meant. This statement also holds true for metals where the commonly used term "grain size" refers, in general, to the mean diameter of an aggregate of grains. It follows that grain growth, or cellular growth, refers to the growth of the average diameter of the aggregate. Attention is called, however, to the difficulties involved in accurately defining this concept of an average grain size in a metal, as pointed out in Chapter 6.

Let us now derive an expression for a soap froth that relates the size of the average cell to the time. In this derivation we shall not limit ourselves to the two-dimensional case, but shall consider that we are working in three dimensions. We shall first assume that the rate of growth of the cells is proportional to the curvature of the cell walls of the average cell at a given instant of time. This is in agreement with the fact that boundaries move as a result of gaseous diffusion caused by a pressure difference from one side of a soap film to the other, that is proportional to the curvature of the boundary. If the symbol D represents the mean diameter of the average-sized soap cell, and c the curvature of the cell walls, we have

$$\frac{dD}{dt} = K'c \qquad \textbf{8.8}$$

where t is the time, and K' a constant of proportionality. Let us also assume that the curvature of the average-sized cell is inversely proportional to its diameter, and rewrite Eq. 8.8 in the following form:

$$\frac{dD}{dt} = \frac{K}{D} \qquad \textbf{8.9}$$

where K is another constant of proportionality. Integration of this equation leads to the following result:

$$D^2 = Kt + c$$

Assuming that D_0 is the size of the average cell at the start of the observation ($t = 0$), an evaluation of the constant of integration gives

$$D^2 = D_0^2 + Kt \qquad \textbf{8.10}$$

Although several broad assumptions have been made in deriving Eq. 8.10, experimental measurements of the growth of cells in a soap froth have shown that this expression fits the observed data quite closely. (Ref. 18) It can therefore be concluded that Eq. 8.10 is essentially correct for the growth of soap cells under the action of surface-tension forces.

If it is assumed that the grain size is very small at the beginning of cell growth, then it is possible to neglect D_0^2 in relation to D^2, with the result that the equation relating the cell size to the time can be expressed in the somewhat simpler form

$$D^2 = Kt$$

or

$$D = kt^{1/2} \qquad 8.11$$

where $k = \sqrt{K}$. According to this relationship, the mean diameter of the cells in a soap froth grows as the square root of the time. Figure 8.30 shows a sketch of this relationship—clearly, as time progresses, the rate of cellular growth diminishes.

Let us consider the case of grain growth in metals. Here, as in soap froth, the driving force for the reaction lies in the surface energy of the grain boundaries. Grain-boundary movements in metals are, in many respects, perfectly analogous to those of the cell walls of soap froth. In both cases, the boundaries move toward their centers of curvature, and the rate of this movement varies with the amount of curvature. On the other hand, while it is known that the growth of soap cells can be explained in terms of a simple diffusion of gas molecules through the walls of the soap film, less is known about the mechanism by which atoms on one side of a grain boundary cross the boundary and join the crystal on the other side.

It was originally proposed (Ref. 19) and generally accepted that the boundary atoms in the crystal on the concave side of the boundary are more tightly bound than the boundary atoms in the crystal in the convex side, because they are more nearly surrounded by neighboring atoms of the same crystal. This tighter binding of the atoms on the concave side of the boundary should make the rate at which atoms jump across the boundary from the convex to the concave crystal greater than that in the opposite direction. The greater the curvature of the boundary, then the greater should be this effect, and the faster the movement of the crystal boundary. However, because detailed knowledge of the structure of metallic grain boundaries is lacking, the exact nature of the transfer mechanism by which atoms cross a boundary is still not known, and it is not possible to explain quantitatively the apparently irrational results obtained when studying the growth of crystals in a metal.

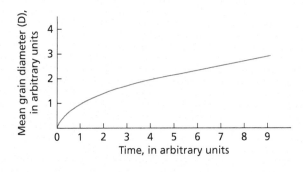

Fig. 8.30 Graphical representation of the ideal grain growth law, $D = kt^{1/2}$

While a quantitative theory capable of explaining grain growth in pure metals and alloys is lacking, much is known about the causes for the apparent abnormal nature of metallic crystal growth. A number of these will now be considered, but first let us reconsider some of the factors that have been brought out in the study of soap froths. If metallic grain growth is assumed to occur as a result of surface-energy considerations and the diffusion of atoms across a grain boundary, then it is to be expected that at any given temperature a grain-growth law of the same form as that found in a soap froth might be observed, namely,

$$D^2 - D_0^2 = Kt \qquad 8.12$$

Also, if the diffusion of atoms across a grain boundary is considered to be an activated process, then it can be shown that the constant K in Eq. 8.12 can be replaced by the expression

$$K = K_0 e^{-Q/RT}$$

where Q is an empirical heat of activation for the process, T is the temperature in degrees Kelvin, and R is the international gas constant. The grain-growth law can therefore be written as a function of both temperature and time in the following manner:

$$D^2 - D_0^2 = K_0 t e^{-Q/RT} \qquad 8.13$$

Most early experimental studies of metallic growth have failed to confirm the above relationship. However, some work has given results that conform quite closely to Eq. 8.13. Figure 8.31 shows part of these results for brass (10 percent zinc and 90 percent copper). Note that in each case, straight lines are obtained when D^2 is plotted against the time. Let us now rearrange Eq. 8.13 as follows:

$$\frac{D^2 - D_0^2}{t} = K_0 e^{-Q/RT}$$

Taking logarithms of both sides of this equation gives

$$\log \frac{D^2 - D_0^2}{t} = \frac{Q}{2.3\, RT} + \log K_0 \qquad 8.14$$

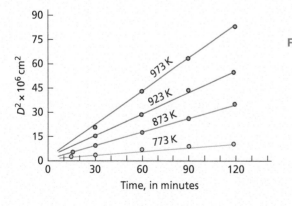

Fig. 8.31 Grain-growth isotherms for alpha-brass (10 percent Zn–90 percent Cu). Notice that the grain diameter squared (D^2) varies directly as the time. (Reprinted from Feltham, P., and Copley, G.J., "Grain-growth in a brasses", *Acta Met.*, **6** 539 (1958) with permission from Elsevier. www.sciencedirect.com/science/journal/00016160)

This relationship tells us that the quantity $\log(D^2 - D_0^2)/t$ should vary directly as the reciprocal of the absolute temperature $(1/T)$ and that the slope of this linear relationship is $Q/2.3R$. Now, the quantity $(D^2 - D_0^2)/t$ equals the slope of a grain-growth isotherm, such as those shown in Fig. 8.32. This figure shows the logarithm of the slopes of the four lines in Fig. 8.31 ($\log[D^2 - D_0^2]/t$) plotted as functions of the reciprocal of the absolute temperature $(1/T)$. The data in question give an excellent straight line from which it can be deduced that the activation energy Q for grain growth in alpha-brass containing 10 percent Zn is 73.6 KJ per gm atom.

Many of the experimental isothermal grain-growth data correspond to empirical equations of the form

$$D - D_0 = kt^n \qquad 8.15$$

where the exponent n is, in most cases, smaller than the value $\frac{1}{2}$ predicted by the grain-growth equation. Frequently a value near 0.3 is found. As discussed below, the rate equation can also be affected by the presence of impurities and foreign particles. Furthermore, the exponent n is not usually constant for a given metal or alloy if the isothermal reaction temperature is changed. As a general rule, the exponent increases with increasing temperature and approaches the value $\frac{1}{2}$.

Finally, it should be emphasized that while Eq. 8.15 applies to the average grain size, it provides no information on the growth of individual grains when the metal contain different size grains. In fact, almost all deformed and recrystallized metals have a grain-size distribution whose spread depends on the amount of deformation prior to annealing. This is nicely documented by the work of Rhines and Patterson (Ref. 20) on the grain-size distribution of recrystallized aluminum. They show that the grain volume or weight of recrystallized aluminum has a distribution that is log-normal and that the spread of the distribution is determined by the amount of prior deformation. The grain-size distribution is broadest after small deformation. Moreover, Rhines and Patterson show that the grain distribution retains its form with increased time after recrystallization, implying that grain growth is more rapid and remains so after small deformation. The authors assert that the metal retains memory of the degree of the prior cold work.

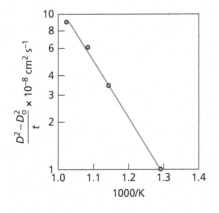

Fig. 8.32 The logarithms of the slope of the isotherms of Fig. 8.31 vary directly as the reciprocal of the absolute temperature. (Reprinted from Feltham, P., and Copley, G. J., "Grain-growth in a brasses," *Acta Met.*, **6** 539 (1958) with permission from Elsevier. www.sciencedirect.com/science/journal/00016160)

8.24 Impurity Atoms in Solid Solution

Of considerable importance in grain growth is that foreign atoms in a lattice can interact with grain boundaries. This interaction is analogous to the interaction between impurity atoms and dislocations, previously mentioned as a factor in recrystallization (that is, nucleus growth). If the size of a foreign atom and that of the parent crystal are different, then there will be an elastic stress field introduced into the lattice by each foreign atom. However, since grain boundaries are regions of lattice misfit, the strain energy of the boundary, as well as that of the lattice surrounding a foreign atom, can be reduced by the migration of the foreign atom to the neighborhood of the grain boundary. In this manner, we can conceive of grain-boundary atmospheres just as we can dislocation atmospheres. That these atmospheres can effectively hinder the motion of grain boundaries has been verified experimentally. Figure 8.33 shows the grain-growth exponent as a function of impurity content for copper containing small percentages of aluminum in solid solution. Notice that as the metal approaches 100-percent purity, the grain-growth exponent increases toward the theoretical value $\frac{1}{2}$. Also, note that at the higher temperature (900 K) the rate of approach is faster. This temperature dependence of the grain-growth exponent can be explained by assuming that the grain-boundary solute atmospheres are broken up by thermal vibrations at high temperatures.

The effect of solutes in retarding grain growth varies with the element concerned. While solid solutions of zinc in copper may behave normally (provided other impurities are eliminated) with respect to the grain-growth law ($n \simeq \frac{1}{2}$), even as small a quantity as 0.01 percent of oxygen will effectively retard grain growth in copper. This difference is probably due to the different magnitudes of strain that various elements produce in the lattice: those elements that distort the lattice structure the most have the largest effect on the rate of grain growth.

8.25 Impurities in the Form of Inclusions

Solute atoms not in solid solution are also capable of interacting with grain boundaries. It has been known for more than sixty years that impurity atoms in the form of second-phase inclusions or particles can inhibit grain growth in metals. The inclusions referred to are frequently found in commercial alloys and so-called commercially pure metals. For the most part, they may consist of very small oxide, sulfide, or silicate particles that were incorporated in the metal during its manufacture. In the present discussion, however, they may be considered to be any second-phase particle that is finely divided and distributed throughout the metal.

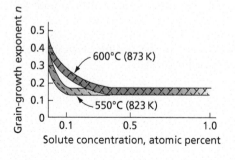

Fig. 8.33 Variation of the grain-growth exponent of copper with the concentration of aluminum in solid solution. (From Weinig, S., and Machlin, E. S., *Trans. AIME*, **209** 843 [1957].)

Let us consider a simplified theory of the interaction between inclusions and grain boundaries, due to Zener. Figure 8.34A is a schematic sketch of an inclusion located in a grain boundary represented as a vertical straight line. For convenience, the particle has been made spherical. As shown in the left-hand sketch, the inclusion and boundary are in a position of mechanical equilibrium. If the boundary is moved to the right, as indicated in Fig. 8.34B, the grain boundary assumes a curved shape in which the boundary (because of its surface tension) strives to maintain itself normal to the surface of the particle. The vectors marked σ in the figure indicate the direction and magnitude of the surface-tension stress at the circular line of contact (in three dimensions) between the grain boundary and the surface of the inclusion. The total length of the line of contact is $2\pi r \cos\theta$, where r is the radius of the spherical particle, and θ is the angle between the equilibrium position of the boundary and the vector σ. The product of the horizontal component of this vector $\sigma \sin\theta$ and the length of the line of contact (between particle and boundary) gives the pull of the boundary on the particle:

$$f = 2\pi r\sigma \cos\theta \sin\theta \qquad 8.16$$

This force, by Newton's Second Law, is also the drag of the particle on the grain boundary; it is a maximum when θ is 45°. Substituting this value gives for the maximum force

$$f = \pi r\sigma \qquad 8.17$$

Notice that the drag of a single particle varies directly as the radius of the particle. Since the volume of each particle varies as the cube of its radius, the effect of second-phase inclusions in hindering grain boundary motion will be greater the smaller and more abundant are the particles (assuming particles of the same shape).

In many cases, second-phase particles tend to dissolve at high temperatures. It is also generally true that second-phase particles tend to coalesce at high temperatures and form fewer large particles. Both of these effects—the decrease in amount of the second phase and the tendency to form larger particles—remove the retarding effect of the inclusions on grain growth in metals. An

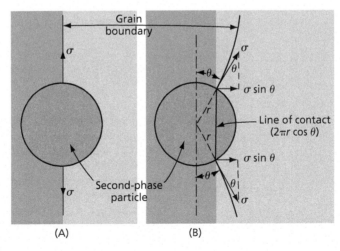

Fig. 8.34 Interaction between a grain boundary and a second-phase inclusion

excellent example is shown in Fig. 8.35 where grain-growth data are plotted on log-log coordinates. A plot of this type for data that conform to an equation of the form

$$D = k(t)^n \qquad 8.18$$

gives a straight line, the slope of which is the grain-growth exponent n. In this alloy of aluminum containing 1.1 percent manganese, a second phase ($MnAl_6$) exists up to a temperature of 898 K, but dissolves at temperatures above 923 K, so that only a single solid solution remains. The effect of the inclusions on the grain growth is evident. At 898 K and below, grain growth is almost completely stopped (n is of the order of 0.02); on the other hand, at 923 K, grain growth occurs very rapidly, with a grain-growth exponent (0.42) close to the theoretical 0.5. The curved upper part of the curve, corresponding to the 923 K data, is the result of a geometrical effect that will be explained in the next section.

Holes or pores in a metal can have the same effect on grain-boundary motion as second-phase inclusions. This is clearly shown in the photograph of Fig. 8.36 where the cross-section of a small metal magnet made of Remalloy (12 percent Co, 17 percent Mo, and 71 percent Fe) can be seen. This specimen was made by the powder metallurgy technique involving the heating of a compressed metal compact to a temperature below the melting point of the metals contained in the compact, but sufficiently high to weld and diffuse the particles together. The sample presented in Fig. 8.36 has a structure consisting of a single homogeneous solid solution. A number of pores are clearly visible in the photograph, and at a number of them it can be seen that crystal boundaries have been held up by the pores. As an example, the nearly horizontal boundary of the very large grain that occupies the upper part of the picture passes through three pores. The motion of this boundary was toward the bottom. Observe how the boundary is curved back toward the pore in each case.

The effect of impurities on grain growth can be summarized as follows. Solute atoms in solid solution can form grain-boundary atmospheres, the presence of which retards the normal surface-tension induced boundary motion. In order for the boundary to move, it must carry its atmosphere along with it. On the other hand, solute atoms in the form of second-phase impurities can also interact with the boundaries. In this case, the boundary must pull itself through the inclusions that lie in its path. In either case, an increase in temperature lowers the retarding effect of the solute atoms and grain growth occurs under conditions more closely resembling the growth of soap cells.

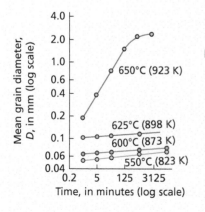

Fig. 8.35 The effect of second-phase inclusions on grain growth in a manganese-aluminum alloy (1.1 percent Mn). Grain growth is severely inhibited at the temperatures below 923 K because of the presence of second-phase precipitate particles ($MnAl_6$). (From Beck, P. A., Holzworth, M. L., and Sperry, P., *Trans. AIME*, **180** 163 [1949].)

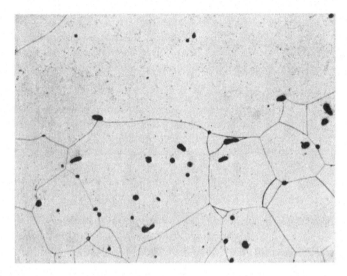

Fig. 8.36 Interaction between pores and grain boundaries

8.26 The Free-Surface Effects

Specimen geometry may play a part in controlling the rate of grain growth. Grain boundaries near any free surface of a metal specimen tend to lie perpendicular to the specimen surface, which has the effect of reducing the net curvature of the boundaries next to the surface. This means that the curvature becomes cylindrical rather than spherical and, in general, cylindrical surfaces move at a slower rate than spherical surfaces with the same radius of curvature. The reason for this difference may be understood in terms of the soap-bubble analogy where it is easily shown that the pressure difference across a cylindrical surface is $2\gamma/R$ and across a spherical surface $4\gamma/R$, where γ is the surface tension of a soap film and R the radius of the curvature.

Mullins (Ref. 21) has pointed out the importance of another phenomenon associated with grain boundaries that meet a free surface. It probably is of greater importance in its effect on grain growth than the reduction of the curvature of boundaries at a surface. The phenomenon in question is *thermal grooving*. At the high temperatures normally associated with annealing, grooves may form on the surfaces where grain boundaries intersect the specimen surface. These channels are a direct result of surface-tension factors. (See Fig. 8.37.) Point *a* in this figure represents the line where three surfaces meet: the grain boundary and the free surfaces to the right and left of point

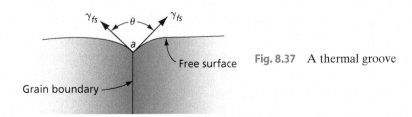

Fig. 8.37 A thermal groove

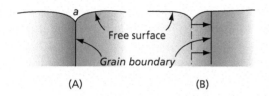

Fig. 8.38 Moving a grain boundary away from its groove increases its surface if the boundary is nearly normal to the free surface

a, respectively. In order to balance the vertical components of the surface tensions in the three surfaces that meet at this line, a groove must form with a dihedral angle θ that satisfies the relation

$$\gamma_b = 2\gamma_{f.s.} \cos \frac{\theta}{2} \qquad 8.19$$

where γ_b is the surface tension of the grain boundary, and $\gamma_{f.s.}$ that of the two free surfaces. Diffusion of atoms over the free surfaces is believed to be the most important factor involved in the transport of atoms out of the grooved regions during the formation of the grooves.

Grain-boundary grooves are important in grain growth because they tend to anchor the ends of the grain boundaries (where they meet the surface), especially if the boundaries are nearly normal to the surface. This anchoring effect can be explained in a very qualitative fashion with the aid of Fig. 8.38. The left-hand figure represents a boundary attached to its groove, while the right-hand figure shows the same boundary moved to the right and freed from its groove. Freeing this boundary from its groove increases the total surface area and, therefore, the total surface energy. To free the boundary from its groove requires work and, as a result, the groove restrains the movement of the boundary.

When the average grain size of a metal specimen is very small compared to the dimensions of the specimen, thermal grooving, or the lack of curvature in the surface grains, has little effect on the overall rate of growth. However, when the grain size approaches the dimensions of the thickness of the specimen, it can be expected that grain-growth rates will be decreased. In this respect, it has been estimated (Ref. 22) from experimental data that when the grain size of metal sheets becomes larger than one-tenth the thickness, the growth rate decreases. It is just this effect that explains the divergence of the curve for the 923 K data in Fig. 8.35 from a straight line at times greater than 625 min. The size of the specimens used in these experiments was small. When the average grain size became larger than approximately 1.8 mm, grain growth was retarded because of the free-surface effect.

8.27 The Limiting Grain Size

In the preceding section it was pointed out that the specimen dimensions can influence the rate of grain growth when the average crystal size approaches the thickness of the specimen. In many cases, this situation can have the effect of putting a practical upper limit on the grain size; that is, the growth may be slowed down to the point where it appears that no further growth is possible. In extreme cases, this free-surface effect can completely stop grain growth. Consider the case of a wire in which the grains have become so large that their boundaries cross the crystal in the manner shown in Fig. 8.39. Boundaries of this nature have no curvature and cannot migrate under the action of surface-tension forces. Further grain growth is then not possible.

Fig. 8.39 One example of a stable grain-boundary configuration

Second-phase inclusions are also known to put an upper limit on the grain size of a metal, or matrix in the case of composite. (Ref. 23) Here it can be considered that the boundary has trapped so many inclusions that the surface-tension force, which is small due to its lack of sufficient curvature, cannot overcome the restraining force of the inclusions.

Metal grain boundaries differ from soap films for they possess only a single surface, whereas the latter have two surfaces. With this in mind, the driving force per unit area for grain-boundary movement (Δp in the analogous soap bubble case) can be written in the following manner:

$$\eta = 2\frac{\gamma}{R} \qquad 8.20$$

where η is the force per unit area, γ the surface tension of the grain boundary, and R the net radius of curvature of the boundary. At the point where the grain boundary is no longer able to pull itself away from its inclusions, the restraining force of the inclusions must equal the above force. This restraining force is equal to that of a single particle times the number of particles per unit area, or

$$\eta = 2\frac{\gamma}{R} = n_s \pi r \gamma \qquad 8.21$$

where n_s is the number of inclusions per unit area and r the radius of the particles. Let us assume that the inclusions are uniformly distributed throughout the metal, then the approximate number of these inclusions that can be expected to be holding back a surface of area A are those whose centers lie inside a volume with an area A and whose thickness equals twice the radius of the particles. This volume $2Ar$ will hold $2n_v Ar$ particles, where n_v is the number of particles per unit volume, and it is concluded that the number of particles per unit area equals $2n_v r$. The number of particles per unit volume, n_v, may be expressed:

$$n_v = \frac{\zeta}{\frac{4}{3}\pi r^3} \qquad 8.22$$

where ζ is the volume fraction of the second phase, and $\frac{4}{3}\pi r^3$, the volume of a single particle. If these last two quantities are substituted into the equation that relates the driving force for boundary movement to the retarding force of the inclusions, Zener's relationship is obtained:

$$\frac{r}{R} = \frac{3}{4}\zeta$$

or

$$R = \frac{4}{3}\frac{r}{\zeta} \qquad 8.23$$

where r is the radius of the inclusions, R the radius of curvature of the average grain, and ζ the volume fraction of inclusions. This relationship assumes spherical particles and a uniform distribution of particles, neither of which can be realized in an actual metal. Nevertheless, it gives us a first approximation of the effect of inclusions on grain growth. Since we can assume that the radius of curvature is directly proportional to the average grain size, this equation shows that the ultimate grain size to be expected in the presence of inclusions is directly dependent on the size of the inclusions.

8.28 Preferred Orientation

Other factors in addition to grain-boundary atmospheres, second-phase inclusions, and the free-surface effects are known to affect the measured rate of grain growth. Among these is a preferred orientation of the crystal structure. By a preferred orientation one signifies a nearly identical orientation in all the crystals of a given sample of metal. When this situation occurs, it has been generally observed that grain-growth rates are reduced.

8.29 Secondary Recrystallization

In a previous section it was pointed out that it is frequently possible to obtain a limiting grain size in a metal. When this grain-growth inhibition occurs as a result of the presence of inclusions, the size effects, or from the development of a strong preferred orientation, a secondary recrystallization is often possible. This secondary recrystallization behaves in much the same manner as the primary one and is usually induced by raising the annealing temperature above that at which the original grain growth occurred. After a nucleation period, some of the crystals start to grow at the expense of their neighbors. The causes for the nucleation are not entirely clear but grain coalescence could well produce this effect. It is relatively easy to understand the growth that occurs once it starts, for the enlarged crystals quickly become grains with a large number of sides as they consume their smaller neighbors. As mentioned earlier, grains with many sides possess boundaries that are concave away from their centers and, consequently, the grains become larger in size. Secondary recrystallization is really a case of exaggerated grain growth occurring as a result of surface-energy considerations, rather than as a result of the strain energy of cold work that is responsible for primary recrystallization.

Figure 8.40 is a good illustration of secondary recrystallization. In the center of the photograph, one sees a large grain with 13 sides, all of which are concave away from the center of the crystal. The specimen shown in the figure is the same as that of Fig. 8.36. The latter also shows a large grain growing at the expense of its neighbors (that occupying the top of the figure).

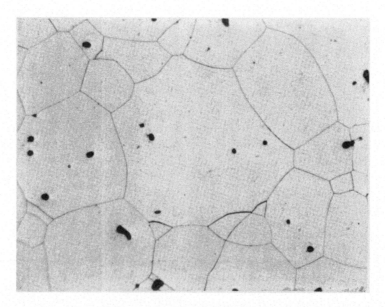

Fig. 8.40 A specimen undergoing secondary recrystallization. Notice that the central grain has thirteen sides

Very large grains, or even single crystals, can sometimes be grown by secondary recrystallization because the number of grains that finally results depends only on the number of secondary nuclei. Growth-inhibiting factors that control the grain growth after primary recrystallization, such as inclusions and free surfaces, do not control the growth of grains in secondary recrystallization. In the latter case, as in primary recrystallization, the nuclei grow until the matrix is completely recrystallized.

8.30 Strain-Induced Boundary Migration

While it is generally conceded that normal grain growth occurs as a result of the surface energy stored in the grain boundaries, it is also possible for crystals to grow as a result of strain energy induced in the lattice by cold work. Strain-induced boundary migration should not be expected in a metal after complete recrystallization unless it has been distorted either by handling after recrystallization, or by residual strains caused by uneven rates of recrystallization in different parts of the specimen.

Strain-induced boundary movements differ from recrystallization in that no new crystals are formed. Rather, the boundaries between pairs of grains move so as to increase the size of one grain of a pair while causing the other to disappear. As the movement occurs, the boundary leaves behind a crystalline region which is lower in its strain energy. In contrast to surface-tension-induced grain-boundary migration, this form of boundary movement occurs in such a manner that the boundary usually moves away from its center of curvature. This movement is shown schematically in Fig. 8.41, where the moving grain boundary is shown with an irregular curved shape. The irregular form of a boundary that grows as a consequence of strain energy can be explained by assuming that the rate of motion is a function of the strain magnitude in the metal. The boundary should move faster into those regions where the distortion has been greatest. One of the interesting aspects of strain-induced boundary migrations is that instead of the boundary of the moving grain lowering its surface energy through the movement, it may actually increase it by increasing its area.

Strain-induced migration of boundaries only occurs after relatively small or moderate amounts of cold work. Too great a degree of deformation will bring about normal recrystallization. On the other hand, this form of boundary movement can be induced in sheet specimens where the normal grain growth has been inhibited by the size effects previously discussed.

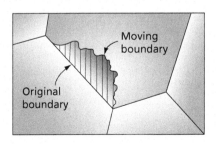

Fig. 8.41 Schematic representation of strain-induced boundary migration. In this case, the boundary moves away from its center of curvature, which is in the opposite direction to the movement in surface-tension-induced boundary migration. (After Beck, P. A., and Sperry, P. R., *Jour. Appl. Phys*, **21** 150 [1950].)

Problems

8.1 In Sec. 8.1, it was stated that a high-purity Cu specimen, after 30 percent deformation at room temperature, was found to have a recoverable strain energy of about 25 J/g-mol. The Burgers vector for Cu is 0.256 nm and the shear modulus about 5.46×10^{10} Pa. Determine the dislocation density of an array of uniformly distributed alternating right- and left-hand screw dislocations in the Cu that might possess this amount of strain energy.

8.2 (a) Using the 283 K data, given in Sec. 8.5 for the Drouard, Washburn and Parker zinc single crystal, determine the value of A in the rate equation, Eq. 8.2.

(b) Next determine the temperature at which the crystal should recover one-fourth of its yield point in 5 seconds and $b = 3.0$ nm.

8.3 A simple equation can be easily derived that relates the density of the excess edge dislocations in a bent crystal to the radius of curvature of the bent region. This expression is simply $r = 1/\beta b$, where r is the radius of curvature, β the excess edge density, and b the Burgers vector. Compute the local radius in a region where β is 10^{12} m/m^3.

8.4 A well known equation, which treats a crystal as an isotropic elastic solid, was developed in the early days of dislocation theory to predict the surface energy of a low-angle tilt boundary. Basically it assumes that the strain field of an individual edge dislocation is neutralized by the fields of the dislocations above and below it, in the boundary, and at a distance from the boundary equal to approximately one-half the spacing between the dislocations in the boundary. The derivation of this equation is given in most texts on dislocations. The equation is

$$\Gamma = \frac{\mu b}{4\pi(1-v)} \alpha(A - \ln \alpha)$$

where Γ is the surface energy, b the Burgers vector, μ the shear modulus, α the angle of tilt of the boundary in radians, A a constant representing the core energy of the dislocation, and v is Poisson's ratio.

(a) Write a computer program that will determine Γ as a function of the angle of tilt α. Use this program to determine the variation of Γ with α from 0.0001 to 1.047 rad in 60 steps. Assume $\mu = 8.6 \times 10^{10}$ Pa, $b = 0.25$ nm, $A = 0.5$, and $v = 0.33$.

(b) Determine the maximum value of the surface energy and divide each value of Γ by the maximum value to obtain a set of relative values of the surface energy. Plot these values as a function of α in degrees and compare the curve that is obtained with Fig. 6.8.

8.5 One of the phenomena during the latter stages of recovery is the coalescence of tilt boundaries into a single tilt boundary with a larger tilt angle. This is accompanied by a loss of surface energy. Compute the fractional loss of surface energy when two tilt boundaries with tilt angles of 0.5° combine to form a 1.0° tilt boundary. (Use the parameters of Prob. 8.4.)

8.6 An investigation of the recovery of a zinc single crystal gave the following data:

Temp., K	Time to Recover 50% of the Yield Point, Hr
283	0.007
273	0.022
263	0.079
253	0.306
243	1.326
233	6.521

(a) Plot these data in the form $\ln(1/\tau)$ vs. $1/T$, where τ is the 50% recovery time and T is the absolute temperature. From the slope of the curve that is obtained determine the activation energy, Q, for the recovery of the yield point. (See Eq. 8.2.)

(b) Determine the corresponding value of the preexponential constant, A, of this equation.

8.7 With the aid of the equation developed in Prob. 8.6, determine the time to recover 50% of the yield stress of the deformed zinc crystal at 213 K and 300 K.

8.8 The data in Fig. 8.16 of Decker and Harker corresponding to the complete recrystallization of pure copper are:

Temp., K	Recrystallization Time, 10^3 S
316	2,300
361	33
375	10
385	7
392	4
408	1.5

(a) Use the above data to determine the activation energy Q and the preexponential constant A in the rate equation for recrystallization, Eq. 8.5.

(b) Determine the recrystallization temperature for the copper; i.e., the temperature corresponding to complete recrystallization in 1 hour.

(c) How long would it take to completely recrystallize the copper at room temperature, 300 K?

8.9 (a) The surface energy of a film of soap and water is about 3×10^{-2} J/m^2. Compute the increase in the pressure on the inside of a soap bubble with a diameter of 6 cm.

(b) In some studies designed to investigate the effect of very rapid cooling on the freezing point of metals, droplets of a liquid metal have been rapidly cooled. The diameter of these droplets were of the order of 50 μm. Consider the case of liquid gold droplets of this diameter with a surface energy of 13.2×10^{-2} J/m^2 and compute the increase in the internal pressure for this case.

8.10 (a) The goal is to write a computer program for the grain-growth law, Eq. 8.13, using data from Fig. 8.31. Express Q in J/mol, R in J/mol-K, and leave D^2 and D_0^2 in units of 10^{-6} cm^2 and the time t in minutes so that the data of Fig. 8.31 can be reproduced. The first step is to determine the value of K_0. To do this, solve for K_0, letting $t = 90$, $T = 973$, $D_0^2 = 2$, and $D^2 = 63$.

(b) Next insert the value of K_0 found above into Eq. 8.13 and write a program that allows the value of the temperature, T, to be substituted into the equation and also contains a (FOR—NEXT) LOOP that varies the time, t, from 0 to 120 min. in steps of 30 min. The purpose is to obtain four values of D^2 at any arbitrary temperature. Check your program against the 973 K (700°C) data in Fig. 8.31 and then use it to determine a set of D^2 values for 1000 K. Note, take $D_0 = 2$.

8.11 If a silver specimen is annealed for a long time under conditions favoring the development of grooves along the lines where internal grain boundaries intersect the outer surface of the sample, a groove angle of 139.5° occurs. If the grain boundary energy for silver is 0.790 J/m^2, what is the energy of the solid-vapor surface?

8.12 Determine, to a first approximation, the limiting grain size in a 2 cm-thick plate of a metal containing a 1 percent volume fraction of a stable spherical precipitate whose average diameter is 600 nm.

References

1. Gordon, P., *Trans. AIME*, 203 1043 (1955).
2. Greenfield, P., and Bever, M. B., *Acta Met.*, **4** 433 (1956).
3. Clarebrough, H. M., Hargreaves, M. E., and West, G. W., *Proc. Roy. Soc.*, London, **232A** 252 [1955].
4. Scholz, F., Driver, J. H., and Woldt, E., *Scripta Mat.*, 40 949 (1999).
5. Drouard, R., Washburn, J., and Parker, E. R., *Trans. AIME*, 197 1226 (1953).
6. Orowan, E., *Communication to the Congres de la Société Française de la Metallurgie d'Octobre*, 1947.
7. Avrami, M. (now M. A. Melvin), *Jour. Chem. Phys.*, **7** 1103 (1939); *ibid.*, **8** 212 (1940); *ibid.*, **9** 177 (1941). Also Johnson, W. A., and Mehl, R. F., *Trans. AIME*, **135** 416 (1939).
8. Cahn, R. W., *Recrystallization, Grain Growth and Texture*, ASM Seminar Series, pp. 99–128, American Society for Metals, Metals Park, Ohio, 1966.
9. Detert, K., and Zieb, J., *Trans. AIME*, **233** 51 (1965).

10. Bailey, J. E. and Hirsch, P. B., *Proc. Roy. Soc.*, A267 11 (1962).
11. Cahn, R. W., *Proc. Roy. Soc.*, A63 323 (1950).
12. Beck, P. A., *Adv. Phys.*, 3 245 (1954).
13. Hu, Hsun, *Recovery and Recrystallization of Metals*, AIME Conference Series, pp. 311–62, Interscience Publishers, New York, 1963.
14. Burk, J. E., and Turnbull, D., *Prog. Met. Phys.*, 3 220 (1952).
15. Nielsen, J. P., *Recrystallization, Grain Growth and Texture*, ASM Seminar Series, pp. 141–64, American Society for Metals, Metals Park, Ohio, 1966.
16. Moldovan, D., Wolf, D. and Phillpot, S. R., *Acta Mat*, **49**, 3521 (2001).
17. Rhines, F. N., *Met. Trans.*, **1** 1105–20 (1970).
18. Fullman, R. L., ASM Seminar, *Metal Interfaces*, p. 179 (1952).
19. Harker, D., and Parker, E. A., *Trans. ASM*, **34** 156 (1945).
20. Rhines, F. N., and Patterson, B. R., *Met. Trans. A*, 13A, 985 (1982).
21. Mullins, W. W., *Acta Met*, **6**, 414 (1958).
22. Beck, P. A., *Phil. Mag. Supplement*, **3** 245 (1954).
23. Moshksar, M. M., Doty, H., and Abbaschian, R., *Intermetallics*, **5**, 393 (1997).

Chapter 9

Solid Solutions

Learning Objectives

Upon completion of Chapter 9, you will be able to:

1. Compare and contrast interstitial solid solutions, substitutional solid solutions, and intermediate phases
2. Understand solubility limits as affected by atomic sizes and electromotive forces following Hume-Rothery rules
3. Understand interactions of solute atoms with dislocations and the formation of dislocation atmosphere
4. Express drag-stress dislocation velocity relationships, including the effects of strain rates and temperature
5. Discuss dislocation velocity and how diffusion rates affect it
6. Define Ludder bands, and describe how and why they form
7. Describe dynamic strain aging, including how its interval is affected by strain rates

9.1 Solid Solutions

When homogeneous mixtures of two or more kinds of atoms occur in the solid state, they are known as *solid solutions*. These solutions are quite common and are equivalent to liquid and gaseous forms, for the proportions of the components can be varied within fixed limits, and the mixtures do not separate naturally. The term *solvent* refers to the more abundant atomic form, and *solute* to the less abundant. These solutions are also usually crystalline.

Solid solutions occur in either of two distinct types. The first is known as a *substitutional solid solution*. In this case, a direct substitution of one type of atom for another occurs so that solute atoms enter the crystal to take positions normally occupied by solvent atoms. Figure 9.1A shows schematically an example containing two kinds of atoms (Cu and Ni). The other type of solid

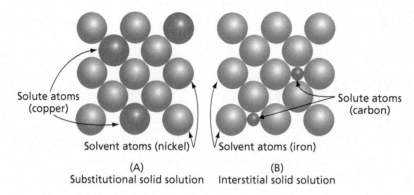

Fig. 9.1 The two basic forms of solid solutions. Note: In the interstitial example on the right, carbon is dissolved interstitially in the body-centered cubic form of iron. For the substitutional solid solution shown on the left, the nickel atoms replace copper atoms

solution is shown in Fig. 9.1B. Here the solute atom (carbon) does not displace a solvent atom, but, rather, enters one of the holes, or interstices, between the solvent (iron) atoms. This type of solution is known as an *interstitial solid solution*.

9.2 Intermediate Phases

In many alloy systems, crystal structures or phases are found that are different from those of the elementary components (pure metals). If these structures occur over a range of compositions, they are, in all respects, solid solutions. However, when the new crystal structures occur with simple whole-number fixed ratios of the component atoms, they are intermetallic compounds with stoichiometric compositions.

The difference between intermediate solid solutions and compounds can be more easily understood by actual examples. When copper and zinc are alloyed to form brass, a number of new structures are formed in different composition ranges. Most of these occur in compositions which have no commercial value whatsoever, but that which occurs at a ratio of approximately one zinc atom to one of copper is found in some useful forms of brass. The crystal structure of this new phase is body-centered cubic, whereas that of copper is face-centered cubic, and zinc is close-packed hexagonal. Because this body-centered cubic structure can exist over a range of compositions (it is the only stable phase at room temperature between 47 and 50 weight percent of zinc), it is not a compound, but a solid solution. This is also sometimes called a nonstoichiometric compound or a nonstoichiometric intermetallic compound. On the other hand, when carbon is added to iron in an amount exceeding a small fraction of one-thousandth of a percent at ambient temperatures, a definite intermetallic compound is observed. This compound, called cementite, has a fixed composition (6.67 weight percent of carbon) and a complex crystal structure (orthorhombic, with 12 iron atoms and 4 carbon atoms per unit cell) which is quite different from that of either iron (body-centered cubic) or carbon (graphite). Unlike bcc iron, cementite is a very hard and brittle phase, strongly influencing properties of steels, as discussed in more detail in chapter 18.

A binary alloy system may contain both stoichiometric and nonstoichiometric compounds. For example, aluminum and nickel alloys contain five nickel aluminide intermediate phases designated as Al_3Ni, Al_3Ni_2, $AlNi$, Al_3Ni_5, and $AlNi_3$. The first intermediate phase, Al_3Ni, has a fixed stoichiometric composition of 75 at. % Al and 25 at. % Ni. The other compounds, on the other hand, are nonstoichiometric. Additional examples of intermetallics will be given in Section 11.14.

9.3 Interstitial Solid Solutions

An examination of Fig. 9.1B shows that solute atoms in interstitial alloys must be small in size. The conditions that determine the solubilities in both interstitial and substitutional alloy systems have been studied in great detail by Hume-Rothery and others. According to their results, extensive interstitial solid solutions occur only if the solute atom has an apparent diameter smaller than 0.59 times that of the solvent. The four most important interstitial solute atoms are carbon, nitrogen, oxygen, and hydrogen, all of which are small in size.

Atomic size is not the only factor that determines whether or not an interstitial solid solution will form. Small interstitial solute atoms dissolve much more readily in transition metals than in other metals. In fact, we find that carbon is so insoluble in most nontransition metals that graphite-clay crucibles are frequently used for melting them. Some of the commercially important transition metals are

Iron	Vanadium	Tungsten
Titanium	Chromium	Thorium
Zirconium	Manganese	Uranium
Nickel	Molybdenum	

The ability of transition elements to dissolve interstitial atoms is believed to be due to their unusual electronic structure. All transition elements possess an incomplete electronic shell inside of the outer, or valence, electron shell. The nontransitional metals, on the other hand, have filled shells below the valency shell.

The extent to which interstitial atoms can dissolve in the transition metals depends on the metal in question, but it is usually small. On the other hand, interstitial atoms can diffuse easily through the lattice of the solvent and their effects on the properties of the solvent are larger than might otherwise be expected. Diffusion, in this case, occurs not by a vacancy mechanism, but by the solute atoms jumping from one interstitial position to another.

9.4 Solubility of Carbon in Body-Centered Cubic Iron

Let us consider a specific interstitial solid solution, carbon in body-centered cubic iron, and confine our attention to temperatures below 1000 K. This restriction is applied so that we do not have to concern ourselves with considerations of the face-centered cubic form of iron, which is stable at temperatures as low as 1000 K in the presence of carbon.

The solubility of carbon in body-centered cubic iron is low. In fact, the number of carbon atoms in an iron crystal is roughly equivalent to the number of vacancies found in crystals. It is therefore possible to deduce an equation for the equilibrium number of carbon atoms in an iron crystal.

There are, however, significant differences between the two cases, as will be shown. The iron-carbon case will now be considered.

The positions marked with an x in Fig. 9.2 are those that carbon atoms take when they enter the iron lattice. They lie either midway between two corner atoms or in the centers of the faces of the cube, where they are midway between two atoms located at the centers of unit cells. In the diamond lattice, the carbon atom has an apparent diameter of 0.1541 nm. However, Fig. 9.2B shows that the lattice constant of iron (length of one edge of the unit cell) is 0.2866 nm. The diameter of the iron atom is, accordingly, 0.2481 nm, and another simple calculation shows that the width of the hole occupied by carbon atoms is only 0.0385 nm. It is quite apparent that the carbon atom does not fit well into the body-centered cubic crystal of iron. The lattice in the vicinity of each solute atom is badly strained, and work must be done in order to introduce the interstitial atom into the crystal.

Let us designate this work by the symbol w_c. If n_c is the number of carbon atoms in an iron crystal, the internal energy of the crystal (in the presence of carbon atoms) will be increased by the amount $n_c w_c$.

Each interstitial atom increases the entropy of the crystal in the same manner that each vacancy does. This form of entropy increment is known as an intrinsic entropy. This *intrinsic entropy* arises from the fact that the introduction of an interstitial atom affects the normal modes of lattice vibrations. The solute atom distorts the orderly array of iron atoms and this, in turn, makes the thermal vibrations of the crystal more random, or irregular. The total entropy contribution from this source is $n_c s_c$, where s_c is the intrinsic entropy per carbon atom.

In the derivation of the equation for the equilibrium number of vacancies in a crystal, the intrinsic entropy of the vacancies $n_v s_v$ was neglected in driving equation 7.40. However, if it had been incorporated in the expression for the free energy associated with the presence of vacancies, the solution of the problem would have been no more difficult. In that case, the final result would have had the form

$$\frac{n_v}{n_o} = e^{-(w_v - Ts_v)/kT} = e^{-g_v/kT} \qquad 9.1$$

and not

$$\frac{n_v}{n_o} = e^{-w_v/kT} \qquad 9.2$$

where g_v is the Gibbs free energy for a single vacancy, n_v is the number of vacancies, n_o is the number of atoms, w_v is the work to form a vacancy, T is the absolute temperature K, and k is the Boltzmann's constant.

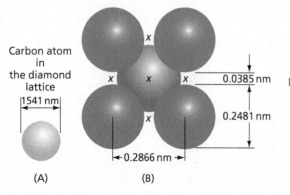

Fig. 9.2 The interstitial positions in the body-centered cubic iron unit cell that may be occupied by carbon atoms

9.4 Solubility of Carbon in Body-Centered Cubic Iron

The preceding, more exact equation, including the intrinsic entropy, can also be expressed as follows:

$$\frac{n_v}{n_o} = e^{S_v/k_e - w_v/kT} \qquad 9.3$$

However, since the exponent S_v/k is a constant, this expression reduces to

$$\frac{n_v}{n_o} = Be^{-w_v/kT} = Be^{-Q_f/RT} \qquad 9.4$$

where B is a constant, Q_f is the enthalpy or energy to introduce a mole of vacancies into the lattice, R is the universal gas constant, and T the temperature in degrees Kelvin.

Now, with reference to carbon atoms in an iron crystal, it should be noted that the mixing of carbon atoms and iron atoms to form a solid solution involves an entropy of mixing. The calculation of this mixing entropy, however, differs from that used for the mixing entropy associated with vacancies in a crystal lattice because vacancies and atoms interchange positions, while carbon atoms do not interchange with iron atoms. In fact, the iron atoms can be assumed to remain fixed in space as far as the movements of carbon atoms are concerned. This is because below 1000 K the jump rate of iron atoms into vacancies becomes negligible compared to the jump rate of carbon atoms between interstitial sites. The entropy of mixing associated with the carbon atoms involves only the distribution of the carbon atoms in the interstitial sites. As may be deduced with the aid of Fig. 9.2, there are three interstitial sites per Fe atom in the bcc iron lattice. Thus, there are two Fe atoms per unit cell as shown in Sec. 1.4, and if one examines the bcc unit cell, it will be found that there are twelve interstitial sites like those shown in Fig. 9.2. Each of these, however, belongs to two cells so that there are six per unit cell. Thus, 6/2 gives 3 sites per Fe atom. The mixing entropy thus involves the distribution of n_c carbon atoms on $3n_{Fe}$ interstitial sites where n_{Fe} is the number of iron atoms. On the assumption that the solid solution of carbon in iron is dilute, the entropy of mixing is thus

$$S_m = k \cdot \ln \frac{(3n_{Fe})!}{n_c!(3n_{Fe} - n_c)!} \qquad 9.5$$

With the aid of Stirling's approximation $(\ln(x)! \approx x \cdot \ln(x) + x)$, Eq. 9.5 becomes

$$S_m = k[3n_{Fe} \ln 3n_{Fe} - 3n_{Fe} - n_c \ln n_c + n_c - (3n_{Fe} - n_c) \ln(3n_{Fe} - n_c) + (3n_{Fe} - n_c)] \qquad 9.6$$

which reduces to

$$S_m = k[3n_{Fe} \ln 3n_{Fe} - n_c \ln n_c - (3n_{Fe} - n_c) \ln(3n_{Fe} - n_c)] \qquad 9.7$$

Let ΔG_c represent the increase in the Gibbs free energy of the system due to the presence of carbon atoms in solid solution in the iron; then

$$\Delta G_c = n_c g_c - TS_m \qquad 9.8$$

where g_c is the free energy associated with a single carbon atom. The carbon concentration corresponding to a minimum G_c can be obtained by setting $\Delta G_c/dn_c = 0$. If the indicated operation is performed, one obtains, with the aid of Stirling's approximation,

$$\Delta G_c/dn_c = g_c + kT[\ln n_c - \ln(3n_{Fe} - n_c)] = 0 \qquad 9.9$$

which leads to

$$\frac{n_c}{(3n_{Fe} - n_c)} = e^{-g_c/kT} \qquad 9.10$$

For a dilute solution, where $n_c \ll n_{Fe}$, this reduces to

$$C = n_c/n_{Fe} = 3e^{-g_c/kT} = 3e^{s_c/k}e^{-w_c/kT} \qquad 9.11$$

where $g_c = w_c - s_c T$, and w_c and s_c are the work to add a carbon atom into the lattice and the intrinsic entropy associated with the addition of this atom, respectively. Equation 9.11 may also be written in the form

$$C = Be^{-Q_c/RT} \qquad 9.12$$

where $B = e^{S_c/R}$ is a constant, $Q_c = N q_0$ the work to introduce a mole of carbon atoms, $R = 8.37$ J per mole-K, T the temperature in degrees Kelvin, N is Avogadro's number, and $S_c = N s_c$.

Now let us consider the physical significance of w_c, which has been defined as the work required to introduce a carbon atom into an interstitial position. This element of work depends upon the source of the carbon atoms, which in alloys of pure iron and carbon can be either graphite crystals or iron-carbide crystals. In either case, the crystals that supply the carbon atoms are assumed to be in intimate contact with the iron. The most important of these two sources of carbon is iron-carbide, for commercial steels are almost universally aggregates of iron and iron-carbide. Graphite rarely appears in steels, even though it represents a more stable phase than iron-carbide. (Iron-carbide is a metastable phase and there is a free-energy decrease when it decomposes to form graphite and iron. However, the rate of this decomposition is extremely slow in the temperature range of normal steel usage, and, for all practical purposes, Fe_3C may be considered a stable phase.)

In metallurgical terminology, iron-carbide is called *cementite*, and the interstitial solid solution of carbon in body-centered cubic iron is known as *ferrite*. These terms will be used in the following sections of the book.

According to Chipman, (Ref. 1) the experimentally determined activation energy Q_c for the transfer of a mole of carbon atoms from cementite to ferrite is 77,300 J per mole. At the same time Chipman gives the corresponding value for graphite to ferrite as 106,300 J per mole. The work to take a carbon atom from graphite and place it interstitially in iron is therefore greater than the work to take it from iron-carbide and place it in iron. This, of course, has an effect on the equllibrium number of carbon atoms in the iron lattice. The solubility of carbon in ferrite is lower when the iron is in equilibrium with graphite than when it is in equilibrium with cementite, because a greater energy is required to remove an atom from graphite and place it in the iron. This will be further presented in the Fe–C phase diagram in Fig. 18.1. The solubility equation

$$C_c = Be^{-Q_c/RT} \qquad 9.13$$

can be evaluated in terms of the data reported in Chipman's paper, which specifically apply to the case of ferrite in equilibrium with cementite (that is, where the interstitial atoms are supplied by iron-carbide). If this is done, we obtain

$$C_c = 11.2e^{-77,300/RT} \qquad 9.14$$

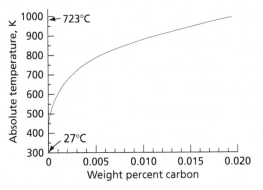

Fig. 9.3 Solubility of carbon in alpha-iron (body-centered cubic iron)

where C_c is the equilibrium concentration of carbon atoms (n_c/n_{Fe}). This equation is easily converted to read the carbon concentration in weight percent, thus

$$C'_c = 240 e^{-77{,}300/RT} \qquad 9.15$$

where $R = 8.314$ J per mole-K.

Figure 9.3 shows a plot of the equilibrium carbon concentration between room temperature and 727°C (1000 K). The curve clearly shows that the solubility of carbon in ferrite is very small. The equilibrium value at room temperature is only 8.5×10^{-12} weight percent, or one part in 8.5×10^{-14} by weight. This is equivalent to an atom fraction of 4×10^{-13}, or a mean separation between solute atoms of 3,000 solvent atoms. Figure 9.3 also shows the temperature dependence of the equilibrium number of carbon atoms. At 727°C the carbon concentration reaches its maximum value, 0.022 percent. In this case, there is about one carbon atom for every 1000 iron atoms, or a separation equivalent to approximately 10 solvent atoms between carbon atoms.

9.5 Substitutional Solid Solutions and the Hume-Rothery Rules

In Figure 9.1A, the copper and nickel atoms are drawn with the same diameters. Actually, the atoms in a crystal of pure copper have an apparent diameter (0.2551 nm) about 2 percent larger than those in a crystal of pure nickel (0.2487 nm). This difference is small and only a slight distortion of the lattice occurs when a copper atom enters a nickel crystal, or vice versa, and it is not surprising that these two elements are able to dissolve in each other and crystallize simultaneously into a face-centered cubic lattice in all proportions. Nickel and copper form an excellent example of an alloy series of complete solubility.

Silver, like copper and nickel, crystallizes in the face-centered cubic structure. It is also chemically similar to copper. However, the solubility of copper in silver, or silver in copper, equals only a fraction of 1 percent at room temperature. There is, thus, a fundamental difference between the copper-nickel systems and the copper-silver systems. This dissimilarity is due primarily to a greater difference in the relative sizes of the atoms in the copper-silver alloys. The apparent diameter of silver is 0.2884 nm, or about 13 percent larger than that of copper. It is thus very close to the limit noted first by Hume-Rothery, who pointed out that an extensive solid solubility of one metal in another only occurs if the diameters of the metals differ by less than 15 percent. This criterion for solubility is known as the *size factor* and is directly related to the strains produced in the lattice of the solvent by the solute atoms.

The size factor is only a necessary condition for a high degree of solubility. It is not a sufficient condition, since other requirements must be satisfied. One of the most important requirements is the relative positions of the elements in the electromotive series. (This series may normally be found in either a general chemistry text or an elementary book on the subject of corrosion.) Two elements, which lie far apart in this series, do not, as a rule, alloy in the normal sense, but combine according to the rules of chemical valence. In this case, the more electropositive element yields its valence electrons to the more electronegative element, with the result that a crystal with ionic bonding is formed. A typical example of this type of crystal is found in NaCl. On the other hand, when metals lie close to each other in the electromotive series, they tend to act as if they were chemically the same, which leads to metallic bonding instead of ionic.

Two other factors are of importance, especially when one considers a completely soluble system. Even if the size factor and electromotive series positions are favorable, such a system is only possible when both components (pure metals) have the same valence, and crystallize in the same lattice form. More detailed discussion of Hume-Rothery rules can be found in Reference 2 listed at the end of the chapter.

9.6 Interaction of Dislocations and Solute Atoms

A dislocation is a crystal defect the presence of which in a crystal means that large numbers of atoms have been displaced from their normal lattice positions. This displacement of the atoms around the center of a dislocation results in a complex two-dimensional strain pattern with the dislocation line at its center, see sections 4.13 and 4.14.

9.7 Dislocation Atmospheres

When a crystal contains both dislocations and solute atoms, interactions may occur. Of particular interest is the interaction between substitutional solutes and dislocations in the edge orientation. If the diameter of a solute atom is either larger or smaller than that of the solvent atom, the lattice of the latter is strained. A large solute atom expands the surrounding lattice, while a small one contracts it. These distortions may be largely relieved if the solute atom finds itself in the proper place close to the center of a dislocation. Thus, the free energy of the crystal will be lowered when a small solute atom is substituted for a larger solvent atom in the compressed region of a dislocation in, or close to, the extra plane of the dislocation. (See Fig. 4.38.) In fact, it can be shown by computations that the stress field of the dislocation attracts small solvent atoms to this area. Similarly, large solute atoms are drawn to lattice positions below the edge (i.e., the expanded region of the dislocation).

Substitutional atoms do not react strongly with dislocations in the screw orientation where the strain field is nearly pure shear. The lattice distortion associated with substitutional atoms can be assumed to be spherical in shape. Figure 9.4 shows that a state of pure shear is equivalent to two equal normal strains (principal strains)—one tensile and one compressive. Lattice strains of this type will not react strongly with the spherical strain associated with substitutional solute atoms.

Now consider the interaction of interstitial atoms and dislocations. Interstitial atoms can react with the edge orientations of dislocations since they normally cause an expansion of the solvent lattice. As a result, interstitial atoms are drawn toward the expanded regions of edge dislocations. Also, because of the nonspherical lattice distortions that interstitial atoms produce in body-centered cubic metals, they are capable of reacting with the screw components of dislocations in this type of

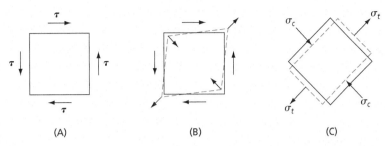

Fig. 9.4 **(A)** Stress vectors on a unit cube corresponding to a state of pure shear. **(B)** Deformation of a unit cube corresponding to shear stresses of Fig. 9.4A. **(C)** A state of pure shear is equivalent to two equal normal stresses, one compressive and one tensile, aligned at 45° to the shear stresses

lattice. An example has already been cited when it was shown that the carbon atom enters a restricted space between a pair of iron atoms and shoves them apart. (See Fig. 9.2.) Since, in the body-centered cubic lattice, the atom pairs referred to always lie along the edges of the unit cells or at the centers of two adjacent unit cells, the distortions (expansions) are in $\langle 100 \rangle$ directions. A nonspherical lattice distortion such as this will react with the shear strain field of a screw dislocation as follows. In those regions close to a screw dislocation where a $\langle 100 \rangle$ direction lies close to the principal tensile strain of the dislocation's strain field (Fig. 9.4), the holes separating iron-atom pairs lying along this direction will be enlarged. Carbon atoms will naturally seek these enlarged interstitial positions.

Solute atoms are drawn toward dislocations as a result of the interactions of their strain fields. However, the rate at which the solute atoms move under these attractive forces is controlled by the rate at which they can diffuse through the lattice. At high temperatures, diffusion rates are rapid, and solute atoms concentrate quickly around dislocations. If the solute atoms have a mutual attraction, the precipitation of a second crystalline phase may start at dislocations. In this case, dislocations act as sinks for solute atoms, and the flow can be assumed to be in one direction—toward the dislocations. Movements of this nature can be expected to continue until the solute concentration in the crystal is depleted (lowered to the point where it is in equilibrium with the newly formed phase). On the other hand, if the solute atoms do not combine to form a new phase, an equilibrium state should develop, with equal numbers of solute atoms entering and leaving a finite volume containing a dislocation. Under these conditions, a steady-state concentration of solute atoms builds up around the dislocation which is higher than that of the surrounding lattice. This excess of solute atoms associated with a dislocation is known as its *atmosphere*. The number of atoms in the atmosphere depends on the temperature. Increasing the temperature tends to tear solute atoms away from dislocations and to increase the entropy of the crystal. Increasing temperatures thus lowers the solute concentrations around dislocations, and at a suffciently high temperature, the concentrations can be lowered to the point where it can be considered that dislocation atmospheres no longer exist.

9.8 The Formation of a Dislocation Atmosphere

A simple theory rationalizing the kinetics of solute atom movements in the stressfield of an edge dislocation was proposed by Cottrell and Bilby (Ref. 3) in 1949. Their theory assumed that a difference in size between the substitutional solute atoms and the solvent atoms should result in strains in the lattice centered at the solute atoms. When a solute atom lies in the stress-field of a dislocation, an interaction energy between the solute atom and the dislocation should occur.

Following Cottrell, (Ref. 4) this interaction energy is computed as follows. It is first assumed that the crystal can be treated as an isotropic elastic solid so that isotropic elasticity theory may be used and the strain associated with the solute atom has spherical symmetry. The effective radial strain of a solute atom is then taken as

$$\varepsilon = \frac{r' - r}{r} \qquad 9.16$$

where r' is the radius of the solute atom and r that of the solvent atom. The volume strain was then computed by multiplying the surface area of a sphere of radius r (i.e., $4\pi r^2$) by Δr, where by Eq. 9.16, $\Delta r = \varepsilon r$. This gives a volume strain of $4\pi\varepsilon r^3$. The interaction energy was then obtained by multiplying the volume strain by the hydrostatic stress of the dislocation at the solute atom. This hydrostatic stress equals

$$-\frac{1}{3}(\sigma_{xx} + \sigma_{yy} + \sigma_{zz}) \qquad 9.17$$

where σ_{xx}, σ_{yy}, and σ_{zz} are the three orthogonal components of the stress expressed in Cartesian coordinates (see Eq. 4.9.) The result is

$$U = -4/3\pi\varepsilon r^3 (\sigma_{xx} + \sigma_{yy} + \sigma_{zz}) \qquad 9.18$$

However, the interaction energy is more conveniently expressed in polar coordinates. See Eq. 4.10. In these coordinates, the interaction energy becomes

$$U = \frac{4(1 + \nu)\mu b \varepsilon r^3 \sin\theta}{3(1 - \nu)R} \qquad 9.19$$

or

$$U = \frac{A \sin\theta}{R} \qquad 9.20$$

where θ and R are defined in Fig. 9.5 and

$$A = \frac{4(1 + \nu)}{3(1 - \nu)}\mu b \varepsilon r^3. \qquad 9.21$$

The parameter A is called the *interaction constant*.

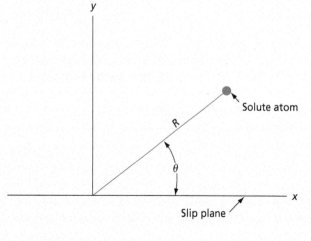

Fig. 9.5 A figure defining the polar coordinate variables R and θ

9.9 The Evaluation of A

The evaluation of the interaction constant A will now be considered for the metal iron where

$$b = 2.48 \times 10^{-10} \text{ m}$$
$$r = 1.24 \times 10^{-10} \text{ m}$$
$$\mu = 7 \times 10^{11} \text{ Pa}$$
$$\nu = 0.33$$

9.22

where b is the Burgers vector, r the solvent atom radius, μ the shear modulus, and ν is Poisson's ratio. Substituting the values from Eqs. 9.22 into Eq. 9.21 yields $A = 8.8 \times 10^{-29} \, \varepsilon \text{ Nm}^2$. Thus, if the (substitutional) solute atom radius is 10 percent larger than the iron atom, A would equal $8.8 \times 10^{-30} \text{ Nm}^2$.

The above calculation gives the interaction energy between a solute atom and an edge dislocation in a substitutional solid solution. The interaction energy between an interstitial solute atom and an edge dislocation in a bcc metal can also be evaluated using Eq. 9.20. However, the calculation of A is more complicated because the strain in the lattice due to the interstitial atom does not have spherical symmetry. It has a tetragonal symmetry. An example of this type of solid solution is carbon steel, which normally contains interstitial carbon atoms dissolved in the bcc iron lattice. The calculation of the interaction constant for this case will not be covered here. The original calculations of Cottrell and Bilby gave $A = 3.0 \times 10^{-20}$ dyne cm^2 (3.0×10^{-29} Nm^2). However, a later and more accurate calculation by Schoeck and Seeger (Ref. 5) used $A = 1.84 \times 10^{-29}$ Nm^2.

9.10 The Drag of Atmospheres on Moving Dislocations

Consider an edge dislocation in a body-centered cubic metal such as iron. Directly above the dislocation line the compressive stress that exists there tends to lower the carbon or nitrogen atom concentration to a value below that existing in the crystal as a whole. At the same time, the tensile stress below the dislocation attracts these interstitial solute atoms. The dislocation atmosphere around an edge dislocation therefore consists of an excess concentration of interstitial atoms below the edge, and a deficiency of these atoms above the edge.

When such a dislocation moves at a high enough temperature so that solute atoms are very mobile, its atmosphere tends to move with it. The movement of a dislocation away from its atmosphere creates an effective stress on the solute atoms that draws them back toward their equilibrium distribution. This motion can only occur by thermally activated jumps of the atoms from one interstitial position to another. As a result, the atmosphere tends to lag behind the dislocation. At the same time, the distribution of the atoms in the atmosphere also changes. This is because the structure of the atmosphere is now influenced by several additional factors. The most important of these is probably that the movement of the dislocation through the crystal tends to bring additional solute atoms into the atmosphere. At the same time, a corresponding number of solute atoms must leave the atmosphere on the side opposite to the direction of motion. In this process, it can be considered that the movement of the dislocation through the crystal tends to realign those solute atoms lying just above its slip plane into positions below its slip plane. The atmosphere associated with a moving dislocation is thus a dynamic concept, but its existence can have a strong influence on the motion of a dislocation.

Now let us consider how the atmosphere affects the motion of a dislocation. The interaction stress between the solute atoms in the atmosphere and the dislocation makes it more difficult to move the dislocation, and this stress has to be overcome in order to advance the dislocation. The drag-stress due to a dislocation atmosphere is therefore one of the important components of the flow-stress of a metal. This stress component is a function of the dislocation velocity. The qualitative nature of its dependence on the velocity can be easily visualized. At both very high and very low velocities the drag-stress has to be very small. At extremely high dislocation velocities the dislocation passes by the solute atoms at such a fast rate that there is insufficient time for the atoms to rearrange themselves. Under these conditions, the solute atoms can be considered as a set of fixed obstacles through which the dislocation moves. An atmosphere of solute atoms should not exist under these conditions. On the other hand, when the dislocation is at rest there exists no net stress between the dislocation and its atmosphere. If the dislocation is now given a small velocity, the center of its atmosphere will move to a position behind the dislocation. The distance of separation between the dislocation and the effective center of its atmosphere increases with increasing velocity. Computations (Ref. 6) indicate that this causes an increase in the drag-stress that is proportional to the dislocation velocity. However, eventually a maximum drag-stress should be reached, since at very large velocities the atmosphere itself becomes less and less well-defined. Figure 9.6A shows the nature of this dependence of the drag-stress on the dislocation velocity.

It is appropriate now to consider the effect of temperature on the drag-stress dislocation velocity relationship. The drag-stress is, in effect, a manifestation of a coupling between a moving dislocation and a set of interstitial atoms that also must move in order to both form and maintain the atmosphere. Accordingly, it can be assumed that the maximum drag-stress corresponds to a direct relationship between the dislocation velocity and the diffusion rate of the solute atoms. Increasing the temperature

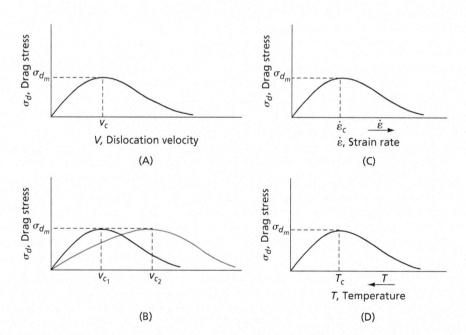

Fig. 9.6 Variation of the drag-stress with, **(A)** the dislocation velocity, **(B)** the dislocation velocity at two different temperatures, **(C)** the strain rate, and **(D)** the temperature at constant strain rate

increases the rate at which the solute atoms move and, as a result, the dislocation velocity corresponding to a maximum drag-stress also has to increase. This is shown in Fig. 9.6B. Note that at the higher temperature the critical velocity v_c is higher. However, the maximum drag-stress σ_{d_m} remains the same. This is in agreement with the theoretical treatment of Cottrell and Jaswon.

By the Orowan equation, we have $\dot{\gamma} = \rho b \bar{v}$, see section 5.21. Therefore, at a constant dislocation density, the average dislocation velocity should be directly proportional to the strain rate. Therefore, it is reasonable to assume that when deformation occurs at a nearly constant value of ρ, the relationship between the strain rate and the drag-stress component of the flow-stress should be similar to that shown in Fig. 9.6C. Finally, because of the interrelationship between strain rate and temperature, it may be deduced that a similar relationship exists between the temperature and the drag-stress when the strain rate is held constant. This is illustrated in Fig. 9.6D. Note that in this case the temperature decreases as the abscissa coordinate increases. This is because a very low temperature at a constant strain rate is equivalent to a high strain rate at a constant temperature. In either case, the solute becomes immobile with respect to the moving dislocation. The results of recent modeling calculations using a phase-field model also indicate that the average velocity of dislocations depends on the ratio of the solute mobility to dislocation mobility. (Ref. 7)

9.11 The Sharp Yield Point and Lüders Bands

When the stress-strain curves of metals deformed in tension are plotted, two basic types of curves are observed, as shown in Fig. 9.7. The curve on the left exhibits what is known as a *sharp yield point*. In this curve, the stress rises with insignificant plastic deformation to point *a*, the upper yield point, shown by point *a* in Figure 9.7A. At this point, the material begins to yield, with a simultaneous drop in the flow-stress required for continued deformation. This new yield stress, point *b*, is called the lower yield point and corresponds to an appreciable plastic strain at an almost constant stress. Eventually the metal starts to harden with an increase in the stress necessary for additional deformation. After this occurs, there is little difference between the appearance of the stress-strain curves for metals with a yield point and those without it.

The sharp yield point is an especially important effect because it occurs in iron and in low-carbon steels. Its existence is of considerable concern to manufacturers who stamp or draw thin sheets of these materials in forming such objects as automobile bodies. The significance of the yield point is this: once plastic deformation starts in a given area, the metal at this point is effectively softened and suffers a relatively large plastic deformation. This deformation then spreads into the material

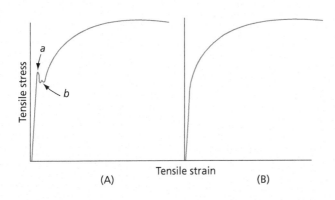

Fig. 9.7 **(A)** Tensile stress-strain curve for a metal exhibiting a sharp yield point. **(B)** Stress-strain curve for a metal that does not exhibit a sharp yield point

adjoining the region, which has yielded because of the stress concentration at the boundary between the deformed and undeformed areas. In general, deformation starts at positions of stress concentration as discrete bands of deformed material, called *Lüders bands*. In the usual tensile test specimen (Fig. 9.8), the fillets are stress raisers and points at which Lüders bands form. The edges of these bands make approximately 50° angles with the stress axis and are designated *Lüders lines*. Lüders bands should not be confused with slip lines. It is quite possible to have hundreds of crystals cooperating to form a Lüders band, each slipping in a complicated fashion on its own slip planes. Once a Lüders band has formed at one fillet of a tensile test specimen, it can then move through the gage length of the specimen. The bands can form simultaneously at both ends of a tensile test specimen, or even, under certain conditions, at a number of positions throughout the specimen gage length. In any case, the deformation starts at localized areas and spreads into undeformed areas. This occurs at an almost constant stress and explains the horizontal part of the stress-strain curve at the lower yield point. In fact, the lower yield stress can be viewed as the stress required to propagate the Lüders bands. The velocity with which a Lüders front moves increases with an increase in the applied stress. Most testing machines tend to deform a specimen at a fixed rate. If two Lüders bands move through a specimen's gage section, the front of each Lüders band will move at approximately half the rate of the front of a single band. This means that the lower yield stress required to move two fronts will be less than that required to move a single front and, in general, an increase or decrease in the number of moving Lüders fronts must be accompanied by a corresponding variation in the lower yield stress. This accounts for the fluctuating lower yield stress shown schematically in Fig. 9.7. Only after the Lüders deformation has covered the entire gage section does the stress-strain curve start to rise again.

In the previously mentioned example of low-carbon-steel sheet used in forming automobile bodies, the effect of using metals containing a yield point is to develop roughened surfaces. These surfaces result from the uneven spread of Lüders bands, which leave striations on the surface commonly called *stretcher strains*. On the other hand, when material without a yield point is used, work hardening instead of softening sets in as soon as plastic deformation starts. This tends to spread deformation uniformly over large areas and to produce smooth surfaces after deformation.

It may not be necessary to exceed all of the yield point strain in order to remove most of the harmful effects of the Lüders phenomenon. Annealed steel sheet is often given a slight reduction

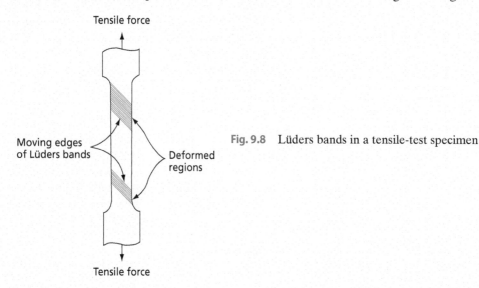

Fig. 9.8   Lüders bands in a tensile-test specimen

in thickness by rolling, which amounts to about a 1 percent strain. This is called a *temper roll*, and it produces a very large number of Lüders band nuclei in the sheet. When the metal is later deformed into a finished product, these small bands grow; but, because of their small size and close proximity to each other, the resulting surface roughening is greatly reduced.

9.12 The Theory of the Sharp Yield Point

It has been proposed by Cottrell (Ref. 4) that the sharp yield point which occurs in certain metals is a result of the interaction between dislocations and solute atoms. According to this theory, the atmosphere of solute atoms that collect around dislocations serves to pin down, or anchor, the dislocations. Additional stress, over that normally required for movement, is needed in order to free a dislocation from its atmosphere. This results in an increase in the stress required to set dislocations in motion, and corresponds to the upper yield-point stress. The lower yield point in the original Cottrell theory then represents the stress to move dislocations that have been freed from their atmospheres. It is important to note that, in general, yield points involve interactions between dislocations and solute atoms at temperatures low enough so that the thermal mobility of the solute atoms is very low. Thus, an applied stress may pull dislocations away from their atmospheres.

All evidence points to the correctness of Cottrell's assumption that the increase in yield stress associated with the upper yield point is caused by the interaction of dislocations with solute atoms. Whether or not the yield point is associated with a simple breaking away of dislocations from their atmospheres is still in doubt. The original Cottrell theory does not satisfactorily explain certain experimental data.

A theory of the yield point originally proposed by Johnston and Gilman (Refs. 8 and 9) to explain yielding in LiF crystals is worth noting. In effect, this theory postulates that in a metal with a very low initial dislocation density, the first increments of plastic strain cause an extremely large relative increase in the dislocation density. To be explicit, suppose that the dislocation density in a metal were to increase directly with the strain at a rate of about 10^{12} m/m^3 per each one percent strain. If the initial dislocation density was only 10^8 m/m^3, the dislocation density would increase by a factor of about 10,000 times in the first one percent of strain. However, in the next one percent of strain, the density would only increase by a factor of two; and for each succeeding one percent increment, the relative increase would be progressively smaller.

The Johnston-Gilman analysis shows how this rapid dislocation multiplication at the beginning of a tensile test can produce a yield drop. The average testing machine tries to deform a tensile specimen at a constant rate. When the dislocation density is very low, the resultant strain is largely elastic, and the stress rises rapidly. Associated with this resultant large stress is a very high dislocation velocity. At the same time, it should be remembered that the dislocation density is increasing rapidly. A high dislocation velocity and an increasing number of mobile dislocations eventually produce a point of instability where the specimen deforms plastically at the same rate as the machine. Beyond this point the load has to fall and, in so doing, lowers the dislocation velocity, with the result that the rate of decrease in the load, while rapid at first, becomes slower and slower. This continued decrease in flow-stress with strain beyond the yield point is normally masked by the work hardening that occurs in the metal. While the Johnston-Gilman theory does not account for the Lüders deformation, it does give a very interesting analysis of the phenomenon of yielding.

In the case of iron and steel, the room-temperature yield point has been shown to be due to either carbon or nitrogen in interstitial solid solution. An important question is how much carbon or nitrogen is needed to form atmospheres around the dislocations. This can be roughly estimated in the following fashion. The number of carbon atoms in an atmosphere is not known with certainty, but can be assumed to be of the order of 1 carbon atom per atomic distance along the dislocation. The iron atom in the body-centered cubic structure has a diameter of about 0.25 nm. In a dislocation 1 cm long, there are 10^7 nm units, or 4×10^7 atomic distances. It can, accordingly, be assumed that there will be 4×10^7 carbon atoms per cm of dislocation.

The density of dislocations in a soft annealed crystal is usually about 10^8 cm^{-2}. The corresponding number in a highly cold-worked metal is 10^{12}, or 10,000 times as many as in unstrained material. In the soft state, the total length of all dislocations in a cubic centimeter is 10^8 cm, and with 4×10^7 carbon atoms per cm, the total number of carbon atoms is 4×10^{15}. Let us compare this with the total number of iron atoms in the same crystal. The length of the body-centered cubic unit cell is 0.286 nm. Thus there are 4.3×10^{22} unit cells in a cubic centimeter. Since the body-centered cubic structure has 2 atoms per unit cell, the number of iron atoms is 8.6×10^{22}, or about 10^{23}. The concentration of carbon atoms required to form an atmosphere of 1 carbon atom per atomic distance along a dislocation is thus:

$$\frac{n_c}{n_c + n_{Fe}} \simeq \frac{n_c}{n_{Fe}} = \frac{4 \times 10^{15}}{10^{23}} = 4 \times 10^{-8} \qquad 9.23$$

or four carbon atoms in every hundred million iron atoms. In a heavily cold-worked crystal, the same ratio would be

$$\frac{n_c}{n_{Fe}} = 4 \times 10^{-4} \qquad 9.24$$

or about 0.04 percent. The significance of the above figures is plain: very little carbon (or nitrogen) is required in order to have sufficient interstitial solute available to form dislocation atmospheres.

9.13 Strain Aging

Figure 9.9A shows a tensile stress-strain curve for a metal with a sharp yield point in which loading was stopped at point c and then the load was removed. During the unloading stage, the stress-strain curve follows a linear path parallel to the original elastic portion of the curve (line ab), and drops to point d. On reloading within a short time (hours), the specimen behaves elastically to approximately point c, and then deforms plastically, and no sharp yield point is observed. On the other hand, if the specimen is not retested for a period of some months and during this period is allowed to age at room temperature, the yield point reappears, as is shown in Fig. 9.9B. The aging period has raised

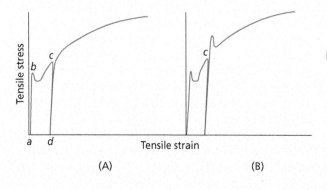

Fig. 9.9 Strain aging. **(A)** Load removed from specimen at point c and specimen reloaded within a short period of time (hours). **(B)** Load removed at point c and specimen reloaded after a long period of time (months)

the stress at which the specimen is strengthened and hardened. This type of phenomenon, where a metal hardens as a result of aging after plastic deformation, is called *strain aging*. The yield point that reappears during strain aging is also associated with the formation of solute atom atmospheres around dislocations. Those dislocation sources that were active in the deformation process just before the specimen was unloaded are pinned down as a result of the aging process. Because solute atoms must diffuse through the lattice in order to accumulate around dislocations, the reappearance of the yield point is a function of time. It also depends on the temperature, in as much as diffusion is a temperature-dependent function. The higher the temperature the faster the rate at which the yield point will reappear. The yield point is not normally observed in iron and steels tested at elevated temperatures (above approximately 400°C). This fact can be explained by the tendency for dislocation atmospheres to be dispersed as a result of the more intense thermal vibrations found at elevated temperatures.

The yield point and strain aging phenomena are most closely associated with iron and low-carbon steel. However, they are also observed in many metals including other body-centered cubic, face-centered cubic, and hexagonal close-packed metals. In many cases the phenomena are not as pronounced as they are in steel.

9.14 The Cottrell-Bilby Theory of Strain Aging

One of the earlier successes in the area of dislocation theory was the Cottrell-Bilby theory of strain aging. (Ref. 3) Actually, this theory does not directly concern the rise in the flow-stress due to aging a strained metal. Rather, it deals with the time- and temperature-dependent growth of solute atmospheres near dislocations. Specifically, they addressed the case of an edge dislocation in bcc iron containing carbon atoms in solid solution. As indicated in Sec. 9.9, the interaction energy between a solute carbon atom and an edge dislocation can be computed with the aid of Eq. 9.20 or

$$U = \frac{A \cdot \sin(\theta)}{R} \qquad 9.25$$

where U is the interaction energy, A the interaction constant, R the distance from the dislocation line to the interstitial atom, and θ the angle between R and the slip plane to the right of the dislocation, as indicated in Fig. 9.5. For fixed values of U, Eq. 9.25 yields a set of circles passing through the dislocation line. A number of these equipotential lines are shown in Fig. 9.10. Note that the equipotential lines corresponding to positive values of U lie above the slip plane while those for negative U values lie below the slip plane. The largest positive or negative values of U correspond to the smallest circles.

From the above it is evident that there is a variation of the interaction energy from one circle to the next in Fig. 9.10. The carbon atoms experience a force, F, due to this energy gradient. A result of this force is that the carbon atoms may attain a drift velocity that can be computed with the aid of the Einstein equation

$$v = (D/kT) \cdot F \qquad 9.26$$

where v is the velocity of the solute atoms, D the diffusion coefficient of the solute, k Boltzmann's constant, and F the force on a solute atom. The concept of the diffusion coefficient, D, is developed in detail in Chapters 11 and 12. In simple terms, the diffusion coefficient gives a measure of the atomic mobility of the solute atoms due to the fact that thermal vibrations can cause them to jump back and forth

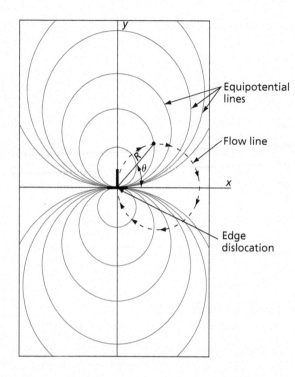

Fig. 9.10 Equipotential lines around an edge dislocation

between adjacent interstitial sites. These jumps become more frequent as the temperature increases. In the absence of an energy gradient, the jumps tend to occur in a statistically random fashion, with the result that the solute concentration remains statistically homogeneous. In other words, the jumps in any one direction tend to be balanced by an equal number of jumps in the opposite direction. However, in the presence of an energy gradient, such as that in the stress field of an edge dislocation, the jumps tend to occur more frequently in the direction of decreasing interaction energy than in the opposite direction. The result is a net flow of solute down the energy gradient. The Einstein equation is able to give a quantitative measure of this directed flow of solute atoms down the energy gradient. If one considers the dimensions of the components of Eq. 9.26, those of D are m²/s, that of $1/kT$ is 1/J, while those of F are J/m. Thus v has the dimensions m/s or those of a velocity. The maximum energy gradient occurs in a direction normal to the equipotential lines in Fig. 9.10 and thus falls along a circular dashed line like that shown in the figure. The dashed circular line in Fig. 9.10 is thus a possible path for the movement of a carbon atom. Cottrell and Bilby (Ref. 3) estimated that the mean gradient of the energy, U, around one of these circles of radius r is of the order of

$$dU/dr = F = A/r^2 \qquad 9.27$$

where dU/dr is the average energy gradient, F the average force, A the interaction constant, and r the radius of the circular path along which the solute atom moves. Thus, the mean velocity of an atom along its path is about

$$v \simeq \frac{AD}{kT \cdot r^2} \qquad 9.28$$

The time to move completely around a path of radius r at this velocity is

$$t = \frac{r^3 \cdot kT}{AD} \qquad 9.29$$

Let t represent the time for a solute atom to move completely around a path of radius r_1. A reasonable assumption is that, in this time interval, all solute atoms lying within the circle of radius r_1 should have moved up to the region underneath the dislocation. However, all of the circular flow lines with radii greater than r_1 should still be active and continue to supply carbon atoms to the region below the dislocation line. In a small time increment, dt, each of these latter flow lines will add to the dislocations a number of solute atoms from a length increment measured from its end point equal to $[(AD/kT)/r^2] \, dt$. The total accumulation of this solute in the time increment dt should thus be

$$2dt \int (AD/kT/r^2) \, dt \qquad 9.30$$

Note that the above equation is multiplied by 2 in order to take into account the fact that for every flow line to the right of the dislocation there is an equivalent one on the left of the dislocation. In order to account for all active flow lines, the above integration should be carried out between the limits $r = (ADt/kT)^{1/3}$ and $r = \infty$. The first of these limits is obtained by solving Eq. 9.27 for r. If Eq. 9.30 is integrated between these limits there is obtained $2(AD/kT)^{2/3}(dt/t^{1/3})$. This represents the area that has supplied solute to the dislocation between t and $t + dt$.

Cottrell and Bilby next computed the number of solute atoms supplied to the dislocation from the depleted area. They deduced that it should equal the number that the area would have contained before the aging process started, assuming that the solute was uniformly distributed. On this basis, they observed that the number of solute atoms which would arrive in a given time t is

$$n(t) = n_o 2 (AD/kT)^{2/3} \int_0^t \frac{dt}{t^{1/3}} = \alpha n_o (ADt/kT)^{2/3} \qquad 9.31$$

where $n(t)$ is the number of solute atoms that have collected under the dislocation per unit length of the dislocation, n_o is the total number of solute atoms per unit volume in the original solid solution, α is a numerical constant equal to approximately 3, A is the interaction constant, D is the diffusion coefficient of the solute atom, k is Boltzmann's constant, T is the absolute temperature, and t is the time of aging.

If one desires to know the fraction of solute, f, that has segregated to the dislocations in the time t, it may be computed by dividing $n(t) \cdot \rho$ by n_o where $n(t)$ is the number of segregated solute atoms on a unit length of dislocation, ρ the total length of dislocation per unit volume, and n_o the initial number of solute atoms per unit volume. Thus,

$$f = n(t)\rho/n_o = \alpha\rho(ADt/kT)^{2/3} \qquad 9.32$$

Note that, if the drift of solute occurs isothermally, all parameters on the right of Eq. 9.32, except for the time, t, should be effectively constant. Thus, Eq. 9.32 predicts that the rate of arrival of solute atoms at the dislocation should vary as $t^{2/3}$. This prediction has been generally verified by experiment for short aging times. When the aging period is long, the $t^{2/3}$ law generally does not hold. A better equation for longer aging times is the Harper (Ref. 10) equation. This is

$$f = 1 - \exp[-\alpha\rho(ADt/kT)^{2/3}]. \qquad 9.33$$

Note that Eq. 9.33 reduces to Eq. 9.32 when t is small, that is, when the exponent is small.

At the present time it is generally felt that the Harper equation should be given the status of an empirical equation even though it was derived by Harper on theoretical grounds, because the basic assumption used in his derivation has been questioned. In any event, the Harper equation is often able to describe very well the dependence on time of the solute flow that leads to atmospheres at dislocations. This is clearly demonstrated in Fig. 9.11, which shows Harper's original results. Note that in this figure $\log_{10}(1 - f)$ plots linearly against $t^{2/3}$ for data obtained at six different temperatures between 294.5 and 324.5 K. These data were obtained by first cold working iron wires containing carbon in solid solution. The cold working was used in order to create a number of new dislocations in the iron. An internal friction method was then used to measure the concentration of carbon remaining in solid solution in the iron wires as they were allowed to age. The assumption was made that the carbon which had accumulated at the dislocations up to a given time, t, had been removed from the solid solution. On this basis, the concentration of carbon still in solid solution should equal $(1 - f)$.

It should be noted that the Cottrell-Bilby and Harper equations (Eqs. 9.32 and 9.33) are only concerned with the flow of the solute to dislocations. That is, they do not measure directly the change in the flow-stress resulting from the accumulation of solute at the dislocations. In order to do this, one must assume that the relative change in the flow-stress $d\sigma/d\sigma_m$ is equal to f, where $d\sigma$ is the incremental increase in the flow-stress and $d\sigma_m$ is the maximum possible increase. This assumption has often proved to yield credible results. Figures 9.12 and 9.13 give examples where some return of the yield point data are plotted using this assumption. These data were obtained by Szkopiak and Miodownik (Ref. 11) using cold-worked bcc niobium wires containing interstitial oxygen atoms in solid solution, a system equivalent to iron with carbon. Figure 9.12 shows their data plotted using the Cottrell-Bilby relation, Eq. 9.32, while Fig. 9.13 shows the same data using the Harper equation. Note that a plot of f against $t^{2/3}$ only gives a reasonable straight line for the initial stages of aging, while the Harper equation plot of $\log_{10}(1 - f)$ against $t^{2/3}$ corresponds to a good linear relation for all of the data.

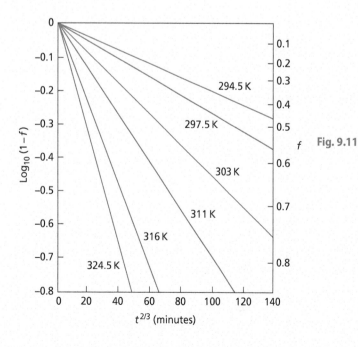

Fig. 9.11 Harper's plot showing that his equation is able to describe the time dependence of the flow of solute to the dislocation atmospheres. Cold-worked iron wire specimens with carbon in solid solution were used

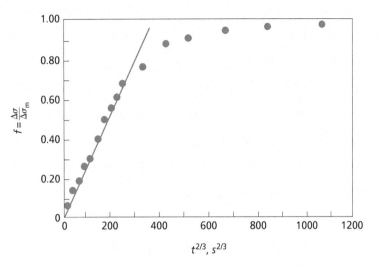

Fig. 9.12 The Cottrell-Bilby strain aging equation was used to plot data from niobium specimens with oxygen atoms in solution. (After Szkopiak, Z. C., and Miodownik, J., *J. Nuclear Materials* **17** 20 (1965).)

9.15 Dynamic Strain Aging

As indicated previously, the higher the temperature the faster the yield point reappears. At a sufficiently high temperature, the interaction between the impurity atoms and the dislocations should occur during deformation. When aging occurs during deformation, the phenomenon has been

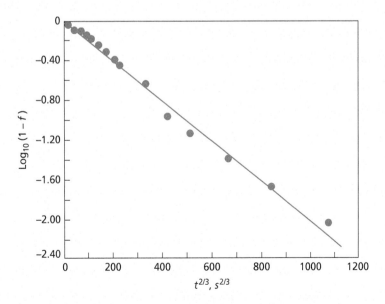

Fig. 9.13 The data in Fig. 9.12 are replotted in this illustration according to the Harper equation. (After Szkopiak, Z. C., and Miodownik, J., *J. Nuclear Materials,* **17** 20 (1965).)

called *dynamic strain aging*. There are many physical manifestations of dynamic strain aging, some of which will now be described.

First, it is important to note that dynamic strain aging tends to occur over a wide temperature range and that the temperature interval in which it is observed depends on the strain rate. Increasing the strain rate raises both the upper and lower temperature limits associated with the dynamic strain aging phenomena. Thus, in steel deformed at a normal cross-head speed of 8.4 μm/s (0.02 in. per min.), dynamic strain aging effects are observed between approximately 370 K and 620 K. However, at a strain rate 10^6 faster, these same limits occur at about 720 and 970 K.

An interesting aspect of dynamic strain aging is that when it occurs the yield stress (or the critical resolved shear stress) of a metal tends to become independent of temperature. This is shown in Fig. 9.14, where the 0.2 percent yield stress of titanium is plotted as a function of temperature. Note that from slightly above 600 K to about 800 K the yield stress is almost constant. At the same time, the flow-stress becomes almost independent of strain rate. In many metals the flow-stress can be related to the strain rate by a simple power law

$$\sigma = A\,(\dot{\varepsilon})^n \qquad 9.34$$

where A is a constant and the exponent n is called the *strain rate sensitivity*. This equation can also be written in the form

$$\frac{\sigma_2}{\sigma_1} = \left(\frac{\dot{\varepsilon}_2}{\dot{\varepsilon}_1}\right)^n \qquad 9.35$$

and taking the logarithms of both sides we have

$$n = \frac{\log_e \frac{\sigma_2}{\sigma_1}}{\log_e \frac{\dot{\varepsilon}_2}{\dot{\varepsilon}_1}} \qquad 9.36$$

In most modern testing machines very rapid changes in the applied strain rate may be made during a tensile test. If this is done and the corresponding values of strain rate and flow-stress

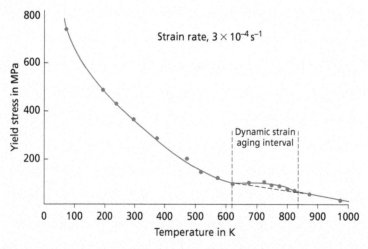

Fig. 9.14 The variation of the yield stress (0.2 percent strain) with the temperature for commercial purity titanium. Note that in the dynamic strain aging interval the flow-stress becomes approximately constant. (Data courtesy of A. T. Santhanam.)

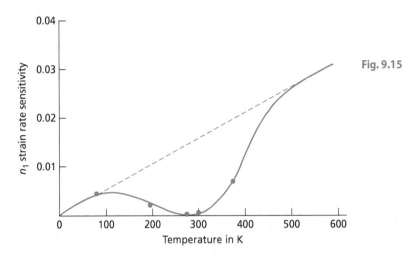

Fig. 9.15 Aluminum alloys are subject to strong dynamic strain aging effects near room temperature. Note that the strain rate sensitivity of 6061 S-T aluminum shows a minimum strain rate sensitivity just below room temperature. (After Lubahn, J. D., *Trans. AIME*, **185** 702 (1949).)

(measured just before and after the rate change) are substituted into this equation, one is able to determine a value of n. When measurements of this type, corresponding to a change in strain rate between two rates differing by a simple ratio, are made on a metal that is not subject to dynamic strain aging, the value of n tends to rise linearly with the absolute temperature. However, when dynamic strain aging is observed, the value of n becomes very low in the temperature range of strain aging, because, in this interval, the flow-stress becomes nearly independent of the strain rate. This is illustrated for the case of an aluminum alloy in Fig. 9.15.

Inside the dynamic strain aging temperature range the plastic flow often tends to become unstable. This is manifested by irregularities in the stress-strain diagram. These discontinuities may be of several types. In some cases the load tends to rise abruptly and then fall. In others the plastic flow is best described as jerky. However, actual sharp load drops are often observed, as indicated schematically in Fig. 9.16. These serrations, as they appeared in aluminum alloys, were first studied in detail by Portevin and LeChatelier, (Ref. 12) and it is now common to call the phenomenon associated with these serrations the *Portevin-LeChatelier effect*.

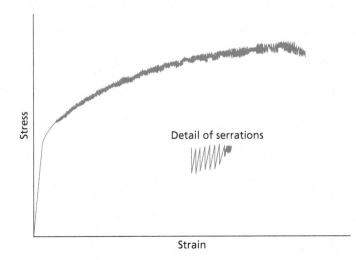

Fig. 9.16 Discontinuous plastic flow is a common aspect of dynamic strain aging. This diagram indicates one form of serrations that may be observed

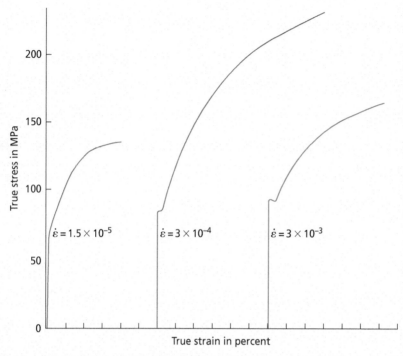

Fig. 9.17 In the temperature range of dynamic strain aging, the work hardening rate may become strain-rate dependent. This figure shows the stress-strain curves for three titanium specimens deformed at 760 K at different rates

One of the most significant aspects of dynamic strain aging, observed primarily in metals containing interstitial solutes, is that the work hardening rate can become abnormally high during dynamic strain aging and, at the same time, it can become strain-rate and temperature dependent. This effect is illustrated for commercially pure titanium in Fig. 9.17. The three stress-strain curves shown in the figure correspond to specimens strained at the same temperature but at three different strain rates. Note that the specimen deformed at the intermediate rate was subject to a much higher degree of work hardening than either of the specimens deformed at a rate 20 times slower or 10 times faster. This illustration suggests that, at the temperature in question, there is a maximum work hardening rate corresponding to a specific strain rate. A similar maximum work hardening rate will be observed if the temperature is raised or lowered, provided that the strain rate is adjusted accordingly. Thus, if the temperature is raised, the strain rate at which maximum work hardening is observed also rises.

Finally, one of the other well-known manifestations of dynamic strain aging is the phenomenon called *blue brittleness* when it occurs in steel. In approximately the center of the temperature range where the other phenomena of dynamic strain aging are observed, it has been found that the elongation, as measured in a tensile test, becomes very small or passes through a minimum on a curve of elongation plotted against the temperature. This subject is considered in Section 21.11.

The various dynamic strain aging phenomena do not appear to the same degree in all metals. However, they are commonly observed and it is probably safe to state that, in general, dynamic strain aging is the rule rather than the exception in metals.

Problems

9.1 At room temperature the stable crystal structure of iron is bcc. However, above 1183 K it becomes fcc. The iron fcc crystal structure is able to dissolve a much larger concentration of carbon than is the bcc structure. A primary reason for this is believed to be that while the fcc structure is more close packed, the octahedral interstitial sites in this lattice are much larger than the sites occupied by the carbon atoms in the bcc structure. The hole at the center of the fcc unit cell is such a site. See Fig. 1.6A. The minimum opening in one of these sites corresponds to the distance between atoms along a $\langle 100 \rangle$ direction. Note that this is the same as in the bcc case. Compare the fcc opening with that of the bcc lattice. For the sake of convenience assume the fcc iron atom diameter is the same as that of the bcc atom at room temperature.

9.2 The data reported by Chipman also give a solubility equation for carbon in alpha iron (bcc iron) when the carbon is supplied by graphite particles in the iron. This equation is

$$C_{cg} = 27.4 \exp\{-106300/RT\}$$

where C_{cg} is the carbon concentration in weight per percent in equilibrium with the graphite, R is the gas constant in J/mol-K, and T is the absolute temperature. Write a computer program for this equation and use it to obtain the weight percent of the carbon as a function of the temperature between 300 and 1000 K at 50°. Plot the data to obtain a curve and compare the curve with that in Fig. 9.3.

9.3 An approximate equation giving the atom fraction of carbon soluble in Austenite, or the fcc cubic form of iron, when the carbon comes from Fe_3C, is

$$C_\gamma = 1.165 \exp\{-28,960/RT\}$$

where Q is in J/mol, R in J/mol-K, and T in degrees K. Write a program that first gives the atom fraction of carbon at a given temperature and then converts the answer to weight percent carbon. With the aid of this program determine the atom fraction and weight percent carbon in Austenite at 1000, 1100, 1200, 1300, 1400, and 1421 K, respectively.

9.4 In polar coordinates the hydrostatic pressure (see Eqs. 4.10 and 9.17) may be written in the form

$$-\frac{1}{3}(\sigma_{rr} + \sigma_{\theta\theta} + \sigma_{zz})$$

where, due to the two-dimensional nature of the stress field of a dislocation, $\sigma_{zz} = \nu(\sigma_{rr} + \sigma_{\theta\theta})$. Using the expressions for the stress components of an edge dislocation, given earlier, derive Eq. 9.19.

9.5 Szkopiak and Miodownik, *J. Nucl. Mat.*, **17** 20 (1965), computed the interaction constant, A, for niobium containing oxygen. They took the shear modulus, μ, to be 3.7×10^{10} Pa, the Burgers vector or atom diameter as 0.285 nm, and ν Poisson's ratio as 0.35. Their equivalent for ε equaled 0.0806. Demonstrate that these values make $A = 6.81 \times 10^{-30}$.

9.6 According to Cottrell and Bilby, Eq. 9.29, which is

$$t = r^3 \frac{kT}{AD}$$

where A is the interaction constant, D the solute diffusion coefficient, k Bolzmann's constant, T the temperature in degrees Kelvin, and r the radius of the circular path along which the solute drifts — gives a measure of the time that it takes the solute to move completely around the path. The nature of one of these paths is shown by the dashed circle in Fig. 9.10. Compute the value of the time for the solute to travel completely around a circle whose $r = 8b$ at 300, 500, 700, and 900 K using the following data relative to iron with interstitial carbon atoms in solution.

$A = 1.84 \times 10^{-29}$ N · m^2

$D = 8 \times 10^{-7} \exp(-82,840/T)$ m^2/s

R = The gas constant

T = Temperature in K

b = 0.248 nm

9.7 The goal of this problem is to determine the equilibrium number of carbon atoms in a cubic meter of iron of a dilute iron-Fe_3C alloy at 600 K as predicted by Eq. 9.14. The atomic volume of iron is 7.1 cm^3 per gram-atom.

9.8 There is an iron carbon alloy with a concentration of 1.8×10^{23} carbon atoms in solid solution per cubic meter. Assume that this material is deformed and then held at 600 K for 10 min. At 600 K the diffusion coefficient for the carbon in the iron is 4.9×10^{-14} m²/s. Assume that the interaction constant, A, is 1.84×10^{-24} and that the constant $\alpha = 3$. Use these data to predict $n(t)$, the number of carbon atoms, per meter of edge dislocation, that should collect below these dislocations in the 10 min. aging period.

9.9 A dilute iron-carbon alloy is quickly cooled from an elevated temperature so that it contains in solid solution about 4×10^{23} carbon atoms per cubic meter. Determine the fraction of this carbon concentration that will precipitate to an array of edge dislocations whose dislocation density is 10^{12} m/m³ in an aging period of 10 hours at 400 K. Note: the diffusion coefficient of the carbon at 400 K is 1.22×10^{-17} m²/s; assume that the interaction constant is 1.84×10^{-29} and that $\alpha = 3$.

9.10 The Cottrell-Bilby and the Harper equations may be written in the form $f = Gt^{2/3}$ and $f = 1 - \exp(-Gt^{2/3})$, where $G = \alpha\rho(AD/kT)^{2/3}$. In the case of a dilute iron carbon alloy at 500 K, $G = 8.44 \times 10^{-4}$. Write a program that will evaluate both forms of f, the fraction precipitated, from $t = 0$ to 36,000 s in steps of 600 s, and plot the results on a graph of f versus $t^{2/3}$ to show the difference in the two concepts of the fraction precipitated.

References

1. Chipman, J., *Met. Trans.*, **3** 55 (1972).
2. Cahn, R. W., and Haasen, P., *Physical Metallurgy,* North-Holland, p. 158 (1983).
3. Cottrell, A. H., and Bilby, B. A., *Proc. Phys. Soc.*, **A62** 49 (1949).
4. Cottrell, A. H., *Dislocations and Plastic Flow in Crystals,* Oxford University Press, London, 1953.
5. Schoeck, G., and Seeger, A., *Acta Met.,* **7** 469 (1959).
6. Cottrell, A. H., and Jaswon, M. A., *Proc. Roy. Soc.,* **A199** 104 (1949).
7. Hu, S. Y., Choi, J., Li, Y. L., and Chen, L. Q., *J. Applied Phys.*, **96** 229 (2004).
8. Johnston, W. G., *J. Appl. Phys.,* **33** 2716 (1962).
9. Johnston, W. G. and Gilman, J. J., *J. Appl. Phys.,* **30** 129 (1959).
10. Harper, S., Phys. Rev., **83** 709 (1951).
11. Szkopiak, Z. C., and Miodownik, J, *J. Nuclear Materials,* **17** 20 (1965).
12. Portevin, A., and LeChatelier, F., *Comp. Rend. Acad. Sci., Paris,* **176** 507 (1923).

Chapter 10

Phases

Learning Objectives

Upon completion of Chapter 10, you will be able to:

1. Describe the differences between phases, solutions (liquid and solid), and phase mixtures
2. Compare and contrast total and partial molar free energies. Derive the relationship between them
3. Understand equality of partial molar free energy of components with phases at equilibrium in multicomponent systems
4. Differentiate between ideal and nonideal solutions. Explain activities and activity coefficients
5. Understand the Gibbs phase rule and relationship between the number of equilibrium phases and degrees of freedom
6. Explain the effects of temperature on free energies and stability regions for polymorphic metals
7. Use graphical determination of partial molar free energies and determination of compositions at equilibrium

10.1 Basic Definitions

The concept of phases is very important in the field of metallurgy. A phase is defined as a *macroscopically homogeneous body of matter*. This is the precise thermodynamic meaning of the word. However, this term is often used more loosely in speaking of a solid or other solution which can have a composition that varies with position and still be designated as a phase. For the present this fact will be ignored and the basic definition will be considered to hold. Let us consider a simple system: a single metallic element, for example, copper, a so-called one-component system. Solid copper conforms to the definition of a phase, and the same is true when it is in the liquid and gaseous forms. However, solid, liquid, and gas have quite different characteristics, and at the freezing point (or at the boiling point) where liquid and solid (or liquid and gas) can coexist respectively, two homogeneous types of matter are present rather than one. It can be concluded that each of the three forms of copper—solid, liquid, and gas—constitutes a separate and distinct phase.

Certain metals, for example, iron and tin, are polymorphic (allotropic) and crystallize in several structures, each stable in a different temperature range. Here each crystal structure defines a separate phase, so that polymorphic metals can exist in more than one solid phase. As an example, consider the phases of iron, as given in Table 10.1. Notice that there are three separate solid phases for iron, each denoted by one of the Greek symbols: alpha (α), gamma (γ), or delta (δ). Actually, there are only two different solid iron phases since the alpha and delta phases are identical; both are body-centered cubic.

At one time, iron was believed to possess a third solid phase, the beta (β) phase, because heating and cooling curves indicated a phase transformation in the temperature range from 773 to 1064 K. It has since been demonstrated that in this temperature range iron changes from the ferromagnetic to the paramagnetic form on heating. Since the crystal structure remains body-centered cubic throughout the region of the magnetic transformation, the β phase is no longer generally recognized as a separate phase. In other words, the magnetic transformation is now considered to occur inside the alpha phase.

Let us now consider alloys instead of pure metals. *Binary alloys*, two-component systems, are mixtures of two metallic elements, while *ternary alloys* are three-component systems: mixtures of three metallic elements. At this time, two terms that have been used a number of times should be clearly defined: the words *system and component*. A *system*, as used in the sense usually employed in thermodynamics, or physical chemistry, is an isolated body of matter. The components of a system are often the metallic elements that make up the system. Pure copper or pure nickel are by themselves one-component systems, while alloys formed by mixing these two elements are two-component systems. Metallic elements are not the only types of components that can be used to form metallurgical systems; it is possible to have systems with components that are pure chemical compounds. The latter type is perhaps of more interest to the physical chemist, but it is also of importance in the field of metallurgy and ceramics. A typical two-component system, with components that are chemical compounds, is formed by mixing common salt, NaCl, and water, H_2O. Another important compound that is usually considered as a component occurs in steels. Steels are normally considered to be two-component systems consisting of iron (an element) and iron carbide (Fe_3C), a compound.

When alloys are formed by mixing selected component metals, the gaseous state, generally speaking, is of little practical interest. In any case, there is only a single phase in the gaseous state, for all gases mix to form uniform solutions. In the liquid state, it sometimes happens, as in alloys formed by adding lead to iron, that the liquid components are not miscible, and then it is possible to have several liquid phases. However, in most alloys of commercial interest, the liquid components dissolve in each other to form a single liquid solution. In the discussion of phases and phase diagrams that follows, our attention will be concentrated on alloys in which the liquid components are miscible in all proportions.

Table 10.1 Phases of Pure Iron

Stable Temperature Range K	Form of Matter	Phase	Identification Symbol of Phase
Above 3013	gaseous	gas	gas
1812 to 3013	liquid	liquid	liquid
1673 to 1812	solid	body-centered cubic	(delta)
1183 to 1673	solid	face-centered cubic	(gamma)
Below 1183	solid	body-centered cubic	(alpha)

The nature of the solid phases that occur in alloys will now be considered. Certain metals, for example, the lead-iron combination mentioned above, do not dissolve appreciably in each other in either the liquid or the solid state. Since this is the case, there are two solid phases, each of which is extremely close to being a pure metal (a component). The solid phases in most alloy systems, however, are usually solid solutions of either of two basic types. First, there are the *terminal solid solutions* (phases based on the crystal structures of the components). Thus, in binary alloys of copper and silver, there is a terminal solid solution in which copper acts as the solvent, with silver the solute, and another in which silver is the solvent and copper the solute. Second, in some binary alloys, crystal structures different from those of either component may form at certain ratios of the two components. These are called *intermediate phases*. Many of them are solid solutions in every sense of the word, for they do not have a fixed composition (ratio of the components) and appear over a range of compositions. A well known intermediate solid solution is the so-called β phase in brass (copper-zinc) which is stable at room temperature in the range 47 percent Zn and 53 percent Cu to 50 percent Zn and 50 percent Cu, all measured in weight percentages. This intermediate phase is not the only one that appears in copper-zinc alloys. There are three other intermediate solid solutions, making in all six copper-zinc solid phases: the four intermediate phases and the two terminal phases.

In some alloys intermediate crystal structures are formed that are best identified as compounds. One intermetallic compound has already been discussed, namely, iron-carbide, Fe_3C.

10.2 The Physical Nature of Phase Mixtures

A brief explanation of the physical significance of "a system of several phases" is in order. To illustrate, consider a simple mixture of two phases. In general, these phases will not be separated into two distinct and separate regions, such as oil floating on water. Rather, the usual metallurgical two-phase system is comparable to an emulsion of oil droplets in a matrix of water. In speaking of such systems, it is common to refer to the phase (water) that surrounds the other phase as the *continuous* or *matrix phase*, and the phase (oil) that is surrounded, the *discontinuous* or *dispersed phase*. It should be noticed, however, that the structure of a two-phase system may be so interconnected that both phases are continuous. An example was cited in Chapter 6, where it was mentioned that, under the proper conditions, a second phase, present in a limited amount, might be able to run through the structure as a continuous network along grain edges. The factors controlling this type of structure are described in Chapter 6.

In the solid state, a metallurgical system of several phases is a mixture of several different types of crystals. If the crystal sizes are small, surface energy effects become important and should rightfully be included in thermodynamic or energy calculation. In the following presentation, for the sake of simplicity, the surface energy will be neglected. It will also be assumed that all systems are removed from electric- and magnetic-field gradients so that their effects on our systems can also be ignored.

10.3 Thermodynamics of Solutions

It can be concluded that the phases in alloy systems are usually solutions—either liquid, solid, or gaseous. it is true that solid phases do sometimes form, with composition ranges so narrow that they are considered as compounds, but it is also possible to think of them as solutions of very limited solubility. In the discussion immediately following, the latter viewpoint is taken and all alloy phases are spoken of as solutions.

In general, the free energy of a solution is a thermodynamic property of the solution, or a variable that depends on the thermodynamic state of the solution (that is, system). More detailed discussion and analysis of thermodynamic properties can be found in thermodynamic textbooks. (Refs. 1 through 5) In a one-component system (a pure substance) in a given phase, the thermodynamic state is determined uniquely if any two of its thermodynamic properties (variables) are known. Among the variables that are classed as properties are the temperature (T), the pressure (P), the volume (V), the enthalpy (H), the entropy (S), and the free energy (G). Thus, in a one-component system of a given mass and phase, if the two variables temperature and pressure are specified, the volume of the system will have a definite fixed value. At the same time, its free energy, enthalpy, and other properties will also have values that are fixed and determinable.

Solutions have additional degrees of freedom compared with pure substances, and it is necessary to specify values for more than just two properties to define the state of a solution. In general, temperature and pressure are employed as two of the required variables, while the composition of the solution furnishes the remainder. The number of independent composition variables is, of course, one less than the number of components. This can be seen if the composition is expressed in atom or mole fractions. In a three-component system, for example, the mole fractions are given by

$$N_A = \frac{n_A}{n_A + n_B + n_C}, \quad N_B = \frac{n_B}{n_A + n_B + n_C}, \quad N_C = \frac{n_C}{n_A + n_B + n_C} \qquad 10.1$$

where N_A, N_B, and N_C are the mole fractions, and n_A, n_B, and n_C are the actual number of moles of the A, B, and C components, respectively. By definition of the mole fraction, we have the condition

$$N_A + N_B + N_C = 1 \qquad 10.2$$

From this expression the value of any one of the mole fractions can be computed once the other two are known. There are only two independent mole fractions in a ternary system.

Most metallurgical processes occur at constant temperature and pressure, and, under these conditions, the state of a solution can be considered to be a function of its composition. Similarly, any of the state functions (properties), such as free energy (G), can be considered a function of only the composition variables. In the case of a three-component system, the total free energy (G) of a solution is written:

$$G = G(n_A, n_B, n_C) \text{ (temperature and pressure constant)} \qquad 10.3$$

where n_A, n_B, and n_C are the number of moles, respectively, of the A, B, and C components.

By partial differentiation, the differential of the free energy of a single solution of three components at constant pressure and temperature is

$$dG = \frac{\partial G}{\partial n_A} \times dn_A + \frac{\partial G}{\partial n_B} \times dn_B + \frac{\partial G}{\partial n_C} \times dn_C \qquad 10.4$$

where the partial derivatives, such as $\partial G/\partial n_A$, represent the change in the free energy when only one of the components is varied by an infinitesimal amount. Thus, for a very small variation of component A, while the amounts of the components B and C in the solution are maintained constant, we have

$$\frac{dG}{dn_A} = \frac{\partial G}{\partial n_A} \qquad 10.5$$

In the present case, the partial derivatives are the partial molar free energies of the solution and are designated by the symbols $\overline{G}_A$, $\overline{G}_B$, and $\overline{G}_C$, so that Eq. 10.5 can also be written

$$dG = \overline{G}_A dn_A + \overline{G}_B dn_B + \overline{G}_C dn_C \qquad 10.6$$

The total free energy of a solution composed of n_A moles of component A, n_B moles of component B, and n_C moles of component C can be obtained by integrating Eq. 10.6. (Ref. 1) This integration can be made quite easily in the following manner.

Let us start with zero quantity of the solution and form it by simultaneously adding infinitesimal quantities of the three components dn_A, dn_B, and dn_C. Each time that we add the infinitesimals, however, let us make the amounts of the components in the infinitesimals have the same ratio as the final numbers of moles of the components n_A, n_B, and n_C, so that

$$\frac{dn_A}{n_A} = \frac{dn_B}{n_B} = \frac{dn_C}{n_C} \qquad 10.7$$

If the solution is formed in this manner, its composition at any instant will be the same as its final composition. In other words, the composition will be constant at all times, and since the partial-molar free energies are functions of only the composition of the solution (at constant temperature and pressure), they will also be constant during the formation of the solution. Integration of

$$dG = \overline{G}_A dn_A + \overline{G}_B dn_B + \overline{G}_C dn_C$$

starting from zero quantity of solution, with the condition that $\overline{G}_A$, $\overline{G}_B$, and $\overline{G}_C$ are constant, gives us

$$G = n_A \overline{G}_A + n_B \overline{G}_B + n_C \overline{G}_C \qquad 10.8$$

where n_A, n_B, and n_C are the number of moles of the three components in the solution.

Let us differentiate Eq. 10.8 completely to obtain

$$dG = n_A d\overline{G}_A + \overline{G}_A dn_A + n_B d\overline{G}_B + \overline{G}_B dn_B + n_C d\overline{G}_C + \overline{G}_C dn_C$$

But we have already seen that the derivative of the free energy is

$$dG = \overline{G}_A dn_A + \overline{G}_B dn_B + \overline{G}_C dn_C$$

and the only way that both of these expressions for the derivative of the total free energy can be true is for

$$n_A d\overline{G}_A + n_B d\overline{G}_B + n_C d\overline{G}_C = 0 \qquad 10.9$$

This equation gives a relationship between the number of moles of each component in a three-component solution and the derivatives of the partial-molar free energies. Similar relationships can be deduced for a two-component solution and for a solution of more than three components. Thus, for two components,

$$n_A d\overline{G}_A + n_B d\overline{G}_B = 0$$

and for four components

$$n_A d\overline{G}_A + n_B d\overline{G}_B + n_C d\overline{G}_C + n_D d\overline{G}_D = 0$$

The significance of these relationships in explaining the phenomena of polyphase systems in equilibrium will be shown in Sec. 10.6.

10.4 Equilibrium between Two Phases

A binary (two-component) system with two phases in equilibrium will now be considered. The total free energy for the first phase, which we shall designate as the alpha (α) phase, is

$$G^\alpha = n_A^\alpha \overline{G}_A^\alpha + n_B^\alpha \overline{G}_B^\alpha$$

while that of the beta (β) phase is

$$G^\beta = n_A^\beta \overline{G}_A^\beta + n_B^\beta \overline{G}_B^\beta \qquad 10.10$$

Let a small quantity (dn_A) of component A be transferred from the alpha phase to the beta phase. As a result of this transfer, the free energy of the alpha phase will be decreased, while that of the beta phase will be increased. The total free-energy change of the system is the sum of these two changes and can be represented by

$$dG = dG^\alpha + dG^\beta = \overline{G}_A^\alpha(-dn_A) + \overline{G}_A^\beta(dn_A)$$

or

$$dG = (\overline{G}_A^\beta - \overline{G}_A^\alpha)\, dn_A \qquad 10.11$$

But we have assumed that the two given phases are at equilibrium. This signifies that the state of the two-phase system is at a minimum with respect to its free energy (total free energy of the two solutions). The variation in the free energy for any infinitesimal change inside the system, such as the shift of a small amount of component A from one phase to the other, must, accordingly, be zero. Therefore,

$$dG = (\overline{G}_A^\beta - \overline{G}_A^\alpha)\, dn_A = 0$$

Since dn_A is not zero, we are able to deduce the following important conclusion:

$$\overline{G}_A^\alpha = \overline{G}_A^\beta \qquad 10.12$$

and in the same manner it can be shown that

$$\overline{G}_B^\alpha = \overline{G}_B^\beta$$

The above quite general results are not restricted to systems of only two components or systems containing only two phases. In fact, it may be shown that, in the general case where there are M components with μ phases in equilibrium, the partial molar free energy of any given component is the same in all phases, or

$$\overline{G}_A^{\alpha} = \overline{G}_A^{\beta} = \overline{G}_A^{\gamma} = \cdots = \overline{G}_A^{\mu}$$
$$\overline{G}_B^{\alpha} = \overline{G}_B^{\beta} = \overline{G}_B^{\gamma} = \cdots = \overline{G}_B^{\mu}$$
$$\overline{G}_C^{\alpha} = \overline{G}_C^{\beta} = \overline{G}_C^{\gamma} = \cdots = \overline{G}_C^{\mu} \qquad 10.13$$
$$\vdots \qquad \cdots \qquad \vdots$$
$$\overline{G}_M^{\alpha} = \overline{G}_M^{\beta} = \overline{G}_M^{\gamma} = \cdots = \overline{G}_M^{\mu}$$

where the superscripts designate the phases (solutions) and the subscripts the components. Further, since the partial molar free energy of any component is the same in all phases, we need no longer use the phase superscripts when specifying partial molar free energies of the components.

10.5 The Number of Phases in an Alloy System

One-Component Systems For a complete understanding of alloy systems, it is necessary to know the conditions determining the number of phases in a system at equilibrium. In a one-component system the conditions are well known. A glance at Table 10.1 shows that in the single-component system (pure iron) under the conditions of constant pressure, two phases can coexist only at those temperatures having a phase change: the boiling point, the melting point, and the temperatures at which solid-state phase changes occur. At all other temperatures only a single phase is stable.

Let us now consider the causes for phase changes in a one-component system. In particular, let us consider a solid-state phase change; the interesting allotropic transformation of white tin to gray tin. The former, or beta (β) phase has a body-centered tetragonal crystal structure that can be thought of as a body-centered cubic structure with one elongated axis. This is the ordinary, or commercial, form of tin and possesses a true metallic luster. The other, or alpha (α) phase, that is gray in color, is the equilibrium phase at temperatures below 286.2 K. Fortunately, from both a practical and a scientific point of view, the transformation from white to gray tin is very slow at all temperatures below 286.2 K. With regard to the practical aspect, it is fortunate that tin does not turn instantly to gray tin once the temperature falls below the equilibrium temperature, because objects made of tin are usually ruined when the change occurs. Gray tin has a diamond cubic-crystal structure, a basically brittle structure, and this fact, coupled with the large volume expansion (about 27 percent) that accompanies the transformation, can cause the metal to disintegrate into a powder. On the other hand, the fact that the change occurs very slowly makes it possible to study the properties of a single element in two crystalline forms over a very wide range of temperatures. In this respect, the specific heat at constant pressure of both the gray and the white forms of tin have been measured from ambient temperatures down to almost absolute zero. A plot of these data is shown in Fig. 10.1. The importance of this information lies in the fact that with it, it is possible to compute the free energy of both solid phases as a function of temperature. Let us see how this is done.

By definition, the Gibbs free energy of a pure substance is

$$G = H - TS \qquad 10.14$$

where G = the free energy in Joules per mole
T = the temperature in degrees Kelvin
H = the enthalpy (Joules per mole)
S = the entropy in Joules per mole degree Kelvin

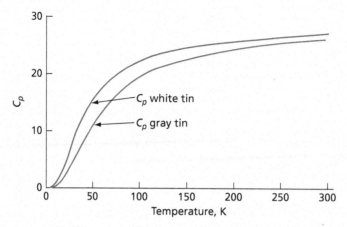

Fig. 10.1 Heat capacity C_p as a function of temperature for both forms of solid tin

To evaluate the free energy of the substance at some temperature T, one needs to know both the enthalpy H and the entropy S. Both of these quantities can be found in terms of the specific heat at constant pressure C_p. Thus, in a reversible process at constant pressure, the heat exchanged between the system and its surroundings equals the enthalpy change of the system, or

$$q = dH \qquad 10.15$$

where q represents a small transfer of heat into or out of the system and dH is the accompanying enthalpy change of the system. But for one mole of a substance by the definition of specific heat

$$q = \int C_p \, dT$$

where C_p is the specific heat at a constant pressure (the number of calories required to raise one mole of the substance 1 K). Thus,

$$dH = C_p \, dT$$

and taking the enthalpy of the system at absolute zero as H_0 the enthalpy at any temperature T is, accordingly,

$$H = H_0 + \int_0^T C_p \, dT \qquad 10.16$$

Simlarly, for a reversible process by the thermodynamic definition of entropy

$$dS = \frac{dq}{T} = \frac{C_p \, dT}{T}$$

Integration of this equation from 0 K to the temperature in question leads to

$$S = S_0 + \int_0^T \frac{C_p \, dT}{T} = \int_0^T \frac{C_p \, dT}{T} \qquad 10.17$$

The term S_0, which is the entropy at absolute zero, may be equated to zero, for at this temperature, assuming perfect crystallinity (a pure substance with no entropy of mixing) and lattice vibrations in their ground levels, there is no disorder of any kind. (This is actually a statement of the third law of thermodynamics, which says that the entropy of a pure crystalline substance is zero at the absolute zero of temperature.)

With the aid of the equations stated in the preceding paragraph, we can now express the free energy of both the alpha and beta phases of tin as functions of the temperature and the specific heat at constant pressure.

$$G^\alpha = H_0^\alpha + \int_0^T C_p^\alpha \, dT - T \int_0^T \frac{C_p^\alpha \, dT}{T}$$

$$G^\beta = H_0^\beta + \int_0^T C_p^\beta \, dT - T \int_0^T \frac{C_p^\beta \, dT}{T}$$

10.18

These equations can be evaluated by means of graphical calculus, using the curves of Fig. 10.1. The results are shown in Fig. 10.2, where the free energies of the two phases are equal at the temperature 286.2 K. At temperatures below this value, gray tin has the lowest free energy, while for temperatures above it, white tin has the lowest free energy. Since the phase with the lowest free energy is always the most stable one, we see that gray tin is the preferred phase below 286.2 K, while white tin is the more stable at temperatures higher than this.

An interesting feature to notice in Fig. 10.2 is that the free energy of both phases decreases with increasing temperature. This is a general property of free-energy curves, which demonstrates the importance of the TS term in the free-energy equation. Figure 10.3 shows another method of interpreting the information given in Fig. 10.2. The curve marked ΔG in this figure represents the difference in free energy between the two solid phases of tin as a function of temperature. This curve is also the difference of the other two curves designated in the figure as ΔH and $T\Delta S$, where

$$\Delta H = H_0^\beta + \int_0^T C_p^\beta \, dT - H_0^\alpha - \int_0^T C_p^\alpha \, dT$$

10.19

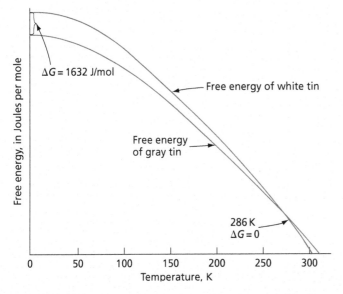

Fig. 10.2 Temperature-free energy curves for two solid forms of tin

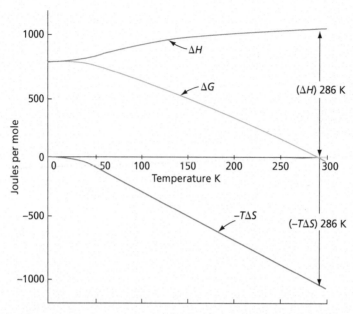

Fig. 10.3 Curves showing the relationship among ΔG, ΔH, and $-T\Delta S$ for the two solid phases of tin, where $\Delta G = \Delta H - T\Delta S$

and

$$T\Delta S = T\int_0^T \frac{C_p^\beta \, dT}{T} - T\int_0^T \frac{C_p^\alpha \, dT}{T}$$

At the temperature of transformation (286.2 K), ΔH and $T\Delta S$ are equal, so that ΔG is zero.

The above allotropic transformation can also be interpreted in the following manner. At low temperatures, the phase with the smaller enthalpy is the stable phase where, by definition, the enthalpy is

$$H = U + PV \qquad 10.20$$

where U is the internal energy, P the pressure, and V the volume. In solids, the PV term is usually very small compared with the internal energy. Thus, the stable phase at lower temperatures is that with the lower internal energy or with the tighter binding of the atoms. On the other hand, at high temperatures the $T\Delta S$ becomes more important, and the phase with the greater entropy becomes the preferred phase. Generally, this is a phase with a looser form of binding or with greater freedom of atom movement. In tin, it appears that, although the diamond cubic lattice (alpha phase) has the lower density, the binding forces between atoms in this structure are of such a nature that the atoms are more restrained in their vibratory motion than they are in the tetragonal, or beta, lattice.

In many cases, the structure of closest packing represents the more closely bound phase, and a more open structure the one with the greatest entropy of vibration. In this classification, where the high-temperature stable phase is body-centered cubic and the low-temperature phase a close-packed structure, such as face-centered cubic or close-packed hexagonal, the elements are Li, Na, Ca, Sr, Ti, Zr, Hf, and Tl.

The element iron presents an unusual example of allotropic-phase changes, in that the stable phase at low temperatures is body-centered cubic that transforms to face-centered cubic at 1183 K. On further heating, a second solid-state transformation to body-centered cubic occurs at 1673 K. The solid finally melts at 1812 K. The free energies of these phases versus temperature are shown in Fig. 10.4.

An explanation of these allotropic reactions that occur in iron has been given by Zener and can be interpreted in the following broad general terms. The two competing solid phases of iron are face-centered cubic (gamma phase) and body-centered cubic (alpha phase). However, the alpha phase has two basic modifications: ferromagnetic and paramagnetic. The ferromagnetic form of the body-centered cubic phase is stable at low temperatures, while the paramagnetic form is stable at high temperatures. The change from ferromagnetism to paramagnetism takes place by what is known as a *second-order transformation*, occurring over a range of temperatures extending from about 773 K to several hundred degrees above this temperature. It would, thus, appear that at low temperatures, the competing, phases in iron are actually ferromagnetic alpha and gamma, where the phase of the lowest internal energy and entropy is the former. The alpha phase is, accordingly, the stable phase at low temperatures, but gives way to the gamma phase with increasing temperatures as the entropy term of the difference in free energy between the phases becomes dominant. This transformation takes place at 1183 K.

At about the same time that the gamma phase becomes the more stable form, the alpha phase loses its ferromagnetism, therefore, for further increases in temperature the competing phases are gamma and paramagnetic alpha. In the paramagnetic form, the alpha phase has a greater internal energy and entropy than it does in the ferromagnetic form. This effect is large enough to reverse the relative positions of the alpha phase and the gamma phase. The gamma phase now becomes the one with the lowest internal energy and entropy and gives way to the alpha phase when the temperature exceeds 1673 K.

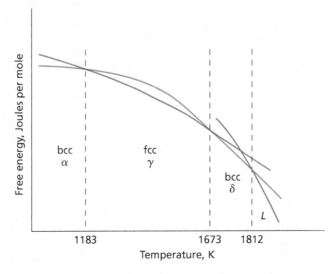

Fig. 10.4 Free energies of bcc, fcc, and liquid iron versus temperature

Two-Component Systems The study of systems of more than one component involves a study of solutions. The simplest type of multicomponent system is a binary one, and the least complex structure in such a system is a single solution. Some of the aspects of single-phase binary systems will now be considered.

Ideal Solutions Assume that a solution is formed between atoms of two kinds, A and B, and that any given atom of either A or B exhibits no preference as to whether its neighbors are of the same kind or of the opposite kind. In this case, there will be no tendency for either A atoms or for B atoms to cluster together or for opposite types of atoms to attract each other. Such a solution is said to be an *ideal solution*, and its free energy per mole is expressed as the sum of the free energy of N_A moles of A atoms, plus N_B moles of B atoms, less a decrease in free energy due to the entropy associated with mixing A atoms and B atoms, where N_A and N_B are the mole, or atom, fractions of A and B, respectively.

$$G = G_A^0 N_A + G_B^0 N_B + T\Delta S_M \qquad 10.21$$

where G_A^0 = free energy per mole of pure A
G_B^0 = free energy per mole of pure B
T = the absolute temperature
ΔS_M = the entropy of mixing

If the previously derived expression for the entropy of mixing, Eq. 7.35, is substituted in Eq. 10.21,

$$G = G_A^0 N_A + G_B^0 N_B + RT(N_A \ln N_A + N_B \ln N_B) \qquad 10.22$$

Rearranging Eq. 10.22 gives us

$$G = N_A(G_A^0 + RT \ln N_A) + N_B(G_B^0 + RT \ln N_B) \qquad 10.23$$

It is customary in chemical thermodynamics to express an extensive property of a solution, such as its free energy, in the following form:

$$G = N_A \overline{G}_A + N_B \overline{G}_B \qquad 10.24$$

where G is the free energy per mole of solution, N_A and N_B are the mole fractions of the components A and B in the solution, and $\overline{G}_A$ and $\overline{G}_B$ are called the *partial-molar free energies* of components A and B, respectively. In the ideal solution discussed in the preceding paragraph, the partial-molar free energies are equal to

$$\overline{G}_A = G_A^0 + RT \ln N_A$$
$$\overline{G}_B = G_B^0 + RT \ln N_B \qquad 10.25$$

These relationships can also be written:

$$\Delta \overline{G}_A = \overline{G}_A - G_A^0 = RT \ln N_A$$
$$\Delta \overline{G}_B = \overline{G}_B - G_B^0 = RT \ln N_B \qquad 10.26$$

The quantities $\Delta \overline{G}_A$ and $\Delta \overline{G}_B$ represent the increase in free energy when one mole of A, or one mole of B, is dissolved at constant temperature in a very large quantity of the solution. These free-energy changes are functions of the composition, and it is necessary to define them in terms of the addition of a quantity of pure A or pure B to a very large volume of the solution in order that the composition of the solution remains unchanged.

Nonideal Solutions In general, most liquid and solid solutions are not ideal and, in solid solutions, it is not to be expected that two atomic forms, chosen at random, will show no preference either for their own or for their opposites. If either of these events occurs, then a larger or smaller free-energy change ($\Delta \overline{G}_A, \Delta \overline{G}_B$) will be observed than that expected in an ideal solution. For a nonideal solution, the change in molar free energy is not given by Eq. 10.26:

$$\Delta \overline{G}_A = RT \ln N_A$$

which holds for an ideal solution. However, in order to retain the form of this simple relationship, we define a quantity "a," known as the activity, in such a fashion that for a nonideal solution

$$\Delta \overline{G}_A = RT \ln a_A$$
$$\Delta \overline{G}_B = RT \ln a_B$$

10.27

where a_A is the activity of component A, and a_B the activity of component B. The activities of the components of a solution are useful indicators of the extent to which a solution departs from an ideal solution. They are functions of the composition of the solution, and Fig. 10.5 shows typical activity curves for two types of alloy systems. In both Figs. 10.5A and 10.5B the straight lines marked N_A and N_B correspond to the atom fractions of the components A and B, respectively. Figure 10.5A shows an alloy system in which activities a_A and a_B are greater than the corresponding atom fractions N_A and N_B at an arbitrary composition, as indicated by the dotted vertical line marked x. The significance of this positive deviation will now be considered. Since both activities are greater than the corresponding mole fractions, we need consider only

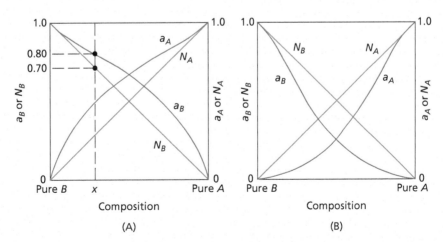

Fig. 10.5 Variation of the activities with concentration: **(A)** positive deviation, and **(B)** negative deviation

one component, which we shall arbitrarily choose as B. According to the figure, N_B is 0.70 while a_B is 0.80. Substituting these values and solving for $\Delta \overline{G}$ leads to the following results for an ideal solution:

$$\Delta \overline{G} = RT \ln 0.70 = -0.356 \, RT$$

For the nonideal solution:

$$\Delta \overline{G} = RT \ln 0.80 = -0.223 \, RT$$

A solution that behaves in the manner shown in Fig. 10.5A has a smaller decrease in free energy as a result of the formation of the solution than if an ideal solution had formed. In situations such as this, the attraction between atoms of the same kind is greater than the attraction between dissimilar atoms. The extreme example would be one in which the two components were completely insoluble in each other, in which case the activities would be equal to unity for all ratios of A and B. The other set of curves in Fig. 10.5B shows the opposite effect: unlike atoms are more strongly attracted to each other than are those of the same kind. Here we find negative deviations of the activity curves from the mole fraction lines.

It is apparent from the above that when the activities of the components of a solution are compared to their respective atom fractions, they indicate roughly the nature of the interactions of the atoms in the solution. For this reason, it is frequently convenient to use quantities known as *activity coefficients*, which are the ratios of the activities to their respective atom fractions. The activity coefficients of a binary solution of A and B atoms are thus defined as

$$\gamma_A = \frac{a_A}{N_A} \quad \text{and} \quad \gamma_B = \frac{a_B}{N_B} \qquad 10.28$$

In the diffusion chapters (Chapters 12 and 13) it will be shown that when a composition gradient exists in a solid solution, diffusion usually occurs in such a way as to produce a uniform composition in the solid solution. The thermodynamic reasons for this will now be considered. Thus, if one has an ideal solution, then by Eqs. 10.25 the free energy of a mole of the solution should be given by

$$G = N_A G_A^0 + N_B G_B^0 + RT(N_A \ln N_A + N_B \ln N_B) \qquad 10.29$$

The first two terms ($N_A G_A^0 + N_B G_B^0$) on the right-hand side of Eq. 10.29 may be considered to represent the free energy of one total mole of the two components if the components are not mixed, or do not mix. In the nonmiscible iron-lead system, these two terms would be all that are required to specify the free energy of one total mole of iron and lead. The other term, $RT(N_A \ln N_A + N_B \ln N_B)$, is the contribution of the entropy of mixing to the free energy of the solution. Notice that this term is directly proportional to the temperature and becomes increasingly important as the temperature is increased.

To illustrate the dependence of the free energy of an ideal solution on its composition, let us consider the following assumed data. Free-energy curves for two hypothetical pure elements A and B, which are assumed to be completely soluble in each other in the solid state, are shown in Fig. 10.6. Notice that in both curves the free energy is shown to decrease with increasing temperature, in agreement with the previously considered free-energy curves of the two solid phases of pure tin (Fig. 10.2). At 500 K, the curves of Fig. 10.6 show that the free energy of the A component (G_A^0) is 6280 J per mole, while that of the B component (G_B^0) is 8370 J per mole. These two values

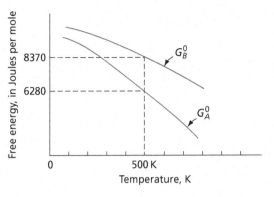

Fig. 10.6 Free energies of the elements are different functions of the absolute temperature

are plotted in Fig. 10.7 at the right- and left-hand sides of the figure, respectively. The dashed line connecting these points represents the terms $N_A G_A^0 + N_B G_B^0 = 6280\,N_A + 8370\,N_B$, or the free energy of the solution less the entropy-of-mixing term. The latter term can be evaluated with the aid of the data given in Table 10.2, where values of $(N_A \ln N_A + N_B \ln N_B)$ are given as a function of the atom fraction N_A (column 2), while values of the product $RT(N_A \ln N_A + N_B \ln N_B)$ are given in column 3 for the temperature 500 K. The data of column 3 are plotted in Fig. 10.7 as the dashed curve at the bottom of the figure. The free energy of the solution proper is shown as the solid curve of Fig. 10.7 and is the sum of the two dashed curves.

At this point let us assume that we have a diffusion couple, one part of which consists of half a mole of pure component A, and the other part of half a mole of pure component B. The average composition of this pair of metals, when expressed in mole fraction units, is $N_A = N_B = 0.5$, while its free energy at 500 K corresponds to point a in Fig. 10.7 and is 7325 J/mol. The same quantity of A and B, when mixed in the form of a homogeneous solid solution, has the free energy of point b in Fig. 10.7 and is 4457 J/mol. Clearly, mixing one-half mole of component A with one-half mole of component B reduces the free energy of the pair of metals by 2868 Joules, the value of the contribution of the entropy of mixing to the free energy of the solution. This same 2868 J per mole is therefore the driving force that is capable of causing the components of the couple to diffuse.

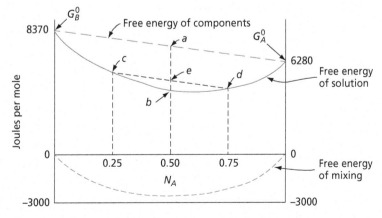

Fig. 10.7 Hypothetical free energy of an ideal solution

Table 10.2 Data for Computing the Entropy of Mixing Contribution to the Free Energy of an Ideal Solution

Atom Fraction, N_A	$(N_A \ln N_A + N_B \ln N_B)$	$RT(N_A \ln N_A + N_B \ln N_B)$ for Temperature 500 K
0.00	0.000	000 J/mol
0.10	−0.325	−1351
0.20	−0.500	−2079
0.30	−0.611	−2540
0.40	−0.673	−2798
0.50	−0.690	−2868
0.60	−0.673	−2798
0.70	−0.611	−2540
0.80	−0.500	−2079
0.90	−0.325	−1351
1.00	−0.000	000

Now suppose that instead of two pure components, the above couple had consisted of two homogeneous solid solutions: one solid solution with a composition $N_A = 0.25$ and the other with the composition $N_A = 0.75$. At 500 K the free energy of the two portions of this couple are shown in Fig. 10.7 as points c and d, respectively. If the average composition of the whole couple is again 0.5, the average free energy of the couple will lie at point e. A single homogeneous solid solution of the same average composition will have the free energy of point b. Here again we see that the homogeneous solid solution has the lower free energy and represents the stable state.

With arguments similar to the above, it is possible to show that any macroscopic nonuniformity of composition in a solution phase represents a state of higher free energy than a homogeneous solution. If the temperature is sufficiently high that the atoms are able to move at an appreciable rate, diffusion will occur that has, as its end result, a homogeneous solid solution. While the arguments used above were based on an average total composition $N_A = 0.5$, any other total or average composition would have given the same final results.

10.6 Two-Component Systems Containing Two Phases

Let us now take up the case of two-component systems that contain not a single phase, but two phases. In either of the two phases, an equation of the form

$$n_A d\overline{G}_A + n_B d\overline{G}_B = 0 \qquad 10.30$$

must be satisfied. Now, if each term is divided by the quantity $n_A + n_B$, we can rewrite this relationship in the form

$$N_A d\overline{G}_A + N_B d\overline{G}_B = 0 \qquad 10.31$$

where $n_A/(n_A + n_B) = N_A$ and $n_B/(n_A + n_B) = N_B$. If the phases are labeled alpha (α) and beta (β), then in the respective phases we have

$$\text{alpha phase: } N_A^\alpha d\overline{G}_A + N_B^\alpha d\overline{G}_B = 0$$
$$\text{beta phase: } N_A^\beta d\overline{G}_A + N_B^\beta d\overline{G}_B = 0$$

10.32

where the phase superscripts of the partial-molar free energies are omitted because, assuming equilibrium, the partial-molar free energy of either component is the same in both phases.

This pair of equations restricts the values of the mole fractions of the components in the solutions. These equations will be referred to as *restrictive equations*. In order to understand how restrictive equations work, let us consider an example.

Imagine that an alloy of copper and silver is formed by melting together an equal amount of each component. After the mixture has been formed, let it be frozen into the solid state and then reheated and held at 1052 K long enough to attain equilibrium. If, at the end of this heating period, the alloy is cooled very rapidly to room temperature, we can assume that the phases which were stable at the elevated temperature will be brought down to room temperature unchanged. Metallographic examination of such an alloy will reveal a structure consisting of two solid-solution phases, and a corresponding chemical analysis of the phases will show that the phases have the following compositions:

Alpha (α) Phase	Beta (β) Phase
$N_{Ag} = 0.86$	$N_{Ag} = 0.05$
$N_{Cu} = 0.14$	$N_{Cu} = 0.95$

Substituting these values in the restrictive equations gives us

$$0.86 d\overline{G}_{Ag} + 0.14 d\overline{G}_{Cu} = 0$$
$$0.05 d\overline{G}_{Ag} + 0.95 d\overline{G}_{Cu} = 0$$

If the first equation is divided by the second, there results

$$\frac{0.86 d\overline{G}_{Ag}}{0.05 d\overline{G}_{Ag}} = \frac{-0.14 d\overline{G}_{Cu}}{-0.95 d\overline{G}_{Cu}}$$

or

$$\frac{0.86}{0.05} = \frac{0.14}{0.95}$$

This impossible result suggests that the only way that the above equations can be true is for both $d\overline{G}_{Ag}$ and $d\overline{G}_{Cu}$ to be zero, which means that at a constant temperature (1052 K) and a constant pressure (one atmosphere), there can be no change in the partial-molar free energies when the two phases are in equilibrium. Further, since the partial-molar free energies are functions of only the composition of the phases, this, in turn, implies that the compositions of the two phases must be constant. No matter what the relative amounts of the two phases happen to be, as long as there is some of both phases present, the composition in atomic percent of the alpha phase will be 86 percent Ag and 14 percent Cu, while that of the beta phase will be 5 percent Ag and 95 percent Cu.

10.7 Graphical Determinations of Partial-Molar Free Energies

A graphical method for determining the partial-molar free energies of a single binary solution is shown in Fig. 10.8. This figure shows the same molar-free energy curve previously given in Fig. 10.7. Suppose that one desires to determine the partial-molar free energies of the two components of the solution when the composition has some arbitrary value, perhaps $N_A = 0.7$. A vertical line through this composition intersects the free-energy curve at point x, thereby determining the total free energy of the solution. Now if a tangent is drawn to the free-energy curve at point x, the ordinate intercepts of this tangent with the sides of the diagram (that is, at compositions $N_A = 0$ and $N_A = 1$) give the partial-molar free energies. The intercept on the left is $\overline{G}_B$ and that on the right $\overline{G}_A$. That these relationships are true is easily shown by the geometry of the figure, which is reproduced for clarity in Fig. 10.8B. Thus, the free energy of the solution is

$$G = \overline{G}_B + N_A(\overline{G}_A - \overline{G}_B)$$

or

$$G = N_A \overline{G}_A + (1 - N_A)\overline{G}_B$$

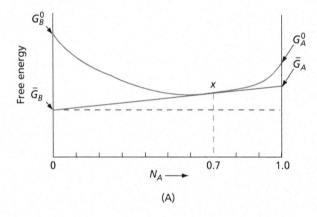

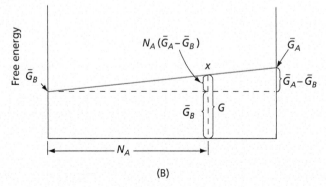

Fig. 10.8 Graphical determination of the partial-molar free energies

but

$$(1 - N_A) = N_B$$

and therefore

$$G = N_A \overline{G}_A + N_B \overline{G}_B$$

which is the basic equation for the free energy of a binary solution.

The fact that at constant pressure and temperature the compositions of the phases are fixed in a binary two-phase mixture at equilibrium can also be shown with the aid of free-energy diagrams of the phases plus the condition that the partial-molar free energies of each component be the same in both phases at equilibrium (that is, $\overline{G}_A^\alpha = \overline{G}_A^\beta$ and $\overline{G}_B^\alpha = \overline{G}_B^\beta$). Figure 10.9 shows hypothetical free-energy curves for two phases of a binary system at a temperature where the free-energy curves intersect. This intersection of the curves is a condition for the existence of two phases at equilibrium. Certainly, if the free-energy-versus-composition curve of one phase lies entirely below that of another phase, the lower one will always be the more stable and there can only be a single equilibrium phase. It is also important to notice that the free-energy curves of the type shown in Fig. 10.9 are functions of temperature. It is entirely possible for a pair of these curves to intersect at one temperature, but at some higher or lower temperature to have no points of intersection. In Chapter 11 more will be said on this subject when phase diagrams are discussed.

Now assume that we have an alloy containing both phases in which the composition of the alpha phase is $N_{A_1}^\alpha$ and that of the beta phase is $N_{A_1}^\beta$. Points a and b on the respective free-energy curves give the free energies of the alpha and beta phases. The partial-molar free energies of the two phases can be obtained by drawing tangents to points a and b. Their interception with the sides of the figure determine the desired quantities. As may be seen by examining Fig. 10.9, the arbitrarily chosen compositions of the two phases result in partial-molar free energies of the components which are not the same in the two phases: $\overline{G}_A^\alpha \neq \overline{G}_A^\beta$ and $\overline{G}_B^\alpha \neq \overline{G}_B^\beta$. Actually the only way that two phases can have identical partial-molar free energies is for both phases to have the same tangent. In other words, as shown in Fig. 10.10, the composition of the two phases must have values corresponding to the intersection of a common tangent with the two free-energy curves. Since only one common tangent can be drawn to the curves of Fig. 10.10, the compositions of the alpha and beta phases must be $N_{A_2}^\alpha$ and $N_{A_2}^\beta$.

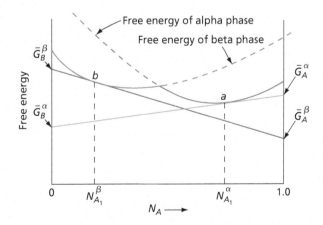

Fig. 10.9 Two phases having compositions of N_{A1}^α and N_{A1}^β, respectively, cannot be in equilibrium

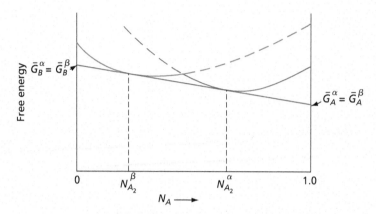

Fig. 10.10 Two phases in equilibrium given by the common tangent

10.8 Two-Component Systems with Three Phases in Equilibrium

In two-component systems, three phases in equilibrium occur only under very restricted conditions. The reason for this is not hard to see, for the partial-molar free energies of the components must be the same in each of the three phases. In a graphical analysis like that used in the two-phase example, this means that we must be able to draw a single straight line that is tangent simultaneously with the free-energy curves of all three phases; see Fig. 10.11. Now it is necessary to notice that each free-energy curve varies with temperature in a manner different from that of the other phases and, in general, there will only be one temperature where it is possible to draw a single straight-line tangent to all three curves. It can be seen, therefore, that three phases in a binary system can only be in equilibrium at one temperature. Furthermore, the compositions of

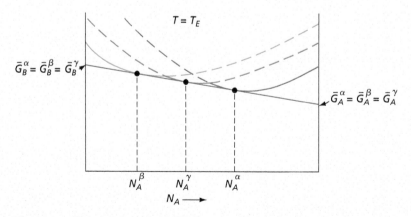

Fig. 10.11 Three phases in equilibrium

Table 10.3 Types of Three-Phase Transformations That Can Occur in Binary Systems

Type of Transformation	Phase A	Nature of the Phase Transformation Phase B	Phase C
Eutectic	Liquid	⇌ Solid Solution	+ Solid Solution
Eutectoid	Solid Solution	⇌ Solid Solution	+ Solid Solution
Peritectic	Liquid Solution	+ Solid Solution	⇌ Solid Solution
Monotectic	Liquid Solution	⇌ Liquid Solution	+ Solid Solution

the phases are fixed by the points where the common tangent touches the free-energy curves. We must conclude that in a binary system, when three phases are in equilibrium at constant pressure, the temperature and the composition of each phase is fixed.

At this point, it is well to recall the phase relations in single-component systems where it was observed that, under the condition of constant pressure, two phases could be in equilibrium at only those fixed temperatures at which phase changes occur. An analogous situation exists in two-component systems, for the temperatures at which three phases can be in equilibrium are also temperatures at which phase changes take place. Associated with each of these three-phase reactions is a definite composition of the entire alloy. At this composition, a single phase can be converted into another of two phases. In certain alloy systems, the single phase is the stable phase at temperatures above the transformation temperature, while two phases are stable below it. In other systems the reverse is true.

As an example of a three-phase reaction, we may take the lead-tin alloy containing 61.9 percent Sn and 38.1 percent Pb (expressed in weight percent). This alloy forms a simple liquid solution at temperatures above 456 K, and at temperatures below 456 K it is stable in the form of a two-phase mixture. Each of the latter is a terminal solid solution, the compositions of which are alpha phase 19.2 percent Sn and beta phase 97.5 percent Sn. A reaction of this type, when a single liquid phase is transformed into two solid phases, is known as an *eutectic reaction*. The temperature at which the reaction takes place is the eutectic temperature, and the composition which undergoes this reaction is the eutectic composition (in the present example 61.9 percent Sn and 38.1 percent Pb).

The eutectic reaction is only one of several well-known three-phase transformations, as can be seen by examining Table 10.3.

In the four transformation equations in Table 10.3, the reactions proceed to the right when heat is removed from the system, and to the left when heat is added to the system. These reactions will be treated separately in Chapter 11.

10.9 The Gibbs Phase Rule

We will now summarize most of the pertinent points of the preceding sections. For this purpose, consider Fig. 10.12, a phase diagram of a one-component system where the variables are the pressure and the temperature. A point such as that marked a, lying within the limits of the gaseous phase, represents a gas whose state is determined by the temperature T_a and the pressure P_a. Changing either or both variables so as to bring the gas to some other arbitrary state (a') is entirely possible.

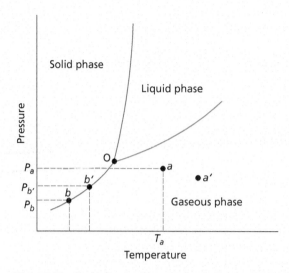

Fig. 10.12 A single-component phase diagram

As long as one stays within the limits of the region indicated as belonging to the gaseous phase, it is evident that there are two degrees of freedom. That is, both the temperature and the pressure may be arbitrarily varied. This statement also applies to the other two single-phase regions, liquid and solid. In summary, when a single-component system exists as a single phase, there are two degrees of freedom. This conclusion is stated in tabular form along the first horizontal line in Table 10.4.

Now consider the case where there are two phases in a single-component system. This can only occur if the state of the system falls along one of the three lines in Fig. 10.12 separating the single-phase fields. For example, point b represents a solid-vapor (gas) mixture in equilibrium at temperature T_b and pressure P_b. If now the temperature is changed to T_b', the pressure has to be changed to exactly P_b'. If this is not done, the two-phase system becomes a single-phase system: either all solid or all gas. It can be concluded that there is only one degree of freedom. This result is tabulated on the second line in Table 10.4.

Table 10.4 The Relative Number of Phases and Degrees of Freedom in One- and Two-Component Systems

Number of Components C	Number of Phases P	Degrees of Freedom F
1	1	2 (T, P)
1	2	1 (T or P)
1	3	0
2	1	3 (T, P, N_A or N_B)
2	2	2 (T, P)
2	3	1 (T or P)
2	4	0

The last possibility for the single-component system is illustrated by the third line in Table 10.4 and corresponds to the case where three phases are in equilibrium in a single-component system. This can only occur at the triple point designated by the symbol 0 in Fig. 10.12. This three-phase equilibrium only occurs at one specific combination of temperature and pressure and there are, accordingly, no degrees of freedom.

Table 10.4 next lists the various possibilities for a two-component system. Note that for a single phase in a two-component system there are three degrees of freedom. These are normally considered to be the temperature, the pressure, and the composition of the phase. In the case of two phases in equilibrium in a binary system there are two degrees of freedom. This conforms to the conclusions of Sec. 10.6, where it was demonstrated that at a chosen temperature and pressure where two phases could be maintained in equilibrium, the compositions of the phases were automatically determined. Of course, it is possible to vary the temperature and pressure, but such variations bring about determinable changes in the composition of the phases.

Next, consider a binary system with three phases in equilibrium. Section 10.8 has shown that here, if the pressure is fixed, three-phase equilibrium can only occur at one temperature. Varying the pressure naturally changes this temperature. However, this result implies that there is only one degree of freedom.

Finally, the case of four phases in a two-component system is equivalent to a triple point in a single-component system. Four-phase equilibrium can only occur at a single combination of temperature and pressure, and the compositions of all phases are fixed. There are, accordingly, no degrees of freedom.

This type of analysis could be extended to systems with larger numbers of components, but it is simpler to invoke the use of the Gibbs phase rule, named after J. Willard Gibbs. This relationship can be deduced from the data in Table 10.4. Note that if along any one of the horizontal lines in the table one adds the number of phases P (given in column 2) to the number of degrees of freedom F (given in column 3), the result is always the number of components, C, plus a factor of 2. Thus we have

$$P + F = C + 2$$

The Gibbs phase rule is often of very great value in determining the factors involved in phase equilibria.

10.10 Ternary Systems

The phase structures of binary alloys are our primary concern here, but the phases in ternary alloys will also be considered. In general, ternary alloys are much more difficult to understand than binary alloys and, as a result, one is usually forced into a study of binary systems, in spite of the fact that many practical problems involve three or more components.

Analogous to the relations already considered for single- and two-component systems, at certain fixed temperatures four-phase transformations may occur in which the compositions of the phases are fixed throughout the reactions. Furthermore, in every case there are certain definite alloy compositions at which the entire alloy undergoes these phase transformations. In these reactions, either a single phase is changed into three other phases, or two phases are changed into two different phases. A typical example of the former type of transformation is a ternary eutectic where a single liquid solution changes into three solid solutions.

The preceding paragraph points out the similarity between four-phase systems in ternary alloys and three-phase systems in binary alloys. In the same manner, three-phase ternary systems are analogous to two-phase binary systems. Thus, at constant temperature and pressure, a three-phase system in a ternary alloy will have fixed compositions of the phases. This fact does not mean that the composition of the alloy as a whole has one value, but that the phases, which may be in any proportion, have fixed compositions.

Three-component alloys may also have structures containing two phases or even a single phase. It is important to notice that in a ternary alloy containing two phases, the compositions of the phases are not fixed. This statement is true of binary systems, but not of ternary systems.

Problems

10.1 List and identify the phases in Fig. 2.27. Explain whether the phase rule can be applied to this structure.

10.2 The three elements zirconium, titanium, and hafnium form an alloy system in which the components are able to form a single solid solution containing the elements mixed in any proportion. Assume that one has such a solid solution containing 0.50 kg titanium, 0.30 kg zirconium, and 1.00 kg hafnium, and determine the mole fraction of each of the components in the alloy. The gram-atomic weights of Ti, Zr, and Hf are 47.9, 91.2, and 178.6 g/mol, respectively.

10.3 Given an alloy containing 70 percent of copper and 30 percent nickel, by weight, determine the composition of this alloy in atomic percent. Note the atomic weights of copper and nickel are 63.546 and 58.71 grams per mole, respectively.

10.4 Consider the activity coefficient vs. composition curves for two alloys in Fig. 10.5. In Fig. 10.5A, a dashed line is drawn at the composition $N_A = 0.70$. As discussed in the text, the activity of the B component, a_B, equals 0.80. The activity a_A is 0.50. Assume the free energy of 1 mole of pure A is 50,000 J/mol and that of pure B is 70,000 J/mol and the temperature is 1000 K.

(a) Compute the free energy of 1 mole of the 0.70 N_A alloy.

(b) Determine the free-energy decrease when a mole of this solution forms from the pure components.

10.5 In Fig. 10.5B, at the composition $N_A = 0.70$ we have

$$a_A = 0.10$$
$$a_B = 0.44$$
$$T = 1000 \text{ K}$$
$$G_A^0 = 50,000 \text{ J/mol}$$
$$G_B^0 = 70,000 \text{ J/mol}$$

(a) Compute the free-energy per mole of the solution with $N_A = 0.30$.

(b) Determine the decrease in free energy if a mole of an ideal solution were to be formed from the pure components.

(c) What is the significance of the difference in the free energies of solution of the two $N_A = 0.70$ alloys in Figs. 10.5A and 10.5B?

10.6 (a) Compute the free-energy decrease associated with forming one mole of an ideal solution at 1000 K, if $G_A^0 = 50{,}000$ J/mol and $G_B^0 = 70{,}000$ J/mol.

(b) Compare your answer with those for the (b) parts in Probs. 10.4 and 10.5 and explain the significance of the differences.

10.7 Compute the activity coefficients corresponding to the four activities involved in Probs. 10.4 and 10.5.

10.8 Consider the hypothetical free-energy diagram of an ideal solution shown in Fig. 10.7. Now assume that a layer of pure metal A is bonded to a layer of pure metal B, that 40 percent by weight of the couple is in the A metal layer and that the atomic weights of metals A and B are 80 and 60 grams per mole, respectively. Further assume that the couple is held long enough at 500 K so that, as a result of diffusion, a homogeneous solid solution is obtained.

(a) What would be the free energy of the couple at the beginning of the diffusion anneal? Give the answer in Joules per mole.

(b) What would be the free energy of the homogeneous solid solution at the end of the diffusion anneal? Give the answer in Joules per mol.

10.9 Assume that the following data hold for an ideal binary solid solution; $T = 1000$ K, $G_A^0 = 7000$ J/mol, and $G_B^0 = 10{,}000$ J/mole. Draw the free energy versus mole fraction diagram for this solid solution (see Fig. 10.8) and determine the partial molar free energies per mole for the solid solution composition containing 0.4 N_A.

10.10 With reference to Fig. 10.10, take $\overline{G}_B^\alpha = 2100$ J/mol, $\overline{G}_A^\beta = 1400$ J/mol, $N_{A_2}^\beta = 0.22$, and $N_{A_2}^\alpha = 0.64$.

(a) Determine the free energy of 1.0 mole of the alloy whose mole fraction is 0.22.

(b) What fraction of this composition will be in the α phase?

(c) Determine the free energy of 1.0 mole of the composition $N_A = 0.50$.

(d) How much of this latter composition will be in the α phase?

10.11 (a) As explained in Sec. 10.10, a ternary alloy may have a ternary eutectic point at which four phases may coexist. Describe the nature of these four phases. How many degrees of freedom are there when four phases coexist in a ternary alloy? What is the meaning of this?

(b) A ternary alloy may also have a three-phase field. How many degrees of freedom will there be in the three-phase field? Explain the significance of this number of degrees of freedom.

(c) In a ternary alloy a single-phase field may also occur. How many degrees of freedom will it contain?

References

1. Darken, L. S., and Gurry, R. W., *Physical Chemistry of Metals*, McGraw-Hill, New York, 1953.
2. Gaskell, D., *Introduction to Thermodynamics of Materials*, McGraw-Hill, New York, 1995.
3. Ragone, D. V., *Thermodynamics of Materials*, Vol. 1, John Wiley and Sons, New York, 1995.
4. Lupis, C. H. P., *Chemical Thermodynamics of Materials*, North-Holland, 1983.
5. DeHoff, R. T., *Thermodynamics in Materials Science*, McGraw-Hill, New York, 1983.

Chapter **11**

Binary Phase Diagrams

Learning Objectives

Upon completion of Chapter 11, you will be able to:
1. Apply phase diagrams to determine equilibrium compositions and temperatures
2. Derive the lever rule and apply it to determine the relative amounts of phases
3. Explain free energy versus composition curves and their utilization to construct equilibrium phase diagrams
4. Understand the influence of negative deviations of activities from ideality and how it may lead to the formation of superlattices
5. Compare and contrast isomorphous alloy systems with eutectic diagrams
6. Discuss microstructural differences between eutectic, hypoeutectic and hypereutectic compositions
7. Compare and contrast eutectic, peritectic, monotectic, eutectoid and peritectoid
8. Identify intermediate (intermetallic) phases and describe their formations via peritectic or congruent transformations
9. Construct ternary diagrams consisting of three simple eutectics and identify freezing pathways through eutectic troughs and ternary eutectics

11.1 Phase Diagrams

Phase diagrams, also called *equilibrium diagrams* or *constitution diagrams*, are a very important tool in the study of alloys. They define the regions of stability of the phases that can occur in an alloy system under the condition of constant pressure (atmospheric). The coordinates of these diagrams are temperature (ordinate) and composition (abscissa). Notice that the expression "alloy system" is used to mean all the possible alloys that can be formed from a given set of components. This use of the word *system* differs from the thermodynamic definition of a system, which refers to a single isolated body of matter. An alloy of one composition is representative of a thermodynamic system, while an alloy system signifies all compositions considered together.

The interrelationships between the phases, the temperature, and the composition in an alloy system are shown by phase diagrams only under equilibrium conditions. These diagrams do not apply directly to metals not at equilibrium. A metal quenched (cooled rapidly) from a higher temperature to a lower one (for example, room temperature) may possess phases or metastable phase and compositions that are more characteristic of the higher temperature than they are of the lower temperature. As discussed in section 14.3, certain alloy compositions can also form non-cystalline and amorphous or glassy structure upon rapid cooling. In time, as a result of thermally activated atomic motion, the quenched specimen may approach its equilibrium low-temperature state. If and when this occurs, the phase relationships in the specimen will conform to the equilibrium diagram. In other words, the phase diagram at any given temperature gives us the proper picture only if sufficient time is allowed for the metal to come to equilibrium.

In the sections that follow, unless specifically stated otherwise, it will always be implied that equilibrium conditions hold. Further, because phase diagrams are of great importance in solving problems in practical metallurgy, we shall conform to common usage and express all compositions in weight percent instead of atomic percent, as has been the case heretofore.

11.2 Isomorphous Alloy Systems

In the discussion that follows, only two-component, or binary, alloy systems will be considered. The simplest of these systems is the *isomorphous*, in which only a single type of crystal structure is observed for all ratios of the components. A typical isomorphous system is represented in the phase diagram of Fig. 11.1. The binary alloy series in question is copper-nickel. These elements combine, as in all alloy systems of this type, to form only a single liquid phase and a single solid phase. Since the gaseous phase is generally not considered, there are only two phases involved in the entire diagram. Thus, the area in the figure above the line marked "liquidus" corresponds to the region of stability of the liquid phase, and the area below the solidus line represents the stable region for the solid phase. Between the liquidus and solidus lines is a two-phase area where both phases can coexist.

The significance of several arbitrary points located on Fig. 11.1 will now be considered. A set of coordinates—a temperature and a composition—is associated with each point. By dropping a vertical line from point x until it touches the axis of abscissae, we find the composition 20 percent Cu. Similarly, a horizontal line through the same point meets the ordinate axis at 500°C.[*] Point x signifies an alloy of 20 percent Cu and 80 percent Ni at a temperature of 500°C. The fact that the given point lies in the region below the solidus line tells us that the equilibrium state of this alloy is the solid phase. The structure implied is, therefore, one of solid-solution crystals, and each crystal will have the same homogeneous composition (20 percent Cu and 80 percent Ni). Under the microscope, such a structure will be identical to a pure metal in appearance. In other respects, however, a metal of the given composition will have different properties. It should be stronger and have a higher resistivity than either pure metal, and it will also have a different surface sheen or color.

Let us turn our attention to the point, y, which falls inside the two-phase region, the area bounded by the liquidus and solidus lines. The indicated temperature in this case is 1200°C, while the composition is 70 percent Cu and 30 percent Ni. This 70 to 30 ratio of the components represents

---

[*]Note: Phase diagram temperatures are normally expressed in °C rather than in K.

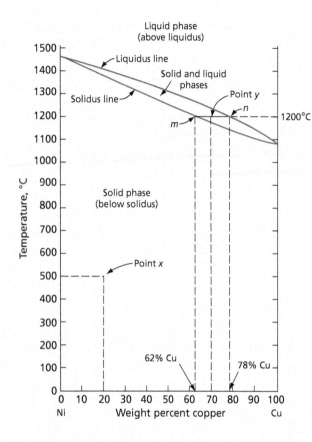

Fig. 11.1 The copper-nickel phase diagram

the average composition of the alloy as a whole. It should be remembered that we are now dealing with a mixture of two phases (liquid plus solid) and that neither possesses the average composition.

To determine the compositions of the liquid and the solid in the given phase mixture, it is only necessary to extend a horizontal line (called a *tie line*) through point y until it intersects the liquidus and solidus lines. These intersections give the desired compositions. In Fig. 11.1, the intersections are at points m and n, respectively. A vertical line dropped from point m to the abscissa axis gives 62 percent Cu, which is the composition of the solid phase. In the same manner, a vertical line dropped from point n shows that the composition of the liquid is 78 percent Cu.

11.3 The Lever Rule

A very important relationship, which applies to any two-phase region of a two-component, or binary, phase diagram, is the so-called *lever rule*. With regard to the lever rule, consider Fig. 11.2, which is an enlarged portion of the copper-nickel diagram of Fig. 11.1. In other words, Fig. 11.2 in just an enlarged view of the upper right-hand corner of Fig. 11.1. Line mn is the same in the new figure as in the old one in Fig. 11.1 and lies on the 1200°C isothermal. One change has been made, however, and that is in the average alloy composition that is now assumed to be at point z, or at 73 percent Cu. Because this new composition still lies on the line mn and therefore in a two-phase region, it must also be a mixture of liquid and solid at the given temperature. Further,

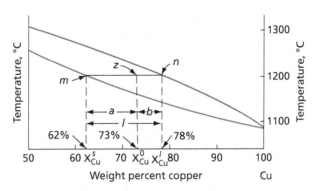

Fig. 11.2 The lever rule

since points *m* and *n* are identical in both Figs. 11.1 and 11.2, it can be concluded that this new alloy must be composed of phases whose compositions are the same as those in the previous example. The quantity that does change, when the composition is shifted along line *mn* from point *y* to *z*, is the relative amounts of liquid and solid. The compositions of the solid and the liquid are fixed as long as the temperature is constant. This conclusion is in perfect agreement with our thermodynamic deductions, Gibbs phase rule, in Chapter 10, where it was shown that at constant temperature and pressure the compositions of the phases in a two-phase mixture are fixed.

While a composition shift of the alloy (taken as a whole) cannot alter the compositions of the phases in a two-phase mixture at constant temperature, it does change the relative amounts of the phases. In order to understand how this occurs, we shall determine the relative amount of liquid and of solid in an alloy of a given composition. For this purpose, take the alloy corresponding to point *z*, which has the average composition $X_{Cu}^o = 73$ percent Cu. In 100 grams of this alloy, 73 grams of them must be copper and the remaining 27 nickel, so that we have

$$\text{Total weight of the alloy} = 100 \text{ grams}$$
$$\text{Total weight of copper} = 73 \text{ grams}$$
$$\text{Total weight of nickel} = 27 \text{ grams}$$

Let *w* represent the weight of the solid phase of the alloy expressed in grams, and $(100 - w)$ the weight of the liquid phase. The amount of copper in the solid form of the alloy equals the weight of this phase times the percentage of copper ($X_{Cu}^s = 62$ percent) which it contains. Similarly, the weight of copper in the liquid phase is equal to the weight of liquid times the percentage of copper in the liquid ($X_{Cu}^l = 78$ percent). Thus,

$$\text{Weight of copper in the solid phase} = 0.62w$$
$$\text{Weight of copper in the liquid phase} = 0.78(100 - w)$$

The total weight of copper in the alloy must equal the sum of its weight in the liquid and in the solid phases, or

$$73 = 0.62w + 0.78(100 - w)$$

Collecting similar terms

$$73 - 78 = (0.62 - 0.78)w$$

and solving for w, the weight of the solid phase,

$$w = \frac{5}{0.16} = 31.25 \text{ grams}$$

Since the total weight of the alloy is 100 grams, the weight percent of the solid phase is 31.25 percent, and the corresponding weight percent of the liquid phase is 68.75 percent.

Now reexamine the equation given above

$$73 - 78 = (0.62 - 0.78)w$$

and divide each side by 100 grams, the total weight of the alloy. If this is done, one obtains

$$0.78 - 0.73 = (0.78 - 0.62)\frac{w}{100}$$

or

$$\frac{w}{100} = \frac{0.78 - 0.73}{0.78 - 0.62}$$

where $w/100$ is the weight percent of the solid phase.

The preceding equation is worthy of careful consideration. In this expression the denominator is the difference in composition of the solid and the liquid phase (expressed in weight percent copper), that is, it is exactly the composition difference between points m and n. On the other hand, the numerator is the difference between the composition of the liquid phase and the average composition (composition difference of points n and z). The weight fraction of the solid phase (expressed in weight percent of the total alloy) is therefore given by

$$\text{weight fraction solid} = \frac{\text{composition of liquid} - \text{average composition}}{\text{composition of liquid} - \text{composition of solid}}$$

$$= \frac{X^l_{Cu} - X^o_{Cu}}{X^l_{Cu} - X^s_{Cu}} \qquad 11.1$$

Similarly, it may be shown that

$$\text{weight fraction liquid} = \frac{\text{average composition} - \text{composition of solid}}{\text{composition of liquid} - \text{composition of solid}}$$

$$= \frac{X^o_{Cu} - X^s_{Cu}}{X^l_{Cu} - X^s_{Cu}} \qquad 11.2$$

The preceding equations represent the lever rule as applied to one specific problem. The same relationships can be expressed in a somewhat simpler form, for in Fig. 11.2, a, b, and l represent composition differences between the points zm, nz, and nm, respectively, so that

$$\text{weight fraction solid} = \frac{b}{l} \qquad 11.3$$

$$\text{weight fraction liquid} = \frac{a}{l} \qquad 11.4$$

The above information will now be summarized.

Given a point such as z in Fig. 11.2, which lies inside a two-phase region of a binary-phase diagram,

(a) Find the composition of the two phases.

Draw an isothermal line (*a tie line*), through the given point. The intersections of the tie line with the boundaries of the two-phase region determine the composition of the phases. (In the present example, points *m* and *n* determine the compositions of the phases, solid and liquid, respectively.)

(b) Find the relative amounts of the two phases.

Determine the three distances, a, b, and l (in units of percent composition), as indicated in Fig. 11.2. The amount of the phase corresponding to point *m* is given by the ratio b/l, while that corresponding to point *n* is given by a/l. Notice carefully that the amount of the phase on the left (*m*) is proportional to the length of the line segment (*b*) lying to the right of the average composition (point *z*), while the phase that lies at the right (*n*) is proportional to the line segment (*a*) lying to the left of point *z*. Note that the compositions of the two phases can be read directly from the phase diagram; however, the amounts of the phases need to be calculated.

11.4 Equilibrium Heating or Cooling of an Isomorphous Alloy

Equilibrium heating or cooling means a very slow rate of temperature change so that at all times equilibrium conditions are maintained in the system under study. Just how slowly one has to heat or cool an alloy in order to keep it effectively in a state of equilibrium depends on the metal under consideration and the nature of the phase changes that occur as the temperature of the alloy is varied. For present purposes, it will arbitrarily be assumed that all temperature changes are made at a slow enough rate that equilibrium is constantly maintained. In Chapter 14 some aspects of nonequilibrium temperature changes will be considered when the more intimate details of freezing are investigated. For the time being, however, our attention will be concentrated on the phase changes that occur as a result of temperature variations.

First, the phase changes that occur when a specific alloy of an isomorphous system has its temperature varied through the freezing range will be considered. For this purpose, the hypothetical phase diagram of Fig. 11.3 with assumed components *A* and *B* will be used. As an arbitrary alloy, consider the composition 70 percent *B* and 30 percent *A*. A vertical line is drawn through this composition in Fig. 11.3. Two segments of this line are drawn as solid lines (*ab* and *cd*), while the third segment between points *b* and *c* is shown as a dashed line. The solid portions fall in single-phase regions: *ab* in the solid-solution area, *cd* in the liquid-solution area. The dashed segment *bc* lies in a two-phase region of liquid-plus-solid phases. Any point on one of the solid sections corresponds to a single homogeneous substance. Points in the length *bc*, on the other hand, do not correspond to a single homogeneous form of matter, but to two—a liquid solution and a solid solution—each with a different composition. Furthermore, the compositions of these phases change as the temperature inside the two-phase region is varied. This can perhaps be best understood by considering the complete cycle of the phase change that occurs as the alloy is cooled from point *d* down to room temperature at point *a*.

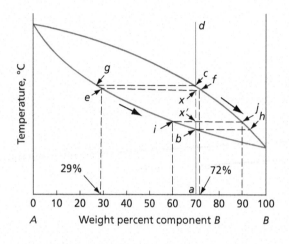

Fig. 11.3 Equilibrium cooling of an isomorphous alloy

At point *d*, the alloy is a homogeneous liquid and, until point *c* is reached, it remains as a simple homogeneous liquid. Point *c*, however, is the boundary of the two-phase region, which signifies that solid begins to form from the liquid at this temperature. The freezing of the alloy, accordingly, commences at the temperature corresponding to point *c*. Conversely, when the alloy has been cooled to point *b*, it must be completely frozen, for at all temperatures below *b* the alloy exists as a single solid-solution phase. Point *b*, therefore, corresponds to the end of the freezing process. An interesting feature of the freezing process is at once apparent: the alloy does not freeze at a constant temperature, but over a temperature range. The temperature range from *b* to *c* is called the *equilibrium freezing range* of the alloy.

The analysis of phase-change phenomena that occur in the freezing range between points *c* and *b* can now be made rather easily with the aid of the two rules set down at the end of the last section. Suppose that the temperature of the alloy has been lowered to a position just below point *c*. The position is designated by the symbol *x* in Fig. 11.3, and at this position

(a) An isothermal line drawn through point *x* determines the composition of the phases by its intersections with the liquidus and solidus lines, respectively.

Liquid phase (point *f*) has the composition 72 percent B.
Solid phase (point *e*) has the composition 29 percent B.

(b) By the lever rule, the amount of the solid phase is:

$$\frac{x-f}{e-f} = \frac{70-72}{29-72} = \frac{2}{43} = 4.7 \text{ percent}$$

and the amount of the liquid phase is 95.3 percent.

The above data show that the given alloy, at a temperature slightly below the temperature at which freezing starts, is still largely a liquid (95.3 percent) and the composition of the liquid (72 percent B) is close to the average composition. On the other hand, the solid phase, which is present in only a small quantity (4.7 percent), possesses a composition differing considerably from the average (29 percent B).

Now let us assume that the alloy is very slowly cooled from point *x* to point *x'*. If this operation is done slowly enough, equilibrium conditions will be maintained at the end of the drop in temperature and we can apply once more the two rules for determining the compositions and amounts of the phases. When this is done, it is found that the composition of the solid phase is now 60 percent B and that of the liquid phase is 90 percent B, while the amount of the liquid and the solid phase is 33 percent and 67 percent, respectively. Comparing these figures with those for the higher temperature (corresponding to point x) shows two important facts. First, as the temperature is lowered in the freezing range, the amount of the solid increases, while the amount of the liquid decreases. This result might well be expected. Second, as the temperature falls, the compositions of both phases change. This fact is not self-evident. It is also interesting to note that the composition shift of both phases occurs in the same direction. Both the liquid and the solid become richer in component B as the temperature drops lower and lower. This apparently anomalous result can only be explained by the fact that as the temperature falls, there is also a corresponding change in the amounts of the phases and the fact that at all times, the average composition of the liquid and the solid is constant.

Another important fact relative to the freezing process in a solid-solution type of alloy is that as the process continues and more and more solid is formed, there must be a continuous composition change taking place in the solid that has already frozen. Thus, at the temperature of point x, the solid has the composition 29 percent B, but at the temperature of x', it has the composition 60 percent B. The only way that this can occur is through the agency of diffusion. Since the composition change is in the direction of increasing concentrations of B atoms, this implies a steady diffusion of B atoms from the liquid toward the center of the solid, and a corresponding diffusion of A atoms in the reverse direction.

The equilibrium freezing of an isomorphous alloy can now be analyzed. On cooling the liquid solution, freezing starts when the liquidus line is reached (point c, Fig. 11.3). The first solid to form has the composition determined by the intersection of an isothermal line drawn through point c with the solidus line (point g). As the alloy is slowly cooled, the composition of the solid moves along the solidus line toward point c, while, simultaneously, the composition of the liquid moves along the liquidus line toward point h. However, at any given instant in the cooling process, such as typified by point x or x', the solid and the liquid must lie at the ends of an isothermal line. In addition, as the freezing process proceeds, the relative amounts of liquid and solid change from an infinitesimal amount of solid in a very large amount of liquid at the start of the process to the final condition of complete solidification.

The preceding discussion has been concerned with the freezing process. The reverse process, melting, in which the alloy is heated from a low temperature, where it is a solid, into the liquid phase, is equally simple to analyze. In this case, point b represents the temperature at which melting commences, point h the composition of the first liquid to form, and point g the composition of the last solid to dissolve.

It is well to note at this point that for all compositions of an isomorphous system, freezing or melting occurs over a range of temperatures. Thus, the melting and freezing points do not coincide as they do in pure metals.

11.5 The Isomorphous Alloy System from the Point of View of Free Energy

In an alloy system such as copper-nickel, the atoms of the components are so similar in nature that it can be assumed, at least as a first approximation, that they form an ideal solution in both liquid and solid phases. The free-energy-composition curves for both phases must, therefore, be similar in nature to those in Fig. 10.7. Consider now the situation that obtains at the melting point of pure nickel. At this temperature (1455°C) the free energies of the liquid and of the solid phases are equal at the composition of pure nickel. Figure 11.1 shows us, however, that for any other alloy composition of copper and nickel at this temperature, the liquid phase is the stable phase and, therefore, its free energy must lie below that of the solid phase. This relationship between the two free-energy curves is shown in Fig. 11.4A. Notice that the two curves intersect only at the composition of pure nickel. At any temperature higher than 1455°C, the two curves will separate so that the liquid-phase-free-energy-composition curve lies entirely below that of the solid phase. On the other hand, decreasing the temperature to a value that is still above the freezing point of pure copper (1084.9°C) has the effect of shifting the curves of the liquid phase upward with respect to that of the solid phase. The two curves now intersect at some intermediate composition, as shown in Fig. 11.4B. This figure represents the free energy of the two solutions at the same temperature (1200°C) as points y and z in Figs. 11.1 and 11.2, respectively. Notice that in this figure the common tangent drawn to the two free-energy curves has points of tangency at the compositions 62 percent Cu and 78 percent Cu. These values are the same as indicated in Figs. 11.1 and 11.2 for the compositions of the phases at 1200°C, which agrees with the results of Chapter 10, where it was shown that the points of contact of the common tangent determine the compositions of the phases in a two-phase mixture.

The complete copper-nickel-phase diagram can be mapped out by considering the relative motion of the two free-energy-composition curves as the temperature is dropped. With decreasing temperature, the effect, as indicated above, is to raise the liquid-phase curve relative to that of the solid phase. As this happens, the point of intersection shifts continuously from the composition of pure nickel at the temperature 1455 °C to the composition of pure copper at 1084.9°C. Concurrent with this movement of the point of intersection toward larger copper concentrations is a similar motion of both points of contact of the common tangent to the two free-energy curves. Below 1084.8 °C the two free-energy-composition curves no longer intersect, and the curve representing the solid phase lies entirely below the liquid curve.

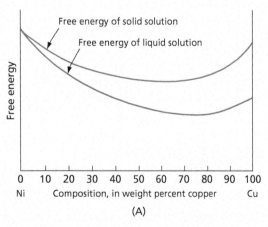 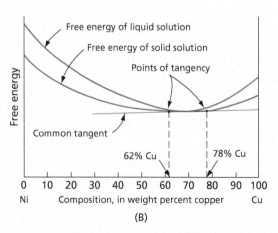

Fig. 11.4 Free-energy-composition curves for the copper-nickel alloy system. **(A)** Freezing point of pure nickel; 1455°C **(B)** 1200°C

11.6 Maxima and Minima

The copper-nickel phase diagram is typical of an alloy system in which the free-energy-composition curves, for a given temperature, of the two solution phases (liquid and solid) intersect at only one composition. Other alloy systems are known which have equivalent curves intersecting at two compositions. When this happens, it is generally observed that in the corresponding phase diagrams liquidus and solidus curves are so shaped as to form either a minimum or a maximum. This can be demonstrated graphically with the aid of Figs. 11.5 and 11.6. Figure 11.5 shows schematically the relationships between the free-energy curves that lead to a minimum configuration. In this case, the curve for the solid has less curvature than that for the liquid. The figure on the left of Fig. 11.5 shows that, with decreasing temperature, intersections of the two free-energy curves occur first at the compositions of pure components (A and B). These intersections then move inward toward the center of the figure and eventually meet at a single point. At the temperature where this occurs (T_c), the two free-energy curves are tangent. Continued lowering of the temperature (T_d) causes the free-energy curves to separate and makes the solid phase the only stable phase at all compositions. In an example such as this, it is possible to draw two common tangents to the free-energy curves. The points at which the tangents contact the free-energy curves determine the limits of the two-phase regions of the equilibrium diagram at any given temperature—for example, T_b. The motion of these points, with varying temperature, maps out the phase diagram shown on the right of Fig. 11.5.

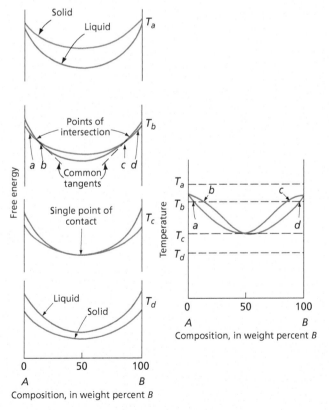

Fig. 11.5 Relationship of the free-energy curves that lead to a minimum

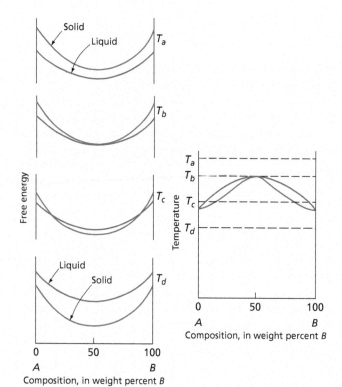

Fig. 11.6 Relationship of the free-energy curves that lead to a maximum

A series of figures, similar to that of Fig. 11.5, is shown in Fig. 11.6. In this case, however, the solid solution is assumed to have the free-energy curve with the greater curvature, so that the free-energy curves, on falling temperature, meet first at a single central point at temperature T_b. This single intersection then splits into two, with the result that the phase diagram has the shape shown on the right of Fig. 11.6.

An important aspect of phase diagrams is shown in Figs. 11.5 and 11.6. When the boundaries of a two-phase region intersect, they meet at a maximum or minimum, and both curves (liquidus and solidus) are tangent to each other and to an isothermal line at the point of intersection. Such points are called *congruent points*. It is characteristic of congruent points that freezing can occur with no change in composition or temperature at these points. Thus, the freezing of an alloy at a congruent point is similar to the freezing of a pure metal. The resulting solid, however, is a solid solution and not a pure component.

It should be further emphasized that the boundaries that define the limits of a two-phase region in an equilibrium diagram can only meet at either congruent points, or at the compositions or pure components. These points, where the phase boundaries meet, are known as *singular points* and, accordingly, in an equilibrium diagram, the single-phase regions, or fields, are always separated by two-phase regions except at singular points.

A number of isomorphous phase diagrams show congruent points that correspond to minima of the liquidus and solidus curves. A typical example is shown in Fig. 11.7, which is the equilibrium diagram for the gold-nickel-alloy system. The significance of the solid line lying below the liquidus and solidus lines will be considered in Sec. 11.8 on miscibility gaps.

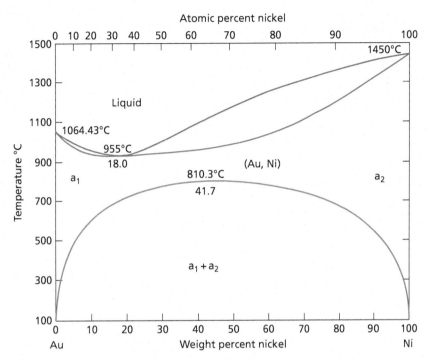

Fig. 11.7 Gold-nickel phase diagram. (From *Binary Alloy Phase Diagrams*, Massalski, T. B., Editor-in-Chief, ASM International, 1986, p. 289. Reprinted with permission of ASM International®. All rights reserved. www.asminternational.org)

Congruent points at which the liquidus and solidus meet at a maxima are not ordinarily found in simple isomorphous equilibrium diagrams. They are observed, however, in several more complicated alloy systems that possess more than a single solid phase. One of the simpler systems is the lithium-magnesium system whose phase diagram is shown in Fig. 11.8. The maximum appears at 601°C and 13 percent Li.

11.7 Superlattices

Copper and gold also freeze to form a continuous series of solid solutions, as shown in Fig. 11.9. The chemical activities of these solid solutions exhibit a negative deviation. This fact is shown in Fig. 11.10, where it can be observed that at 500°C the activities of both gold and copper are smaller than the corresponding mole fractions. Negative deviations of the activities are generally considered as evidence that the components of a binary system possess a definite attraction for each other or at least a preference for opposite atomic forms as neighbors. An interesting result of this effect is the formation, in this system, of ordered structures in which gold and copper atoms alternate in lattice positions in such a way as to form the maximum number of gold-copper atomic bonds and the minimum number of copper-copper and gold-gold bonds. At higher temperatures, thermally induced atomic movements are too rapid to permit the grouping together of a large number of atoms in stable ordered structures. Two opposing factors, namely, the attraction of unlike forms for each other

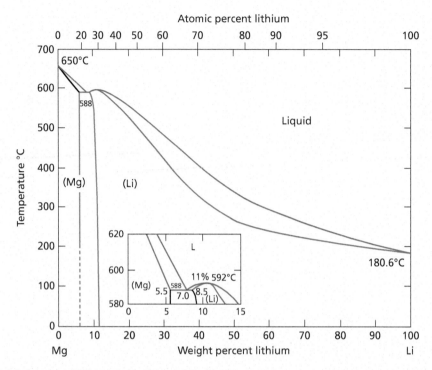

Fig. 11.8 Magnesium-lithium phase diagram. (From *Binary Alloy Phase Diagrams*, Massalski, T. B., Editor-in-Chief, ASM International, 1986, p. 1487. Reprinted with permission of ASM International®. All rights reserved www.asminternational.org.)

and the disrupting influence of thermal motion, therefore, lead to a condition known as *short-range order*. In this situation, gold atoms have a statistically greater number of copper atoms as neighbors than would be expected if the two atomic forms were arranged on the lattice sites of a crystal in entirely random fashion. The effectiveness of thermal motion in destroying an extensive periodic arrangement of the gold and copper atoms, of course, decreases with falling temperature, so that at low temperatures and at the proper compositions gold and copper atoms can arrange themselves in stable configurations that extend through large regions of a crystal. When this occurs, a state of long-range order is said to exist and the resulting structure is called a *superlattice* or *a superstructure*.

An ordered region of a crystal is known as a *domain*. The maximum theoretical size of a domain is determined by the size of the crystal or grain in which the domain lies. Usually, however, a given metal grain will contain a number of domains, and the relationship between grains and domains is indicated in Fig. 11.11. A portion of two schematic ordered crystals is shown based on the assumption of equal numbers of black (*A*) and white (*B*) atoms. The grain at the upper left of the figure contains three domains, while that at the lower right has two. At the domain boundaries, which are outlined by dashed lines, *A* atoms face *A* atoms, and *B* atoms face *B* atoms. Inside of each domain, each *A* atom and each *B* atom is surrounded by atoms of the opposite kind. At the juncture between two domains, the sequence of the *A* and *B* atoms is reversed and it is common practice to call the domains *antiphase domains* and the boundaries *antiphase boundaries*. In the copper-gold system, the transformation from short-range order to long-range order produces supperlattices within three basic composition ranges. One of these regions surrounds the composition corresponding to equal

11.7 Superlattices

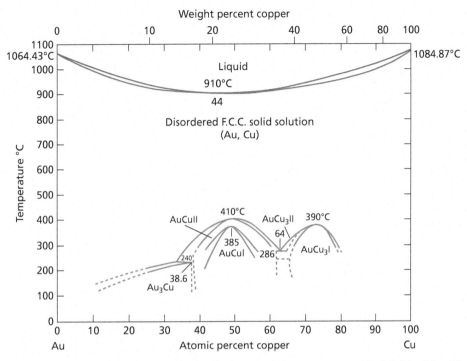

Fig. 11.9 Copper-gold phase diagram. (From *Bulletin of Alloy Phase Diagrams*, Vol. 8, No. 5, by Okamoto, H., Chakrabarti, D. J., Laughlin, D. E., and Massalski, T. B., 1987, p. 454. Reprinted with permission of ASM International®. All rights reserved www.asminternational.org)

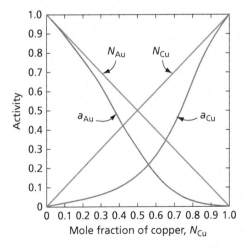

Fig. 11.10 Activities of copper and gold in solid alloys at 500°C. (After Oriani, R. A., *Acta Met.*, **2** 608 [1954].)

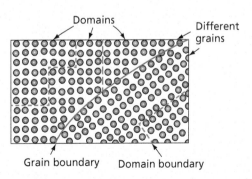

Fig. 11.11 Ordered domains in two different grains. Ordering based on equal numbers of (A) black atoms and (B) white atoms

numbers of gold and copper atoms, and the others have a 3 to 1 ratio of either copper to gold or gold to copper. Each superlattice is a phase in the usual sense and is stable inside a definite range of temperature and composition. Five different superstructures have been identified in the copper-gold system: two corresponding to the composition CuAu, one to the composition Au_3Cu and two for $AuCu_3$. The boundaries delineating the two-phase region surrounding each superlattice phase meet at congruent maxima. The congruent maximum of the Cu_3Au phase appears at 390°C, while the maxima of CuAu lie at 410° and 385°C. Notice that with respect to the CuAu phases, the lower temperature phase is formed on cooling from the higher termperature phase. Thus, when an alloy with a composition approximately equal to AuCu cools down from a temperature near the solidus, it first transforms from the high-temperature short-range ordered phase (disordered face-centered cubic phase) to the phase designated in Fig. 11.9 as AuCuII (orthorhombic). On further cooling, this latter phase transforms into AuCuI (tetragonal). Similarly, the gold-rich $AuCu_3$ composition transforms to $AuCu_3$I (orthorhombic) at high temperatures, then to $AuCu_3$ (face-centered cubic) at lower temperatures. The structure of the Au_3Cu phase is also based on the face-centered cubic lattice.

The unit cells of two of the five superlattices of the copper-gold system are shown in Fig. 11.12. The left drawing shows the structure of Cu_3AuI. It is merely a face-centered cubic unit cell with copper atoms at the face centers and gold atoms at the corners. It is easily proved that this configuration corresponds to the stoichiometric ratio Cu_3Au. There are six face-centered copper atoms, each belonging to two unit cells, or a total of three copper atoms per unit cell. On the other hand, there are eight corner atoms, each belonging to eight unit cells, or one gold atom per unit cell. A perfect lattice, composed of unit cells of this nature, has a ratio of three copper atoms to each gold atom. Actually, as the phase diagram shows, the Cu_3Au structure, as well as the CuAu structures, is capable of existing over a range of compositions. This range is somewhat limited because, as one deviates from the strict stoichiometric ratios, the perfection of the order decreases and with it the stability of the superlattices.

The right diagram in Fig. 11.12 represents the unit cell of the lower-temperature CuAu phase (CuAuI, or tetragonal phase). This structure is also a modification of the face-centered cubic lattice with alternating (001) planes completely filled with gold and copper atoms. The tetragonality of this phase is directly related to the alternating stacking of planes of gold atoms and planes of copper atoms. Such a structure has axes of equal length in a plane containing atoms of the same type, but an axis of different length in the direction perpendicular to this plane. The unit cell is thus distorted from a cube into a tetragon. The other phase, CuAuII, has a somewhat more complicated structure that will not be described, but it can be considered an intermediate step between the short-range-ordered, high-temperature face-centered cubic structure and the low-temperature-ordered tetragonal structure.

Ordered phases are found in a number of alloy systems. (Ref. 1) One of these that has been studied extensively occurs in the copper-zinc system and corresponds to an equal number of copper and zinc atoms. This phase will be considered briefly later in this chapter.

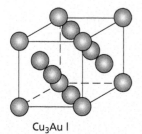
Cu_3Au I

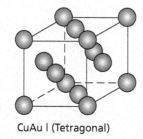
CuAu I (Tetragonal)

Fig. 11.12 Unit cells of two of the five known ordered phases in copper-gold alloys. (● gold atoms, ● copper atoms)

11.8 Miscibility Gaps

Gold and nickel, like copper and nickel and copper and gold, also form an alloy system that freezes into solid solutions in all proportions. See the phase diagram in Fig. 11.7. This system serves as an example of a completely soluble system in which the constituents tend to segregate as the temperature is lowered. Consider now the curved line that occupies the lower central region of the phase diagram. At all temperatures below 810.3°C and inside the indicated line, two phases are stable, α_1 and α_2. The first, α_1, is a phase based on the gold lattice with nickel in solution, and the other, α_2, nickel with gold in solution. Both phases are face-centered cubic, but differ in lattice parameters, densities, color, and other physical properties.

The two-phase field, where α_1 and α_2 are both stable, constitutes an example of what is commonly called a *miscibility gap*. A necessary condition for the formation of a miscibility gap in the solid state is that both components should crystallize in the same lattice form.

In Chapter 10, it was pointed out that many alloy systems have solvus lines that show increasing solubility of the solute with rising temperature. It is quite possible to consider the boundary of the miscibility gap as two solvus lines that meet at high temperatures to form a single boundary separating the two-phase field from the surrounding single-phase fields.

The gold-nickel system is particularly significant because it demonstrates that there is still much to learn about solid-state reactions. When a binary alloy is formed between A and B atoms, there are two possible types of atomic bonds between nearest neighbors: bonds between atoms of the same kind (*A-A* or *B-B* bonds), and bonds between unlike atoms (*A-B* bonds). Associated with each bond between a pair of atoms is a chemical bonding energy which may be written as ε_{AA} or ε_{BB} for pairs of like atoms, and ε_{AB} for a pair of unlike atoms. The total energy of the alloy may, of course, be written as the sum of the energies of all the bonds between neighboring atoms; the lower this energy the more stable the metal. If now the bonding energy between unlike atoms is the same as the average bonding energy between like atoms $\frac{1}{2}(\varepsilon_{AA} + \varepsilon_{BB})$, then there is no essential difference between the bonds, and the solution should be a random solid solution. When ε_{AB} is lower than the average bonding energy of like atoms, short-range order at higher temperatures and long-range order at lower temperatures are to be expected. On the other hand, segregation and precipitation are usually associated with the condition where ε_{AB} is greater than $\frac{1}{2}(\varepsilon_{AA} + \varepsilon_{BB})$.

The fact that gold-nickel alloys exhibit a miscibility gap would generally be construed as evidence that the bonding energy for a gold-nickel pair is larger than the average of the bond energies for gold-gold and nickel-nickel pairs because this effect is expected in segregation. Thermodynamic measurements of solid solutions of these alloys at temperatures above the miscibility gap show that the activities of both gold and nickel exhibit positive deviations. This fact also normally indicates that unlike pairs have a higher bond energy and that gold and nickel atoms prefer to segregate. However, X-ray diffraction measurements show a small but definite short-range order to exist in the solid solutions above the miscibility gap. This apparent contradiction is strong evidence that there is more than one factor involved in determining the type of structure that results. The simple, so-called quasi-chemical theory, which pictures the development of segregation, or ordering, as the result of only the magnitudes of the interatomic bonding energies, is not sufficient to explain the observed results in the gold-nickel system. Other factors certainly must be involved in determining the nature of the solid-state reactions that occur in solid-solution alloys. In the gold-nickel system, the ambiguous results have been explained as being primarily due to the large difference in size of the gold and nickel atoms. (Ref. 2) This difference amounts to about 15 percent and is at the Hume-Rothery limiting value for extensive solubility. When a random solid solution is formed of gold and nickel atoms, the fit between atoms is rather poor and the lattice is badly strained. One way that the strain energy associated with the misfit of gold and nickel atoms can be relieved is by causing the atoms to assume an ordered arrangement. Less strain is produced when gold and nickel atoms alternate in a crystal than when gold or nickel atoms cluster together. By this means, it is possible to explain the short-range order that exists at high temperatures. A still greater decrease in the strain energy of the lattice is possible if it breaks down to form crystals of the gold-rich and nickel-rich phases. Notice that, in this event, it is postulated that separate crystals of the two phases are formed with conventional grain boundaries between them, and that we are not talking about coherent clusters of gold atoms or nickel atoms existing in the original solid-solution crystals. Clustering in the coherent sense would raise, not lower, the strain energy. The nucleation of the segregated phases in a solid solution in which clustering does not occur is, of course, a difficult process, and the precipitation of the phases inside the miscibility gap of the gold-nickel system is a very slow and time-consuming process that is apparently only nucleated at the grain boundaries of the matrix phase.

Miscibility gaps are not only found in solid solutions, but also often in the liquid regions of phase diagrams. A particular liquid miscibility gap will be discussed in a later section.

11.9 Eutectic Systems

The copper-silver phase diagram, Fig. 11.13, can be taken as representative of eutectic systems. In systems of this type, there is always a specific alloy, known as the *eutectic composition*, that freezes at a lower temperature than all other compositions. Under conditions approaching equilibrium (slow-cooling), it freezes at a single temperature like a pure metal. In other respects, the

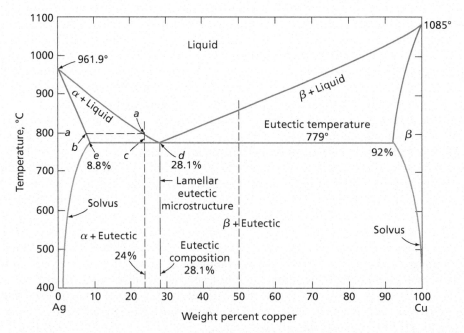

Fig. 11.13 Copper-silver phase diagram. (From *Constitution of Binary Alloys*, by Hansen, M., and Anderko, K. Copyright, 1958. McGraw-Hill Book Co., Inc., New York, p. 18. Used by permission.)

solidification reaction of this composition is quite different from that of a pure metal since it freezes to form a mixture of two different solid phases. Hence, at the eutectic temperature, two solids form simultaneously from a single liquid phase. A transformation, where one phase is converted into two other phases, requires that three phases be in equilibrium. In Chapter 10, it was shown that, assuming constant pressure, three phases can only be in equilibrium at an invariant point, that is, at a constant composition (in this case the eutectic composition) and at a constant temperature (eutectic temperature). The eutectic temperature and composition determine a point on the phase diagram called the *eutectic point*, which occurs in the copper-silver system at 28.1 percent Cu and 779.4°C. The eutectic transformation can be represented as:

$$\text{L (with eutectic composition)} \underset{\text{heating}}{\overset{\text{cooling}}{\rightleftarrows}} \alpha \text{ (terminal solid solution with Ag as solvent)} + \beta \text{ (terminal solid solution with Cu as solvent)}$$

$$T = T_{\text{eutectic}} \qquad \qquad 11.5$$

At this time, it is perhaps well to compare Fig. 11.7, showing the gold-nickel equilibrium diagram and its miscibility gap, with the copper-silver diagram of Fig. 11.13. The close relationship between the two systems is evident. The copper-silver system corresponds to one in which the miscibility gap and the solidus lines intersect. The eutectic point, therefore, is equivalent to the minimum point in the gold-nickel system. In the gold-nickel system, however, an alloy with the composition of the minimum first solidifies as a single homogeneous solid solution and then, on further cooling, breaks down into two solid phases as it passes into the miscibility gap. An alloy possessing the eutectic composition of the copper-silver system, on the other hand, freezes directly into a two-phase mixture.

The fundamental requirement for a miscibility gap is a tendency for atoms of the same kind to segregate in the solid state. The same is also true of an eutectic system. A true miscibility gap, like that in the gold-nickel system, can only occur if the component metals are very similar chemically and crystallize in the same lattice form, since the components must be capable of dissolving in each other at high temperatures. In an eutectic system, the components do not have to crystallize in the same structure nor do they necessarily have to be chemically similar. If the two atomic forms are quite different chemically, however, then intermediate crystal structures are liable to form in the alloy system. Eutectics can still be observed in such systems, but they will not form between terminal phases, as shown in Fig. 11.13.

11.10 The Microstructures of Eutectic Systems

Any composition of an isomorphous system at equilibrium and in the solid state consists of a single homogeneous group of solid-solution crystals. When viewed under the microscope, such a structure does not differ essentially from that of a pure metal. It is usually difficult, therefore, to tell much about the composition of these single-phased alloys from a study of their microstructure alone. On the other hand, the appearance under the microscope of an alloy from the two-phase field (solid) of an eutectic system is very characteristic of its composition.

In discussing the microstructure and other aspects of the alloys of an eutectic system, it is customary to classify them with respect to the side of the eutectic composition on which they fall. Compositions lying to the left of the eutectic point are designated as *hypoeutectic*, while those to the right are called *hypereutectic*. These designations are easily remembered if one recalls the fact that *hypo-* and *hyper-* are Greek prefixes signifying below and above. Thus, reading the copper-silver equilibrium diagram in the usual way from left to right (increasing copper content), it is found that alloys with less than 28.1 percent Cu (eutectic composition) fall in the hypoeutectic class, whereas those containing more than 28.1 percent Cu belong to the hypereutectic group.

Figure 11.14 shows the microstructure of an alloy of 24 percent Cu and 76 percent Ag. Since this composition contains less copper than 28.1 percent (eutectic composition), the alloy is hypoeutectic. Two distinct structures are visible in the photograph: a more or less continuous gray area, inside of which is found a number of oval-shaped white areas. Notice that both the white and gray regions have their own characteristic appearance. Copper-silver alloys that contain between 8.8 percent and 28.1 percent Cu and are slowly cooled from the liquid phase, will show varying amounts of these two structures. The amount of the gray structure increases from zero to 100 percent as the composition of the alloy is changed from 8.8 percent to 28.1 percent. Figure 11.15 is another photomicrograph of the same alloy, but in this case the magnification is higher. What appears as a rough gray structure at lower magnification is shown by Fig. 11.15 to be an aggregate of small platelets of one phase in a matrix of another phase. This is the eutectic structure of the copper-silver system. The small platelets are composed of a copper-rich phase, while the continuous matrix is a silver-rich phase. This alternating platelet structure is commonly called lamellar eutectic structure. The two phases forming the eutectic have colors that are characteristic of the element which is present in each in the greater amount. The copper-rich area is tinted red, while the silver-rich is white, so that both phases are clearly visible in a polished specimen that has not been etched. (In the present photographs, the specimens were etched

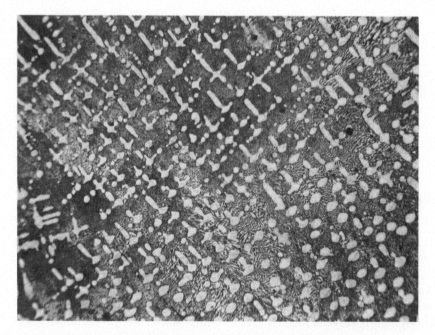

Fig. 11.14 A hypoeutectic structure from the copper-silver phase diagram containing approximately 24 percent copper. Lighter oval regions are proeutectic alpha dendrites, while the gray background is the eutectic 2-phase structure of $\alpha + \beta$

11.10 The Microstructures of Eutectic Systems

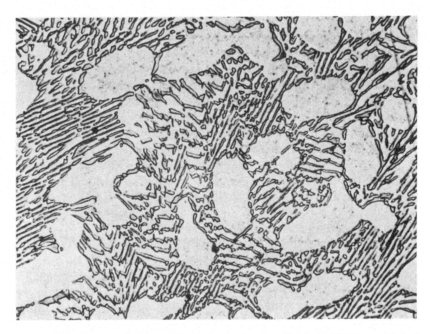

Fig. 11.15 The microstructure of Fig. 11.14 shown at a greater magnification. (White matrix is the alpha or silver-rich phase. Dark small platelets are the beta or copper-rich phase. The eutectic structure is thus composed of beta platelets in an alpha matrix)

to increase the contrast.) Careful study of Fig. 11.15 shows that the large white oval areas that have no copper particles are continuous with the white areas of the eutectic regions. These are, accordingly, extended regions of the silver-rich phase.

From the above, it can be concluded that hypoeutectic alloys in this alloy system possess a microstructure consisting of a mixture of the lamellar eutectic structure and regions containing only the silver-rich phase. The important thing to notice is that in a photograph of an eutectic alloy, such as Fig. 11.14, the features of the structure that stand out are not the phases themselves, but the contrast between the eutectic structure and the regions in which only a single phase exists. It is customary to call these parts of the microstructure that have a clearly identifiable appearance under the microscope, the *constituents* of the structure. Unfortunately, the term *constituent* is frequently confused with the term *component*. The words actually have quite different meanings. The components of an alloy system are the pure elements (or compounds) from which the alloys are formed. In the present case, they are pure copper and pure silver. The constituents, on the other hand, are the things that we see as clearly definable features of the microstructure. They may be either phases, such as the white regions in Fig. 11.14, or mixtures of phases, the gray eutectic regions of Fig. 11.14.

The microstructure in Fig. 11.14 is characteristic of an eutectic alloy that has been cooled (slowly) from the liquid phase. Cold-working and annealing this alloy will not change the amounts of the phases that appear in the microstructure, but such a treatment can well change the shape and distribution of the phases. In other words, Fig. 11.14 shows the structure of a hypoeutectic copper-silver alloy in the cast state, and it follows that what we see in this photograph is a function of the freezing process of the eutectic alloy.

Some of the details of the freezing process in alloys of an eutectic system will now be considered. For this purpose, we shall assume that the alloys are cooled from the liquid state to a temperature slightly below the eutectic temperature. At this temperature all compositions will be completely solidified, but considerations of the effect on the microstructures of the solvus lines, which show a decrease in solubility of the phases with decreasing temperatures, are eliminated. Actually, the change in the microstructure caused by the decreasing solubilities of the phases is, in general, slight and will be discussed briefly after we have explained the nature of the equilibrium freezing process.

A vertical line is shown in Fig. 11.13 that passes through the composition 24 percent Cu. This line represents the average composition of the alloy whose microstructure is shown in Fig. 11.14. At temperatures above point a, the alloy is in the liquid phase. Cooling to a temperature below point a causes it to enter a two-phase region where the stable phases are liquid and the alpha solid. The latter is a solid solution of copper in silver. Freezing of the given composition, accordingly, starts with the formation of crystals that are almost completely silver in composition (point b). Until the alloy reaches the eutectic temperature, the freezing process is similar to that of an isomorphous alloy with the liquid and solid moving along the liquidus and solidus lines respectively. The silver-rich crystals that form in this manner grow as many-branched skeletons called *dendrites* whose nature will be explained in Chapter 14. It suffices for the present to point out that when the alloy has cooled to just above the eutectic temperature (point c), it contains a number of skeleton crystals immersed in a liquid phase. Using the previously stated rules for analyzing a two-phase alloy at a given temperature, it is evident that the composition of the liquid at this temperature must correspond to the eutectic composition (point d). The solid phase, on the other hand, has a composition 8.8 percent Cu (point e). Since the average composition falls at a point about four-fifths of the way (24 − 8.8 percent) from the composition of the solid phase to that of the liquid phase (28.1 − 8.8 percent), it is to be expected, by the lever rule, that the ratio of liquid to solid will be approximately 4 to 1.

Immediately above the eutectic temperature the alloy is a mixture of liquid and solid. The solid is in the form of dendrites surrounded by the liquid phase of the eutectic composition. Further cooling of the alloy freezes this liquid at the eutectic temperature as given by Eq. 11.5. Such a freezing process results in the formation of the eutectic structure: a mixture of two solid phases (copper platelets in a matrix of silver) that freezes between the branches of the silver-rich crystals. A cross-section of a structure formed in this way has the appearance shown in Fig. 11.14. Since just above the eutectic temperature four-fifths of the alloy is liquid eutectic, it is to be expected that four-fifths of the completely solidified alloy will consist of eutectic solid. The remaining fifth of the structure corresponds to the silver-rich dendrites that form during the freezing process at temperatures above the eutectic temperature. In Fig. 11.14 this part of the structure is the clear oval-shaped areas. The explanation of their shape is readily apparent if one considers that the plane of the photograph represents a cross-section of the arms of the skeleton crystal.

In a hypoeutectic alloy of the type shown in Fig. 11.14 or 11.15, the silver phase appears in two locations: in the eutectic, where it is present with the copper particles, and in the dendrite arms, where it is the only phase. The copper-rich phase, on the other hand, only appears in the eutectic structure. It is common practice to differentiate between the two forms of the silver-rich phase and to designate the dendritic silver-rich regions as "primary." The remaining silver-rich regions are designated *eutectic*. The primary regions are also commonly called the *proeutectic constituent* since they form at temperatures above the eutectic temperature.

All hypoeutectic compositions between 8.8 percent Cu and 28.1 percent Cu solidify in a manner similar to the 24 percent composition that has just been discussed. The ratio of the eutectic to the primary alpha phase in the alloys after they have passed through the eutectic temperature, of course, varies with the composition of the metal. The amount of eutectic in the microstructure increases directly with increasing copper content as we move from the composition 8.8 percent Cu to 28.1 percent Cu. At this latter value the structure will be entirely eutectic; at the former value it will consist of only primary alpha crystals.

At compositions below 8.8 percent Cu all alloys freeze, under equilibrium conditions, in an isomorphous manner and are single-phased (as long as they are not lowered below the solvus line). When cooled below the solvus line, these compositions (0 to 8.8 percent Cu) become supersaturated and precipitate beta-phase (copper-rich) particles. This means that they are theoretically capable of age-hardening and the precipitation processes described in Chapter 16 apply. Notice that this phenomenon does not develop the eutectic structure that can only be formed when a liquid of the eutectic composition is solidified.

Let us now focus our attention on hypereutectic copper-silver alloys. Figure 11.16 shows a typical microstructure of one of these compositions (50 percent Cu–50 percent Ag). A vertical line is drawn through the appropriate composition in Fig. 11.13 to show where this alloy falls on the phase diagram. When hypereutectic alloys (28.1 percent Cu to 92 percent Cu) are cooled from the liquid state, the copper-rich beta phase forms from the liquid until the eutectic temperature is reached. This depletion of the copper content of the liquid shifts the liquid composition (along the liquidus line) toward the eutectic composition. After the eutectic temperature is passed, the resulting microstructure is a mixture of primary beta plus eutectic. The primary beta appears in Fig. 11.16 as oval-shaped dark areas. The structure of the eutectic in this alloy is the same as in the hypoeutectic alloys—copper particles in a matrix of silver. One very interesting feature can be seen by comparing this microstructure with the comparable hypoeutectic structure of Fig. 11.15. In both cases, the silver phase is the continuous phase, while the copper phase is discontinuous. Thus, the alpha, or

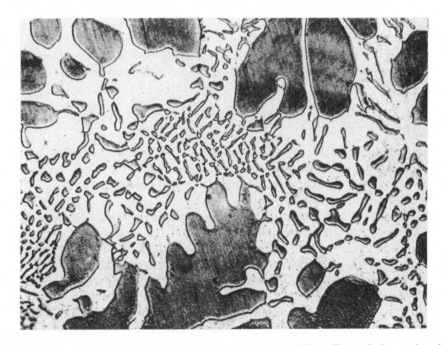

Fig. 11.16 Hypereutectic copper-silver structure consisting of proeutectic beta (large dark areas) and eutectic

silver-rich, phase of the eutectic is continuous with the primary silver dendrite arms in the hypoeutectic structure, but in the hypereutectic structure the primary copper-rich phase is not continuous with the copper phase in the eutectic. Similar results, where one phase tends to surround the other phase, are found in other eutectic systems besides the copper-silver system.

The amount of the eutectic structure in hypereutectic alloys varies directly with the change in copper concentration as one moves from 28.1 percent Cu to 92 percent Cu; decreasing from 100 percent at the former composition to 0 percent at the latter. Above 92 percent Cu to 100 percent Cu, all compositions freeze in an isomorphous manner to form single-phased (beta or copper-rich) structures. The latter, on cooling to room temperature, become supersaturated in silver and are subject to precipitation of the alpha phase. This effect is exactly analogous to that observed in the silver-rich alloys containing less than 8.8 percent Cu.

The effect of cooling the various compositions lying between 8.8 percent Cu and 92 percent Cu from the eutectic temperature to room temperature still remains to be described. In general, all of these alloys will contain part of the eutectic structure and are, therefore, two-phase structures. According to the phase diagram, both the alpha phase and the beta phase have decreasing solubilities with temperature and, on slow-cooling, both phases tend to approach the pure state, which means that copper will diffuse out of the silver-rich phase and silver will diffuse out of the copper-rich phase. With sufficient slow-cooling, no new particles of either phase should form since it will be easier, for example, for the silver atoms leaving the beta phase to enter the alpha phase which is already present, than it will be to nucleate additional alpha particles. The net effect of the decreasing solubility of the phases (with falling temperature) upon the appearance of the microstructure is therefore small.

A final word should be said about the significance of the eutectic point. Hypoeutectic compositions, when cooled from the liquid phase, start to solidify as silver-rich crystals. The liquidus line to the left of the eutectic point can, therefore, be considered as the locus of temperatures at which various liquid compositions will start to freeze out the alpha phase. Similarly, the liquidus line to the right of the eutectic point represents the locus of temperatures at which the beta phase will form from the liquid phase. The eutectic point, which occurs at the intersection of the two liquidus lines, is, therefore, the point (with respect to temperature and composition) at which the liquid phase can change simultaneously into both alpha and beta phases as shown in Equation 11.5.

11.11 The Peritectic Transformation

The eutectic reaction, in which a liquid transforms into two solid phases, is just one of the possible three-phase reactions that can occur in binary systems. Another that appears frequently involves a reaction between a liquid and a solid that forms a new and different solid phase. This three-phase transformation occurs at a peritectic point.

Figure 11.17 shows the constitution diagram of the iron-nickel system. A peritectic point appears at the upper left-hand corner. An enlarged view of this region is given in Fig. 11.18.

Before studying the peritectic reaction, some of the basic features of the alloy system in question should be considered. Iron and nickel have apparent atomic diameters that are almost identical (Fe, 0.2476 nm and Ni, 0.2486 nm). Since both iron and nickel belong to group VIII of the periodic table, these two elements are chemically similar. Both crystallize in the face-centered cubic system; nickel is face-centered cubic at all temperatures but iron only in the range 912°C to 1394°C. Conditions are thus ideal for the formation of a simple isomorphous system except for the fact that the stable crystalline form of iron is body-centered cubic at temperatures above 1394°C and below 912°C. It is

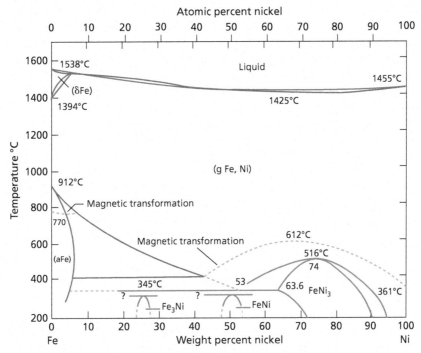

Fig. 11.17 Iron-nickel phase diagram. (From *Binary Alloy Phase Diagrams*, Massalski, T. B., Editor-in-Chief, ASM International, 1986, p. 1086. Reprinted with permission of ASM International®. All rights reserved. www.asminternational.org)

not surprising, therefore, that iron-nickel alloys are face-centered cubic except for two small body-centered cubic fields at the upper and lower left-hand corners of the phase diagram.

Figure 11.17 also shows that a superlattice transformation, based on the composition $FeNi_3$, occurs in this system. The boundaries delineating this face-centered cubic order-disorder transformation appear at the lower right-hand side of the phase diagram.

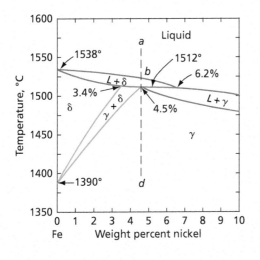

Fig. 11.18 The peritectic region of the iron-nickel phase diagram

The addition of nickel to iron increases the stability of the face-centered cubic phase. As a consequence, the temperature range in which this crystalline phase is preferred expands with increasing nickel content, and the boundaries separating the body-centered cubic fields from the face-centered field slope upward and downward respectively (with increasing nickel content).

With reference to Fig. 11.18, let us consider only that part of the diagram that lies above 1394°C. According to normal terminology, the high-temperature form of the body-centered cubic phase that appears in this temperature range is called the delta (δ) phase, whereas the face-centered cubic phase is designated the gamma (γ) phase.

In the indicated temperature interval, when the composition of the alloy is very close to pure iron, delta is the stable phase. With increased nickel content, the stable structure becomes the gamma phase. Liquid alloys of very low-nickel concentrations (<3.4 percent Ni) freeze directly to the body-centered cubic phase, whereas those containing more than 6.2 percent Ni freeze to form the face-centered cubic phase.

The composition interval 3.4 percent Ni to 6.2 percent Ni represents a transition interval in which the product of the freezing reaction shifts from the delta to the gamma phase. The focal point of this part of the phase diagram is the peritectic point, which occurs at 4.5 percent Ni (the peritectic composition) and 1512°C (the peritectic temperature).

With the aid of the dashed line *ad* in Fig. 11.18, one can follow the freezing reaction of an alloy of peritectic composition. Solidification starts when the temperature of the liquid phase reaches point *b*. Between this point and the peritectic temperature, the alloy moves through a two-phase field of liquid and delta phases. Solidification, therefore, commences with the formation of body-centered cubic dendrites that are low in nickel. The liquid, accordingly, is enriched in nickel. At a temperature immediately above the peritectic temperature (1512 °C), the rules for analyzing a two-phase mixture give us:

The composition of the phases:

           delta phase   3.4 percent Ni
           liquid phase   6.2 percent Ni

The amount of each phase (by lever rule):

           delta phase   61 percent
           liquid phase   39 percent

Directly above the peritectic temperature, an alloy of the peritectic composition has a structure composed of solid delta-phase crystals in a matrix of liquid phase. On the other hand, the phase diagram shows that just below the peritectic temperature, the given alloy lies in a single-phase field (gamma), signifying a simple homogeneous solid-solution phase. Cooling through the peritectic temperature combines the delta and liquid phases to form the gamma phase. This is the iron-nickel peritectic transformation. In this particular system, the peritectic point is the direct consequence of the fact that the liquid phase freezes to form two different crystalline forms in adjacent composition ranges, and of the fact that one of the phases (gamma) is the more stable at lower temperatures and, therefore, displaces the other.

Notice that the peritectic reaction, like the eutectic, involves a fixed ratio of the reacting phases: 61 percent delta (3.4 percent Ni) combines with 39 percent liquid (6.2 percent Ni) to form the gamma phase (4.5 percent Ni). If an alloy at the peritectic temperature does not contain this exact ratio of liquid phase to delta phase, the reaction cannot be complete and some of the phase that is in excess will remain after the peritectic temperature is passed. Compositions in the interval are 3.4 percent to 4.5 percent Ni, which lie to the left of the peritectic composition, contain an excess of the delta phase (more than 61 percent) immediately above 1512°C. On passing through the peritectic temperature, they enter a two-phase field: delta phase and gamma phase. Similarly, alloys to the right of the peritectic point, lying between 4.5 percent Ni and 6.2 percent Ni, after passing through the peritectic temperature, enter a two-phase field: gamma phase and liquid phase.

11.12 Monotectics

Monotectics represent another form of three-phase transformation in which a liquid phase transforms into a solid phase and a liquid phase of different composition. Monotectic transformations are associated with miscibility gaps in the liquid state. A reaction of this type occurs in the copper-lead system at 955°C and 37.4 percent Pb, as can be seen in Fig. 11.19. Notice the similarity between this monotectic and the eutectic of the copper-silver system (Fig. 11.13). The liquid miscibility gap lies just to the right of the monotectic point.

Note that the copper-lead phase diagram also possesses an eutectic point at 327.5°C and 99.9 percent Pb (0.1 percent Cu). Because it lies so close to the composition of pure lead, it is not possible to show it on the scale of the present diagram. Finally, this system is representative of one composed of elements that do not mix in the solid state; at room temperature the solubility

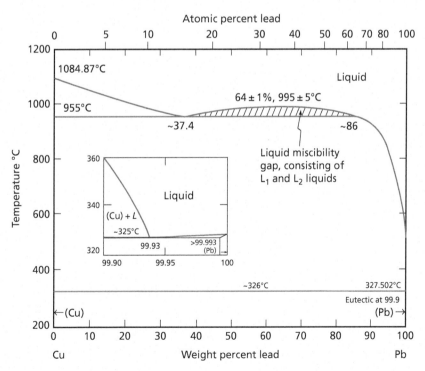

Fig. 11.19 Copper-lead phase diagram. (From *Binary Alloy Phase Diagrams*, Massalski, T. B., Editor-in-Chief, ASM International, 1986, p. 946. Reprinted with permission of AjSM International®. All rights reserved. www.asminternational.org)

of copper in lead is less than 0.007 percent, while the solubility of lead in copper is of the order of 0.002 to 0.005 percent Pb. The terminal solid solutions are, accordingly, nearly pure elements.

Liquid miscibility gap, or Liquid Phase Separation (LPS), is directly related to the effects of temperature and composition on the Gibbs free energy of the system. For these monotectic alloys, when temperature is lowered, the entropy of mixing cannot overcome the positive heat of mixing, and phase separation occurs. This can be visualized by considering a generalized monotectic phase diagram, Fig. 11.20, where the liquid miscibility gap is shown by the dome-shaped region containing L_1 and L_2. The monotectic point is shown by T_M, and there is an eutectic point designated by T_E. The critical temperature, T_C, represents the temperature above which the liquid is fully miscible for all compositions.

Note that the shape of the dome, location and compositional range vary with the alloy system; for example, compare Fig. 11.19 with Fig. 11.20. Regardless, the monotectic phase diagram represents equilibrium between components, and Gibbs phase rule and lever rules apply as before. To utilize the diagram, consider the cooling of an alloy with composition X_0 of approximately 35% B from point designated by a. The alloy at this point is a single homogenous liquid phase and will remain so until the temperature is lowered to point b, at the boundary of the two-phase region. Slightly below this temperature, liquid L_2 begins to form. As temperature continues to decrease, the compositions of both L_1 and L_2 liquids follow their respective boundaries. At temperature T_1, the equilibrium compositions of the two liquids are shown by points c and d, which represents

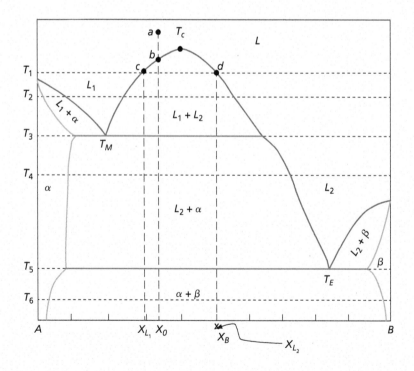

Fig. 11.20 A hypothetical monotectic phase diagram of components A and B, showing dome shaped liquid miscibility gap of $L_1 + L_2$ (from N. Derimow and R. Abbaschian, "Liquid Phase Separation in High-Entropy Alloys—A Review," *Entropy*, 20(11), 890, 2018)

approximately 30 and 50% B, respectively. The lever rule can be applied to determine the relative amounts of each liquid as

$$\%L_1 = 100 \times (50-35)/(50-30) = 75\%$$
$$\%L_2 = 100 \times (35-30)/(50-30) = 25\%$$

The free energy versus composition diagrams for the liquid, and component phases at this temperature are shown in Fig. 11.21A. The diagrams are similar to those shown in Figs. 11.4 and 11.5, except the curve for the liquid has a "W" shape, with a maximum in the middle and two minima on either side. The common tangent to this curve gives equilibrium compositions for liquids L_1 and L_2, as given in the diagram. Note that the dome represents free energy of the liquid had it not phase separated to the mixture of two liquids L_1 and L_2. Note that the free energy of the system is lowered by the LPS from point represented by m in Fig. 11.21A to n, which is a mixture of o and p. Moreover, as can be seen from Figs. 11.20 and 11.21A, the equilibrium phases depend on the composition: L_1 when B content is less than that of point c, $L_1 + L_2$ for compositions between c and d, and L_2 for all compositions higher than d.

At lower temperature T_2 as shown in Fig. 11.21B, the free energy curves shift up but with different amounts, and a portion of the curve for α now falls below the free energy of the liquid. This also allows for drawing a common tangent with the curve for the liquid, resulting in equilibrium phases consisting of α, $\alpha + L_1$, L_1, $L_1 + L_2$ and L_2 as B concentration is increased. At the monotectic temperature T_M, the last of liquid L_1 disappears, and the alloy enters $\alpha + L_2$ region. Below this temperature, the phase diagram is similar to the eutectic system discussed earlier.

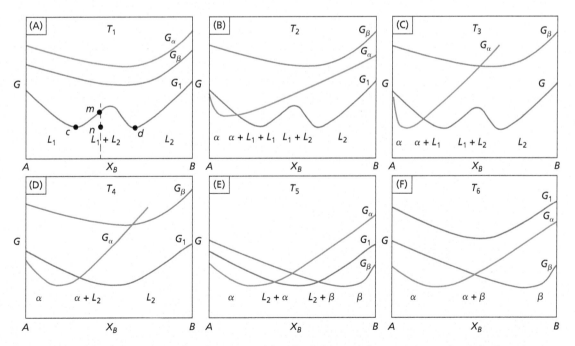

Fig. 11.21 Gibbs free energy as function composition B of phases and temperatures shown in the monotectic diagram Fig. 11.20 (from N. Derimow and R. Abbaschian, "Liquid Phase Separation in High-Entropy Alloys—A Review," *Entropy*, 20(11), 890, 2018).

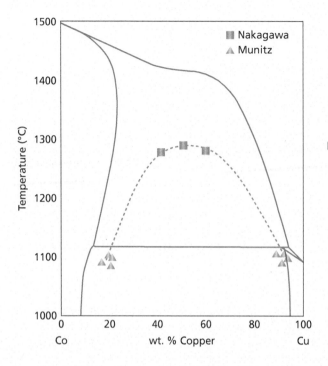

Fig. 11.22 The Cu-Co phase diagram showing metastable miscibility gap. (Reproduced from A. Munitz, S. P. Elder-Randall, and R. Abbaschian, *Metallurgical Transactions A*, 1992, Vol. 23A, 1817–1827)

Liquid Phase Separation is widely observed in many binary alloys. A partial listing is given in a table in the review paper by Derimow and Abbaschian (Ref. 3). Note that LPS may also form under metastable conditions upon supercooling of melts, examples of which are Fe-Cu and Co-Cu alloys. (Refs. 4 and 5) Under normal conditions, both alloys solidify with the formation of dendrites of Fe or Co in a copper matrix, respectively, similar to that discussed previously for eutectic systems. However, it has been shown that when some of these liquids are supercooled to a certain degree, they enter a metastable liquid immiscibility region that is located beneath stable liquidus curves.

Finally, it should be mentioned that liquid immiscibility is not limited to binary alloys. It can also take place in alloys that contain three or more components. It has additionally been observed in other, newer alloys, such as High Entropy Alloys (HEA), that have been explored in recent years. An example of this is a CoCrCu alloy, which contains 1/3 Mole of each component as shown in Fig. 11.22 (Ref. 3).

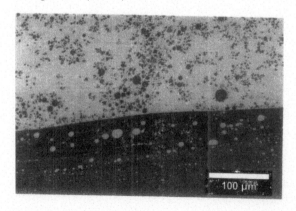

Fig. 11.23 Backscattered electron image displaying emulsion of CoCr-rich (darker) and Cu-rich (lighter) liquids in an as-cast alloy of CoCrCu (Derimow N, Abbaschian R. Liquid Phase Separation in High-Entropy Alloys-A Review. Entropy (Basel). 2018 Nov 20;20(11):890. doi: 10.3390/e20110890. PMID: 33266614; PMCID: PMC7512472.)

11.13 Other Three-Phase Reactions

The three basic types of three-phase reactions (eutectics, peritectics, and monotectics) that we have studied so far involve transformations between the liquid and solid phases. They are, therefore, associated with the freezing or melting processes in alloys. Several other important three-phase reactions involve only changes between solid phases. The most important of these occurs at eutectoid and peritectoid points. At an eutectoid point, a solid phase decomposes into two other solid phases when it is cooled. The reverse is true at a peritectoid point where, on cooling, two solid phases combine to form a single solid phase. The similarity between the eutectoid and peritectoid transformations on the one hand and the eutectic and peritectic transformations on the other hand is quite obvious. A summary of important phase diagram reactions is given in Table 11.1, where α, β, and γ are separate solid phases, and L_1 and L_2 are immiscible liquids.

Table 11.1 Summary of Important Equilibrium Phase Transformations

Transformation Name	Equilibrium Reaction	Phase Diagram Appearance
Congruent	$L \underset{\text{heating}}{\overset{\text{cooling}}{\rightleftharpoons}} \alpha$	
Eutectic	$L \underset{\text{heating}}{\overset{\text{cooling}}{\rightleftharpoons}} \alpha + \beta$	
Eutectoid	$\alpha \underset{\text{heating}}{\overset{\text{cooling}}{\rightleftharpoons}} \beta + \gamma$	
Monotectic	$L_1 \underset{\text{heating}}{\overset{\text{cooling}}{\rightleftharpoons}} L_2 + \alpha$	
Peritectic	$L + \alpha \underset{\text{heating}}{\overset{\text{cooling}}{\rightleftharpoons}} \beta$	
Peritectoid	$\alpha + \beta \underset{\text{heating}}{\overset{\text{cooling}}{\rightleftharpoons}} \gamma$	

A very important eutectoid reaction occurs in the iron-carbon system and will be considered in considerable detail at a later point.

11.14 Intermediate Phases

Intermediate phases, also known as *intermetallic phases*, are the rule rather than the exception in phase diagrams. Equilibrium diagrams given to this point were selected for their simplicity in order to demonstrate certain basic principles. Except for ordered or superlattice phases, no intermediate phases have been shown. Several important features of alloy systems that contain intermediate phases will now be treated briefly.

The silver-magnesium phase diagram (Fig. 11.24) is of interest in the study of intermediate phases since it shows two basic ways in which the single-phase fields associated with intermediate phases are formed. This alloy system possesses a total of five solid phases, of which two are terminal phases (alpha—fcc based on the silver lattice, and delta—cph based on the magnesium lattice). Both the Ag_3Mg, beta prime and epsilon phases (the intermediate phases) are stable over a range of composition and are, therefore, examples of true solid solutions.

Figure 11.24 shows that the β' phase is centered about a composition having equal numbers of magnesium and silver atoms. In order that this fact might be clearly visible, the diagram has been

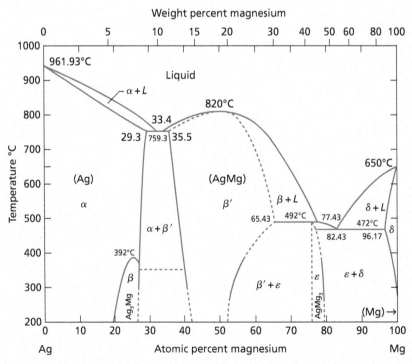

Fig. 11.24 Silver-magnesium phase diagram. (From *Binary Alloy Phase Diagrams*, Massalski, T. B., Editor-in-Chief, ASM International, 1986, p. 42. Reprinted with permission of ASM International®. All rights reserved www.asminternational.org)

plotted in atomic percent. This phase is a superlattice based on the body-centered cubic structure. With exactly equal numbers of silver and magnesium atoms, a space lattice results, in which the corner atoms of the unit cell are of one form, while the atom at the center of the cell is of the other type. An example of this type of structure in a different system is shown in Fig. 11.25. The β' phase in the present alloy system is interesting in that the ordered structure is stable to the melting point. In order to indicate that the given structure is not a simple random solid solution, the symbol β' rather than β has been used.

The β' phase field is bounded at its upper end by a maximum having coordinates of 820°C and 50 atomic percent Mg. This body-centered cubic phase, therefore, forms on freezing by passing through a typical maximum configuration of liquidus and solidus lines. On the other hand, the upper limits of the single-phase field of the epsilon phase terminate in a typical peritectic point at 492°C and 75 atomic percent Mg. Formation of this latter phase (which has a complex, not completely resolved, crystalline structure), accordingly, results from a peritectic reaction.

The preceding paragraph has defined two basic ways that intermediate phases are formed during freezing—either by a transformation at a congruent maximum or through a peritectic reaction. Both types of reaction are common in phase diagrams. Intermediate phases may also form as a result of a transformation that takes place in the solid state. The formation of superlattices from solid solutions possessing a state of short-range order at congruent maxima has already been discussed. New solid phases may also form at peritectoid points that are the solid-state equivalent of peritectic points.

Returning now to the silver-magnesium system, Fig. 11.24 also shows that this system, in addition to the peritectic point and the congruent point, also possesses two eutectic points. The first lies at 59.3°C and 33.4 atomic percent Mg and corresponds to a reaction in which a liquid transforms into an eutectic mixture of the α and β' phases. The other eutectic point, at 472°C and 82.43 atomic percent Mg, yields an eutectic structure which is a mixture of the ε and δ phases. Notice that, in each case, the eutectic structures are composed of different sets of phases.

The intermediate phases in the silver-magnesium system are solid solutions stable over relatively wide composition ranges. Thus, at 200°C, the single-phase field of the β' phase extends from approximately 42 atomic percent Mg to 52 atomic percent Mg, while that of the ε phase covers the composition range from 75 atomic percent Mg to 79 atomic percent Mg. Many intermediate phases, on the other hand, have single-phase fields that are vertical lines. Such phases are commonly classed as *intermetallic compounds*.

Intermediate phases of the compound type are formed during freezing in the same manner as the solid-solution phases; they may form either at congruent maxima or at peritectic points. The equilibrium diagram of the magnesium-nickel system, Fig. 11.26, which is also plotted in atomic percent, shows both a congruent point (1147 °C and 67 percent Ni) and a peritectic point (750°C and 33.3 percent Ni). In this respect, it is analogous to the silver-magnesium system and, like the

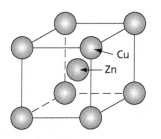

Fig. 11.25 The unit cell of the β' phase of the copper-zinc system. This type of structure is also found in the compound cesium-chloride and is usually called the CaCl structure

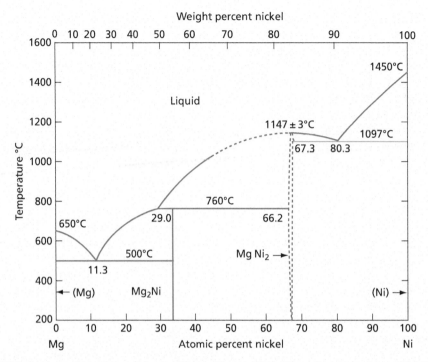

Fig. 11.26 Nickel-magnesium phase diagram. (From *Binary Alloy Phase Diagrams*, Massalski, T. B., Editor-in-Chief, ASM International, 1986, p. 1529. Reprinted with permission of ASM International®. All rights reserved. www.asminternational.org)

former, it also possesses two eutectic points. The only basic difference between the two systems lies in the absence of solubility in the phases of this latter system.

In the magnesium-nickel system, all phases, including the terminal phases, have very limited single-phase fields. The compound $MgNi_2$ crystallizes in a hexagonal lattice. This phase may have a small compositional range at high temperatures, as indicated by the dashed lines. The other intermetallic phase appears at a ratio of the components equivalent to 1 nickel atom to 2 magnesium atoms or Mg_2Ni.

The first compound ($MgNi_2$) freezes or melts at a congruent maximum point. The vertical dashed lines representing this compound can be considered to divide the phase diagram into two independent parts. Each of these parts is a phase diagram in itself, as is shown in Fig. 11.27, where only the left section of the complete nickel-magnesium diagram is located. This partial diagram can be considered the nickel-$MgNi_2$ phase diagram: a system with one component an element and the other a compound.

11.15 The Copper-Zinc Phase Diagram

The equilibrium diagram of the copper-zinc system is shown in Fig. 11.28. Since copper-zinc alloys comprise the commercially important group of alloys known as brasses, the diagram is important for this reason alone. It is also significant because it is representative of a group of binary-equilibrium

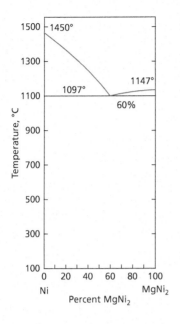

Fig. 11.27 The nickel-magnesium diagram can be divided at the composition of the compound MgNi$_2$ into two simpler diagrams. This figure is the Ni–MgNi$_2$ diagram and corresponds to the left-hand portion of the Ni–Mg diagram

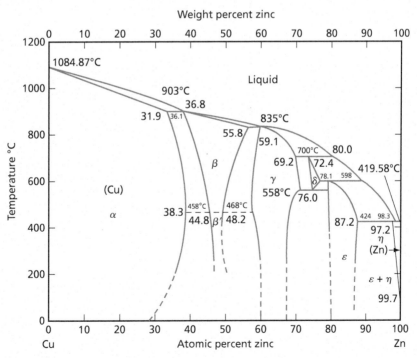

Fig. 11.28 Copper-zinc phase diagram. (From *Binary Alloy Phase Diagrams*, Massalski, T. B., Editor-in-Chief, ASM International, 1986, p. 981. Reprinted with permission of ASM International®. All rights reserved. www.asminternational.org)

diagrams formed when one of the noble metals (Au, Ag, Cu) is alloyed with such elements as zinc and silicon. Attention is called to the fact that Fig. 11.28 is plotted in atomic percent. A weight-percent scale is given across the top of this diagram and it can be seen that copper-zinc compositions are almost identical when expressed in either weight or atomic percentages.

The seven solid phases in the copper-zinc system are classified as follows:

Terminal phases:

> alpha (α) fcc based on the copper lattice
> eta (η) cph based on the zinc lattice

Intermediate phases:

> beta (β) disordered body-centered cubic
> beta prime (β') ordered body-centered cubic
> gamma (γ) cubic, low symmetry
> delta (δ) body-centered cubic
> epsilon (ε) close-packed hexagonal

Except for the alpha and beta-prime phases, all single-phase fields terminate at their high-temperature extremities in peritectic points. There are, accordingly, five peritectic points in Fig. 11.28. The delta phase differs from the others in that it is stable over a rather limited temperature range (700°C to 558°C). Notice that the delta-phase field terminates at its lower end in an eutectoid point.

Another significant feature of this phase diagram is the order-disorder transformation that occurs in the body-centered cubic phase ($\beta - \beta'$). Near room temperature, the β' field extends from about 48 percent to 50 percent Zn. The stoichiometric composition CuZn falls at the edge of the β' field (50 percent Zn). The ordered body-centered cubic phase is, therefore, one that is based on a ratio of approximately one zinc to one copper atom. In this respect, it is similar to the copper-gold ordered phase CuAu, but here the structure is body-centered, whereas the CuAu phase is based on the face-centered cubic lattice. Ordering with equal numbers of two atomic forms in a body-centered cubic lattice produces a structure in which each atom is completely surrounded by atoms of the opposite kind (see Fig. 11.25). An example in the silver-magnesium system has already been mentioned. The atomic arrangement can be visualized by imagining that in each unit cell of the crystal, the corner atoms are copper atoms, while those at the cell centers are zinc atoms. That such an arrangement corresponds to the formula CuZn is easily proved: since one-eighth of each corner atom (copper) belongs to a given cell, and there are eight corner atoms, the corner atoms contribute one copper atom to each cell. Similarly, there is one zinc atom at the center of each cell which belongs to this cell alone.

The transformation is indicated in the phase diagram by a single line running from 456°C to 468°C. Some work by Rhines and Newkirk has indicated that the β and β' fields are separated by a normal two-phase field ($\beta + \beta'$), but the details of this region are still not completely worked out, which is due primarily to experimental difficulties connected with the rapidity of the transformation. The details of the transformation cannot be studied by quenching specimens from the two-phase region ($\beta + \beta'$), since any normal quench will not suppress the complete transformation to β'.

11.16 Ternary Phase Diagrams

The two-dimensional phase diagrams discussed in the preceding sections cannot be used to describe temperature-phase relations in alloys which contain three or more components. As discussed in Sec. 10.9, there are three independent variables, or degrees of freedom, for binary alloys. By fixing pressure for a given system, the remaining two variables can be plotted in a two-dimensional space. For a three-component alloy system, on the other hand, the number of independent variables is four (T, P, and two compositions). As such, a three-dimensional construction is required to describe temperature and phase relations at constant pressure.

Ternary diagrams are commonly constructed by plotting the compositions along the sides of an equilateral triangle base, called a Gibbs triangle, within which the composition is represented by a point, as shown in Fig. 11.29. The corners of the triangle represent the pure constituents, whereas its sides give compositions for the binary combinations. The temperature for ternary diagrams is given along the vertical axis. As such, each side of this three-dimensional structure represents one of the contributing binary phase diagrams. As an example, a ternary phase diagram involving three binary eutectic alloys *A-B*, *A-C*, and *B-C* is shown in Fig. 11.30. There are three such curved liquidus surfaces for the ternary shown in the figure, one of which is delineated by hatch lines. The intersections of the liquidus surfaces form curved lines, called *eutectic troughs*, which eventually lead to the ternary eutectic point *E*. The ternary eutectic point in this diagram represents the lowest temperature at which a liquid will exist in the alloy at a predefined constant pressure. It is also the only point where four phases (α, β, γ, and liquid) are at equilibrium at the constant pressure in which the diagram is drawn.

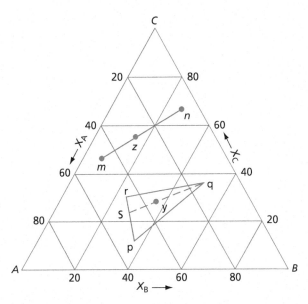

Fig. 11.29 Gibbs triangle for ternary alloy compositions

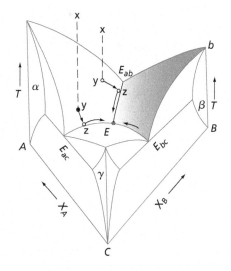

Fig. 11.30 A ternary phase diagram with three eutectic binaries between *A*, *B*, and *C*

Utilization of the ternary diagram to determine the equilibrium phases in alloys is similar to the process discussed in Sec. 11.2. To illustrate it, let's consider a liquid alloy with an overall composition given by point X in Fig. 11.30. Upon cooling the liquid, the first solid will form when the temperature reaches the liquidus surface at point Y. The composition of this solid will be determined by a tie line that connects point Y to a corresponding point on the solidus surface. (Note that these solidus surfaces are not shown in the figure, only their intersections with the sides of the diagram can be seen.) Upon further cooling, the liquid composition changes, following the contours of the liquidus surface, until it reaches the eutectic trough at point Z. The liquid composition then moves down the trough until the ternary eutectic E is reached.

The solidus points in the ternary diagrams also form curved surfaces. This can be easily surmised considering that each point on the liquidus surface is at equilibrium with a solid on the solidus surface. The compositions of these phases are given by a horizontal tie line connecting the two surfaces. Since drawing solidus surfaces and tie lines make the diagrams rather complicated, ternary diagrams are often presented by a series of isothermal sections. Each isothermal section represents a horizontal cross-section of the three-dimensional diagram at a given temperature. Figure 11.31 shows three such isothermal sections for the ternary diagram shown in Fig. 11.30. The first section, Fig. 11.31A, is drawn at a temperature below the A-B eutectic temperature but above the ternary eutectic temperature. A few tie lines in the two-phase fields are indicated with dotted lines. Note that no tie lines are necessary in a three-phase field $\alpha + \beta +$ liquid since the composition of each phase is uniquely defined by the corresponding corner of the triangular region.

The isothermal section shown in Fig. 11.31B is drawn at a temperature below all binary eutectic temperatures but above the ternary eutectic temperature. In this temperature, only alloys with overall composition within the central triangle will be all liquid, while those alloys poor in A, B, or C will be completely solid. All other compositions will consist of a mixture of solid plus liquid. When the temperature falls below the ternary eutectic, as shown in Fig. 11.31C, the alloy is completely solid, and it will contain one, two, or three phases dependent upon the overall composition.

The isothermal sections can also be utilized to determine the fraction of phases present in a ternary alloy by applying the lever rule discussed in Sec. 11.3. To illustrate this, consider an alloy with overall composition of 30 wt. % A, 15 wt. % B, and 55 wt. % C, as shown by point z in Fig. 11.29. Let's assume that the alloy contains two phases α and β with compositions given by points m and n, respectively. The fraction of the α phase in the alloy is equal to the length of line zn divided by the length of line mn. The fraction of the β phase would be zm/mn. A similar procedure can be used to determine the phase fractions in alloys containing three phases. For this case, a triangle is first formed between the compositions of the three phases. In order to satisfy mass balance for the components, the triangle will surround the overall alloy composition. An example is shown in Fig. 11.29 for an alloy composition given by point Y, which is considered to contain three phases given by points p, q, and r. The line drawn from a corner through y to point s can then be used to calculate the fractions.

The above-mentioned ternary diagram is a relatively simple one since it only deals with three solid phases. The diagrams can become quite complicated if the binaries include intermediate phases or other solid-state reactions. Examples of these systems can be found in phase diagram handbooks or other similar sources.

For alloys which contain four or more components, the phase relations cannot be described by a single diagram since more than a three-dimensional space would be required. These alloys are generally described by multiple quasi-binary diagrams in which the concentration of one component is changed at a time.

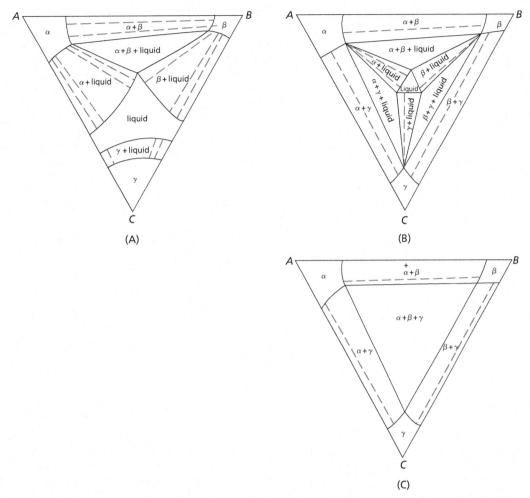

Fig. 11.31 Three isothermal sections for the ternary diagram shown in Fig. 11.30. A few of the tie lines are shown by the dotted lines. **(A)** At a temperature below *A-B* eutectic temperature but above eutectic temperatures of the ternary, *A-C* and *B-C*; **(B)** at a temperature below the three binary eutectics but above the ternary eutectic, and **(C)** at a temperature below the ternary eutectic temperature

Problems

11.1

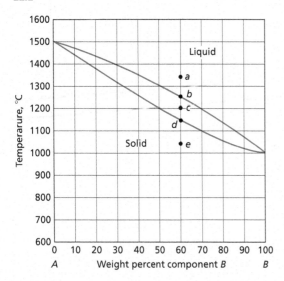

The points *a, b, c, d,* and *e* on the 60 percent component *B* line give some temperatures through which a slowly cooled alloy containing 60 percent *B* will pass on freezing from a liquid to a solid. Identify the phase or phases and the amount of each that should exist in the microstructure at each indicated point.

11.2 A gold-nickel alloy containing 60 percent nickel is heated to 1100°C and allowed to come to equilibrium. Determine the amount and composition of the liquid and solid phases when equilibrium is attained.

11.3 (a) A copper-75 percent silver alloy is slowly cooled from the liquid state to 900°C and allowed to come to equilibrium. Estimate the amount and composition of both the liquid and solid phases.

(b) Make a sketch of the 900°C equilibrium structure of the alloy.

(c) Now assume the alloy is slowly cooled to just below the eutectic temperature. What are the weight percentages and compositions of the phases and constituents at this point?

11.4 (a) Sterling silver is an alloy of silver with 7.5 percent copper. Describe the structure that one should expect if a specimen of sterling silver were to be heated from room temperature to 782°C and allowed to come to equilibrium at this latter temperature.

(b) If the specimen of sterling silver equilibriated at 782 °C is now cooled very slowly to 400°C, what would be the nature of the microstructure? Give the amount and composition of each phase.

(c) Finally assume the specimen is cooled very rapidly (quenched) from 782°C to 400°C. Describe the structure that one might expect.

11.5 The alloy that formerly was used in U.S. silver coins contained 10 percent copper.

(a) If one were to heat one of these coins to 782°C, what would be the expected effect on the microstructure?

(b) Would one successfully be able to mechanically work the metal in the coin at 782°C? Explain.

11.6 Consider the iron-nickel peritectic transformation in Fig. 11.18.

(a) What are the compositions and weight percentages of the phases just above the peritectic temperature (1512°C)?

(b) Answer this same question with regard to a temperature just below the peritectic temperature.

11.7 Given that the rate of diffusion of nickel in iron is very much greater in the liquid state than in the solid state, what effect should this have on the ease of obtaining an equilibrium microstructure (i.e., one that is homogeneous) when an alloy containing the peritectic composition 4.5 percent nickel is cooled through the peritectic temperature?

11.8 Answer the following questions with regard to an alloy of copper with 64 percent lead that is slowly cooled in a crucible without being stirred from 1100°C to room temperature.

(a) What is the nature of the alloy at 1100°C?

(b) Now consider the alloy at a temperature just above the monotectic temperature at 955°C where it is composed of two liquids of different compositions. Do the liquids have the same density? If not, what would you expect to happen?

(c) One of these liquids has the composition of the monotectic point. What should happen to this liquid as it passes through the monotectic temperature of 955°C? Describe the physical nature of the contents of the crucible after the alloy is cooled below 955°C and until it reaches the eutectic temperature at 326°C.

11.9 (a) What phases are in equilibrium at the peritectic temperature of the iron-nickel alloy system?

(b) What relationship exists between the partial-molal free energies of these phases at this temperature?

(c) Sketch the relationship that must exist between the free energy versus composition curves of these phases at 1512°C.

11.10 Make two additional sketches for the free-energy-composition curves of the iron-nickel system corresponding to a temperature about 25°C above and 25°C below 1512°C.

11.11 Label all regions of the Cu-Zn diagram in Fig. 11.23 and identify transformations that take place at each horizontal line.

11.12 For the three-phase alloy y shown in Fig. 11.25, Calculate the weight fractions of the constituent phases. The alloy composition is given by Point y, and those of the constituent phases by p, q, and r.

References

1. For example, see "Alloy Phase Diagrams," *ASM Handbook*, Vol. 3, ASM International, 1992.
2. Averbach, B. L., Flinn, P. A., and Cohen, M., *Acta Met.*, **2** 92 (1954).
3. Derimow, N. and Abbaschian, R., "Liquid Phase Separation in High-Entropy Alloys – A Review," *Entropy*, 20(11), 890 (2018).
4. Munitz, A., Elder-Randall, S. P. and Abbaschian, R., "Supercooling Effects in Cu-10 wt.% Co Alloys Solidified at Different Cooling Rates," *Metallurgical Transactions A*, 1992, Vol. 23A, (1817–1827).
5. S. P. Elder, A. Munitz, and G. J. Abbaschian, "Metastable Liquid Immiscibility in Fe-Cu and Co-Cu Alloys," *Materials Science Forum*, 1989, Vol. 50, 137–150.

Chapter **12**

Diffusion in Substitutional Solid Solutions

Learning Objectives

Upon completion of Chapter 12, you will be able to:

1. Understand the mechanism of atomic movement and the meaning of activation energy in substitutional solid solutions
2. Describe Kirkindal shift and how its velocity relates to diffusivities across a welded couple boundary
3. Understand Fick's laws and how they relate to atomic flux with diffusivity and concentration gradients
4. Explain how vacancies are created or annihilated across a welded diffusion couple
5. Derive Darken's equations relating intrinsic diffusivities to interdiffusion coefficient in a binary alloy
6. Determine interdiffusion and intrinsic coefficients using the Gruber, and Matano-Boltzmann methods
7. Understand the influence of temperature and high diffusivity paths on interdiffusion
8. Differentiate between tracer diffusion, self-diffusion and chemical diffusion coefficients
9. Compare and contrast diffusion in isomorphous alloy systems with non-isomorphic alloy systems involving intermediate phases

The field of diffusion studies in metals is of great practical, as well as theoretical importance. By *diffusion* one means the movements of atoms within a solution. In general, our interests lie in those atomic movements that occur in solid solutions. This chapter will be devoted in particular to the study of diffusion in substitutional solid solutions and the following chapter will be concerned with atomic movements in interstitial solid solutions.

12.1 Diffusion in an Ideal Solution

Consider a solid solution composed of two forms of atoms A and B, respectively. Let us arbitrarily designate the A component as the solute and the B component as the solvent, and consider that the solution is ideal. This, of course, implies that there is no preferential interaction between solute and solvent atoms or that the two forms act inside the crystal as though they were a single chemical species.

Experimental work has shown that the atoms in face-centered cubic, body-centered cubic, and hexagonal metals move about in the crystal lattice as a result of vacancy motion. Let it now be assumed that the jumps are entirely random; that is, the probability of jumping is the same for all of the atoms surrounding a given vacancy. This statement implies that the jump rate does not depend on the concentration.

Figure 12.1 represents a single crystal bar composed of a solid solution of A and B atoms in which the composition of the solute varies continuously along the length of the bar, but is uniform over the cross-section. For the sake of simplifying the argument, the crystal structure of the bar is assumed to be simple cubic with a $\langle 100 \rangle$ direction along the axis of the bar. It is further assumed that the concentration is greatest at the right end of the bar and least at the left end, and that the macroscopic concentration gradient dn_A/dx applies on an atomic scale so that the difference in composition between two adjacent transverse atomic planes is:

$$(a)\frac{dn_A}{dx} \qquad 12.1$$

where a is the interatomic, or lattice spacing (see Fig. 12.2). Let the mean time of stay of an atom in a lattice side be τ. The average frequency with which the atoms jump is therefore $1/\tau$. In the simple cubic lattice pictured in Fig. 12.2, any given atom, such as that indicated by the symbol x, can jump in six different directions: right or left, up or down, or into or out of the plane of the paper. The exchange of A atoms between two adjacent transverse atomic planes, such as those designated X and Y in Fig. 12.2, will now be considered. Of the six possible jumps that an A atom can make in either of these planes, only one will carry it over to the other indicated plane, so that the average frequency with which an A atom jumps from X to Y is $1/6\tau$. The number of these atoms that will jump per second from plane X to plane Y equals the total number of the atoms in plane X times the average frequency with which an atom

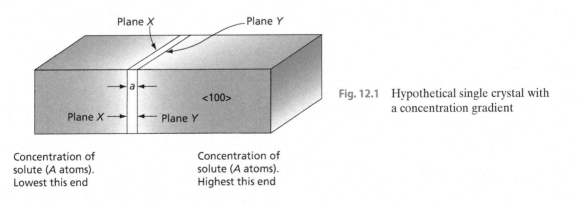

Fig. 12.1 Hypothetical single crystal with a concentration gradient

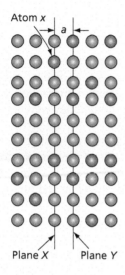

Fig. 12.2 Atomistic view of a section of the hypothetical crystal of Fig. 12.1

jumps from plane X to plane Y. The number of solute atoms in plane X equals the number of solute atoms per unit volume (the concentration n_A) times the volume of the atoms in plane X, (Aa), so that the flux of solute atoms from plane X to plane Y is

$$J_{X \to Y} = \frac{1}{6\tau}(n_A a) \qquad 12.2$$

where $J_{X \to Y}$ = flux of solute atoms from plane X to plane Y per unit cross-section
τ = mean time of stay of a solute atom at a lattice site
n_A = number of A atoms per unit volume
a = lattice constant of crystal

The concentration of A atoms in plane Y may be written:

$$(n_A)_Y = n_A + (a)\frac{dn_A}{dx} \qquad 12.3$$

where n_A is the concentration at plane X, and a is the lattice constant, or distance between planes X and Y. The rate at which A atoms move from plane Y to X is thus

$$J_{Y \to X} = \left[n_A + (a)\frac{dn_A}{dx}\right]\frac{a}{6\tau} \qquad 12.4$$

where $J_{Y \to X}$ represents the flux of A atoms from plane Y to plane X. Because the flux of solute atoms from right to left is not the same as that from left to right, there is a net flux (designated by the symbol J) which can be expressed mathematically as follows:

$$J = J_{X \to Y} - J_{Y \to X} = \frac{a}{6\tau}(n_A) - \left[n_A + (a)\frac{dn_A}{dx}\right]\frac{a}{6\tau} \qquad 12.5$$

or

$$J = -\frac{a^2}{6\tau} \cdot \frac{dn_A}{dx} \qquad \text{12.6}$$

since the cross-sectional area was chosen to be a unit area. Notice that in Eq. 12.6, the flux (J) of A atoms is negative when the concentration gradient is positive (concentration of A atoms increases from left to right in Fig. 12.2). This result is general for diffusion in an ideal solution; the diffusion flux is down the concentration gradient. Notice that if one considers the flow of B atoms instead of A atoms, the net flux will be from left to right, in agreement with a decreasing concentration of the B component as one moves from left to right. Again, the flux (in this case of B atoms) is down the concentration gradient.

Let us now make the substitution

$$D = \frac{a^2}{6\tau} \qquad \text{12.7}$$

in the equation for the net flux, which gives:

$$J = -D\frac{dn_A}{dx} \qquad \text{12.8}$$

This equation is identical with that first proposed by Adolf Fick in 1855 on theoretical grounds for diffusion in solutions. In this equation, called *Fick's first law*, J is the flux, or quantity per second, of diffusing matter passing normally through a unit area under the action of a concentration gradient dn_A/dx. The factor D is known as the *diffusivity*, or the *diffusion coefficient*.

As originally conceived, the diffusivity D in Fick's first law was assumed to be constant for measurements made at a fixed temperature. However, only in the case of a solution composed of two gases has it been verified experimentally that the diffusivity approaches a constant value at a fixed temperature. For example, it has been shown that for binary mixtures of the gases oxygen and hydrogen, which have rather large molecular mass differences (16 to 1) and a correspondingly large difference in their arithmetic mean speed (4 to 1), the diffusivity (D) changes by less than 5 percent when the atom fraction $n_1/(n_1 + n_2)$ changes from 0.25 to 0.75. When the molecular masses are more nearly equal, the variation in the diffusivity with composition becomes even smaller and harder to detect in gaseous solutions.

In contrast to gaseous solutions, the diffusivity in both liquid and solid solutions is seldom constant. An example of this variation of the diffusion coefficient with composition can be seen in Fig. 12.3, taken from the work of Million and Kucera. (Ref. 1) The figure shows the variation of the diffusivity of cobalt in nickel-cobalt alloys at four different temperatures. Ni and Co form an isomorphous phase diagram, similar to the one shown in Fig. 11.1. In other words, Co and Ni form a single solid solution phase at all compositions within the temperature range indicated in Fig. 12.3. It should be noted that a solid-state transition form α to ε phase exists around 70 at. % Co in the alloys, but this transformation is at a much lower temperature than those indicated in Fig. 12.3. The measurements show that at 1450°C, the diffusivity decreases linearly as cobalt content in the alloy is increased. Since log scale in used for the Y-axis, actually, diffusivity decreases by a factor of about 4 as Co content changes from zero percent to 100 percent. At other temperatures, the diffusivity still depends on the Co content, but its variation is not linear. The data show none ideal behavior with

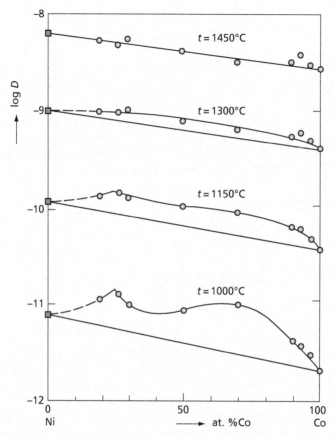

Fig. 12.3 Concentration dependence of log D at 1450°C, 1300°C, 1150°C, and 1000°C in Ni–Co alloys. (From Acta Metallurgica Volume 17, Issue 3, B. Million and J. Kucera, Concentration dependence of diffusion of cobalt in nickel-cobalt alloys, p. 342, Copyright 1969, with permission from Elsevier. https://www.sciencedirect.com/science/article/abs/pii/000161606990073X)

two maxima around 25 and 75 percent Co. Although the exact reason or reasons for the existence of the maxima do not exist, they may be attributable to short-range ordering or clustering in the alloys around these compositions. Based on the above results and similar measurements done by other investigators, (Ref. 2) it can be concluded that in metallic solid solutions, the diffusivity is usually not constant and depends on the composition. Of course, it also strongly depends on the temperature as discussed later.

12.2 The Kirkendall Effect

An experiment will now be discussed which shows that, in a binary solid solution, each of the two atomic forms can move with a different velocity. In the original experiment, as performed by Smigelskas and Kirkendall, (Ref. 3) the diffusion of copper and zinc atoms was studied in the

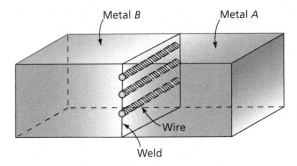

Fig. 12.4 Kirkendall diffusion couple

composition range where zinc dissolves in copper and the alloy still retains the face-centered cubic crystal structure characteristic of copper (alpha-brass range). Since their original work, many other investigators have found similar results using a large number of different binary alloys. Figure 12.4 is a schematic representation of a Kirkendall diffusion couple: a three-dimensional view of a block of metal formed by welding together two metals of different compositions. In the plane of the weld, shown in the center of Fig. 12.4, a number of fine wires (usually of some refractory metal that will not dissolve in the alloy system to be studied) are incorporated in the diffusion couple. These wires serve as markers with which to study the diffusion process. For the sake of the argument, let us assume that the metals separated by the plane of the weld are originally pure metal A and pure metal B; that on the right side of the weld is pure A, while that on the left is pure B. In order that a total amount of diffusion, which is large enough to be experimentally measurable, can be obtained in a specimen of this type, it is necessary that it be heated to a temperature close to the melting point of the metals comprising the bar, and maintained there for a relatively long time, usually of the order of days, for diffusion in solids is much slower than in gases or liquids. Upon cooling the specimen to room temperature, it is placed in a lathe and thin layers parallel to the weld interface are removed from the bar. Each layer is then analyzed chemically and the results plotted to give a curve showing the composition of the bar as a function of distance along the bar. Such a curve is shown schematically in Fig. 12.5, from which it is easily deduced that there has been a flow of B atoms from the left side of the bar toward the right, and a corresponding flow of A atoms in the other direction.

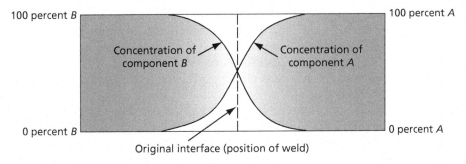

Fig. 12.5 Curves showing concentration as a function of distance along a diffusion couple. Curves of this type are usually called *penetration curves*

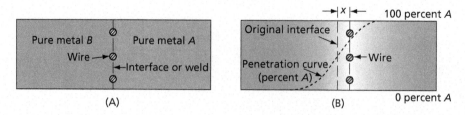

Fig. 12.6 Marker movement in a Kirkendall diffusion couple

Curves such as those shown in Fig. 12.5 were obtained for diffusion couples many years before the Smigelskas and Kirkendall experiment was performed. The original feature to their work was the incorporation of marker wires between the members of the diffusion couple. The interesting result which they obtained was that the wires moved during the diffusion process. The nature of this movement is shown in Fig. 12.6, where the left figure represents the diffusion couple before the isothermal treatment (anneal), and the right, the same bar after diffusion has occurred. The latter figure shows that the wires have moved to the right through the distance x. This distance, while small, is measurable, and for those cases where the markers have been placed at the weld between two different metals, the distance has been found to vary as the square root of the time during which the specimen was maintained at the diffusion temperature.

The only way to explain the movement of the wires during the diffusion process (in Fig. 12.6) is for the A atoms to diffuse faster than the B atoms. In this present example, it would be expected that more A atoms than B atoms must pass through the cross-section (defined by the wires) per unit time, causing a net flow of mass through the wires from right to left. Measurement of the position of the wires with respect to one end of the bar will show the movement of the wires.

The Kirkendall effect can be taken as a confirmation of the vacancy mechanism of diffusion. Through the years a number of mechanisms have been proposed to explain the movement of atoms in a crystal lattice. These can be grouped roughly into two classifications: those involving the motion of a single atom at a time, and those involving the cooperative movement of two or more atoms. As examples of the former, we have diffusion by the vacancy mechanism, as shown in Fig. 7.1, and diffusion of interstitial atoms (such as carbon in the iron lattice where the carbon atoms jump from one interstitial position to an adjacent one). While interstitial diffusion is recognized as the proper mechanism to explain the movement of small interstitial atoms through a crystal lattice, it is generally conceded that a mechanism of diffusion which involves placing large atoms (which normally enter into substitutional solutions) into interstitial positions is not feasible on energy considerations. The distortion of the lattice caused by placing one of these atoms in an interstitial position is very large, requiring a very large activation energy. Of these two possibilities based on the motion of individual atoms, the vacancy mechanism is much to be preferred in explaining diffusion in substitutional solid solutions.

The simplest conceivable cooperative movement of atoms is a direct interchange, as is illustrated schematically in Fig. 12.7. Here two adjacent atoms jump past each other and exchange positions. However, this involves the outward displacement of the atoms surrounding the jumping pair during the period of the transfer. Theoretical computations by Huntington and Seitz of the energy required to make a direct intercharge indicated that it is much larger than that required for the jump of an atom into a vacancy in metallic copper, and it is commonly believed that this

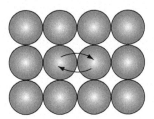

Fig. 12.7 Direct interchange diffusion mechanism

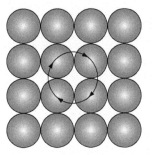

Fig. 12.8 Zener ring mechanism for diffusion

conclusion can be applied to other metals. For this reason the direct interchange is usually ruled out as an important mechanism in the diffusion of metals.

Another possible mechanism that explains diffusion in substitutional solid solutions is the Zener ring mechanism. In this case, it is assumed that thermal vibrations are sufficient to cause a number of atoms, which form a natural ring in a crystal, to jump simultaneously and in synchronism in such a manner that each atom in the ring advances one position around the ring. This mechanism is illustrated in Fig. 12.8, where a ring of four atoms is shown and the arrows indicate the motion of the atoms during a jump. Zener has suggested, on the basis of theoretical computations, that a ring of four atoms might be the preferred diffusion mechanism in body-centered cubic metals because their structure is more open than that of close-packed metals (face-centered cubic and close-packed hexagonal). The more open structure would require less lattice distortion during the jump. It has also been proposed as a more probable mechanism than the direct interchange because the lattice distortion that occurs during the jump is smaller, requiring less energy for the movement. However, the principal objection to the acceptance of the ring mechanism, even in body-centered cubic metals, is that it has been shown conclusively by diffusion experiments involving couples composed of a number of different body-centered cubic metals, that a Kirkendall effect occurs. It is possible to explain different rates for the diffusion of two atoms A and B by a vacancy mechanism; the rate at which the two atoms jump into the vacancies need only be different. In a ring mechanism, or in a direct exchange, the rate at which A atoms move from left to right must equal the rate at which B atoms move from right to left. Thus, in those alloy systems where a Kirkendall effect has been observed, one must rule out interchange mechanisms.

In summation, the vacancy mechanism is generally conceded to be the correct mechanism for diffusion in face-centered cubic metals. First, because of the various methods that have been proposed to explain the movements of atoms in this type of crystal, it requires the least thermal energy to activate it. Second, the Kirkendall effect has now been observed in a great many diffusion experiments involving couples composed of face-centered cubic metals. In body-centered cubic metals, while calculations have indicated that diffusion by a ring mechanism might require less thermal energy than vacancy diffusion, the discovery of a Kirkendall effect in body-centered cubic metals indicates that these metals also diffuse by a vacancy mechanism. Finally, while the picture is not quite as clear in the case of hexagonal metals, results of diffusion experiments are in accord with a vacancy mechanism. In this regard, it should be mentioned that, due to the asymmetry of the hexagonal lattice, the rate of diffusion is not the same in all directions through the lattice. Thus, diffusion in the basal plane occurs at a different rate from diffusion in a direction perpendicular to the basal plane. If a vacancy mechanism is assumed for hexagonal metals, this implies that the jump rate of an atom into vacancies lying in the same basal plane as itself will differ from the rate at which it jumps into vacancies lying in the basal planes directly above or directly below it.

12.3 Pore Formation

The Kirkendall experiment demonstrates that the rate at which the two types of atoms of a binary solution diffuse is not the same. Experimental measurements have shown that the element with the lower melting point diffuses faster. Thus, in alpha-brass (a mixture of copper and zinc atoms) zinc atoms move at the faster rate. On the other hand, diffusion couples formed of copper and nickel show that copper moves faster than nickel, in agreement with the fact that copper melts at a lower temperature than nickel.

In the case of a copper-nickel couple, shown schematically in Fig. 12.9, there is a greater flow of copper atoms toward the nickel side than nickel atoms toward the copper side. The right-hand side of the specimen suffers a loss of mass because it loses more atoms than it gains, while the left-hand side gains in mass. As a result of this mass transfer, the right- and left-hand portions of the specimen shrink and expand respectively. In cubic metals, these volume changes would be isotropic (Ref. 4) except for the fact that in a diffusion couple of any considerable size the diffusion zone containing the regions that undergo the volume changes may be only a small part of the total specimen. Shrinkage or expansion in a direction perpendicular to the weld interface occurs without an appreciable restraint from the rest of the specimen, but dimensional changes parallel to the weld interface are resisted by the bulk of the metal that lies outside of the diffusion zone. The net effect of this restraining action is twofold: the dimensional changes are essentially one-dimensional (along the axis of the bar lying perpendicular to the weld interface) and a state of stress is set up in the diffusion zone. The region to the right of the weld that suffers a loss of mass is placed under a two-dimensional tensile stress, while that on the side that gains mass is placed in a state of two-dimensional compressive stress. These stress fields may bring about plastic flow, with the resulting structural changes normally associated with plastic deformation and high temperatures, formation of substructures, recrystallization, and grain growth.

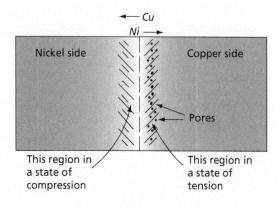

Fig. 12.9 In a copper-nickel diffusion couple the copper diffuses faster into the nickel than the nickel does into the copper

In addition to the effects mentioned in the previous paragraph, another phenomenon is encountered when two diffusing atom forms move with greatly different rates during diffusion, and it is directly connected with the vacancy motion associated with diffusion in metals. It has frequently been observed that voids, or pores, form in that region of the diffusion zone from which there is a flow of mass. Since every time an atom makes a jump, a vacancy moves in the opposite direction, an unequal flow in the two types of atoms must result in an equivalent flow of vacancies in the reverse direction to the net flow of atoms. Thus, in the copper-nickel couple shown in Fig. 12.9, more copper atoms leave the area just to the right of the weld interface than nickel atoms arrive to take their place, and a net flow of vacancies occurs from the nickel-rich side of the bar toward the copper-rich side. This movement reduces the equilibrium number of vacancies on the nickel-rich side of the diffusion range, while increasing it on the copper side. The degree of super-saturation and undersaturation is, however, believed to be small. On the side of a diffusion couple that loses mass, it has been estimated by several workers (Refs. 5 and 6) that the vacancy supersaturation is of the order of 1 percent, or less, of the equilibrium number of vacancies. Since the concentration of vacancies is not greatly affected by the conditions of flow, and the number is, at best, a small fraction of the number of atoms, vacancies must be created on the side of a couple that gains mass, and absorbed on the side that loses mass. Various sources and sinks have been proposed for the maintenance of this flux of vacancies. Grain boundaries and exterior surfaces are probable positions for both the creation and elimination of vacancies, but it is now generally agreed that the experimental facts agree best on the concept that the most important mechanism for the creation of vacancies is *dislocation climb*, and that dislocation climb and void formation account for most of the absorbed vacancies, positive climb being associated with the removal of vacancies and negative climb with the creation of vacancies.

The formation of voids, as a result of the vacancy current that accompanies the unequal mass flow in a diffusion couple, is influenced by several factors. It is generally believed that voids are heterogeneously nucleated, that is, they form on impurity particles. The tensile stress that exists in the region of the specimen where the voids form is also recognized as a contributing factor in the development of voids. If this tensile stress is counteracted by a hydrostatic compressive stress placed on the sample during the diffusion anneal, the voids can be prevented from forming.

12.4 Darken's Equations

The Kirkendall effect shows that in a diffusion couple composed of two metals, the atoms of the components move at different rates, and the flux of atoms through a cross-section defined by markers is not the same for both atom forms. It is, thus, more logical to think in terms of two diffusivities D_A and D_B corresponding to the movement of the A and B atoms, respectively. These quantities may be defined by the following relationships.

$$J_A = -D_A \frac{\partial n_A}{\partial x} \quad \text{and} \quad J_B = -D_B \frac{\partial n_B}{\partial x} \qquad 12.9$$

where J_A and J_B = the fluxes (number of atoms per second passing through a unit cross-section area) of the A and B atoms, respectively

D_A and D_B = the diffusivities of the A and B atoms

n_A and n_B = the numbers of A and B atoms per unit volume, respectively

The diffusivities D_A and D_B are known as the intrinsic diffusivities and are functions of composition and therefore of position along a diffusion couple.

Darken's equations, (Ref. 7) which make it possible to determine the intrinsic diffusivities experimentally, will now be derived. Several important assumptions are made in this derivation. First, it is assumed that all of the volume expansion and contraction during diffusion, due to unequal mass flow, occurs only in the direction perpendicular to the weld interface. (The cross-section area does not change during the diffusion.) As pointed out earlier, this condition is effectively realized when the diffusion couple is large in size compared to the diffusion zone. It is also assumed that the total number of atoms per unit volume is a constant or that

$$n_A + n_B = \text{constant}$$

This is equivalent to a requirement that the volume per atom be independent of concentration, which is an assumption that has not been realized experimentally, but deviations from this relationship can generally be compensated for by computation. The following derivation is also based on the condition that porosity does not occur in the specimen during the diffusion process.

Let us first consider the velocity of the Kirkendall markers through space. In Fig. 12.10, the cross-section at the distance x from the left end of a bar, with a square cross-section of unit area, represents the position of the markers at the time t_0, while the cross-section at the distance x' represents their position after a time interval dt. At the same time the origin of coordinates (left end of the bar) is assumed to be far enough away from the weld that its composition is not affected by the diffusion. The velocity v of the Kirkendall markers may be written

$$v = \frac{x - x'}{dt} \qquad 12.10$$

The marker velocity is also equal in magnitude, but opposite in direction to the volume of matter flowing past the markers per second divided by the cross-section area A of the bar at the markers.

$$v = \frac{\text{volume}}{\text{seconds}} \times \frac{1}{\text{unit area}}$$

The volume of matter that flows through the markers per second is equal to the net flux of atoms (net number of atoms per second) passing the wires times the volume per atom, or

$$\frac{\text{volume}}{\text{seconds}} = \frac{J_{\text{net}}}{n_A + n_B}$$

where $1/(n_A + n_B)$ is the volume per atom by the definition of n_A and n_B. The net flux equals the sum of the fluxes of the A and B atoms, or

$$J_{\text{net}} = J_A + J_B = -D_A \frac{\partial n_A}{\partial x} - D_B \frac{\partial n_B}{\partial x} \qquad 12.11$$

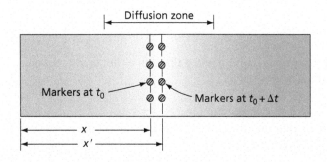

Fig. 12.10 Marker movement is measured from a point outside the zone of diffusion (one end of bar)

12.4 Darken's Equations

Substituting the above expressions into the equation for the marker velocity gives us the following equation:

$$v = -\frac{\text{volume}}{\text{seconds}} \times \frac{1}{\text{area}} = +\frac{\left(D_A \frac{\partial n_A}{\partial x} + D_B \frac{\partial n_B}{\partial x}\right)}{(n_A + n_B)} \quad 12.12$$

or

$$v = +\frac{\left(D_A \frac{\partial n_A}{\partial x} + D_B \frac{\partial n_B}{\partial x}\right)}{n_A + n_B} \quad 12.13$$

Remembering that $n_A + n_B$ is a constant, and that by definition

$$N_A = \frac{n_A}{n_A + n_B}, \quad N_B = \frac{n_B}{n_A + n_B}$$

$$N_B = 1 - N_A \quad \text{and} \quad \frac{\partial N_B}{\partial x} = -\frac{\partial N_A}{\partial x}$$

where N_A and N_B are the atom fractions of the A and B atoms, we may write the velocity of the markers in the following manner:

$$v = (D_A - D_B)\frac{\partial N_A}{\partial x} \quad 12.14$$

In principle it is possible to insert an imaginary set of markers on any cross-section along a diffusion couple. Equation 12.14 may therefore be used to compute the velocity with which any lattice plane normal to the diffusion flux moves. However, since D_A and D_B are normally functions of composition, v is a function of the composition of the specimen at the cross-section where v is measured. This means that the value of v is normally a function of position along the diffusion couple.

Equation 12.14 is one of our desired relationships, for it relates the two intrinsic diffusivities to the marker velocity and the concentration gradient ($\partial N_A/\partial x$), two quantities that can be experimentally determined. However, another equation is also needed in order that we may solve it and Eq. 12.14 simultaneously for the two unknowns D_A and D_B. This needed equation can be obtained by considering the rate at which the number of one of the atomic forms (A or B) changes inside a small-volume element.

Figure 12.11 also represents the diffusion couple of Fig. 12.10. In this case, however, the (unit) cross-section (designated mm) at a distance x from the left end of the bar represents a cross-section fixed in space; that is, fixed with respect to the end of the bar, which is assumed outside of the diffusion

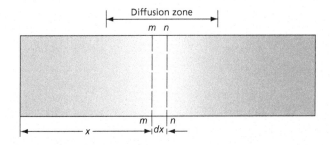

Fig. 12.11 The cross-sections mm and nn are assumed to be fixed in space; that is, they do not move with respect to the left end of the bar

zone. The cross-section at x (mm), accordingly, is not an instantaneous position of the markers as was the case in Fig. 12.10. A second (unit) cross-section of the bar (nn) is also shown, the distance of which from the end of the bar is $x + dx$. The two cross-sections define a volume element $1 \cdot dx$.

Let us consider the rate at which the number of A atoms changes inside this volume. This quantity equals the difference in the number of A atoms moving into and out of the volume per second, or to the difference in the flux of A atoms across the surface at x and at $x + dx$. Because a cross-section fixed in the lattice moves relative to our cross-sections fixed in space (at x and $x + dx$), the flux of A atoms through one of these space-fixed boundaries is due to two effects. First, because the metal is moving with a velocity v, a number ($n_A v$) of A atoms is carried each second through the cross-section at x, where n_A is the number of A atoms per unit volume, and v is the volume of metal flowing past the cross-section at x in a second. To this flux must be added the usual diffusion flux, so that the total number of atoms per second crossing the boundary at x is:

$$(J_A)_x = -D_A \frac{\partial n_A}{\partial x} + n_A v$$

The flux of A atoms through the cross-section at $x + dx$ is

$$(J_A)_{x+dx} = (J_A)_x + \frac{\partial (J_A)_x}{\partial x} \cdot dx$$

The rate at which the number of atoms inside the volume dx changes is therefore

$$(J_A)_x - (J_A)_{x+dx} = \frac{\partial}{\partial x} \left[D_A \frac{\partial n_A}{\partial x} - n_A v \right] \cdot dx$$

or the rate at which the number of A atoms per unit volume changes is

$$\frac{(J_A)_x - (J_A)_{x+dx}}{dx} = \frac{\partial n_A}{\partial t} = \frac{\partial}{\partial x} \left[D_A \frac{\partial n_A}{\partial x} - n_A v \right]$$

This equation may also be written:

$$\frac{\partial N_A}{\partial t} = \frac{\partial}{\partial x} \left[D_A \frac{\partial N_A}{\partial x} - N_A v \right] \qquad 12.15$$

because $n_A + n_B$ is assumed constant and the division of each term of the previous equation by this quantity makes it possible to convert concentration units from numbers of A atoms per unit volume to atom fractions. The expression 12.14 is now substituted into Eq. 12.15, and with the aid of the relationship

$$N_A = 1 - N_B$$

one obtains the final relationship

$$\frac{\partial N_A}{\partial t} = \frac{\partial}{\partial x} [N_B D_A + N_A D_B] \frac{\partial N_A}{\partial x} \qquad 12.16$$

This equation is in the form of Fick's second law, which is generally written

$$\frac{\partial N_A}{\partial t} = \frac{\partial}{\partial x} \tilde{D} \frac{\partial N_A}{\partial x} \qquad 12.17$$

where the quantity $\tilde{D}$ is seen to be equal to $[N_B D_A + N_A D_B]$. In fact, it is just this relationship that we have been seeking:

$$\tilde{D} = N_B D_A + N_A D_B \qquad 12.18$$

for the quantity $\tilde{D}$ may be evaluated experimentally, as can the atom fractions N_A and N_B. This is the second of Darken's equations for the determination of the intrinsic diffusivities D_A and D_B.

12.5 Fick's Second Law

Fick's second law, Eq. 12.17, is the basic equation for the experimental study of isothermal diffusion. Solutions of this second-order partial differential equation have been derived corresponding to the boundary conditions found in many types of diffusion samples. Most metallurgical specimens, such as the diffusion couple shown in Fig. 12.5, involve only a (net) one-dimensional flow of atoms and the assumption that the specimen is long enough in the direction of diffusion (so that the diffusion process does not change the composition at the ends of the specimen). There are two standard methods of measuring the diffusion coefficient when using this type of specimen. In one case, the diffusivity is assumed constant, and in the other, it is taken as a function of composition. The former method, known as the *Grube method* (also called Grube-Jedele), is strictly applicable only to those cases in which the diffusivity varies very slightly with composition. However, it may be applied to diffusion in alloy systems where the diffusivity varies moderately with composition if the two halves of a couple are made of metals differing slightly in composition. Thus, the couple in Fig. 12.4, instead of consisting of a bar of pure A welded to another of pure B, might have consisted of an alloy of perhaps 60 percent of A and 40 percent of B welded to one of 55 percent of A and 45 percent of B. Our preceding analysis would hold just as well for this couple as for the couple composed of pure metals. Over a small range of composition such as this, the diffusivity is essentially constant and the measurement effectively gives an average value of the diffusivity for the interval.

If the diffusivity $\tilde{D}$ is assumed to be a constant, then Fick's second law can be written

$$\frac{\partial N_A}{\partial t} = \frac{\partial}{\partial x} \tilde{D} \frac{\partial N_A}{\partial x} = \tilde{D} \frac{\partial^2 N_A}{\partial x^2} \qquad 12.19$$

The solution of this equation for the case of a diffusion couple consisting originally of two alloys of the elements A and B, one having the composition N_{A_1} (atom fraction) and the other the composition N_{A_2} at the start of the diffusion process, is:

$$N_A = N_{A_1} + \frac{(N_{A_2} - N_{A_1})}{2}\left[1 + \text{erf}\frac{x}{2\sqrt{\tilde{D}t}}\right] \quad \text{for} \quad -\infty < x < \infty \qquad 12.20$$

where N_A is the composition or atom fraction at a distance x (in cm) from the weld interface, t is the time in seconds, and $\tilde{D}$ is the diffusivity. The symbol erf $x/2\sqrt{\tilde{D}t}$ represents the error function, or probability integral with the argument $y = x/2\sqrt{\tilde{D}t}$. This function is defined by the equation

$$\text{erf}(y) = \frac{2}{\sqrt{\pi}}\int_0^y e^{-y^2} dy \qquad 12.21$$

Table 12.1 Error Function Values

y	erf(y)
0	0
0.2	0.2227
0.4	0.42839
0.477	0.50006
0.6	0.60386
0.8	0.74210
1.0	0.84270
1.4	0.95229
2.0	0.99532
3.0	0.99998

and is tabulated in many mathematical tables (Ref. 8) in much the same manner as trigonometric and other frequently used functions. Table 12.1 gives a few values of the error function for some corresponding values of the argument y. In applying this table, the error function becomes negative when y (or x in $x/2\sqrt{\tilde{D}t}$) is negative.

Figure 12.12 shows the theoretical penetration curve (distance versus composition curve) obtained when the solution of Fick's equation is plotted as a function of the variable $x/2\sqrt{\tilde{D}t}$, using the data of Table 12.1. Note that this curve is obtained under the assumption that $\tilde{D}$ is constant, or that $\tilde{D}$ varies only slightly within the composition interval N_{A_1} to N_{A_2} (original compositions of the two halves of the diffusion couple).

The curve in Fig. 12.12 shows that, under the stated conditions, the concentration is a single-valued function of the variable $x/2\sqrt{\tilde{D}t}$. Thus, if a diffusion couple has been maintained at some fixed temperature for a given period of time (t) so that diffusion can occur, then a single determination of the composition at an arbitrary distance (x) from the weld permits the determination of the diffusivity $\tilde{D}$. Thus, suppose that a diffusion couple is formed by the welding together of two alloys of the elements A and B having the compositions N_{A_1}, 40 percent of element A (the alloy to the right

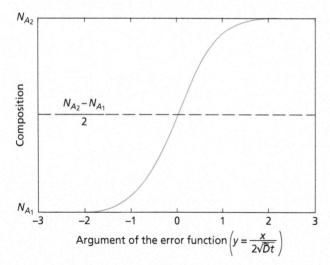

Fig. 12.12 Theoretical penetration curve, Grube method

of the weld), and N_{A_2}, 50 percent of element A (left of the weld). Let the couple be heated quickly to some temperature T_1 and held there 40 hr (144,000 s), and let it be assumed that after cooling to room temperature, chemical analysis shows that at a distance 2×10^{-3} m to the right of the weld the composition N_A is 42.5 percent A, we have, on substituting the assumed data in Eq. 12.20,

$$0.425 = 0.400 + \frac{(0.500 - 0.400)}{2}\left[1 - \text{erf}\frac{0.002}{2\sqrt{\tilde{D}(144{,}000)}}\right]$$

or

$$\text{erf}\frac{0.001}{\sqrt{144{,}000\tilde{D}}} = 0.500$$

Table 12.1 shows, however, that when the error function equals 0.500 (i.e., 0.50006), the value of the argument $y = (x/2\sqrt{\tilde{D}t})$ is 0.477, and so

$$\tilde{D} = 3.04 \times 10^{-11}, \text{m}^2/\text{s}$$

In general, the Grube method is based on the evaluation of the error function in terms of the composition of the specimen at an arbitrary point and then, with the aid of the error-function table, the argument y is found. This, in turn, permits the value of the diffusivity to be determined.

In the numerical computation just considered, a composition 42.5 percent A was assumed at a distance of 2×10^{-3} m from the weld when the diffusion time was 40 hr. With the assumption of a constant diffusivity ($\tilde{D}$) and the same diffusion temperature, let us compute the length of time needed to obtain the same composition, 42.5 percent A, at twice the distance from the weld interface. Since the composition N_A remains the same, this problem requires that the argument of the probability integral has the same value as in the last example. That is,

$$\frac{x_1}{2\sqrt{\tilde{D}t_1}} = \frac{x_2}{2\sqrt{\tilde{D}t_2}}$$

or

$$\frac{x_1^2}{t_1} = \frac{x_2^2}{t_2} \qquad\qquad 12.22$$

where $x_1 = 2 \times 10^{-3}$ m
$x_2 = 4 \times 10^{-3}$ m
$t_1 = 40$ hr
t_2 = time to reach a composition of 42.5 percent at 4×10^{-3} m from weld interface

Substituting these values into Eq. 12.22 and solving for t_2 gives a value for the latter of 160 hr. The equation $x_1^2/t_1 = x_2^2/t_2$ is often used to indicate the relationship between distance and time during isothermal diffusion even when $\tilde{D}$ is not a constant. In this latter case, it still serves as a rough but very convenient approximation.

12.6 The Matano Method

The second well-known method of analyzing experimental data from metallurgical diffusion samples was devised by Matano, (Ref. 9) who first proposed it in 1933. It is based on a solution of Fick's second law, originally proposed by Boltzmann in 1894. In this method, the diffusivity is assumed to be a function of concentration, which requires the solution of Fick's second law in the form:

$$\frac{\partial N_A}{\partial t} = \frac{\partial}{\partial x} \tilde{D}(N_A) \frac{\partial N_A}{\partial x} \qquad 12.23$$

This mathematical operation is much more difficult than when $\tilde{D}$ is constant and, as a consequence, the Matano-Boltzmann method of determining the diffusivity $\tilde{D}$ uses graphical integration. The first step in this procedure, after the diffusion anneal and chemical analysis of the specimen, is to plot a curve of concentration versus distance along the bar measured from a suitable point of reference, say from one end of the couple. For the purpose of simplifying the following discussion, it will be assumed that the number of atoms per unit volume $(n_A + n_B)$ is constant. The second step is to determine that cross-section of the bar through which there have been equal total fluxes of the two atomic forms (A and B). This cross-section is known as the Matano interface and lies at the position where areas M and N in Fig. 12.13 are equal. The position of the Matano interface is determined by graphical integration, but, in general, it has also been experimentally determined that, in the absence of porosity, the Matano interface lies at the position of the original weld. (This is not the position of the weld after diffusion has occurred since, as we have seen, markers placed at the weld move during diffusion.) Once the Matano interface has been located, it serves as the origin of the x coordinate. In agreement with the normal sign convention, distances to the right of the interface are considered positive, while those to the left of it are negative. With the coordinate system thus defined, the Boltzmann solution of Fick's equation is

$$\tilde{D} = -\frac{1}{2t} \frac{\partial x}{\partial N_A} \int_{N_{A_1}}^{N_A} x\, dN_A \qquad 12.24$$

where t is the time of diffusion, N_A is the concentration in atomic units at a distance x measured from the Matano interface, and N_{A_1} is the concentration of one side of the diffusion couple at a point well removed from the interface where the composition is constant and not affected by the diffusion process.

In a manner similar to that used to explain the Grube method, we shall take arbitrarily assumed diffusion data and solve them by the Matano method. Table 12.2 represents this concentration-distance data corresponding to no actual alloy system, but it is representative, in a broad general way, of diffusion data obtained in actual experiments. The diffusion couple is assumed to be formed from pure A and pure B.

Figure 12.14 shows the penetration curve obtained when the data of Table 12.2 are plotted. Let us reconsider the Boltzmann solution of Fick's second law:

$$\tilde{D} = -\frac{1}{2t} \frac{\partial x}{\partial N_A} \int_{N_{A_1}}^{N_A} x\, dN_A \qquad 12.25$$

Suppose that we desire to know the diffusivity at a particular concentration, which we shall arbitrarily take as 0.375. This concentration corresponds to the point marked C in Fig. 12.14. In order to

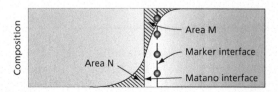

Fig. 12.13 The Matano interface lies at the position where area M equals area N

12.6 The Matano Method

Table 12.2 Assumed Diffusion Data to Illustrate the Matano Method

Composition Atomic Percent Metal A	Distance from the Matano Interface, mm
100.00	5.08
93.75	3.14
87.50	1.93
81.25	1.03
75.00	0.51
68.75	0.18
62.50	−0.07
56.25	−0.27
50.00	−0.39
43.75	−0.52
37.50	−0.62
31.25	−0.72
25.00	−0.87
18.75	−1.07
12.50	−1.35
6.25	−1.82
0.00	−2.92

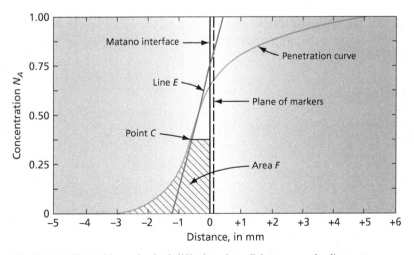

Fig. 12.14 Plot of hypothetical diffusion data (Matano method)

compute the diffusivity at this point, we must first evaluate two quantities with the aid of Fig. 12.14. The first of these is the derivative $\partial x/\partial N_A$, the reciprocal of the slope of the penetration curve at point C. The tangent to the curve at this point is shown in the figure as line E and its slope is 610 m^{-1}. The other quantity is the integral, the integration limits of which are $N_{A_1} = 0$ and $N_A = 0.375$. The indicated integral corresponds to the cross-hatched area (F) of Fig. 12.14. Evaluation of this area by a graphical method (Simpson's Rule) yields the value 4.66×10^{-4} m. The diffusivity at the composition 0.375 is, accordingly,

$$\tilde{D}(0.375) = \frac{1}{2t}\left(\frac{1}{\text{slope at 0.375}}\right) \times (\text{area from } N_A = 0 \text{ to } N_A = 0.375)$$

If the diffusion time is now assumed to be 50 hr (180,000 s), the complete evaluation of the diffusivity is:

$$\tilde{D}(0.375) = \frac{1}{2(180,000)} \times \frac{1}{6.10} \times 4.66 \times 10^{-4} = 2.1 \times 10^{-12} \text{ m}^2/\text{s}$$

Computations similar to the above may be made to determine the diffusivity $\tilde{D}(N_A)$ at any concentration not too close to the terminal compositions ($N_A = 0$ and $N_A = 1$). In any case, the slope at the desired concentration and the area (between the penetration curve, the vertical line representing the position of the Matano interface and within the composition limits zero to N_A) must be determined. Since the desired area approaches zero as the composition approaches one of the terminal compositions, the accuracy of the determination falls off as N_A becomes very close to either zero or one.

Figure 12.15 shows the variation of $\tilde{D}(N_A)$ with the concentration N_A for the data of Fig. 12.14, as determined by computation at several compositions. Notice that large values of $\tilde{D}$ are obtained as one approaches the concentration $N_A = 1$. There is also a minimum in this curve in the middle of the concentration range. A diffusivity-concentration curve of this form has been reported (Refs. 10 and 11) for the diffusion of zirconium and uranium. However, the curves of Fig. 12.3 are more typical of the diffusion data reported to date.

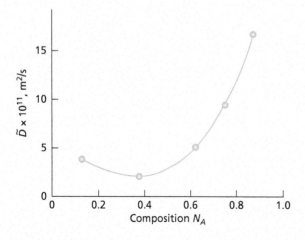

Fig. 12.15 Variation of the interdiffusion coefficient $\tilde{D}$ with composition. (Data of Table 12.1)

12.7 Determination of the Intrinsic Diffusivities

The determination of the intrinsic diffusivities will now be illustrated with the use of the assumed data of Table 12.2. First we must derive an expression for the marker velocity v in terms of the marker displacement and the time of diffusion t. Experimentally, it has been determined that the markers move in such a manner that the ratio of their displacement squared to the time of diffusion is a constant. Thus

$$\frac{x^2}{t} = k$$

where k is a constant. The marker velocity is, accordingly,

$$v = \frac{\partial x}{\partial t} = \frac{k}{2x}$$

but k equals x^2/t, so that

$$v = \frac{x}{2t} \qquad \qquad 12.26$$

In Fig. 12.14, an arbitrarily assumed position of the marker interface is shown at a distance $x = 0.0001$ m from the Matano interface. The diffusion time t taken for the data is 50 hr, or 180,000 s. These numbers correspond to a marker velocity

$$v = \frac{0.0001}{2(180,000)} = 2.78 \times 10^{-10} \text{ m/s}$$

At the position of the markers we also have $N_A = 0.65$ and $N_B = 0.35$ and

$$\tilde{D}_M = 5.5 \times 10^{-12}; \qquad \frac{\partial N_A}{\partial x} = 244 \text{ m}^{-1}$$

The value of $\tilde{D}_M$ is obtained from Fig. 12.15, while dN_A/dx is the slope of the penetration curve in Fig. 12.14 at the position of the markers, and N_A and N_B are the atom fractions of A and B, respectively, at the position of the markers. The above values can now be substituted into the Darken equations:

$$\tilde{D} = N_B D_A + N_A D_B$$

$$v = (D_A - D_B)\frac{\partial N_A}{\partial x}$$

yielding

$$5.5 \times 10^{-12} = 0.35\, D_A + 0.65\, D_B$$

$$2.78 \times 10^{-10} = (D_A - D_B)244$$

The solution of this pair of simultaneous equations has as a result

$$D_A = 6.24 \times 10^{-12}$$
$$D_B = 5.10 \times 10^{-12}$$

These values tell us that the flux of A atoms through the marker interface from right to left is approximately 1.2 times that of the flux of B atoms moving from left to right.

The preceding section has demonstrated that it is possible to determine experimentally the intrinsic diffusivities of a binary diffusion system (D_A and D_B). These quantities are valuable because they measure the speed with which the individual atomic forms move during diffusion. However, there has been, to date, very little success in the development of a theory capable of predicting the numerical values of the intrinsic diffusivities starting from a consideration of atomic processes. While it is generally agreed that diffusion in metallic substitutional solid solutions is the result of the movement of vacancies, the factors that control jump rates into vacancies of the two different atomic forms are complex and not completely understood. Thus, in our previous derivation of Fick's first law, several simplifying assumptions were made that do not hold for real metallic substitutional solid solutions. First, it was assumed that the solution was ideal, but, as we have seen, most metallic solutions are not ideal, and in a nonideal solution the diffusion rates are influenced by a tendency for like atoms either to group together, or to avoid each other. Second, it was assumed that the rate of jumping was independent of composition, that is, whether the jump was made by an A or a B atom. The assumption that the rate of jumping is independent of the composition is certainly not true, as may be judged by the fact that measured diffusivities vary widely with composition.

As discussed above, theoretical interpretation of substitutional diffusion is difficult because of the number of variables that must be considered. For this reason, the major effort in diffusion studies has been directed toward the investigation of diffusion in relatively simple systems that are more amenable to interpretation. In general, these consist of the study of diffusion in very dilute substitutional solid solutions or the study of diffusion using radioactive tracers.

12.8 Self-Diffusion in Pure Metals

In self-diffusion studies, one investigates the diffusion of a solute consisting of a radioactive isotope in a solvent that is a nonradioactive isotope of the same metal. In such a system, both atomic forms are identical except for the small mass difference between the isotopes. The principal effect of this mass difference is to cause the solute isotope to vibrate about its rest point in the lattice with a frequency slightly different from that of the solvent isotope, giving the two isotopes a slightly different jump rate. This difference is easily calculated because the vibration frequency is proportional to the reciprocal of the square root of the mass, and since the jump rate into vacancies is proportional to the vibration frequency, we have

$$\frac{1}{\tau^*} = \sqrt{\frac{m}{m^*}}\left(\frac{1}{\tau}\right) \qquad 12.27$$

where $1/\tau$ and $1/\tau^*$ are the jump-rate frequencies of normal and radioactive isotopes respectively (τ and τ^* are the mean times of stay of the respective atoms at lattice positions, and m and m^* are the masses of the two isotopes [m^* radioactive]).

12.8 Self-Diffusion in Pure Metals

Except for the mass difference, solute and solvent are chemically identical and the solid solution is truly ideal. Considerations of the effect of the departure of a solution from ideality may thus be neglected. Furthermore, the mass correction is usually small so that, to a good approximation, we may assume that the intrinsic diffusivity of the radioactive isotope is the same as that of the nonradioactive isotope. When the intrinsic diffusivities are equal, the interdiffusion coefficient equals the intrinsic diffusivities, as may be seen by considering the Darken equation:

$$\tilde{D} = N_B D_A + N_A D_B = (N_A + N_B)D = D$$

where $\tilde{D}$ is the interdiffusion coefficient (also called diffusivity), $D = D_A = D_B$, the intrinsic diffusivity of either the radioactive or nonradioactive isotopes, and $(N_A + N_B)$ is unity by the definition of atom fractions. Because the intrinsic coefficients do not depend on composition, it is also true that the interdiffusion coefficient does not depend on composition. Therefore experimental determinations of self-diffusivities may be made by using the simpler Grube method.

Because self-diffusion in pure metals occurs in an ideal solution and with a diffusivity that is independent of concentration, experimentally determined self-diffusion coefficients of pure metals are usually of high accuracy. Furthermore, because the diffusion process occurs in a relatively simple system, the measured diffusivities are capable of theoretical interpretation. The assumption made in our derivation of Fick's first law are those actually observed in self-diffusion experiments, and Eq. 12.7 is correct for self-diffusion in a simple cubic system: that is,

$$D = \frac{a^2}{6\tau}$$

where D is the diffusivity, a is the lattice constant, and τ the mean time of stay of an atom in a lattice position. While only polonium is believed to crystallize in a simple cubic lattice, similar relationships can be derived for other metallic lattices. As examples, we have for face-centered cubic metals

$$D_{\text{FCC}} = \frac{a^2}{12\tau} \qquad 12.28$$

and body-centered cubic metals

$$D_{\text{BCC}} = \frac{a^2}{8\tau} \qquad 12.29$$

and, in general, for any lattice,

$$D = \frac{\alpha a^2}{\tau} \qquad 12.30$$

where α is a dimensionless constant depending on the structure.

In the chapter on vacancies, it was shown (Eq. 7.46) that

$$r_a = Ae^{-(H_m + H_f)/RT}$$

where r_a, the number of jumps made per second by an atom in a pure metal crystal, is identical to our quantity $1/\tau$; H_f is the enthalpy change or work to form a mole of vacancies; H_m is the enthalpy change or energy barrier that must be overcome to move a mole of atoms into vacancies; R is the universal gas constant (8.31 J/mol-K), and T the absolute temperature in degrees Kelvin.

In the above expression, the coefficient A may be replaced by Zv and the equation rewritten in the form

$$r = Zve^{-H_m/RT}e^{-H_f/RT} \qquad 12.31$$

where Z is the lattice coordination number and v the lattice vibration frequency. This relationship may be interpreted as follows. The jump rate of atoms into vacancies (r_a) varies directly as (1) the number of atoms (Z) that are next to a vacancy; (2) the frequency (v) or number of times per second that an atom moves toward a vacancy; (3) the probability ($e^{-H_m/RT}$) that an atom will have sufficient energy to make a jump; (4) the concentration of vacancies in the lattice ($e^{-H_f/RT}$). Equation 12.31 neglects entropy changes associated with the formation and movement of vacancies and should be more correctly written:

$$\frac{1}{\tau} = Zve^{-(\Delta G_m + \Delta G_f)/RT} \qquad 12.32$$

where ΔG_m and ΔG_f are the free-energy changes associated with the movement and formation of vacancies respectively. These quantities may be expressed as

$$\Delta G_m = \Delta H_m - T\Delta S_m$$
$$\Delta G_f = \Delta H_f - T\Delta S_f$$

where ΔS_m is the entropy change per mole resulting from the strain of the lattice during the jumps, and ΔS_f the increase in entropy of the lattice due to the introduction of a mole of vacancies. Thus, the self-diffusion coefficient is:

$$D = \alpha a^2 Zve^{-(\Delta G_m + \Delta G_f)/RT} \qquad 12.33$$

In the body-centered cubic lattice, α is 1/8, while Z is 8, and, similarly, in the face-centered cubic lattice, α is 1/12, so that for both forms of cubic crystals

$$D = a^2 ve^{-(\Delta G_m + \Delta G_f)/RT}$$
$$D = a^2 ve^{(\Delta S_m + \Delta S_f)/R}e^{-(H_m + H_f)/RT} \qquad 12.34$$

This expression will be discussed further when we consider the temperature dependence of the diffusion coefficient. For the present, it suffices to point out that considerable success has been obtained in the theoretical interpretation of the various factors in this equation. A detailed discussion, such as is necessary to treat these quantities adequately, is beyond our scope and for this reason the reader is referred to advanced papers covering this subject. (Refs. 12 and 13)

12.9 Temperature Dependence of the Diffusion Coefficient

It has already been seen that the diffusion coefficient is a function of composition. It is also a function of temperature. The nature of this temperature dependence is shown clearly in the equation for the self-diffusion coefficient stated in the previous section,

$$D = a^2 ve^{(\Delta S_m + \Delta S_f)/R}e^{-(H_m + H_f)/RT}$$

Suppose that we set

$$D_0 = a^2 v e^{(\Delta S_m + \Delta S_f)/R} \qquad 12.35$$

and

$$Q = \Delta H_m + \Delta H_f \qquad 12.36$$

where D_0 and Q are constants since all the quantities from which they are formed are effectively constant. The quantity Q is the activation energy of diffusion, and D_0 is called the *frequency factor*. The self-diffusion coefficient may now be written in the simplified form

$$D = D_0 e^{-Q/RT} \qquad 12.37$$

In this form, the equation can be applied directly to the study of experimental data.

Taking the common logarithm $\{2.3 \log_{10}(x) = \ln(x)\}$ of both sides of Eq. 12.37 gives us

$$\log D = -\frac{Q}{2.3RT} + \log D_0 \qquad 12.38$$

This is an equation in the form

$$y = mx + b$$

where the dependent variable is $\log D$, the independent variable $1/T$, the ordinate intercept $\log D_0$, and the slope $-Q/(2.3\ R)$.

In light of the above, it is evident that if logarithms of experimental values of a self-diffusion coefficient yield a straight line when plotted against the reciprocal of the absolute temperature, then the data conform to the equation

$$D = D_0 e^{-Q/RT} \qquad 12.39$$

The slope of the experimentally determined straight line determines the activation energy Q since $m = -Q/2.3R$ or $Q = -2.3Rm$. At the same time, the intercept of the line with the ordinate designated by b yields the frequency factor D_0, since $b = \log D_0$ or $D_0 = 10^b$.

The above method of determining experimental activation energies and frequency factors can be illustrated with the use of some representative data given in Table 12.3.

The data of Table 12.3 are plotted in Fig. 12.16. The slope of the resulting straight line is -8000, or

$$m = -\frac{Q}{2.3R} = -8000$$

Table 12.3 Assumed Data to Show the Temperature Dependence of Self-Diffusion

Temperature K	Self-Diffusion Coefficient D	$\frac{1}{T}$	log D
700	1.9×10^{-15}	1.43×10^{-3}	-14.72
800	5.0×10^{-14}	1.25×10^{-3}	-13.30
900	6.58×10^{-13}	1.11×10^{-3}	-12.12
1000	5.00×10^{-12}	1.00×10^{-3}	-11.30
1100	2.68×10^{-11}	0.91×10^{-3}	-10.57

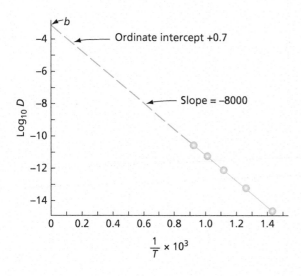

Fig. 12.16 Experimental diffusion data plotted to obtain the activation energy Q and frequency factor D_0

Solving for Q and remembering that R is 8.314 J/mol-K, gives

$$Q = 2.3(8.314)8000 = 153{,}000 \text{ J/mol}$$

The ordinate intercept of the experimental curve has the value -3.3. The value of D_0 is, accordingly,

$$D_0 = 10^b = 10^{-3.3} = 5 \times 10^{-4} \frac{m^2}{s}$$

The experimentally determined equation for the self-diffusion coefficient is, accordingly,

$$D = 5 \times 10^{-4} e^{-153{,}000/RT} \text{ m}^2 \text{ s}^{-1}$$

The preceding discussion has been concerned only with the temperature dependence of self-diffusion coefficients. However, experimentally determined values of chemical interdiffusion coefficients $\tilde{D}$, and of their component intrinsic diffusivities D_A and D_B, also show the same form of temperature dependence. Thus, speaking in general, all diffusion coefficients tend to follow an empirical activation law, so that we have for self-diffusion

$$D^* = D_0^* e^{-Q/RT} \qquad \text{12.40}$$

and for chemical diffusion

$$\tilde{D} = \tilde{D}_0 e^{-Q/RT} \qquad \text{12.41}$$

and

$$D_A = D_{A_0} e^{-Q_A/RT}$$
$$D_B = D_{B_0} e^{-Q_B/RT} \qquad \text{12.42}$$

where $\tilde{D} = N_B D_A + N_A D_B$.

Although the activation energy in the case of self-diffusion has a significance capable of explanation in terms of atomic processes, the meanings of the activation energies Q, Q_A, and Q_B for chemical diffusion when the solute concentration is high are vague and not clearly understood, particularly when other extrinsic factors also contribute to the diffusion process. (Refs. 14 and 15) Therefore, except when the solute concentration is very low, these activation energies should only be considered as empirical constants.

12.10 Chemical Diffusion at Low-Solute Concentration

The chemical interdiffusion coefficient $\tilde{D}$ also assumes a simple form when the solute concentration becomes very small, as can be seen by referring again to Darken's equation, Eq. 12.18:

$$\tilde{D} = N_B D_A + N_A D_B$$

Suppose that component B is taken as the solute. Let us assume that at all points in the diffusion couple the concentration of B is very low. Then

$$N_A \simeq 1 \qquad N_B \simeq 0$$

and

$$\tilde{D} \simeq D_B \qquad \qquad 12.43$$

Thus, at very low solute concentrations, the chemical interdiffusion coefficient approaches the intrinsic diffusivity of the solute.

If the solute concentration is much smaller than the solubility limit, solute atoms can be considered to be uniformly and widely dispersed throughout the lattice of the solvent. Considerations relative to the interaction between individual solute atoms can be neglected and each solute atom assumed to have an equivalent set of surroundings. All neighbors of each solute atom will be solvent atoms. Under these conditions, it is possible to give a theoretical interpretation to the frequency factor D_{B_0} and the activation energy for diffusion Q_B. In fact, for cubic crystals we can write an expression for D_B that is entirely equivalent to that which we have previously derived for the self-diffusion coefficient. See Eq. 12.34.

$$D_B = a^2 \nu e^{(\Delta S_{Bm} + \Delta S_{Bf})/R} e^{-(\Delta H_{Bm} + \Delta H_{Bf})/RT}$$

In this expression, ν is the vibration frequency of the solute atom in the solvent lattice, ΔS_{Bm} the entropy change per mole associated with the jumping of solute atoms into vacancies, H_{Bm} the energy barrier per mole for the jumping of solute atoms, ΔS_{Bf} the entropy increase of the lattice for the formation of a mole of vacancies adjacent to solute atoms, and H_{Bf} the work to form a mole of vacancies in positions next to solute atoms. Notice that, while the above expression has the same form as the self-diffusion equation, most of the quantities on the right-hand side of the equation have somewhat different meanings.

The frequency factor D_{B_0} and activation energy Q_B for chemical diffusion at low solute concentrations are

$$D_{B_0} = a^2 \nu e^{(\Delta S_{Bm} + \Delta S_{Bf})/R} \qquad \qquad 12.44$$
$$Q_B = H_{Bm} + H_{Bf}$$

Table 12.4 Solute Diffusion in Dilute Nickel-Base Substitutional Solutions*

Solute	Frequency Factor, D_{B_0}, m²/s	Activation Energy, Q_B, J/mol	Diffusivity $D_B = D_{B_0} e^{-Q'/RT}$ at 1470 K
Mn	7.50×10^{-4}	280,000	8.42×10^{-14}
Al	1.87×10^{-4}	268,000	5.60×10^{-14}
Ti	0.86×10^{-4}	255,000	7.47×10^{-14}
W	11.10×10^{-4}	321,000	0.44×10^{-14}
Ni	1.27×10^{-4}	279,000	1.55×10^{-14}

*Values for Mn, Al, Ti, and W from the data of Swalin and Martin, *Trans. AIME*, **206** 567 (1956).

Table 12.4 contains experimental data for the diffusion of several different solutes (at low concentrations) in nickel. The chemical diffusion data shown in this table were obtained with the use of diffusion couples consisting of a plate of pure nickel welded to another composed of an alloy of nickel containing the indicated element in an amount of the order of 1 atomic percent. For the systems studied, these couples conform to the condition that the solute concentration be low.

Column one of Table 12.4 indicates the diffusing solute atom. The symbol Ni in this column indicates that the values on the lowest line correspond to self-diffusion in pure nickel. Columns two and three give values of the frequency factor D_{B_0} and activation energy Q_B, respectively, while column four lists computed values of the diffusivity D_B for the temperature 1470 K.

Table 12.4 shows that the diffusivities of the elements Mn, Al, Ti, and W in dilute solid solution in nickel differ, but not in large measure, from the nickel self-diffusion coefficient.

12.11 The Study of Chemical Diffusion Using Radioactive Tracers

Consider the diffusion couple shown in Fig. 12.17, consisting of two halves with the same chemical composition, but with a fraction of the B atoms in the right-hand member radioactive. If such a couple is heated and the atoms allowed to diffuse, there should be no detectable change in chemical composition throughout the length of the bar. However, there will be a redistribution of the radioactive atoms. This change in concentration of tracer atoms along the axis of the bar can be determined in the following manner. After the diffusion anneal, the specimen is placed in a lathe and thin layers of equal thickness removed parallel to the weld interface. Measurement of the radioactive radiation intensity from each set of lathe turnings indicates the concentration of the radioactive B atoms in the corresponding layer. A plot of these intensity values as a function of position along the bar is equivalent to a normal penetration curve. In a specimen of this type, one measures the diffusion of B atoms in a homogeneous alloy of atoms A and B. Since the specimen is chemically homogeneous, the composition is everywhere the same and there can be no variation of the diffusivity with composition, which means that the penetration curve can be analyzed for the diffusivity by the Grube method. The diffusion coefficient found in this manner is

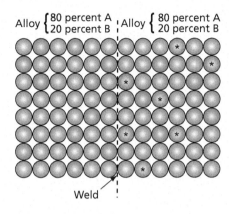

Fig. 12.17 Schematic representation of a diffusion couple using the tracer technique
◯ atom A, ◯ atom B, ◉ radioactive B

like a self-diffusion coefficient, but indicates the rate at which B atoms diffuse in an alloy of A and B atoms rather than in a matrix of pure B atoms.

In some binary systems it is possible to find radioactive isotopes of both of the component elements (A and B) which are suitable for use as tracers. When this is possible, measurements of the tracer diffusion coefficients for both elements can be made over the complete range of solubility. It is customary to give these quantities the symbols D_A^* and D_B^*. (The tracer diffusivities are denoted by asterisks to differentiate them from the intrinsic diffusion coefficients D_A and D_B.) The coefficients D_A^* and D_B^*, like the intrinsic diffusivities D_A and D_B, are functions of composition, which means that the rate of diffusion of either the A or the B atoms in a homogeneous crystal containing both atomic forms is not the same as it is in a pure metal of either component. We now have two different types of diffusion coefficients describing the diffusion process of the two atomic forms—tracer and intrinsic—in a substitutional solid solution. It is not unreasonable to wonder if they are related. A relationship does exist and was first derived by Darken (Ref. 7) and has been fully verified by experiment. According to Darken, the relationship between the intrinsic diffusivities and the self-diffusion coefficients are:

$$D_A = D_A^*\left(1 + N_A \frac{\partial \ln \gamma_A}{\partial N_A}\right)$$

$$D_B = D_B^*\left(1 + N_B \frac{\partial \ln \gamma_B}{\partial N_B}\right)$$

12.45

where D_A and D_B are the intrinsic diffusion coefficients, D_A^* and D_B^* are the tracer-diffusion coefficients, γ_A and γ_B are the activity coefficients of the two components, and N_A and N_B are the atom fractions of the two components. With the aid of the Gibbs-Duhem equation (a well-known physical chemistry relationship),

$$N_A \partial \ln \gamma_A = -N_B \partial \ln \gamma_B$$

12.46

and from the fact that $\partial N_A = -\partial N_B$ we have

$$1 + \frac{N_A \partial \ln \gamma_A}{\partial N_A} = 1 + \frac{(-N_B \partial \ln \gamma_B)}{-\partial N_B} = 1 + \frac{N_B \partial \ln \gamma_B}{\partial N_B}$$

12.47

The two factors that give the intrinsic diffusivities, when multiplied by the respective tracer-diffusion coefficients, are actually equal and it is customary to call this quantity the *thermodynamic factor*.

Let us consider the significance of the thermodynamic factor. In an ideal solution, the activity (a_A) equals the concentration of the solution N_A, and the activity coefficient γ_A (which is the ratio of these two quantities), is one. Since the logarithm of unity is zero, the thermodynamic factor becomes

$$1 + \frac{N_A \partial \ln \gamma_A}{\partial N_A} = (1 + 0) = 1$$

and the intrinsic-diffusion coefficient equals the tracer-diffusion coefficient. The tracer-diffusion coefficient can thus be considered a measure of the rate at which atoms would diffuse in an ideal solution, and the thermodynamic factor can be considered a correction that takes into account the departure of the crystal from ideality.

It is possible to express the chemical diffusivity in terms of the tracer-diffusion coefficients, for by Darken's equation, Eq. 12.18:

$$\tilde{D} = N_B D_A + N_A D_B$$

When one expresses the intrinsic diffusivities D_A and D_B in terms of the self-diffusion coefficients, we have

$$\tilde{D} = N_B D_A^* \left(1 + N_A \frac{\partial \ln \gamma_A}{\partial N_A}\right) + N_A D_B^* \left(1 + N_B \frac{\partial \ln \gamma_B}{\partial N_A}\right)$$

or

$$\tilde{D} = (N_B D_A^* + N_A D_B^*)\left(1 + N_A \frac{\partial \ln \gamma_A}{\partial N_A}\right) \qquad 12.48$$

since the two forms of the thermodynamic factor are equal.

Figures 12.18 through 12.20 contain experimental data for gold-nickel diffusion at 1173 K. At this temperature, gold and nickel dissolve completely in each other and form a completely soluble series of alloys. The significance of this experimental information is that it gives experimental confirmation of the Darken relationships.

In Fig. 12.18, the tracer-diffusion coefficients are plotted as a function of composition, and the very large variation of D_A^* and D_B^* is apparent. Notice that the tracer-diffusion rate of nickel atoms in pure gold is about 1000 times larger than nickel atoms in pure nickel. Figure 12.19 shows the thermodynamic factor

$$\left(1 + N_{Ni} \frac{\partial \ln \gamma_{Ni}}{\partial N_{Ni}}\right) \qquad 12.49$$

as a function of composition for gold-nickel alloys at 1173 K. The variation of the interdiffusion coefficient as a function of composition is given in Fig. 12.20. Two curves are shown: the one marked D (calculated) is derived from the self-diffusion coefficients (Fig. 12.18) and the thermodynamic factor (Fig. 12.19), the other marked $\tilde{D}$ (observed) is obtained from direct chemical-diffusion measurements using Matano analysis. Quite good agreement is found between the calculated and observed curves. The slight divergence between the two curves at high nickel concentrations can be explained on the basis of experimental errors.

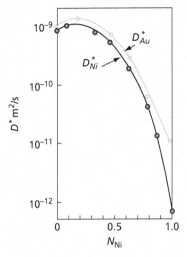

Fig. 12.18 Self-diffusion coefficients of Au and Ni in gold-nickel alloys at 1173 K. (From *Acta Metallurgica* Volume 5, Issue 1, J. E. Reynolds, B. L. Averbach, and Morris Cohen, Self-diffusion and interdiffusion in gold-nickel alloys, p. 29, Copyright 1957, with permission from Elsevier. https://www.sciencedirect.com/science/article/abs/pii/0001616057901529)

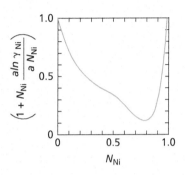

Fig. 12.19 Thermodynamic factor for interdiffusion at 1173 K. (From *Acta Metallurgica* Volume 5, Issue 1, J. E. Reynolds, B. L. Averbach, and Morris Cohen, Self-diffusion and interdiffusion in gold-nickel alloys, p. 29, Copyright 1957, with permission from Elsevier. https://www.sciencedirect.com/science/article/abs/pii/0001616057901529)

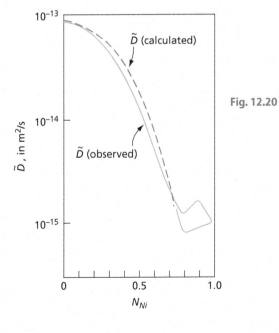

Fig. 12.20 Calculated and observed interdiffusion coefficients in gold-nickel alloys at 1173 K. (From *Acta Metallurgica* Volume 5, Issue 1, J. E. Reynolds, B. L. Averbach, and Morris Cohen, Self-diffusion and interdiffusion in gold-nickel alloys, p. 29, Copyright 1957, with permission from Elsevier. https://www.sciencedirect.com/science/article/abs/pii/0001616057901529)

12.12 Diffusion along Grain Boundaries and Free Surfaces

Atom movements in solids are not restricted to the interiors of crystals. It is a well known fact that diffusion processes also occur on the surfaces of metallic specimens and along the boundaries between crystals. In order to reduce complexity, these two forms of diffusion have purposely been neglected in our previous discussions.

Experimental measurements have shown that the surface and grain-boundary forms of diffusion also obey activation, or Arhennius-type laws, so that it is possible to write their temperature dependence in the form

$$D_s = D_{s_0} e^{-Q_s/RT}$$
$$D_b = D_{b_0} e^{-Q_b/RT}$$

12.50

where D_s and D_b are the surface and grain-boundary diffusivities, D_{s_0} and D_{b_0} are constants (frequency factors), and Q_s and Q_b are the experimental activation energies for surface and grain-boundary diffusion.

Sufficient evidence has been accumulated to conclude that diffusion is more rapid along grain boundaries than in the interiors of crystals, and that free-surface diffusion rates are larger than either of the other two. These observations are understandable because of the progressively more open structure found at grain boundaries and on exterior surfaces. It is quite reasonable that atom movements should occur most easily on free metallic surfaces, with more difficulty in boundary regions, and least easily in the interior of crystals.

Because of the very rapid movements of atoms on free surfaces, surface diffusion plays an important role in a large number of metallurgical phenomena. However, grain-boundary diffusion is of more immediate concern because, in the average metallic specimen, the grain-boundary area is many times larger than the surface area. Furthermore, grain boundaries form a network that passes through the entire specimen. It is this latter property that often causes large errors to appear in the measurement of lattice, or volume diffusion, coefficients. When the diffusivity of a metal is measured with polycrystalline samples, the results are usually representative of the combined effect of volume and grain-boundary diffusion. What is obtained, therefore, is an apparent diffusivity, D_{ap}, which may not correspond to either the volume- or the grain-boundary-diffusion coefficient. However, under certain conditions, the grain-boundary component may be small, so that the apparent diffusivity equals the volume diffusivity. On the other hand, if the conditions are right, the grain-boundary component may be so large that the apparent diffusivity diverges considerably from the lattice diffusivity. Let us investigate these conditions.

Diffusion in a polycrystalline specimen cannot be described as a simple summation of diffusion through the crystals and along the boundaries. Diffusion in the boundaries tends to progress more rapidly than diffusion through the crystals, but this effect is counteracted because as the concentration of solute atoms builds up in the boundaries, a steady loss of atoms occurs from the boundaries into the metal on either side of the boundary. The nature of this process can be visualized with the aid of Fig. 12.21, which represents a diffusion couple composed of two pure metals A and B. Both halves of the couple are assumed to be polycrystalline, but, for the sake of convenience, grain boundaries are indicated only on the right side of the couple. A number of short arrows are in the figure to represent the nature of the movement of A atoms into the B matrix. That group of parallel arrows perpendicular to the weld interface represents the volume component of diffusion. Other arrows parallel to the grain boundaries indicate the movement of atoms along boundaries, and a third set of arrows perpendicular to the boundaries represent the diffusion from the boundaries into the crystals.

In the usual diffusion experiment, one removes thin layers of metal parallel to the weld interface (like that indicated in Fig. 12.21 by two vertical dashed lines separated by the distance dx). These layers are chemically analyzed to obtain the penetration curve. The concentration of A atoms in the layer dx depends on how many A atoms reach this layer by direct-volume diffusion, and how many reach the layer by the short-circuiting grain-boundary paths. The problem is quite complex, but for a given ratio of the grain-boundary and lattice diffusivities (D_b/D_l), the relative number of A atoms that reach the layer dx by traveling along grain boundaries and by direct lattice diffusion is a function of the grain size. The smaller the grain size, the greater the total grain-boundary area available for boundary diffusion and, therefore, the greater the importance of boundaries in the diffusion process.

The importance of grain-boundary-diffusion phenomena in diffusion measurements is also a function of temperature. In Fig. 12.22, two sets of self-diffusion data are plotted for silver specimens: the one at the upper right for grain-boundary measurements and the one at the lower left for lattice measurements (single-crystal specimens). Both groups of points plot as straight lines on the log $D - 1/T_{\text{Abs}}$ coordinate system. For grain-boundary self-diffusion the equation of the line is:

$$D_b = 2.5 \times 10^{-6} e^{-84{,}500/RT},\ \text{m}^2/\text{s}$$

For volume or lattice self-diffusion,

$$D_l = 89.5 \times 10^{-6} e^{-192{,}000/RT},\ \text{m}^2/\text{s}$$

Notice that in this example the activation energy for grain-boundary diffusion is only about half that for volume diffusion. This fact is significant for two reasons: first, it emphasizes the fact that diffusion is easier along grain boundaries and, second, it shows that grain-boundary and volume diffusions have a different temperature dependence. Volume diffusion, when compared with boundary diffusion, is more sensitive to temperature change. Thus, as the temperature is raised, the rate of diffusion through the lattice increases more rapidly than the rate of diffusion along the boundaries. Conversely, as the

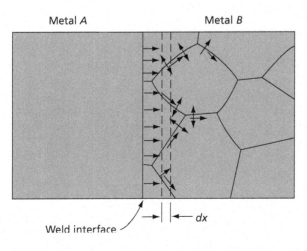

Fig. 12.21 The combined effect of grain boundary and volume diffusions

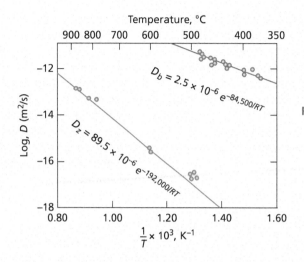

Fig. 12.22 Lattice and grain-boundary diffusion in silver. (From data of Hoffman, R. E., and Turnbull, D. J., *Jour. Appl. Phys.*, **22** 634 [1951])

temperature is lowered, the rate of diffusion along the boundaries decreases less rapidly. The net effect is that at very high temperatures diffusion through the lattice tends to overpower the grain-boundary component, but at low temperatures diffusion at the boundaries becomes more and more important in determining the total, or apparent, diffusivity.

The above facts are illustrated in Fig. 12.23, where the curves of Fig. 12.22 are redrawn as dashed lines, and another curve (solid line) is also shown that corresponds to self-diffusion measurements made on fine-grain polycrystalline silver specimens (grain size 35 microns before diffusion anneal). This latter curve consists of two segments, one part with the equation

$$D_{ap} = 2.3 \times 10^{-9} e^{-110,000/RT}, \text{m}^2/\text{s}$$

for temperatures below about 973 K, and the other coinciding with the single-crystal, or volume-diffusion, data for temperatures above about 973 K. It can be concluded that at temperatures above this value the volume-diffusion component is overpowering in silver specimens with a 35-micron grain size. Below this temperature, the grain-boundary component is a factor in determining the measured diffusivity.

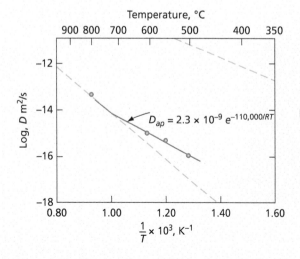

Fig. 12.23 Diffusion in polycrystalline silver. (From data of Hoffman, R. E., and Turnbull, D. J., *Jour. Appl. Phys.*, **22** 634 [1951])

The above results have general implications: (a) diffusivities determined with polycrystalline specimens are more liable to be representative of lattice diffusion if they are measured at high temperatures and (b) the reliability of the data can be increased by controlling the grain size of the specimens. The larger the grain size, the smaller the grain-boundary contribution to the diffusivity. Thus, for accurate measurements of lattice diffusivities using polycrystalline specimens, high temperatures and large-grained specimens should be used.

12.13 Fick's First Law in Terms of a Mobility and an Effective Force

Fick's first law is often expressed by using a different set of variables. This form of the equation will now be developed by considering the diffusion of one component in an ideal binary solution. Fick's first law, Eq. 12.8, in this case may be written

$$J_A = -D_A^* \frac{\partial n_A}{\partial x} \qquad 12.51$$

where J_A is the number of A atoms passing an interface of unit area per second, n_A is the number of A atoms per unit volume, and D_A^* is the diffusivity of A in an ideal solution. Note that this diffusion coefficient is equivalent to that measured with radioactive tracers in a solution of constant chemical composition (see Sec. 12.11). This expression may be reexpressed in the form

$$J'_A = -D_A^* (n_A + n_B) \frac{\partial N_A}{\partial x} \qquad 12.52$$

where J'_A equals J_A/A and represents the flux of A atoms per cm², $(n_A + n_B) \times N_A$ equals n_A by the definition of the mole fraction, and $(n_A + n_B)$ is assumed to be a constant.

By Eq. 10.25, $\overline{G}_A$, the partial-molal free energy of the A component in an ideal solution is

$$\overline{G}_A = G_A^\circ + RT \ln N_A$$

where G_A° is the free energy of a mole of pure A at the temperature T and N_A is the mole fraction of A. Taking the derivative of $\overline{G}_A$ with respect to x, the distance along a diffusion couple, and noting that by definition $\partial G_A^\circ/dx$ is zero, we have

$$\frac{\partial \overline{G}_A}{\partial x} = \frac{RT}{N_A} \frac{\partial N_A}{\partial x}$$

Solving this expression for $\partial N_A/\partial x$ and substituting the result into Fick's equation and noting that $(n_A + n_B)N_A$ equals n_A (the number of atoms of A per unit volume), gives us

$$J'_A = -\frac{D_A^*}{RT} n_A \frac{\partial \overline{G}_A}{\partial x} \qquad 12.53$$

Now let us make the substitution $B = D_A^*/RT$ and we obtain the result

$$J'_A = -n_A B \frac{\partial \overline{G}_A}{\partial x} \qquad 12.54$$

where J'_A is the flux per m², B is a parameter called the mobility, n_A is the concentration of A atoms in number of atoms per m², and $\partial \overline{G}_A/\partial x$ is the partial derivative of the partial-molal free energy of the A component in the solution with respect to the distance x. Both $\partial \overline{G}_A/\partial x$ and B have a physical significance worth noting. The partial-molal free energy $\overline{G}_A$ has the dimensions of an energy, and its derivative with respect to the distance x can be considered as an effective "force" causing diffusion to occur in this direction. B has the dimensions of velocity divided by force. In terms of its dimensions, Eq. 12.54 may be written

$$\frac{\text{Number of atoms}}{\text{m}^2 \times \text{s}} = \frac{\text{Number of atoms}}{\text{m}^3} \times \frac{\text{m}}{\text{s} \times \text{force}} \times \text{force}$$

While the above relationship was derived for the special case of diffusion in an ideal solution, the result is of general use and may be applied to diffusion in a nonideal solution where the partial-molal free energy is given (see Eq. 10.27) by

$$\overline{G}_A = G_A^\circ + RT \ln a_A$$

where a_A is the activity of the A component in the solution. Let us now consider the derivative of this partial-molal free energy with respect to x, the distance along a diffusion couple. Since by the definition of the activity coefficient γ_A (see Eq. 10.28), $a_A = \gamma_A N_A$, we have

$$\frac{\partial \overline{G}_A}{\partial x} = RT \frac{\partial}{\partial x} (\ln \gamma_A N_A) = RT \frac{\partial}{\partial N_A} (\ln \gamma_A N_A) \frac{\partial N_A}{\partial x}$$

or

$$\frac{\partial \overline{G}_A}{\partial x} = RT \left(\frac{1}{N_A} + \frac{\partial \ln \gamma_A}{\partial N_A} \right) \frac{\partial N_A}{\partial x} = \frac{RT}{N_A} \left(1 + N_A \frac{\partial \ln \gamma_A}{\partial N_A} \right) \frac{\partial N_A}{\partial x}$$

Substituting this relationship back into the equation

$$J'_A = -\frac{D_A^*}{RT} n_A \frac{\partial \overline{G}_A}{\partial x}$$

and simplifying, using the relationship $N_A = n_A/(n_A + n_B)$, we obtain

$$J'_A = -\left(1 + N_A \frac{\partial \ln \gamma_A}{\partial N_A} \right) D_A^* \frac{\partial n_A}{\partial x} \qquad 12.55$$

where $(1 + N_A \partial \ln \gamma_A/\partial N_A)D_A^*$ is the intrinsic diffusion coefficient D_A as defined in Sec. 12.11. We have therefore derived the Darken relationship $D_A = (1 + N_A \partial \ln \gamma_A/\partial N_A)D_A^*$ given in this section.

12.14 Diffusion in Non-Isomorphic Alloy Systems

When a thin sheet of copper is welded to a thin sheet of nickel and the diffusion couple thus formed is annealed at an elevated temperature, the composition should vary across the couple in a manner like that shown in Fig. 12.12. In other words, the composition should change smoothly and continuously from pure nickel on the one side of the couple to pure copper on the other side.

A penetration curve of this type is obtained when an alloy system contains only a single solid phase. The corresponding penetration curve in an alloy system containing a number of different solid phases, such as the copper-zinc system, is basically different. For example, when a copper-zinc diffusion couple is annealed at a temperature of the order of 400°C, a layered structure is formed. Each layer in this couple corresponds to one of the five solid phases that can exist at this temperature. In order to explain the nature of such layered structures, a hypothetical alloy system will be considered whose phase diagram is shown in Fig. 12.24. This system has a single intermediate phase, the β phase, in addition to the terminal α and γ phases. A diffusion couple formed by welding a layer of pure metal A to a layer of pure metal B, after a diffusion anneal at a temperature T_1, can show three distinct layers corresponding, respectively, to the $\alpha, \beta,$ and γ phases, as shown in Fig. 12.25. If the composition-distance curve is determined for this couple, it also should have the form shown in Fig. 12.25. Note that where the penetration curve crosses either of the boundaries separating a pair of phases, there is a sharp discontinuity in the composition. These sudden changes in composition are equal to the composition differences across the two-phase $\alpha + \beta$ and $\beta + \gamma$ fields. Thus, consider the boundary between the alpha and beta phases. Note that as the interface is approached from the alpha-phase side, the composition curve, corresponding to the B component, rises to the point marked a. This point represents the same composition as point a in the phase diagram of Fig. 12.24. On the other side of the boundary, the composition in the beta phase is given by point b, corresponding to point b on the phase diagram. Similarly, at the beta-gamma boundary, the composition changes from c to d, corresponding to points c and d on the phase diagram. In brief, the diffusion couple in an alloy system of this type is like a constant temperature section cut across the phase diagram in which each of the single-phase fields appears with a finite width, but in which the two-phase fields are represented by surfaces.

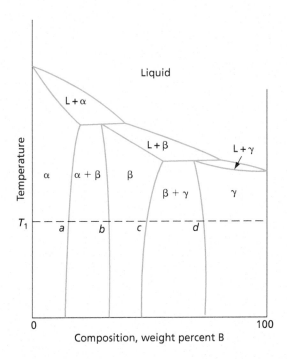

Fig. 12.24 The equilibrium diagram of a hypothetical alloy system with three solid phases

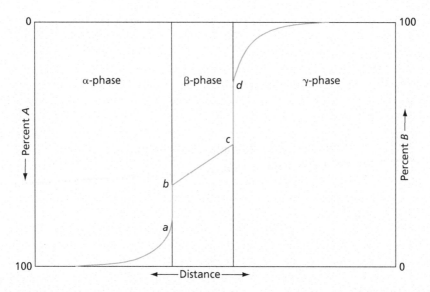

Fig. 12.25 A diffusion couple formed by welding a layer of pure metal A to one of pure metal B after an anneal at T_1 (see Fig. 12.24) will show a layered structure. Each layer in this structure corresponds to one of the phases in the equilibrium diagram. A curve showing the variation of the composition (B component) across the phase diagram is also shown in this figure

The reason why the two-phase fields appear as surfaces in a diffusion couple is not hard to understand. Two important factors need to be considered. First, for diffusion to occur, a concentration gradient must exist across the couple. More accurately, an activity gradient has to exist. If this gradient is missing at any position across the couple the flux of A or B atoms past this position will stop. According to Eq. 10.27, we may write for the B component

$$\Delta \overline{G}_B = \overline{G}_B - G_B^\circ = RT \ln a_B \qquad 12.56$$

where $\overline{G}_B$ is the partial molal free energy of the B component, G_B° is the free energy per mole of pure B at the temperature in question, and a_B is the activity of the B component. This equation shows that at constant temperature a variation in a_B with distance across the couple corresponds to a variation in the partial-molal free energy with distance. Thus, continuous diffusion across the couple can only occur if the partial-molal free energy decreases continuously with distance. Now consider Fig. 12.26, which shows the hypothetical free energy composition curves for the three phases, α, β, and γ, at the temperature T_1. As discussed earlier in Section 10.8, a line drawn so that it is tangent to a pair of these free-energy curves defines the limits of a two-phase field. This is indicated in Fig. 12.26. Also by Sec. 10.8, the intersection with the sides of the figure of a tangent drawn to any point on a free-energy composition curve gives the partial-molal free energy of the phase in question. By drawing a series of tangents to this set of three free-energy curves and selecting the minimum value of the partial-molal free energy corresponding to a given composition, it is possible to determine a curve corresponding to the variation of the partial-molal free energy of the B component with composition across the whole phase diagram when the temperature is T_1. Such a curve is shown in Fig. 12.27. Note that starting with pure B, the partial-molal free energy of the B component falls continuously with decreasing

12.14 Diffusion in Non-Isomorphic Alloy Systems   395

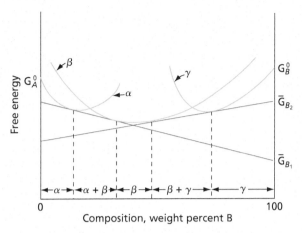

Fig. 12.26 The free energy versus composition curves for all three phases corresponding to the temperature T_1 are shown in the diagram

concentration of B until the two-phase $\gamma + \beta$ field is reached. In this region the partial-molal free energy, defined by the common tangent to the gamma and beta energy curves in Fig. 12.27, is constant and equal to $\overline{G}_{B_1}$. A similar constant partial-molal free-energy interval is also obtained in the alpha-beta two-phase region. Here the partial molal free energy is $\overline{G}_{B_1}$.

The significance of the above can be easily stated. The only way that a gradient in the partial-molal free energy can be obtained in an interval where the partial-molal free energy wants to

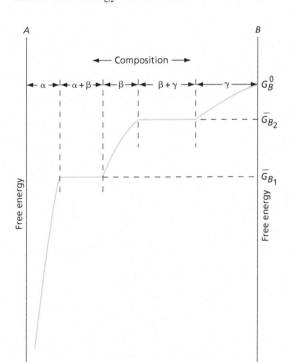

Fig. 12.27 The variation of $\overline{G}_B$ with composition across the phase diagram at T_1

remain constant is for the thickness of the interval to vanish. In other words, in the diffusion couple the two-phase regions have to appear as surfaces; that is, regions of zero thickness. If this did not occur, there could be no net diffusion flux across the couple. Finally, note that by Fig. 12.27 the partial-molal free energy of the β and γ phases are equal at the interface. While a composition difference exists at the boundary, this is not true of the free energy. At the interface there is a change in slope of the partial-molal free energy versus distance curve, but there is no discontinuity in $\overline{G}_B$.

The width of the single-phase layers in a diffusion couple will now be considered briefly. It is quite possible for a phase or phases to appear to be missing in a diffusion couple. Thus, as shown by Bückle, when a diffusion couple is formed from copper and zinc and annealed for about a half hour at 380°C, a layered structure, analogous to that shown schematically in Fig. 12.28, is obtained. Note that the β' phase layer is so thin that it is not visible at a magnification of about 150×. The thickness of a diffusion layer is determined by the relative velocities with which its two boundaries move. These boundary movements are controlled by diffusion and are therefore analogous to the diffusion-controlled planar interface growth phenomenon to be discussed in Sec. 15.7. As an example, we will now consider the growth of the β-phase layer in Fig. 12.25. At the start of the anneal this layer has a zero thickness. With time it should grow in size. However, its net growth depends on the relative velocities with which its two boundaries move. The two boundaries will not normally move with the same velocity because the growth-controlling variables are not the same at the two boundaries. By analogy with the planar growth equation developed in Sec. 15.7, we may write, with respect to the boundary between the beta and alpha phases,

$$(n_B^b - n_B^a) A dx_{\alpha\beta} = AD_\alpha \left(\frac{dn_B^a}{dx}\right) dt - AD_\beta \left(\frac{dn_B^b}{dx}\right) dt \qquad 12.57$$

where n_B^b and n_B^a are the compositions (number of B atoms per unit volume) of the two phases at the boundary, dn_B^a/dx and dn_B^b/dx are the concentration gradients in the alpha and beta phases at the boundary, and D_α and D_β are the corresponding diffusion coefficients. The left-hand side of this equation corresponds to the number of B atoms required to move the boundary through a distance $dx_{\alpha\beta}$, while the right-hand side represents the net flux that will give this number. As will be

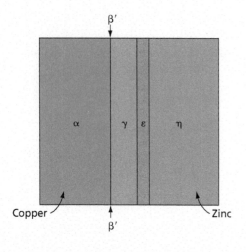

Fig. 12.28 A copper-zinc diffusion couple annealed for a short time at a temperature of about 380°C does not show a visible layer of the β' phase. (After Bückle, H., *Symposium on Solid State Diffusion*, p. 170. North Holland Publishing Co. Copyright, Presses Universitaires de France, 1959.)

discussed in Sec. 15.7, the boundary is assumed to move only as the result of carbon atoms diffusing into cementite. In this case one has to consider both the flux of B atoms up to the interface in the β phase, as well as the flux of B atoms away from the interface in the α phase. The Eq. 12.57 may be simplified to

$$\frac{dx_{\alpha\beta}}{dt} = \frac{1}{(n_B^b - n_B^a)} \left[D_\alpha \left(\frac{dn_B^a}{dx} \right) - D_\beta \left(\frac{dn_B^b}{dx} \right) \right] \qquad 12.58$$

and a corresponding growth velocity expression written for the β to γ interface,

$$\frac{dx_{\gamma\beta}}{dt} = \frac{1}{(n_B^c - n_B^d)} \left[D_\gamma \left(\frac{dn_B^c}{dx} \right) - D_\beta \left(\frac{dn_B^d}{dx} \right) \right] \qquad 12.59$$

Note that the growth velocities of the two interfaces of the β phase depend on a number of parameters. These include the concentrations of the phases at the boundaries, the diffusion coefficients, and the concentration gradients. Since the diffusion coefficients are normally functions of the composition, the solution of layer-growth problems is usually very difficult. Also, it is quite possible to conceive of growth conditions that will not allow a phase to develop a layer of visible thickness. In fact, by suitable choices of diffusion coefficients, it is possible to make either boundary move in either direction.

The above analysis has assumed a condition of dynamic equilibrium and implies a diffusion couple of such a size that its outer layers do not change their compositions. In a diffusion couple of a finite size, the situation can be quite different. Thus, if in the above example the copper and zinc layers were very thin and present in a ratio of 48 percent zinc to 52 percent copper, a sufficiently long anneal should result in a specimen containing only a single phase. This would be the β' phase, or the phase corresponding to the average composition.

A word should be said about some practical examples of layered structures. A typical example is furnished by galvanized iron. When steel is dipped in molten zinc, the zinc diffuses into the iron and a layered structure is formed that contains four phases in addition to the base metal (steel). The outermost of these phases is a liquid. On cooling, this liquid passes through an eutectic point so that the outermost layer is basically an eutectic. Hot-dipped tin plate also has a layered alloy structure. In fact, in most cases where one metal is plated on another under conditions where diffusion can occur, it will be found that layered structures tend to develop.

Problems

12.1 (a) What is the number of atoms in a cubic meter of copper? The gram-atomic weight of copper is 63.54 grams per mole and the atomic volume of copper is 7.09 cm^3 per gram-atom.

(b) Next, compute the number of copper atoms per m^3 given that the lattice constant, a, of copper is 0.36153 nm and that there are 4 atoms per unit cell in a face-centered cubic crystal.

12.2 A diffusion couple, made by welding a thin one-centimeter square slab of pure metal A to a similar slab of pure metal B, was given a diffusion anneal at an elevated temperature and then cooled to room temperature. On chemically analyzing successive layers of the specimen, cut parallel to the weld interface, it was observed that, at one position, over a distance of 5000 nm, the atom fraction of metal A, N_A, changed from 0.30 to 0.35. Assume that the number of atoms per m^3 of both pure metals is 9×10^{28}. First determine the concentration gradient dn_A/dx. Then if the diffusion coefficient, at the point in question and annealing temperature, was 2×10^{-14} m^2/s, determine the number of A atoms per second that would pass through this cross-section at the annealing temperature.

12.3 On the assumption that the self-diffusion coefficient of a simple cubic metal whose lattice constant, a, equals 0.300 nm is given by the equation

$$D = 10^{-4}e^{-200,000/RT}, \text{ m}^2/\text{s},$$

determine the value of the diffusion coefficient at 1,200 K and use this to determine the mean time of stay, τ, of an atom at a lattice site.

12.4 It is determined by experiment that the Kirkendall markers placed at the interface of a diffusion couple, formed by welding a thin plate of metal A to a similar plate of metal B, move with a velocity of 4.5×10^{-12} m/s toward the A component when the concentration $N_A = 0.38$ and the concentration gradient, dn_A/dx, is 2.5×10^2 per m. The chemical diffusion coefficient $\tilde{D}$ under these conditions is 3.25×10^{-14} m^2/s. Determine the values of the intrinsic diffusivities of the two components.

12.5 This problem and the next require an error function (probability integral) table. However, it is suggested that one develop such a table using the following procedure. The error function is based upon integrating the exponential e^{-y^2} between the limits $y = 0$ and $y = y$. A convenient way of doing this is to integrate the first six terms of the series expansion for e^{-y^2}, which are

$$e^{-y^2} = 1 - y^2 + y^4/2! - y^6/3! + y^8/4! - y^{10}/5! + \cdots.$$

This integration yields the following expression for the error function, accurate to 5 significant figures for y from 0 to 0.80.

$$\frac{2}{\sqrt{\pi}} \int_0^y e^{-y^2} dy$$

$$= \frac{2}{\sqrt{\pi}} \left[y - \frac{y^3}{3} + \frac{y^5}{10} - \frac{y^7}{42} + \frac{y^9}{216} - \frac{y^{11}}{1320} + \cdots \right]$$

A simple computer program should now be written around this latter equation and the error function evaluated in steps of 0.01 from $y = 0$ to $y = 0.80$.

A thin plate of a binary alloy of composition $N_{A1} = 0.245$ and $N_{B1} = 0.755$ was welded to a similar plate of composition $N_{A2} = 0.255$ and $N_{B2} = 0.745$ so as to form a diffusion couple. After a diffusion anneal for 200 hours at 1300 K, the composition at a plane 2×10^{-4} m from the original weld interface, on the side whose starting composition was $N_A = 0.245$, was observed to be $N_A = 0.248$. Compute the diffusivity using the Grube method.

12.6 Using the diffusivity determined in Prob. 12.5 and the error function equation associated with this problem, write a computer program to determine the composition of the specimen as a function of the time at a point 2×10^{-4} m from the weld interface and on the side of the couple with the starting composition $N_A = 0.245$. Determine this composition by varying the time from 3 to 70 hours in 2-hour intervals. Plot your results in a curve of composition versus time. What is the significance of the composition limit that is approached at long times?

12.7 With the aid of the data available in Table 12.3 and Fig. 12.14 determine $\tilde{D}$, the interdiffusion coefficient, at the composition $N_A = 0.625$ using the following directions:

(a) Draw a tangent to the penetration curve at $N_A = 0.625$ in order to determine $\partial x/\partial N_A$ at this composition.

(b) With the aid of Simpson's Rule, graphically integrate the data in Table 12.3 from $N_A = 0$ to $N_A = 0.625$.

(c) Using the results of the above, evaluate the Matano equation for D letting the time, t, equal 50 hours. Compare your answer with Fig. 12.15.

12.8 According to Eq. 12.28 the self diffusion coefficient for a face-centered cubic metal may be expressed by the following equation: $\tilde{D} = a^2/12\tau$ where a is the lattice parameter and τ is the mean time of stay of an atom in a lattice site. Derive this relationship using a procedure similar to that in Sec. 12.1 for the case of a simple cubic lattice.

12.9 With the aid of the data in Table 12.4, determine a value for the self-diffusion coefficient of nickel at a temperature of 1173 K and compare it with the value shown in Fig. 12.18.

12.10 Determine the interdiffusion coefficient, $\tilde{D}$, for nickel in a 50 atomic percent Ni nickel-gold alloy at 1173 K using the data plotted in Figs. 12.18 and 12.19.

References

1. Million, B., and Kucera, J., *Acta Met,* **17**, pp. 339–344 (1969).
2. For compilation of data, see *Smithells Metals Reference Book*, 6th edition, Chapter 13, Butterworths, London, 1983.
3. Smigelskas, A. D., and Kirkendall, E. O., *Trans. AIME*, **171** 130 (1947).
4. Balluffi, R. W., and Seigle, L. I., *Acta Met.*, **3** 170 (1955).
5. Barnes, R. S., and Mazey, D. J., *Acta Met.*, **6** 1 (1958).
6. Balluffi, R. W., *Acta Met.*, **2** 194 (1954).
7. Darken, L. S., *Trans. Met. Soc. AIME,* **175** 184 (1948).
8. *Transport Phenomena in Materials Processing*, D. R. Poirier and G. H. Geiger, TMS Publication, 1994, p. 303.
9. Matano, C., *Jap. J. Phys.* **8** 109 (1933).
10. Adda, Y., and Philbert, J., *La Diffusion dans Les Metaux*. Eindhoven, Holland Bibliothèque Technique Philips, 1957.
11. Walsoe De Reca, E., and Pampillo, C., *Acta Met.*, **15** 1263 (1967).
12. LeClaire, A. D., *Prog. in Metal Phys.*, **1** 306 (1949); **4** 305 (1953); *Phil. Mag.*, **3** 921 (1958).
13. Shewmon, P. G., *Diffusion in Solids*, McGraw-Hill, New York, 1963, p. 134.
14. Herzig, C., Mishin, Y., and Divinski, S., *Met. and Mat. Trans. A*, 33A 765 (2002).
15. Glicksman, M. E., Diffusion in Solids: Field Theory, Solid-state Principles, and Applications, John Wiley, 2000, p. 257.

Chapter 13

Interstitial Diffusion

> **Learning Objectives**
>
> Upon completion of Chapter 13, you will be able to:
> 1. Understand mechanism of interstitial diffusion
> 2. Explain Snoek relaxation in bcc metals
> 3. Differentiate between elastic and anelastic strains
> 4. Derive fundamental equations for relaxation time and internal friction measurement techniques
> 5. Describe activation coefficient for interstitial diffusion and Arrhenius relations of diffusivity with reciprocal of temperature
> 6. Compare and contrast diffusivities of substitutional and interstitial elements

In Chapter 12, the diffusion of atoms in substitutional solid solutions was considered. In this chapter, we shall be concerned with diffusion in interstitial solid solutions. In the former case, atoms move as a result of jumps into vacancies; in the present case, diffusion occurs by solute atoms jumping from one interstitial site into a neighboring one. Interstitial diffusion is basically simpler, since the presence of vacancies is not required for the solute atoms to move. The following expression for the diffusivity of the solute atoms in a dilute substitutional solid solution was presented in Chapter 12

$$D = \alpha a^2 Z \nu e^{-(\Delta G_m + \Delta G_f)/RT} \qquad 13.1$$

where a is the lattice parameter of the crystal, α is a geometrical factor that depends upon the crystal, Z is the coordination number, ν is the vibration frequency of a solute atom in a substitutional site, ΔG_f is the free-energy change per mole associated with the formation of vacancies, and ΔG_m is the free-energy per mole required for solute atoms to jump over their energy barriers into vacancies. A similar expression can be written for interstitial diffusivity:

$$D = \alpha a^2 p \nu e^{-\Delta G_m/RT} \qquad 13.2$$

In this case, p is the number of nearest interstitial sites, α and a have the same meanings as above, v is the vibration frequency of a solute atom in an interstitial site, and ΔG_m is the free-energy per mole needed for solute atoms to jump between interstitial sites. This expression, in contrast to the previous one, contains only one free-energy term: the direct result of the fact that interstitial diffusion is not dependent upon the presence of vacancies. Because a free-energy change is capable of being expressed in the form

$$\Delta G = \Delta H - T\Delta S \qquad 13.3$$

the expression for the interstitial diffusivity can be written:

$$D = \alpha a^2 p v e^{+\Delta S_m/R} e^{-\Delta H_m/RT} \qquad 13.4$$

where ΔS_m is the entropy change of the lattice (per mole solute atoms), and ΔH_m is the work (per mole of solute atoms) associated with bringing solute atoms to the saddle point during a jump between interstitial positions.

13.1 Measurement of Interstitial Diffusivities

Interstitial diffusion is often studied, especially when it occurs at high temperatures, with the same experimental techniques (Matano, Grube, etc.) used for the study of diffusion in substitutional solid solutions. On the other hand, a great deal of success has been achieved in the investigation of interstitial diffusion, especially in body-centered cubic metals, with an entirely different technique. This technique has the advantage that it can be used at very low temperatures where normal methods of studying diffusion are inoperative because of very slow diffusion rates. The method, internal-friction measurement, will be discussed in the next section and considerable time will be spent on it, not only because it is an important tool for the study of diffusion, but also because the general field of internal friction in metals is of importance in the study of metallurgical phenomena. While space does not permit a discussion of the use of internal-friction methods to areas other than the study of diffusion, an appreciation of the advantage of this type of technique can be gained by investigating its application to diffusion studies.

Before taking up the subject of internal-friction measurements, it should be pointed out that, as in the case of substitutional alloys, experimental measurements of interstitial diffusion coefficients conform to an equation of the type

$$D = D_0 e^{-Q/RT} \qquad 13.5$$

where D is the diffusivity or diffusion coefficient, D_0 is a constant known as the frequency factor, and Q is the experimental activation energy for diffusion. Comparison of this expression with the theoretical one given in Eq. 13.4 shows that

$$Q = \Delta H_m \qquad \text{and} \qquad D_0 = \alpha a^2 p v e^{\Delta S_m/R} \qquad 13.6$$

Excellent agreement has been found between the quantities ΔH_m and ΔS_m (which may be computed theoretically for basic atomic considerations) and the experimentally determined quantities Q and D_0. When the diffusion takes place by a single mechanism, the Arrhenius Plot of $\ln D$ versus

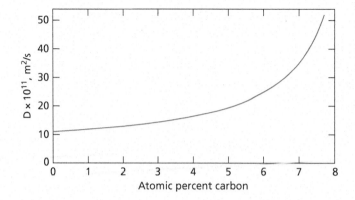

Fig. 13.1 The diffusion coefficient is also a function of composition in interstitial systems. Diffusion of carbon in face-centered cubic iron at 1127°C. (From Wells, C., Batz, W., and Mehl, R. F., *Trans. AIME*, 188 553 [1950])

$1/T$ is linear, with a constant slope of $-Q/R$. It is well to note, however, that we are speaking of dilute interstitial solid solutions. When the concentration of the solute becomes appreciable, so that large numbers of interstitial sites are occupied, solute atoms interact, or at least interfere, with each other's jumps. As was found to be the case in substitutional solid solutions, interstitial diffusivities are, in general, functions of composition. For example, see Fig. 13.1.

As indicated earlier, most interstitial diffusion coefficient measurements conform to Eq. 13.5. However, there are other measurements involving interstitial diffusion in body-centered cubic metals that show this not to be the case. An example is shown in Fig. 13.2 for diffusion of carbon in bcc iron. The temperature designated by T_c represents the Curie temperature. In this Arrhenius plot, the diffusivity deviates from linearity toward faster diffusion at high temperatures. Similar positive deviation from linearity have been observed for diffusion of nitrogen in bcc iron as well as diffusion of carbon, nitrogen, and oxygen in the group V metals molybdenum, niobium, vanadium, and tantalum. Such positive deviations in these bcc metals have been attributed to other factors contributing to the overall diffusivity of the interstitials. (Refs. 1 through 7)

13.2 The Snoek Effect

The study of interstitial diffusion by internal-friction methods usually makes use of an effect first explained by Snoek. (Ref. 8) In a body-centered cubic metal like iron, interstitial atoms, such as carbon or nitrogen, take positions either at the centers of the cube edges or at the centers of the cube faces (see Fig. 13.3). Both positions are crystallographically equivalent, as may be deduced from Fig. 13.3. An interstitial atom at either x or w would lie between two iron atoms aligned in a $\langle 100 \rangle$ direction (the iron atoms on either side of the w position lie at the center of the unit cell shown in the figure and the center of the next unit cell in front of this cell [not shown in the figure]). It has been previously mentioned (Chapter 9, Fig. 9.2B) that the space available for the solute atom between two iron atoms is smaller than the diameter of the solute atom. The occupancy of one of these positions (such as that designated by x in Fig. 13.3) by a solute atom pushes apart the two solvent atoms a and b. An atom at x or w thus increases the length of the

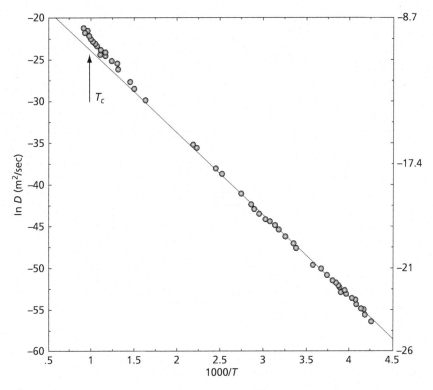

Fig. 13.2 Arrhenius plot of ln D vs 1/T for the diffusion of C in bcc Fe. (Carbon Diffusivity in B.C.C. Iron, McLellan, R.B., and Wasz, M.L., Dept of Mechanical Engineering and Materials Science, William Marsh Rice University, Houston TX. *J. Phys. Chem. Solids*, Vol 54, No.5, pp. 583–586, Copyright 1993. With permission from Elsevier. https://www.sciencedirect.com/science/article/abs/pii/002236979390236K)

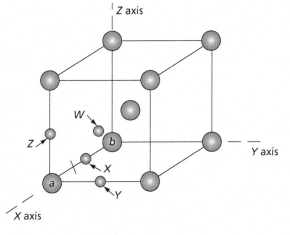

Fig. 13.3 The nature of the sites that interstitial carbon atoms occupy in the body-centered cubic iron lattice

crystal in the [100] direction. Similarly, an atom at y or z increases the length of the crystal in the [010] or [001] directions.

For the sake of convenience, let us define the axis of an interstitial site as that direction along which the solvent atoms (at either side of the interstitial site) are spread when it is occupied by an interstitial solute atom.

When a body-centered cubic crystal containing interstitial atoms is in an unstressed state, a statistically equal number of solute atoms will be found in the three types of sites, the axes of which are parallel to the [100], [010], and [001] directions, respectively. Now, if an external force is applied to the crystal so as to produce a state of tensile stress parallel to the [100] axis, it will have the effect of straining the lattice in such a manner that those sites with axes parallel to [100] will have their openings enlarged, while those with axes normal to the stress ([010] and [001]) will have their openings diminished. The effect of a stress, therefore, is to give the solute atoms a somewhat greater preference for interstitial sites with axes that are parallel to the stress. After the application of the stress, the number of solute atoms tends to increase in these preferred sites, so that an equal division of solute atoms among the three types of sites ceases to exist.

When the applied stress is small, so that the elastic strain is small (order of 10^{-5} or smaller), the number of excess solute atoms per unit volume that eventually find themselves in interstitial positions, with axes parallel to the tensile-stress axis, is directly proportional to the stress. Thus,

$$\Delta n_p = K s_n \qquad 13.7$$

where Δn_p is the additional number of solute atoms in preferred positions, K is a constant of proportionality, and s_n is the tensile stress. Each of the additional solute atoms in one of the preferred positions adds a small increment to the length of the specimen in the direction of the tensile stress. The total strain of the metal thus consists of two parts: the normal elastic strain, ε_{el}, and the anelastic strain, ε_{an}, which is caused by the movement of solute atoms into sites with axes parallel to the stress axis

$$\varepsilon = \varepsilon_{el} + \varepsilon_{an} \qquad 13.8$$

When a stress is suddenly applied, the elastic component of the stress can be considered to develop instantly. The anelastic strain, however, is time dependent and does not appear instantly. The sudden application of a stress to a crystal places the solute atoms in a nonequilibrium distribution, for equilibrium now corresponds to an excess of solute atoms, Δn_p, in sites with axes parallel to the stress. The attainment of the equilibrium distribution occurs as a result of the normal thermal movements of solute atoms. The net effect of the stress is to cause a slightly greater number of jumps to go into the preferred positions than come out of them. However, when equilibrium is attained, the number of jumps per second into and out of the preferred positions will be the same. Obviously, the number of excess atoms in preferred positions and the anelastic strain must both be a maximum at equilibrium.

The rate at which the number of additional atoms in preferred interstitial sites grows depends directly on the number of the excess sites that are still unoccupied at any instant. The rate is greatest, therefore, at the instant that the stress is applied, because at this time the number of excess atoms in preferred sites is zero. As time goes on, however, and the number of excess atoms approaches its maximum value, the rate becomes progressively smaller and smaller. As in all physical problems where the rate of change depends on the number present, an exponential law can be expected to govern the time dependence of the number of additional interstitial atoms. In the present case, this law is:

$$\Delta n_p = \Delta n_{p(\max)}[1 - e^{-t/\tau_\sigma}] \quad 13.9$$

where Δn_p is the number of excess solute atoms at any instant, $\Delta n_{p(\max)}$ the maximum number that may be attained under a given tensile stress, t is the time, and τ_σ is a constant known as *relaxation time* at constant stress.

Since the anelastic strain is directly proportional to the number of excess atoms in preferred sites, it can also be expressed by an exponential relation

$$\varepsilon_{an} = \varepsilon_{an(\max)}[1 - e^{-t/\tau_\sigma}] \quad 13.10$$

where ε_{an} and $\varepsilon_{an(\max)}$ are the respective instantaneous and maximum (equilibrium values) of the anelastic strain. The relationship between the elastic and anelastic strains is shown in Fig. 13.4.

The effect of removing the stress after the anelastic strain has reached its maximum value is also shown in Fig. 13.4. If the stress is suddenly removed, the elastic strain is recovered instantly, while the anelastic component is time dependent. For the condition of stress removal, the anelastic strain follows a law of the form:

$$\varepsilon_{an} = \varepsilon_{an(\max)}\, e^{-t/\tau_\sigma} \quad 13.11$$

where ε_{an} is the anelastic strain at any instant, $\varepsilon_{an(\max)}$ is the anelastic strain at the instant of the removal of the stress, and t and τ_σ have the same meanings as in the previous equations.

The significance of τ_σ can be seen if the time t is set equal to τ_σ and then substituted into Eq. 13.11

$$\varepsilon_{an} = \varepsilon_{an(\max)} e^{-t_\sigma/t_\sigma} = \frac{1}{e}\varepsilon_{an(\max)} \quad 13.12$$

The relaxation time τ_σ is thus the time that it takes the anelastic strain to fall to $1/e$ of its original value. If τ_σ is large, the strain relaxes very slowly, and if τ_σ is small, the strain relaxes quickly. The rate at which the strain relaxes is consequently an inverse function of the relaxation time. It is also an inverse function of the mean time of stay of an atom in an interstitial position τ, for small values of τ correspond to large jump rates, $1/\tau$, and rapid strain relaxation. These two essentially different

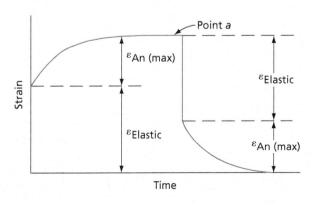

Fig. 13.4 Relationship between elastic and anelastic strains. (After A. S. Nowick, Reference 9)

time concepts (relaxation time and mean time of stay of an atom in an interstitial position) are directly related and, in the case of the body-centered cubic lattice, it can be shown that

$$\tau = \frac{3}{2}\tau_\sigma \qquad 13.13$$

This relationship will now be derived following Nowick. (Ref. 9) For this purpose, let us rewrite the expression for the anelastic strain, Eq. 13.10:

$$\varepsilon_{an} = \varepsilon_{an(max)} - \varepsilon_{an(max)} e^{-t/\tau_\sigma} \qquad 13.14$$

The derivative of this latter is

$$\frac{d\varepsilon_{an}}{dt} = \frac{\varepsilon_{an(max)}}{\tau_\sigma} e^{-t/\tau_\sigma} = \frac{\varepsilon_{an(max)} - \varepsilon_{an}}{\tau_\sigma} \qquad 13.15$$

As stated earlier, this equation shows that the time rate of change of the anelastic strain equals the difference between the maximum attainable anelastic strain (under a given applied stress) and the instantaneous value of the anelastic strain. A similar relationship holds for Δn_p, the number of excess carbon atoms per unit volume in preferred sites, so that

$$\frac{d(\Delta n_p)}{dt} = \frac{\Delta n_{p(max)} - \Delta n_p}{\tau_\sigma} \qquad 13.16$$

where if the stress is assumed to be applied along one of the three $\langle 100 \rangle$ axes of the iron crystal, say the z axis, $\Delta n_p = n_z - n/3$, and $\Delta n_{p(max)} = n_{z(max)} - n/3$. This is based on the assumption that under zero stress, the carbon atoms will be uniformly distributed in the three possible $\langle 100 \rangle$ sites. Therefore, we may also write

$$\frac{dn_z}{dt} = \dot{n}_z = \frac{\left(n_{z(max)} - \frac{n}{3}\right) - \left(n_z - \frac{n}{3}\right)}{\tau_\sigma} \qquad 13.17$$

An expression can also be written for $\dot{n}_z$ in terms of the difference in the rates at which carbon atoms enter and leave z sites. This equation is

$$\dot{n}_z = n_x\left(\frac{1}{\tau_{xz}}\right) + n_y\left(\frac{1}{\tau_{yz}}\right) - n_z\left(\frac{1}{\tau_{zx}}\right) - n_z\left(\frac{1}{\tau_{zy}}\right) \qquad 13.18$$

where n_x, n_y, and n_z are the numbers of carbon atoms per unit volume in x, y, and z sites, respectively, and $1/\tau_{xz}$, $1/\tau_{yz}$, $1/\tau_{zx}$, and $1/\tau_{zy}$ are the jump frequencies of the carbon atoms between the types of sites indicated by the subscripts. Thus, $1/\tau_{xz}$ is the jump rate of a carbon atom from an x to a z site, and $1/\tau_{zx}$ is the jump frequency in the opposite direction.

Under an applied constant stress, the jump frequencies will be different from those obtained when there is no stress. This is because a stress applied along the z axis, as assumed above, lowers the energy barrier for a jump into a z site from an x or y site, while it raises the energy barrier for a jump of the reversed kind. This is indicated schematically in Fig. 13.5 for the case of atoms jumping between x and z sites. Due to the symmetry of the lattice, an identical curve would be obtained for interchanges between y and z sites. Note that as a result of the applied stress, the energy level of the x site is higher by an amount u than the energy level of the z site. In effect, this makes the

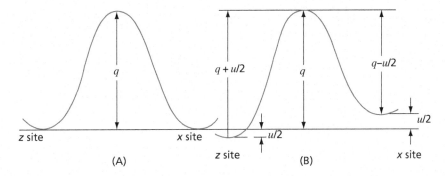

Fig. 13.5 The effect of an applied stress along a z-axis direction in a bcc metal on the energy barrier for the jumping of carbon atoms between x and z sites. **(A)** The energy barrier when the stress is zero. **(B)** The barrier when the applied stress is finite

energy barrier for a jump from an x to a z site $(q - u/2)$, and for a jump from a z to an x site $(q + u/2)$. The jump frequency in the absence of a stress is

$$\frac{1}{\tau} = \frac{1}{\tau_0} e^{-q/kT} \qquad 13.19$$

and by the symmetry of the lattice we have

$$\frac{1}{\tau_{xz}} = \frac{1}{\tau_{yz}} = \frac{1}{2\tau} = \frac{1}{2\tau_0} e^{-q/kT} \qquad 13.20$$

where the factor $\frac{1}{2}$ is introduced because an atom in an x site, for example, will normally make half of its jumps into z sites and the other half into y sites. On the other hand, in the presence of the stress along the z axis

$$\frac{1}{\tau_{xz}} = \frac{1}{2\tau_0} e^{-(q-u/2)/kT} \qquad 13.21$$

and

$$\frac{1}{\tau_{zx}} = \frac{1}{2\tau_0} e^{-(q+u/2)/kT} \qquad 13.22$$

and by the symmetry of the lattice

$$\frac{1}{\tau_{yz}} = \frac{1}{2\tau_0} e^{-(q-u/2)/kT} \qquad 13.23$$

and

$$\frac{1}{\tau_{zy}} = \frac{1}{2\tau_0} e^{-(q+u/2)/kT} \qquad 13.24$$

Substituting these relationships into the equation for $\dot{n}_z$ gives us

$$\dot{n}_z = (n_x + n_y)\frac{e^{-(q-u/2)/kT}}{2\tau_0} - \frac{2n_z}{2\tau_0} e^{-(q+u/2)/kT} \qquad 13.25$$

or

$$\dot{n}_z = \frac{e^{-q/kT}}{2\tau_0}[(n_x + n_y)e^{u/2kT} - 2n_z e^{-u/2kT}] \qquad 13.26$$

Because u, the difference in the energy levels due to the stress, is normally very small and therefore $u/2 \ll kT$, we may make the substitutions

$$e^{u/2kT} = 1 + \frac{u}{2kT} \quad \text{and} \quad e^{-u/2kT} = 1 - \frac{u}{2kT}$$

so that

$$\dot{n}_z = \frac{e^{-q/kT}}{2\tau_0}\left[(n_x + n_y)\left(1 + \frac{u}{2kT}\right) - 2n_z\left(1 - \frac{u}{2kT}\right)\right] \qquad 13.27$$

Since $n_x + n_y + n_z = n$, where n is the total number of carbon atoms per cm^3 and $e^{-q/kT}/\tau_0 = 1/\tau$, the jump frequency in the absence of a stress, this equation can also be written

$$\dot{n}_z = \frac{1}{2\tau}\left[(n - 3n_z) + (n + n_z)\frac{u}{2kT}\right]$$

or

$$\dot{n}_z = \frac{3}{2\tau}\left[\left(\frac{n + n_z}{3}\right)\frac{u}{2kT} - \left(n_z - \frac{n}{3}\right)\right] \qquad 13.28$$

and since $n_z u$ is a small quantity and n_z is not greatly different from $n/3$, we may make the approximation in the first term inside the brackets that $n_z = n/3$, so that

$$\dot{n}_z = \frac{3}{2\tau}\left[\frac{2}{9}\frac{nu}{kT} - \left(n_z - \frac{n}{3}\right)\right] \qquad 13.29$$

Now let us assume that the stress is applied for a very long time. In this case, $n_z \to n_{z(\max)}$ and $\dot{n}_z \to 0$. Therefore, in the limit when $t \to \infty$, we find that

$$\frac{2}{9}\frac{nu}{kT} = \left(n_{z(\max)} - \frac{n}{3}\right) \qquad 13.30$$

This equation states that the energy level difference u is directly proportional to the final excess number of carbon atoms in z sites. In addition, we have

$$\dot{n}_z = \frac{3}{2\tau}\left[\left(n_{z(\max)} - \frac{n}{3}\right) - \left(n_z - \frac{n}{3}\right)\right] \qquad 13.31$$

which by Eq. 13.17 shows that

$$\frac{1}{\tau_\sigma} = \frac{3}{2\tau} \qquad 13.32$$

or

$$\tau = \frac{3\tau_\sigma}{2} \qquad 13.33$$

In addition, Eq. 13.32 is significant in that an experimentally determined value (relaxation time τ_σ) is capable of directly yielding the value of a theoretically important quantity (the mean time of stay of a solute atom in an interstitial position τ). In addition, once the value of τ is determined, it is possible to compute directly the diffusion coefficient for interstitial diffusion from the relation

$$D = \frac{\alpha a^2}{\tau} \qquad 13.34$$

This equation is identical in form to that used in Chapter 12 in the discussion of substitutional diffusion. In the present case, the quantity a is the lattice constant of the solvent, τ is now the mean time of stay of a solute atom in an interstitial site, and α is a constant that is determined by the geometry of the lattice and the nature of the diffusion process (in this case, jumping of solute atoms between interstitial positions). In interstitial diffusion, the constant α equals $\frac{1}{24}$ in the body-centered cubic lattice, and $\frac{1}{12}$ in the face-centered cubic lattice. Since interstitial diffusion has been most thoroughly investigated in body-centered cubic lattices, the present discussion will be confined to this lattice. Here α is $\frac{1}{24}$ and $\tau = \frac{3}{2}\tau_\sigma$. Therefore,

$$D = \frac{a^2}{24\tau} = \frac{a^2}{36\tau_\sigma} \qquad 13.35$$

Substitution of the experimentally determined relaxation time τ_σ in Eq. 13.35 yields the diffusivity directly.

13.3 Experimental Determination of the Relaxation Time

If the relaxation time is very long (of the order of minutes to hours), it can be determined by the elastic after-effect method. In this case, a stress is applied to a suitable specimen and maintained until the anelastic component of the stress has effectively reached its equilibrium value. This corresponds to a point such as a in Fig. 13.4. After this condition has been reached, the stress is quickly removed and the strain measured as a function of time. The data obtained correspond to the curve at the lower right-hand side of Fig. 13.4 from which the relaxation time can be obtained either by determining the time required for the anelastic part of the strain to fall $1/e$ of its value, or by more elaborate methods (Ref. 10) of handling the data.

The elastic after-effect method is not particularly suitable for use when the relaxation time is of the order of seconds instead of minutes. In this case, a convenient and frequently used method for determining relaxation time makes use of a torsion pendulum of the type (Ref. 11) shown schematically in Fig. 13.6. The specimen, in the form of a wire, is clamped into two pin vises; the upper one rigidly fixed in the apparatus, the lower one connected to the inertia bar and free to rotate with it. Small iron blocks are located at each end of the inertia bar in order to make possible a smooth release of the pendulum when it is set into oscillation. In placing the pendulum in motion, it is

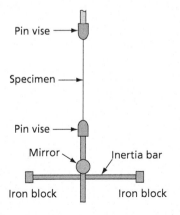

Fig. 13.6 Torsion pendulum. (After Ké, T. S., *Phys. Rev.*, **71** 553 [1947])

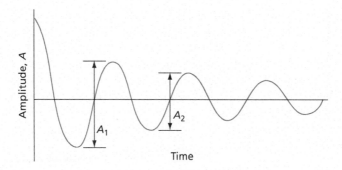

Fig. 13.7 Damped vibration of a torsion pendulum

given an initial twist about its axis so that the iron blocks are brought into contact with two small electromagnets that hold it in the twisted position until the current is broken in the magnet circuit. When this is done, the pendulum is free to oscillate. A mirror placed on the bar connecting the lower pin vise to the inertia bar makes it possible to follow the oscillations of the pendulum when a light beam is reflected from its surface onto a translucent scale. Figure 13.7 shows a typical trace of the pendulum amplitude as a function of time. Note that the amplitude decreases with increasing time because of vibrational energy losses inside the wire (neglecting air-friction effects on the torsion bar). Such an energy loss is said to be due to internal friction in the metal. There are many sources of internal friction in metals. We are interested in that due to the presence of interstitial solute atoms in body-centered cubic metals.

Three possible cases will be considered. First, let us assume that the period of the torsion pendulum is very small compared with the relaxation time of the metal. In this case, the length of time during a cycle that the wire is subject to a given type of stress is very much shorter than the mean time of stay of a solute atom in an interstitial position. Stated slightly differently, the stress alternates so rapidly that it is impossible for the solute atoms to follow the stress changes. The anelastic component of the strain can be taken as zero and the pendulum considered to vibrate in a completely elastic manner. A stress-strain diagram for this case (Fig. 13.8) is a straight line with a slope equal to the modulus of elasticity. In anelastic work this slope is normally designated as M_U and is called the *unrelaxed modulus*. The line is drawn so that it passes through the origin of

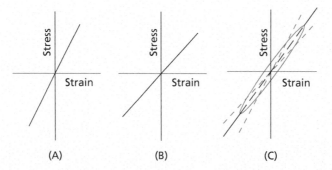

Fig. 13.8 Stress-strain curves for a torsion pendulum: **(A)** Pendulum period much shorter than the relaxation time, **(B)** Pendulum period much longer than the relaxation time, and **(C)** Pendulum period approximately equal to the relaxation time

coordinates, in agreement with the fact that a pendulum oscillates to either side of its rest point. The curve of Fig. 13.8 can also be considered to represent the stress-strain diagram for a complete oscillation of the stress.

The second possibility corresponds to the other extreme where the period of the pendulum is very much larger than the relaxation time. In this case, the solute atoms have no difficulty in following the stress alternations and it can be assumed that a state of equilibrium is constantly maintained. At all times, the anelastic strain will have its maximum, or equilibrium, value corresponding to the instantaneous value of the stress. Both the anelastic and elastic components of the strain vary directly as the stress and, therefore, the total strain varies linearly with the stress. However, in contrast to the case where the period of the pendulum is very short, the strain at each value of the stress will now be of greater magnitude due to the finite value of the anelastic strain. Figure 13.8B shows that the stress-strain curve for this second case has a smaller slope than the previous one, signifying that the modulus of elasticity (ratio of stress to strain) is smaller. The modulus of elasticity, measured under these conditions, is known as the *relaxed modulus* and is given the symbol M_R.

The third case is the intermediate one where the period of the torsion pendulum approximately equals the relaxation time. Here the stress cycles are slow enough that the anelastic strain assumes finite values, but the stress variation is not slow enough for a state of equilibrium to be effectively reached. Under these conditions, the anelastic strain does not vary linearly with the stress, and the total strain contains a nonlinear component (anelastic part). The stress-strain curve for a complete cycle of stress now assumes the form of an ellipse (Fig. 13.8C) with the major axis lying between the lines corresponding to the stress-strain curves for very short and very long periods. The area inside this (*hysteresis*) loop has the dimensions of work and represents the energy lost in the specimen per unit volume during a complete cycle. In the two other examples (Figs. 13.8A and 13.8B), the area inside the stress-strain loop is zero. It appears, therefore, that the energy loss per cycle is a function of the period of oscillation, having a zero value when the period is either very long or very short, and a finite value when the period has intermediate values.

The energy loss per cycle in a torsion pendulum can be determined directly from a plot of the pendulum amplitude as a function of time. In Fig. 13.7, A_1 and A_2 represent two adjacent vibration amplitudes and the assumption is made that the difference in the magnitude of these two amplitudes is small, a condition usually met in work of the type being discussed.

Now in an oscillating system, the energy of vibration is proportional to the square of the vibration amplitude, so that the vibrational energy, when the pendulum has amplitude A_1, is proportional to A_1^2, and when it has the amplitude A_2, the energy corresponds to A_2^2, and the fractional loss of energy per cycle is

$$\frac{\Delta E}{E} = \frac{A_2^2 - A_1^2}{A_1^2} \qquad 13.36$$

where E is the energy of the pendulum, ΔE is the loss of energy during a cycle, and A_1 and A_2 are the amplitudes at the beginning and end of the cycle.

Factoring the expression on the right-hand side of Eq. 13.36 leads to

$$\frac{\Delta E}{E} = \frac{(A_1 - A_2)(A_1 + A_2)}{A_1^2}$$

but since it has been assumed that the difference between A_1 and A_2 is small, we can write $A_1 - A_2 = \Delta A$ and $A_1 + A_2 = 2A_1$. Therefore,

$$\frac{\Delta E}{E} = \frac{2\Delta A}{A} \qquad 13.37$$

This equation states that the fractional energy loss per cycle is twice the fractional amplitude loss per cycle. This latter quantity is easily determined experimentally.

The measure of the internal friction given above, $\Delta E/E$, is usually known as the *specific damping capacity*. This quantity is often used by engineers to express the energy-absorbing properties of materials of construction.

A more commonly used measure of the internal friction in problems of the type currently being discussed is the *logarithmic decrement*, which is the natural logarithm of the ratio of successive amplitudes of vibration. Thus,

$$\delta = \ln \frac{A_1}{A_2} \qquad 13.38$$

where δ is the logarithmic decrement, and A_1 and A_2 are two successive vibration amplitudes.

Provided that the damping is small, we may write (Ref. 10)

$$\delta = \frac{1}{2} \frac{\Delta E}{E} = \frac{\Delta A}{A} \qquad 13.39$$

Still another method of expressing the internal friction in a metal undergoing cyclic strain uses the phase angle α by which the strain lags behind the stress. The tangent of this angle can also be taken as an index of the internal-energy loss. Again, if the damping is small, (Ref. 10) it can be shown that

$$\tan \alpha = \frac{1}{\pi} \ln \frac{A_1}{A_2} = \frac{\delta}{\pi} \qquad 13.40$$

This quantity, $\tan \alpha$, is often written as Q^{-1} and called the *internal friction*. This is in analogy with the damping, or energy loss, in an electrical system.

The energy loss per cycle $\Delta E/E$ is a smoothly varying function of the period, or frequency, of the pendulum capable of being stated mathematically. When expressed in terms of the angular frequency of the pendulum ω

$$\omega = 2\pi\nu = \frac{2\pi}{\tau_p} \qquad 13.41$$

where ν = pendulum frequency in cycles per second and τ_p = pendulum period in seconds, this expression assumes the following simple and convenient form

$$\frac{\Delta E}{E} = 2\left(\frac{\Delta E}{E}\right)_{max} \left[\frac{\omega \tau_R}{1 + \omega^2 \tau_R^2}\right] \qquad 13.42$$

where $\frac{\Delta E}{E}$ = fraction of energy lost per cycle, $\left(\frac{\Delta E}{E}\right)_{max}$ = maximum fractional energy loss, ω = angular frequency of pendulum, τ_R = relaxation time for interstitial diffusion.

The relaxation time τ_R, as measured using a torsion pendulum, is not exactly the same as τ_σ, the relaxation time as measured in an elastic after-effect experiment, or other experiments carried out at constant stress. The two quantities are, however, simply related to each other by the expression (Ref. 10)

$$\tau_R = \tau_\sigma \left(\frac{M_R}{M_U}\right)^{1/2} \qquad 13.43$$

where M_R is the relaxed modulus and M_U is the unrelaxed modulus. In the case of the interstitial diffusion of carbon in alpha iron, M_R and M_U are very nearly equal, so we may assume $\tau_R = \tau_\sigma$ and also that the mean time of stay of a carbon atom in an interstitial site is $\tau = \frac{3}{2}\tau_R$.

Equation 13.43 is symmetrical in both ω and τ_R, so that the variation of the fractional energy loss $\Delta E/E$ or δ is the same no matter whether we hold τ_R constant and vary ω or hold ω constant and vary τ_R. In either case, the given function yields a curve symmetrical with respect to the point of maximum-energy loss, if δ is plotted as a function of the log ω, or of the log τ_R. Figure 13.9 shows the shape of this curve and, as indicated in the figure, the maximum logarithmic decrement occurs when the following relationship holds:

$$\omega = \frac{1}{\tau_R} \qquad 13.44$$

Equation 13.44 states that the energy loss is a maximum when the angular frequency of the pendulum equals the reciprocal of the relaxation time. This fact is important because it gives a direct relationship between the experimentally measurable frequency of the pendulum and the relaxation time.

The relaxation time is only a function of temperature, while the frequency of the pendulum is a function of its geometry. In theory, there are two basic methods of determining the point of maximum energy loss: the frequency of the pendulum may be varied while the temperature is maintained constant (thereby keeping the relaxation-time constant) or the frequency of the pendulum may be kept constant while the temperature is varied. The latter method is more often used as it is usually more convenient. In this case we have

$$D = D_0 e^{-Q/RT} = \frac{a^2}{36\tau_R} \qquad 13.45$$

Solving this equation for relaxation time yields:

$$\tau_R = \frac{a^2}{36 D_0 e^{-Q/RT}} = \tau_{R_0} e^{+Q/RT} \qquad 13.46$$

where τ_{R_0} is a constant equal to $a^2/36 D_0$. From this relationship it is clear that the $\log_{10} \tau_R$ varies as $1/T$, where T is the absolute temperature. A plot of the fractional energy loss as a function of $1/T$ should give a curve of the type shown in Fig. 13.9. A set of five of these curves for iron, containing carbon in solid solution, is shown in Fig. 13.10. Each curve was obtained by adjusting the torsion pendulum to operate at a different frequency. The maximum energy loss of each curve occurs at a different temperature. The frequencies corresponding to the curves of Fig. 13.10 are listed in the first column of Table 13.1, the angular frequencies in column 2; the temperatures of the energy-loss maxima in column 3; the reciprocal of the temperature in column 4; and the logarithms of the relaxation times in column 5, as computed with the aid of the equation

$$\tau_R = \frac{1}{\omega} \qquad 13.47$$

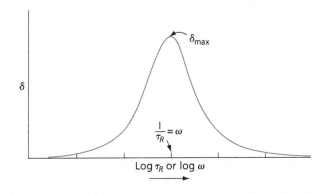

Fig. 13.9 Theoretical relationship between the fractional energy loss per cycle and either log τ_R or log ω. At δ_{max}, $\frac{1}{\tau_R}$ equals ω

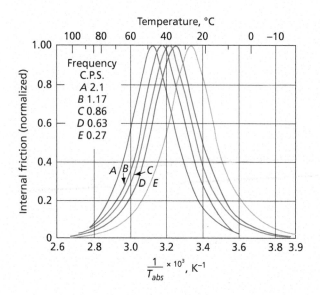

Fig. 13.10 Internal friction as a function of temperature for Fe with C in solid solution at five different pendulum frequencies. (Reprinted figure with permission from Wert, C., and Zener, C., *Phys Rev.*, **76** 1169 [1949]. Copyright 1949 by the American Physical Society)

τ_{R_0} and Q may be obtained from these data by plotting the $\log_{10} \tau_R$ as a function of $1/T$. This plotting is done in Fig. 13.11, where the slope of the line is

$$m = \frac{Q}{2.3R} = 4320$$

Note that the factor 2.3 enters the above equation because we are using logarithms to the base 10. The activation energy Q is accordingly

$$Q = 8.314(4320) = 82{,}600 \text{ J/mol} \qquad 13.48$$

The value of the constant τ_{R_0} is obtained from the ordinate intercept in Fig. 13.11, which has a value of

$$\log_{10} \tau_{R_0} = -14.54$$

so that

$$\tau_{R_0} = 2.92 \times 10^{-15} \text{ sec}$$

Table 13.1 Data Corresponding to the Five Curves Shown in Figure 13.10*

Frequency cps	Angular Frequency	Absolute Temperature T	$\frac{1}{T}$, K^{-1}	$\log_{10} \tau_R$
2.1	13.2	320	3.125×10^{-3}	-1.120
1.17	7.35	314	3.178×10^{-3}	-0.866
0.86	5.39	311	3.22×10^{-3}	-0.731
0.63	3.95	309	3.25×10^{-3}	-0.595
0.27	1.69	300	3.35×10^{-3}	-0.227

*Reprinted figure with permission from Wert, C., and Zener, C., *Phys Rev.*, **76** 1169 [1949]. Copyright 1949 by the American Physical Society.

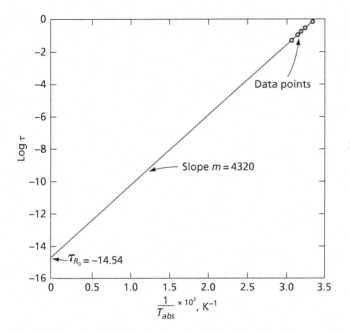

Fig. 13.11 Variation of log τ with $1/T_{abs}$ for data in Fig. 13.10. (Data of Wert, C., and Zener, C., *Phys. Rev.*, **76** 1169 [1949])

The mean time of stay of a carbon atom in an interstitial position is, accordingly,

$$\tau = \frac{3}{2}\tau_R = \frac{3}{2}\tau_{R_0}e^{Q/RT} = \left(\frac{3}{2}\right) 2.92 \times 10^{-15} e^{82,600/RT} \qquad 13.49$$

and the diffusivity of carbon in body-centered cubic iron (alpha-iron), taking the lattice constant a for iron as 0.286 nm, is:

$$D = \frac{a^2}{36\tau_R} = 8.0 \times 10^{-7} e^{-82,600/RT}, \text{ m}^2/\text{s} \qquad 13.50$$

Additional details on the use of internal friction techniques to study interstitial diffusivity in iron and other materials can be found in References 12-15.

13.4 Experimental Data

A compilation of certain experimentally determined interstitial-diffusivity equations is given in Table 13.2. The value given in the table for the diffusion of carbon in iron is not the same as that given above, for it is a later and somewhat more accurate value. In general, it is noted that in the elements listed, the activation energies of interstitial diffusion are somewhat smaller than those of substitutional diffusion; compare diffusivity values for interstitials with niobium self-diffusion and others given in Table 12.4. It should be noted that while Snoek effects in bcc metals, particularly carbon in bcc iron, have been well documented, Snoek relaxation in iron-carbon bct martensite is not clear. However, a recent Monte Carlo simulation and mean-field thermodynamic kinetic modeling show that Snoek relaxation in bct martensite does occur. (Ref 16)

Table 13.2 Diffusivity Equations for Interstitial Diffusion in Certain Body-Centered Cubic Metals

Solvent Metal	C	Diffusing Element N	O	Nb
Iron*△	$2.0 \times 10^{-6} e^{-84,100/RT}$	$1.00 \times 10^{-7} e^{-74,100/RT}$		
Vanadium†		$50.21 \times 10^{-7} e^{-151,000/RT}$	$26.61 \times 10^{-7} e^{-125,000/RT}$	
Tantalum†		$5.21 \times 10^{-7} e^{-158,000/RT}$	$10.50 \times 10^{-7} e^{-110,000/RT}$	
Niobium†		$25.62 \times 10^{-7} e^{-152,000/RT}$	$7.31 \times 10^{-7} e^{-110,000/RT}$	
Niobium††	$1 \times 10^{-6} e^{-42,000/RT}$	$6.3 \times 10^{-6} e^{-161,000/RT}$	$4.2 \times 10^{-7} e^{-107,000/RT}$	$5.2 \times 10^{-5} e^{-395,000/RT}$

* Wert, C., *Phys. Rev.*, **79** 601 (1950)
△ Wert, C., and Zener, C., *Phys. Rev.*, **76** 1169 (1949)
† Boratto, F., *Univ. of Florida Thesis* (1977)
†† See Ref. 17.

13.5 Anelastic Measurements at Constant Strain

In addition to measurements made at constant stress, as exemplified by the technique of the elastic after-effect (zero stress) and vibrational studies, of which the torsion pendulum is only one of a number of techniques, there is a third basic type of measurement. This involves the measurement of the relaxation of the stress under the condition of a constant strain. In a typical experiment of this type, a specimen may be loaded in a universal (constant deformation rate) testing machine to some predetermined stress, whereupon the machine is stopped. Because the specimen is subjected to a stress, it continues to deform anelastically. For the purpose of this discussion it will be assumed that the testing machine and its grips are very *hard* elastically. Under this assumption, the ends of the specimen will be fixed in space so that the total strain in the specimen is constant, or

$$\varepsilon = \varepsilon_{an} + \varepsilon_{el} = 0 \qquad 13.51$$

Thus as the specimen increases its length anelastically there must be a corresponding decrease in its elastic strain. This will, of course, relax the stress in the specimen in a manner like that shown schematically in Fig. 13.12. This curve is similar in its general form to the elastic after-effect curve and follows

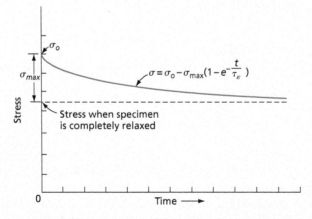

Fig. 13.12 The relaxation of the stress at constant strain

a similar law. It is, accordingly, possible to measure a relaxation time at constant strain, designated τ_ε. It is not the same as τ_σ, the relaxation time at constant stress, but it is related to it by the equation

$$\frac{\tau_\sigma}{\tau_\varepsilon} = \frac{M_U}{M_R} \qquad 13.52$$

where M_U is the unrelaxed modulus and M_R is the relaxed modulus. Both of these relaxation times are related to that obtained in a torsion pendulum as follows (Ref. 18)

$$\tau_R = (\tau_\sigma \tau_\varepsilon)^{1/2} \qquad 13.53$$

This shows that τ_R, obtained in a torsion pendulum, is the geometric mean of the other two relaxation times.

Problems

13.1 For interstitial diffusion in a body-centered cubic lattice, prove that α, in the diffusion equation $D = \alpha a^2/\tau$, equals 1/24.

13.2 (a) Determine the mean time of stay of an oxygen atom in an interstitial site in niobium at 300 K. See Table 13.2 for the necessary diffusion parameters and take the lattice parameter, a, for niobium as 0.3301 nm.

(b) Do the same for a temperature of 400 K.

13.3 In a special torsion pendulum used for elastic after-effect measurements, it is not practical to measure the relaxation time if it is less than 1 minute. On this basis, determine the maximum temperature at which the apparatus should be used for determining the relaxation time of nitrogen atoms in iron. Take $a = 0.28664$ nm and see Table 13.2 for the diffusion equation of iron.

13.4 Compare the relaxation times, τ_σ, of vanadium and tantalum, due to the presence of oxygen atoms in solid solution, at 400 K. The lattice parameters of vanadium and tantalum are, respectively, equal to 0.3029 and 0.3303 nm.

13.5 (a) Determine the mean time of stay, τ, of a nitrogen atom in a tantalum interstitial site at 400 K.

(b) At what temperature would τ equal 1.0 second?

13.6 A torsion pendulum with a tantalum wire has a pendulum frequency, ν, of 0.82 Hz. When the logarithmic decrement, δ, is measured with this pendulum, the peak in the plot of δ versus $1/T$ occurs at 415.5 K. Show by using Eq. 13.45, with $T = 415.5$ K, that this peak corresponds to the presence of oxygen in the tantalum.

13.7 A torsion pendulum with a vanadium wire containing nitrogen and a period of 2.00 seconds shows a decrease in the pendulum amplitude, A, of 10 percent in 100 cycles of oscillation at 350 K. Compute:

(a) The specific damping capacity.

(b) The logarithmic decrement.

(c) Tan α, where α is the phase angle by which the strain lags the stress.

13.8 Write a computer program suitable for obtaining data to make a plot of the logarithmic decrement, δ, versus the reciprocal of the absolute temperature, $1/T$ using Eq. 13.42, in order to obtain a "bell-shaped" Lorentz curve. In determining the data, vary $1/T$ from 1.4×10^{-3} to 2.1×10^{-3} in steps of 1.0×10^{-5} and assume that the pendulum frequency is 0.9 Hz and the pendulum wire is made of niobium containing nitrogen in solid solution. With the aid of this curve and Eq. 13.45, determine the relaxation time, τ_R.

References

1. McLellan, R. B., and Wasz, M. L., *J. Phys. Chem. Solids*, **54** 583 (1993).
2. Wuttig, M., *Scripta Metal.*, **5** 33 (1971).
3. Kimura, H., and Yoshioko, K., *Mater. Sci. Eng.*, **26** 171 (1976).
4. McLellan, R. B., Rudee, M. L., and Ishibachi, T., *Trans. Metall. Soc. AIME*, **233** 1938 (1965).
5. Homan, C. G., *Acta Metall.*, **12** 1071 (1964).
6. Farraro, R., and McLellan, R. B., *Mater. Sci. Eng.*, **39** 47 (1979).
7. Murch, G. E., and Thorn, R. J., *J. Phys. Chem. Solids*, **38** 789 (1977).
8. Snoek, J., *Jour. Physica*, **6** 591 (1939).
9. Nowick, A. S., *Prog. in Metal Phys.*, **41** (1953).
10. Nowick, A. S., and Berry, B. S., *Anelastic Relaxation in Crystalline Solids*, Academic Press, New York and London, 1972.
11. Ké, T. S., *Phys. Rev.*, **71** 553 (1947).
12. Berry, B. S., *Acta Meta.*, **10** 271 (1962).
13. Buzzichelli, G., and Baldi, G., *Scripta Met.*, **5** 943 (1971).
14. Numakura, H., Kashiwazaki, K., Yokoyama, H., and Koiwa, M., *J. Alloys and Compounds*, **310** 344 (2000).
15. Golovin, I. S., *Mat. Sci. Eng.*, **A442** 92 (2006).
16. Maugis, P., and Huang, L., *J. Alloys and Compounds*, **907** 164502 (2022).
17. Mehrer, H., *Diffusion in Solids, Fundamentals, Methods, Diffusion-Controlled Processes*, **155** 313 (2007).
18. Zener, Clarence M., *Elasticity and Anelasticity of Metals*, The University of Chicago Press, Chicago, Ill., 1948.

Chapter **14**

Solidification of Metals

Learning Objectives

Upon completion of Chapter 14, you will be able to:

1. Differentiate between structures of liquids, solids and metallic glasses
2. Explain the fundamentals of liquid-solid transformation and the differences between melting and solidification
3. Compare and contrast activation energies for atomic attachment and detachment across solid-liquid interfaces
4. Differentiate between the nature of faceted and non-faceted solid-liquid interfaces, including their growth by continuous and lateral growth mechanisms
5. Understand alloy solidification, solute redistribution at the interface, and calculate microsegregation and coring using the Scheil equation
6. Describe solute build-up at the interface and the development of constitutional supercooling
7. Explain how a planar interface becomes unstable and how it leads to cellular or dendritic growth
8. Understand the mechanism of eutectic solidification and formation of lamellar structures

Most commercial metal objects are frozen from a liquid phase into either their final shapes, called *castings*, or intermediate forms, called *ingots*, which are then worked into the final product. A large number of metal objects are also made by hot pressing and sintering of alloy particles, which are produced via liquid-solid transformation. Since the properties of the end result are determined, in large measure, by the nature of the solidification process, the factors involved in the transformation from liquid to solid are of the utmost practical importance. Before the subject of solidification is considered, however, the nature of the liquid state will be treated briefly.

14.1 The Liquid Phase

For most cases of practical importance, liquid metal alloys occur as a single homogeneous liquid phase. To simplify the present discussion, we shall treat a special case—a pure metallic element with only a single solid phase. Figure 14.1 is a schematic phase diagram of a single-component system.

From what we have previously learned, we know that the solid is a crystalline phase in which the atoms are aligned in space in definite patterns over long distances. The regularity of crystal lattices makes it easy to study their structures with the aid of X-ray diffraction and electron microscopy; therefore, a great deal is known about the internal arrangements of atoms in metal crystals. At the same time, the uniformity of the structure of crystals makes it possible to employ mathematics in the study of their properties. On the other hand, the gas phase represents the other extreme from the solid phase where the structure is one of almost complete randomness, or disorder, instead of almost complete order. Here, in most cases, the atoms can be assumed to be removed far enough from each other that metallic gases can be treated as ideal gases. The physical properties of metallic gases, like those of metallic solids, are therefore capable of mathematical analysis.

While the solid crystalline phase is pictured as a completely ordered arrangement of atoms (neglecting defects such as dislocations and vacancies) and the gas phase as a state of random disorder (ideal gas), no simple picture has as yet been devised to represent the structure of the liquid phase. The principal trouble is that the liquid phase possesses neither the long-range order of the solid nor the lack of interaction between atoms characteristic of the gas phase. It is, therefore, essentially an indeterminant structure. Actually, in a liquid, the average separation between atoms is very close to that in the solid. This fact is shown by the small change in density on melting, which for close-packed metals amounts to 2 to 6 percent only, where part of this density change is probably associated with the formation of additional structural defects in the liquid phase. A compilation of physical properties of pure metals at elevated temperatures and those of liquid metals can be found in Reference 1. Further, the latent heat of fusion released when a metal melts is relatively small, being only $\frac{1}{25}$ to $\frac{1}{40}$ of the latent heat of vaporization. Since the latter is a measure of the energy required to place the atoms in the gas phase, where it can be assumed that they are so far apart that they do not interact, it serves as a good measure of the energy binding atoms together. Both the small size of the energy released when a metal melts and the fact that the atoms in the liquid and solid phases

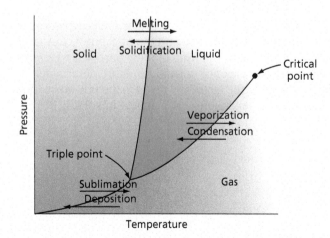

Fig. 14.1 Pressure-temperature phase diagram for a single-component system

have almost identical separations lead to the conclusion that the bonding of atoms in the solid and liquid phases are probably similar. X-ray diffraction studies of liquid metals tend to confirm this assumption. Time does not permit us to delve into X-ray-diffraction techniques as applied to liquid metals, but the results of these investigations are interpreted as showing the following: liquid metals possess a structure in which the atoms, over short distances, are arranged in an ordered fashion and have a coordination number approximately the same as the solid. The X-ray results also show that liquid metals do not possess long-range order. A plausible picture of the liquid structure might be the following: the atoms over short distances approach arrangements close to those exhibited in crystals, but due to the presence of many structural defects, the exact nature of which is unknown, the long-range order typical of a crystal cannot be achieved. In a number of cases, it has been suggested that the liquid phase is essentially the same as that of the solid, with the incorporation into the structure of a large number of structure defects, such as vacancies, interstitial atoms, or dislocations. However, these models have not been completely satisfactory and it must be admitted that the exact nature of the defects that exist in the liquid phase are not known. The fact that the atoms in the liquid have an extremely high mobility modifies this essentially static picture of the liquid phase. Diffusion measurements of liquids, made at temperatures just above the melting point, show that atomic movements in liquids are several orders of magnitude (Ref. 2) (several powers of 10) more rapid than in solids just below the melting point. The rapid diffusion rates in liquids are undoubtedly a result of structural defects characteristic in the liquid phase. Since the nature of the liquid state is unresolved, a simple picture of the way atoms move in a liquid cannot be presented. This is in contrast to solids, where the evidence strongly favors the vacancy-motion concept. One thing is clear, however: the energy barriers for atom movements in liquid diffusion are very small. Since diffusion in the liquid state must be associated with a corresponding motion of structural defects and the motion is very rapid, we are led to the conclusion that the liquid phase is one of everchanging structure. The local order existing at any one position in space changes continually with time. This represents a basic difference between liquid and solid phases. In a solid, diffusion by vacancy movement does not, in general, alter the structure of the crystal, whereas in the liquid the local structure is one of constant change resulting from atomic motion. Another important consequence of the ease of atom movements in the liquid phase is the development of the property that is most characteristic of the liquid phase: fluidity, or the inability of a liquid to support shear stresses of even very low magnitudes.

A surprising fact is that the liquid phases of most metals are very similar in their properties, even though their properties may be quite different in the solid state (Ref. 3). Thus, on melting, the close-packed metals are inclined to lower their coordination slightly, but metals of loose packing, such as gallium and bismuth, usually increase their coordination. The ultimate result is that liquid metals tend to have the same number of nearest neighbors. The recent X-ray analysis shows that 8–11 nearest neighbors surround each atom in a liquid (Ref. 4). The analysis further shows that a metallic liquid consists of clusters of atoms which have crystal-like structure. The clusters are in turn surrounded by free atoms and free electrons. Various behavior of metals, such as heat capacity, heat of fusion, and diffusion, can be explained based on the assumption that clusters rotate around the center of their mass. Similarly, the electrical and thermal conductivities of the liquid phases of metals also tend toward common values.

The above facts may be interpreted as demonstrating that the properties of the liquid phases of metals are more dependent on the structural defects characteristic of liquids than they are upon the nature of the bonding forces between the atoms. On the other hand, in solid crystalline phases, the physical properties are very dependent on the nature of the bonding forces between atoms, since these interatomic forces determine the types of crystals that form and, as a consequence, the properties of the solids.

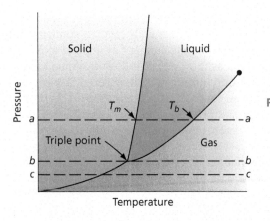

Fig. 14.2 Same as Fig. 14.1, but showing the isobars considered in Fig. 14.3

From the standpoint of the free energy of the phases and the reasons for the existences of transformations between solid, liquid, and gases, consider the following. Of the three phases, the solid crystalline phase possesses the lowest internal energy and the highest degree of order or lowest entropy. The liquid represents a phase with a slightly higher internal energy as measured by the heat of fusion, and a slightly larger entropy corresponding to its more random structure. Finally, the gas phase has the highest internal energy and the greatest entropy or disorder. It should be noted that at extremely high pressures and temperatures, beyond those of the critical point D in Fig. 14.1, the liquid and vapor phases are indistinguishable from each other. The free-energy curves for these three phases may be plotted in a manner similar to those for the two solid phases of tin and, in each case, the slope will be steeper, in a more negative sense, the greater the inherent entropy of the phase. Thus, the free energy of the gas phase falls most rapidly with increasing temperature, the liquid phase next, and the solid phase least rapidly. The free energies of the three phases are also functions of the pressure, so that their positions relative to each other are not the same at different pressures. This fact is illustrated in Fig. 14.3, where the free-energy curves are drawn schematically for three isobaric lines (constant pressure) aa, bb, and cc of Fig. 14.2, which is based on Fig. 14.1. In the left-hand figure, corresponding to the isobaric aa, it can be seen that at

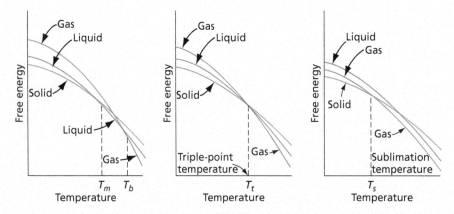

Fig. 14.3 Free-energy curves for the phases in a one-component system at three different pressures corresponding to isobars aa, bb, and cc, respectively, of Fig. 14.2

temperatures below the melting point, T_m, the solid phase has the lowest free energy, but at T_m the free-energy curve of the liquid phase crosses that of the solid, and the liquid phase becomes the more stable until T_b, the boiling temperature, is reached. At this point, the free-energy curve of the gas phase crosses that of the liquid phase and for all higher temperatures the gas phase is the most stable. The middle figure corresponds to a constant-pressure line that passes through the triple point. Along this isobaric line (*bb*) the three free-energy curves cross at a single point (the triple point) where all three phases may coexist. At temperatures below the triple point, the solid phase has the lowest free energy, and at all temperatures above it the gas phase has the least free energy. At this particular pressure, the liquid phase is only stable at one temperature—the triple-point temperature. Finally, on the right in Fig. 14.3 is shown the relative positions of the three free-energy curves when the pressure is very low (that is, at *cc*). In this case, the free-energy curve of the liquid phase lies above that of the gas phase, so that the liquid phase is not stable at any temperature.

14.2 Nucleation

The solidification of metals occurs by nucleation and growth. The same is also true of melting, but there is one important difference. Nucleation of the solid phase during freezing is a much more difficult process than the formation of nuclei of the liquid phase during melting. As a result, metals do not superheat to any appreciable extent before they are liquefied, whereas some supercooling occurs almost every time a metal is frozen. Further, under appropriate conditions, liquid metals can be cooled to temperatures far below their equilibrium freezing points before solidification begins. This fact is illustrated in Table 14.1, which gives the maximum supercooling observed for a number of metals. It is not known whether any of the values listed in Table 14.1 correspond to freezing by homogeneous nucleation, since heterogeneous nucleation, where crystals are formed on foreign nucleating sites, reduces the magnitudes of the attainable supercooling. Moreover, the level of the

Table 14.1 Maximum Supercooling of Some Pure Metal Liquids

	Substance	Max. Supercooling, K
(1)	Mercury	88
(2)	Cadmium	110
(3)	Lead	153
(4)	Aluminum	160
(5)	Tin	187
(6)	Silver	227
(7)	Gold	230
(8)	Copper	236
(9)	Iron	286
(10)	Manganese	308
(11)	Nickel	365
(12)	Platinum	370
(13)	Niobium	525

(1 to 5) After Perepezko, J. H., *Mat. Sci. Eng.*, **65** 125 (1984).

(6 to 13) After Flemings, M. C., and Shiohara, Y., *Mat. Sci. Eng.*, **65** 157 (1984).

attainable supercooling depends on the potency of the heterogeneous nucleation sites. We can conclude that the supercooling corresponding to homogeneous nucleation is at least as large, if not larger, than the values in Table 14.1.

The difficulty associated with the formation of homogeneous nuclei of the solid crystalline phase is typical not only of pure metals, but also it occurs in alloys, as can be seen in Fig. 14.4, which shows it is possible to supercool copper-nickel alloys approximately 300°C (300 K) at all compositions. The very large magnitudes of the supercooling shown in Table 14.1 and Fig. 14.4 are only observed under rigidly controlled experimental conditions designed to retard the occurrence of potent heterogeneous nucleants by emulsification (Ref. 5) or containerless melting and solidification either in low gravity environment of space or using electromagnetic levitation (Refs. 6 and 7). In the usual metals of commerce, the formation of nuclei are aided by the presence in the liquid melt of accidental particles of impurities or by the mold surfaces. When nucleation occurs in this manner, supercooling is reduced and can only amount to a few degrees. The subject of nucleation in commercial metals will be discussed in more detail when the freezing of ingots is discussed. Let us first, however, briefly treat the subject of nucleation during melting—the reverse of freezing.

In melting most metals do not normally superheat, although superheating on the order of 0.1 K has been reported for gallium (Ref. 8).

An explanation for this fact has been ordered by Hollomon and Turnbull (Ref. 9) based on the assumption that nucleation in melting occurs on the surfaces of solids. In this case, the nucleus, consisting of a region of the liquid phase, will be enveloped by two basically different surfaces. On one side is a liquid-solid interface and on the other, a gas-liquid interface. The total surface energy of this pair of surfaces is usually smaller than that of a single solid-gas interface. This conclusion is obtained from the experimental fact that when a solid and a liquid are in contact at the melting point, the liquid normally wets the surface of the solid so that, as proved in college physics textbooks,

$$\gamma_{gs} > \gamma_{sl} + \gamma_{gl} \qquad 14.1$$

where γ_{gs}, γ_{sl} and γ_{gl} are the surface tensions or surface energies of the gas-solid, solid-liquid, and gas-liquid interfaces, respectively. What Eq. 14.1 really signifies is that the surface energy associated with liquid nuclei can aid rather than hinder the formation of the nucleus. Thus, let Fig. 14.5A represent

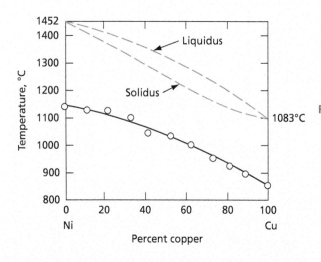

Fig. 14.4 Supercooling of Cu-Ni alloys as a function of composition (Cech, R. E., and Turnbull, D., Trans. AIME, **191** 242 [1951])

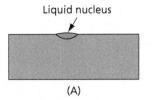
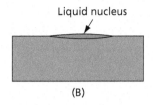

Fig. 14.5 The formation of a liquid nucleus during melting

a liquid nucleus at an early stage of development, and Fig. 14.5B the same nucleus at a slightly later time. Spreading of the liquid nucleus along the surface, as shown in these figures, decreases the area on the gas-solid interface, while increasing those of the gas-liquid and liquid-solid interface. As a result, there is a decrease in surface energy, and since any rise of the temperature above the equilibrium melting temperature causes a decrease in the volume free energies, it can be seen that both surface and volume free energies favor melting for even the slightest amount of superheating. There is also another factor: the diffusion rate also increases with rising temperature. The rate of a melting reaction should therefore tend to increase with superheating because of this fact.

14.3 Metallic Glasses

In recent years, techniques have been developed to very rapidly cool or quench certain liquid metallic alloys in order to produce an amorphous or glassy structure. Since some of these metallic glasses have very unusual and unique properties, the subject of metallic glasses is of considerable interest. To better understand these materials, it is worth considering the basic nature of the glassy state. For this purpose, we will consider the commercial inorganic glasses, which may be considered as alloys with inorganic oxides as components. A representative bottle glass, for example, might contain 71 percent SiO_2, 15 percent Na_2O, 12 percent CaO, and 2 percent other oxides.

Normally when a composition like the above is cooled from the liquid state, it may solidify in two alternative ways. If the rate of cooling is below some critical rate, the liquid may freeze to form a crystalline solid. On the other hand, if the rate of cooling is faster than the critical rate, it may pass through the freezing range without crystallizing so that it becomes a supercooled liquid that then transforms into a glass at a lower temperature. The critical cooling rate, below which crystallization occurs and above which a glass is formed, is quite low in a common inorganic glass; that is, $\leq 10^{-1}$ K/s. This means it is relatively easy to obtain glass structures with these compositions. For metallic alloys, on the other hand, the formation of a glass structure is quite difficult and requires cooling rates normally in excess of 10^5 K/s.

At the glass transition point, where a supercooled liquid changes into a glass, discontinuities may appear in plots of certain important properties made against the temperature. For example, the upper curve of Fig. 14.6A shows schematically the specific volume, $V(m^3/mol)$, of a liquid that is cooled fast enough to produce a glass. There is an *inflection* or a change of slope at the point marked I_g, which denotes the glass transition point. On the other hand, in the lower curve, corresponding to a cooling rate slow enough to allow crystallization to occur, a sharp decrease in volume occurs when the liquid freezes at T_{mp}. Note that the slope of this latter cooling curve is less after crystallization than it was before. Conversely, there is no change in the slope of curve of the more rapidly cooled material as the melt is cooled past the freezing point and becomes supercooled. It is also significant that the glass transition is not associated with a volume change as is the case when a more slowly cooled melt

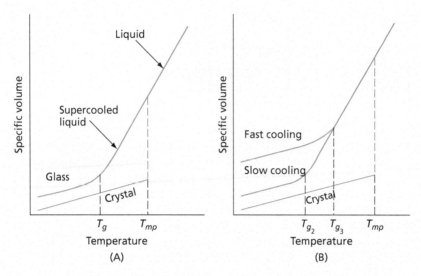

Fig. 14.6 Specific volume of liquid, glass, and crystal versus temperature

freezes. At temperatures below the glass transition the slopes of both the glass and crystallized solid curves are nearly identical. The volume of the glass, as may be seen in Fig. 14.6A, is greater than that of the crystalline solid at all temperatures where the two solid forms can coexist. Furthermore, the volume difference as well as the glass transition temperature, T_g, depend on the cooling rate, as shown in Fig. 14.6B. This volume difference, which can be attributed to a more open structure in the glass, is normally not large; it is in the range of a few percent for silicate glasses. In metallic glasses it is usually of the order of 0.5 percent (Ref. 10). In a glass, long-range order does not exist as in a crystalline solid. It is generally agreed (Ref. 11), however, that the glass structure has limited short-range order.

The transition on cooling from a supercooled liquid to a glass is believed to be caused by the very rapid rise of the viscosity of a supercooled liquid when the temperature is decreased. Although a liquid is normally incapable of supporting a shear stress, a shear stress applied to a liquid can still cause it to flow. This shear of the liquid is resisted by a frictional force, and generally one may write:

$$\tau = \eta \, d\gamma/dt \qquad 14.2$$

where τ is the shear stress, $d\gamma/dt$ is the shear rate of the liquid, and η is a constant of proportionality known as the viscosity. The classical viscosity unit is the poise, which can be viewed as the shear stress required to move two parallel layers of fluid, 1 cm apart, past each other with a velocity of 1 cm/sec. In the international system of units, the unit of dynamic viscosity is a *pascal second* (Pa-s). One Pa-s equals 10 poise.

As mentioned previously, the viscosity of a typical glass melt increases rapidly as the temperature is decreased. With the aid of empirical observations combined with theoretical deductions, an approximate equation has been developed (Ref. 12) that is capable of describing with some accuracy the temperature dependence of the viscosity. This is known as the *Vogel-Futcher-Tammann equation (VFT equation)* and has the form:

$$\log \eta = A + B/(T - T_0) \qquad 14.3$$

where η is the viscosity; T is the absolute temperature; and A, B, and T_0 are constants. These constants may be evaluated for a given material by using corresponding values of η and T for three points on the viscosity-temperature curve. Three convenient points to use for this purpose are:

1. **The Glass Transition Point.** At the glass transition point the viscosity is normally 10^{12} Pa-s. The glass transition temperature can be measured in terms of the inflection point on a dilatometric curve, which gives the variation in length of a glass specimen as a function of its temperature. Such a curve will be similar to that shown schematically in Fig. 14.6, where the glass transition occurs at the inflection point marked T_g.
2. **The Softening Point (Littleton Point).** This point is measured (Ref. 12) using a glass fiber with a diameter between 1 and 6.5 mm with a length of 27.9 cm. The fiber is hung vertically from its upper end inside a furnace where it is heated at a rate between 5 and 10 K/min. The creep rate is measured by following the movement of the lower end of the fiber. On the basis of empirical measurements it has been determined that when the creep rate of the fiber is 1 mm/min, $\eta = 10^{5.5}$ Pa-s. Thus, the softening point is determined by measuring the temperature at which a creep rate of 1 mm/min is attained and by assuming that the viscosity equals $10^{5.5}$ Pa-s at this temperature.
3. **The Sinking Point.** This point was developed (Ref. 13) by Dietzel and Bruckner. It involves the use of a small 80 percent platinum 20 percent rhodium rod with a diameter of 0.5 mm, a length of 20 cm, and a weight of 0.746 gm. The temperature at which this rod sinks to a depth of 2 cm in a time of 2.0 min. has been found to correspond to a viscosity of $10^{3.22}$ Pa-s. In essence, this technique involves varying the temperature in small steps until the desired sinking rate is observed.

Scholze (Ref. 12) has given the following viscosity-temperature data (for the three fixed points listed above) for a specific glass; Jena thermometer Glass 16^{III}.

	Temp., K	η, Pa-s
(1) Glass Transition Point	828	10^{12}
(2) Softening Point	988	$10^{5.5}$
(3) Sinking Point	1246	$10^{3.22}$

Substituting these data in Eq. 14.3 gives three simultaneous equations, which allows the three constants in Eq. 14.3 to be evaluated. The results are $A = -2.866$, $B = 3830.3$, and $T_0 = 561.8$ K so that Eq. 14.3 becomes, for this glass,

$$\log \eta = -2.386 + 3830.3/(T - 561.8) \qquad \textbf{14.4}$$

The variation of log η, with temperature, T, computed with this equation is shown in Fig. 14.7. The positions of the glass transition and the softening and sinking points are also marked in Fig. 14.7. It is interesting that glass is normally mechanically worked between a viscosity of 10^5 and $10^{2.5}$ Pa-s, or roughly between the softening and sinking points. In addition, glass is annealed to remove internal stresses near the glass transition point.

As mentioned earlier, at the glass transition point, significant changes occur in the physical properties of a material cooled so as to yield a glass. For example, Fig. 14.6 shows that the coefficient of volume expansion (or linear expansion) decreases sharply below the glass transition point. Figure 14.8 shows that the specific heat (C_p) of a crystalline solid decreases smoothly and continuously

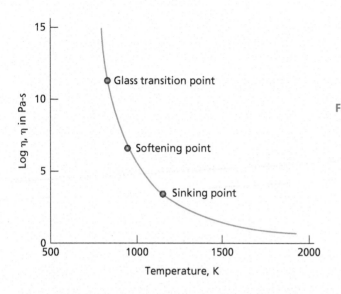

Fig. 14.7 The relation between the viscosity, η, and the temperature, T, in a specific glass (Jena Thermometer Glass). (Based on data in Sholze, H., *Glas*, p. 125, Springer Verlag, Berlin-Heidelberg-New York, 1979)

on cooling. The same substance, when cooled so as to give a glass, will have a larger specific heat as it becomes a supercooled liquid. Also, as the temperature decreases, the difference between the specific heats of the supercooled liquid and the crystalline solid increases until the glass transition point is reached. However, at the transition point, the C_p of the supercooled melt falls and becomes approximately equal to that of the crystalline solid. Below the glass transition point, the specific heats of glass and solid are approximately the same.

As suggested above, the changes that occur in the supercooled liquid as it becomes a glass are believed to be directly associated with the viscosity of the melt. Below the glass point, viscous flow of the melt becomes so slow that it begins to behave as if it were elastic, and the diffusive motion of atoms past each other becomes so slow that material becomes, as described by Turnbull (Ref. 14), *configurationally frozen*. That is, the supercooled liquid state existing at the glass transition becomes to all intents and purposes frozen in place at temperatures below T_g.

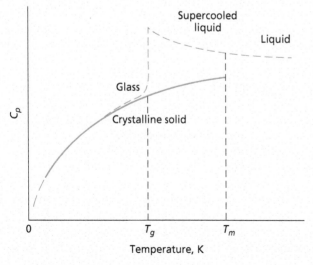

Fig. 14.8 A schematic drawing that shows the difference between the variation with temperature of the specific heat (C_p) of a glass and a crystalline solid

It is also true that the viscosity is closely related to the rate at which a shear stress can be relaxed. Accordingly, if a viscous flow produced by a shear stress is interrupted, the shear stress, τ, existing at this time will normally decrease at a rate proportional to the shear stress remaining or:

$$d\tau/dt = -\tau/t_r \qquad 14.5$$

where τ is the shear stress, t is the time, and t_r is a constant called the *relaxation time*. As originally proposed by Maxwell, t_r is closely related to the viscosity by the equation:

$$\eta = Gt_r \qquad 14.6$$

where G is the shear modulus.

The shear modulus of a typical glass is of the order of 2×10^{10} Pa-s. Assuming that this value does not change significantly for small temperature changes near T_g and that $\eta = 10^{12}$ Pa-s at T_g, then at T_g, t_r should be about 50 s or approximately 1 min. Thus, in terms of the data in Fig. 14.7, a shear stress can be effectively relaxed in about 1 min. at 550 K. Lowering the temperature to 500 K, according to Fig. 14.7, increases η to about 10^{16} Pa-s and the relaxation time to approximately 5×10^5 s or 140 hours or nearly 6 days. Thus, below the glass transition temperature the glass relaxes so slowly that it may be considered to be elastic for all practical purposes.

In summary, above T_g, a supercooled liquid that is being cooled at a fixed rate is able to maintain a state of internal equilibrium. Its free energy, however, is higher than that of the crystalline solid and it is metastable relative to the crystalline solid. Because atomic diffusion is so slow below T_g, it is no longer possible to maintain internal equilibrium in a supercooled liquid when it is cooled below T_g. However, it should be remembered that T_g is normally measured in glass samples subjected to a fixed rate of cooling and that the glass transition point normally depends on this cooling rate. The slower this cooling rate, the lower the temperature below which internal equilibrium can no longer be maintained; that is, the lower T_g. In practice, a standard rate of 4 K/s is commonly used so that T_g will have a fixed reference value.

As already suggested, the rate of cooling also has an important bearing on whether or not a given substance can be converted into a glass. Normally, a critical cooling rate exists below which a sufficient degree of crystallization will occur so that it cannot be considered a glass. Above this critical rate, of course, a glass is obtained. This critical cooling rate is a measure of the ease with which one can attain a glass structure on cooling. In a typical inorganic glass, crystallization occurs with difficulty and the critical cooling rate is very slow and of the order of 10^{-1} K/s.

Thus, in theory it should be possible to attain a glass structure in any metal provided that it can be cooled fast enough from the melt so that crystallization does not occur. Unfortunately, metal liquids tend to have very low viscosities and to crystallize very rapidly. According to Turnbull (Ref. 14) there have been no reports of the successful quenching of pure metal melts to glasses. On the other hand, metallic glasses have been successfully obtained with a number of metallic alloys. These alloys, for the most part, have had compositions lying near deep eutectic points, whose temperatures were very low. The reason for this is not difficult to understand. A schematic two-component phase diagram is shown in Fig. 14.9. It is important to remember that, to obtain a glass, crystallization should be prevented. The liquid must be cooled through the freezing range and past the glass transition temperature before a significant number of crystals have been able to form. Once the supercooled melt is below T_g, crystallization is in theory no longer possible. According to Turnbull (Ref. 14), T_g is not strongly dependent upon composition. With this in mind, compare the freezing of the two compositions 1 and 2 in Fig. 14.9. Composition 1 passes through the liquidus line at a high temperature designated as a in Fig. 14.9. At this temperature the atomic mobility is very high so that crystallization of the α phase should occur

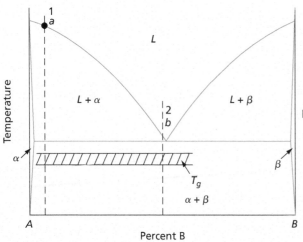

Fig. 14.9 A binary phase diagram with a single deep eutectic

easily in the melt. Furthermore, the freezing range is large since it extends roughly from a to T_g. This means that the melt has to be cooled over a very large temperature range in which it is possible for α crystals to form and grow. On the other hand, at Composition 2 the liquidus lies at a much lower temperature so that the thermal energy available for crystal growth is smaller and at the same time the temperature interval between the liquidus and T_g is short. Practically speaking, this means that a glass structure can be obtained with this latter composition using a slower rate of cooling. It should be emphasized, however, that the above-mentioned practical reasoning may not hold true for all cases. For example, consider the glass-forming ability of binary alloy $Cu_{100-x}Zr_x$ ($x = 34, 36, 38.2,$ and 40 at. %) series. It has been recently shown (Ref. 15) that the best glass former $Cu_{64}Zr_{36}$ corresponds neither to the largest supercooling region nor to the highest reduced glass transition temperature. Obviously, there are other factors (Ref. 16) that have a bearing on the ease of glass formation. Large differences in the sizes of the atoms of the components may tend to inhibit crystallization and make the formation of a glass easier. Another consideration is the nature of the atomic binding between the components.

The early works on production of metallic glasses had concentrated mainly on binary alloy compositions, using very fast cooling rates of 10^5 k/s or higher. For a comprehensive description of some of these works, see the 1967 Campbell Memorial Lecture of Pol Duwez (Ref. 17). Because of the high cooling rate requirement, these metallic glasses were generally in the form of powders, wires, and ribbons since the cooling rate during solidification is roughly proportional to the inverse square of sample thickness. As an example, Gilman (Ref. 18) has described the continuous casting of specific ferrous alloys to produce glass ribbons of great length at speeds ranging from 7.5 to 30 m/s. The thickness of these ribbons was between 25 and 130 μm and their width was between 0.08 and 50 mm. One set of alloys was based on the Fe–Be system, with other elements added to enhance either processing conditions or some of the physical aspects of the materials in the ribbons. These ribbons may be produced by projecting a narrow stream of liquid alloy onto a rotating cold copper wheel. Under the proper conditions, the jet spreads out as it contacts the wheel and solidifies into a continuous ribbon.

In the past three decades, it has been shown that bulk metallic glasses can be formed from a variety of multicomponent bulk glass-forming alloys at rather low cooling rates. For example, $Pd_{40}Cu_{30}Ni_{10}P_{20}$ requires a critical cooling rate of 0.10 k/s to form metallic glass (Ref. 19). As such, the alloy can be produced by normal casting techniques with thicknesses approaching 100 mm. Other bulk metallic galls alloys consist of Mg-, lanthanide-, Zr-, Ti-, Fe-, Co-, Pd-Cu, and Ni-based systems. A comprehensive review of the other bulk metallic glass produced can be found in a paper by Inoue (Ref. 20), who attributes the enhanced glass-forming ability of these alloys to the stabilization of the supercooled liquid. Inoue also suggests the following three empirical requirements to form bulk metallic glasses: (1) multicomponent alloy systems consisting of more than three elements, (2) atomic size ratio difference larger than 12 percent among the three main constituents, and (3) negative heat of mixing of the three main constituents.

It is significant that although metallic glasses are formed from supercooled liquids, as are the inorganic glasses, and that the former as well as the latter can be considered amorphous solids, the metallic glasses still possess many of the aspects of a true metal. Thus, they can have the high conductivity and characteristic metallic sheen of a metal. Therefore, it is difficult to visually differentiate between a metallic glass and a crystalline metal.

Metallic glasses exhibit many unique physical and mechanical properties, such as high strength coupled with high ductility, high corrosion resistance, and soft ferromagnetism in Fe-Ni-Co-based metallic glasses. These glasses are among the most easily magnetized of all the ferromagnetic materials, while at the same time possessing high saturation magnetizations. As far as their mechanical behavior is concerned, metallic glasses deform by homogeneous shear above their glass transition temperatures. In this type of flow, every atom or molecule responds to the applied shear stress and participates in the deformation process. Homogeneous shear flow is the common deformation mechanism in liquids, and in this respect metallic glasses above their transition temperatures behave like liquids (provided that they do not crystallize).

Below the glass transition temperature, however, the deformation of metallic glasses is highly inhomogeneous, and occurs as a result of the formation of highly localized shear bands. Each shear band is normally accompanied by an extensive local offset. For example, Masumato and Maddin (Ref. 21) found 20 nm thick shear bands in the alloy Pd_{80}-Si_{20}, with shear offsets of about 200 nm. Consequently, the formation of multiple shear bands can produce extensive or macroscopic ductility. Pampillo and Chen (Ref. 22) have shown that a 40 percent reduction in height could be obtained from shear bands upon compressing small cylinders of Pd_{75}-Cu_6-$Si_{16.5}$. The existence of shear bands in metallic glasses is considered to indicate the absence of work hardening in these materials. In general, in materials that undergo work hardening, shear bands may move laterally into undeformed regions only at low strains. On the other hand, in metallic glasses, the absence of work hardening allows continuing deformation along existing shear bands. Because of their lack of work hardening, the tensile strengths of metallic glasses are normally approximately equal to their yield strengths.

Another attractive property of metallic glasses is higher corrosion resistance than observed in polycrystalline materials. A factor contributing heavily to this is the superior chemical homogeneity of metallic glasses. Another consideration is their general lack of grain boundaries and defects such as dislocations. For these same reasons, metallic glasses are more resistant to pitting attack in acidic solutions (Refs. 19 and 23). For example, the pitting resistance of Fe_{70}-P_{13}-C_7 in sulfuric acid solutions containing chloride ions is considerably better than that of 18Cr-8Ni stainless steel.

14.4 Atomic Movement at S/L Interface

It has been shown (Ref. 24) that the movement of the boundary separating a liquid from a solid crystalline phase, under a temperature gradient normal to the boundary, may be considered the resultant of two different atomic movements. Thus, at the interface, those atoms that leave the liquid and join the solid determine a rate of attachment, while those that travel in the opposite direction determine a rate of detachment. Whether the boundary moves so as to increase or

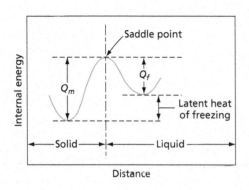

Fig. 14.10 The relationship between the activation energies for attachment and detachment

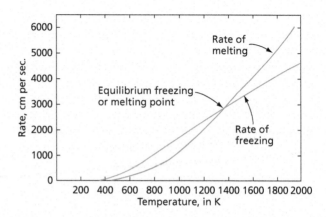

Fig. 14.11 Rates of freezing and melting for copper. (Chalmers B., *Trans. AIME*, **200** 519 [1954])

decrease the amount of solid depends on whether or not the rate of attachment is larger than the rate of detachment. This point of view is equivalent to viewing the movement of the interface as a two-directional diffusion problem in which

$$R_f = R_{f_0} e^{-Q_f/RT}$$
$$R_m = R_{m_0} e^{-Q_m/RT}$$

14.7

where R_f and R_m are the rates of attachment and detachment respectively, R_{f_0} and R_{m_0} are approximate constants, Q_f and Q_m are activation energies, and R and T have their usual significance.

The activation energy Q_f represents the energy to take an atom that lies on the liquid side of the boundary to the saddle point, as shown in Fig. 14.10. Similarly, Q_m represents the energy required to bring an atom on the solid side of the interface up to the saddle point. In Fig. 14.10, the potential energy wells shown on each side of the saddle point differ by the latent heat of fusion. An atom in the solid possesses a lower energy than one in the liquid, but it should be noted that it also possesses a lower entropy.

The rates of attachment, R_f, and detachment, R_m, may be expressed either as atoms per second crossing the boundary or as velocities of the boundary in centimeters per second. The constants R_{f_0} and R_{m_0} depend on a number of factors that can be estimated. This has been done for copper by Chalmers, as may be seen in Fig. 14.11, which shows the curves for the freezing rate and the melting rate as functions of temperature. Notice that they intersect at the equilibrium melting point of copper. Another interesting feature of Fig. 14.11 is that on either side of the true melting point the curves diverge sharply. The difference in the individual rates determines the actual rate with which the interface moves, which signifies that the observed growth rate, or melting rate, increases as the temperature deviates from the equilibrium freezing point.

14.5 The Heats of Fusion and Vaporization

The heat of vaporization can be considered a measure of the energy to remove an atom from a solid surface and place it in the vapor. In other words, this is the energy required to break the bonds binding the atom to its neighbors at the surface. Atoms leaving the surface to enter the vapor are generally believed to do so at jogs in ledges on the surface. Such a ledge with jog is shown in Fig. 14.12, where the shaded circles represent the atoms of an incomplete plane. The atom, marked A, in the uppermost row of the incomplete plane lying next to the jog is in nearest neighbor contact with three other atoms in the incomplete plane as well as with three atoms in the complete plane below it. The total number of atomic bonds that are broken when this atom leaves the surface is therefore six. Furthermore, when an atom is removed from a jog, the atom behind it assumes an equivalent position so that the removal or addition of atoms to a jog is a repeatable process. Another possibility is the loss or addition of a single atom sitting on a close-packed plane, such as the atom B in Fig. 14.12. These are bound to the surface by only three bonds so that their vaporization requires much less energy than the removal of an atom from a jog. However, such an event is not repeatable as is the loss of an atom from a jog. A possibility does exist, nevertheless, that an atom at an A position may leave the jog and diffuse to a B position, and then leave the surface. This dual event still requires breaking six bonds; three when it leaves the jog and three when it leaves the surface by evaporation. This sequence is therefore, equivalent to the direct removal of an atom from a jog. Consequently, the heat of vaporization of a single atom is generally assumed to be equivalent to breaking six atomic bonds.

The relations involved in the evaporation of atoms from a surface can be reversed when one considers growth from the vapor. An individual atom may arrive at a position such as B in Fig. 14.12, with the formation of three solid bonds, and then move by surface diffusion to a jog where three more bonds are created.

Solidification to a faceted interface from a melt is normally treated as analogous to growth from the vapor; that is, by the addition of atoms from the liquid to jogs or by the deposition of atoms on close-packed surfaces followed by surface diffusion to jogs. However, a noticeable difference exists in the size of the bonds formed when an atom leaves the melt to join a solid, compared to those formed when it leaves the vapor to join a solid. The size of the latter bonds is naturally much smaller, in agreement with the large difference between the heats of fusion and vaporization. In a pure metal, the heat of fusion is normally about 1/20th the heat of vaporization. Thus, in pure mercury at 234 K the heat of fusion is 11.8 J/gr-mol, while the heat of vaporization is 309 J/gr-mol.

The breaking of the bonds associated with removing an atom from the surface is reasonably easy to model for vaporization. Here an atom is removed from intimate contact with its neighbors and placed into a gas where its interaction energy with other gaseous atoms may be considered negligible. On the other hand, when an atom leaves the surface to enter the melt, the picture is not well defined. Atomic binding in the liquid is not well understood. It is, moreover, only slightly less than that in the solid as implied by the small size of the heat of fusion. In spite of this difficulty, the freezing or melting process has been reasonably well modeled using concepts developed to explain the growth or loss of a solid in contact with its vapor. Thus, a metal is assumed to solidify primarily by the movement of jogs along ledges. As an atom attaches itself to a solid at a jog, six bonds are assumed to be formed with each bond representing one-sixth of the heat of fusion.

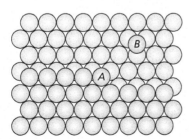

Fig. 14.12 Atoms leaving the surface of a crystal to join the vapor phase are normally assumed to do so at a jog in a surface ledge, as shown at atom (A)

The major difference in the solidification of different materials will, however, depend on the population of the jogs and the nature of the liquid-solid interface.

14.6 The Nature of the Liquid-Solid Interface

It is now realized that the interface between the liquid and the solid can vary considerably in structure depending on the nature of the material solidifying and the amount of the supercooling at the interface. However, to reduce the complexity of the problem one usually considers only the two extreme cases. These are (1) the diffuse and (2) the atomically smooth interfaces. In the diffuse interface, the change from the liquid to solid is assumed to occur over a number of atomic layers in which the liquid structure changes gradually to that of the solid (Ref. 25). In other words, the ordering of atoms from the non-crystalline liquid state to the crystalline solid state is gradual. This type of interface is represented very roughly in Fig. 14.13A, where the solid lies at the left of the figure and liquid at its right. This two-dimensional cut through the interface indicates the

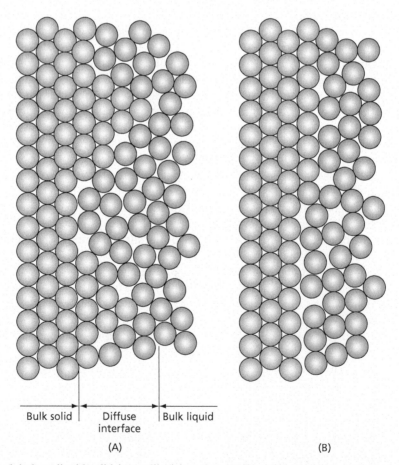

Fig. 14.13   Two models for a liquid-solid (crystalline) interface. **(A)** The diffuse interface. **(B)** The atomically smooth interface

presence of numerous closely spaced ledges. In three dimensions one can also assume that many closely spaced jogs occur on these ledges. The significant feature of this model is that it implies that this interface has a very high accommodation factor for the liquid atoms. Thus, growth occurs by the addition of atoms continuously at every atomic site and the interface advances normal to itself. This type of growth mechanism is called *continuous* or *normal growth*. A two-dimensional cut of an atomically flat interface is shown in Fig. 14.13B. While this interface is basically closed-packed, it is also assumed to have a few ledges with jogs. These ledges spread parallel to the interface, that is, laterally, upon the attachment of atoms to them. As will be described later, the interface remains stationary except during the passage of the ledges.

We shall now return to a consideration of the constants R_{f_0} and R_{m_0}, which appear in the rate equations (Eqs. 14.7). One of the factors that determine the magnitude of these constants is known as the *accommodation factor*, the chance that an atom in a given phase, either the liquid or the solid, can find a position on the other side of the boundary where it can attach itself. For the movement of atoms from the solid to the liquid, the accommodation factor should be more or less independent of the chemical nature of the atoms composing the liquid. This follows from the fact that the liquid phases of metals have very similar structures. On the other hand, different crystalline structures present entirely different types of surfaces toward the liquid phase, so that the accommodation factor for the motion of atoms from the liquid toward the solid varies with the nature of the solid. It is also true that the motion of atoms from the liquid to the solid depends on the indices of the particular crystalline plane that faces the liquid. The less close-packed the plane, the easier it is for liquid atoms to attach themselves to the crystal. This fact may be explained with the aid of Fig. 14.14, which represents a face-centered cubic structure, which shows the atomic arrangements on {100} and {111} surfaces. The holes, or pockets, in the surface available for the accommodation of a liquid atom as it joins the crystal are larger for the less close-packed {100} plane than for the more closely packed {111} plane. As a result of this difference, for a given amount of supercooling there is a difference in the growth velocity of the two crystallographic planes: the least closely packed plane grows with the greater velocity.

The fact that low-atomic-density planes grow faster does not mean that a growing crystal assumes faces that are planes of this type. On the contrary, the tendency is for the crystal to assume faces that are close-packed, or slow-growing. The reason for this is easy to understand, for the low-density planes of fast growth tend to grow themselves out of existence, leaving behind only close-packed surfaces. This effect is shown schematically in Fig. 14.15(A), where several stages in the growth of the faces of a crystal are shown. Figure 14.15 (B) shows a photograph of an un-cut yellow diamond grown under high pressure-high temperature conditions, using a split-sphere apparatus in Reza Abbaschain's laboratory (Refs. 26–28). The yellow color is due to ppm level nitrogen substitutional impurity.

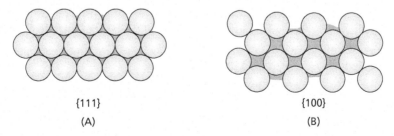

Fig. 14.14 Planes of looser packing, such as {100}, are better able to accommodate an atom that leaves the liquid to join the solid than a closer packed plane, such as {111}. Illustrated planes correspond to a face-centered cubic lattice. (After Chalmers, B., *Trans. AIME*, **200** 519 [1954])

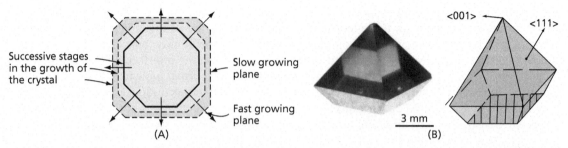

Fig. 14.15 **(A)** A crystal growing in a liquid tends to develop faces that are slow-growing (close-packed). **(B)** Faceted faces on an un-cut laboratory grown yellow diamond. (Photographed by Steven Herrera, used by permission)

It has already been indicated that the rate with which the interface between liquid and solid moves during solidification is a function of the amount of supercooling. In addition, the interface motion is also affected by the sign of the temperature gradient in front of the interface. Thus, the crystal growth process, when the temperature gradient in advance of the interface is a rising one, is different from that when the gradient is a falling one. These two cases will be considered later.

14.7 Continuous Growth

An ideal diffuse liquid-solid interface is predicted to undergo continuous growth at a rate varying linearly with the amount of undercooling at the interface. This may be rationalized as follows:

The jump rate of the atoms in a pure metal liquid, v_l, is related to the liquid diffusion coefficient by the expression:

$$v_l = 6D_l/\lambda^2 \qquad 14.8$$

where D_l is the liquid diffusion coefficient and λ is the mean jump distance. It is now assumed that D_l can be expressed by:

$$D_l = \lambda^2 v_0 \exp[-\Delta G_l/RT] \qquad 14.9$$

where v_0 is the atomic vibration frequency. Thus,

$$v_l = 6v_0 \exp[-\Delta G_l/RT] \qquad 14.10$$

and ΔG_l is the free energy of activation for liquid diffusion. A relation can now be written for the rate at which atoms jump from the liquid to the solid

$$v_{ls} = v_l/6 = v_0 \exp[-\Delta G_l/RT] \qquad 14.11$$

In deriving Eq. 14.11, it is assumed that the atomic migration and diffusion at or near the interface is the same as that in the bulk liquid. This may not be true for a diffuse interface for which the properties are expected to change gradually from that of the bulk liquid to that of the bulk solid. In other words, within the interfacial region, the portion near the solid is more like a solid while the portion near the liquid is more like a liquid. Therefore, the diffusivity is expected to vary and depend on the location within the interface, and particularily to be different from D_1.

The corresponding atomic jump rate from solid to liquid, v_{sl}, involves the breaking of the atomic bonds holding the atoms to the solid surface. In other words, the free energy of melting, ΔG_m, must be added to ΔG_l. Accordingly,

$$v_{sl} = v_0 \exp[-(\Delta G_l + \Delta G_m)/RT] \qquad 14.12$$

The net rate of interchange of atoms between solid and liquid is therefore

$$v_{net} = v_{ls} - v_{sl} = v_0 \exp[-\Delta G_l/RT] \cdot \{1 - \exp[-\Delta G_m/RT]\} \qquad 14.13$$

Near the equilibrium freezing point, ΔG_m is small compared to RT so that $\Delta G_m/RT$ is small compared to unity and Eq. 14.13 can be reduced, with the aid of Eq. 14.8, to

$$v_{net} = v_0[\exp(-\Delta G_l/RT)] \cdot (\Delta G_m/RT) = D_l \Delta G_m/\lambda^2 RT \qquad 14.14$$

Now consider ΔG_m. At T_m, the equilibrium freezing temperature, $\Delta G_m = 0$ and

$$\Delta G_m = \Delta H - T_m \Delta S = 0 \qquad 14.15$$

where ΔH and ΔS are the enthalpy and entropy of freezing. However, if under-cooling should occur:

$$\Delta G_m = \Delta H - T \Delta S \neq 0 \qquad 14.16$$

where T is the temperature at which freezing occurs. Now letting $T = T_m - \Delta T$, we have

$$\Delta G_m = \Delta H - T_m \Delta S + \Delta T \Delta S \qquad 14.17$$

Since most bulk metals tend to freeze at temperatures very close to T_m, and ΔH and ΔS are not strongly temperature dependent, we may assume that the first two terms on the right-hand side of Eq. 14.17 cancel (see Eq. 14.15), and remembering that $\Delta S = \Delta H/T_m$, one obtains

$$\Delta G_m = \Delta H \Delta T/T_m \qquad 14.18$$

Substituting this last relation into Eq. 14.14 gives

$$v_{net} = D_l \Delta H \Delta T/(\lambda^2 R T_m) \qquad 14.19$$

If it is now assumed that λ can be equated to the lattice parameter of the solid, a, and that the rate of advance of the interface, V, equals a times v_{net} then:

$$V = a v_{net} = \beta D_l \Delta H_m \Delta T/(a R T_{m2}) \qquad 14.20$$

where V is the growth rate, ΔH_m the freezing enthalpy, D_l the liquid diffusion coefficient at T_m, T_m the equilibrium freezing point temperature, a the lattice parameter of the solid, ΔT the amount of undercooling, and the coefficient β is introduced to correct for the various assumptions made in deriving the rate equation (Ref. 29). Equation 14.20 may now be simplified to read:

$$V = B \Delta T \qquad 14.21$$

which shows that the solidification rate of a diffuse interface is a linear function of the interface supercooling.

14.8 Lateral Growth

The diffuse interface represented by Fig. 14.13A is replete with many ledges containing jogs at which an atom moving from the liquid may attach itself to the solid and attain the six bonds of the heat of fusion. Such a surface has a very high accommodation factor and growth occurs readily. This is not the case when the surface may be described as atomically flat (faceted), as is the case with many metals. Examples include Bi, Ga, Si, and intermetallic compounds such as Nb_5Si_3 and $NbSi_2$.

Atomically flat interfaces are believed to grow primarily by the movement of ledges over these surfaces. This type of growth is known as *lateral growth*. The existence of ledges on this type of interface can be explained in several ways. One explanation involves surface nucleation and growth of two-dimensional discs as indicated in Figs. 14.16A and 14.16B. Another explanation is related to the surface discontinuity caused by the intersection of a twin boundary with the surface. However, of more importance is the fact that when a screw dislocation emerges at the surface of a crystal, a ledge is automatically produced. This may be seen in the simple screw dislocation models of Figs. 4.13, 4.14, and 14.16C. As indicated in Fig. 14.17, these ledges are capable of moving across a crystal surface as atoms join or leave the solid at jogs lying along the ledges. Since one end of the ledge in Fig. 14.17 is anchored at a screw dislocation, the movement of the ledge across the surface tends to make the ledge take the shape of a spiral whose center is at the point of emergence of the dislocation line, as shown in Fig. 14.16C. As the spiral ledge makes a complete rotation about the screw dislocation, an extra plane is either added or removed from the crystal depending upon whether the crystal is growing or decreasing in size. Support for believing that growth spirals

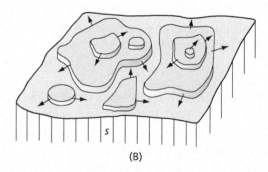
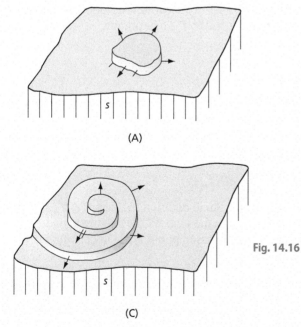

Fig. 14.16 Schematic drawings showing the interfacial processes for the lateral growth mechanisms. **(A)** Mononuclear. **(B)** Polynuclear. **(C)** Spiral growth. (Note the negative curvature of the clusters and/or islands is just a drawing artifact.)

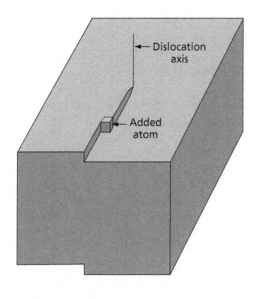

Fig. 14.17 A crystal may grow by the addition of atoms to a jog on the ledge that forms when a screw dislocation intersects a crystal surface

associated with screw dislocations are important in crystal growth comes from the fact that most crystals in bulk are known to contain significant numbers of growth dislocations.

Lateral growth by screw dislocation ledges has been treated theoretically and is predicted to be controlled by an equation of the form

$$V = B_2(\Delta T)^2 \qquad 14.22$$

where B_2 is a constant and ΔT is the amount of undercooling.

For lateral growth by two-dimensional nucleation, the growth rate is an exponential function of the reciprocal of the supercooling (Refs. 30 and 31). The rate equation depends on whether a single nucleus forms per layer (mononuclear) or many nuclei form simultaneously. The two respective equations are

$$V = B_3 A \exp(-B_4/\Delta T)$$
$$V = B_5 \exp(-B_4/3\Delta T) \qquad 14.23$$

where B_3, B_4, and B_5 are constants, and A is the surface area.

14.9 Stable Interface Freezing

The foregoing discussions dealt with the atomistic mechanisms of growth and the nature of the interface on an atomic scale. Another factor that has a strong effect on the shape of the interface on a macroscopic scale is the removal of the heat of fusion from the interface. Let it be assumed that the temperature gradient rises as we move from an interface into a pure liquid, and that the temperature gradient is linear and perpendicular to the interface. Under these conditions, the interface is believed to maintain a stable planar shape and move forward as a unit. If by chance a close-packed plane (one of slow growth) lies perpendicular to the heat-flow direction, a strictly planar interface should, in theory, develop. However, the chance of finding a close-packed plane in exactly the right position is very small, and one has to consider those cases where a crystallographic plane of high-atomic density is almost parallel to the interface, as well as those situations where the interface does not approximate any high-density plane.

Consider the first of these possibilities. Here the interface tends to develop a series of close-packed planar steps. A schematic example of such a surface is shown in Fig. 14.18. Since each of the facets has an inclination in the heat-flow direction, the temperature cannot be uniform over its entire area. Because of the rising temperature in advance of the interface, those portions of each facet that are most advanced will be in contact with hotter liquid than those portions to the rear. Since, for a given crystallographic plane, the rate of growth is a function of the degree of supercooling, it is not possible for the facets to maintain a strictly crystallographic surface and grow with a constant velocity. As a consequence, the steps assume a curved shape. The most advanced, or hotter, portion of each facet corresponds to a lower indice, or higher accommodation factor surface, while the most retarded, or cooler, portion corresponds to a slow-growing, or lower, accommodation factor surface. In this manner, it should be possible to obtain an interface that grows at a constant speed.

The structure described in the preceding paragraph is that which attains when close-packed planes are nearly parallel to the interface. When there are no close-packed planes approximately parallel to the interface, a simpler interface, consisting of a random (crystallographic) planar surface, may be more stable. Such an interface is similar to one obtained when a close-packed crystallographic plane is strictly perpendicular to the heat flow direction, but differs in that it is not a close-packed plane, but a plane with high, or irrational, indices.

Figure 14.19 shows shape and morphology of primary graphite flakes grown out of a hypereutectic Ni-C melt containing 3 wt.% C. The alloy was melted using electromagnetic levitation and superheated to 1800°C and then cooled either within the levitation coil by purified Ar/He gas mixture. The Ni-C phase diagram contains a simple eutectic reaction, similar to the phase diagram shown in Fig 11.13, except that the solubility of carbon in nickel is very limited, less than 1 wt.%, and there is no solubility of Ni in carbon. The eutectic temperature is at 1326°C and its eutectic composition is around 1.9 wt.% C. As such, for this hypereutectic alloy, the first solid to form is pure graphite. The deeply etched graphite flake of Fig. 14.19A shows evidence of macrosteps and ledges formed on the flat flake surface, which is parallel to the basal {0001} plane of graphite with hexagonal crystal structure. The flakes are surrounded by the prismatic planes which grow faster

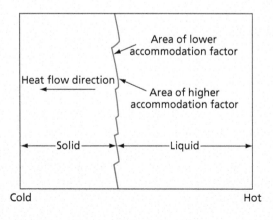

Fig. 14.18 A stable interface when a close-packed lattice plane is almost parallel to the interface

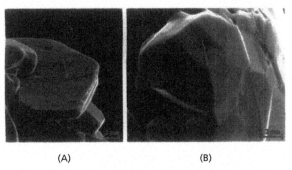

Fig. 14.19 Growth of graphite basal plane by **(A)** 2-D nucleation and migration of ledges and **(B)** spiral growth. The arrows show the evidence of ledge-assisted and spiral growth on the graphite interfaces. (From Amini, S., Kalaantari, H., Mojgani, S. and Abbaschian, R., "Graphite crystals grown within electromagnetically levitated metallic droplets," *Acta Materialia*, 2012, Vol. 60, 7123–7131. Reprinted with permission of Elsevier)

than the basal plane at the cooling rates imposed during solidification. Fig. 14.19B shows evidence of spiral growth assisted by a dislocation. The basal planes grew either by 2-D nucleation and lateral step growth mechanism schematically shown in Figs. 14.16A and B, or by dislocation-assisted growth depicted in Fig. 14.16C. Note that with increasing the cooling rate, the growth rate of basal planes increases exponentially and approaches those of the prismatic planes, leading to the formation of spherical shaped graphite (Refs. 32 and 33).

14.10 Dendritic Growth in Pure Metals

A very important type of crystalline growth occurs when the liquid-solid interface moves into a supercooled liquid whose temperature falls, or decreases, in advance of the interface. One of the most important ways that a temperature gradient of this nature can form is as follows.

Suppose that Fig. 14.20 represents a region containing a liquid-solid interface and that the heat is flowing away from the interface in both directions; that is, the heat is being removed through both the solid and the supercooled liquid. Because of the heat of fusion released at the interface, the temperature of the interface usually rises above that of both the liquid and the solid. Under these conditions, the temperature drops as one moves from the interface into the solid, because this is the heat-flow direction. It also falls off into the liquid because there is a natural flow of heat from the interface into the supercooled liquid. The resulting temperature contour is shown in Fig. 14.20 and is known in general as a *temperature inversion* (Ref. 34).

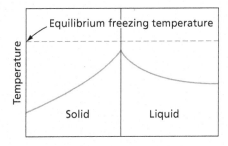

Fig. 14.20 Temperature inversion during freezing. (After Chalmers, B., *Trans. AIME*, **200** 519 [1954])

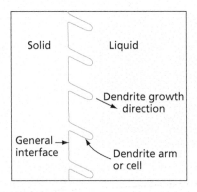

Fig. 14.21 Schematic representation of the first stage of dendritic growth. A temperature inversion is assumed to exist at the interface; that is, the temperature in the liquid drops in advance of the interface

When the temperature falls in the liquid in advance of the interface, the latter becomes unstable and, in the presence of any small perturbation, cells may grow out from the general interface into the liquid. The resulting structure may also become quite complicated, with secondary branches forming on the primary cell, and possibly with tertiary branches forming on the secondary ones. The resulting branched crystal often has the appearance of a miniature pine tree and is, accordingly, called a *dendrite* after the Greek word *dendrites* meaning "of a tree."

The reasons for the branched growth of a crystal into a liquid whose temperature falls in advance of the solid are not hard to understand. Whenever a small section of the interface (perturbation) finds itself ahead of the surrounding surface, it will be in contact with liquid metal at a lower temperature. Its growth velocity will be increased relative to the surrounding surface, which is in contact with liquid at a higher temperature, and the formation of a cell is only to be expected. Associated with the development of each cell is the release of a quantity of heat (latent heat of fusion). This heat raises the temperature of the liquid adjacent to any given cell and retards the formation of other similar projections on the general interface in the immediate vicinity of a given projection. The net result is that a number of cells of almost equal spacing are formed that grow parallel to each other in the fashion shown in Fig. 14.21. The direction in which these cells grow is crystallographic and is known as the *dendritic growth direction*. The direction of dendritic growth depends on the crystal structure of a metal, as may be seen in Table 14.2.

The branches, or cells, shown in Fig. 14.21 are first order, or primary, in nature. How secondary branches may form from primary ones will now be considered. For this purpose consider Fig. 14.22, where section *aa* represents the general interface. Notice that in this figure the direction

Table 14.2 Dendritic Growth Directions in Various Crystal Structures*

Crystal Structure	Dendritic Growth Direction
Face-centered cubic	$\langle 100 \rangle$
Body-centered cubic	$\langle 100 \rangle$
Hexagonal close-packed	$\langle 10\bar{1}0 \rangle$
Body-centered tetragonal (tin)	$\langle 110 \rangle$

*Chalmers, B., *Trans. AIME*, **200** 519 (1954).

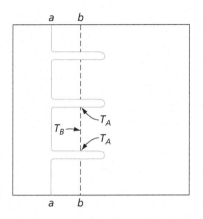

Fig. 14.22 Secondary dendrite arms form because there is a falling temperature gradient starting at a point close to a primary arm and moving to a point midway between the primary arms. Thus, $T_B < T_A$

of dendritic growth is assumed to be normal to the general interface. This is done to simplify the presentation. Once the cells have formed, growth at the general interface will be slow because here the supercooling is small and the latent heat of fusion associated with the formation of the cells tends to further decrease its magnitude. At section bb, on the other hand, the average temperature of the liquid is, by definition, lower than at aa. However, even on this section at points in the liquid close to the cell walls, the temperature will be higher than midway between the cells ($T_A > T_B$) because of the latent heat released at the cells. There is, therefore, a decreasing temperature gradient not only in front of the primary cells, but also in directions perpendicular to the primary branches. This temperature gradient is responsible for the formation of secondary branches, which form at more or less regular intervals along the primary branches, as shown in Fig. 14.23, which shows a 3-D image of Ni-2 wt%Al dendrites, exposed in a shrinkage porosity, as described in Section 14.19. Initiation of some secondary arms on the primary arms can be clearly seen. Tertiary arms are also visible on the upper left corner. Since the secondary branches form for the same basic reasons as the primary branches, their directions of rapid growth are along directions equivalent crystallographically to those taken by primary arms. In the case of cubic metals, dendrite arms may form along all of the $\langle 100 \rangle$ directions and the arms are perpendicular to each other. Similarly, tertiary branches will form from the secondary branches if the space is available for their growth.

Fig. 14.23 In a cubic crystal, the dendrite arms form along $\langle 100 \rangle$ directions. Primary and secondary arms are thus normal to each other. (Photograph courtesy of Shaahin Amini, 19 Covington, Mission Viejo, CA 92692)

Fig. 14.24 Dendrite tip and side branching in (left) succinonitrile (SCN), a body-centered cubic crystal; (right) pivalic anhydride (PVA), a face-centered cubic crystal. (Reprinted from *Journal of Crystal Growth* 264, M.E. Glicksman and A.O. Lupulescu, Dendritic crystal growth in pure materials, 541–549, Fig. 1, Copyright 2004, with permission from Elsevier. https://www.sciencedirect.com/science/article/abs/pii/S0022024803022449)

The shape of the dendrite tip and formation of secondary branches during growth of transparent materials are shown in Fig. 14.24. The materials, succinonitrile and pivalic anhydride were grown under microgravity conditions by M. E. Glicksman and his colleagues. For additional information, consult References 35–37. The dendrite arms usually grow in thickness as they grow in length, with the eventual result that the arms grow together to form a single nearly homogeneous crystal.

Dendritic growth, as outlined above, will occur in the freezing of pure metals when the interface is allowed to move forward into sufficiently supercooled liquid. In metals of relatively low purity, however, it is almost impossible to obtain enough thermal supercooling, so the entire freezing process is not dendritic unless heat is constantly removed from the liquid. In the absence of efficient heat removal from the liquid, the temperature rises very rapidly to its melting point because of the heat release at the interface. This rapid temperature rise, which follows nucleation of a supercooled liquid, is called *recalescence*.

In the absence of cooling from outside, a very large supercooling (of the order of 100 K) is required in pure metals for complete dendritic freezing. The amount of solid that forms from a supercooled liquid without heat removal by the surroundings can be estimated from:

$$f_s = C_p \cdot \Delta T / \Delta H_m \qquad 14.24$$

where f_s is the fraction solidified, C_p is the average specific heat of the solid and liquid, ΔH_m is the heat of fusion, and ΔT is the supercooling.

14.11 Freezing in Alloys with Planar Interface

Several examples of freezing in simple alloy systems will now be discussed. Consider an isomorphous-alloy system with the phase diagram shown in Fig. 14.25A. In this system at the temperature T_1, solid of composition (a) is in equilibrium with liquid of composition (b). Let it now be assumed that Fig. 14.25B represents a volume of liquid of composition (b) placed in a long tubular mold and that heat is only removed from the left end of the mold so that the heat flow is linear and from right to left. Under these conditions, freezing will start at the left end of the liquid and the small volume element (dx) may be taken to represent the first solid to form. For simplicity, let's assume that diffusion in the solid is negligible and no convective mixing takes place in the liquid. The freezing of this volume element will take place at temperature T_1 and the composition of the solid will correspond to point (a) in Fig. 14.25A. If this layer of solid is frozen in a relatively short period of time with negligible solid-state diffusion, we can assume that it is formed from the liquid layer adjacent to the interface and not from the entire volume of the liquid. Since the solid contains a higher ratio of A to B atoms than does the liquid layer from which it forms, the latter is depleted in A atoms

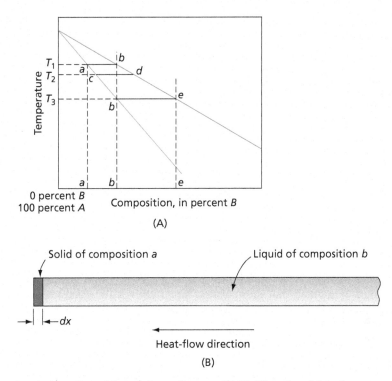

Fig. 14.25 **(A)** Part of an isomorphous binary phase diagram. **(B)** Unidirectional freezing

and enriched in B atoms. In other words, solute atoms B are rejected into the liquid as the interface moves, and forms a solute build-up layer. This composition change in the liquid just ahead of the interface increases as more solid is formed, but the increase eventually stops when a steady-state condition is attained. Concurrent with the change in composition of the liquid next to the interface is a similar change in the composition of the solid that can form from this layer of liquid. Thus, if the composition of the liquid corresponds to point d in Fig. 14.25A, then the solid that freezes must have the composition of point c. It is generally assumed that solid and liquid across the interface are at *local equilibrium*. It is also important to recognize that the indicated freezing process can only occur if the temperature at the interface is lowered from T_1 to T_2. The concentration of an excess of B atoms in the liquid adjacent to the interface reaches a maximum when the liquid next to the solid attains the composition of point e. At this instant the liquid is able to freeze solid of composition b. When this occurs, the steady state has been attained and the solid, which is formed from the liquid layer enriched in B atoms, is the same as the liquid drawn into this layer. At this time, the temperature at the interface must be T_3, as shown in Fig. 14.25A. The composition as a function of distance along the mold is shown in Fig. 14.26. This figure corresponds to the time when the steady-state freezing process has just been attained. A similar illustration is found in Fig. 14.26B, but pertains to a later time when the interface has moved farther to the right. Notice that, in both cases, the composition of the solid rises from its original value a to that of the original liquid b. At the interface there is also a sudden rise in composition to the value e as one passes from solid into liquid. Following this sudden rise, the composition in the liquid decreases exponentially back to the value of the original liquid, which is b. The shape of the composition-distance curve in this exponential region is a function

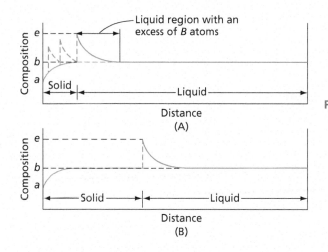

Fig. 14.26 Composition-distance curves corresponding to two different stages in the one-dimensional freezing problem of Fig. 14.25B

of the rate of freezing and the atomic diffusion rates in the liquid. It is important to note that in the above discussion it is implied that there are no liquid convection currents. In the presence of such currents it is not possible to attain as large a concentration of B atoms in advance of the interface. There are other factors that may also limit the magnitude of the composition change in the liquid at the interface. However, for our purposes we can assume that the distance-composition curves shown in Fig. 14.26 are representative of what can actually happen under similar conditions during the freezing of an alloy. The equations describing the composition-distance curves in the solid and liquid can be found elsewhere (Ref. 38).

14.12 The Scheil Equation

The one-dimensional freezing problem has also been considered by several authors (Refs. 39 through 41) for freezing under normal conditions. From this has come a significant relationship, the Scheil equation or the nonequilibrium lever rule. Several important assumptions are made in its derivation. First, diffusion of the solute does not occur in the solid; second, in the liquid the diffusion of the solute is rapid and complete; and third, the solid and liquid across the interface are in local equilibrium. The first assumption requires that the composition of the freezing solid vary continuously as the freezing front advances and that the solid retain this compositional variation after the front has passed. On the other hand, because of complete diffusion in the liquid, the liquid composition is always uniform, although this uniform composition changes as freezing continues. The physical bases for these assumptions about the diffusion in the liquid and the solid come from the fact that the diffusion rate is normally several orders of magnitude greater in the liquid than in the solid and the mass transport in the liquid is further aided by convective mixing. The Scheil equation is normally derived using an idealized eutectic phase diagram, such as that in Fig. 14.27, where the liquidus and solidus are assumed to be straight lines. At any temperature, T_i, the linear liquidus and solidus lines give the solid and liquid compositions. These compositions are assumed to be in equilibrium at the freezing front, linearly related to each other by the equation $C_s = kC_l$, where k is the equilibrium redistribution coefficient and C_s and C_l are the solute concentrations by weight in the solid and liquid, respectively. Finally, it is assumed that any difference between the molar volumes of the liquid and solid can be ignored.

14.12 The Scheil Equation

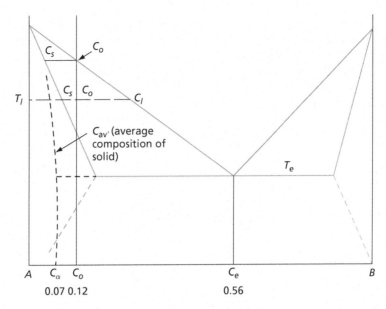

Fig. 14.27 Idealized eutectic phase diagram with the liquidus and solidus drawn as straight lines. The average composition of the solid, using the Scheil equation, is shown to the left of the solidus line

Now let f_s represent the fraction by weight of the solid in the alloy. The corresponding liquid fraction is $(1 - f_s)$. When a small increment of the solid, df_s, freezes, a quantity of the solute equal to $C_s df_s$ is transferred from the liquid to the solid. However, since the concentration of the solute in the solid is less than that in the liquid, the liquid will have its solute concentration increased by an amount:

$$dC_l = (C_l - C_s)df_s/(1 - f_s) \qquad 14.25$$

Using the relation $C_s/C_l = k$ from the phase diagram, this equation reduces to:

$$dC_l/C_l = (1 - k) \cdot df_s/(1 - f_s) \qquad 14.26$$

Equation 14.26 is now integrated using the boundary conditions existing at the initiation of freezing; namely $f_s = 0$ and $C_l = C_o$. This yields:

$$C_l = C_o(1 - f_s)^{(k-1)} \qquad 14.27$$

or in terms of the solute concentration in the solid

$$C_s = kC_o(1 - f_s)^{(k-1)} \qquad 14.28$$

Equation 14.28 gives C_s, the concentration of the solid that freezes at the interface. This concentration increases as the interface advances along the mold. The average composition of the solid also changes as the freezing progresses. A schematic curve showing the variation of this average composition is given by the curved dashed line to the left of the solidus line in Fig. 14.27. Note that in this non-equilibrium freezing process, solidification continues to a much lower temperature than it does in equilibrium freezing, which is at the solidus temperature (see Fig. 14.27).

Normally the freezing continues down to the eutectic temperature. When the eutectic temperature is reached, the remaining liquid will have the eutectic composition, and it freezes to form eutectic solid. Thus, the solid resulting from this non-equilibrium freezing normally will contain eutectic in its microstructure. The variation of the solid composition, C_s, with fraction solidified, f_s, in this non-equilibrium freezing is shown in Fig. 14.28. The amount of eutectic can be calculated using Eq. 14.27 and letting C_l equal C_e, where C_e is the composition of the eutectic. It should be noted that as dendrites move with velocity R, upward as depicted in the schematic of Fig. 14.28C, they also thicken sidewise behind the tip. The thickening arms eventually reach the neighboring arms and freeze the remaining liquid in between.

It should be noted that in most solidification processes, the extent of the segregation is usually less than that calculated using the Scheil equation. This is because of solid-state diffusion and the depression of the temperature at the dendrite tip (Ref. 42). Moreover, in practice, the actual redistribution coefficient deviates from the equilibrium value calculated from the phase diagram

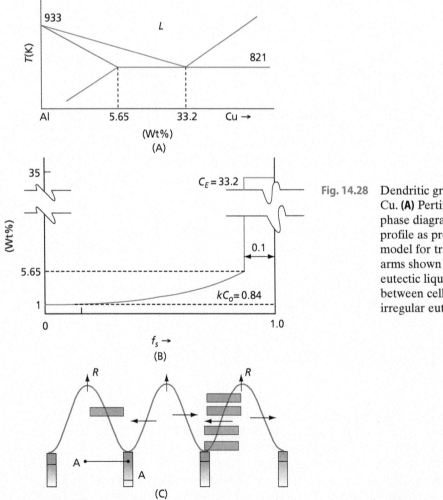

Fig. 14.28 Dendritic growth of Al-4.9 pct Cu. **(A)** Pertinent part of the phase diagram. **(B)** Composition profile as predicted by the Scheil model for trace A-A on dendrite arms shown in **(C)**. Note that the eutectic liquid ($f_e = 0.1$) freezes between cell walls, often as irregular eutectic

and becomes dependent on the solidification velocity. At sufficiently rapid solidification rates, the value of *k* becomes equal to unity since the solute atoms do not have time to move away from the interface and they become entrapped by the advancing interface (Ref. 43). The effect of solidification velocity on *k* can be estimated with the following equation:

$$k' = (k + \beta R)/(1 + \beta R) \qquad 14.29$$

where k' is the actual redistribution coefficient, R is the solidification rate, k is the equilibrium redistribution coefficient, and β is a constant to be determined experimentally.

14.13 Dendritic Freezing in Alloys

Dendritic freezing is a common phenomenon in many alloy systems. Here the supercooling that furnishes the driving force for dendritic growth is normally of a different type. The form discussed in Sec. 14.10 is called *thermal supercooling* and can also be a factor in the freezing of alloys, but constitutional supercooling, which we shall now consider, following in general the proposals of Rutter and Chalmers (Ref. 44), is of much greater importance. *Constitutional supercooling* results when a solid freezes with a composition different from that of the liquid from which it forms.

For the sake of the present argument, let us assume that we can neglect considerations of convection currents and other complicating factors and that the steady-state composition variation shown in Fig. 14.26 holds. In most practical examples of freezing, liquid metal is poured into a mold cavity and freezes as a consequence of heat losses through the mold shell. As a result, the temperature is always lowest at the mold walls and rises toward the center of the mold. Solidification, accordingly, starts at the walls and proceeds inward. Applying these considerations to the present problem signifies that we should concentrate our attention on a freezing process in which the temperature rises in advance of the liquid-solid interface. This situation is illustrated in Fig. 14.29, where the temperature of the liquid is assumed to rise linearly with distance from the interface. A second curve in the figure shows the freezing point of the liquid alloy as a function of distance from the interface. This varies with distance from the interface due to the change in composition of the liquid as one moves away from the interface. At the interface, the freezing temperature, as shown in Fig. 14.25, is T_3, but away from the interface it at first rises rapidly and then levels off to the temperature T_1, the temperature at which the bulk of the liquid will begin to freeze. As shown in Fig. 14.29, the temperature of the liquid and the freezing point of the liquid intersect at two points: at the interface, and at the distance x from the interface. The pertinent point, however, is

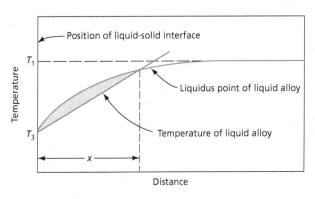

Fig. 14.29 Inside the distance (x), the temperature of the alloy is lower than its freezing point. This is known as *constitutional supercooling*

that within the distance x the liquid lies at a temperature below its freezing point. Inside this range it is effectively supercooled in spite of the fact that the temperature gradient is positive. This is the direct result of the concentration gradient in the liquid alloy in front of the interface.

Whether or not true dendritic freezing, with its associated shooting out of branches of primary and higher order, occurs when the liquid in advance of the interface is constitutionally supercooled depends on the amount of the supercooling. In large commercial castings, the supercooled layer (distance x in Fig. 14.29) is usually large because of a low temperature gradient ahead of the interface and/or a low freezing rate, and dendritic freezing is important. On the other hand, if the supercooled layer is thin, the growth of fully developed dendrites is not possible because of the limited depth of the supercooled layer into which they can grow. In this case, the instability of the interface may result in the formation of a surface composed of more or less oval projections of the type shown in Fig. 14.30. Because the movement of an interface of this type is coupled with the forward motion of the narrow supercooled region, its shape is stable. This leads to a very interesting result. In order for the surface to maintain its shape, freezing must occur uniformly over the entire surface. However, the solid lying on the surface at the centers of the projections, which are farthest to the right, lies at a temperature (T_1) higher than that at the cusps (T_2), which are farthest to the left. Note that T_2 can go down to as low as eutectic temperature. Coupled with this temperature difference is a difference in the composition of the liquid that freezes at the two positions. That which solidifies at the cusps has a higher concentration of solute than that at the center of the projections. The result of this freezing process is the formation of a cellular structure in which the cell walls (horizontal lines in Fig. 14.30) are defined as regions of high-solute concentration. Figure 14.31 shows an actual photograph taken normal to the interface of one of these cellular structures. Notice that what one sees in this figure is the surface of a single crystal which is not uniform in composition; the dark lines represent areas of increased solute concentration.

The cellular structure discussed above is noteworthy, for it shows how a nonuniform solute distribution can be obtained on a scale smaller than the crystal size. In general, this type of phenomenon is called *micro-segregation*. This is only one aspect of the segregation problem that occurs during the freezing of alloys. In a later part of this chapter the important subject of segregation in alloy castings will be considered in more detail.

As may be seen in Fig. 14.29, the curves showing the liquidus point and the temperature gradient are drawn so that they only coincide at two points; this leads to constitutional supercooling. A planar interface under this condition is unstable and leads to the formation of cells or dendrites. On the other hand, if the temperature in the liquid rises faster (corresponding to a steeper temperature gradient) than that of the liquidus point, the planar interface would be stable and no dendrites will form.

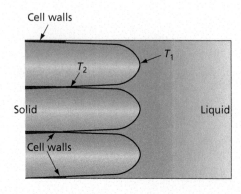

Fig. 14.30 When the region of constitutional supercooling is narrow, a cellular structure may form as the result of the movement of a stable interface of the type shown on the left

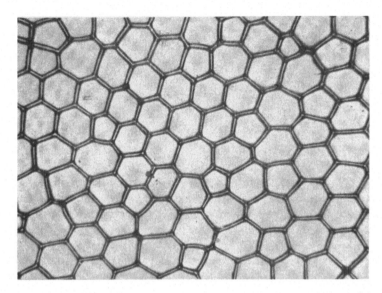

Fig. 14.31 Cellular structure in tin as viewed normal to the interface. 100X. (Rutter, J. W., ASM Seminar, *Liquid Metals and Solidification*, 1958, p. 243. Reprinted with permission of ASM International®. All rights reserved. www.asminternational.org)

A planar interface is desirable for growing single crystals. The constitutional supercooling criterion for the stability of a planar interface during steady-state solidification can be written (Ref. 45) as:

$$G_L/R = m_l C_o (1 - k)/k D_l \qquad 14.30$$

where G_L is the temperature gradient in the liquid, R is the solidification rate, m_l is the slope of the liquidus, C_o is the alloy composition, k is the redistribution coefficient, and D_L is the liquid diffusivity. For most solidification processes, Eq. 14.30 can be used to determine whether or not an interface will remain stable. However, for some solidification processes, particularly for rapid solidification, the criterion can lead to erroneous results. This is because the constitutional supercooling criterion ignores the temperature gradient in the solid and the solid-liquid surface energy. The constitutional supercooling criterion also ignores the influence of surface energy on the stabilization of the interface. For these cases it would be more accurate to use the morphological stability criterion developed by Mullins and Sekerka (Ref. 46), which will not be considered here.

14.14 Freezing of Ingots

The casting of ingots is a very important step in the manufacture of wrought products: items such as plates and beams which are plastically worked into their final shape. The size of these castings depends on the type of metal and its eventual use. In the manufacture of steel, large ingots, weighing 6 to 8 metric tons, are common.

When an ingot is frozen, three separate phases of the freezing process may occur, with each phase developing a characteristic arrangement of grain sizes and shapes. The basic structures are illustrated in Fig. 14.31. In a narrow band following the contour of the mold lies the "chill zone," consisting of small equiaxed (equal-sized) grains which usually have random orientations. Inside of this outer zone, the grains become larger in size, elongated in shape, and with their lengths parallel to the heat-flow direction (normal to the mold walls). These grains have a very strong preferred orientation with a direction of dendritic growth parallel to their long axis. Because of the shape of the crystals in this zone, it is customary to call it the *columnar zone*. The last zone lies at the center of the ingot and represents the last metal to freeze. In this region the grains are again equiaxed and of random orientation, but are much larger than the grains in the chill zone.

The factors (Ref. 47) instrumental in developing the above structures will now be considered. First take the case of a pure metal. When liquid metal is poured into a mold, the mold walls, which are at a much lower temperature (usually room temperature) than the liquid, rapidly cool the layer of liquid with which they are in contact. As a result, the temperature of the liquid metal for a short distance away from the mold walls drops below the equilibrium freezing temperature. Because of the rapidity with which the liquid temperature falls, a considerable magnitude of supercooling usually results. When nucleation (usually heterogeneous) occurs, its rate will be relatively rapid, with the result that the average size of the grain of this solid will be small. Because the crystals form independently, their orientation will be random. Finally, because their growth is limited by similar neighboring crystals nucleated at approximately identical times, their sizes will be nearly uniform and the structure is said to be *equiaxed*.

The crystals in the chill zone develop through both nucleation and growth. Crystal nuclei form in the liquid and grow in size until they make contact with neighboring crystals. The columnar zone forms in a different manner. Here crystal growth predominates and very little nucleation is observed. As soon as nucleation starts in the chill zone, the temperature in this region begins to rise again toward the equilibrium freezing temperature. This is the natural result of the release of the latent heat of fusion.

The columnar zone is thus composed of crystals that start at the chill zone and grow side by side in one direction, the heat-flow direction. In a pure metal, these crystals may continue to grow in this manner to the center of the ingot. Their growth stops when they meet the grains growing out from the opposite wall. The central equiaxed zone shown in Fig. 14.32 is not found in pure metal ingots. It is, however, a phenomenon commonly observed in the freezing of alloys. While it would appear that dendritic growth occurs only briefly during the freezing of pure metals, it has been proposed by Walton and Chalmers (Ref. 47) that the preferred orientation found in the columnar zone results

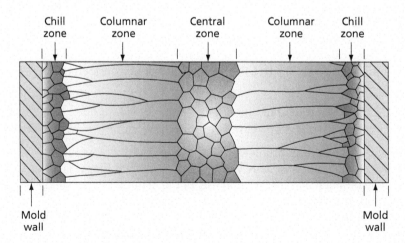

Fig. 14.32 A section through a large ingot, and the three basic zones of freezing that may be found in a casting

from the dendritic freezing phase. At the start of dendritic freezing, those crystals on the interface (chill zone) which possess a direction of rapid dendritic growth nearly normal to the interface will, in general, shoot out their spikes more rapidly than their less favorably oriented neighbors. The growth of these latter is also adversely affected by the heat of fusion released by the more rapidly growing crystals. In this manner, certain crystals are suppressed while others continue to grow in size, with the end result that only those crystals survive that have dendritic-growth directions nearly parallel to the heat-flow direction. The general nature of the development of the preferred orientation is shown schematically in Fig. 14.32. Notice that as the columnar grains grow in length there is also an increase in their diameter. This is, above all, the result of the elimination of less favorably oriented crystals. Still, there is other evidence that grain coarsening in ingots can arise from other causes, such as the grain-boundary movements that would be due to grain-boundary surface-energy factors.

When a pure metal freezes in a mold, the supercooling promoting dendritic growth can only be of the thermal form. In alloys, on the other hand, in addition to the thermal supercooling, one has the possibility of producing constitutional supercooling. When this occurs, dendritic freezing is also observed and with it a corresponding development of a preferred orientation. The resulting preferred orientation is of the same type as that discussed above with a dendritic-growth direction in each columnar grain parallel to the heat-flow direction. The mechanism by which dendritic growth in alloy castings eliminates the less favorably oriented crystals differs (Ref. 47) somewhat from that in pure metals, but need not concern us here.

One important result of constitutional supercooling in alloys is that it favors the development of the central equiaxed zone. As the solid-liquid interfaces that start from opposite sides of a mold approach each other at the center of a large ingot, their respective zones of constitutional supercooling eventually overlap. This is shown schematically in Fig. 14.33. The left-hand figure represents an early stage in the freezing process, while the right-hand figure represents a later stage when the zones have come together. At this point, a very large supercooling can occur in the liquid near the center of the ingot. There are two basic causes for this. First, the concentration of solute in the liquid just ahead of the interface tends to increase with increasing growth of the columnar zone, thus requiring lower and lower temperatures at the interface for continued solidification. Second, the temperature at the center of the ingot tends to approach that of the interfaces as the latter get closer together. This effectively flattens out the contour of the temperature-distance

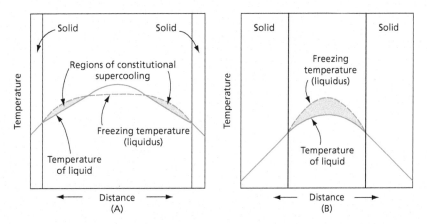

Fig. 14.33 The development of the constitutional supercooled region at the center of an alloy ingot that produces the central equiaxed zone of the ingot

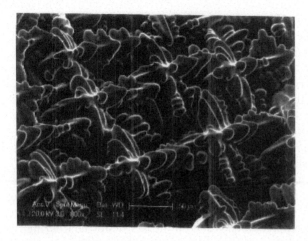

Fig. 14.34 A dendritic crystal colony formed by three-dimensional dendritic growth. Tertiary arms are shown on some of the secondary arms. (Photograph courtesy of Shaahin Amini, 19 Covington, Mission Viejo, CA 92692)

curve in the liquid. When a central equiaxed zone appears in an ingot, it is indicative of the fact that constitutional supercooling may have occurred in this region and that it was able to develop to the point where nucleation could occur in the liquid at the center of the casting. Crystallization in this central zone may occur, therefore, through the appearance and growth of new crystals, and not through the continued growth of the elongated crystals of the columnar zone.

One other point is worth noting, relative to the formation of the central zone of an ingot. When nuclei form in this region, the heat of fusion that they release raises the temperature in the immediate surroundings of each nucleus. Each new crystal is thus surrounded by a temperature field and, as a result, begins its growth dendritically. However, since they form freely in the liquid, dendrite arms will shoot out along all the directions of dendrite growth. In a cubic crystal, arms form along all of the six $\langle 100 \rangle$ directions.

Dendritic growth in the central zone only continues until the heat of fusion removes the constitutional supercooling. When this happens, freezing is completed by the filling in of the spaces between the dendrite arms and between neighboring crystals. In the abscence of complete filling, shrinkage porosity will form as shown in Fig. 14.34.

It should be mentioned that grain nucleation, as a result of constitutional supercooling, is just one way that the equiaxed grains can form at the center of an ingot. Another way they can form is by grain multiplication, which will be described in the next section.

14.15 The Grain Size of Castings

A detailed description of the factors in freezing that determine the final grain size of a casting is beyond our scope. Cole and Bolling (Ref. 48) have summarized many of them in a paper that has particular reference to techniques used to obtain small grain sizes, primarily in the equiaxed central zone of a casting. Under the right conditions, this zone can become quite large. There are a number of significant reasons why it is desirable to produce castings with a very fine grain size. For example, it has been found that superplasticity occurs primarily as a result of a very finely divided structure.

There are two basic ways of increasing the number of grains in a casting. One involves increasing the basic nucleation rate, while the other depends on breaking up the dendrites as they grow so as to form additional crystal seeds. This latter is called *grain multiplication*. Rapid cooling, as in die casting and casting into small cold molds, tends to increase the nucleation rate. It can also increase grain multiplication because of the enhanced convection currents associated with this type of casting. The nucleation rate can also be increased by adding catalysts (inoculants) to the melt. These inoculants are normally small, second-phase particles that increase the rate of heterogeneous nucleation. For example, boron and titanium are often added to aluminum alloys to increase nucleation. These elements can be considered to react with the aluminum in the alloy to form particles that act as catalysts.

Of the two basic methods of decreasing the grain size, the one of most general application is grain multiplication. This process probably occurs in most casting operations. As the dendrite arms grow and branch out, there is a continual release of the heat of fusion. This heat release, under the right conditions, can produce local melting and cause some dendrite arms to neck and pinch off. When this occurs, a small crystal fragment is formed. Agitation of the liquid metal due to convection currents tends to remove these crystallites from the liquid-solid interface and distribute them throughout the liquid. Many of them may eventually melt, but others grow into new, randomly-oriented crystals.

Artificial means can be employed to increase grain multiplication. In general, these involve fragmenting the growing dendrites at the growing liquid-solid interface and causing them to be redistributed throughout the melt by a forced fluid flow. Forced vibrations, alternating magnetic fields, ultrasonic vibrations, and a number of other methods may be used for this purpose. An effective method (Ref. 48) involves the oscillation of a mold through several revolutions, followed by a change in the direction of rotation. This not only promotes the detachment of dendrite arms, but also causes them to be rather uniformly distributed throughout the liquid as a result of the agitation of the liquid.

14.16 Segregation

The liquids that are frozen to form industrial alloys usually contain, in addition to solute elements added intentionally for their beneficial effects, many impurity elements that find their way into the liquid metal by a number of different routes. Thus, impurity elements, present in the ores from which the basic metals were obtained, are frequently only partly eliminated during the smelting and refining operations. The refractory brick linings of furnaces used in melting or refining (purifying) and the gases in the furnace atmospheres may be other sources. In the latter case, elements enter the liquid metal in the form of dissolved gases. The various elements dissolved in commercial liquid metals can and often do react with each other to form compounds (oxides, silicates, sulfides, etc.). In many cases, the latter may be less dense than the liquid, and rise to the surface and join the slag which floats on top of the liquid metal. On the other hand, it is quite possible for small impurity particles (compounds) to exist in the liquid. Certain of the latter undoubtedly form nucleation centers for heterogeneous nucleation. This fact has been used to some extent to control the grain size of castings by inoculating liquid metals with elements that combine to form nucleation catalysts. Increasing the number of nucleation centers will, of course, produce a finer grain size in the solidified casting.

When an alloy is frozen, a more or less general rule which applies is that solute elements, whether present as alloying elements or as impurities, are more soluble in the liquid state than in the solid state. This fact usually leads to a segregation of the solute elements in the finished casting. There are two basic ways of looking at the resulting nonuniformity of the solute. First, because the liquid becomes progressively richer in the solute as freezing progresses, the solute concentrations in a casting tend to rise in those regions that solidify last (center of the ingot). This and similar long-range composition fluctuations fall in the classification of macrosegregation. In general, macrosegregation refers to the change in the average composition of the metal as one moves from place to place in an ingot. Segregation of this form is not always caused by the selective freezing out of high-melting-point constituents. Gravitational effects are often a factor in producing macrosegregation, especially during the formation of the central equiaxed zone. The crystals that form freely in the liquid often have a different density from that of the liquid. As a result, they may either rise toward the surface of the casting, or settle toward the bottom. An extreme example (Ref. 49) of gravity-induced segregation occurs in an alloy system somewhat different from the solid-solution alloys presently being considered. Thus, in the lead-antimony system, an eutectic forms at 11.1 percent Sb and 252°C. When alloys containing more than 11.1 percent Sb (say 20 percent) are frozen, crystals of almost pure antimony form from the liquid until the composition of the latter reaches the eutectic composition, at which time the eutectic mixture begins to freeze. Since antimony crystals have a lower density than the liquid from which they originate, they tend to rise toward the surface. Slow cooling of this alloy, therefore, results in a structure whose lower fraction is composed almost entirely of eutectic solid, with an upper portion containing the primary-alpha antimony crystals with some eutectic solid filling in the gaps between the primary crystals.

In castings not only do we find composition variations (macrosegregation) over large distances, but it is also possible to have localized composition variations on a scale smaller than the crystal size. This is called *microsegregation* and one form has already been described; the composition segregation associated with the cellular structure resulting from the combined movement of the liquid-solid interface and a very narrow zone of constitutional supercooling. A much more frequent form of microsegregation, commonly known as *coring*, is caused by dendritic freezing in alloys. The original dendrite arms, which shoot out into the liquid metal, freeze as relatively pure metal. The liquid surrounding these arms is, accordingly, enriched in solutes and normally, when this liquid solidifies, the spaces between the arms become regions high in solute concentration.

Dendritic segregation, or coring, is very common in alloy castings solidified under normal conditions. It can occur with a solute concentration as low as 0.01 percent under the proper conditions. Rapid solidification, however, can drastically reduce segregation and in some cases lead to its total elimination. When a casting is sectioned and the surface prepared for metallographic examination, the exposed surface will usually be a planar section cutting through the forest of dendrite arms. Since the composition at the center of an arm differs from that at points midway between arms, the dendrite arms can be revealed by etching with a suitable metallographic etch. Dendritic segregation in a copper-tin alloy is shown in Fig. 14.35.

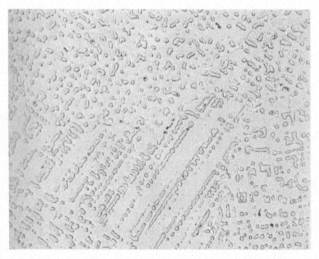

Fig. 14.35 Coring, or dendritic segregation, in a copper-tin alloy. Several different grains are shown. Note that the dendrite arms have different orientations in each crystal. 200X

14.17 Homogenization

The segregation of the solute due to coring can usually be removed if the equilibrium structure of the alloy is a single phase. This is accomplished by a heat treatment known as a *homogenizing anneal*. In this process, the metal is heated to a high temperature for a sufficient time to allow diffusion to homogenize the structure.

Before considering the kinetics of homogenization, the freezing process that leads to coring will be re-examined. Fig. 14.36 shows one corner of an isomorphous phase diagram. If a metal of composition x is frozen under equilibrium conditions, the solid will follow the solidus curve from point b to point c. This curve represents both the composition of the solid that is forming at any instant and the average composition of all the solid that has previously formed. In normal freezing there is insufficient time for diffusion to homogenize the structure, and coring results. In this case, the solidus line from point b to point f represents only the composition of the solid that can form at any given temperature. Since the metal that has solidified earlier is always richer in component A than the solid that is just freezing, the average composition of the solid has to lie to the left of the solidus line. The average composition of the solid during normal freezing, therefore, follows a line such as be in Fig. 14.36. At the same time, because there is always some diffusion, the composition of the first solid to form (at the center of the dendrite arms) also changes and follows a path such as bd.

Note that one of the significant features of normal freezing is that it tends to occur over a greater temperature range than required for equilibrium freezing. Thus, point d, which represents the end of the normal freezing process in Fig. 14.36, lies well below point c, corresponding to the completion of equilibrium freezing.

At the end of normal freezing the composition will vary over the range from point b to f in the phase diagram of Fig. 14.36. For the purpose of deriving an equation, it will now be assumed that the dendrites can be considered as a set of parallel plates whose composition varies linearly and periodically through this composition range, as indicated in Fig. 14.37. The horizontal dashed line in this figure represents the average composition of the alloy n_{Bav}. At the center of each dendrite arm, such as at point m, the composition is less than the average by the amount Δn_B, and midway between the

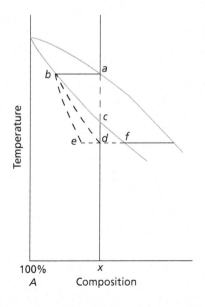

Fig. 14.36 During nonequilibrium freezing, the average composition of the solid follows a path such as bd

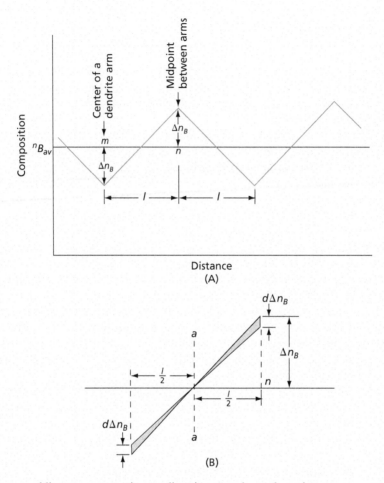

Fig. 14.37 An assumed linear concentration gradient in a cored metal specimen

dendrite arms, as at point n, it is greater than the average by Δn_B. The distance from the center of a dendrite arm to a point midway between a pair of arms is assumed to equal l. This quantity is equal to one-half the dendrite arm spacing, the distance between the centers of the two arms.

In a time interval dt, the concentration at point m may be assumed to increase by Δn_B, while at point n it should decrease by the same amount. In order for this to happen, a number of B atoms equal to

$$d(\Delta n_B)\tfrac{1}{2}A \qquad 14.31$$

has to pass through the cross-section aa, where $d(\Delta n_B)l/2$ is the area of the small cross-hatched triangles in Fig. 14.37, and A is the cross-section area of the specimen. This number can be equated to $JAdt$, where J is the flux in atoms/s · m². Since the concentration gradient is assumed to be linear and equal to $\Delta n_B/(l/2)$, we may write:

$$d\Delta n_B \cdot l/2 = J \cdot dt = -D \cdot \Delta n_B/(l/2) \cdot dt \qquad 14.32$$

where D is the diffusion coefficient. This equation reduces to

$$\frac{d(\Delta n_B)}{dt} = -\frac{4D(\Delta n_B)}{l^2} \qquad 14.33$$

Note that the rate of change of Δn_B is proportional to Δn_B. This means that Δn_B must vary exponentially with the time, or

$$\Delta n_B = \Delta n_{B_0} e^{-(4D/l^2)t}$$

where Δn_{B_0} is the concentration differential when the time is zero. The quantity $l^2/4D$ is equivalent to a relaxation time τ, so that we may also write

$$\Delta n_B = \Delta n_{B_0} e^{-t/\tau} \qquad 14.34$$

A somewhat better approximation is to assume that the concentration varies sinusoidally with the distance rather than linearly. Figure 14.38 shows this type of variation, which corresponds (Ref. 50) to a relaxation time $\tau = l^2/\pi^2 D$ instead of $l^2/4D$.

As demonstrated in Sec. 13.3, the relaxation time is equal to the time required to make a function that depends exponentially on the time decrease in value by a factor $1/e$. In a typical cored specimen, the dendrite arm spacing may be of the order of 10^{-4} m or 10^{-2} cm. The diffusion coefficient (see Table 12.4) at elevated temperatures is, in many cases, equal to about 10^{-13} m²/s. Using these values gives a relaxation time

$$\tau = \frac{l^2}{\pi^2 D} = \frac{10^{-8}}{\pi^2 10^{-13}} \approx 10^4 \text{ s} \approx 3 \text{ hours}$$

This result clearly shows that for a relatively coarse dendrite arm spacing, a very long annealing time may be required in order to approach complete homogenization.

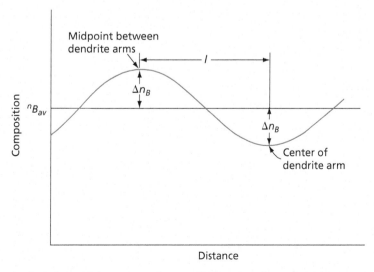

Fig. 14.38 A hypothetical sinusoidal concentration profile in a cored metal specimen

When only a single phase is present, it may not be of major importance whether or not complete homogenization is attained. However, if coring yields a second nonequilibrium phase, this may not be true. This is particularly the case in a number of precipitation hardening alloys. As an example, consider the alloy of aluminum with 4 percent copper. If equilibrium is maintained during freezing, this alloy will freeze as a homogeneous solid solution. In practice, equilibrium is never attained and the metal becomes badly cored. The aluminum side of the aluminum-copper phase diagram is shown in Fig. 14.39. During freezing, the average composition of the solid of a 4 percent copper alloy should move along a line such as bc. Point c lies on the eutectic isotherm. When the alloy is cooled to this temperature it is still a mixture of solid and a small quantity of liquid lying between the dendrite arms. The liquid has the eutectic composition and will freeze to form an eutectic composed of the alpha and theta phases. The amount of eutectic may be computed with Eq. 14.27. Sometimes when an eutectic forms and the primary phase (in this example the alpha phase) occurs as a widely distributed set of closely spaced dendrite arms, the corresponding phase (α) in the eutectic may form preferentially on the existing primary dendrites. This leaves the second phase (θ) as the only remaining visible phase in the eutectic. When this occurs, the eutectic is said to be a *divorced eutectic*.

In any case, precipitation hardening alloys, such as the aluminum–4 percent copper alloy, can develop a two-phase structure as a result of coring. This is undesirable for several reasons. First, it ties up a large fraction of the hardening agent (the copper) in the precipitate where it is not available for strengthening the metal during a precipitation hardening treatment. Second, as will be shown in Chapter 21 on fracture, precipitates can have a harmful effect on the ductility of a metal. Homogenizing a cored precipitation hardening alloy of this type will increase not only its strength, but also its ductility. This is clearly demonstrated in Fig. 14.40, where the tensile and yield strengths, as well as the elongation, of a commercial aluminum alloy are plotted as a function of the volume fraction of the second phase in the alloy.

Considerable attention is currently being devoted to attempts to develop casting techniques that will reduce the dendrite arm spacing. This is because the relaxation time for homogenization varies as the square of the dendrite arm spacing. The closer the dendrite arms, the easier it is to remove the second phase by homogenization. The gains to be achieved in precipitation hardening alloys by removing the second phase are clearly shown in Fig. 14.40. The most important factor in determining

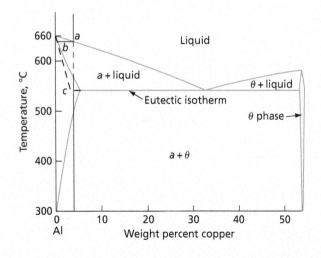

Fig. 14.39 Coring in an eutectic alloy system

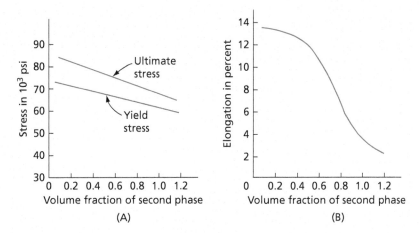

Fig. 14.40 **(A)** The ultimate stress and the yield stress of age-hardened 7075 aluminum alloy as a function of the volume fraction of second phase. **(B)** The elongation of age-hardened 7075 aluminum alloy as a function of the volume fraction of second phase. (From Singh, S. N., and Flemings, M. C., *TMS-AIME*, **245** 1811 [1969])

the dendrite arm spacing is the rate of cooling. Very rapid cooling can produce a dendrite arm spacing of a few microns. For most alloys, the dendrite arm spacing (DAS) is related to the cooling rate, ε, by

$$DAS = k\varepsilon^{-n} \qquad 14.35$$

where k is a constant and n is in the range of 1/3 to 1/2. Other factors that lead to a reduction in the time required for homogenization cannot be covered here. Information about them is given in several papers (Refs. 51 and 52).

Finally, on the subject of homogenization, it should be mentioned that care should be exercised in selecting the temperature for the homogenization anneal. Partial melting or liquation of the metal may occur if it is heated to just under the solidus curve of an equilibrium diagram. Coring normally allows part of the structure to freeze at a temperature below that corresponding to the completion of freezing under equilibrium conditions. This part of the structure will melt if the metal is reheated to just below the solidus. In some cases, serious damage to a metal object may result from this cause.

14.18 Inverse Segregation

Dendritic freezing in the columnar zone of ingots may sometimes lead to a phenomenon known as *inverse segregation*. In the normal freezing of ingots, the center and top portions of the ingot, which freeze last, become richer in solutes than the outer portions of the ingot, which freeze first. In some alloys, however, the dendrite arms may extend for long distances before the spaces between them are filled in by the freezing operation. Under the proper conditons, channels between the dendrite arms offer a path by which liquid from the center of the casting can work its way back toward the surface. The latter phenomenon is promoted by the fact that as solidification progresses, the casting as a whole can shrink away from the mold walls, causing a suction that pulls the liquid toward the surface. Other important factors are the internal pressure that develops in an ingot due to gas evolution during freezing and convective currents developed in the liquid. "Tin sweat," which occurs in tin bronzes (copper-tin alloys), is an example. When the liquid contains a relatively large concentration of dissolved hydrogen, the release of this gas near the end of freezing forces the liquid enriched in tin through interdendritic pores to the surface, where it coats the normally yellow-bronze-colored surface with a fine layer of white alloy (containing approximately 25 percent Sn).

14.19 Porosity

Other than shrinkage cracks resulting from unequal cooling rates in different sections of castings, there are two basic causes for porosity in cast metals: gas evolution during freezing, and the shrinkage in volume that accompanies the solidification of most metals.

Gases differ from other solutes in that their solubility in metals depends markedly on the applied pressure. Many gases with which metals come in contact are diatomic: O_2, N_2, H_2, etc. Providing that the solubility is small, it is often possible to express the relationship between the pressure and the solubility of diatomic gases in a simple form, known as Sievert's law.

$$c_g = k\sqrt{p} \qquad 14.36$$

In Eq. 14.36, c_g is the solubility of the dissolved gas, p is the gas pressure, and k is a constant. Some experimental data corresponding to hydrogen dissolved in pure magnesium are shown in Fig. 14.41. The melting point of pure magnesium is 650°C (923 K). Notice that the data plots well as straight lines (on c_g and $\sqrt{p}$ coordinates) both above and below the melting point. This signifies that Sievert's law is probably valid for the solution of hydrogen in both liquid and solid magnesium.

The solubility of gases in metals is also a function of temperature and, in most cases, the solubility increases rapidly with temperature. If the maximum solubility is small, it is often possible to express the equilibrium concentration of a gas in a metal, at a constant gas pressure, as an exponential function of the type previously derived for the solubility of carbon in iron; namely

$$c_g = Be^{-Q/RT} \qquad 14.37$$

where c_g is the concentration of the gas in the metal, B is a constant, Q is the work to introduce a mole of gas atoms into the metal, and R and T have their usual significance.

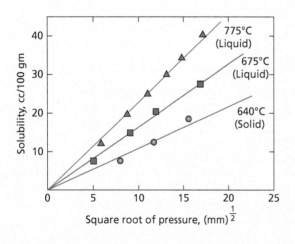

Fig. 14.41 Solubility of hydrogen in both liquid and solid magnesium as a function of the partial pressure of hydrogen. (From Koene, J., and Metcalfe, A.G., *Trans. ASM*, **51** 1072 [1959] Reprinted with permission of ASM International®. All rights reserved. www.asminternational.org)

The solubility of hydrogen as a function of temperature in copper at a constant pressure of one atmosphere is given in Fig. 14.42. This curve shows not only the rapid increase in solubility with increasing temperature, but also a large decrease in solubility when the metal transforms from a liquid to a solid at 1083°C. Similar changes in the solubility of gases in metals occur at the freezing points of most metals. It can be concluded that, in general, when a metal freezes, its ability to hold gases is strongly decreased.

Let us consider these factors in relation to the freezing process. In almost all cases, the solubility of dissolved gases undergoes a large drop when metals freeze. For this reason, gaseous solutes segregate during freezing in a similar fashion to other solutes. Segregation signifies, therefore, localized increases in the gas concentration of a liquid. Gas bubbles may form in the liquid in those regions where the concentration of gas rises above the equilibrium concentration (saturation value). Homogeneous nucleation of bubbles, however, requires a high degree of supersaturation for the same reason as the nucleation of solid crystals in a liquid. In both cases, a surface with its inherent surface energy is formed between the old and new phases. As a result, gas bubbles usually form heterogeneously and, in most cases, the nucleation centers lie on the liquid-solid interface. Nucleation at the interface is also furthered by the build-up in concentration of the gaseous solutes at both the general interface and in between the dendrite arms as a result of segregation effects.

A very important factor in gas-bubble formation is the local pressure in the liquid. As previously mentioned, the equilibrium concentration of a gas in a liquid depends on the pressure. In metals with a given quantity of absorbed gas, gas-bubble nucleation is therefore promoted by a lowering of the pressure, and hindered by pressure increases. Thus, if liquid metals are frozen under sufficiently high pressures, gas evolution may be prevented. In this respect, die castings are a form of casting in which small metal parts of close dimensional tolerances are manufactured by forcing liquid metals (under very high pressures) into steel dies. Gas porosity is effectively eliminated in this casting method. On the other hand, it frequently happens that the normal shrinkage that occurs when a metal freezes causes pressure drops in the liquid. This situation may occur when the entire outer surface of a casting freezes over, leaving the liquid in the center surrounded by a shell of solid metal. A similar effect may also happen in localized areas of complicated castings wherein a small volume of liquid finds itself enclosed by solid. In either case, a vacuum may develop and, if it does, gas-bubble nucleation will be promoted.

Gas-bubble formation is a nucleation and growth phenomenon similar in many respects to other nucleation and growth phenomena previously considered. In particular, as mentioned above, the formation of the interface makes it hard to nucleate bubbles. However, once a bubble has

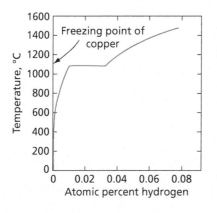

Fig. 14.42 Solubility of hydrogen in copper at one atmosphere pressure. (From *Constitution of Binary Alloys*, by Hansen, M., and Anderko, K. Copyright, 1958. McGraw-Hill Book Co., Inc., New York, p. 587. Used by permission.)

formed and grown to a size greater than its critical radius, its growth becomes progressively easier. This fact is easily shown in the following manner. A gas bubble in a liquid is analogous to a soap bubble, except that it consists of a single liquid-gas interface separating a gas phase from a liquid phase. By analogy with the soap bubble, we may write

$$p_g - p_l = \frac{2\gamma}{r} \qquad 14.38$$

where γ is the surface energy in dyne/cm, p_g is the internal pressure in the gas bubble in dyne/cm^2, p_l is the pressure in the liquid in dyne/cm^2, and r is the radius of curvature of the bubble in centimeters. As the bubble grows in size, its radius of curvature increases. As a result, the pressure differential between the gas in the bubble and the pressure in the liquid falls. This implies (by Sievert's law) that the bubble, as it grows in size, is able to be in equilibrium with a decreasing concentration of gas atoms in the surrounding liquid. Continued growth of gas bubbles is also furthered by the solute concentration gradient, which develops in the surrounding liquid as the gas bubble absorbs more and more gas atoms. The concentration falls toward the bubble, causing a diffusion of gas atoms from the surrounding liquid toward the bubble.

The effect of gas-bubble formation on the final solid metal structure during freezing depends upon whether or not the bubbles are trapped into the solid structure, and upon the shape of the entrapped cavities. Rapidly growing bubbles have the tendency to free themselves from the interface and rise toward the upper surface of the casting. If this surface is frozen over, a large cavity may form at the top of the ingot. If the upper surface is unfrozen, the gas is eliminated from the casting. In general, since gas bubbles that form near the top of an ingot grow under a lower hydrostatic head, they will be larger in size and more inclined to escape. This action is further increased by the sweeping action of bubbles that rise from points at greater depths in the mold.

When the growth rate is very slow, bubbles may be trapped by the solid growing around them. In this case, the cavity formed in the solid is approximately spherical in shape and is commonly called a *blowhole*. See Fig. 14.43A. In some cases, blowholes may not be particularly harmful to the properties of the metal. This is especially true in those metals that receive a high degree of hot-working during their manufacture, providing that the blowholes are formed at some distance below the surface of the ingot. This is required in order that the internal surfaces of the blowholes do not become oxidized through contact with oxygen from the surrounding air. If the surface of the blowhole is not oxidized, it is often possible, under the extreme pressures of hot-rolling, to collapse the blowholes and to weld their sides together. If the welds are successful, the blowholes are eliminated from the metal and any gas remaining in the cavities is reabsorbed into the metal (now a solid). Conversely, if the surfaces of the collapsed blowhole do not weld, a long slitlike defect is formed in the metal, which is called a *seam*.

Wormhole porosity is another form of cavity defect that may be found in castings. These are long tubular cavities produced when gas is evolved at an intermediate rate, so that gas bubbles grow in length at the same rate that the liquid-solid interface moves. The nature of wormhole porosity is shown in Fig. 14.43B, where it may be observed that the holes are elongated in the heat-flow direction. Wormhole porosity is easily observed in a nonmetallic ingot familiar to almost everyone: a large cake of ice frozen in a commercial icehouse from water containing dissolved air. The central whitish areas are usually regions of wormhole porosity.

Porosity of a much finer and more irregular shape than that described above occurs when shrinkage effects and gas evolution combine. Thus, in regions completely enclosed by solid metal which still contain some interdendritic liquid, pores can develop, which more or less parallel the dendrite arms.

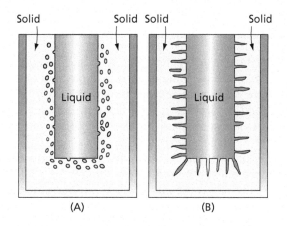

Fig. 14.43 Two forms of gas porosity: **(A)** blowholes and **(B)** wormhole porosity, both of which are macroscopic in size as contrasted to dendritic porosity, which is microscopic

If the pores form primarily as a result of shrinkage, they will be very fine in cross-section and very numerous. On the other hand, if gas evolution occurs while the pores are forming, the cavities will be larger and less numerous.

Another form of interdendritic porosity may occur as a result of shrinkage effects alone. When dendrites form during freezing, a network of small channels is formed between the dendrite arms. When the liquid at the end of the network (roots of the dendrites) freezes, shrinkage takes place, which sucks in the liquid. If the channels are narrow and extensive, as is the case when the dendrites are fully developed, the liquid flow is restricted. As a result, small pores form near the roots of the dendrite arms.

While considering the subject of gas evolution, it is worth noticing that controlled-gas evolution may sometimes be used to advantage. Thus, a very large percentage of the steel manufactured in this country is poured into ingots while it still contains a relatively high but controlled amount of gas. This gas content is adjusted so that when the ingot freezes enough, blowholes are trapped in the metal to compensate for the normal shrinkage of the metal as it freezes. In subsequent manufacturing stages (hot-rolling), the blowholes are eliminated as discussed earlier. This form of ingot casting greatly simplifies the problems associated with the casting of large steel ingots and is called *rimming*. Rimming is associated with a relatively high concentration of oxygen in the liquid steel that combines with some of the carbon dissolved in the steel to form carbon monoxide on freezing. This is the gas primarily associated with the formation of the blowholes. A high-oxygen content in the steel unfortunately leads to a relatively high concentration of oxides, silicates, and similar particles in the finished product. In many instances, nonmetallic inclusions may not be particularly detrimental in a steel. However, the higher the demands on the steel, the more detrimental are nonmetallic inclusions to the steel properties. Thus, high concentrations cannot be tolerated in highly stressed machinery steels. Furthermore, for steels containing more than about 0.3 percent carbon, it becomes very difficult to weld blowholes during the rolling processes, and high-quality steels and steels containing more than 0.3 percent carbon are usually deoxidized before casting. These degassed steels freeze without gas evolution and are said to be *killed*. The principal factor to be avoided in a killed-steel ingot is the formation of a large shrinkage cavity along the central axis of the casting, known as a *pipe*. A properly designed killed-steel ingot mold must feed liquid metal into the central regions of the casting as freezing progresses. This is accomplished by placing risers (liquid metal resevoirs) or hot tops at critical locations of the casting. The casting of ingots by this method is more expensive because the molds are complicated and more costly.

14.20 Eutectic Freezing

For solidification of a pure metal with a planar interface at a supercooling, ΔT, the free-energy decrease is given by Eq. 14.18, which shows the energy decrease to be linearly proportional to the supercooling. A similar equation applies for the volume free-energy change for solidification of an eutectic liquid. In this case, however, two phases, α and β, form upon solidification, and the total free-energy change includes not only the volume free-energy change, but also the surface energy created between these phases, thus requiring an additional energy expenditure. In a simple "plate-like" lamellar eutectic, the boundary area created is proportional to $2/\lambda$ per m^3, where λ is the eutectic spacing; see Fig. 14.44. The corresponding free-energy increment per mole is $2\gamma_{\alpha\beta} V_m/\lambda$, where $\gamma_{\alpha\beta}$ is the surface energy of the α-β boundary and V_m is the molar volume of the eutectic. This additional free energy that is needed to form the solid requires an additional driving force for freezing. Thus, the freezing of a lamellar eutectic cannot take place at its equilibrium freezing temperature, T_e, and the liquid must be cooled below T_e for freezing to occur. Accordingly, for a given lamellar spacing, λ, the free-energy change for freezing becomes:

$$\Delta G(\lambda) = -\Delta G(\lambda_\infty) + 2\gamma_{\alpha\beta} V_m/\lambda \qquad 14.39$$

where $\Delta G(\lambda_\infty)$ is the bulk free-energy change when λ is so very large that the interfacial energy is negligible. $\Delta G(\lambda_\infty)$ is, of course, a function of the amount of the undercooling, ΔT_0, and may be represented, similar to Eq. 14.18, by the approximate relationship:

$$\Delta G(\lambda_\infty) = \Delta H \Delta T_0/T_e \qquad 14.40$$

where ΔH is the change in the enthalpy, ΔT_0 is the undercooling, and T_e is the equilibrium freezing temperature. Consequently:

$$\Delta G(\lambda) = -\Delta H \Delta T_0/T_e + 2\gamma_{\alpha\beta} V_m/\lambda \qquad 14.41$$

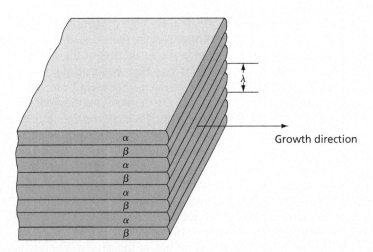

Fig. 14.44 Simple plate-like lamellar eutectic with an interarm spacing λ

Setting $\Delta G(\lambda)$ equal to zero and solving for λ yields:

$$\lambda_{min} = 2\gamma_{\alpha\beta}V_m T_e/\Delta H \Delta T_0 \qquad 14.42$$

where λ_{min} represents a minimum achievable lamellar spacing. For a given amount of undercooling, ΔT_0, a spacing greater than λ_{min} results in a negative free-energy change, $\Delta G(\lambda)$, while a value of $\lambda < \lambda_{min}$ corresponds to a positive free-energy change. In this latter case freezing cannot occur, so theoretically λ cannot be smaller that λ_{min}. But by Eq. 14.42, $\Delta H \Delta T_0/T_e = 2\gamma_{\alpha\beta}V_m/\lambda_{min}$, and one may write:

$$\Delta G(\lambda) = -\frac{2\gamma_{\alpha\beta}V_m}{\lambda_{min}}\left(1 - \frac{\lambda_{min}}{\lambda}\right)$$

or

$$\Delta G(\lambda) = -\frac{\Delta H \Delta T_0}{T_e}\left(1 - \frac{\lambda_{min}}{\lambda}\right) = -\Delta G(\lambda_\infty)\left\{1 - \frac{\lambda_{min}}{\lambda}\right\} \qquad 14.43$$

As shown in Fig. 14.45, when a coupled lamellar eutectic is forming, the liquid concentration ahead of the interface varies sinusoidally (Ref. 53). When $\lambda = \lambda_{min}$, it is apparent that, for any given value of ΔT_0, $\Delta G(\lambda) = 0$. Thus, if $\lambda = \lambda_{min}$, the solidification process should be infinitely slow and the difference between the solute concentrations in the liquid in front of an α and a β lamella ($C = C_1^\alpha - C_1^\beta$) should become vanishingly small. In other words, the concentration of the solute in the liquid should be the same all along the liquid-solid interface and equal to C_e, the eutectic composition. On the other hand, when $\lambda = \lambda_\infty$, the value of $\Delta G(\lambda)$ will be a maximum and the concentration difference, ΔC_1, should be a maximum. On the basis of arguments such as this, it has been assumed that $\Delta C_1 \propto \Delta G(\lambda)$ and that

$$\Delta C_1 = \Delta C_{1_{max}}\left(1 - \frac{\lambda_{min}}{\lambda}\right) \qquad 14.44$$

It is generally believed that the growth of the eutectic is controlled by the diffusion of the solute in the liquid lying in front of the two-phased eutectic solid. This is based primarily on the assumption that the interfaces between the liquid and the α and β phases are very mobile. If the growth rate of the solid, R, does depend on the solute diffusion, then it should be proportional to $D \cdot dC/dl$, where D is the liquid diffusion coefficient and dC/dl is the effective concentration gradient of the solute in the liquid. The problem is complex but it is possible to obtain a reasonable model for the growth process by simply writing:

$$R = kD \cdot \Delta C_1/\lambda \qquad 14.45$$

where R is the growth rate, k is a constant of proportionality, D is the average diffusivity, ΔC_1 is the difference between the solute concentration in front of an α lamella and that in front of a β lamella, and λ is the interlamellar spacing as defined in Fig. 14.45. If one substitutes for ΔC_1 in this growth rate equation using the relation in Eq. 14.44, and assumes that, by Eq. 14.42, $\lambda_{min} = B/\Delta T_0$ where $B = 2\gamma_{\alpha\beta}V_m T_e/\Delta H$, it follows that:

$$R = kD \cdot \Delta C_{1_{max}}(1 - B/\lambda \Delta T_0) \qquad 14.46$$

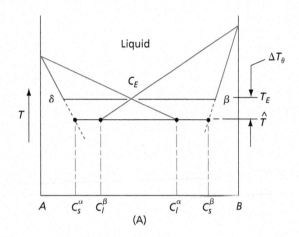

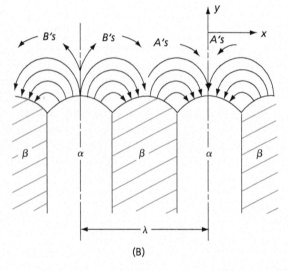

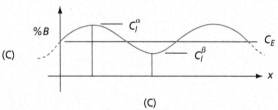

Fig. 14.45 The factors involved in the freezing of a coupled lamellar eutectic. **(A)** The freezing process in relation to the phase diagram. **(B)** A atoms are rejected by advancing β plates and B atoms are rejected by the α plates. **(C)** The liquid concentration ahead of the advancing interface varies sinusoidally

For this latter equation, ΔC_1 can be estimated by extrapolating the α and β liquidus lines below the eutectic temperature as shown in Fig. 14.45A. On the basis of this figure one can assume that $\Delta C_1 = k_1 \Delta T_0$ where k_1 is a constant of proportionality. On substituting this relationship in Eq. 14.46 we have:

$$R = \frac{\Delta T_0}{A\lambda}\left\{1 - \frac{B}{\lambda \Delta T_0}\right\} \qquad 14.47$$

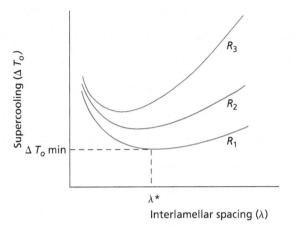

Fig. 14.46 Curves of undercooling, ΔT_o, versus interlamellar spacing, λ, for three different values of the growth rate, R

where $A = 1/kk_1D$. Solving for the undercooling, ΔT_0, in Eq. 14.47 yields

$$\Delta T_0 = AR\lambda + \frac{B}{\lambda} \qquad 14.48$$

Equation 14.48 involves three variables—ΔT_0, λ, and R—but only two are independent, so its solution requires an additional condition. This problem can be handled by assuming several constant values for the growth velocity, R, and then plotting corresponding curves of ΔT_0 versus λ, as shown schematically in Fig. 14.46 for three assumed growth velocities. Note that each of these curves passes through a minimum. In addition, experimental data generally imply that the eutectic growth rate tends to seek a minimum undercooling.

If it is now assumed that growth occurs in such a manner that undercooling is a minimum, then one has the condition that $d\Delta T_0/d\lambda = 0$ and by Eq. 14.48, assuming R constant:

$$\frac{d\Delta T_0}{d\lambda} = AR - \frac{B}{\lambda^2} = 0 \qquad 14.49$$

$$\lambda = (B/AR)^{1/2} \qquad 14.50$$

If the expression for λ in Eq. 14.50 is substituted in Eq. 14.48, the following expression for ΔT_0 will be obtained:

$$\Delta T_0 = 2(ABR)^{1/2} \qquad 14.51$$

Equations 14.50 and 14.51 are significant because they predict (1) that the lamellar spacing of the eutectic should vary inversely as the square root of the growth rate, and (2) that the undercooling should vary directly as the square root of the growth rate. Experimental support for these predictions are shown in the following two figures. Figure 14.47, from Flemings (Ref. 51) shows a plot in which λ plots linearly against $R^{-1/2}$ and uses experimental lead-tin data taken from several sources. On the other hand, Fig. 14.48, from Hunt and Chilton (Ref. 54) is also based on lead-tin data and shows an example where ΔT_o varies directly as $R^{1/2}$.

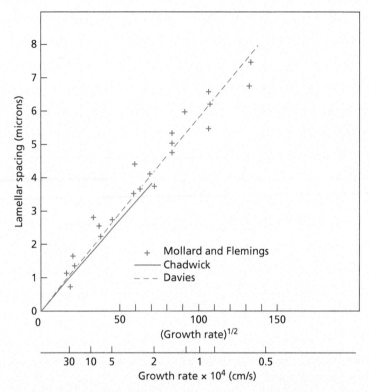

Fig. 14.47 Lamellar spacing as a function of growth rate; tin-lead composites. [Results of Chadwick and Davies are for eutectic composition. Points from Mollard and Flemings are for off-eutectic alloys. (From Flemings, M.C., *Solidification processing*, McGraw-Hill, New York, 1974, p. 101. Used with permission of the author)]

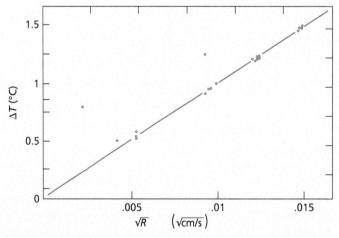

Fig. 14.48 Interface undercooling for tin-lead eutectic. [From Hunt, J. D., and Chilton, J. R., *J. Inst. Metals*, **92** 21 (1963–64)]

Problems

14.1 (a) Suppose that 0.100 kg of liquid copper is supercooled 250 K, where it is allowed to nucleate and solidify adiabatically (no heat is lost to the surroundings). Calculate how much copper will solidify until the temperature recalesces to its melting point of 1356 K. The specific heats of solid and liquid copper in the temperature range of interest are, respectively:

$$C_{ps} = 22.64 + 5.86 \times 10^{-3}T, \text{ J/K} \cdot \text{mol}$$
$$C_{pl} = 31.4, \text{ J/K} \cdot \text{mol}$$

where T is the temperature in degrees Kelvin. The heat of fusion of copper is 13.20 kJ/mol and its molar weight is 0.0635 kg/mol.

(b) How much supercooling would be necessary in order to solidify the entire sample adiabatically?

14.2 (a) Consider the face-centered cubic crystal structure and the planes (100), (110), and (111). On the basis of the degree of close-packing associated with each plane, rank these planes in order of their growth velocity during freezing.

(b) Also rank the (100), (110), and (111) planes for the body-centered cubic lattice.

14.3 The dendrite arm spacing (DAS) is dependent on the cooling rate, ε, by an equation of the form $DAS = k\varepsilon^{-n}$. Suppose for Al-4.9% Cu, DAS was measured as 100 and 10 μm at the cooling rates 0.1 and 60 K/s, respectively.

(a) Determine k and n for the alloy.

(b) What cooling rate is necessary to reduce the arm spacing to 1.0 μm?

(c) Melt spinning and powder atomization are two techniques which are commonly used to achieve rapid cooling rates. Briefly describe what these techniques are.

14.4 An Al-5% Cu ingot is unidirectionally solidified under the conditions of no diffusion in the solid, complete diffusion in the liquid, and local equilibrium at the interface, so that the Scheil equation applies.

(a) Calculate the composition of the liquid when the ingot is 50 percent solid. What is the average composition of the solid?

(b) What is the interface temperature at this point?

(c) How much eutectic and second phase θ will have formed when the ingot is completely solidified?

(d) Plot the composition profile in the solidified ingot.

14.5 Some eutectics grow as rod eutectics while others grow as lamellar eutectics, as shown in the accompanying figure. Suppose that for an eutectic, the weight fraction of the α phase is f_α and for the β phase it is f_β and the densities of α and β are the same.

(a) Calculate f_α for the rod eutectic as a function of λ and r_α, where λ is the rod spacing and r_α is the radius of the α rods.

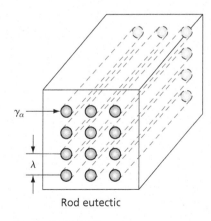

Rod eutectic

Fig. for Prob. 14.5

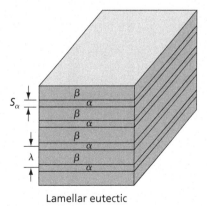

Lamellar eutectic

(b) Calculate f_α for the lamellar eutectic as a function of λ and S_α, where λ is the lamellar spacing and S_α is the thickness of the α plates.

(c) Calculate the α/β interface energy per unit volume of each eutectic, using $\gamma_{\alpha/\beta}$ for the interface energy per unit area.

(d) At what f_α would both eutectics have the same interfacial energy?

(e) What type of eutectic structure do you expect for an alloy in which the volume fraction of α is less than that calculated in part (d)? Justify your answer.

14.6 Assume that an alloy of nickel with 2 wt. percent aluminum is cast and, by metallographic examination, a dendrite arm spacing of 50 μ is observed. It is also determined that the composition difference between the center of an arm and the midpoint between two arms is 1 percent. Estimate the time required for homogenization if the annealing temperature is to be 1400°C and the composition difference is to be reduced to one-tenth of its original value. *Note*: see Table 12.5.

14.7 In a precipitation hardening alloy, coring usually results in the precipitation of an eutectic between the dendrite arms. With regard to aluminum-copper alloys, Singh and Flemings (Ref. 55) have assumed that this eutectic is divorced (consists only of the θ phase) and have derived an equation to describe the kinetics of homogenization that is similar to that discussed in the text for homogenization of a cored solid solution alloy. This equation, which applies to a solution treatment close to the solvus, is

$$g = g_0 e^{-\pi^2 Dt/4l_0^2}$$

where g is the volume fraction of the θ phase at time t, g_0 is the volume fraction of θ when t is zero, D is the diffusion coefficient, and l_0 is equal to half the dendrite arm spacing, as shown in the schematic figure on the next page. (Note that the concentration inside the precipitate is constant since it equals that of the θ phase. In the alpha phase, the concentration of copper is assumed to vary sinusoidally with distance and to be a minimum at the midpoint corresponding to the centers of the dendrites.)

(a) Assuming that at the annealing temperature D is 10^{-14} m²/s and that the original volume fraction of the θ phase is 1.0 percent, what solution time would be required to obtain a volume fraction of 0.01 percent if the dendrite arm spacing is 100 μ?

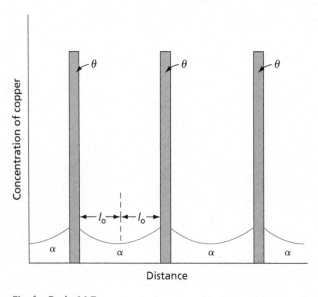

Fig. for Prob. 14.7

(b) What would be the corresponding time if the dendrite arm spacing was 10 μ?

14.8 Care must be taken in melting certain copper alloys to prevent absorption of hydrogen from the furnace gases or as a result of reaction with water vapor. The release of this gas on freezing can result in an open, porous structure of little value. As an exercise, compute the volume of hydrogen that should be released in 10 cc of copper melted in a hydrogen atmosphere at a pressure of one atmosphere when the metal is frozen. The atomic volume of copper is 7.09 cm^3 per gm atom. Assume the ideal gas law and refer to Fig. 14.41.

References

1. *Smithells Metals Reference Book*, 6th edition, Brandes, E. A., Editor, Butterworths, 1983, pp. 14–1 through 14–13.
2. Nachtrieb, N. H., ASM Seminar (1958), *Liquid Metals and Solidification*, p. 49.
3. Paskin, A., *Advances in Physics*, 16: 62, 1967, 223–240.
4. Fredriksson, H., and Fredriksson, E., *Mat. Sci. Eng. A*, 413–414, 2005, 455–459.
5. Wilde, G., Sebright, J. L., and Perepezko, J. H., *Acta Mat.*, **54**, 2006, 4759–4769.
6. Gokhale, A.B. and Abbaschian, R., "Containerless Processing Using Electromagnetic Levitation," in Space Commercialization: Platforms and Processing, Edited by F. Shahrokhi, et al., AIAA, March, 1990, 151–178.
7. Herlach, D. M., Cochrane, R. F., Egry, I., Fecht, H.-J. & Greer, A. L. Containerless processing in the study of metallic melts and their solidification. *Int. Mat. Rev.*, **38** 273 (1993).
8. Abbaschian, G. J., and Ravitz, S. F., *Crystal Growth*, **28** 16 (1975).
9. Hollomon, J. H., and Turnbull, D., *Prog. in Metal Phys.*, **4** 333 (1953). Pergamon Press, Inc., New York.
10. Turnbull, D., *Scripta Met.*, **11** 113 (1977).
11. Gilman, J. H., *Metal Prog.*, **42** (July 1979).
12. Scholze, H., *Glas*, p. 125, Springer Verlag, Berlin-Heidelberg-New York, 1977.
13. Dietzel, A., and Bruckner, R., *Glastechn. Ber.*, **30** 73 1957.
14. Turnbull, D., *Met. Trans. A*, **12A** 695 (1981).
15. Xu, D., Lohwongwatana, B., Duan, G., and Johnson W. L., *Acta Mat.*, **52**, 2004, 2621–2624.
16. Polk, D. E., and Giessen, B. C., *Metallic Glasses*, p. 1, ASM, Metals Park, Ohio, 1976.
17. Duwez, P., *ASM Trans.*, **60** 607 (1967).
18. Gilman, J. J., *Met. Prog.*, **42** 42 (July 1979).
19. Inoue, A., and Nishiyama, N., *Mater. Sci. Eng.*, A226–228, 1997, 401.
20. Inoue, A., *Acta Mater.*, 48, 2000, 279–306.
21. Masumato, T., and Maddin, R., *Acta Met.*, **19** 725 (1971).
22. Pampillo, C. A., and Chen, H. S., *Mat. Sci. and Eng.*, **13** 181 (1974).
23. Polk, D. E., and Giessen, B. C., *Metallic Glasses*, p. 30, ASM, Metals Park, Ohio, 1976.
24. Chalmers, B., *Trans. AIME*, **200** 519 (1954).
25. Abbaschian, R., and Kurz, W., in *Solidification Processes and Microstructures*, Edited by M. Rappaz, C. Bekermann, and R. Tvivedi, TMS publication, 2004, pp. 319–324.
26. Abbaschian, R., Clarke, C., "Recent progress in growth of diamond crystals" published in *Innovative Superhard Materials and Sustainable Coatings for Advanced Manufacturing: Proceedings of the NATO Advanced Research Workshop on Innovative Superhard Materials and Sustainable Coating,* Kiev, Ukraine, 12–15 May 2004, edited by Lee, J., and Novikov, N., (2005), Chapter 13, pp. 193–202.
27. Abbaschian, R., Zhu, H., Clarke, C., "High pressure–high temperature growth of diamond crystals using split sphere apparatus", *Diamond & Related Materials*, **14**, 2005, pp. 1916–1919.
28. Shigley, J.E., Abbaschian, R. and Clarke, C., *Gem and Gemology*, **38**, 4, 2002, p. 301.

29. Cahn, J. W., Hillig, W. B., and Sears, G. W., *Acta Metallurgica*, **12** 142 (1964).
30. Peteves, S. D., and Abbaschian, G. J., *J. Crystal Growth*, **79** 775 (1986).
31. Peteves, S. D., and Abbaschian, R., *Met. Trans. A*, **22A**, 1271 (1991).
32. Amini, S. and Abbaschian, R., "Nucleation and growth kinetics of graphene layers from a molten phase," *Journal of Carbon*, 2013, **51**, 110–123.
33. Amini, S., Kalaantari, H., Mojgani, S., and Abbaschian, R., "Graphite crystals grown within electromagnetically levitated metallic droplets," *Acta Materialia*, 2012, **60**, 7123–7131.
34. Weinberg, F., and Chalmers, B., *Canadian Jour. of Phys.*, **29** 382 (1951); **30** 488 (1952).
35. Glicksman, M. E., *Mat. Sci. Eng.*, **65**, 1984, pp. 45–55.
36. Glicksman, M. E., and Lupulescu, A. O., *J. Crystal Growth*, **264**, 2004, pp. 541–549.
37. Lacombe, J. C., Koss, M. B., and Glicksman, M. E., *Met. Mat. Trans. A*, **38A**, 2007, pp. 116–125.
38. Tiller, W. A., et al., *Acta Met.*, **1** 428 (1953).
39. Gulliver, G. M., *Metallic Alloys*, Charles Griffin and Co., Ltd, London, 1922.
40. Scheil, E., *Z. Metallk.*, **34** 70 (1942).
41. Pfann, W. G., *Trans. AIME*, **194** 747 (1952).
42. Sarreal, J., and Abbaschian, G. J., *Met. Trans. A.*, **17A** 2036 (1986).
43. Aziz, M. J., *J. Appl. Phys.*, **53** 1158 (1982).
44. Rutter, J. W., and Chalmers, B., *Canadian Jour. of Phys.*, **31** 15 (1953).
45. Chalmers, B., *Principles of Solidification*, John Wiley and Sons, New York, 1964.
46. Mullins, W. W., and Sekerka, R. F., *J. Appl. Phys.*, **444** (1964).
47. Walton, D., and Chalmers, B., *Trans. AIME*, **215** 447 (1959).
48. Cole, G. S., and Bolling, G. F., *Ultrafine Grain Metals*. Ed. Burke, J. J., and Weiss, V., p. 31, Syracuse University Press, Syracuse, N.Y., 1970.
49. Brick, R. M., and Phillips, A., *Structure and Properties of Alloys*. McGraw-Hill Book Company, Inc., New York, 1949.
50. Shewmon, P. G., *Transformations in Metals*, p. 41, McGraw-Hill Book Co., New York, 1969.
51. Flemings, M. C., *Solidification Processing*, p. 148, McGraw-Hill, New York, 1974.
52. Bower, T. F., Singh, S. N., and Flemings, M. C., *Met. Trans.*, **1** 191 (1970).
53. Jackson, K. A., and Hunt, J. D., *Trans. Met. Soc. AIME*, **236**, 1966, 1129.
54. Hunt, J. D., and Chilton, J. R., *J. Inst. Metals*, **92** 21 (1963–64).
55. Singh, S. N., and Flemings, M. C., *TMS-AIME*, **245** 1803 (1969).

Chapter **15**

Nucleation and Growth Kinetics

Learning Objectives

Upon completion of Chapter 15, you will be able to:

1. Understand nucleation and growth of liquid from a vapor phase, and express the energies involved as a function of nucleus radius and supersaturation
2. Compare and contrast the Volmer-Weber theory of nucleation and embryo distribution with Becker-Doring
3. State the similarities and differences between freezing and solid-solid transformations
4. Explain the free energies of homogeneous and heterogenous nucleation
5. Describe Zener's model of one-dimensional growth and specify the assumptions
6. Derive equations showing that interface position varies with the square root of time while growth velocity varies inversely with the square root of time
7. Differentiate between interface-controlled growth and diffusion-controlled growth
8. Describe the influence of temperature on dissolution and growth of precipitates

Structural changes in metallic systems usually take place by nucleation and growth whether a change in state, a phase change within one of the three states, or a simple structural rearrangement within a single phase is involved. A number of examples have already been discussed and others will be covered in later chapters. In Sec. 5.2, the subject of the nucleation of dislocations was considered. In the broadest sense, the nucleation of dislocation loops and their growth during straining represent a structural change. Nucleation and growth during recrystallization are discussed in Sec. 8.13; in solidification in Sec. 14.2 and 14.3; in precipitation hardening in Sec. 16.3 and 16.4; in the formation of deformation twins in Sec. 17.4 and 17.6; in martensite transformations in Sec. 17.18 and 17.19; and in the eutectoid transformation of steel to form pearlite in Sec. 18.3 and 18.4.

Because metallurgical transformations depend so strongly on the processes of nucleation and growth, this chapter will consider in greater detail the conditions controlling the formation of nuclei, their rate of formation, and the rate of their growth.

15.1 Nucleation of a Liquid from the Vapor

The simplest case that can be considered is that involving the formation of liquid droplets in a vapor phase. There are a number of reasons why this is so. In an ordinary nucleation process, a small region of a new phase is created within another phase. Associated with such an event is the formation of a boundary separating the two phases. The nature of this boundary is often difficult to define when the newly formed particle is very small and contains very few atoms. Fortunately, the conditions controlling the formation of nuclei are not normally sensitive to the properties of the particles when they are extremely small. There is always a critical particle size above which the particles become stable, and below which they are unstable. Below this critical size the particles are normally called *embryos*, or *clusters*, and above it, *nuclei*. When an embryo has grown to approximately the critical size, it is usually of sufficient size that one can treat the particle as macroscopic. Thus, in the region of primary interest, it is often possible to ignore the complications associated with very small groups of atoms and consider the nucleation problem from a simple macroscopic point of view. It is worth noting that the formation of a solid crystalline nucleus from a vapor or a liquid is more complicated than the formation of a liquid droplet from its vapor. This has to do with the shape of the particle as it grows. In a crystal the surface energy may be a function of the orientation of the surface, and a crystal nucleus of minimum surface energy may have a complicated shape. Furthermore, the growth of a crystal nucleus is also a more complicated process than the growth of a liquid droplet. In a solid, the atoms have to fit into the fixed pattern of the crystal lattice. The problem of attachment of atoms to the crystal surface has to consider this fact, and when the crystal grows under a relatively low driving force, the growth may occur by the addition of atoms to steps on the crystal surface.

Another simplifying factor in considering the nucleation of a liquid from a vapor that also applies to the nucleation of crystals from either the vapor or liquid phases is that the nucleation process does not involve strain energy. In solid-solid reactions, the newly formed phase often does not fit well into the matrix surrounding it, with the result that both the nuclei and the matrix usually become strained.

Finally, with regard to vapor-liquid transformations, it may be mentioned that the vapor phase itself is capable of being modeled in a relatively simple fashion using the concept of the ideal gas and the kinetic theory of gaseous collisions.

The nucleation of liquid droplets in a vapor will not be treated in detail. However, the basic principles and primary results will be covered. For a more extensive treatment, one is referred to standard texts on transformation (Refs. 1 and 2).

There are two basic ways that a drop of liquid can form in a vapor. It can form at a small foreign particle, such as a dust particle, in the vapor. In this case the formation of the liquid droplet is made easier by the presence of the foreign particle and one has a typical case of *heterogeneous nucleation*. On the other hand, a droplet may result from a concentration fluctuation of the atoms or molecules of the vapor. This would correspond to *homogeneous nucleation*.

Homogeneous nucleation, in general, is difficult to obtain. For example, in the freezing of a liquid to form a solid, which also involves nucleation and growth, heterogeneous nucleation is the rule rather than the exception. The nuclei of the solid form at mold walls or impurity particles in the liquid itself. Thus, very pure water does not freeze homogeneously until supercooled to temperatures of the order of $-40°C$ (233 K).

15.1 Nucleation of a Liquid from the Vapor

Homogeneous nucleation is difficult because, when a second phase particle forms, a surface is created between the particle and the matrix phase. This requires positive energy and opposes the loss of energy associated with creating the volume of the particle.

Accordingly, we may write the following equation to describe the formation of a liquid droplet like that shown in Fig. 15.1:

$$\Delta G = \Delta G_v + \Delta G_\gamma \qquad 15.1$$

where ΔG is the free energy associated with the formation of the droplet, ΔG_v is the free energy associated with the volume of the droplet, and ΔG_γ is the energy of the surface that is created. Since the surface energy depends on the area of the nucleus and the volume energy depends on its volume we may write:

$$\Delta G = A_1 r^3 + A_2 r^2 \qquad 15.2$$

where A_1 and A_2 are constants and r is the droplet radius. Figure 15.2 is a plot of Eq. 15.2. At small radii, the surface free energy, $(A_2 r^2)$, is larger than the volume free energy, $(A_1 r^3)$, and the total free energy is positive. The situation changes, however, as the radius grows in size, so that with larger radii the free energy becomes negative. The radius, r_0, is known as the *critical radius*. Below this value a droplet lowers its free energy by decreasing its size, so that droplets with radii smaller than r_0 tend to disappear. On the other hand, droplets with radii larger than r_0 undergo a decrease in free energy with increasing radius. For this reason they are stable and should continue to grow.

Homogeneous nucleation requires that thermal fluctuations produce droplets large enough to exceed r_0; otherwise the second phase cannot nucleate. Now we may also write Eq. 15.2 in the form:

$$\Delta G = 4/3 \pi r^3 \frac{\Delta g^{vl}}{v_l} + 4\pi r^2 \gamma \qquad 15.3$$

where Δg^{vl} is the chemical free-energy change per atom associated with the transfer of atoms from the vapor to the liquid phases, v_l is the volume of an atom in the liquid phase, r is the radius of the droplet, and γ is the specific surface free energy. Two important parameters are associated with the maximum in Fig. 15.2. The first is the critical particle radius, r_0. A relation for r_0 may be obtained by setting the derivative of Eq. 15.3 to zero and solving for r_0. This yields

$$r_0 = -2\gamma v_l / \Delta g^{vl} \qquad 15.4$$

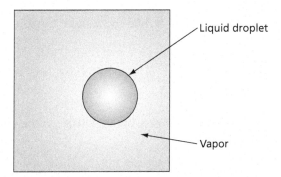

Fig. 15.1 A liquid droplet in a vapor

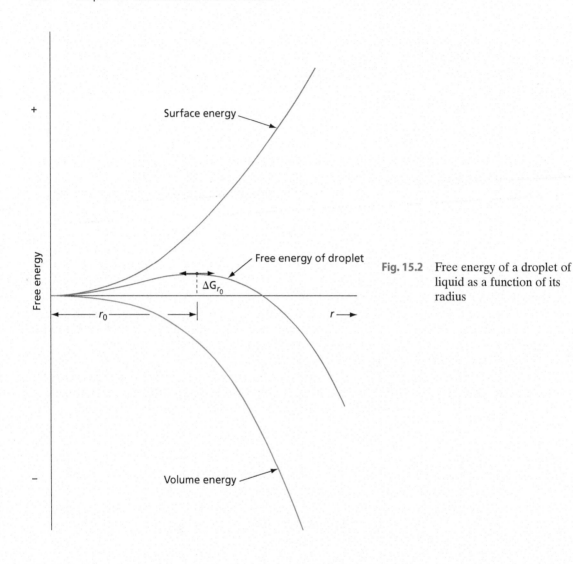

Fig. 15.2 Free energy of a droplet of liquid as a function of its radius

where r_0 is the critical radius, γ the specific surface energy, v_l the volume of a liquid atom, and Δg^{vl} the free-energy difference between an atom in the vapor and liquid phases. Note that in Fig. 15.2, Δg^{vl} is assumed negative. On this basis, r_0 will be positive.

The second parameter is ΔG_{r_0}. If Eq. 15.4 is solved for $\Delta g^{vl}/v_l$ and the result is substituted into Eq. 15.3, one obtains

$$\Delta G_{r_0} = 4\pi\gamma r_0^2/3 \qquad 15.5$$

where ΔG_{r_0} is the free energy of the nucleus at the maximum.

Equation 15.3 may also be written with the number of atoms, n, in the particle as the independent variable (i.e., instead of r the particle radius):

$$\Delta G_n = \Delta g^{vl} n + \eta\gamma n^{2/3} \qquad 15.6$$

where $n = (4/3)\pi r^3/v_l$ and η is a shape factor, which, if one assumes a sphere, equals $(3v_l)^{2/3}$. A diagram showing the variation of all the terms in this equation, with n the number of atoms, is shown in Fig. 15.3. Note that the free energy of the particle, ΔG_n, passes through a positive maximum and then decreases and eventually becomes negative.

In Fig. 15.3 it is assumed that the vapor is supersaturated. This is equivalent to assuming that it lies at a position on a temperature-pressure diagram like point a in Fig. 15.4. At this temperature and pressure, liquid is the stable phase and the chemical free energy change per atom, Δg^{vl}, is negative. On the other hand, if the vapor lies at point c, the vapor phase is the stable phase and Δg^{vl} is positive, and the free energy of the particle is always positive and increases monotonically with the number of atoms, as shown in Fig. 15.5. It is therefore evident that, in this case, there is no tendency for liquid droplets to continue to grow and to form large stable particles. This does not mean, however, that vapor atoms cannot come together and combine to form a distribution of very small embryos. In the vapor, the random motion of the atoms causes local fluctuations in concentration or density. Certain of these fluctuations may bring a sufficient number of atoms close enough together so that, in a small volume, the atoms may tend to arrange themselves in a structure more characteristic of the liquid phase than of the vapor phase. Such a fluctuation is called a *heterophase fluctuation*. These are basically the sources from which embryos may be considered to derive. In a stable phase, such as at point c in Fig. 15.4, the probability of observing an embryo of a given size containing n atoms may

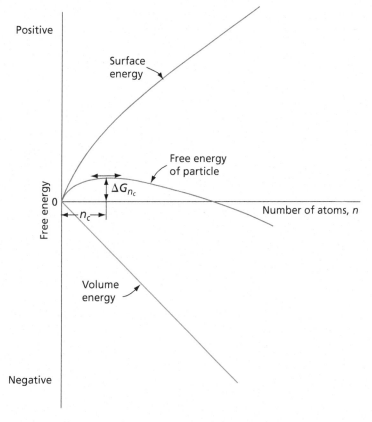

Fig. 15.3 Variation of the free energy per atom in a precipitate particle with the number of atoms in the particle

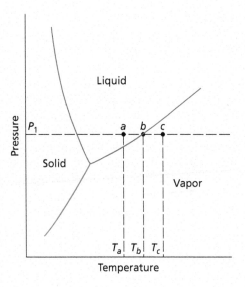

Fig. 15.4 A one-component (temperature-pressure) phase diagram

be shown to be proportional to $e^{-\Delta G_n/kT}$, where ΔG_n is the free energy increase associated with the formation of the particle. As a consequence, the numbers of particles containing n atoms is given by

$$N_n = Ce^{-\Delta G_n/kT} \qquad 15.7$$

where N_n is the number of these particles, ΔG_n is the free energy required to form a particle, C is a slowly varying function of n, and k and T have their usual significance. To a reasonable approximation, C may be assumed equal to N, the total number of atoms under consideration, provided that the number of particles is small compared to N. We may therefore write

$$N_n = Ne^{-\Delta G_n/kT} \qquad 15.8$$

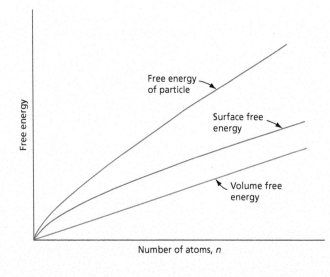

Fig. 15.5 Variation of the total surface and volume free energies with n for a particle at a temperature and pressure corresponding to point c in Fig. 15.4

Attention is now called to the similarity of this equation to Eq. 7.40, which gives the equilibrium number of vacancies in a crystal. Equation 15.8 could be derived in an analogous fashion to that used in obtaining the vacancy equation (Ref. 1). To do this one needs to consider the entropy of mixing associated with the division of atoms between the particles of various sizes.

Let us now consider the variation of N_n with n, as predicted by Eq. 15.8, when the vapor phase is stable. For convenience, we shall consider that the vapor is very close to being in equilibrium with the liquid phase, so that it lies nearly at point b in Fig. 15.4. In this case, the change in chemical free energy per atom, on going from the vapor to the liquid, is approximately zero, so that we may ignore the term containing Δg^{vl} in the equation for the free energy of the particle, which becomes

$$\Delta G_n = \eta \gamma n^{2/3} \qquad 15.9$$

and

$$N_n = N e^{-\eta \gamma n^{2/3}/kT}$$

This latter expression can be easily evaluated using characteristic approximate values for the various parameters. Those that are given roughly correspond to those for tin. Thus, let us assume that γ, the surface free energy, is 0.5 J/m², the boiling temperature is 2550 K, and the diameter of the atom in the liquid phase is 0.3 nm. In this case η, the geometrical factor $\{(4\pi)^{1/3}(3v_l)^{2/3}\}$, is about 4.3×10^{-19} m², and $\Delta G_n = 2.15 \times 10^{-19} n^{2/3}$ J. The corresponding variation of ΔG_n with n is shown as the left graph in Fig. 15.6. The right diagram corresponds to the variation of N_n with n, assuming that the total number of atoms in the assembly is that of one mole, or about 10^{24}. Note that in this latter diagram semilogarithmic coordinates have been used. The normal procedure in demonstrating a curve of this type is to use simple linear coordinates. However, as may be readily seen in Fig. 15.6, N_n varies by over 20 orders of magnitude in the interval shown in the diagram. It is very difficult to represent this fact using a linear system of coordinates.

While Fig. 15.6 is basically qualitative, it does show several very interesting features. First, the number of embryos is very large when the size of the embryo is small. Second, as the size of the embryo increases or the value of n becomes larger, the number of embryos falls very rapidly. For the example given in Fig. 15.6, while there should be about 10^{10} embryos containing about

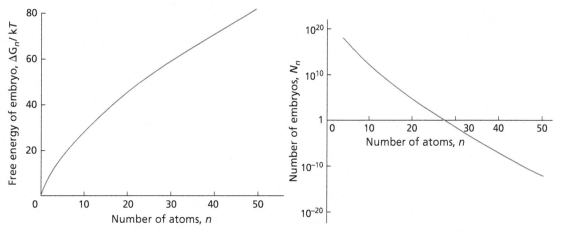

Fig. 15.6 The free energy of an embryo and the number of embryos per mole as a function of n, the number of atoms in an embryo, for a hypothetical metal vapor just above the boiling point of the metal

12 atoms in the assembly, there should be only one embryo containing about 25 atoms. For any embryo larger than this there is only a fractional chance of it existing. The significant point is that in a stable assembly, embryos varying through a wide range of sizes can exist. However, this situation is dynamic. Individual embryos are constantly growing and shrinking in size as they either add or lose atoms. Nevertheless, the distribution implied in Fig. 15.6 is stable. Furthermore, it should be mentioned that there is no tendency for the larger embryos to grow to form nuclei.

Now let us consider the supersaturated vapor corresponding to point a in Fig. 15.4. Here the system is metastable. It wants to transform to the liquid state, but the liquid has to be nucleated. The basic problem is how to represent the distribution of particles according to their sizes, as was done for the stable state in Fig. 15.6. At this point we shall introduce the symbol Z_n to represent the number of particles containing n atoms in a metastable assembly. It is equivalent to N_n in the stable assembly. We know that the larger embryos have to grow to form nuclei. This implies, in effect, that there is a constant net rate of particle growth. In the stable state, since the number of embryos of any one size is fixed, as many of these embryos leave a size class as enter it. In the metastable state, more embryos of a given size class will increase their size by gaining an atom than will decrease their size by losing an atom in a given interval of time. The net result is a steady progression of embryos that increase their sizes to eventually become nuclei. While this implies a basic difference in the variation of N_n with n for the two cases, the earlier theories assumed that in both cases the same relation holds, or that

$$Z_n = Ne^{-\Delta G_n/kT} \qquad 15.10$$

where for the supersaturated vapor the variation of ΔG_n with n is that given in Fig. 15.3. The fact that above n_c this figure shows that ΔG_n decreases and eventually becomes negative implies that given sufficient time, the number of particles of very large size should approach infinity. This, of course, implies that the entire assembly would be effectively in the liquid state. However, in most nucleation problems of interest, one normally starts with a phase that has just been brought into the supersaturated condition. In the present case this would signify a vapor that was originally at point T_c in Fig. 15.4 whose temperature had been lowered to T_a. For this case, it was originally proposed by Volmer and Weber (Ref. 3) that the distribution of embryos might correspond to the solid curve shown in Fig. 15.7. This curve is equivalent to the assumption that the distribution function is

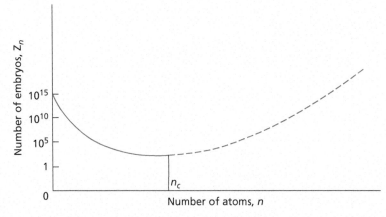

Fig. 15.7 The number of embryos as a function of the number of atoms in an embryo for a system at point a in Fig. 15.4 according to the Volmer-Weber theory

determined by the variation of ΔG_n (shown in Fig. 15.3) up to the critical embryo size at n_c. Above this value of n it is assumed that the particles grow rapidly and are effectively removed from the problem. Thus, a nucleus can be considered to be formed whenever an embryo of the critical size captures an additional atom. The rate of nucleation would, therefore, be proportional to the number of critical-size embryos, Z_{n_c}, and the rate of condensation of vapor atoms on these embryos. The former quantity is given by $Ne^{-\Delta G_n/kT}$, while the latter is proportional to the surface area of the critical-size embryo and the probability per unit area per unit time of a vapor atom condensing on the surface. Accordingly, we have

$$I = q_0 O_c N e^{-\Delta G_n/kT} \qquad 15.11$$

where I is the number of stable nuclei created per second, N is the total number of atoms in the assembly, O_c is the area of the critical embryo, q_0 is the probability per unit time per unit area of capturing a vapor atom, and ΔG_n is the free-energy change associated with forming an embryo of critical size. It should be noted that q_0 can be evaluated in terms of the kinetic theory collision factor $p/(2\pi mkT)^{1/2}$, where p is the pressure and m is the atomic mass.

A basic precept of the Volmer-Weber theory is that after a nucleus has formed it no longer has to be considered with respect to the formation of the nuclei that follow it. Of course, the atoms that have joined to make this nucleus are no longer available for the formation of other nuclei. In effect, the theory assumes that additional atoms are continually added to the system to compensate for those removed from the system. A theory of this type is therefore designated a quasi-steady-state theory. This kind of theory conforms best to the experimental conditions when the number of nuclei is small compared to the total number of atoms.

15.2 The Becker-Döring Theory

The Volmer-Weber theory, which served as the basis of the preceding discussion, assumes that once a nucleus of the critical size obtains an additional atom, it always grows into a stable nucleus. This assumption is not strictly true. The addition of an atom, or even several atoms, to critical nucleus will certainly tend to make it become more stable. However, this increase in stability has to be small. This is because, at n_c, the free energy reaches a maximum and, at the same time, $d\Delta G_n/dn$ passes through zero. Therefore, an embryo that has grown slightly beyond the critical size always has a nearly equal chance of shrinking back and becoming smaller. In formulating their nucleation theory, Becker and Döring (Ref. 4) recognized this fact. They also postulated that in the quasi steady state, the number of embryos of a given size should remain effectively constant, although, individually, embryos might either grow or shrink in size. The distribution function Z_n, giving the number of embryos corresponding to a given number of atoms, was assumed to be equal to the number of embryos predicted by the Volmer-Weber theory for very small-sized embryos, and for very large ones, Z_n was assumed to approach zero. Furthermore, unlike the Volmer-Weber theory, Z_n was not assumed to go to zero just above n_c. The difference in the assumptions with regard to Z_n in the two theories is illustrated in simple schematic fashion in Fig. 15.8. Finally, with regard to Z_n, the Becker-Döring theory (Ref. 4) does not require an exact specification for the distribution function near the critical nucleus size.

The Becker-Döring theory postulates that the nucleation rate I must equal the difference in the two rates at which embryos of a given size containing n atoms grow into embryos containing

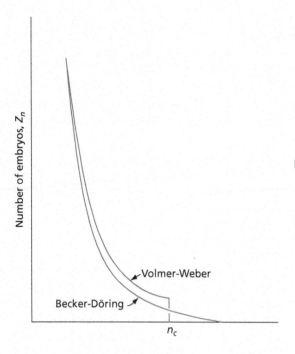

Fig. 15.8 A qualitative comparison of the Becker-Döring and Volmer-Weber distribution functions

$n + 1$ atoms, and the inverse rate corresponding to the reversion of embryos containing $n + 1$ atoms into those containing n atoms. This follows from the assumption that Z_n is independent of time. We may express this relation in the form

$$I = i_{n \to n+1} - i_{n+1 \to n} \qquad 15.12$$

where I is the number of nuclei formed per sec per m^3, $i_{n \to n+1}$ is the rate of conversion of embryos of size n to size $n + 1$, and $i_{n+1 \to n}$ is the corresponding opposite rate. For the case of a liquid nucleating in a vapor, this equation may also be written

$$I = q_0 O_n Z_n - q_{n+1} O_{n+1} Z_{n+1} \qquad 15.13$$

where Z_n is the number of embryos containing n atoms, q_0 is the probability per m^2 per sec of an atom jumping from the vapor to the liquid, O_n is the surface area of a droplet containing n atoms, and the corresponding symbols with a $n + 1$ subscript relate to an embryo containing $n + 1$ atoms that is assumed to lose an atom. It should be noted that whereas q_0 can be assumed to be independent of n, since the chance of a vapor atom striking a droplet should be nearly independent of the free energy of the droplet, this is not true of an atom that leaves the droplet to return to the vapor. In this case, the flux of atoms from the embryo should depend on the free energy of the droplet and thus on its size. As may be seen in Fig. 15.2, the free energy of an atom in an embryo becomes more positive with increasing size of the particle. This means that the rate of evaporation of atoms from a unit area of the surface of an embryo should increase as it becomes larger. An equation similar to the above can be written for all values of the number of atoms in the embryo, as, for example,

$$I = q_0 O_{n+1} Z_{n+1} - q_{n+2} O_{n+2} Z_{n+2} \qquad 15.14$$

It is thus apparent that there is a large set of these related equations. With the aid of the proper assumptions (Refs. 1 and 5), these relationships may be solved to yield the following expression for the nucleation current:

$$I = \frac{q_0 O_c N}{n_c} \left(\frac{\Delta G_{n_c}}{3\pi kT}\right)^{1/2} e^{-\Delta G_{n_c}/kT} \qquad 15.15$$

where the subscript c refers to quantities measured at the critical embryo size. It should be noted that the above equation only differs from the Volmer-Weber equation, given in the preceding section, by the pre-exponential term

$$\frac{1}{n_c}\left(\frac{\Delta G_{n_c}}{3\pi kT}\right)^{1/2}$$

It will be shown in the next section that for the case of freezing, ΔG_{nc} varies approximately as the square of the supercooling ΔT, or the temperature difference between the temperature of transformation and the equilibrium or freezing temperature. The factor T also appears in the denominator of the above expression. In spite of these facts, the variation of this pre-exponential factor with temperature, in the range of temperatures involved in nucleation problems, is much smaller than the variation due to the exponential term. In addition, the accuracy of experimental determinations of nucleation rate data is not very high. Therefore, the basic Volmer-Weber relationship may still be considered to adequately represent the nucleation current. The principal advantage of the Becker-Döring theory is it treats the nucleation problem on a basis that is physically more satisfying.

15.3 Freezing

Let us now consider the freezing of a pure metal. While a rigorous treatment would require that one should consider the problem of the shape of the embryo and nucleus as well as the problems associated with the attachment of atoms to the surface of a solid, these factors apparently do not strongly influence the rate of nucleation. One can therefore assume, to a reasonable approximation, that the solid embryos take a simple spherical shape and that atoms attach themselves to this particle at all points on its surface. In this case, the nucleation rate should depend on the number of critical nuclei, N_{n_c}, and the rate at which an atom can attach itself to the embryo. The first quantity, in analogy with the vapor-liquid reaction, should be given by $Ne^{-\Delta G_{n_c}/kT}$, where N is the total number of atoms in the assembly and ΔG_{n_c} is the free-energy change associated with the formation of a critical embryo. The rate of jumping of atoms across the boundary between the liquid and the solid may be expressed as the product of an atomic vibration frequency ν and the probability that an atom will make a successful jump. In making this jump there will normally be an energy barrier that the atom must overcome. Let us designate the magnitude of the free energy associated with this barrier as Δg_a. The nucleation rate or nucleation current then becomes

$$I = \nu e^{-\Delta g_a/kT}(Ne^{-\Delta G_{n_c}/kT}) \qquad 15.16$$

In analogy with the equation for the free energy associated with the formation of a liquid embryo in a vapor, we may set down the following expression for ΔG_n:

$$\Delta G_n = n\Delta g^{ls} + \eta \gamma_{ls} n^{2/3} \qquad 15.17$$

where n is the number of atoms in an embryo, Δg^{ls} is the free-energy difference between an atom in the liquid and solid phases, η is a shape factor, and γ_{ls} is the surface free energy of the liquid-solid boundary. It is now possible to evaluate Δg^{ls} approximately in terms of the heat of fusion. Thus at the freezing point

$$\Delta g^{ls} = \Delta h^{ls} - T_0 \Delta s^{ls} = 0 \qquad 15.18$$

where Δh^{ls} and Δs^{ls} are the heat of fusion and entropy of fusion per atom, and T_0 is the freezing point temperature. Solving this relation for Δs^{ls} gives $\Delta s^{ls} = \Delta h^{ls}/T_0$. Now both Δs^{ls} and Δh^{ls} are not strongly varying functions of the temperature near the freezing point and may be considered to be constant. Therefore, near the freezing point we have

$$\Delta g^{ls} = \Delta h^{ls} - \frac{T \Delta h^{ls}}{T_0} = \frac{\Delta h^{ls} \Delta T}{T_0} \qquad 15.19$$

In this last expression, ΔT is the temperature increment measured with respect to the freezing point and therefore represents the degree of supercooling.

The free energy associated with the formation of a nucleus may now be written

$$\Delta G_{n_c} = \frac{4\eta^3 \gamma_{ls}^3 T_0^2}{27(\Delta h^{ls})^2 \Delta T^2} = \frac{A}{\Delta T^2} \qquad 15.20$$

where $A = \dfrac{4\eta^3 \gamma_{ls}^3 T_0^2}{27(\Delta h^{ls})^2}$

This indicates that the free energy needed to form a critical embryo varies inversely as the square of the degree of supercooling. Thus not only does the size of the critical nucleus decrease with the supercooling, but so does the free energy required to form a critical nucleus. Let us now investigate the effect of some of these factors on the rate of nucleation. The nucleation rate, when expressed in terms of this simplified expression for ΔG_{n_c}, becomes

$$I = \nu N e^{-(\Delta g_a + A/\Delta T^2)/kT} \qquad 15.21$$

The most important part of this relation is the exponent, because it controls the nucleation rate. As written above, the exponent has two terms that do not depend upon the temperature in the same way. The first term, $\Delta g_a/kT$, represents the effect of the boundary on the jumping of atoms toward the nucleus. To a reasonable approximation, Δg_a may be assumed to be temperature independent and, therefore, this term should vary inversely with the absolute temperature. The nature of this variation is shown schematically in Fig. 15.9, where T_0 represents the melting temperature. In this diagram, the values of the exponential terms have been plotted using a logarithmic scale. Note that $\Delta g_a/kT$ increases continuously with decreasing temperature. On the other hand, the behavior of the second term, $A/\Delta T^2 kT$, is quite different. At the melting temperature, where ΔT is zero, this quantity is infinite. This implies that at the melting temperature, the nucleation rate must be zero. With decreasing temperature this term falls rapidly and eventually becomes much smaller than the first term. Since the nucleation rate is controlled by the sum of these two terms, it is a maximum where their sum is a minimum. This point is indicated on the diagram. Above this temperature, the rate of nucleation is effectively controlled by the energy barrier associated with forming the critical embryo; below it the nucleation rate is controlled by the jump rate of the atoms across the boundary from the liquid to the solid.

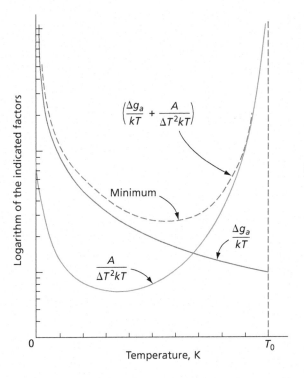

Fig. 15.9 Variation with temperature of the various terms involved in the exponent of the nucleation rate equation

Figure 15.9 is instructive in another fashion: it shows a very rapid rise of the exponent as one approaches the melting point. This is in excellent agreement with the fact that homogeneous nucleation is very difficult to achieve near the melting point. Conversely, it is also in agreement with the fact that in a very pure metal it is often possible to achieve a very high degree of supercooling.

15.4 Solid-State Reactions

Reactions that occur between liquid and solid or solid and solid are said to occur in condensed systems. In the preceding section we have considered the simplest possible type of condensed system transformation: the freezing of a pure metal. In a more general case, such as in the precipitation of a new solid phase from a single-phase solid solution, one may have to consider that the reaction may be largely controlled by diffusion. This situation would occur when solute, present in a relatively small concentration, has to diffuse through the matrix in order for the embryo to grow. In 1940 Becker (Ref. 6) proposed that when this occurs, the nucleation rate might be written

$$I = Ae^{-Q_d/kT}e^{-\Delta G_{n_c}/kT} \qquad 15.22$$

where A is a constant, Q_d is the activation energy for the diffusion of the solute, and ΔG_{nc} is the free energy required to form the critical embryo.

In condensed systems involving solid-state reactions, strain energy is usually an important factor. The formation of a new phase in a solid normally involves some form of deformation. An example will be discussed later in the martensite transformation sections. At that point it will be observed that most martensite transformations are believed to involve a plane-strain deformation. This corresponds to a shear parallel to the habit plane of the martensite plate and an expansion or contraction normal to this plane. Equivalent deformations can be expected in most solid-solid transformations. Strains in both the matrix and the newly formed particles will result from these deformations.

Like the surface energy, the strain energy also opposes the formation of a nucleus. It is normally assumed that the strain energy is proportional to the volume of the embryo and, therefore, to the number of atoms in the embryo. If this is true, the free energy associated with the embryo becomes

$$\Delta G_n = n(\Delta g^{\alpha\beta} + \Delta g_s) + \eta\gamma n^{2/3} \qquad 15.23$$

where Δg_s is the strain free energy per atom, $\Delta g^{\alpha\beta}$ is the free-energy difference between an atom in the matrix (alpha phase) and in the embryo (beta phase), η is the shape factor, n is the number of atoms in the embryo, and γ is the specific interface free energy. It should be noted that $\Delta g^{\alpha\beta}$ is negative, whereas both Δg_s and γ are positive. If Δg_s is larger in absolute magnitude than $\Delta g^{\alpha\beta}$, ΔG_n must increase in magnitude with increasing number of atoms in the embryo. Under these conditions it will be impossible to form a nucleus. In general, it is to be expected that Δg_s will be largest for coherent embryos. When the interface between the embryo and the matrix is coherent, there exists a perfect match between planes and directions across the interface separating the two structures. These crystallographic features may, of course, suffer a change in direction as they cross the interface. The general features of a coherent interface may be understood with the aid of Fig. 15.10. The upper diagram in this figure represents a supersaturated solid solution of atoms A and B. The A atoms are assumed to be the solvent, and the B the solute. It should be noted that the critical nucleus for coherent precipitates will normally have an approximately spherical shape. On the other hand, if the precipitate is softer than the matrix or it is elastically anisotropic, platelets will form.

For the purpose of simplification, we will assume that the solubilities of A in B and of B in A are both small enough at the temperature of precipitation so that they can be assumed to equal zero. Let it also be assumed that the precipitate is not a compound, such as Fe_3C, but the β phase: the crystal structure of the B atoms. In the present case, the B atoms will be attracted to each

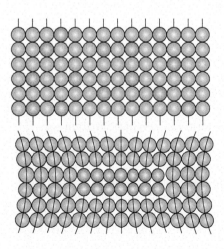

Fig. 15.10 Coherency. The upper figure represents a supersaturated solid solution of B atoms (green circles) in a matrix of A atoms (purple circles). The lower figure shows a coherent precipitate particle formed by clustering of the B atoms

other, and the first stage in the formation of a precipitate particle will be the formation of a cluster of B atoms. The lattice planes in this cluster will, in general, be continuous with the planes of the matrix, and the cluster is said to be a *coherent particle*. If, as shown in Fig. 15.10, the diameter of the solute atom differs from that of the solvent, then the matrix and the nucleus will both be strained by the presence of the latter. The strain associated with a nucleus will enlarge as its size increases, but its size cannot increase indefinitely. It is possible that the particle may break away from the lattice of the matrix, and when this occurs, a surface, or grain boundary, forms between the two phases. Such a loss of coherency would greatly reduce the state of strain associated with the precipitate particles. There exists an alternate possibility, however, that now appears to be more probable. This is that the strains associated with coherency are lost or reduced when a new phase that is noncoherent or less coherent forms. This point of view is consistent with the fact that multiple precipitate structures are observed in many alloys.

It is most important that it be recognized that the above discussion is far from complete. The subject of nucleation in the solid state is very complex. In this regard it should be mentioned that there also exists a third type of interface between the matrix and the second phase. This is a semicoherent boundary that is basically a coherent interface containing a grid of dislocations in the boundary. In a coherent boundary, the mismatch between the two crystal structures is small enough to be accommodated by elastic strains. In the semicoherent interface the mismatch is accommodated by dislocations. An extremely simple model for this type of boundary is furnished by the small-angle grain boundary shown in Fig. 6.2. Another model is that of the incoherent twin boundary of Fig. 17.16. Because of space limitations, the complexity of the total problem can only be suggested. In a given transformation, a coherent boundary should be preferred if the mismatch is extremely small, whereas with increasing degree of mismatch the semicoherent boundary might possess the smaller total energy. This statement has to be tempered by the fact that the size of the particle also has a bearing on the question. With growing particle size, the total nonchemical energy of the particle (strain plus surface energies) may be reduced if a coherent boundary is replaced by a semicoherent boundary.

At this point we shall ignore further considerations of the semicoherent boundary and consider only coherent and incoherent boundaries, primarily because these two types of boundaries have been treated in somewhat more detail and are perhaps easier to define. In the case of the coherent boundary, it has been shown (Ref. 1) that the strain energy of the embryo is not very dependent on the shape of the particle. In other words, the strain energy associated with the formation of a spherical embryo is not very different from that associated with a platelike or even needle-shaped embryo. This is not the case for an incoherent embryo. Here the shape may be very important, as has been demonstrated by Nabarro (Ref. 7), who attacked the problem from the point of view of isotropic elasticity. His computations also involved another basic assumption: that the matrix was much stiffer elastically than the particles, so that the strain energy of the particles could be neglected in comparison to that of the matrix. In effect, Nabarro considered the strain energy associated with expanding a cavity in the matrix. This expansion would be analogous to that resulting from the pumping of an incompressible fluid into a hole. The shapes of these cavities were assumed to correspond to various ellipsoids of revolution. His results are shown schematically in Fig. 15.11, plotted as a function of the ratio r_1/r_2 of the semi-axes of the ellipsoid of revolution, where the three semi-axes were taken to be r_1, r_2, and r_2. If the ratio r_1/r_2 is very small, the ellipsoid approximates a disc. If r_1 and r_2 are equal, the ellipsoid is a sphere; whereas if r_1 is much larger than r_2, the shape of the ellipsoid approaches a needle. As may be seen in the diagram, the strain energy is a maximum for a sphere and is least for a disc. Furthermore, as the thickness of the disc

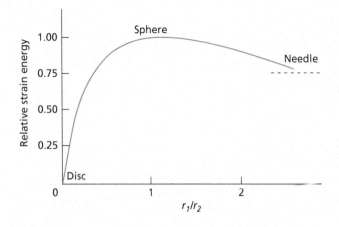

Fig. 15.11 The strain energy of an incoherent nucleus as a function of its shape. (After Nabarro, F. R. N., *Proc. Phys. Soc.*, **52** 90 [1940])

approaches zero, the strain energy also approaches zero. It should be pointed out that the above description is an oversimplification of the process. Incoherent nucleation may involve vacancies which by absorption or emission can relieve internal strains.

In evaluating the results of Nabarro for incoherent embryos, one has to recognize that the total surface energy of the embryo should be least for a sphere. Thus, the effect of shape on the two energy factors is opposing. Because the strain energy is a function of the volume of the particle, and the surface energy is dependent on the surface area, one should expect that the tendency to form discs should become more important as the particle grows in size. It should be mentioned that many precipitates certainly assume a plate-like morphology, and this is in good agreement with Nabarro's predictions. However, the tendency for a plate-like habit to develop can also be the result of other factors, such as the nature of the mechanism that controls the growth of the particle. Thus, for example, a plate may form if it is easier for atoms to attach themselves to the edge of the particle than to its flat surfaces. For more detailed analysis on the influence of crystallography on the shape and kinetics of homogeneous nucleation in solids, see a series of papers by LeGoues et al. (Refs. 8 through 11).

15.5 Heterogeneous Nucleation

In nature most nucleation occurs heterogeneously, as indicated previously at several points in this book. In the preceding sections we have shown that homogeneous nucleation is a rather difficult process. At this time a simple problem will be considered that illustrates one reason why heterogeneous nucleation may be a much easier process to invoke. In freezing, the container or mold walls often offer preferred sites for nucleation. This problem can be placed on a quantitative basis if several simplifying assumptions are made. First, let us assume that solid embryos form on the mold walls as spherical caps, as suggested in Fig. 15.12. Second, let us assume that at the positions where the surfaces of the caps make contact with the mold walls, there exists a state of quasi-equilibrium between the surface forces, as indicated in Fig. 15.13. This balance between the surface forces is assumed to take place in the direction parallel to the mold surface. In the direction normal to the mold surface, the surface tensions are not balanced, implying a net traction acting on the surface.

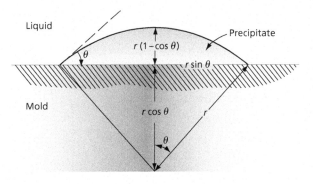

Fig. 15.12 A hypothetical spherical cap embryo

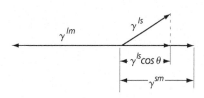

Fig. 15.13 The surface forces associated with the spherical cap embryo

With sufficient time to allow diffusion to occur in the mold wall, a complete balance between both sets of components might be obtained. The equation relating the surface forces parallel to the mold walls is

$$\gamma^{lm} = \gamma^{sm} + \gamma^{ls} \cos \theta \qquad 15.24$$

where γ^{lm}, γ^{sm}, and γ^{ls} are the surface tensions between liquid and mold wall, solid and mold wall, and liquid and solid, respectively, and θ is the angle of contact of the embryo surface with the mold wall. Note that θ is a function of only the three surface tensions. This signifies that no matter how large the size of the particle, the angle of contact should be the same. In effect, this says that as the embryo grows, its shape remains invariant as a spherical cap. This point is illustrated in Fig. 15.14.

We can now write an equation giving the free energy of an embryo. In this case we shall express the equation in terms of the radius of the surface of the spherical cap rather than in terms of the number of atoms in the embryo. The equation in question is

$$\Delta G^{het} = V_c \Delta g_v + A_{ls}\gamma^{ls} + A_{sm}(\gamma^{sm} - \gamma^{lm}) \qquad 15.25$$

where ΔG^{het} is the free energy associated with the heterogeneously nucleated embryo, V_c the volume of the cap-shaped embryo, A_{ls} is the area of the cap that faces the liquid, A_{sm} is the area of the interface between the embryo and the mold wall, Δg_v is the free energy per unit volume associated with the freezing process, and γ^{ls}, γ^{sm}, and γ^{lm} are the surface energies as defined above. Note that A_{sm} is multiplied by the difference between γ_{sm} and γ^{lm}. This is because the surface

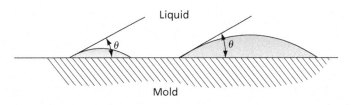

Fig. 15.14 As the embryo grows, its shape is invariant

formed between the embryo and the mold replaces an equivalent area of liquid and mold interface.

Now consider Fig. 15.12, which corresponds to a cross-section taken through the center of the embryo. Note that the height of the cap is equal to $r(1 - \cos\theta)$ and that the radius of the circular area, corresponding to the interface between the cap and the mold wall, is $r(\sin\theta)$. We may therefore set down the following relationships:

$$A_{ls} = 2\pi r^2(1 - \cos\theta)$$
$$A_{sm} = \pi r^2 \sin^2\theta \qquad \textbf{15.26}$$
$$V_c = 1/3\pi r^3(2 - 3\cos\theta + \cos^3\theta)$$

If all of the above relationships and Eq. 15.24 are substituted into the equation for the free energy of the embryo and the resulting expression is simplified, one obtains

$$\Delta G^{het} = \frac{4}{3}\pi r^3 \frac{(2 - 3\cos\theta + \cos^3\theta)}{4}\Delta g_v + 4\pi r^2 \gamma^{ls}\frac{(2 - 3\cos\theta + \cos^3\theta)}{4}$$

or

$$\Delta G^{het} = (V_{sph}\Delta g_v + A_{sph}\gamma^{ls})\frac{(2 - 3\cos\theta + \cos^3\theta)}{4} \qquad \textbf{15.27}$$

where V_{sph} and A_{sph} represent the volume and area of a sphere, respectively. In this last equation the quantity inside the first set of parentheses on the right side is the same as the free energy associated with the formation of a spherical embryo by homogeneous nucleation. Therefore, for the present example, we may equate the free energy of the heterogeneously nucleated embryo to the free energy of an equivalent homogeneously nucleated one. This relation is as follows:

$$\Delta G^{het} = \Delta G^{hom}\frac{(2 - 3\cos\theta + \cos^3\theta)}{4} \qquad \textbf{15.28}$$

where ΔG^{hom} is the free energy of a spherical embryo of radius equal to the radius of the cap of the heterogeneously nucleated particle. This expression states that the free energy to form a nucleus heterogeneously at the mold wall varies directly as the homogeneous free energy modified by a factor that is a function only of the angle of contact θ between the mold wall and the embryo. Since this angle is determined by the relative surface energies of the three surfaces in question, it is obvious that the free energy of heterogeneous nucleation is directly dependent on these surface energies. Also, since Eq. 15.28 holds for all values of r, it must hold when r is equal to r_c, the critical radius. This, in turn, signifies that

$$\Delta G^{het}_c = \Delta G^{hom}_c \frac{(2 - 3\cos\theta + \cos^3\theta)}{4} \qquad \textbf{15.29}$$

The factor $(2 - 3\cos\theta + \cos^3\theta)/4$ that converts the homogeneous nucleation free energy to the free energy of heterogeneous nucleation is plotted as a function of θ in Fig. 15.15. It is significant that this factor is very small for even rather large values of the contact angle. Thus, when θ is 10 degrees, the multiplying factor is of the order of about 10^{-4}. When θ is 30 degrees, it is only about 0.02, and at 90 degrees, or at one-half the limit of applicability of

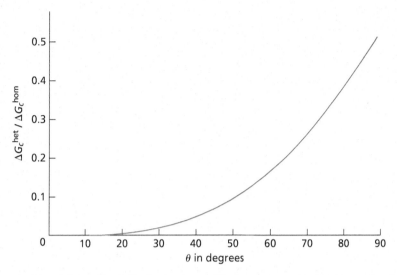

Fig. 15.15 The variation of the ratio of the free energy of heterogeneous nucleation to that for homogeneous nucleation as a function of θ, the angle of contact between the embryo and the mold wall

Eq. 15.29, it is still only equal to one-half. The significance of this large decrease in the free energy to form a critical nucleus cannot be overestimated. The effect has to be very large. In this regard, let us consider the nucleation rate equation for the case of heterogeneous nucleation. The number of critical embryos should be given by the equation

$$N^m e^{-\Delta G_c^{het}/kT} \qquad \textbf{15.30}$$

where N^m is the number of atoms in the liquid phase facing the mold wall. This is a reasonable assumption, because only the atoms in contact with the mold can form an embryo at the surface. Note that N^m is a much smaller number than N, the number of atoms in the assembly, that appears in the corresponding equation for the embryos undergoing homogeneous nucleation. On the other hand, the exponent in Eq. 15.30 can easily be enough smaller to more than compensate for this difference in the pre-exponential term. By analogy to our previous expressions for the nucleation rate, we may now write for the case of heterogeneous nucleation:

$$I = \nu N^m (e^{-\Delta g_a/kT})(e^{-\Delta G_c^{het}/kT}) \qquad \textbf{15.31}$$

where ν is a frequency and Δg_a is the free energy associated with the jump of an atom across the interface between the liquid and solid embryo.

15.6 Growth Kinetics

Growth kinetics become important once an embryo has exceeded the critical size and become a stable nucleus. In some cases this may occur at an extremely early stage in the development of the particle. As was the case with nucleation kinetics, there are also many facets to the study of growth kinetics, and only a couple of very simple examples will be considered. The primary purpose of these is to show the nature of the approach used in studying growth. First, it should be noted that reactions involving a large heat of transformation, such as occurs in freezing, present a special problem. Here the growth rate may be largely determined by the rate at which it is possible to remove the heat of fusion. The simplest growth theories, accordingly, ignore this type of problem and correspond to transformations in which the heat of transformation is very small, so that they may be assumed to be isothermal. Certain solid state reactions probably come closest to satisfying this condition. Thus, for example, consider the phase changes in iron (Ref. 12). The heat of fusion of iron is 15,360 J per mole, while the heat of transformation from the delta to the gamma phase is 690 J per mole, and that for the transformation from gamma to alpha is 900 J per mole.

Let us assume that we are concerned with a transformation in a pure substance that occurs within the solid state and that a particle has grown sufficiently so that it has become a stable nucleus. It is further assumed that the particle is spherical in shape, that no volume change is involved as atoms leave the alpha phase and join the beta phase in the particle, and that surface energy or capillarity effects can be neglected. This means that we shall ignore strain energy. Finally, it is assumed that growth occurs continuously or without the need for steps on the surface where the atoms can successfully attach themselves. The role of growth involving the movement of steps has received considerable attention in recent years. The subject is too lengthy to consider here, so the reader is referred to other sources (Refs. 1 and 13). In general, it is felt that step-wise growth is probably most important when the driving force for growth is very small.

Under the conditions stated above, it is reasonable to assume that the curve relating the free energy of an atom as it moves across the boundary from the alpha phase matrix onto the beta phase precipitate has the form given in Fig. 15.16. Note that in crossing the boundary, the atom has to pass over an energy barrier equal to Δg_a, and when it joins the beta phase it has a lower energy than it had in the alpha phase. The corresponding difference in free energy is indicated on the diagram by $\Delta g^{\alpha\beta}$. This difference in energy per atom is assumed to be equal to the chemical free-energy difference between an atom in the α and β phases.

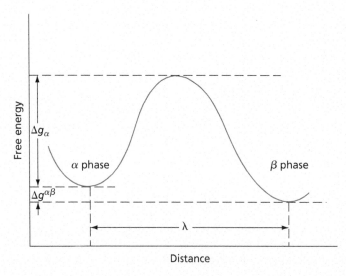

Fig. 15.16 The energy barrier associated with the growth of a precipitate during a solid-state phase change in a polymorphic pure metal

We may now write an expression for the net rate of atom transfer from the matrix to the particle as equal to the difference in the rate of atom movement toward and away from the particle. This is

$$I = S\nu e^{-\Delta g_a/kT} - S\nu e^{-(\Delta g_a + \Delta g^{\beta\alpha})/kT} \qquad 15.32$$

where S is the number of atoms facing the surface, ν is an atomic vibration frequency, and I is the net number of atoms per second leaving the matrix to join the beta phase. This expression reduces to

$$I = S\nu e^{-\Delta g_a/kT}(1 - e^{-\Delta g^{\beta\alpha}/kT}) \qquad 15.33$$

Let us now assume that when the atoms jump they move through an average distance λ. The velocity of the boundary will therefore be given by

$$v = \frac{\lambda I}{S} \qquad 15.34$$

where I/S represents the average number of jumps per second per atom facing the boundary, and λ is the distance corresponding to each of these jumps. The velocity may now be expressed in terms of the equation for I as

$$v = \lambda \nu e^{-\Delta g_a/kT}(1 - e^{-\Delta g^{\beta\alpha}/kT}) \qquad 15.35$$

Now consider $\Delta g^{\beta\alpha}$. This quantity represents the decrease in free energy when an atom joins the nucleus, ignoring surface and strain energy effects. In analogy with Δg^{ls}, this free energy may be expected to vary directly as the degree of supercooling. For a very small supercooling, $\Delta g^{\beta\alpha}$ will be small, and for a sufficiently small supercooling, we may assume $\Delta g^{\beta\alpha} \ll kT$. If this is true, then the exponential term $e^{-\Delta g^{\beta\alpha}}$ may be assumed to be approximately equal to the first two terms in its series expansion, or

$$e^{-\Delta g^{\beta\alpha}/kT} \approx 1 - \Delta g^{\beta\alpha}/kT \qquad 15.36$$

Under these conditions, the growth velocity becomes

$$v \approx \lambda \nu \left(\frac{\Delta g^{\beta\alpha}}{kT}\right) e^{-\Delta g_a/kT} \qquad 15.37$$

Since $\Delta g^{\beta\alpha}$, as indicated above, should vary approximately as ΔT (the supercooling), for small supercooling the growth velocity should be roughly proportional to the degree of supercooling. On the other hand, if the degree of supercooling is large, $\Delta g^{\beta\alpha}$ may become greater than kT. This is also encouraged by the fact that large supercooling normally requires that the temperature be lowered to very small values. When this happens, the exponential term becomes very small and the quantity $(1 - e^{-\Delta g^{\beta\alpha}/kT})$ may be equated to unity. The velocity equation may then be written

$$v \approx \lambda \nu e^{-\Delta g_a/kT} \qquad 15.38$$

This last expression shows that as T becomes very small, the growth velocity again approaches zero. Since it is also zero at the transformation temperature, the velocity must be a maximum

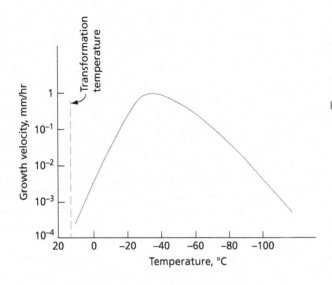

Fig. 15.17 The growth rate in the transformation of beta tin to alpha tin (on cooling). The equilibrium transformation temperature is 13°C. (Data of Becker, J. H., *J. Appl. Phys.*, **29** 1110 [1958])

at some intermediate temperature. This has actually been experimentally verified for the transformation of beta tin to alpha tin on cooling, as shown in Fig. 15.17. Note that just below the transformation temperature at 13°C, the growth rate is very small but increases with increasing supercooling; then it eventually decreases again as the temperature becomes very low. This general trend of the dependence of the growth rate on temperature is similar to that which was shown to be characteristic of nucleation.

15.7 Diffusion Controlled Growth

In the preceding section growth was considered under conditions where a new phase grows out of another phase by a simple transfer of atoms of a single component. We shall now treat the case where the transformation not only involves the formation of a new phase, but where this phase also possesses a composition different from that of the old phase. A relatively simple example of this type of reaction is the precipitation of iron-carbide particles from a supersaturated solid solution of carbon in alpha iron. This reaction has been studied in detail by Wert (Ref. 14) using torsion pendulum measurements. Since the carbide particles contain 6.7 percent carbon, whereas the matrix from which they form always has a small fraction of a percent of carbon, it is obvious that carbon atoms have to diffuse over relatively long distances in order for the carbide particles to grow.

As a result of Wert's observations on the precipitation of carbides in dilute solutions of alpha iron, Zener (Ref. 15) proposed a simple theory for this type of growth. For our present purposes, however, let us first consider Zener's theory in terms of a precipitate that is not spherical, but which grows as a plate in the direction normal to its surface. This assumption of one-dimensional growth greatly simplifies the problem. Now consider Fig. 15.18A. In this diagram the shaded region represents the precipitate after it has grown to a thickness equal to x. Figure 15.18B shows a schematic plot of the concentration of the solute B as a function of distance in which the concentration is expressed in atoms per m^3. Note that as in the case of iron-carbide precipitating in iron, the concentration of the solute in the precipitate is higher than that in the matrix. This concentration is shown as n_B^β. The assumption is also made that the time for the atoms to cross the boundary between the matrix and

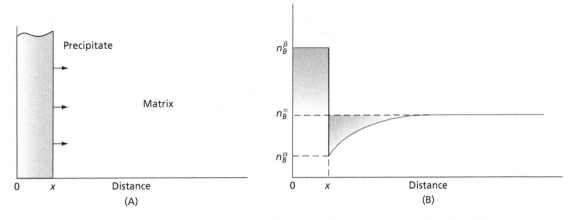

Fig. 15.18 A planar precipitate growing under conditions where growth is controlled by diffusion. The righthand drawing shows schematically how the composition varies with distance

the precipitate is small compared to the time for the solute to diffuse up to the precipitate. Under this condition, the concentration of B in that part of the matrix immediately facing the precipitate may be assumed to equal the equilibrium concentration. This is the concentration of B in the alpha phase that would be in equilibrium with the beta phase and is designated n_B^α in the diagram. Note that, in effect, one considers that "local" equilibrium exists at the boundary. The figure also shows that as one moves away from the interface, the concentration of B rises in the matrix to the value marked n_B^∞. This latter concentration is assumed to be that which the matrix possessed before the precipitation started. The drop in concentration near the interface is the result of the short-range jumping of atoms from the matrix into the precipitate.

Now let us assume that in a small time increment dt, the boundary of the precipitate moves forward into the matrix through a distance dx. This has the effect of converting a volume of material, equal to $A\,dx$, from a concentration n_B^α to a concentration n_B^β. In order for this to occur, $(n_B^\beta - n_B^\alpha) A\,dx$ atoms of B would have to diffuse up to the interface and cross over it. By Fick's first law, this number of atoms should also equal

$$-J\,dt = D\left(\frac{dn_B^\alpha}{dx}\right)dt \qquad 15.39$$

where J is the flux or number of B atoms crossing unit area per second, D is the diffusion coefficient which is assumed to be independent of concentration, and dn_B^α/dx is the concentration gradient of the B component in the matrix at the interface. Equating these two quantities, we have

$$\left(n_B^\beta - n_B^\alpha\right)dx = D\left(\frac{dn_B^\alpha}{dx}\right)dt \qquad 15.40$$

or solving for the interface velocity v

$$v = \frac{dx}{dt} = \frac{D}{\left(n_B^\beta - n_B^\alpha\right)}\frac{dn_B^\alpha}{dx} \qquad 15.41$$

Zener has given an approximate solution to this equation based on the drawing of Fig. 15.19. In this diagram the concentration distance curve, to the right of the boundary in the matrix in

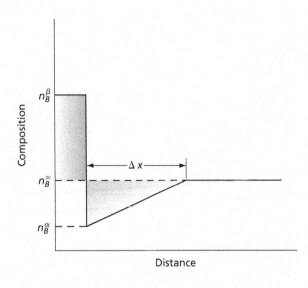

Fig. 15.19 Zener's approximation for the composition-distance curve

Fig. 15.18B, is assumed to be a straight line. The slope of this straight line is determined by the fact that the two shaded regions shown in the diagram should have equal areas. This is because the square section to the left of the boundary represents the B atoms that have joined the precipitate, while the triangular shaped area to the right of the boundary corresponds to the atoms that have left the matrix to enter the precipitate. Equating these areas gives us

$$\frac{1}{2}\Delta n_B^\alpha \Delta x = \left(n_B^\beta - n_B^\infty\right)x \qquad 15.42$$

where $\Delta n_B^\alpha = n_B^\infty - n_B^\alpha$. This yields

$$\Delta x = \frac{2\left(n_B^\beta n_B^\infty\right)x}{\Delta n_B^\alpha} \qquad 15.43$$

and the slope of the straight line concentration gradient is accordingly

$$\frac{\Delta n_B^\alpha}{\Delta x} = \frac{(\Delta n_B^\alpha)^2}{2\left(n_B^\beta - n_B^\alpha\right)x} = \frac{\left(n_B^\infty - n_B^{\alpha 2}\right)}{2\left(n_B^\beta - n_B^\infty\right)x} \qquad 15.44$$

Substituting this approximate slope into the velocity equation yields

$$v = \frac{dx}{dt} = \frac{D\left(n_B^\infty - n_B^\alpha\right)^2}{\left(n_B^\beta - n_B^\alpha\right) \cdot \left(n_B^\beta - n_B^\infty\right)x} \qquad 15.45$$

Integration of this differential equation leads to the following relation for the position of the boundary as a function of time:

$$x = \alpha_1^* \sqrt{Dt} \qquad 15.46$$

where

$$\alpha_1^* = \frac{\left(n_B^\infty - n_B^\alpha\right)}{\sqrt{\left(n_B^\beta - n_B^\alpha\right)}\sqrt{\left(n_B^\beta - n_B^\infty\right)}} \qquad 15.47$$

The subscript 1 of the parameter α_1^* indicates that the approximate solution is for one-dimensional growth. Differentiation of Eq. 15.46 gives the growth velocity in the simplified form

$$v = \frac{dx}{dt} = \frac{\alpha_1^*}{2}\sqrt{D/t} \qquad 15.48$$

This equation indicates that for one-dimensional growth, the interface position varies as $\sqrt{Dt}$ and its velocity as $\sqrt{D/t}$.

These results have a very general application and, as Zener has shown by dimensional analysis, whenever the growth is controlled by a simple diffusion process of the type indicated above, the interface position varies with the square root of the time, and growth velocity varies inversely as the square root of the time. Thus, in the case of three-dimensional or spherical growth, we may therefore write

$$x = \alpha_3^*\sqrt{Dt} \qquad 15.49$$

However, in this case Zener indicates that the parameter α_3^* depends to some degree upon the original concentration of the solute in the alpha phase.

To obtain a better appreciation of the roles of diffusion and the composition gradients in diffusion-controlled growth, let us consider the one-dimensional growth equation of Zener relative to the iron-carbon diagram. This diagram is redrawn in Fig. 15.20 with concentrations expressed in atomic percent. For the purposes of this discussion, we will assume that we can use concentrations in atomic percent to express the various values of n_B in the Zener equation where they will actually correspond to numbers of atoms per cm³. This will not affect our general conclusions. Suppose an alloy of iron containing 0.08 atomic percent of iron is heated to 727°C (1000 K),

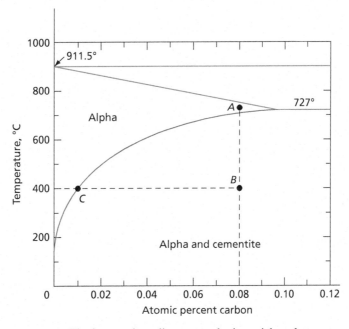

Fig. 15.20 The iron-carbon diagram at the iron-rich end

shown by point A in Fig. 15.20, where it is kept for a long enough period to give a homogeneous Fe-C solid solution. The alloy is then quenched to 400°C (673 K), shown by point B in the figure. The metal will be supersaturated and will contain 0.08 atomic percent carbon, or $n_B^\infty = 0.0008$. At the same time, by the phase diagram, the equilibrium concentration n_B^α is only 0.000011, point C in the figure. Since cementite contains a ratio of one carbon atom to three iron atoms, n_B^β is 0.25. Substituting these values in the expression for α_1^*, Eq. 15.47, gives us

$$\alpha_1^* = \frac{n_B^\infty - n_B^\alpha}{\sqrt{n_B^\beta - n_B^\alpha}\sqrt{n_B^\beta - n_B^\infty}}$$

$$= \frac{0.0008 - 0.000011}{\sqrt{0.25 - 0.000011}\sqrt{0.25 - 0.0008}} \approx 3.2 \times 10^{-3}$$

Any further lowering of the temperature to which the metal is quenched will not markedly change the value of α_1^*. This is because the equilibrium concentration is already small at 400°C, and any further decrease in n_B^α can have only a small effect on the numerator in the above equation. On the other hand, v also depends on $\sqrt{D}$ where, as shown in Sec. 13.5, the diffusion coefficient for carbon in iron may be represented by

$$D = D_0 e^{-82,900/RT} \qquad 15.50$$

Thus we may write

$$v = \alpha_1^* \left(\frac{D_0}{t}\right)^{1/2} e^{-41,430/RT} \qquad 15.51$$

Below 400°C the term $e^{-41,430/RT}$ becomes increasingly more important and, as T becomes smaller, the growth velocity can be considered to decrease primarily because of the decreasing diffusion rate of the carbon atoms.

At elevated temperatures, the roles of diffusion and concentration gradient are effectively reversed. Near 700°C (973 K), for example, a given change in temperature has a stronger effect on the concentration gradient than it does on $D^{1/2}$.

The result of these effects is a growth rate that maximizes at an intermediate temperature. It is small at high temperatures because the concentration gradient tends to diminish, and it is small at low temperatures because of the lowering of the rate of diffusion.

15.8 Interference of Growing Precipitate Particles

According to the Zener theory discussed in the preceding section, the growth velocity for simple geometries should vary inversely as $\sqrt{t}$. This means that with increasing time, the rate of boundary migration must decrease and become very small. This decrease in growth velocity is due to the fact that as the particle grows, it continues to absorb solute atoms from the matrix surrounding it, and the concentration gradient next to the particle decreases. In this theory, each particle is assumed to lie in a matrix of infinite extent. In an actual specimen there will be many particles drawing solute atoms from the matrix, and the distances between them will be finite. In the beginning, as the particles start to grow, there will be no effective competition between particles for the solute atoms, and the Zener assumption is in agreement with the facts. This is indicated schematically for the case of two parallel plate-shaped precipitates in Fig. 15.21A. However, with continued growth, the regions from which

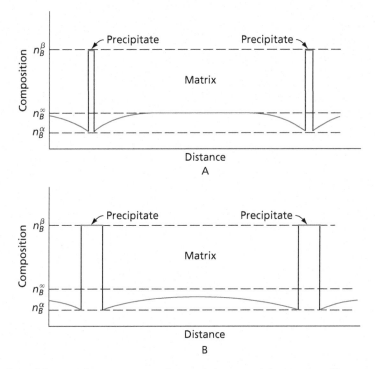

Fig. 15.21 Depletion of the matrix can occur as the precipitate particles grow, as illustrated by this schematic diagram showing planar precipitate particles

the particles are drawing solute atoms will tend to overlap. When this occurs, the maximum value of the solute concentration in the matrix has to fall below n_B^∞. This, in turn, must influence the effective value of the concentration gradient at the surface of the particles and, in general, should act to further decrease the rate of growth. The resultant change in the concentration profile is shown schematically in Fig. 15.21B.

15.9 Interface Controlled Growth

There is still another basic possibility under which growth can occur. In this case the precipitate, as in the preceding sections, is assumed to differ in composition from the matrix. However, here the growth rate is controlled by the mechanism that allows the solute atoms to cross over from the matrix to the precipitate. Thus, suppose that the time required for an atom to jump across the interface is very long compared to that required for it to diffuse to the interface. In this case, as shown schematically in Fig. 15.22, the solute concentration in the matrix will remain effectively constant throughout the matrix. However, with continued growth of the precipitate, the level of the concentration in the matrix has to fall. This means that the driving force for growth also has to fall, since it is directly related to the degree of supersaturation. Attention is called to the difference between this case and that discussed earlier involving the growth of a precipitate where there was no change in composition between the precipitate and the matrix.

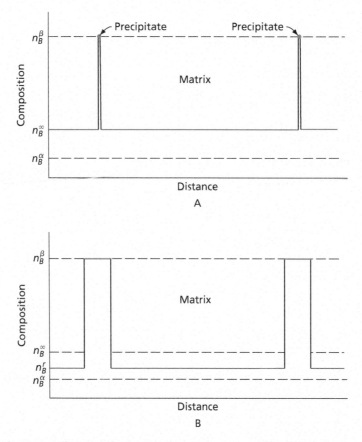

Fig. 15.22 In interface controlled growth, the transfer of atoms across the interface may be so slow that diffusion effectively removes the concentration gradient in the matrix

Returning to Fig. 15.22, attention is called to the fact that one-dimensional growth is again assumed, and two plate-like precipitates are indicated. Part A of this diagram is assumed to correspond to an early period in the growth of the precipitates, while part B corresponds to a later time. Note that as a result of the growth of the precipitates, the concentration level in the matrix drops from n_B^∞ to n_B^r. As before, the equilibrium concentration of solute in the matrix is taken as n_B^α and the concentration in the precipitate or beta phase as n_B^β.

We have already considered an example of growth controlled by an interface reaction: that involved in a phase transformation of a pure metal in Sec. 15.6. In some respects the present growth phenomenon is analogous to that already discussed. Thus, with reference to Fig. 15.16, a B atom, on leaving the matrix, has to pass over an energy barrier (Δg_a) in order to join the precipitate. After this has occurred, its energy level will have been lowered by the amount $\Delta g_B^{\alpha\beta}$. However, there is one important difference: a composition difference exists across the boundary. Thus the number of atoms capable of jumping, that face the surface, is different on the two sides of the boundary. In other words, there now exist two factors, S_1 and S_2, instead of a single factor S. However, at equilibrium the jump rate across the boundary has to be equal in the two directions. At the same time, the free energy of a jumping atom is also the same on both sides of the boundary so that $\Delta g^{\alpha\beta} = 0$.

Therefore, a reasonable assumption for a reaction occurring near equilibrium is that, effectively, $S_1 = S_2$. With this and the previous assumptions that the strain and surface energies may be neglected, we may write a velocity equation of the form

$$v = \frac{\gamma \nu \Delta g_B^{\alpha\beta}}{kT} e^{-\Delta g_a/kT} \qquad 15.52$$

where γ is a factor proportional to the jump distance of an atom, and ν has the same significance as in Eq. 15.33; $\Delta g_B^{\alpha\beta}$ is the free-energy difference between a B atom in the alpha and beta phases; and Δg_a is the energy barrier at the interface that a B atom has to pass over in order to join the precipitate. The energy difference, $\Delta g_B^{\alpha\beta}$, which is assumed to be much smaller than kT, can now be evaluated in terms of the difference in the partial-molal free energy of a B atom in the alpha and beta phases. This follows from the definition of the partial-molal free energy (see Sec. 10.7). Thus we have

$$\Delta g_B^{\alpha\beta} = \frac{1}{N} \{(\overline{G}_B^{\alpha})_t - (\overline{G}_B^{\beta})\} \qquad 15.53$$

where N is Avogadro's number, $(\overline{G}_B^{\alpha})_t$ is the partial-molal free energy of B in the alpha phase, and $(\overline{G}_B^{\beta})$ is the partial-molal free energy of B in the beta phase. Since the partial-molal free energies are expressed per mole, their difference is divided by N to yield a free-energy difference per atom. The subscript t on the term $(\overline{G}_B^{\alpha})_t$ signifies that this quantity changes with time. Substitution of the above relation into the velocity equation gives us

$$v = \frac{\gamma \nu e^{-\Delta g_a/kT}}{NkT} \{(\overline{G}_B^{\alpha})_t - (\overline{G}_B^{\beta})\} \qquad 15.54$$

which may also be written

$$v = \frac{\gamma \nu e^{-\Delta g_a/kT}}{RT} \{(\overline{G}_B^{\alpha})_t - (\overline{G}_B^{\alpha})_e\} \qquad 15.55$$

where $(\overline{G}_B^{\alpha})_e$ is the partial-molal free energy of B in the alpha phase when the latter is at equilibrium with the beta phase. The above substitution is possible since, at equilibrium, $(\overline{G}_B^{\alpha}) = (\overline{G}_B^{\beta})$. The expression can be further simplified if the alpha phase can be assumed to be an ideal solution since, by Eq. 10.25, we have

$$\overline{G}_B^{\alpha} = G_B^0 + RT \ln N_B^{\alpha} \qquad 15.56$$

which, on substitution into the velocity equation, yields

$$v = \gamma \nu e^{-\Delta g_a/kT} \{\ln(N_B^{\alpha})_t - \ln(N_B^{\alpha})_e\} \qquad 15.57$$

where $(N_B^{\alpha})_t$ and $(N_B^{\alpha})_e$ are the mole fractions of the B component in the alpha phase at time t and at equilibrium, respectively. Furthermore, as the concentration of B in the alpha phase approaches the equilibrium concentration, Eq. 15.57 may be approximated by

$$v = \gamma \nu e^{-\Delta g_a/kT} \{(N_B^{\alpha})_t - (N_B^{\alpha})_e\} \qquad 15.58$$

It is interesting to compare this relation for the interface-controlled growth of a precipitate with that derived earlier for the growth rate when it is controlled by diffusion. Thus, in the case of one-dimensional, diffusion-controlled growth, it was found that

$$v = \alpha_1^* \sqrt{D/t} \qquad 15.59$$

where α_1^* is a function of the compositions of the phases but, within limits, may be assumed to be effectively constant. In comparing these two rate equations, the basic difference between the composition-dependent terms $\{(N_B^\alpha)_t - (N_B^\alpha)_e\}$ and α_1^* should be emphasized. Until the solute becomes depleted, α_1^* can be considered a constant, whereas the expression $\{(N_B^\alpha)_t - (N_B^\alpha)_e\}$ is not a constant, since $(N_B^\alpha)_t$ decreases with increasing growth of the precipitate. This difference is due to the fact that in diffusion-controlled growth, local equilibrium is attained at the interface. This is not true in interface-controlled growth. In other words, local equilibrium at the interface is assumed during diffusion-controlled growth, but not during interface-controlled growth. More detailed discussion of the effect of interfacial kinetics on the interface motion can be found in an article by W. C. Johnson (Ref. 16).

Next, it should be noted that in a precipitation reaction, interface growth rates must generally be much slower than diffusion-controlled growth rates. In a broad sense, the precipitation process involves two mechanisms working in series. An atom has to diffuse up to the interface in order to jump across it. Only when the mean time for an atom to jump across the interface is very long will the interface reaction be the controlling one. Finally, it is evident that both types of reactions have rates that decrease with time. In the diffusion-controlled velocity equation, this is revealed by the fact that $v \propto t^{-1/2}$. In the interface-controlled velocity equation, the time dependence appears in the term $(N_B^\alpha)_t$, which decreases as the solute leaves the matrix.

15.10 Transformations That Occur on Heating

Only phase transformations that occur as a result of cooling have been considered to this point. The corresponding reactions that take place on heating are also significant. There are important differences between the kinetics of the reactions that occur on cooling and those that occur on heating. Thus, consider the melting of a pure metal. Appreciable superheating is almost impossible to achieve whereas, under the proper conditions, supercooling can be readily observed. The causes for this difference in behavior are probably related to the fact that it is very easy to heterogeneously nucleate a liquid droplet at an exterior surface or along a grain boundary. In Fig. 15.17 it was shown that the growth rate in the cooling phase transformation from white (β) to grey (α) tin followed a typical nucleation and growth temperature dependence. That is, the growth rate is small just below the transformation temperature, but with decreasing temperature it rises and passes through a maximum and then decreases again. The growth rate of the reversed phase transformation does not show this trend but rises steeply and continuously with increasing temperature, as shown in Fig. 15.23. At very low temperatures, the reaction rate, whether it is nucleation or growth, becomes small because diffusion rates always decrease rapidly with decreasing temperatures. On heating, the diffusion rate is always a steadily increasing function of the temperature. Therefore, it is generally true that both the rate of nucleation and the rate of growth will rise continuously with temperatures above a transformation temperature. In other words, the effect of temperature on the diffusion rate always tends to accelerate the reaction as the temperature is raised above the transformation temperature.

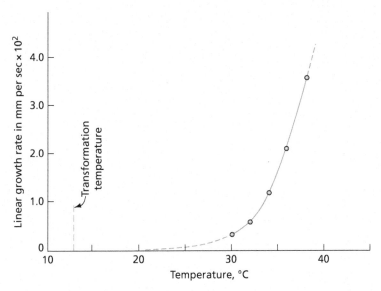

Fig. 15.23 The growth rate in the transformation from alpha tin to beta tin (on heating). (After Burgers, W. G., and Groen, L. J., *Disc. Faraday Soc.*, **23** 183 [1957])

15.11 Dissolution of a Precipitate

Another form of reaction associated with heating is the dissolution of a precipitate. In principle this is the reverse of the precipitation process. The kinetics are usually somewhat different, and this will be discussed briefly later. For the present, let us consider the nature of the process. Figure 15.24 shows the aluminum-rich end of the aluminum-copper phase diagram. Suppose that one has an alloy containing about 4 percent copper that is nearly in equilibrium at 200°C (473 K). The structure should consist of small θ particles in a matrix of almost pure aluminum. Such a structure is shown in Fig. 15.25. Since these particles are readily visible under an optical microscope, the metal is actually in the overaged condition. To develop this structure it was necessary to age the alloy for 200 hours at 200°C (473 K). Aging at room temperature will not produce a structure of this type within any reasonable time interval, since the reaction rates at room temperature are too slow to produce the equilibrium structure. As can be deduced from Fig. 15.24, heating this specimen to 540°C (813 K) will increase the solubility of copper in the aluminum to where it can be completely dissolved. Figure 15.26 shows some measurements made by Batz, Tanzilli, and Heckel (Ref. 17) that reveal some very interesting data about the solution process of this alloy at 540°C (813 K). While the original average particle size was larger than that shown in Fig. 15.25, this has no real bearing on the results. The abscissa of Fig. 15.26 gives the size of the particles in microns, while the ordinate gives the number per unit volume in a given size class. Each solid line represents the size distribution of the particles. Note that with increasing time these curves are displaced to the left with little change in shape. This implies that the size distribution curve is essentially unaltered in shape as the particles dissolve.

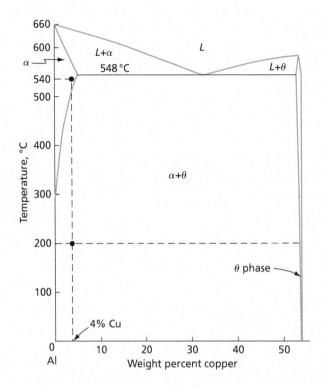

Fig. 15.24 The aluminum end of the aluminum-copper phase diagram

Aaron and Kotler (Ref. 18) have pointed out a number of practical reasons why the kinetics of dissolution should be better understood. In those cases where the presence of the second-phase particles results in an alloy of superior properties, it is desirable to know how it might be possible to develop a precipitate that will dissolve very slowly if the metal is heated. On the other hand, when the presence of a precipitate is harmful to the properties of an alloy, it is advantageous to know how to decrease the life of the precipitate so that the metal may be more easily homogenized.

The dissolution of a precipitate differs in one fundamental respect from the growth of a precipitate. The process does not require nucleation since the particles that are to be dissolved are already present. The problem, therefore, is essentially one of reversed growth, although the kinetics are usually different. In a one-dimensional analysis, it is possible to show that the kinetics of growth and dissolution are both governed by a $t^{1/2}$ law. Thus, Aaron (Ref. 19), using an analysis similar to that of Zener in Sec. 15.6, has shown that the thickness of a precipitate should vary as

$$x(t) = x_o - k\sqrt{Dt} \qquad 15.60$$

where $x(t)$ is the thickness of the planar precipitate at any time t, x_o is its original thickness, k is a constant, and D is the diffusion coefficient of the solute.

In the case of a spherical precipitate, the growth and dissolution kinetics are not generally so simply related. This subject is beyond the scope of this book, and for further information, one is referred to the review paper by Aaron and Kotler (Ref. 20).

15.11 Dissolution of a Precipitate

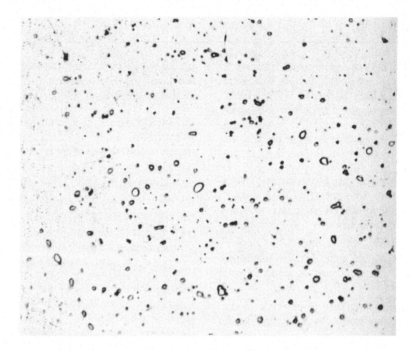

Fig. 15.25 Photomicrograph showing θ or $CuAl_2$ particles in an aluminum-4 percent copper alloy. Magnification 1500× (From Batz, D. L., Tanzilli, R. A., and Heckel, R. W., *Met. Trans.*, **1** 1651 [1970]. Photograph courtesy of R. W. Heckel. With kind permission from Springer Science and Business Media.) Photograph courtesy of R. W. Heckel

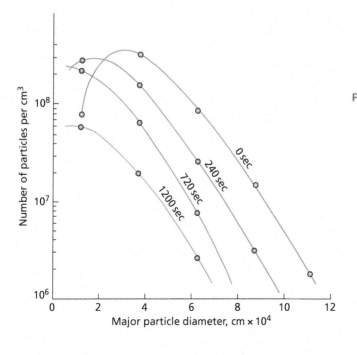

Fig. 15.26 Variation of the distribution of the θ precipitate particles with time at the solution temperature 540°C in an aluminum-4 percent copper alloy. (From Batz, D. L., Tanzilli, R. A., and Heckel, R. W., *Met. Trans.*, **1** 1651 [1970]. With kind permission from Springer Science and Business Media)

Problems

15.1 The surface energy of a pure metal liquid, γ, is 600 dynes/cm^2; the volume of an atom of this metal in the liquid is 2.7×10^{-25} cm^3; and the free-energy difference between an atom in the vapor and liquid, Δg^{vl}, is -2.37 J. Under these conditions, what would be the critical radius of a droplet, r_o, in nm and the free energy of the droplet, ΔG_{ro}, in J?

15.2 The following data concern liquid magnesium at a temperature close to its boiling point of 1380 K: the surface energy of the liquid-vapor interface is 0.440 J/m^2; the density of the liquid phase is 1.50×10^3 kg/m^3; and the atomic weight is 0.02432 kg/mol.

(a) First determine v_l, the volume of the liquid per atom. To do this, use the density and the atomic weight.

(b) Now compute the number of embryos in the vapor at a temperature just above the boiling point, that contain 10 magnesium atoms.

15.3 Derive Eq. 15.20 of the text, starting with Eqs. 15.17 and 15.19.

15.4 The following data are relative to the freezing of the pure metal copper: the melting point of copper is 1356 K; its latent heat of fusion is 2.117×10^5 J/kg; its atomic weight is 0.06355 kg/mol; the surface energy of the liquid-solid interface is 0.177 J/m^2; and the density is 8.35 kg/m^3.

(a) Compute Δh^{ls}, the heat of fusion per atom, in J.

(b) Evaluate the coefficients of both n and $n^{2/3}$ in Eq. 15.15.

(c) Determine the number of atoms in a critical nucleus when the supercooling is 5, 50, and 200 degrees.

15.5 At this point use the data from Prob. 15.3 to evaluate the constant, A, in Eq. 15.20, where $A = 4\eta^3 \gamma_{ls}^3 T_0^2 / 27(\Delta h^{ls})^2$. Use Eq. 15.18 in the next two problems.

15.6 (a) Calculate the free energy, ΔH_{nc}, needed to create a critical nucleus in copper when the amount of supercooling is 5 degrees.

(b) How many times larger than kT is ΔG_{nc}?

(c) Do the answers in parts (a) and (b) of this question imply that homogeneous nucleation is probable just below the melting point?

15.7 (a) Determine the free energy, ΔG_{nc}, associated with a critical nucleus at 1256 K, where 1256 K corresponds to 100 degrees of supercooling.

(b) Now compute ΔG_{nc} if the supercooling is increased to 264 degrees, which by Table 14.1 corresponds to the maximum amount of supercooling that has been observed in copper.

(c) Now discuss the results obtained in parts (a) and (b) of this question with respect to the probability of observing homogeneous nucleation.

15.8 Consider the nucleation rate, I, on freezing. It is normally assumed that the vibration rate of the atoms in solids is about 10^{13} Hertz. Assume that in the case of copper, the energy barrier, Δg_a, that an atom passes over in going from a liquid to the solid is about 4.8×10^{-20} J.

(a) What is the nucleation rate per mole of copper when the supercooling is 264 degrees?

(b) Discuss the significance of the answer to part (a) of this question.

15.9 In the case of pure silver, according to Table 14.1, the maximum supercooling that has been observed is 227 degrees. Determine the homogeneous nucleation rate in silver at this degree of supercooling using the following data. Note, to determine v_s, use the volume of the unit cell of silver divided by the number of atoms per unit cell in this metal. In the following table the lattice parameter, a, is given for 1233 K.

$a = 4.17 \times 10^{-10}$ m
$T_m = 1234.9$ K $\nu = 10^{13}$ Hz
$\gamma_{ls} = 0.123$ J/m^2 $N_0 = 6.023 \times 10^{23}$ atoms/mol
$\Delta H_{ls} = 11{,}960$ J/mol $\Delta g_a = 6.4 \times 10^{-20}$ J

15.10 The accompanying diagram is for a hypothetical embryo of silver growing against an arbitrary mold wall. With the aid of this diagram,

(a) Compute the angle of contact, Θ, of the embryo with the mold wall.

(b) Determine the magnitude of the factor that may be used to convert the homogeneous free energy needed to obtain a nucleus into that of the corresponding heterogeneous free energy.

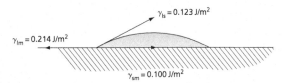

Fig. for Prob. 15.10

15.11 Under the conditions of Prob. 8:

(a) First determine the homogeneous nucleation rate when silver is supercooled by 20 degrees.

(b) Next determine the heterogeneous nucleation rate when silver is supercooled 20 degrees. Use the data in Prob. 8 to determine I in nuclei per m² per second; that is, express N^m in terms of atoms per m².

(c) Discuss the difference between the heterogeneous and the homogeneous nucleation rates in terms of freezing at 20 degrees of supercooling.

Note that the assumed angle of contact, Θ, from Problem 15.10 is relatively large and of the order 20°.

15.12 Although the growth rate equations in Sec. 15.6 ignore several important factors, such as the strain energy and the surface energy, it is interesting to apply this simple theory to an analysis of the growth of the alpha phase from the gamma phase in pure iron. For this purpose, take the data given in the following table and make the indicated computations. Assume that $\lambda = a$ and that Δg_a is independent of the temperature.

$$a = 2.866 \times 10^{-10} \text{ m}$$

$$\nu = 10^{13} \text{ Hz}$$

Transformation temperature, $T_0 = 1184.5$ K

$$\Delta H^{\gamma\alpha} = 900 \text{ J/mol}$$

$$\Delta g_a = 2.08 \times 10^{-19} \text{ J}$$

(a) First take Eq. 15.35 and evaluate all the parameters in this equation so that it becomes a simplified relation between v and ΔT and T. Next write a simple computer program to evaluate v between 1184.5 and 884.5 K in steps of 10 degrees.

(b) Plot the results from part (a) to show the variation of v with T. Express v in units of m/s.

15.13 (a) Suppose that an iron specimen containing 0.09 atomic percent carbon is equilibriated at 720°C (993 K) and then rapidly quenched to 300°C (573 K). Determine the length of the time needed for one side of a plate-shaped carbide precipitate to grow out by 10^3 nm.

(b) How wide a layer of the matrix next to a plate would have its carbon concentration lowered from 0.09 percent carbon to that corresponding to n_α^e to form a layer of cementite 10^3 nm thick?

(c) How long would it take to increase one side of the plate by 10 nm?

References

1. Christian, J. W., *The Theory of Transformation in Metals and Alloys*, 3rd edition, Elsevier, 2002.
2. Machlin, E. S., *An Introduction to Aspects of Thermodynamics and Kinetics Relevant to Materials Science*, 3rd edition, Elsevier, 2007.
3. Volmer, M., and Weber, A., *Z. Phys. Chem.*, **119** 227 (1926).
4. Becker, R., and Döring, W., *Ann. Phys.*, **24** 719 (1935).
5. Rasmussen, D. H., Appleby, M. R., Leedom, G. L., and Babu, S. V., *J. Crystal Growth*, 64, 1983, 229–238.
6. Becker, R., *Proc. Phys. Soc.*, **52** 71 (1940).

7. Nabarro, F. R. N., *Proc. Phys. Soc.*, **52** 90 (1940); and *Proc. Roy. Soc.*, **A175** 519 (1940).
8. LeGoues, F. K., Lee, Y. W., and Aaronson, H. I., *Acta Metall.*, 32 (10), 1984, 1837–1843.
9. LeGoues, F. K., Aaronson, H. I., and Lee, Y. W., *Acta Metall.*, 32 (10), 1984, 1845–1853.
10. LeGoues, F. K., and Aaronson, H. I., *Acta Metall.*, 32 (10), 1984, 1855–1864.
11. LeGoues, F. K., Wright, R. N., Lee, Y. W., and Aaronson, H. I., *Acta Metall.* 32 (10), 1984, 1865–1870.
12. Darken, L. S., and Gurry, R. W., *Physical Chemistry of Metals*, p. 397, McGraw-Hill Book Company, New York, 1953.
13. Fine, M. E., *Introduction to Phase Transformations in Condensed Systems*, The Macmillan Company, New York, 1965.
14. Wert, C., *J. Appl. Phys.*, **20** 943 (1949).
15. Zener, C., *J. Appl. Phys.*, **20** 950 (1949).
16. Johnson, W. C., *Met. and Mat. Trans. A*, 29, 1998, 2021–2032.
17. Batz, D. L., Tanzilli, R. A., and Heckel, R. W., *Met. Trans.*, **1** 1651 (1970).
18. Aaron, H. B., and Kotler, G. R., *Met. Trans.*, **2** 393 (1971).
19. Aaron, H. B., *Materials Sci. J.*, **2** 192 (1968).
20. Aaron, H. B., and Kotler, G. R., *Met. Trans.*, **2** 393 (1971).

Chapter **16**

Precipitation Hardening

Learning Objectives

Upon completion of Chapter 16, you will be able to:

1. Describe the processing steps involved in precipitation hardening of an alloy, and justify each step
2. Relate the free energy changes resulting from volume, surface, and strain energies on the shape and size of precipitates during a precipitation process
3. Discuss effects of temperature and composition on the precipitation hardening of Al-Cu alloys
4. Differentiate between GP zones, θ'' θ' and θ precipitates, and their effects on hardening
5. Understand differences between peak aging and overaged alloy
6. Describe interphase precipitate formation in alloy steels containing carbide formers
7. Explain grain boundary precipitate formation and formation of solute depleted zones

The equation for the equilibrium concentration of carbon in body-centered cubic iron as a function of temperature (see Eq. 9.13), $C = Be^{-Q/RT}$, is not typical only of the iron-carbon system. Similar relationships are observed for nitrogen, or hydrogen in interstitial solution in iron, or for all three elements (carbon, nitrogen, and hydrogen) in interstitial solution in other metals, providing that the maximum solubility is under approximately 1 atomic percent. This 1-percent limitation is usually unimportant in interstitial solid solutions as they are almost always very dilute.

The above law may also be used to predict the maximum solubility in substitutional solid solutions. Here again it is quite accurate as long as the maximum solubility at high temperature does not exceed 1 percent. In those cases where the solubility exceeds 1 percent, the equation still gives a curve that is a useful approximation of experimentally observed solubility relations. Since there are many binary alloys (two-component) with limited solubilities of the component metals in each other, curves like that in Fig. 16.1 are quite common. This figure, incidentally, is the same as Fig. 9.3 in Chapter 9 and shows the solubility of carbon in alpha-iron.

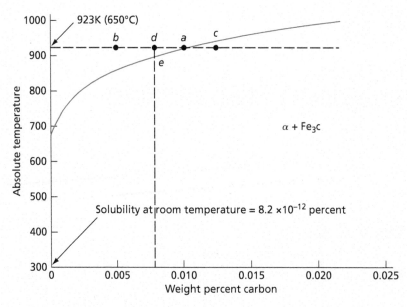

Fig. 16.1 Solubility of carbon in alpha iron

Solubility relationships of the type shown in Fig. 16.1 have great practical significance, for they make possible the precipitation hardening, or age hardening, of metals—an extremely important method of hardening of metals. This type of hardening is used most often in commercial strengthening of nonferrous alloys, especially aluminum and magnesium alloys. For the sake of convenience, the age hardening process in iron alloyed with small quantities of carbon will be taken up here. The basic principles developed, however, are applicable to other more general alloy systems.

16.1 The Significance of the Solvus Curve

Curves of the type shown in Fig. 16.1 are called *solvus lines*. The significance of this line will now be investigated. For this purpose, consider a specific temperature: 923 K (650°C). Assume that at this temperature a grain of ferrite is in contact with a grain of cementite, as shown in Fig. 16.2. In a system of this type, it is possible for carbon atoms to leave the solid solutions (ferrite) and enter the cementite. When this happens, however, three iron atoms of the solution must also join the cementite lattice in order to maintain the strict stoichiometric ratio characteristic of iron-carbide: three atoms of iron to one of carbon. Similarly, when a carbon atom leaves Fe_3C to enter the solution, three iron atoms must leave the compound. In the very dilute solutions presently considered (approximately 0.01 percent carbon), the effect on the concentration of the solid solution of the addition or removal of solvent atoms (iron) simultaneously with the carbon is negligible. It can, therefore, be assumed that concentration changes are due only to transfers of carbon atoms between the two phases. It must be kept in mind, however, that as carbon enters iron carbide, the latter phase grows in volume, and that the composition of this phase does not change. On the other hand, when carbon enters the solid solution (ferrite), the composition of the latter changes.

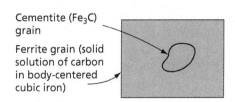

Fig. 16.2 Grain of cementite in contact with a ferrite grain

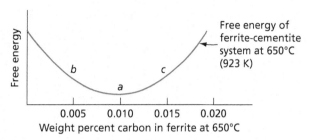

Fig. 16.3 Hypothetical free-energy curves at 600°C for the ferrite-cementite system of Fig. 16.2 as a function of carbon concentration in the ferrite

Figure 16.3 shows a hypothetical plot of the free energy of the ferrite plus cementite system as a function of the carbon concentration in the ferrite. Point *a* on this curve represents the composition at which the free energy of the ferrite-cementite system is a minimum. It is also the same composition (0.010 percent) that one obtains from the solvus curve at 923 K (650°C). The ferrite in a ferrite-cementite aggregate will thus seek the composition of the solvus curve at any given temperature. If for any reason the solid solution should have the composition *b* in Figs. 16.1 and 16.3, carbon will leave the cementite to increase the concentration of the solid solution. The free energy of the system is, of course, lowered by this spontaneous reaction. On the other hand, if the ferrite is supersaturated with carbon and has a composition such as that of point *c*, more cementite will form. This spontaneous reaction also lowers the carbon concentration of the ferrite (and the free energy of the system).

16.2 The Solution Treatment

Consider a specific dilute iron-carbon alloy, one with a total carbon content of 0.008 percent. If this alloy is in equilibrium at room temperature, nearly all of the carbon will be in the form of cementite because the solubility of carbon in ferrite at 300 K is only 8.2×10^{-12} percent (see Fig. 16.1). Suppose that this same alloy is heated to 923 K, indicated by point *d* in Fig. 16.1. At this temperature the equilibrium concentration of carbon in the solid solution is 0.010 percent, which is more than the total amount of carbon in the metal. The cementite stable at room temperature is no longer stable at 923 K and dissolves by yielding its carbon atoms to the solid solution. Because the equilibrium concentration is greater than the total carbon content of the alloy, the cementite must disappear completely if the alloy is maintained for a sufficiently long time at the elevated temperature. The alloy that originally contained two phases (cementite and ferrite) is thus converted to a single phase (ferrite). However, the solid solution attained by maintaining the specimen at 923 K is not a saturated solution because it contains less than the equilibrium concentration. On the other hand, it cannot lower its free energy and assume the equilibrium concentration for a mixture of ferrite and cementite because there is no extra carbon available for this purpose.

Let us study the effect of rapid cooling on the above alloy after it has been transformed into a homogeneous solid solution at 923 K. Very rapid cooling of heated metal specimens can be accomplished by immersing them in a liquid cooling medium, for example, water. This operation is generally known as a *quench*. In the present case, a very rapid quench will prevent an appreciable diffusion of the carbon atoms, so it can be assumed that the solid solution that existed at 923 K is brought down to room temperature essentially unchanged. The alloy, which was slightly unsaturated at the higher temperature, will now be extremely supersaturated. Its 0.008 percent carbon in solution is roughly 10^9 times greater than the equilibrium value (8.2×10^{-12} percent). The alloy is, accordingly, in a very unstable condition. Precipitation of this excess carbon through the formation of cementite will markedly lower the free energy and can be expected to occur spontaneously. The conditions under which this may occur will be discussed in the next section, but at this point it should be mentioned that the first stage in a precipitation-hardening heat treatment has just been outlined. A suitable alloy is heated to a temperature at which a second phase (usually present in small quantities) dissolves in the more abundant phase. The metal is left at this temperature until a homogeneous solid solution is attained, and then it is quenched to a lower temperature to create a supersaturated condition. This heat-treating cycle is known as the *solution treatment*, while the second stage, which is to be discussed, is called the *aging treatment*.

16.3 The Aging Treatment

The precipitation of cementite from supersaturated ferrite occurs by a nucleation and growth process. First, it is necessary to nucleate the cementite crystals. Following nucleation, cementite particles grow in size as a result of the diffusion of carbon from the surrounding ferrite toward the particles. This is called *growth*. No precipitation can occur until nucleation begins, but once it has started, the solid solution can lose its carbon in two ways, either to the growing particles already formed or in the formation of additional nuclei. In other words, nucleation may continue simultaneously with the growth of particles previously formed. The progress of precipitation at a given temperature is shown in Fig. 16.4, where the amount of precipitate, in percentage of the maximum, is plotted as a function of time. Logarithmic units are used for the time scale because spontaneous reactions of this nature usually start rapidly and finish slowly. In general, precipitation does not begin immediately, but requires a finite time t_0 before it is detectable. This time interval is called the *incubation period* and represents the time necessary to form stable visible nuclei. The curve also shows that the precipitation process finishes very slowly, an effect that is to be expected in light of the continued loss of solute from the solution.

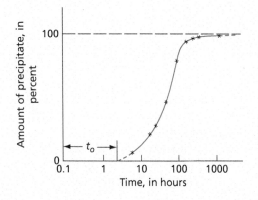

Fig. 16.4 Amount of precipitate as a function of time in an iron-carbon alloy (0.018 percent C) allowed to precipitate from a supersaturated solution at 349 K. (Data of Wert, C., ASM Seminar, *Thermodynamics in Physical Metallurgy*, 1950)

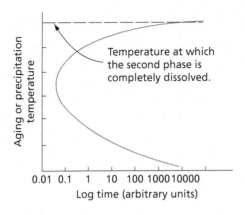

Fig. 16.5 Time for 100 percent of the precipitate to form in a supersaturated alloy

The speed at which precipitation occurs varies with temperature. This is shown qualitatively in Fig. 16.5. At very low temperatures, long times are required to complete the precipitation because the diffusion rate is very slow. Here the rate of the reaction is controlled by the rate at which atoms can migrate. The rate of precipitation is also very slow at temperatures just below the solvus line (point e, Fig. 16.1). In this case, the solution is only slightly oversaturated and the free-energy decrease resulting from precipitation is very small. Nucleation is, accordingly, slow, and precipitation is controlled by the rate at which nuclei can form. The high diffusion rates that exist at these temperatures can do little if nuclei do not form. At intermediate temperatures, between the two above-mentioned extremes, the precipitation rate increases to a maximum, so that the time to complete the precipitation is very short. In this range, the combination of moderate diffusion and nucleation rates makes precipitation rapid.

The most important effect of the precipitation of the second phase (cementite) is that the matrix (ferrite) is hardened. Figure 16.6 shows a typical hardening curve for a dilute iron-carbon alloy. To obtain curves of this nature, a number of specimens are first given a solution heat treatment in order to convert their structures into supersaturated solid solutions. Immediately following the quench of the solution heat treatment, the specimens are placed in a suitable furnace maintained at an intermediate constant temperature (a temperature above room temperature, but below the solvus temperature). They are then removed from the furnace at regular intervals, cooled to room temperature, and tested for hardness. The data obtained in this manner are then plotted to show the effect of time on hardness. The significant feature of this curve is the maximum it possesses. Holding, or aging, the specimens for too long a period at a given temperature causes them to lose their hardness. This effect is known as *overaging*.

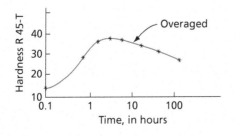

Fig. 16.6 Change in hardness during the aging treatment. Alloy is iron plus 0.015 percent C, and aging temperature 90°C. (Data from Wert, C., ASM Seminar, *Thermodynamics in Physical Metallurgy*, 1950)

The shape of the aging curve is primarily a function of two variables: the temperature at which aging occurs and the composition of the metal. Let us consider the first of these variables. Figure 16.7 shows three curves, each corresponding to aging at different temperatures. The curve marked T_1 represents aging at too low a temperature. In this case, atomic motion is so slow that no appreciable precipitation occurs and hardening occurs slowly. A further lowering of the aging temperature below T_1 will effectively stop all precipitation and prevent hardening. Use is made of this fact in the aircraft industry when aluminum-alloy rivets, which normally harden at room temperature, are kept in deep-freeze refrigerators until they are driven. In this way the rivets, which have been previously solution-treated in the supersaturated condition, are prevented from aging until they are incorporated in the article being fabricated.

Temperature T_2 corresponds to an optimum temperature, a temperature at which maximum hardening occurs within a reasonable length of time. This temperature lies above T_1, but below T_3. At T_3, hardening occurs quickly due to rapid diffusion. However, softening effects also are accelerated, resulting in a lower maximum hardness. The mechanism explaining these effects will be discussed in Sec. 16.9.

The effect of changing the composition, the other variable that influences the aging curve, will now be considered. For low-solute concentrations, the degree of supersaturation is small at the end of the solution treatment, and the free energy of the system is, at best, only slightly higher than that of the equilibrium concentration. Under these conditions it is difficult to nucleate the second phase, and hardening occurs slowly at constant temperatures. Furthermore, the maximum hardness that can be obtained will be small because the total amount of precipitate is not large and, in general, the smaller the amount of precipitate, the smaller the maximum hardness. On the other hand, increasing the total solute concentration makes possible a greater maximum hardness at a given aging temperature. The more solute that is available, the more the precipitate and the greater the hardness. In addition, at higher solute concentrations the maximum hardness will be attained in a shorter time, for both the rate of nucleation and the rate of growth are increased. Nucleation rates rise because of a greater difference in free energy between the supersaturated and equilibrium states, while growth rates are increased because of the greater amount of solute available for the formation of the precipitate. These effects are, however, limited to the extent that it is possible to dissolve the solute in the solvent during the solution treatment. Thus, in Fig. 16.1, the maximum solubility of carbon in iron (0.022 percent) occurs at 1000 K (727°C). Compositions containing carbon in excess of 0.022 percent can still only dissolve this amount in the ferrite; the remainder remains in the form of cementite. Low-carbon-steel compositions containing more than the maximum amount

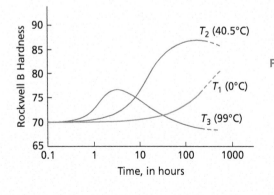

Fig. 16.7 Effect of temperature on the aging curves during precipitation hardening. Curves are for a 0.06 percent C steel. (After Davenport, E. S., and Bain, E. C., *Trans. ASM,* **23** 1047 [1935])

of carbon soluble in iron (0.022 percent) are still able to undergo precipitation hardening as long as the solution-treating temperature is not raised above 1000 K. Figure 16.7 shows a number of aging curves for a steel of 0.06 percent carbon.

Finally, it should be pointed out that for certain alloys, the quenching rate, *pre-aging*, can also markedly influence the precipitation kinetics and morphology (Refs. 1 through 3). Such influence has been attributed to the interaction and association of vacancies with solute atoms. For example, Vyhnal and Radcliffe (Ref. 1) have shown that the carbide precipitate morphology and intensity of precipitation in Fe–0.019 wt. % carbon aged at 25 and 100°C depend on the quenching rate above a critical range. The authors attribute this pre-aging influence to the retardation of the normal loss of vacancies during quenching because of the strong attraction between vacancies and interstitial carbon atoms. With increased quenching rate, a greater fraction of the supersaturated vacancies are retained as stable carbon-vacancy pairs. The influence of the pairs on the precipitation rate and morphology depends on the aging temperature. At low aging temperatures, the pairs serve as preferred nucleation sites, resulting in a high density of fine carbide precipitate and hardness. At high aging temperature, on the other hand, the vacancy-carbon pairs enhance growth by relieving coherency strain.

16.4 Development of Precipitates

How the precipitates form and grow during precipitation is a subject of considerable technical importance. The precipitation hardening process becomes very complex, particularly in commercial alloys, which may contain more than one solute that contributes to the age hardening. Fortunately, however, a great deal of research has been performed on simple binary alloy systems containing only a single solute. Among these systems are alloys formed by adding either copper, silver, or zinc to aluminum. In the first of these systems, that is Al-Cu, the solute atom (copper) has a diameter about 12 percent larger than the aluminum atom. In the other two systems, the solute atom differs in size from that of the solvent by only about one percent.

A significant feature of precipitation hardening, even in the binary systems, is that the precipitating phase normally does not originate in its final stable form. In other words, the precipitate often passes through several stages before a final stable structure appears. Thus, for example, an alloy of aluminum containing 4 percent copper may pass through three different intermediate precipitation stages before the final stable Θ phase ($CuAl_2$) is attained. The first of these stages involves local clustering of the solute atoms to form what are commonly called *Guinier-Preston* or *GP zones*. These zones or clusters are favored by a low aging temperature, a small atomic size misfit, and a high degree of solute supersaturation. The shape that these clusters or zones take is strongly influenced by the amount of the misfit that occurs when the solute atom is placed in the parent lattice. When this is small, as in aluminum-silver alloys, where the silver atoms are almost the same size as the aluminum atoms and diameter of two atoms differ by less than 1 percent, the GP zones tend to be nearly spherical in shape, as observed by Nicholson and Nutting in 1961(Ref. 4). These zones were later shown to have faceted characteristics in alloys containing 17.6 wt.% Ag (Ref. 5). An example of faceted GP zones in a sample aged 160 °C is shown in Fig. 16.8. The proportion of the faceted zones were found to decrease with the increase in the aging temperature up to 350 °C. In the aluminum-copper system, on the other hand, the misfit between the solute and the matrix lattice is much larger since there is about a 12 percent difference between the sizes of the copper and aluminum atoms. This larger misfit signifies a much larger lattice strain when GP zones form inside the matrix. As a consequence, the GP zones tend to form as very thin,

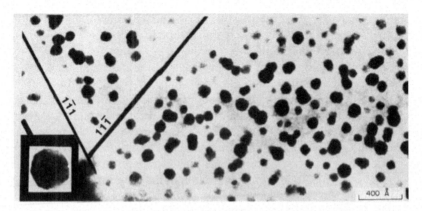

Fig. 16.8 Electron microscopy image of Guinier-Preston zones in an aluminum 17.6 wt.% silver alloy. The small inset clearly shows faceted nature of one zone. (Reprinted from "Faceting of GP zones in an Al-Ag alloy", K.B. Alexander, F.K. Legoues, H.I. Aaronson, D.E. Laughlin, *Acta Met.*, Volume 32, Issue 12, December 1984, Pages 2241-2249. https://doi.org/10.1016/0001-6160(84)90166-4

two-dimensional plates oriented parallel to the {100} aluminum lattice planes. Thus, in the Al-Ag alloy where the strain energy associated with a zone is small, the clusters tend to take a shape that minimizes the surface energy, while in the Al-Cu alloy where the strain energy is large, the zones assume a shape that tends to minimize their volume and the strain energy of the zones.

The plate-like Al-Cu GP zones may have a diameter or length as large as 10 nm (100 Å) but a thickness of only a few tenths of a nanometer (several Å). This means that the thickness of the GP zones in the copper-aluminum alloy is of the order of a few layers of copper atoms. Figure 16.9 shows a high-angle annular dark-field mode scanning transmission electron microscope (HAADF-STEM) micrograph of GP zones for Al–4 wt. % Cu aged at ambient temperature for 504 h. GP zones can be seen in the figure, with reported diameter of 2.5 nm and thickness of 0.256 nm.

If an aluminum-copper alloy is *naturally aged*, that is, aged at room temperature, all of the hardening that it undergoes comes from only the formation of Guinier-Preston zones. The zones begin to appear very soon after the start of aging. The rate at which they appear is initially very rapid but decreases with increasing time. The number of zones eventually approaches a maximum as the metal becomes fully aged. In other words, room-temperature aging of aluminum-copper alloys involves only a single stage and the stable θ phase is not attained. Furthermore, age hardening involving only GP zones occurs, in Al-Cu alloys, not only at room temperature but also at any aging temperatures below roughly 100°C (373 K). In the past, this type of low temperature aging has been called *cold aging*. Since the equilibrium $CuAl_2$ or θ phase does not appear as a result of cold aging, the precipitate in this case should have a metastable character. Supporting this point of view is the fact that hardening due to cold aging in an aluminum alloy can be removed by reheating to about 200°C (473 K), at which temperature the zones dissolve. This phenomenon is known as *reversion*. Softening involving the use of reversion is employed in some commercial operations.

It is believed that vacancies play a very important role in the formation of GP zones, not only in the Al-Cu system but also in the more general sense (Ref. 6). The rates at which the GP zones begin to form, at low and moderate temperatures, are very high. In order for the zones to form this rapidly, the diffusion coefficient of the solute atoms would have to be several orders of magnitude higher

Fig. 16.9 HAADF-STEM micrograph of Al-4wt.%Cu aged for 504 hours at ambient temperature. The electron beam orientation was [001] of the aluminum lattice, and the GP zones are parallel to the (100) and (010) planes of Al. (Reprinted from "A multidisciplinary approach to study precipitation kinetics and hardening in an Al-4Cu (wt.%) alloy", Rodríguez-Veiga, A., Bellona, B., Papadimitriou, I., Esteban-Manzanares, G., Sabirov, I., and Lorca, J., *Journal of Alloys and Compounds*, 757 (2018) 504–519. http://www.elsevier.com/locate/jalcom)

than that obtained by extrapolating the known high-temperature solute diffusion coefficient to the aging temperatures. This accelerated rate of zone formation may be explained in terms of a vacancy supersaturation retained in the aging specimen as a result of its having been rapidly quenched from the higher solution treatment temperature, since the equilibrium vacancy concentration is very much higher at the solution temperature. As shown in Chapter 7, because the diffusion coefficient involves the product of the jump rate of the atoms into vacancies times the concentration of the vacancies, a vacancy supersaturation is expected to increase the diffusion rate at the aging temperature.

16.5 Aging of Al-Cu Alloys at Temperatures above 100°C (373 K)

The formation of Guinier-Preston zones no longer constitutes the only stage of precipitation hardening in an Al-Cu alloy if the aging temperature is raised to 100°C (373 K) or above. One may now observe several intermediate stages, followed by the appearance of the final stable θ or $CuAl_2$ phase. The effect of aging above 100°C is clearly shown in Fig 16.10, where a set of isothermal aging curves of an aluminum alloy, 2014-T4, are shown. In this figure, the yield stress of the metal is plotted as a function of the aging time in hours. This alloy contains, in addition to 4.4 percent Cu, 0.8 percent Si, 0.8 percent Mn, and 0.5 percent Mg. The principal hardening agent of this alloy is its copper and while it is not strictly a binary Al-Cu alloy, its aging behavior is largely representative of the binary Al-Cu alloys. Attention is also called to the fact that the temperatures in this diagram are expressed in degrees Fahrenheit. Note that the room temperature curve, corresponding to hardening due only to the formation of GP zones, shows only a slight rise in the yield stress for aging times in excess of 100 hours. The yield stress up to an aging time of 10,000 hours also shows no indication of decreasing. On the other hand, all of the curves corresponding to specimens aged at temperatures above

212°F (100°C) show a yield stress that quickly rises to a maximum and then falls rapidly at longer aging times. Aging at temperatures above room temperature is normally called *artificial aging*.

A significant feature of the artificially aged curves in Fig. 16.10 is that they show it is possible to attain a higher strength (or hardness) by aging above 100°C. This gain of strength can be retained by controlling the time that the metal is held at the higher aging temperature. It is also worth noting that the maximum increase in strength, in this set of curves, occurs in the 300°F (149°C) curve. Also note that as the aging temperature is increased, the maximum in these curves moves to a shorter time and decreases in height. While the aluminum 2014 alloy curves in Fig. 16.10 readily show the difference between aging at room temperature and aging above 100°C, it is easier to obtain useful information about the development of the intermediate stages of aging from an isothermal aging curve of a binary Al-Cu alloy. Figure 16.11 shows such a curve for a binary alloy of 4.0 percent Cu that was aged for various times at 130°C. In this figure the diamond pyramid hardness (DPH) is plotted against the time of aging measured in days. The initial rise of the curve that begins after the start of aging and extends to a time of one hour, as well as the plateau that follows this rise, are associated with the development of Guinier-Preston zones. However, X-ray diffraction studies have revealed that the second rise is accompanied by the formation of a new structure. Originally, this intermediate structure was called GP(2), but later authors have tended to identify it by the symbol θ'' since it has the characteristics of a three-dimensional ordered phase. It also consists of plates that lie along aluminum {100} planes, but these plates now have a thickness of several atomic layers. It is interesting to note that the sizes or diameters of the θ'' plates are larger than those of the GP zones. In this specific alloy they may become at least four to five times larger in diameter than the GP zones. As indicated in Fig. 16.11, the GP zone and θ'' structures can be seen to overlap each other for a short part of the isothermal aging curve. A greater amount of overlap is apparent between the next intermediate structure, θ', and the θ'' structure. Note that the maximum hardness (or strength) of the alloy is attained when the θ'' constituent is close to its maximum amount. It should be pointed out that the peak aged microstructure consists of both θ'' and θ' precipitates (Refs. 7 through 9). The θ' precipitate has a composition that is the same or very close to that of the stable θ ($CuAl_2$) phase and its crystal structure is also tetragonal like that of the θ phase. The primary difference between these two structures is that θ' precipitates are partially coherent with the lattice of the aluminum matrix, while the θ precipitates are incoherent. Figure 16.11 shows that the growth of the θ' particles is accompanied by a loss in hardness. This is a result of the fact that as the particles

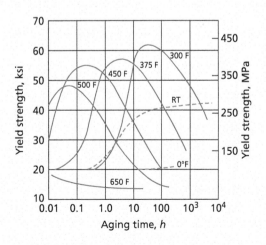

Fig. 16.10 Representative isothermal aging curves of the aluminum alloy 2014-T4. (From Hatch, J. E., Ed., *ALUMINUM Properties and Physical Metallurgy*, American Society for Metals, Metals Park, Ohio, 1984. Reprinted with permission of ASM International®. All rights reserved. www.asminternational.org)

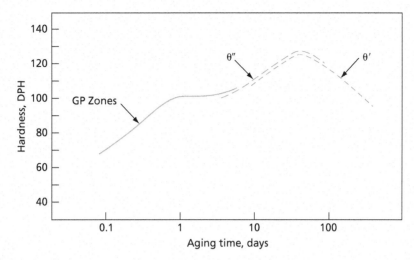

Fig. 16.11 Isothermal aging curve, Al-4 pct Cu at 130°C. (After Silcock, J. M., Heal, T. J., and Hardy, H. K., *J. Inst. Met.*, **82** 239 [1953–4])

grow in size, they decrease in number and there is a general loss in the strength of the coherency strain. Eventually, with increasing aging time, the stable and incoherent θ phase replaces the coherent θ' precipitate and the alloy becomes softened to well below its maximum strength (Ref. 8).

It is now possible to draw a small diagram, as shown in Fig. 16.12, giving the stages in the development of the precipitates in the Al-Cu precipitation hardening alloys.

1. Supersaturated alpha phase.
2. Guinier-Preston zones. This is the final precipitate below 100°C.
3. The θ'' intermediate precipitate structure.
4. The θ' intermediate precipitate structure.
5. The stable θ phase, $CuAl_2$, in a matrix of the equilibrium alpha phase.

The interrelations of these structures are not well understood (Ref. 10). That is, it has not always been possible to determine whether the various precipitate structures evolve directly from

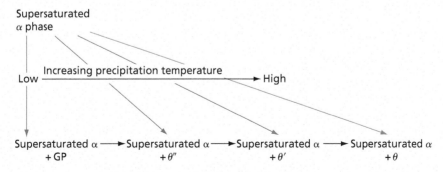

Fig. 16.12 Precipitation sequence in Al-Cu alloys

each other in sequence or whether they are separately nucleated. In the case of the intermediate structure θ'', which replaces the GP zones, the size of this precipitate particle is much greater than that of the GP zones. As a consequence, many of the GP zones must dissolve and release their solute atoms in order to form one of the θ'' particles. Thus, the question of whether a θ'' platelet forms from a GP zone or not is somewhat academic. In addition, it was shown earlier by Guinier that depending on the degree of supersaturation, the final stable θ precipitate may form as a result of a direct transformation from the θ' structure or nucleate directly from the matrix solid solution. Also, it has been deduced as a result of the work of Silcock, Heal, and Hardy (Ref. 7) that, given the correct conditions, the first structure that appears may be either GP zones, θ'', or θ'. This may be interpreted as evidence that all these structures are capable of being independently nucleated.

16.6 Precipitation Sequences in Other Aluminum Alloys

The aluminum-silver system has been studied by a number of investigators. These have shown (Ref. 10) that for concentrations less than about 20 percent Ag the basic sequence of stages appears to be as follows:

(1) Supersaturated $\alpha \rightarrow$ (2) spherical clusters (as in Fig. 16.8) $\rightarrow$ (3) ordered clusters $\rightarrow$ (4) $\gamma' \rightarrow$ (5) stable γ phase and equilibrium α

Studies of the precipitation sequences have been made on a number of other aluminum alloy systems. For the most part, these have concerned alloy systems on which important commercial precipitation hardening aluminum alloys are based. They include the following: (1) aluminum-copper-magnesium alloys, (2) aluminum-magnesium-silicon alloys, (3) aluminum-zinc-magnesium alloys, and (4) aluminum-zinc-magnesium-copper alloys. Three of the above involve ternary systems and the other is a quaternary system. As a consequence, the studies have been complex and the results in many cases not well defined. Therefore, they will not be considered here. Information on this subject may be obtained in standard texts (Refs. 10 and 11).

In the aluminum precipitation hardening alloys, the precipitate particles normally only become visible in an optical microscope when the metal is in an advanced stage of overaging. However, some of the intermediate stages can often be resolved in an electron microscope. Thus, for example, the GP zones of the Al-Ag system can be observed in the electron micrograph of Fig. 16.8.

As evidenced by the photographs in Figs. 16.8 and 16.9, the transmission electron microscope is a useful tool for the study of precipitation phenomena. In addition to making possible the direct observation of very small precipitate structures, it is sometimes possible to use selected area diffraction to deduce the nature of the crystal structure in the precipitate particles. This requires that the particles be large enough to be able to give a diffraction pattern. Very small particles cannot be studied in this manner. For their analysis, the best techniques still rely heavily on X-ray diffraction. Since the problems associated with precipitation are complex, it is often necessary to invoke other techniques in addition to those mentioned above. These involve changes in physical properties such as hardness and electrical resistivity, induced by the various phenomena of precipitation. With regard to electrical resistivity, the orderly motion of electrons through a crystal, which constitutes an electric current, is badly disturbed when an extremely small and uniformly distributed precipitate, such as the GP zones, forms. Thus, when precipitation starts there is an initial large rise in resistivity, which then decreases as the average particle size increases during the progress of the precipitation phenomena.

Another type of physical measurement that can give valuable information about precipitation phenomena is *internal friction*. The use of the torsion pendulum to determine interstitial diffusion coefficients is described in Chapter 13. The torsion pendulum, or other more sophisticated internal friction measuring devices, may also be used to measure the concentration of an interstitial solute in solid solution in an alloy. To a high degree of accuracy, the internal friction in a dilute interstitial solid solution alloy specimen is due only to the solute atoms in solution. Furthermore, the magnitude of the internal friction, which can be measured by a parameter such as δ_{max}, the maximum logarithmic decrement (see Fig. 13.9), is a direct measure of the amount of solute in solution. Thus, when an interstitial solid solution alloy is undergoing precipitation, one can follow the amount of the solute being transferred from the solution to the precipitate particles by measuring the decrease in δ_{max} that accompanies this transfer. As a specific example, consider a supersaturated ferrite solid solution, obtained by quenching an iron specimen containing 0.01 percent carbon from 950 K to room temperature. The transfer of carbon atoms from the solid solution to the cementite precipitate particles that occurs on subsequent aging of the specimen at room temperature can be followed by measurements of δ_{max} since the cementite precipitate depends only on the concentration of the carbon in the ferrite and is, accordingly, independent of the carbon contained in the cementite particles. It is interesting to note that the early data of Wert, used in plotting Fig. 16.1, were obtained in this way. In the alloy systems, where this internal friction method is applicable, it forms an excellent means for studying precipitation effects.

It should be noted that precipitation or age hardening is not limited to the binary Al-Cu or Al-Ag alloys discussed in the previous sections. The hardening is also found in many other aluminum binary, ternary or higher order alloys (Refs. 12 and 13). Precipitation hardening is also found to be beneficial in many other structural materials such as copper-, cobalt-, nickel-, titanium-, iron-base alloys (Refs. 14 and 15). Moreover, the precipitation process has also been widely used in recent years in magnesium alloys, particularly for weight reduction in automotive and aerospace applications. The density of magnesium is approximately 2/3 of aluminum and 1/4 of steels. Magnesium also has low melting temperature and high specific heat. Processing of many casts or wrought magnesium alloys involves age hardening via (1) solution and homogenization treatment within a- Mg single-phase region, (2) quenching to form a supersaturated solid solution, followed by (3) age hardening at low temperatures to obtain finely distributed GP zones, metastable, or stable precipitates depending on the composition and processing conditions (Refs. 16 and 17).

16.7 Homogeneous Versus Heterogeneous Nucleation of Precipitates

A precipitate particle can be nucleated, as in other types of phase transformations, either heterogeneously or homogeneously. Heterogeneous nucleation of precipitates is largely associated with dislocations, dislocation nodes (intersections of two or more dislocations), impurity particles, and discontinuities on grain boundaries. Homogeneous nucleation, on the other hand, involves the spontaneous formation of nuclei as a result of composition fluctuations in the matrix that are large enough to cause a second-phase particle to form in an otherwise perfect crystal.

As mentioned in Chapter 15, homogeneous nucleation is usually difficult to achieve and heterogeneous nucleation normally is favored. However, homogeneous nucleation of Guinier-Preston zones is believed to be possible in a basically continuous crystal lattice, provided a critical vacancy concentration exists. In other words, nucleation, as a result of the formation of vacancy-solute atom clusters, is consistent with experimental data under certain combinations of solution temperatures and quenching rates. It has also been proposed that at low temperatures (Ref. 18) (below 100°C) the critical nucleus size for a GP zone may be so small that the incubation time for a nucleus effectively vanishes. This would account for the rapid growth rates of the zones at these temperatures. GP zones may also form by *spinoidal* transformation, for which there is no barrier for nucleation.

Not all alloy systems exhibit the precipitation phenomena found in precipitation hardening aluminum alloys. For these other systems the formal theory for the nucleation of solid-state precipitates is still very important. This theory was covered in Sec. 15.4. A topic that was not covered in Sec. 15.4 was the effect of temperature on the homogeneous nucleation process. Homogeneous nucleation requires that thermal fluctuations produce particles large enough to exceed r_0, the critical radius (see Fig. 15.2). Otherwise, as explained in Chapter 15, the second phase cannot nucleate. It is customary to call a nucleus that is subcritical ($r < r_0$) in size an *embryo*. Let us now consider how the size of a stable nucleus varies with temperature. With reference to the particle energy equation, Eq. 15.23, it can be assumed as a first approximation that the surface-energy term does not change with temperature. On the other hand, the volume energy varies with temperature, becoming larger at low temperatures because the degree of supersaturation increases (solubility decreases) as the temperature falls. This effect is shown qualitatively in Fig. 16.13, where it can be seen that the critical radius decreases with falling temperature. At temperatures just below the solvus line, r_0 is very large (approaching infinity in size). The rate of homogeneous nucleation is, therefore, very small at this temperature. As the temperature is lowered, the critical radius rapidly decreases in size and so does the energy necessary to form a critical embryo. This latter is measured by the height (ΔG_{r_0}) of the energy curve maxima in Fig. 16.13. Homogeneous nucleation is thus made easier by a drop in temperature, which explains why ice must be supercooled to such a low temperature if heterogeneous nucleation is precluded. On the other hand, in precipitation reactions, nucleation also depends on the ability of the solute to diffuse through the lattice of the solvent. This becomes the controlling factor at very low temperatures. When diffusion effectively becomes negligible, so does precipitation.

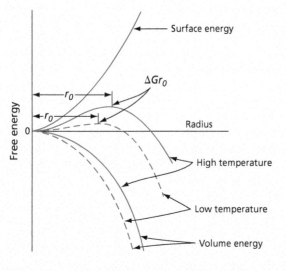

Fig. 16.13 Effect of temperature of precipitation on the free energy of a precipitate particle as a function of its radius

16.8 Interphase Precipitation

A different type of precipitation, called *interphase precipitation*, will now be considered. It is of considerable importance in the field of high-strength low-alloy (HSLA) steels. These steels are not simple binary alloys, since the simplest of them contains at least one other component that normally interacts strongly with the carbon in the alloy. The most important of these elements are vanadium, titanium, niobium, chromium, molybdenum, and tungsten. All have a strong affinity for carbon and, as a result, are commonly known as *carbide formers*. Thus, in a typical low alloy that contains vanadium in addition to carbon, vanadium carbide, VC, will tend to form instead of iron carbide, Fe_3C.

In order to understand the complete nature of interphase precipitation, it is perhaps simplest to first consider the binary iron-carbon diagram, which is usually presented showing the phase relations between iron and the intermediate (metastable) phase iron-carbide, Fe_3C, normally called *cementite*. The complete iron-iron carbide diagram is shown in Fig. 18.1. Note that Fig. 16.1 is merely an expanded view of the lower left-hand corner of Fig. 18.1. A more extensive view of this part of the phase diagram is given in Fig. 16.14. This extends up into the higher temperature region where the γ or face-centered cubic phase, known as *austenite*, is stable. Now consider an alloy of iron with 0.15 percent carbon that has been heated up to point *a* in Fig. 16.14 and left at this temperature long enough to form a homogeneous

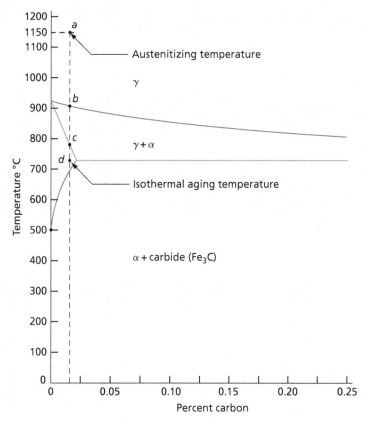

Fig. 16.14 A partial view of the iron-carbon diagram

austenitic solid solution. On slow cooling this alloy enters the two-phase $\alpha + \gamma$ region at point b and and leaves it at point c. Just below this latter point it enters the alpha (bcc) solid solution phase or ferrite region. Thus, on slow cooling the transformation from γ to α occurs, as might be expected, between points b and c. This transformation requires time or slow cooling. Very rapid cooling or quenching a specimen from a to a point such as d and then holding it at point d will tend to suppress the γ to α transformation during cooling and allow it to occur isothermally at the temperature of d. In the HSLA steels where interphase precipitation occurs, the binary iron-carbon diagram is not strictly applicable because the carbide that forms is not cementite but one of the alloy carbides such as vanadium carbide (VC), niobium carbide (NbC), or titanium carbide (TiC). However, in these steels, a rapid quench from the austenite region to the ferrite region followed by a period of isothermal aging will still cause the γ to α transformation to occur isothermally. During this transformation there will be a precipitation of the carbide that does not occur in the simple binary iron-carbon alloy.

Now consider a steel containing 0.15 percent carbon and 0.75 percent vanadium. Such a steel would be heated high into the austenite region to a temperature of about 1150°C and held at this temperature until a uniform austenitic structure was obtained. The next step in the procedure would consist of a rapid cool or quench to a temperature between 700 and 850°C. The specimen then would be allowed to age at this latter temperature. During this aging period an isothermal transformation from the gamma phase to the alpha allotriomorphs occurs.* Also during this transformation the precipitation of the alloy carbide, VC, occurs (Ref. 4). This precipitate is nucleated on the interface between the ferrite and the austenite. The boundary between the austenite and the ferrite is frequently planar, but this is not always true. The transformation of the austenite to the ferrite occurs primarily by the movement of ledges that sweep along the boundary as suggested in Fig. 16.15. In the general case, these ledges can have a height of between 5 and 50 nm depending on the isothermal transformation temperature and the composition of the specimen. As a ledge moves across the interface, a band of metal is transformed from austenite to ferrite, with a thickness equal to the ledge height. The ledge front is believed to move too rapidly for the alloy carbide precipitates to nucleate on the front itself. However, this is not true of the gamma-alpha interface, which lies behind the moving ledge front. This interface remains effectively stationary until the next ledge sweeps across so that the precipitates have time to form on this interphase boundary. They nucleate on the interface and then grow into the alpha phase. The precipitates continue to grow as the ledge front moves away from them with the result that the sizes of the particles increase as the front continues to move. The transformation of austenite into ferrite allotriomorphs is a repeating process so

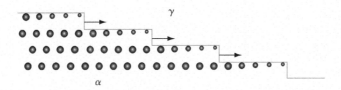

Fig. 16.15 A mechanism for the nucleation and growth of carbides on the interface between the gamma and alpha phases. (From Honeycombe, R. W. K., *Met. Trans. A* **7A** 915 [1976] with kind permission from Springer Science and Business Media)

---

*Ferrite allotriomorphs form at the existing austenite grain boundaries by heterogeneous nucleation. Those allotriomorphs form layers following the gamma grain boundaries. Idiomorphic ferrites, on the other hand, form inside austenite heterogeneously, mostly on non-metallic inclusions in the steel.

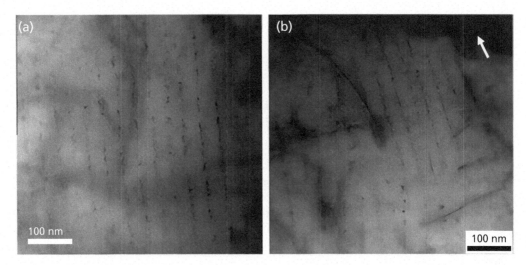

Fig. 16.16 Scanning transmission electron microscopy bright field images of NbC interphase precipitates: **(a)** planar type aged at 923 K , **(b)** non-planar type aged at 1023 K. The arrow shows a grain boundary (Reprinted from "Interphase precipitation in niobium-microalloyed steels", Okamoto, R., Borgenstam, A., Agren, J., *Acta Materialia*, Volume 58, Issue 14 (2010), Pages 4783–4790. https://doi.org/10.1016/j.actamat.2010.05.014

that as the transformation continues additional ledges appear in a regular fashion and move across the interface. A schematic figure due to Honeycombe that shows this process appears in Fig. 16.15. As mentioned earlier, the boundary between the austenite and the ferrite is frequently planar, but this is not always true. As such, the interphase precipitates can be arranged the precipitates could be arranged in periodical fashion but on planer or non-planar sheets, depending on the alloying and aging conditions. An example is shown if STEM images in Fig 16.16 for niobium-microalloyed steel (containing 0.052 wt.% C, 0.05 wt.% Nb, with other minor Si, Mn, P, S and Al additions) aged for short holding times of 10 seconds between 873-1123K. Niobium is strong carbide former and forms NbC precipitates at the interphase boundaries.

16.9 Theories of Hardening

The crystallographic nature of the precipitate particles that form during the various stages of precipitation are now much better understood than they were a few years ago. However, the exact nature of the hardening process is still not completely resolved. It appears that there are at least several hardening mechanisms, and that what is predominant in one alloy is not necessarily important in another. In general, however, it may be said that an increase in hardness is synonymous with an increased difficulty of moving dislocations. Either a dislocation must cut through the precipitate particles in its path, or it must move between them. In either case, it can be shown that a stress increase is needed to move the dislocations through a lattice containing precipitate particles. Thus, the Orowan mechanism illustrated in Fig. 16.17 was proposed to explain the interaction of dislocations with precipitate particles that have grown large enough for dislocation segments to be able to bend and pass between adjacent

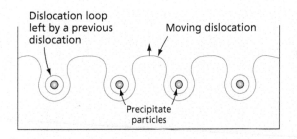

Fig. 16.17 Orowan's mechanism for the movement of dislocations through a crystal containing precipitate particles

particles. It is, accordingly, applicable to the later stages of aging. In this mechanism the dislocation is assumed to form expanding loops around the precipitate particles, which cancel as in a Frank-Read source. This cancellation permits the dislocation to continue its motion, but leaves a dislocation ring around the particle whose stress field adds resistance to the motion of the next dislocation.

It is believed by many that an important factor in the interaction between precipitate particles and dislocations is the presence of stress fields surrounding precipitate particles. This is especially true when the precipitate particle is coherent with the matrix.

Overaging is softening resulting from prolonged aging (see Fig. 16.6). In some age-hardening alloys, it appears concurrently with the loss of coherency by the precipitate. In any case, it may be stated that it is connected with the continued growth of precipitate particles. Growth will continue as long as the metal is maintained at a fixed temperature. This does not mean that all particles continue to grow, while others (the smaller ones) disappear. It merely means that certain particles (the larger ones) continue to grow, while others (the smaller ones), disappear. As aging progresses, the size of the average particle increases, but the number of particles decreases. Maximum hardening is associated with an optimum small-particle size and a corresponding large number of particles, while overaging is associated with few relatively large particles.

The growth of precipitate particles is directly related to the surface tension at the interface between the matrix and the particles. Because of the boundary-surface energy, the free energy per atom in a large precipitate particle is lower than that in a small particle. This free-energy difference is the driving force that causes the dissolution of small particles and the growth of large ones. A simple derivation of this relationship will now be given.

First, assume that the precipitate has a spherical shape, like that shown in Fig. 16.18. The energy of a particle can then be expressed by Eq. 15.2, assuming that the strain energy is small enough to be ignored. Thus,

$$\Delta G = A_1 r^3 + A_2 r^2 \qquad 16.1$$

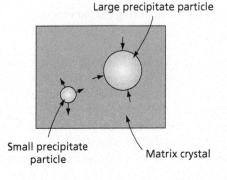

Fig. 16.18 Growth of precipitate particles. Small arrows at surfaces of particles indicate the *net* direction of the flow of solute atoms

This is the total energy of a given amount of precipitate. The energy per unit volume of the precipitate is the above quantity divided by the volume of the particle, or

$$\Delta G' = \frac{\Delta G}{\frac{4}{3}\pi r^3} = A'_1 + \frac{A'_2}{r} \qquad 16.2$$

where $\Delta G'$ is the energy per unit volume,

$$A'_1 = A_1 / \tfrac{4}{3}\pi \quad \text{and} \quad A'_2 = A_2 / \tfrac{4}{3}\pi$$

The energy per atom of the precipitate is proportional to the energy per unit volume, so that

$$\Delta G_a \approx A'_1 + \frac{A'_2}{r} \qquad 16.3$$

where ΔG_a is the free-energy change per atom. This quantity varies inversely as the radius of the particle. The larger the radius, the more negative the free energy of the second-phase atom and, therefore, the more stable it is in the precipitate. Conversely, the smaller the radius, the less stable it is. Under conditions such as these, solute atoms tend to leave smaller particles and enter the matrix, while at the same time they leave the matrix to enter the larger particles. Diffusion of solute through the matrix makes it possible for this process to continue.

16.10 Additional Factors in Precipitation Hardening

In many alloys, precipitation-hardening phenomena are made even more complicated by the fact that nucleation occurs both homogeneously and heterogeneously. Preferred locations for heterogeneous nucleation in these alloys are grain boundaries and slip planes. Since heterogeneous signifies easier nucleation, precipitation tends to occur more rapidly at these locations. This introduces a time lag between the aging responses in areas undergoing heterogeneous and homogeneous nucleations, and overaging frequently occurs at the grain boundaries long before precipitation in the matrix has had a chance to develop fully. Another effect of rapid precipitation at grain boundaries is that precipitate particles may grow large in size and, as a result, deplete the solute from the areas adjacent to the boundaries. A band of metal, free of precipitation, is then left on each side of the boundary (see Fig. 16.19A). This effect can be greatly magnified if the alloy is heated to the solution-treating temperature and slowly cooled. On slow cooling, nucleation starts at temperatures just below the solvus line at points of easy nucleation, as for example, grain boundaries. At the same time, homogeneous nucleation is prevented because of the negligible rate for this form of nucleation at temperatures close to the solvus line. On continued slow cooling, the grain-boundary precipitate continues to grow through diffusion of solute from the matrix to the precipitate. At the same time, the solute concentration in the matrix is continually reduced and the solution never becomes greatly supersaturated. In this manner, nearly all of the solute finds its way to the second phase in the boundaries and general precipitation in the matrix does not occur. Figure 16.19B is representative of a slowly cooled alloy with grain-boundary precipitation.

Heterogeneous nucleation on slip planes is frequently induced by quenching stresses that develop when the metal is rapidly cooled from the solution-treating temperature. Relief of these stresses through plastic deformation by slip leaves large numbers of dislocation segments along slip planes, which act as loci for heterogeneous nucleation.

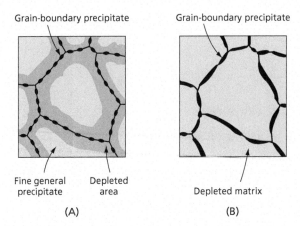

Fig. 16.19 Heterogeneous nucleation at grain boundaries. **(A)** Moderate rate of cooling may result in both heterogeneous nucleation at grain boundaries and homogeneous nucleation in the centers of the grains. **(B)** Very slow cooling may result in the precipitate occurring only at grain boundaries

Precipitation phenomena are sometimes further complicated by the occurrence of recrystallization during the formation of the precipitate. When this happens, it is the matrix that recrystallizes; matrix atoms regroup to form new crystals.

Finally, it should be pointed out that precipitate particles are not always spherical in shape. Frequently, the precipitate has a plate-like, or even a needle-like, form. In many cases plate- or needle-shaped precipitate particles grow in such a manner that they are aligned along specific crystallographic planes or directions of the matrix crystals. Interesting geometrical patterns may result from this type of precipitate growth. It is customary to call such formations *Widmanstätten structures*. Figure 16.20 shows schematically a typical Widmanstätten structure.

Fig. 16.20 Schematic representation of a Widmanstätten structure. Short, dark lines represent plate-shaped precipitate particles that are aligned on specific crystallographic planes of the crystals of the matrix

Problems

16.1 Write a computer program based on Eq. 9.15 and obtain the plot in Fig. 16.1. Now consider that a thin sheet specimen of an alloy of iron containing 0.018 percent carbon is heated to 800 K and held there long enough to come to equilibrium before being quenched in iced brine.

(a) Estimate from your figure the amount of carbon in solid solution just after the quench.

(b) If the specimen is maintained in the iced brine for a very long time, would you expect the amount of carbon in solid solution to remain the same as that existing just after the quench?

16.2 The solubility of the interstitials oxygen and nitrogen in the refractory metals vanadium, niobium, and tantalum is of considerable interest because they can seriously affect the mechanical properties of these metals. According to Bunn, P. and Wert, C., *TMS-AIME*, **230** 936 (1964), the solubility of nitrogen in tantalum is given, in weight percent, by the following equation:

$$C_w = 2.71 \exp(-22{,}600/RT)$$

(a) Plot the solubility curve for N in Ta within the limits $T = 300$ K and $T = 2000$ K.

(b) From a comparison of this curve with that for carbon in iron, deduced in Prob. 16.1, compare the solubility of carbon in iron with that of nitrogen in tantalum at 1000 K.

16.3 A thin sheet specimen of an iron-carbon alloy with 0.016 weight percent carbon is heated to 1000 K, where it is allowed to come to equilibrium. Following this, it is slow cooled to 850 K. It is then quenched in iced brine and aged at 313 K for 100 hours. Describe the resulting microstructure.

16.4 Consider the aluminum end of the aluminum-copper phase diagram in the figure accompanying this problem.

(a) Outline the complete program for obtaining a maximum hardness in an alloy of aluminum with 4 percent copper if the aging temperature is 403 K.
(b) When the maximum hardness is obtained, what would be the nature of the precipitate? Explain.

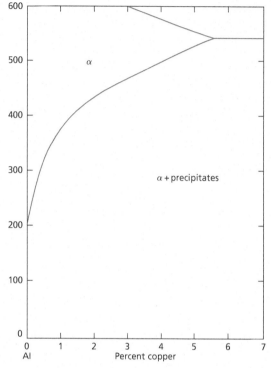

The aluminum-rich end of the copper-aluminum phase diagram that is of interest with respect to precipitation hardening

16.5 Assume that a spherical precipitate particle forms in an age hardening alloy and that the volume free-energy change associated with the formation of the particle is 60 MJ/m^3. The energy of the interface between the particle and the matrix is 0.40 J/m^2. With the aid of Eq. 15.3 plot the free energy of a particle as a function of its radius from $r = 0$ to $r = 3 \times 10^{-8}$ m. With the aid of this diagram determine r_0 and ΔG_{r_0} for the conditions of this problem.

16.6 Assume that the precipitate in Prob. 16.5 has a total volume fraction of 1.5 percent and that the particles are all of the same size, with a radius equal to twice the critical radius.

(a) Compute the number of particles per cubic meter.
(b) Compute the total change in free energy due to the formation of all the precipitate particles in a cubic meter.

16.7 The Orowan mechanism, which explains the hardening effect of precipitate particles when the precipitates have grown to a size where they are no longer coherent with the matrix, is shown in Fig. 16.17. The strengthening arises from the fact that the dislocations must bow out between two neighboring precipitate particles. This bowing out is opposed by the line tension of the dislocations. Ashby has deduced an equation giving the stress needed to move dislocations under conditions conforming to the Orowan model. An equation giving this stress, which assumes that the precipitate particles are spherical and that the system involves iron containing carbon, may be found in Leslie, W. C., *The Physical Metallurgy of Steels*, McGraw Hill Book Company, New York, 1980, p. 198. It is:

$$\sigma(\text{MPa}) = 5.9 f^{1/2}/X \cdot \ln(X/b)$$

where $\sigma(\text{MPa})$ is the stress, f is the volume fraction of the precipitate, X is the mean linear intercept diameter of the precipitate particles and b is the Burgers vector. In the case of iron, $b = 2.5 \times 10^{-4}$ μm. In this equation X is expressed in μm units. Plot a curve showing the dependence of the stress on the particle diameter, X, for the case of a constant volume fraction of precipitate, $f = 0.001$. Let X vary from 0.001 to 0.100 μm.

16.8 In Prob. 16.7 the volume fraction of the precipitate was taken to be 0.001.

(a) Assuming that the precipitate consists of cementite and that the densities of cementite and ferrite are nearly equal, estimate the carbon concentration in the iron.

(b) Would it be possible to get this much carbon in solution at 1000 K? In other words, could one reasonably expect to obtain a 0.001 volume fraction of cementite precipitate on aging just above 300 K after a rapid quench from 1000 K?

16.9 According to Leslie in the reference cited in Prob. 16.7, the particles formed during interphase precipitation in a microalloyed HSLA steel have a diameter of about 5 nm. Assume that the microalloying element is vanadium and that the interphase precipitate is vanadium carbide, VC.

(a) What value of f, the volume fraction of these precipitates, would give an increase in strength of 200 MPa?

(b) Estimate the weight percent of the vanadium incorporated in the precipitate.

References

1. Vyhnal, R. F., and Radcliffe, S. V., *Acta Met.*, 20, 1972, 435–445.
2. Dahmen, U., Pelton, A. R., Witcomb, M. J., and Westmacott, K. H., in *Solid-Solid Phase Transformations*, Ed. H. I. Aaronson, Published by AIME, 1981, pp. 637–641.
3. Pelton, A. R., and Westmocott, K. H., ibid., pp. 643–647.
4. Batte, A. D., and Honeycombe, R. W. K., *J. Iron Steel Inst.*, 211, 1973, 284.
5. Alexander, K. B., Legoues, F.K., Aaronson, H.I., and Laughlin, D.E., *Acta Met.*, 32, 12, 1984, 2241–2249.
6. Hatch, J. E., Ed., *ALUMINUM Properties and Physical Metallurgy*, p. 141, American Society for Metals, Metals Park, Ohio, 1984.
7. Silcock, J. M., Heal, T. J., and Hardy, H. K., *J. Inst. Met.* 82, 1953–54, 239–248.
8. Laird, C., and Aaronson, H. I., *Acta Met.*, 14, 1966, 171–185.
9. Sehitoglu, Foglesong, T., and Maier, H. J., *Met. and Mat. Trans. A*, 36, 2005, 749–761.
10. Christian, J. W., *The Theory of Transformations in Metals and Alloys*, p. 650, Pergamon Press, Oxford, 1965.
11. Hatch, J. E., Ed., *ALUMINUM Properties and Physical Metallurgy*, American Society for Metals, Metals Park, Ohio, 1984.
12. Fine, M. E., *Metall. Trans. A*, **6**, 1975, 625–630.
13. Sigli, C., et al., *Comptes Rendus Physique*, **19**, 8, 2018, 688–709.
14. Dey, G. K., Tewari, R., Rao., P. et al., *Metall. Trans. A*, **24**, 1993, 2709–2719.
15. ASM Handbook, Volume 4A, 2013.
16. Nie, J. F., *Metall. Mater. Trans. A*, **43**, 2012, 3891–3939.
17. Decker, R. F., et al., *JOM*, **71**, 2019, 2219–2226.
18. Turnbull, D., *Solid State Phys.*, **3**, 226 (1956).

Chapter 17

Deformation Twinning and Martensite Reactions

Learning Objectives

Upon completion of Chapter 17, you will be able to:

1. Differentiate between deformation types: twinning and slip
2. Compare and contrast coherent and incoherent twin boundaries
3. Describe nucleation and growth of twins and of martensite
4. Use Bain distortion model to convert fcc to bcc
5. Describe crystallographic features of martensitic transformation
6. Provide justification for the irrational nature of martensitic habit plane
7. Discuss thermoelasticity, stress induced martensite, and shape memory effects

We shall now consider two apparently unrelated types of phenomena that actually have much in common. One, deformation or mechanical twinning, is a mode of plastic deformation, while the other, comprising martensite reactions, is a basic type of phase transformation. Twinning, like slip, occurs as the result of applied stresses. Sometimes stress may be influential in partly triggering a martensitic transformation, but this is an effect of secondary importance. Martensite reactions occur in metals that undergo phase transformations, and the driving force for a martensite reaction is the chemical free-energy difference between two phases.

The similarity between martensite reactions and twinning lies in the analogous way twins and martensite crystals form, for, in both cases, the atoms inside finite crystalline volumes of the parent phase are realigned as new crystal lattices. In twins this realignment reproduces the original crystal structure, but with a new orientation. In a martensite plate, not only is a new orientation produced, but also a basically different crystalline structure. Thus, when martensite forms in steel quickly cooled to room temperature, the face-centered cubic phase, which is stable at elevated temperatures, is converted into small crystalline units of a body-centered tetragonal phase. On the other hand, when twinning occurs in a metal such as zinc, both the parent crystal and the twinned volumes still have the close-packed hexagonal zinc structure. In both twinning and martensite transformations, each realigned volume of material suffers a change in shape that distorts the surrounding matrix. The changes in shape are quite similar, so that martensitic plates and deformation twins look alike, taking the form of small lenses, or plates. Actually, as shall presently be seen, it is quite possible to convert large volumes of the parent structure into elements of one of the new structures.

17.1 Deformation Twinning

The deformation accompanying mechanical twinning is simpler than that associated with martensitic reactions because there is no change in crystal structure, merely a reorientation of the lattice. The change of shape associated with deformation twinning is a simple shear, as shown in Fig. 17.1A, where, for simplicity, it is assumed that the twin traverses the entire crystal. The difference between twinning and slip should be carefully recognized since, in both cases, the lattice is sheared. In slip, the deformation occurs on individual lattice planes, as indicated in Fig. 17.1B. When measured on a single slip plane, this shear may be many times larger than the lattice spacing and depends on the number of dislocations emitted by the dislocation source. The shear associated with deformation twinning, on the other hand, is uniformly distributed over a volume rather than localized on a discrete number of slip planes. Here, in contrast to slip, the atoms move only a fraction of an interatomic spacing relative to each other. The total shear deformation due to twinning is also small, so that normally slip is a much more important primary mode of plastic deformation. Nevertheless, the importance of mechanical twinning is becoming increasingly more apparent in explaining certain elusive mechanical properties of many metals. For example, when a metal twins, the lattice inside the twin is frequently realigned into an

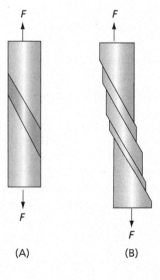

Fig. 17.1 The difference between the shears associated with twinning **(A)**, and slip **(B)**

orientation where the slip planes are more favorably aligned with respect to the applied stress. Under certain conditions, a heavily twinned metal can be more easily deformed than one free of twins. On the other hand, lattice realignment, if confined to a limited number of twins, can induce fracture by permitting very large deformations to occur inside the confined limits of twins. Twins are also of importance in recrystallization phenomena, for the intersections of twins are preferred positions for the nucleation of new grains during annealing.

A good insight into the mechanics of twinning can be gained by studying the simple diagrams of Fig. 17.2. The twinning represented in these sketches is only schematic and does not refer to twinning in any real crystal. The upper diagram represents a crystal structure composed of atoms assumed to have the shape of oblate spheroids. The lower diagram represents the same crystal after it has undergone a shearing action that produced a twin. The twin is formed by the rotation of each atom in the deformed area about an axis through its center and perpendicular to the plane of the paper. Three atoms are marked with the symbols a, b, and c in both diagrams to show their relative positions before and after the shear. Notice that individual atoms are shifted very little with respect to their neighbors. While it is in no way implied that the atom movements in a real crystal are the same as those shown in Fig. 17.2, it is true that in all cases the movement of an atom relative to its neighbors is very small. The two parts of Fig. 17.2 show another important characteristic of twinning: the lattice of the twin is a mirror image of the parent lattice. The lattices of twin and parent are symmetrically oriented across a symmetry

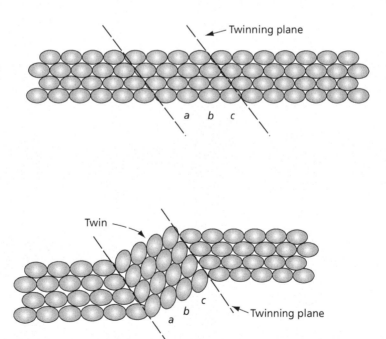

Fig. 17.2 Schematic representation showing how a twin may be produced by a simple movement of atoms

plane called the *twinning plane*. The several ways that this symmetry can be attained will be discussed in the next section.

17.2 Formal Crystallographic Theory of Twinning

The formal crystallographic theory of twinning in metals has been summarized by Cahn (Ref. 1) and the following treatment is largely based on his presentation.

Let us assume that shearing forces are applied to a single crystal specimen, as indicated on the left in Fig. 17.3, and that, as a result of this applied force, the crystal is sheared. The resulting shape of the crystal is shown in the second drawing. Furthermore, assume that after deformation, the sheared section still has the structure and symmetry of the original crystal. In other words, this region is to retain, after shearing, all the crystallographic properties of the metal of which it is composed, so that the size and shape of the unit cell must be unchanged. According to crystallographic theory, the size and shape of the unit cell will be unchanged only if it is possible to find three noncoplanar, rational lattice vectors in the original crystal that have the same lengths and mutual angles after shearing.

Figure 17.4 shows an edge view of a volume that has been sheared. The top surface is displaced a distance e to the right with respect to the bottom surface, and crystallographic planes with orientations perpendicular to the plane of the paper, such as B, C, and D, are rotated to new positions. Original positions of the planes, in each case, are indicated by dotted lines, and the final positions by solid lines. The only plane whose dimensions have not changed as a

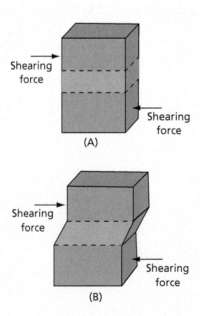

Fig. 17.3 **(A)** shows a single crystal in a position where it is subject to shearing forces. **(B)** shows a twinning shear that has occurred as a result of the force

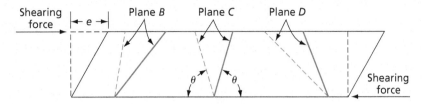

Fig. 17.4 The edge view of a small rectangular volume that has suffered a shear deformation as a result of the applied shearing forces. The dashed lines indicate the original shape of the volume; the solid lines its final shape. The figure is given to illustrate the change in shape of several planes as a result of the shear. The original positions of the planes are given with dashed lines. All three planes have orientations perpendicular to the paper

consequence of the shear is plane C. Plane D has been shortened and plane B has been lengthened. The reason that C has not suffered a change in shape is because it makes the same angle θ with the base both before and after the shear. This plane is unique and, except for the plane defining the top and bottom surfaces of the sheared section, it is the only plane whose shape is not changed by the shear.

The next three-dimensional figure, Fig. 17.5, shows several other arbitrarily chosen planes with more general orientations than those shown in Fig. 17.4. Again the positions of the planes before and after the shear are given (a change in shape is quite evident in each case).

From the above, it is evident that there are only two crystallographic planes in a shearing action that do not change their shape and size as a consequence of the shear. The first is the plane defining the upper and lower surfaces of the sheared volume. This plane contains the shear direction. In the case of shear by twinning, it is known either as the twinning plane or as the first undistorted plane, and is given the symbol K_1. The other plane, designated C in Fig. 17.4, intersects K_1 in a line that is perpendicular to the direction of shear and makes equal angles with K_1 before and after shear occurs. This is called the second undistorted plane and its crystallographic designation is K_2. In Fig. 17.6, the relationship between K_1 and K_2 is graphically illustrated. Several other quantities can also be defined in terms of this figure. The shear direction is shown with an arrow and labeled with its customary designation η_1. The plane that lies perpendicular to K_1 and contains the shear

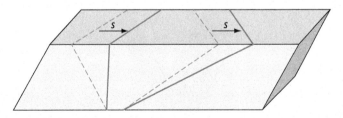

Fig. 17.5 A three-dimensional figure of a small sheared rectangular volume showing the distortion of two arbitrarily chosen planes caused by the shear. Dashed lines are the original positions of the planes, whereas the solid lines are the sheared positions. (The direction of the shear is indicated in each case by the arrow labeled S)

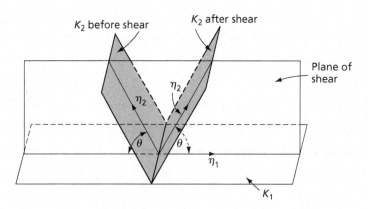

Fig. 17.6 The spatial relationships are given between K_1, K_2, η_1, η_2, and the plane of shear

direction is known as the *plane of shear*. The intersection of the plane of shear with the second undistorted plane K_2 is also an important direction. It is designated with an arrow and is given the symbol η_2. Notice that there are two positions for η_2 corresponding to the positions of K_2 before and after shearing.

Because all other planes suffer a change in size or shape during a shear, it is only in planes K_1 and K_2 that vectors can be found which will not be distorted by a twinning deformation. The basic problem is to find the possible combinations of three noncoplanar lattice vectors, lying in K_1 and K_2, which retain their mutual angles and magnitudes after the shear.

Let ε in Fig. 17.7 be any vector in plane K_1; then there is only one vector in K_2 that makes the same angle with ε before and after the shear. This is η_2, which is perpendicular to the intersection of K_1 and K_2. Since ε is any vector in K_1, η_2 must therefore make the same angle before and after shear with all vectors lying in K_1. Finally, if the assumption is made that K_1 is a rational plane (so that it contains rational directions) and that η_2 is a rational direction, then the conditions for the unit cell to have the same size and shape before and after shearing will have been realized.

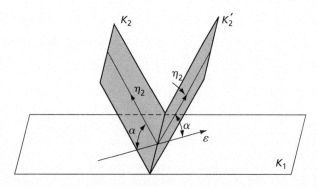

Fig. 17.7 In this figure, which corresponds to a shear of the first kind, notice that η_2 makes equal angles with an arbitrary vector ε, which lies in K_1, both before and after the shear. The sheared position of the second undistorted plane is designated K_2'

17.2 Formal Crystallographic Theory of Twinning

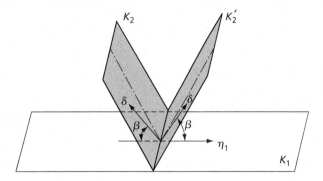

Fig. 17.8 This diagram represents a shear of the second kind. In this case δ represents an arbitrary vector in K_2 which makes the same angle with η_1 before and after shearing. The sheared position of the second undistorted plane is designated K_2'

In the same manner, it is easily seen that η_1 is the only direction in plane K_1 that makes the same angle with arbitrary vectors in K_2 before and after shearing. This follows from the fact that η_1 is also perpendicular to the intersection of K_1 and K_2. The relationship between δ (an arbitrary vector in K_2) with η_1 is shown in Fig. 17.8. It can, therefore, be concluded that another condition for preservation of the lattice structure during twinning is that η_1 and K_2 be rational.

It follows from the above that there are three ways that a crystal lattice can be sheared while still retaining its crystal structure and symmetry:

1. When K_1 is a rational plane and η_2 a rational direction. (A twin of the first kind)
2. When K_2 is a rational plane and η_1 a rational direction. (A twin of the second kind)
3. When all four elements $K_1, K_2, \eta_1,$ and η_2 are rational. (A compound twin)

The lattice rotations that occur as a result of twinning are different, depending upon whether a twin of the first or the second kind is formed.

In a twin of the first kind, the lattice in the sheared section is related to that in the unsheared section by a rotation of 180° about the normal to K_1. This effect is shown in Fig. 17.9. The corresponding 180° rotation in a twin of the second kind is about η_1 as an axis. This is illustrated in Fig. 17.10. On a stereographic projection, the two rotations would appear

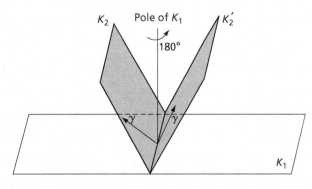

Fig. 17.9 The lattice rotation in a twin of the first kind is 180° about the normal to the twinning plane (K_1). Notice the rotation of the vector γ

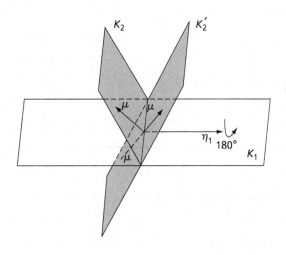

Fig. 17.10 In a twin of the second kind, the rotation is 180° about the shear direction η_1. In the drawing, the nature of this rotation is illustrated by projecting the sheared position of K_2 below the twinning plane (K_1). The 180° rotation of the vector μ is thus emphasized

as shown in Fig. 17.11. In a compound twin, due to symmetry considerations, both types of rotation lead to the same final orientation.

The following is a summary of the basic nomenclature relating to the theory:

K_1 = the twinning plane, or the first undistorted plane
K_2 = the second undistorted plane
η_1 = the shear direction
η_2 = the direction defined by the intersection of the plane of shear with K_2

The plane of shear is the plane that is mutually perpendicular to K_1 and K_2 and contains η_1 and η_2.

The composition plane is the plane separating the sheared and unsheared regions. In actual twins, it is usually close to K_1.

γ is the shear and equals the shear strain.

A twin that occurs inside another twin is called a *second-order twin*.

Twins of the second kind are relatively rare in metals and occur only in crystals of low symmetry. Cahn (Ref. 1) has identified one such twin in uranium. They are also found in salts and minerals

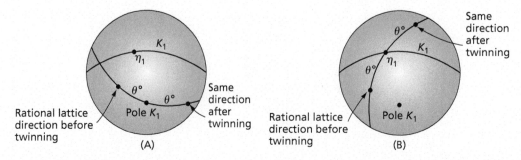

Fig. 17.11 Comparison of the lattice rotations in twins of the first and second kinds. In **(A)**, which corresponds to a twin of the first kind, a rational lattice direction is rotated 180° about the pole of K_1. In **(B)**, the rotation is 180° about η_1 for a twin of the second kind

of orthorhombic or lower symmetry. In metals of high symmetry, with cubic or close-packed hexagonal lattices, the twins generally belong to the compound classification.

From the above theory, it becomes apparent that the orientation of the lattice after twinning is given by a rotation of 180° about either η_1, or the normal to K_1, depending on the kind of twinning involved. On the other hand, the size and sense of shear are determined by the second undistorted plane K_2. It should also be noted that the shear in twinning, as contrasted to slip, is polar, that is, it can occur in only one direction.

The relationship between the shear and the second undistorted plane can be illustrated by a simple example. Consider the case of hexagonal metals. The common form of twinning in these metals is a compound twin in which we have

$$K_1 = (10\bar{1}2) \qquad K_2 = (\bar{1}012)$$

$$\eta_1 = [\bar{1}011] \qquad \eta_2 = [10\bar{1}1]$$

In zinc the c/a ratio is 1.856, while in magnesium it is 1.624. Because of this difference there is a 46.98° angle between the basal plane and the $\{10\bar{1}2\}$ planes in zinc, and a 43.15° angle in magnesium. Figure 17.12 shows how this affects twinning in these two metals.

The parts of Fig. 17.12 are drawn so that the plane of the paper is the plane of shear. Symmetry conditions require that the second undistorted plane, K_2, be rotated clockwise in the case of zinc, and counterclockwise in magnesium. This has the effect, in the former case, of lengthening the crystal inside the twinned volume in a direction parallel to the basal plane. In magnesium on the other hand, the effect is reversed and the crystal is shortened in this direction. Figure 17.13 illustrates the fact that a tensile stress parallel to the basal plane in zinc favors $\{10\bar{1}2\}$ twinning. In magnesium, the sense of shear is such that twinning is favored by compression and not tension in this direction.

Mechanical twins have been observed in all the basic metal crystal structures. This includes both face-centered and body-centered cubic crystals. In some face-centered pure cubic metals, deformation twins are observed only at very low temperatures and at very large strains. Before Blewitt, Coltman, and Redman (Ref. 2) observed twins in copper deformed at 4.2 K in 1957, it was generally

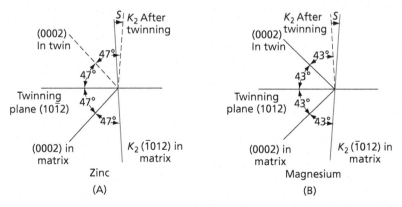

Fig. 17.12   The cause for the difference in shear sense for $\{10\bar{1}2\}$ twinning in zinc and magnesium. Notice that K_2, before twinning, lies to the left of the vertical in zinc, but in magnesium it is on the right. Symmetry conditions require that K_2 rotate in different directions for these two metals. The angles given in these figures have been expressed in whole numbers for the sake of simplicity

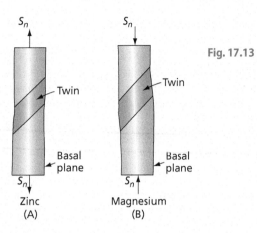

Fig. 17.13 The formation of a $\{10\bar{1}2\}$ twin in zinc increases the length of the crystal in a direction parallel to the basal plane, but decreases it in magnesium. A tensile stress applied parallel to the basal plane will twin a zinc crystal, but a compressive stress is needed in magnesium. In these figures, the basal plane of the parent crystal lies parallel to the stress axes and perpendicular to the plane of the paper

believed that fcc metals probably did not twin. However, since that time, mechanical twinning has been observed in many fcc metals. Twins in face-centered cubic metals form on $\{111\}$ planes and the shear direction is $\langle 11\bar{2} \rangle$.

In body-centered cubic metals, the twinning plane, K_1, is $\{112\}$, while η_1 is $\langle 11\bar{1} \rangle$. Twins in body-centered cubic metals are also primarily observed at low temperatures where they may be a significant factor in the deformation and fracture mechanics. Twinning is more important in hexagonal metals and in metals of low symmetry, such as tetragonal β-tin, orthorhombic uranium, and rhombohedral arsenic, antimony, and bismuth. A comprehensive review of twinning in various structures can be found in an article by Christian and Mahajan (Ref. 3).

17.3 Twin Boundaries

Let us now consider the interface between a twin and the parent crystal. The atomic arrangement at a twin boundary in a face-centered cubic metal is shown in Fig. 17.14. This diagram assumes that the twin interface is exactly parallel to the twinning plane K_1. In this structure, the two lattices (twin and parent) match perfectly at the interface. Atoms on either side of the boundary have the normal interatomic separation expected in a face-centered cubic lattice. The stacking sequence of the close-packed planes is

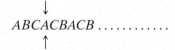

The interfacial energy of the boundary is very small. In the case of copper, it has been determined (Ref. 4) to be 0.044 J/m², which is very much smaller than the surface energy of a copper-copper grain boundary (Ref. 5), which is 0.646 J/m². Barrett (Ref. 6) has drawn diagrams similar to that of Fig. 17.14 for $\{10\bar{1}2\}$ twins in hexagonal metals and $\{112\}$ twins in body-centered cubic metals. He shows that in both cases a reasonable match between twin and parent lattices can be made across K_1. However, atoms on both sides of the interface are displaced small distances from their normal lattice positions. Because atomic bonds are strained in these twin interfaces, they must possess higher interfacial energies than the $\{111\}$ twin boundary of face-centered cubic metals. However, these energies

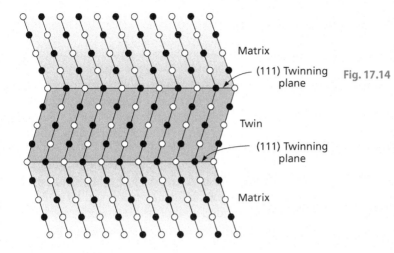

Fig. 17.14 Atomic arrangement at the twinning plane in a face-centered cubic metal. Black and white circles represent atoms on different levels (planes). (After Barrett, C. S., ASM Seminar, *Cold Working of Metals*, 1949, Cleveland, Ohio, p. 65)

are still much smaller than those of normal grain boundaries. The fact that deformation twins invariably form on planes of low indices may be explained in terms of the surface energy associated with the interface between twins and parent crystals. In general, the higher the indices, the poorer the fit at the interface, the higher the surface energy, and the lower the probability of twin formation.

A twin boundary that parallels the twinning plane is said to be a coherent boundary. Most twins start as thin narrow plates that become more and more lens-shaped as they grow. The average twin boundary is, accordingly, incoherent.

In a coherent boundary it is usually quite possible to match the two lattices without assuming the presence of dislocations in the boundaries. (See Fig. 17.14.) In an actual boundary where it is not coherent, it is generally accepted that a dislocation array is necessary in order to adjust the mismatch between lattices of parent and twin. This is demonstrated in Fig. 17.15, which shows schematically segments of both a coherent boundary and an incoherent boundary.

17.4 Twin Nucleation and Growth

Twins form as the result of the shear-stress component of an applied stress that is parallel to the twinning plane and lies in the twinning direction μ_1. The normal stress component (normal to the twinning plane) is unimportant in twin formation, a conclusion based on work (Ref. 7) in which zinc crystals were deformed in tension while subjected to hydrostatic pressure. In these experiments, no measurable difference was found in the tensile stress required to form twins under hydrostatic pressures varying from 1 to 5000 atmospheres. Since the hydrostatic pressure has no

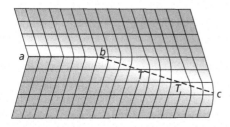

Fig. 17.15 The difference between a coherent twin boundary *ab* and an incoherent twin boundary *bc*. Notice the dislocations in the incoherent boundary. (After Siems, R., and Haasen, P., *Zeits. für Metallkunde*, **49** 213 [1958])

shear component on the twinning plane, but does have a normal component, we must conclude that twins form only as a result of shear stress.

It has been estimated (Ref. 8) that the theoretical shear stress necessary to homogeneously nucleate deformation twins in zinc crystals lies between 40 and 120 kg per mm^2 (56,000 to 168,000 psi). Experimentally measured values of the shear stress to form twins lie much lower than the theoretical values, ranging from 0.5 to 3.5 kg per mm^2. This is strong evidence for the belief that twins are nucleated heterogeneously. A recent direct observation using in-situ high resolution transmission electron microscopy seems to also indicate that the strain field of matrix dislocations could prompt the twinning nucleation in HCP crystals (Ref. 9).

A great deal of evidence suggests that nucleation centers for twinning are positions of highly localized strain in the lattice. Confirmation for this assumption is given by the fact that twins appear to form primarily in metals that have suffered previous deformation by slip (Ref. 8). It would further appear that the slip process must become impeded so that barriers are formed that prevent the motion of dislocations in certain restricted areas.

Because the localized stress fields (of twin nucleation centers) can be formed in a number of different ways, depending on the geometry and orientation of the specimen, as well as the nature of the applied stress, it is probable that there can be no universal critical resolved shear stress for twinning as there is for slip. This explains the experimentally observed wide range of shear stresses required for twinning (that is, 0.5 to 3.5 kg per mm^2 in zinc).

There is an important factor relative to whether or not there is a critical resolved shear stress for twinning that apparently has been ignored. Normally, when one measures the stress required to nucleate a twin, an individual event is observed, whereas the critical resolved shear stress for slip represents the statistical average of many events. That is, when a crystal begins to deform macroscopically by slip, a very large number of dislocations are nucleated and begin to move. If one were to measure the average stress required to nucleate many thousands of small twins, it is possible that for a given set of deformation conditions a relatively fixed value of the stress would be obtained as it is in slip.

The formation of a twin involves not only the creation of a pair of surfaces, but also the accommodation of the twinning shear by the lattice surrounding the twin. As the twin grows, it normally becomes more lenticular and its surface acquires an array of twinning dislocations. As may be seen by considering Fig. 17.15, the movement of the dislocations in an incoherent twin boundary can result in growth of the twin. Thus, a horizontal movement of the dislocations in Fig. 17.15 to the right will move the incoherent boundary to the right, increasing the size of the twin, which is assumed to be at the bottom of the figure.

It is theoretically possible to conceive of the array of twinning dislocations lying in the boundary of a lenticular twin as a single dislocation in the form of a helix. This is shown schematically in Fig. 17.16. Such a dislocation can result from the dissociation of a regular lattice dislocation.

In the case of a body-centered cubic crystal, Cottrell and Bilby have proposed (Ref. 10) that a $\frac{1}{2}[111]$ total slip dislocation should be able to dissociate into a glissile twinning partial dislocation of Burgers vector $\frac{1}{6}[11\bar{1}]$ and a sessile partial dislocation $\frac{1}{3}[112]$. This reaction may be expressed by the equation

$$\tfrac{1}{2}[111] \rightarrow \tfrac{1}{3}[112] + \tfrac{1}{6}[11\bar{1}] \qquad 17.3$$

In the figure, line xy represents the total $\frac{1}{2}[111]$ dislocation, which is assumed to be dissociated between points a and b. The straight dashed line between these two points represents the sessile $\frac{1}{3}[112]$ partial, while the helical dislocation, with $b = \frac{1}{6}[11\bar{1}]$, is the twinning dislocation. This latter is assumed to lie on the $(\bar{1}21)$ plane. The key to this entire picture is the fact that the $(\bar{1}21)$ plane spirals

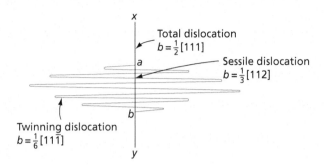

Fig. 17.16 The pole mechanism of Cottrell and Bilby for twinning in a body-centered cubic crystal. The helical twinning dislocation lies on the [$\bar{1}$21] plane, which spirals around the dislocation *xy*. (After Cottrell, A. H. and Bilby, B. A., *Phil. Mag.*, **42** 573 [1951])

about the line *xy*. This occurs because all dislocations comprising the line (that is, line *xa* and *by* whose Burgers vector is $\frac{1}{2}$[111], and line *ab* with Burgers vector $\frac{1}{3}$[112]) have a component of their Burgers vector, normal to the ($\bar{1}$21) plane, equal to the interplanar spacing of this plane. In other words, as far as the ($\bar{1}$21) plane is concerned, line *xy* is a screw dislocation about which this plane spirals.

The schematic twin in Fig. 17.16 can grow in two ways. The twinning dislocation can increase its number of spirals. This would correspond to a growth in thickness. The other possibility is that the spirals themselves could expand, thereby increasing the length of the twin lamellae. Figure 17.16 only represents a part of the Cottrell-Bilby twinning mechanism. In their original paper, they also considered the nature of the original dislocation dissociation that leads to this result. For the details, see the original paper (Ref. 10).

While there are some objections to this type of mechanism (Ref. 11), called a *pole mechanism*, it does indicate how a dislocation can be nucleated and grow. Pole mechanisms have also been proposed for other crystal structures, of which a notable example is that of Venables (Ref. 12) for face-centered cubic crystals.

It is highly probable that if the pole mechanisms do work in metals, any given twin probably grows as a result of a number of pole mechanisms acting together. There is no reason why this should not be possible, and there is a good reason for assuming this; namely the macroscopic size of the average twin is such that it must intersect a large number of slip dislocations, each of which should constitute a pole.

Twins grow in size by increasing both their length and thickness. Large volumes of twinned material often form by the coalescence of separately nucleated twinned areas. The speed of twin growth is primarily a function of two variables that are not entirely independent. The first is the speed of loading that directly influences the rate of growth. The other is the stress required to nucleate twins. If twins nucleate at very low stresses, the stress required for their growth will be of the same order of magnitude as the nucleating stress. In this case, a small submicroscopic twin may form and grow more or less uniformly with an increasing stress until its growth is impeded by some means or other. On the other hand, if twins form under conditions that result in very high stress levels before nucleation, the stress for growth may be much smaller than that for nucleation. When this happens, twins grow at very rapid rates as soon as they are nucleated. Several interesting phenomena are associated with this rapid growth. First, the rapid deformations that ensue set up shock waves in the metal that can be heard as audible clicks. The crackling sound audible when a bar of tin is bent is due to deformation twinning. The other effect is visible in tensile tests of crystal specimens that undergo twinning during loading. The rapid formation of twins results in sudden increments in tensile strain. In a rigid testing machine, this causes the load to suddenly drop, giving the stress-strain curve a saw-tooth appearance in the region of twinning. A stress-strain curve of this nature is shown in Fig. 17.17.

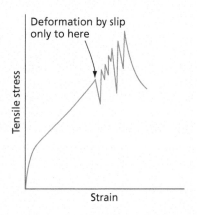

Fig. 17.17 Single-crystal tensile stress-strain curve showing discontinuous strain increments due to twinning. (After Schmid and Boas, *Kristallplastizität*, Julius Springer, Berlin, 1935.)

17.5 Accommodation of the Twinning Shear

While the shear on a lattice plane is smaller in twinning than in slip, this shear becomes macroscopic when the twin attains an appreciable width. The shear of a very thin twin might possibly be accommodated elastically in the matrix surrounding a twin. However, as a twin grows in size, the accommodation normally requires plastic deformation in the surrounding metal. If the matrix is not able to accommodate the twinning shear of a twin of finite size, then a hole may develop at the point where the accommodation was not obtained. A well-known example occurs at twin intersections in iron and is known as a Rose's channel. This phenomenon was originally reported by Rose in 1868. Figure 17.18 shows one way of forming such a channel.

If a twin lies entirely inside a grain, it will usually be lenticular and thus taper to a sharp edge at its periphery. This growth form may be assumed to be related to the shear accommodation process, since it should be easier to accommodate the twinning shear by slip in the matrix if the twin grows as a wedge. However, the twin shape is also undoubtedly influenced by the nature of the dislocation array that forms on the twin boundary. The more tapered the twin profile, the greater will be the separation between the dislocations lying in the boundary.

The sharp leading edge of a twin is lost when a growing twin intersects a free surface. If a twin meets a single free surface, it assumes the shape of a half-lens (Fig. 17.19), and if it crosses a

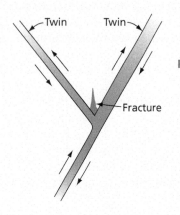

Fig. 17.18 A crack induced in an iron crystal at a twin intersection. Similar cracks were first reported by Rose in 1868. (After Priestner, R., *Deformation Twinning*, AIME Conf. Series, vol. 25, p. 321, Gordon and Breach Science Publishers, New York, 1964)

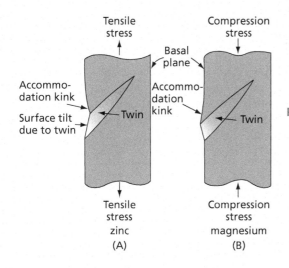

Fig. 17.19 Surface tilts and accommodation kinks resulting from intersections of half-lens twins with the surface in two hexagonal metals

complete crystal, the twin acquires flat sides parallel to the twinning plane K_1 (Fig. 17.1A). In this last case, the twin can be accommodated in the lattice without appreciable distortion of the latter. This is not true for half-lens-shaped twins where the lattice around the twins is forced to accommodate the shearing strain. In the case of a half-lens, one commonly finds accommodation kinks adjacent to the twins. The diagrams of Fig. 17.19 show the nature of simple accommodation kinks in the hexagonal metals zinc and magnesium, as viewed on a surface normal to the basal plane and to the twins. In each case, the basal planes in the crystal adjacent to the twin are bent or kinked so as to allow the lattice of the parent to follow the shear strain of the twin. In some cases, serrated twin interfaces at the crystal surfaces have also been reported (Ref. 13).

In summary, because the twinning shear has to be accommodated, slip normally has to play a very active role in twinning. From a broad viewpoint, we should therefore consider twinning deformation in terms of the coupling between slip and twinning modes. It is highly probable that twinning would not be a significant practical means of plastic deformation in metals such as titanium and zirconium if there was a lack of this coupling. Generally speaking, this implies that twinning also involves slip. The reverse, of course, is not true.

17.6 The Significance of Twinning in Plastic Deformation

The existence of deformation twinning and its effects on mechanical properties has often been ignored in dislocation dynamics studies. This is legitimate if deformation occurs by slip alone. However, in many cases slip and twinning occur simultaneously and the resulting mechanical properties are strongly influenced by the twinning, as for example in the important hcp metals, such as titanium, zirconium, and magnesium and in the commercially significant fcc alloys, such as brasses, bronzes, cupro-nickels, beryllium coppers, and austenitic stainless steels (Ref. 3 and Refs. 13 through 18).

A metal that deforms largely by deformation twinning is iron alloyed with 25 at. % beryllium. This alloy has been extensively studied by Bolling and Richmond (Ref. 19). These authors designated deformation that occurs repeatedly by twinning *continual mechanical twinning* and state that it occurs when the flow-stress for twinning is less than that for gross slip. They have also shown that, at a constant strain rate, when twinning dominates over slip the yield stress tends to rise when the temperature is increased instead of falling as is normally the case in metals deforming by slip in the absence of dynamic strain aging. Second, at a given temperature, if the strain rate is increased, the flow stress tends to decrease. This is also an inverse effect to that normally experienced when deformation occurs primarily by slip. A third observation was that, when twinning dominates, increasing the temperature at constant strain-rate or decreasing the strain-rate at constant temperature can increase the work hardening rate. These are the reverse of the effects normally expected when slip controls the deformation.

17.7 The Effect of Twinning on Face-Centered Cubic Stress-Strain Curves

It was long believed that twinning did not occur in fcc metals. However, it has been amply demonstrated in recent years that twinning is a significant factor in the deformation of these metals. In them the critical resolved shear stress for slip is normally smaller than that to form twins. As a result, an fcc metal usually begins its deformation by slip. The work hardening resulting from this slip deformation increases the flow-stress, and only after the latter reaches the twinning stress does twinning begin.

An important factor is that the twinning stress in fcc metals depends strongly on the stacking fault energy of these metals. This is because the surface energy of a twin boundary is closely related to that of a stacking fault and a large part of the work to form a twin goes into creating its boundaries. Thus, the lower the stacking fault energy of an fcc metal, the lower, in general, is the stress needed to nucleate twins.

In fcc metals, the stacking fault energy decreases with increasing concentration of a solute in solid solution so that twinning becomes more and more significant the higher the solute concentration. Consequently, twinning is a significant feature in a large number of commercially important and highly alloyed fcc metals. These include the brasses, bronzes, cupro-nickels, and many others.

In the low stacking fault energy fcc metals, it is possible to consider deformation twinning and slip as competitive forms of plastic deformation. However, twinning differs from slip in a number of significant ways. Thus, the movement of a total slip dislocation leaves the lattice behind it in perfect registry. Accordingly, there is no change in crystal orientation due to the movement of slip dislocations. This is not true in twinning because the crystal structure inside a twin has an orientation that differs from that of the parent crystal in which the twin formed. Although this orientation is twin-related to the matrix, and it can be shown that some slip dislocations can easily pass into the twin from the matrix or vice versa, experimental evidence strongly suggests that a twin boundary can frequently act like a grain boundary. Thus, twins can often be considered to reduce the grain size by supplying additional grain boundaries.

Because slip is normally the more important of the two types of deformation, the effect of twinning on metal plasticity is probably best considered in terms of how it perturbs the stress-strain behavior expected from slip alone. In this regard, probably the most significant aspect of twinning is its effect on the work hardening rate. This is particularly true in the low stacking fault energy fcc alloys where the dissociation of the total dislocations into widely extended dislocations should have a strong effect on work hardening.

There are two basic ways that twinning can influence the work hardening rate. When twinning starts, the twins in a grain are initially aligned parallel to the primary slip plane. This is because

17.7 The Effect of Twinning on Face-Centered Cubic Stress-Strain Curves

the slip and twinning planes in the fcc lattice are both {111}, and the most highly stressed twinning plane is usually the same as the primary slip plane. In this case the twins do not have a significant effect on the movement of slip dislocations. As a result, in a given increment of deformation, the twins add a small contribution to the strain but not to the stress. Thus, as suggested by Vohringer (Ref. 20), we may write

$$d\varepsilon/d\sigma = (d\varepsilon_s + d\varepsilon_t)/d\sigma_s \qquad 17.4$$

where $d\varepsilon/d\sigma$ is the reciprocal of the work hardening rate, $d\varepsilon_s$ is the strain contribution due to the slip, $d\varepsilon_t$ is the strain increment due to twinning, and $d\sigma_s$ is the rise in the stress due to the slip increment. Vohringer chose the reciprocal of the work hardening rate instead of the work hardening rate, $d\sigma/d\varepsilon$, in his equation because it has some advantage in presenting the experimental data. Equation 17.4 clearly implies that after twinning starts, the reciprocal of the work hardening rate should increase (meaning that the work hardening rate decreases). This has been confirmed (Ref. 18) by experimental data as may be seen in Fig. 17.20, where the reciprocal of the work hardening rate, $d\varepsilon/d\sigma$, is plotted against the true stress, σ, for a Cu–4.9 at. % Sn specimen deformed at room temperature, 298 K. Note that the slope of this curve rises sharply as it passes from the region marked Stage 1, where deformation occurs by slip alone, into Stage 2, where twins form parallel to the active slip plane. This increase in slope of the reciprocal of the work hardening rate at the transition from Stage 2 to Stage 3 signifies a decrease in work hardening rate after the transition so that with continued deformation the work hardening increases at a reduced rate. This slower but nevertheless continuous rise in the flow stress eventually causes twins to form on twinning planes that intersect the primary twinning or slip plane. When this happens, an effective decrease in grain size occurs and the work hardening rate again increases.

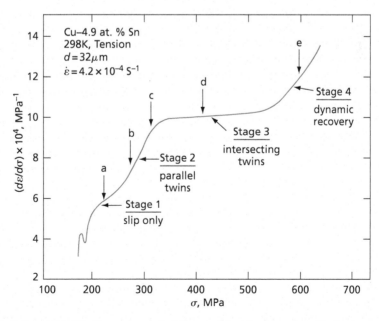

Fig. 17.20 The variation of $d\varepsilon/d\sigma$, the reciprocal of the work hardening rate, with the true stress, σ, for Cu–4.9Sn at 298 K

17.8 Martensite

When the temperature of a metal capable of undergoing a martensite reaction is lowered, it eventually passes through an equilibrium temperature separating the stability ranges of two different phases. Below this temperature, the free energy of the metal is lowered if the metal changes its phase from that stable at high temperatures to that stable at low temperatures. This free-energy difference is the primary driving force for a martensite reaction.

The phase change that occurs in a martensite transformation is brought about by the movement of the interface separating parent phase from the product. As the interface moves, atoms in the lattice structure of the parent phase are realigned into the lattice of the martensite phase. The nature of the individual atomic movements in the region constituting the interface are not known, just as they are not known in deformation twinning. Nevertheless, it is undoubtedly true that the displacement of atoms, relative to their neighbors, is small in magnitude and probably more complicated than those in deformation twinning. Because of the manner in which martensite forms, no composition change occurs as the parent lattice is converted into the product phase and diffusion in either the parent phase or product phase is not required for the reaction to continue. Martensite reactions are, accordingly, commonly referred to as *diffusionless phase transformations*.

The atomic realignments associated with martensite reactions produce shape deformations just as they do in mechanical twinning. Because the new lattice that is formed has a symmetry different from that of the parent phase, the deformation is, of necessity, more complicated. In mechanical twinning we have seen that the distortion is a simple shear parallel to the twinning plane or the symmetry plane between twin and parent crystal. As previously pointed out, the twinning plane is an undistorted plane; all directions in this plane are unchanged by twinning with respect to both their magnitudes and their angular separations. The *habit plane* or plane on which martensite plates form is also usually assumed to be an undistorted plane. The macroscopic shape deformation in the formation of a martensite plate is believed to be a shear parallel to the habit plane plus a simple (uniaxial) tensile or compressive strain perpendicular to the habit plane. A strain of this nature, known as an *invariant plane strain*, is the most general that can occur while still maintaining the invariance of the habit plane. Neither a shear parallel to the habit plane, nor an extension or contraction perpendicular to it will change the positions or magnitudes of vectors lying in the habit plane.

Data showing magnitudes of the shear and normal components of the invariant plane strains associated with martensite reactions are difficult to measure and are relatively scarce. Some of the available information is given in Table 17.1. The normal component of the strain in most martensite transformations is also smaller and harder to measure than the shear component, which probably explains the lack of data for this component in Table 17.1. Because the shape deformation is primarily a shear, martensite plates deform the lattice of the matrix like deformation twins. Individual martensite plates, formed in the interior of a crystal, are therefore lens-shaped, and if a martensite plate crosses a crystal, its boundaries are flat and parallel to the habit plane.

While the atomic movements involved in martensite reactions are small compared to those in slip, there is considerable variation in their magnitudes from one alloy to the next. Both the crystallographic features of a martensite transformation and the kinetics of the reaction are more easily studied in an alloy involving the smallest possible atomic displacement. Two well known martensite transformations fit this category: those in gold-cadmium alloys and that in an indium-thallium alloy. The very important martensite reactions that occur in the hardening of carbon and alloy steels involve

Table 17.1 Habit Planes and Macroscopic Distortions in Martensite Transformations*

System	Phase Changes†	Habit Plane	Shear Direction	Shear Component of Strain	Normal Component of Strain
Fe–Ni (30% Ni)	fcc to bcc	(9, 23, 33)	[156] ± 2°	0.20	0.05
Fe–C (1.35% C)	fcc to bct	(225)	[$\bar{1}12$]	0.19	0.09
Fe–Ni–C (22% Ni, 0.8% C)	fcc to bct	(3, 10, 15)	[$\bar{1}32$] (approx.)	0.19	
Pure Ti	bcc to hcp	(8, 9, 12)	[$11\bar{1}$] (very approx.)	0.22	
Ti–Mo (11% Mo)	bcc to hcp	4° from [$\bar{3}44$] 4° from (8, 9, 12)	[$14\bar{7}$] (approx).	0.28 ± 0.05	
Au–Cd (47.5% Cd)	bcc to orthorhombic	[0.696, −0.686, 0.213]	[0.660, 0.729, 0.183]	0.05	
In–Tl (18–20% Tl)	fcc to fct	[0.013, 0.993, 1]	[$01\bar{1}$] (approx.)	0.024	

*Most of the data are abstracted from a comprehensive summary by Bilby, B. A., and Christian, J. W., *The Mechanism of Phase Transformations in Metals*. The Institute of Metals, London, 1956. For detailed references and explanation of the data in the table, the reader is referred to this summary.

†The abbreviations signify: bct = body-centered tetragonal
fct = face-centered tetragonal

relatively large atomic displacements. In the following sections we shall consider an example of both an alloy in which the atomic movements are small (indium-thallium) and another (iron-nickel, 70%-30%) in which they are large. In this way, a better understanding of the various phenomena involved in martensitic transformations can be obtained. An overview of martensitic transformations and their microstructure can be found in two comprehensive review articles by Wayman (Refs. 21 and 22).

17.9 The Bain Distortion

A face-centered cubic lattice can also be considered as a body-centered tetragonal lattice; see Fig. 17.21 in which (in the upper diagram) a tetragonal unit cell is delineated in the face-centered cubic structure. In a tetragonal structure, the crystallographic axes lie at right angles to each other, as in a cubic structure, but one lattice constant c differs in magnitude from the other two. These latter dimensions of the unit cell are usually designated by the symbol a. When considered as a body-centered tetragonal structure, the normal face-centered structure has a

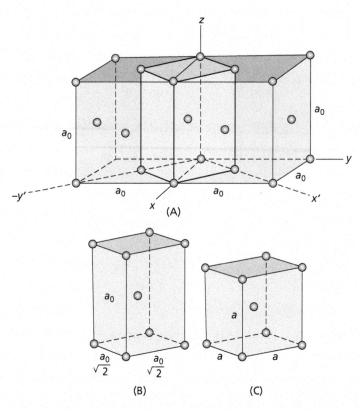

Fig. 17.21 Bain distortion for a face-centered cubic lattice transforming to a body-centered cubic lattice. The body-centered tetragonal cell is outlined in the face-centered cubic structure in **(A)**, and shown alone in **(B)**. The Bain distortion converts **(B)** to **(C)**. (After Wechsler, M. S., Lieberman, D. S., and Read, T. A., *Trans. AIME*, **197** 1503 [1953])

c/a ratio of $\simeq 1.4$. Similarly, a body-centered cubic structure can be considered a body-centered tetragonal structure with a c/a ratio of unity. In 1924, Bain (Ref. 23) suggested that a body-centered cubic lattice could be obtained from a face-centered cubic structure by a compression parallel to the c axis and an expansion along the two a axes. Any simple homogeneous pure distortion of this nature, which converts one lattice into another by an expansion or contraction along the crystallographic axes, belongs to a class known as *Bain distortions*.

The Bain distortion indicated above converts a face-centered cubic lattice into a body-centered lattice with a minimum of atomic movements. However, there is no undistorted plane associated with this Bain distortion (as we shall presently see), so that the invariant plane strain associated with martensitic transformations cannot be explained by a Bain distortion. The Bain distortion does, however, give an approximate measure of the magnitude of atomic movements involved in a transformation. Thus, in the iron-nickel alloy (30 percent Ni) of Table 17.1, which undergoes a face-centered cubic to body-centered cubic transformation, the c axis is shortened by a factor of approximately 0.8, while the a axes are increased in length by a factor of about 1.14. On the other hand, in the indium-thallium alloy, where a face-centered cubic alloy transforms to a face-centered tetragonal structure, as shown in Fig. 17.22, the c (vertical axis) becomes $1.0238\,a_0$, where a_0 is the

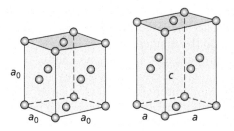

Fig. 17.22 Bain distortion in indium-thallium (18-20 percent Tl) alloy. Face-centered cubic transforms to face-centered tetragonal. In tetragonal structure, $c = 1.0238a_0$ and $a = 0.9881a_0$, where a_0 is cubic lattice constant. (Distortion of lattice is greatly exaggerated in these figures.)

cubic lattice constant, while the a axes decrease to 0.9881 a_0. Clearly, much larger deformations are involved in the iron-nickel transformation than in the indium-thallium (18-20 percent Tl). This fact is reflected in the magnitudes of the macroscopic shears of the martensite transformations given in Table 17.1: 0.20 and 0.024, respectively.

17.10 The Martensite Transformation in an Indium-Thallium Alloy

We shall now consider in some detail the martensite transformation in the indium-thallium alloy, where the small size of the observed shear makes for easier understanding of the observed phenomena. Studies made of single crystals of this alloy are of special interest. Provided that the crystals are carefully annealed and not bent or damaged, they undergo martensitic transformations involving the motion of a single interface between the cubic (parent phase) and the tetragonal (product phase) (Refs. 24 and 25). This transformation does not occur by the formation of lens-shaped plates, or even parallel-sided plates, but by the motion of a single planar boundary that crosses from one side of the crystal to the other. On cooling, an interface first appears at one end of a specimen and, with continued cooling, moves down through the entire length of the crystal. Because of dimensional changes accompanying the reaction, its progress can be followed (Ref. 26) with the aid of a simple dilatometer that permits measurements of the specimen length to be made as a function of temperature. A typical set of data is shown in Fig. 17.23. Notice that on cooling, the length of the specimen begins to change at approximately 345 K, signifying that the martensite transformation

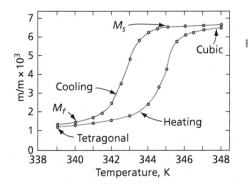

Fig. 17.23 The temperature dependence of the martensite transformation in the indium-thallium (18 percent Tl) alloy. Transformation followed by measurement of change in length of specimen. (From data of Burkart, M. W., and Read, T. A., *Trans. AIME*, **197** 1516 [1953])

started at this temperature. It is customary in all martensite transformations to designate, by the symbol M_s, the temperature corresponding to the start of the transformation. The curve also shows that the specimen did not transform completely until the temperature was lowered to about 340 K. This latter temperature is designated as the M_f, or martensite finish temperature. In order to move the interface from one end of the specimen to the other, it was necessary to drop the temperature 5 K below the M_s temperature. During this temperature interval between M_s and M_f, the interface, or habit plane does not move steadily and smoothly, but rather in a jerky fashion. It moves very rapidly for a short distance in a direction normal to itself and then stops until a further decrease in temperature gives it enough driving force to move forward again. The irregular motion of the interface is not apparent in the dilatometer measurements of Fig. 17.23, but may be readily observed by watching the movement of the interface under a microscope. The necessity for an ever-increasing driving force to continue the reaction is an unusual phenomenon because it implies the existence of a volume relaxation (Ref. 27), or an energy term opposing the transformation which is proportional to the volume of metal transformed. An explanation for the effect can be given in terms of the intersection of the interface with obstacles. It is now believed that martensitic and deformation twin interfaces can be composed of dislocation arrays. It is also recognized that a moving screw dislocation, which cuts through another screw dislocation, acquires a jog or discontinuity that produces a row of vacancies or interstitials as the moving dislocation continues its advance. While the geometry of the dislocation arrays in the crystals and in the interfaces is not known, the screw-dislocation-intersection picture gives us a tentative picture (Ref. 28) of how a moving interface can develop a resistance to its motion proportional to the distance through which it moves. The screw components in an interface should cut other screw dislocations in a number proportional to the distance through which the interface moves: each intersection adding its own contribution to the total force holding back the boundary.

17.11 Reversibility of the Martensite Transformation

The indium-thallium transformation is reversible for, on reheating, the specimen reverts not only to the cubic phase, but also to its original single-crystal orientation. If recooled, the original interface reappears and the cycle can be completely reproduced (provided the specimen is not heated to too high a temperature or held for too long a time between cycles).

The reverse transformation is shown in Fig. 17.23 by the return of the specimen to its original dimensions. It does not, however, retransform in the same temperature interval as that in which the forward transformation occurred, but in one averaging about 2 K higher. A hysteresis in the temperature dependence of the transformation is characteristic of most martensite transformations.

17.12 Athermal Transformation

In the single-crystal experiments discussed in the preceding paragraph, we have seen that the martensite transformation in indium-thallium occurs as a result of temperature changes that increase the driving force for the reaction (free energy). The progress of transformation is, accordingly, not time dependent. In theory, the faster one cools the specimen, the faster the interface moves, with equal amounts of transformation resulting at equal temperatures. Time does, however, have a secondary,

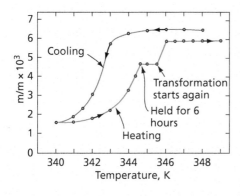

Fig. 17.24 Stabilization in the martensite transformation in an indium-thallium (18 percent Tl) alloy. (From data of Burkart, M. W., and Read, T. A., *Trans AIME*, **197** 1516 [1953])

though negative, effect on this transformation, for isothermal holding of the specimen at any temperature between the start and end of the transformation tends to stabilize the interface against further movement. This effect is demonstrated in Fig. 17.24, where the indicated transformation curve corresponds to a specimen whose heating cycle was interrupted at 345 K and held at this temperature 6 hr. Not only did the holding of the specimen at this temperature not induce additional transformation, but it also stabilized the interface. In order to make the interface move again, it was necessary to increase the driving force by approximately 1 K. An equivalent phenomenon occurs on cooling, so that it can be concluded that the formation of martensite in the present alloy depends primarily on temperature. A transformation of this type is said to be athermal, in contrast to one that will occur at constant temperature (isothermal transformation). Although isothermal formation of martensite is observed in some alloys, martensitic transformations tend to be predominantly athermal.

17.13 Phenomenological Crystallographic Theory of Martensite Formation

The nature of the atomic mechanisms that convert one crystal structure into the other in the narrow confines that we define as the interface are not known. Wechsler, Lieberman, and Read (Ref. 29) have shown, nevertheless, that the crystallographic features of martensite transformations can be completely explained in terms of the following three basic deformations:

1. A *Bain distortion*, which forms the product lattice from the parent lattice, but which, in general, does not yield an undistorted plane that can be associated with the habit plane of the deformation.
2. A *shear deformation*, which maintains the lattice symmetry (does not change the crystal structure) and, in combination with the Bain distortion, produces an undistorted plane. In most cases, this undistorted plane possesses a different orientation in space in the parent and product lattices.
3. A *rotation of the transformed lattice*, so that the undistorted plane has the same orientation in space in both the parent and product crystals.

No attempt is made in this theory to give physical significance to the order of the steps listed here, and the entire theory is best viewed as an analytical explanation of how one lattice can be formed from the other. As such, the Wechsler-Lieberman-Read theory is a phenomenological crystallographic theory of the martensite transformation. This approach to explaining martensitic transformation in metals has been successful in most instances (Refs. 21 and 22). In those cases where the theory appears to work best the martensite plates have a typically lenticular shape. It should be noted that the lens shape is generally associated with a minimum strain energy and is similar to that observed in deformation twinning. However, martensite has also been observed to take several other more intricate shapes. Among the other morphologies are the segmented plates observed in the alloy Fe-8Cr-1C, where "sideplates" are frequently seen to extend out from the principal interface of the martensite plates. In another case, the martensite lenses appear to have a midrib, which is believed to be a region where there is an increased concentration of transformation twins. Still another variant from the typical lenticular morphology is observed in Fe-Ni-C alloys. In this case the martensite plates are parallel sided and not lens shaped. This is known as *thin plate* martensite. Martensite can also be nucleated at the surface and not penetrate deeply into the specimen in some martensitic materials, giving rise to the term *surface* martensite. Alloys of Fe-Ni-C can also exhibit a morphology that has been described as *butterfly* martensite.

The indium-thallium transformation will now be considered in terms of the Wechsler, Lieberman, and Read theory. The Bain distortion for the indium-thallium transformation is shown in Fig. 17.22, where the very small actual deformation is magnified in order to show the effect more clearly. The original structure is cubic, while the final is tetragonal, so that the Bain distortion consists of an extension (2ε) along one axis (the c axis) and a contraction ($-\varepsilon$) along the other two axes. Simple geometrical reasoning will show that the given distortion does not possess an invariant plane. Consider first the set of lines drawn on the front face of the cube in Fig. 17.25. Lines *ab* and *cd* represent those directions on this surface, the lengths of which do not change in the transformation. In the cube shown on the left and in the prism shown on the right, these lengths are the same. All other directions on this surface, such as *ef* and *gh*, have their lengths increased and shortened respectively by the distortion. Now assume that the distortion actually possesses an invariant plane that passes through the front surfaces of each of the two geometrical figures. To be an invariant plane, it must contain one of the two lines whose lengths do not change. For our purpose, assume that this line is *ab* and that the trace of the plane on the right side of each geometrical figure is *bm*, where *bm* is a direction on these latter faces whose length is also not changed by the deformation. The plane defined by *abm* possesses the characteristic that before and after the distortion the two distances *ab* and *bm* are unchanged. However, the angle α between these two directions

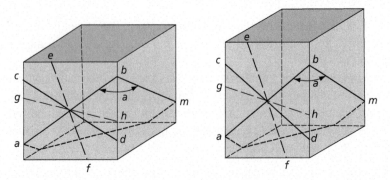

Fig. 17.25 In the cubic to tetragonal transformation, the Bain distortion does not possess an undistorted plane

is obviously changed by the distortion. Plane *abm* cannot be an undistorted plane, for not only must the magnitudes of vectors be maintained, but also the angles between the vectors must be unchanged. If we now take into account the symmetry of the Bain distortion, it becomes quite evident that there is no undistorted plane associated with it.

The directions that are undistorted in the above Bain distortion form a cone whose positions in the cubic and tetragonal structures are indicated in Fig. 17.26. Notice that the vertex angle of the cone decreases during the Bain distortion as the cube is drawn into the right prism. In each of the two sketches of Fig. 17.26, a plane is drawn intersecting the Bain cone in two lines designated as *op* and *oq*. The lengths of these lines are unchanged by the distortion since they lie on the Bain cone. Their angular separation, however, changes. The given plane is drawn to represent the habit, or interface, plane of the transformation in its positions before and after the Bain distortion, and lies very close to the (011) plane of both the cubic and tetragonal structures. Because the angle *poq* changes during the distortion, this plane is not an undistorted plane at the end of the Bain distortion.

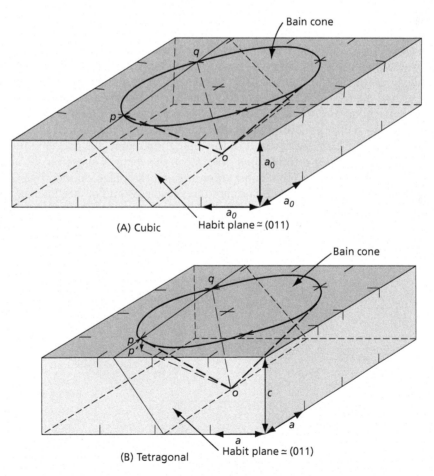

Fig. 17.26 The Bain cone in the cubic and tetragonal structures corresponding to the indium-thallium transformation. The distortion shown is exaggerated in order to emphasize its nature

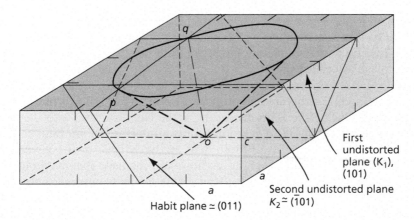

Fig. 17.27 Positions of the two undistorted planes, K_1 and K_2, in the indium-thallium martensite transformation

We shall now turn our attention specifically to the tetragonal structure that is indicated again in Fig. 17.27, where the position of the (101) plane which also passes through line qo is shown. Let us also consider that this is the first undistorted plane (K_1) of a shear (in the same sense as in twinning). A second plane, which lies at almost 90° to K_1, is also given in the diagram. This plane belongs to the same zone as (101) and ($\bar{1}01$) and lies very close to the latter plane. This second plane contains line po and is the second undistorted plane (K_2) of the desired shear. The shearing action is indicated in Fig. 17.28. As a result of this operation, point p is translated to p', with no change in the length of op as it rotates to op'. However, the shear is chosen so as to make angle $p'oq$ equal to angle poq in the original cubic structure, and the plane defined by $p'oq$ is now an undistorted plane. This follows from the fact that the vectors po and qo are unchanged in length and angular relation to each other as a result of the two deformations, which satisfies the theorem (Ref. 30), "If two (noncollinear) vectors in plane are undistorted (unchanged in length) and the

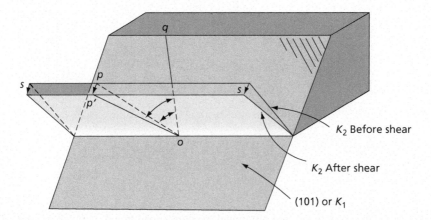

Fig. 17.28 Nature of the shear in the indium-thallium martensitic transformation

angle between them is unchanged by a transformation, all vectors in that plane are undistorted (and all angles are unchanged), that is, the plane is one of zero distortion." Actually, line $p'o$ does not lie strictly in plane poq, so that this line and line oq define a new plane, which is the true habit plane. In other words, plane poq is not the habit plane. However, the angle between plane $p'oq$ and poq is so small that we shall consider $p'o$ to lie in poq and that this latter is the true habit plane. The position of line $p'o$ is, accordingly, plotted in the lower diagram of Fig. 17.26. The plane containing both $p'oq$ and poq of this figure can now be taken as representing the approximate position of the undistorted plane after both the Bain distortion and the shear have occurred. A careful comparison of this sketch with the one above it shows that the undistorted plane in the cubic structure and in the tetragonal structure have different orientations in space. For this reason, $p'oq$ does not yet satisfy the requirements of a habit plane. It must not only be undistorted by the transformation, but it must also be unrotated. A rotation of the tetragonal structure must now be added, which brings plane $p'oq$ into coincidence with plane poq in such a manner that lines op' and oq of the tetragonal phase fall on lines op and oq of the cubic phase, respectively. No effort will be made to demonstrate this rotation with diagrams, for the primary purpose of this discussion has only been to demonstrate the need for considering the three basic steps (Bain distortion, shear, and rotation).

There remains the problem of explaining the nature of the arbitrary shearing process that is needed during the second step of the deformation. In martensitic deformation, the required shear can occur in one of two basic ways: either as a result of slip or as a result of mechanical twinning. Assuming that the deformation is confined to the product lattice, Fig. 17.29 shows how this may accomplished by slip. Glide occurs more homogeneously distributed over a group of slip planes parallel to K_1. In the case of mechanical twinning, what usually happens is that the shear (strain) required to produce the habit plane of the deformation does not equal the shear (strain) that occurs when the product lattice undergoes mechanical twinning. In the present example, mechanical twinning, not slip, furnishes the shear. The twinning occurs in the tetragonal lattice of indium-thallium on {110} planes and with a shear of about 0.06. Notice that the indicated K_1 (101) is a twinning plane. The shear needed on this same plane in order to make angle $p'oq$ equal to poq is one-third of the twinning shear. Clearly, if the entire lattice is twinned after suffering a Bain distortion, it will not be possible to produce the desired undistorted habit plane. This difficulty is resolved in the tetragonal lattice for only part of the structure: twins, taking the form of bands parallel to the twinning plane. Thus, the desired macroscopic shear will be obtained if the net width of the twinned areas is one-third that of the untwinned. This

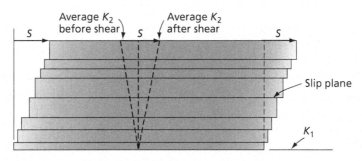

Fig. 17.29 Martensitic shear assuming that deformation occurs by slip

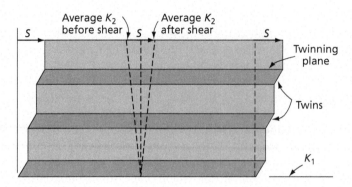

Fig. 17.30 Martensitic shear assuming that deformation occurs by deformation twinning. The required shear is met when only a fraction of the structure twins in a series of bands parallel to the twinning plane

is shown in Fig. 17.30, and the resulting transformed structure is shown in Fig. 17.31. It is very important to note the difference between the two shears involved in the above discussion. In one case, we have considered the shear parallel to the tetragonal twinning planes. The sense of this shear is shown on the right face of the specimen shown in Fig. 17.31. The other shear is the macroscopic shear of the total transformation. This is the net deformation that we observe when we take the resultant of the three distortions (Bain, shear, and rotation) and represents motion parallel to the habit or interface plane. The latter (plus a distortion normal to the habit plane) is what one observes when one considers the specimen as a whole. As a result of the macroscopic shear, the specimen as a whole is tilted at the habit plane. This is demonstrated in the sketch of Fig. 17.31, which also shows the sense of this shear with the aid of vectors located on the front face of the specimen. The shear associated with twinning is on a very much finer scale than the macroscopic shear and is visible on the surface in the tetragonal structure only where it appears as a series of fine parallel surface tilts. Figure 17.32 shows an actual photograph of an indium-thallium alloy illustrating the fine twinned structure in the tetragonal phase. Finally, it should be noted that the twinning that occurs in the tetragonal region is on an extremely fine scale. This is necessary in order to minimize the distortion at the habit plane. Additional consideration shows that the habit plane on a microscopic scale is not a plane of zero distortion, for where each one of the tetragonal bands intersects the interface (Fig. 17.31), the lattice

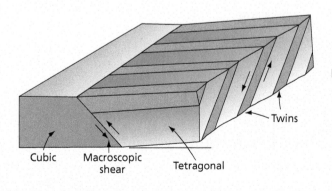

Fig. 17.31 The nature of the twinned tetragonal structure in the indium-thallium martensite transformation

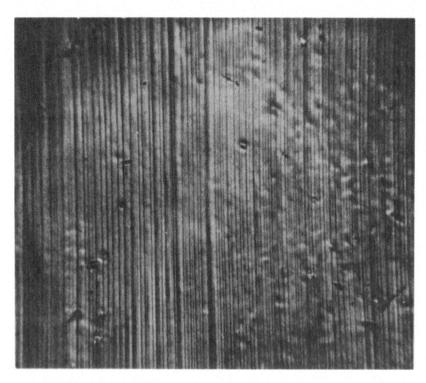

Fig. 17.32 Twinned structure (tetragonal) in an indium-thallium alloy after transformation. (Burkart, M. W., and Read, T. A., *Trans. AIME*, **197** 1516 [1953])

is distorted as the tetragonal structure tries to match the cubic lattice on the other side of the boundary. The habit plane is actually a plane of zero distortion only on a macroscopic scale. The distortion at the interface corresponding to any one of the tetragonal bands is compensated by that of the neighboring bands, which are in a twinned relationship to the first band and distort the interface in the opposite sense. We may thus conclude that the finer the tetragonal twinned bands, the less intense the distortion in the interface.

17.14 Irrational Nature of the Habit Plane

One of the nice features of the Wechsler, Lieberman, and Read (Ref. 29) theory is that it shows quite clearly why the habit plane in martensitic transformations is usually irrational. This characteristic is in sharp contrast to the twinning plane in deformation twinning, which is almost invariably a plane of low rational indices. Let us now reconsider line oq in Fig. 17.27. This line is actually defined as the intersection of the first undistorted (K_1) plane of the shear (twinning plane) and the Bain cone. In the figure it is a rational $[\bar{1}\bar{1}1]$ direction. However, line op is not a rational direction, but diverges slightly from the $[1\bar{1}1]$ direction. It would coincide with $[1\bar{1}1]$ if the shear were exactly equal to the simple twinning shear on the {101} plane, because then the second undistorted plane K_2 (Fig. 17.27) would be the $(\bar{1}01)$ plane. The required shear is only one-third the twinning shear, as we have seen, and the K_2 plane, therefore, makes a small angle with $(\bar{1}01)$. Since op is not a rational direction, neither is op' its position after shear. The final result is that since the habit plane is determined by op' and oq, it is not a rational plane [that is, (011)]. The divergence of the habit plane from a rational plane in the indium-thallium alloy is, however, almost negligible, for it lies only 26′ away from the (011) plane (Ref. 24) and has the indices (0.013, 0.993, 1). The figures illustrating the theory have, of course, been made considerably out of scale for the specific purpose of making the distortions more readily apparent. Distortions of the order used in the figure are, however, observed when larger atom displacements occur in the transformations. The indices of the habit planes in these alloys, accordingly, diverge widely from simple low-indice planes. Additional details on the crystallography of habit planes can be found in References 31 and 32.

It is now necessary to say a word about the multiplicity of the orientations that can be obtained in a martensitic transformation. The number of possible habit planes can be quite large in most reactions. In the indium-thallium alloy there are 24 possible habit planes, all of which lie very close to {110} planes (within 26′), which signifies that there are four habit plane orientations closely clustered about each of the six {110} planes.

17.15 The Iron-Nickel Martensitic Transformation

While the above discussion of the martensitic transformation in indium-thallium alloys has been somewhat lengthy, it has touched on many of the important aspects of martensitic transformations. The reversible characteristic of the transformation is duplicated in most other alloys that undergo martensitic transformations. The athermal nature of the transformation is also a typical martensitic property, and the same holds true for the stabilization phenomenon, the irrational nature of the habit plane, and the complexity of the deformation. There remains now, among other things, a consideration of the effect on the transformation of an increase in the deformation associated with the transformation. For this purpose, let us return to the previously mentioned iron-nickel alloy (70 percent Fe–30 percent Ni) (Ref. 33). Figure 17.33 shows the temperature dependence of the transformation of this alloy (athermal characteristics). In this particular case, the transformation was followed by resistivity measurements. This is possible because the resistance of the body-centered cubic product is lower than that of the face-centered cubic parent phase and the resistance of a mixture of the phases is proportional to the amount of martensite formed. The diagram shows a much larger hysteresis effect than is found in the indium-thallium alloy. In indium-thallium

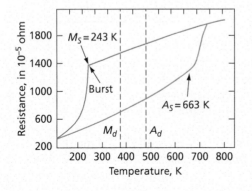

Fig. 17.33 The martensitic transformation in an iron-nickel (29.5 percent Ni) alloy. (From Kaufman, L., and Cohen, M., *Trans. AIME*, **206** 1393 [1956])

the reverse transformation occurs only a few degrees above the forward transformation. In the iron-nickel alloy, martensite begins to form at $M_s = 243$ K. The reverse transformation commences at a temperature designated A_s (austenite start temperature) which is 420 K higher than M_s. The 420 K temperature difference is indicative of the large driving force needed in this alloy to nucleate the transformations.

The difficulty associated with the formation of martensite plates in the iron-nickel alloy appears quite plainly in another respect: the size of the plates that form. These are extremely small and typically ellipsoidal, or lens-shaped. At temperatures slightly below M_s, an interesting phenomenon occurs in these alloys: a large fraction (about 25 percent) of the austenite transforms to martensite in a burst (Ref. 34). It occurs so rapidly that a shock wave results which Machlin and Cohen report as capable of occasionally shattering the Dewar flask that contained the transforming specimen suspended by a thread in a refrigerating liquid. This is excellent evidence of the large driving force for the reaction. The plates that form in the burst (Fig. 17.34) are still small in size, but have a somewhat greater thickness than those formed prior to the burst.

The burst phenomenon can be looked upon as an autocatalytic effect. At an early state in the transformation, a point is reached where the formation of a few more plates triggers the formation of a large number of additional plates.

The progress of the athermal transformation in this alloy is quite different from that in the indium-thallium alloy. In the latter, the transformation was accomplished by the movement of a single interface through the length of the entire specimen and a decreasing temperature was required in order to drive the interface. In the iron-nickel specimens, martensite plates, even in a single crystal of the parent phase, form and grow very rapidly to their final size. The transformation

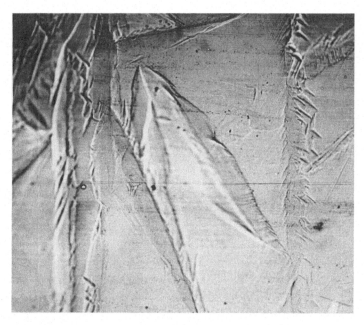

Fig. 17.34 Martensite plates that formed in a burst in an iron-nickel specimen. 500×. (Courtesy of John F. Breedis)

can only proceed with the nucleation of additional plates, which is accomplished by lowering the temperature and increasing the driving force.

The limited growth of the martensite plates in this alloy is undoubtedly associated with the magnitudes of the strains connected with their growth. These latter are large enough to cause plastic flow to occur in the matrix surrounding the martensite plate, and it has been proposed (Ref. 34) that this deformation causes the plates to lose their coherency with the parent lattice. On this basis, it is proposed that growth stops with the loss of coherency. Finally, it should be mentioned that the orientation relationship between the martensite and the parent austenite phase in Fe-Ni alloys depends on the alloy composition (Ref. 35).

17.16 Isothermal Formation of Martensite

Martensite has been observed to form isothermally (Ref. 36), as well as athermally, in 30 percent iron-nickel specimens. In either case, reaction proceeds as the result of the nucleation of additional plates, and not through the growth of existing plates. Curves showing the amount of isothermal martensite formed as a function of time are given in Fig. 17.35. These curves also serve another useful purpose because they demonstrate the interrelation between the athermal and isothermal transformations. The intersection of each isothermal curve with the ordinate axis gives the amount of martensite formed on the quench to the holding temperature.

17.17 Stabilization

The phenomenon of stabilization is also observed in iron-nickel alloys. The mechanism differs from that observed in the indium-thallium alloy, but the effect is the same; isothermal holding of the specimens at a temperature between M_s and M_f stabilizes the transformation so that additional supercooling is required in order to start the transformation again. In the indium-thallium

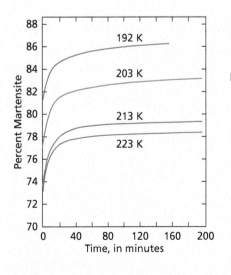

Fig. 17.35 Isothermal transformation subsequent to a direct quench to temperature shown. Notice that the amount of athermal martensite that forms prior to the isothermal transformation increases with decreasing holding temperature. (Fe-29 percent Ni). (From data of Machlin, E. S., and Cohen, Morris, *Trans. AIME*, **194** 489 [1952])

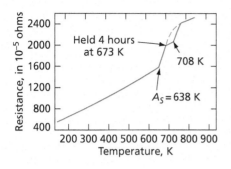

Fig. 17.36 Stabilization during reversed transformation (martensite to austenite) in Fe-29.7 percent Ni. Specimen held at 673 K for 4 hr. Transformation following holding period did not start again until temperature reached 708 K. (Kaufman, L., and Cohen, M., *Trans. AIME*, **206** 1393 [1956].)

specimens, stabilization occurs as a result of a retardation of the movement of the interface, whereas in the iron-nickel alloy, stabilization is manifested by an increased difficulty in the nucleation of additional plates. In order for the reaction to continue, an additional increment of driving force is needed to nucleate more plates. Stabilization is also observed during the reverse transformation when martensite is reacted to form austenite. In this case, no observable isothermal reaction occurs, so that the stabilization effect appears clearly in a resistivity-temperature curve. Such a curve is shown in Fig. 17.36.

17.18 Nucleation of Martensite Plates

The nucleation of martensite plates is a subject of great interest and also of considerable controversy. The available experimental evidence favors the belief that martensite nuclei form heterogeneously, as in most other phase transformations. Theoretical calculations also support the concept of the heterogeneous nucleation of martensite nuclei. In this regard, a straightforward calculation (Ref. 37) gave a value of about 8×10^{-16} J for the energy required in a nucleation event. This is about 10^5 times greater than the value of kT at temperatures where martensite is known to nucleate. Consequently, it is so large that it is unlikely that martensite could be nucleated homogeneously by heterophase fluctuations.

The exact nature of the lattice defects, at which martensite nucleates heterogeneously, is still an unsolved problem. Experimental information about these sites is still inadequate. Over the years, a number of theoretical models for these sites have also been proposed but, while they may adequately describe the martensitic behavior of a specific alloy, they lack generality.

It has often been suggested that dislocations and grain boundaries may act as favorable nucleation sites. Kajiwara (Ref. 38), has described some interesting results that tend to clarify the role of dislocations and grain boundaries in martensite nucleation. He used ten different ferrous alloy compositions of the types Fe-Ni, Fe-Ni-C, and Fe-Cr-C and concluded that while dislocations themselves were not favorable sites for the nucleation of martensite, their existence in the metal could aid in the accommodation of the shape strain of the martensite. On the other hand, it was observed that if the dislocation density in the austenite is too high, the accommodation of the shape strain becomes difficult due to the higher level of work hardening so that martensite nucleation tends to be suppressed. Certain grain boundaries were found to act as nucleation sites. However, the specific crystallographic nature of these boundaries was not determined. Finally, it was concluded, quantitatively, in agreement with the qualitative observations of other investigators, that decreasing the grain size of the parent austenite lowers the M_s temperature. This relationship between the austenite grain size and the martensite start temperature can be rationalized simply in terms of the fact that decreasing the austenite grain size makes it more difficult to accommodate the shape change of a martensite plate in an austenite grain. Therefore, a greater driving force, available at a lower temperature, is required for nucleation.

No simple theory of martensite nucleation is yet available (Ref. 39). This is probably due to the fact that martensite can form under two extremely different sets of conditions. First, it can form athermally, which means that it can form in a very small fraction of a second provided the temperature is lowered sufficiently to activate those nuclei that will respond in this manner. This type of nucleation apparently does not need thermal activation. The ability of martensite to form without the aid of thermal energy is also demonstrated by the fact that martensite plates can be nucleated in some alloys at temperatures as low as 4 K, where the energy of thermal vibrations is vanishingly small. On the other hand, martensite is also capable of forming at constant temperature. The fact that in this case it has a nucleation rate that is time dependent shows that thermal energy can also be a factor in the nucleation of martensite.

17.19 Growth of Martensite Plates

The growth of martensite plates does not have the twofold nature of nucleation, for here it appears that thermal activation is not a factor. The evidence leading to this conclusion includes the fact that plates grow to their final sizes with great rapidity. Measurements show that the growth velocity is of the order of one-third the velocity of sound in the matrix. To this must be added the fact that the growth velocity is independent of temperature. If the growth of martensite plates occurred as the result of atoms jumping over an energy barrier from the parent phase to the product phase, then the jump rate should be a decreasing function of the temperature, and at some finite temperature a noticeable dropping off in the martensitic growth rate should become apparent. This does not occur. Martensite has been observed to form with great rapidity even at 4 K.

17.20 The Effect of Stress

Because the formation of martensite plates involves a change of shape in a finite volume of matter, applied stress can influence the reaction. This is entirely analogous to the formation of deformation twins by stress. However, because the formation of a martensite plate involves both shear and normal components of strain, the dependence on stress is more complicated than in the case of twinning. Nevertheless, theoretical predictions of the effect of various stress patterns on the formation of martensite have been found to agree quite well with experiment (Ref. 40). We shall not discuss this theory in detail, but it is important to note that the M_s temperature can be either raised or lowered as a result of applied stresses. This may be understood in terms of the following, perhaps oversimplified, picture. Suppose, as shown in Fig. 17.37, the macroscopic strain associated with the formation of a martensite plate is a pure shear (normal component zero). Then if the sense of an applied shear stress is the same as the strain, the stress should aid the formation of the plate. A lower driving force for the reaction is to be expected and M_s should be raised. Similarly, if the shear-stress vector is reversed, the formation of the plate becomes more difficult and the temperature at which the plate forms should be lowered. It is important to note that, in respect to the latter,

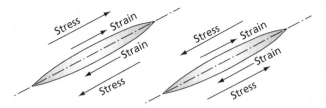

Fig. 17.37 The effect of external stress on martensite plate formation

a simple applied shear stress may not necessarily lower M_s because of the multiplicity of the habit planes on which martensite can form. While the indicated plate might not be favorably oriented to the stress, it is quite probable that there are other plate orientations in the crystal that are.

17.21 The Effect of Plastic Deformation

Plastic deformation of the matrix also has an effect on the formation of martensite, which is primarily to increase the sizes of internal strains and make the nucleation of martensite easier. As a result, martensite can form when the metal is plastically deformed at temperatures well above the M_s temperature. The amount of martensite thus formed decreases, however, as the temperature is raised, and it is common practice to designate the highest temperature at which martensite may be formed by deformation as the M_d temperature. In reversible martensitic transformations, plastic deformation usually has a similar effect on the reverse transformation. The temperature at which the reverse transformation starts is lowered by plastic deformation. The M_d and A_d (austenite start temperature for plastic deformation) are shown as vertical dashed lines in Fig. 17.33. Notice that plastic deformation bring the start of the forward and reverse transformations much closer together—within approximately 100 K. The corresponding difference between M_s and A_s is 420 K.

17.22 Thermoelastic Martensite Transformations

The formation of a martensite plate normally involves a change in shape of a finite volume of material. Because the martensite platelet is coherent with the parent crystal matrix, a state of strain is set up in the martensite particle and the matrix. The strain in the lattice of the parent is called an *accommodation strain*. This accommodation strain can be either elastic or plastic or a combination of both. Whether or not the shape strain is accommodated in the parent elastically or with a sizeable amount of plastic deformation is an important consideration. If the strain in the matrix is elastic, the boundaries between the martensite and the parent phase are normally able to move both easily and reversibly. Thus, if martensite is formed as a result of lowering the specimen temperature, and the accommodation strain in the parent phase is elastic, then on reheating the specimen, its martensite platelets simply shrink back and disappear. Following this, if the specimen is cooled again so as to reform the martensite, the same martensite platelets will normally reappear as on the first quench. Furthermore, this cycle can often be repeated again and again.

On the other hand, if plastic deformation is induced in the matrix during a martensitic transformation, then the boundaries between the martensite and the parent tend to become locked in place by the dislocation structure that results from the growth of the martensite. In this case, on reheating the specimen, the austenite is forced to nucleate inside the martensitic structure.

This means that it is normally difficult to transform the martensite back to the parent phase so that a considerably higher temperature is required to start the reversed transformation. In other words, in a metal where the accommodation strain is largely plastic the temperature difference between M_s and A_s is normally large. An example appears in Fig. 17.33, which shows the martensitic transformation cycle of a Fe-29.5% Ni alloy where the shape change is accommodated plastically. Note that the temperature difference between M_s and A_s is 420 K. A typical metal where the accommodation strain is largely elastic is Au-50% Cd. Here, the formation of austenite on reheating does not require nucleation; it merely forms as a result of the reversed motion of the martensite boundaries. Consequently, the difference in temperature between M_s and A_s is very much smaller, being only about 16 K. The area inside the Au-50% Cd hysteresis loop is also very much smaller than that in Fig. 17.33, signifying a much smaller energy loss during a complete cycle from austenite to martensite to austenite.

Thermoelasticity during a structural phase transformation was originally predicted by Kurdjumov in 1948. Some eleven years later, he suggested that during a thermoelastic transformation an equilibrium should exist between the chemical and non-chemical forces whether a martensite plate grows or shrinks; the nature of this equilibrium depending on the temperature and changing as the temperature is varied. An historical review of Kurdjumov's martensitic theory and laws can be found in an article by O. P. Maksimova (Ref. 41). The thermodynamics of the thermoelastic martensite transformation have been recently treated by Ortin and Planes (Ref. 42). Following their approach and notation, the equilibrium condition for the forward transformation from the parent to the martensite phase may be written:

$$\Delta G^{P-M} = -\Delta G_{ch}^{P-M} + \Delta G_{nch}^{P-M} = 0 \qquad 17.5$$

where ΔG^{P-M} is the Gibbs free energy per mole of the moving plates, ΔG_{ch}^{P-M} is the chemical component of the Gibbs free energy, and ΔG_{nch}^{P-M} is the nonchemical component. The nonchemical part of the free energy can consist of a number of parts; the most important are the elastic energy ΔG_{el}^{P-M} and the work done against frictional forces, which is designated E_{fr}^{P-M}. Thus, we have

$$\Delta G_{nch}^{P-M} = \Delta G_{el}^{P-M} + E_{fr}^{P-M} \qquad 17.6$$

Both the elastic and the frictional terms in Eq. 17.6 are composed of several parts. The elastic part of the free energy is primarily due to the energy stored in the interfaces between the martensite and the parent and to the elastic strains due to the creation of the martensite. In turn, the frictional losses include the work done in moving the martensite interfaces, the energy losses associated with the creation of internal defects induced by the transformation (such as stacking faults and twin boundaries), and the partial plastic accommodation arising from volume and shape changes associated with the transformation. In general, with respect to thermoelastic transformation, the most important of these frictional losses are those due to the motion of the interfaces. These frictional losses result in a thermal hysteresis loss. However, this hysteresis loss is much smaller than that of the non-thermoelastic transformation exemplified by the Fe-29.5% Ni alloy whose curve is shown in Fig. 17.33.

In light of the above, we may now write the thermodynamic equilibrium equation for the forward direction of the transformation in the form:

$$-\Delta G_{ch}^{P-M} + \Delta G_{el}^{P-M} + E_{fr}^{P-M} = 0 \qquad 17.7$$

It is important to note that Eq. 17.7 is concerned only with the local equilibrium or thermodynamic balance at a single-parent martensite boundary lying in a specimen that is only partially transformed at a specified temperature. In this equation, the first term is assumed to be a constant throughout a chemically homogeneous material. However, the latter two terms are not. As a result, thermodynamic equilibrium is not reached at the same time throughout the specimen and the specimen does not transform homogeneously. Also at the local level, one can assume that locally, there is a counterbalance at every temperature between the chemical free energy driving the transformation on the one hand, and the stored elastic strain energy and the irreversible frictional work that has been done, on the other. This simply means that during cooling, when a given temperature is reached, the boundary of a plate moves ahead through a distance such that just enough chemical energy is released to balance the elastic energy that is created plus the energy that is lost to friction by the movement.

Since Eq. 17.7 applies to all the interfaces that are moving at a given temperature, it is possible to write a similar equation for a specimen that is undergoing a transformation at this temperature by integrating the contributions from all of the moving interfaces in the specimen. During the reverse transformation that occurs on heating, where the martensite is reverted to the parent phase, the stored elastic energy is recovered and aids the reverse transformation. It will now be assumed that the temperature during a reversed transformation is above T_0, the equilibrium temperature between the product and the martensite where there is no difficulty in either nucleating the martensite plates or causing them to grow. This, of course, is the ideal condition and at T_0, the martensite transformation should be ideally reversible and without any hystersis during a complete transformation cycle. The equilibrium condition for the reversed transformation assuming that $T > T_0$ may now be written:

$$-\Delta G_{ch}^{M-P} - \Delta G_{el}^{M-P} + E_{fr}^{M-P} = 0 \qquad 17.8$$

Note that in Eq. 17.8 both the chemical and elastic free energies are negative while the frictional loss is positive. This means that in the reversed transformation both the chemical and elastic forces act to overcome the frictional forces.

17.23 Elastic Deformation of Thermoelastic Alloys

Thermoelastic martensitic transformations have been observed in a number of nonferrous alloys. Among these are Cu-Zn, Ag-Cd, Au-Cd, Cu-Zn-Al, and Cu-Al-Ni. Normally the alloys of Au, Ag, and Cu that show this behavior have compositions that place them in the beta phase. However, a thermoelastic martensitic transformation has also been observed in a ferrous alloy, of iron with platinum, with a composition near Fe_3Pt. These alloys tend to exhibit interesting behaviors when subjected to stress, and they can often be deformed elastically through relatively large strains. This has been observed both in the martensitic phase and in the austenitic phase above the martensite start temperature, M_s.

17.24 Stress-Induced Martensite (SIM)

If a stress is applied to a thermoelastic alloy at a temperature above its thermal M_s temperature, in most cases the alloy is able to transform to martensite. This mechanically induced martensite reverses back to the austenitic phase upon removing the stress. Figure 17.38, from Schroeder and Wayman, shows a typical stress-strain curve obtained with a Cu-39.8% Zn specimen that was loaded and then unloaded at −77°C (196 K). At 196 K this alloy is well above its thermal M_s, which

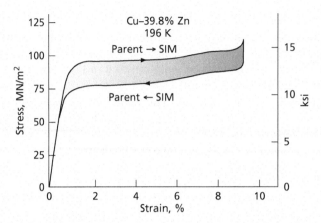

Fig. 17.38 A stress-strain curve loop resulting from loading and unloading a superelastic alloy above M_s. (Reprinted from *Acta Metallurgica*, **27**, Schroeder, T. A., and Wayman, C. M., Pseudoelastic effects in Cu–Zn single crystals, 405, Copyright 1979, with permission from Elsevier. https://www.sciencedirect.com/science/article/abs/pii/0001616079900336)

lies approximately at 149 K. Note that while the strain in Fig. 17.38 was essentially completely recovered so that the deformation cycle was elastic, the stress and the strain were not linearly related. This type of behavior is known as *pseudoelasticity*. A significant feature of the data in Fig. 17.38 is the large magnitude of the elastic strain that was observed: approximately 9 percent. The role that temperature can play in the stress-induced martensite transformation is shown in Fig. 17.39. In this diagram the stresses at which M_s and A_s were observed are plotted against the

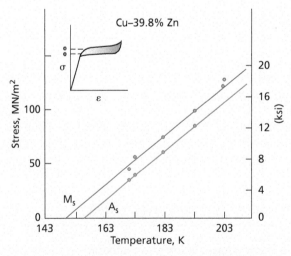

Fig. 17.39 Temperature dependence of the stresses at M_s and A_s. (Reprinted from *Acta Metallurgica*, **27**, Schroeder, T. A., and Wayman, C. M., Pseudoelastic effects in Cu Zn single crystals, 405, Copyright 1979, with permission from Elsevier. https://www.sciencedirect.com/science/article/abs/pii/0001616079900336)

temperature at which the specimens were put through a loading and unloading cycle. Note that the stresses required to obtain both M_s and A_s varied linearly with the temperature. The higher the temperature, the greater the stress required to start either the transformation to martensite or the reversion of the stress-induced martensite (SIM) to parent phase.

17.25 The Shape-Memory Effect

If an alloy that is ordered in both the parent and martensitic phases and capable of undergoing a thermoelastic martensitic transformation is deformed in a tensile test below M_s, one normally obtains a stress-strain curve like that shown schematically in Fig. 17.40. Note that unlike the SIM stress-strain curve in Fig. 17.38, where the specimen was deformed above M_s, the strain is not recovered on unloading. However, heating the specimen into the austenite region allows this strain to be recovered and the specimen to regain its original shape. This is known as the *shape-memory effect*. The ability of the martensite to undergo strains of the order of 6 to 8 percent is now reasonably understood. It is largely due to the 24-fold multiplicity of the orientations that the martensite may take when it is cooled between M_s and M_f. Thus, it is possible for a single austenite grain to be transformed into a region containing 24 martensite orientations. In general, one of these martensite variants will have a shape deformation that is the most favorable for the deformation of the specimen; that is, the one that makes the maximum contribution to the tensile deformation of the specimen. This orientation allows the stress to do the maximum amount of work. As the specimen is deformed, the remaining 23 variants will be converted into this preferred orientation. This conversion can occur either by movements of the interfaces between the martensite variants or as result of twinning that rotates the structure of a given variant into that of the variant best aligned for the deformation. Thus, in effect, the deformation converts the multivariant martensite structure into the equivalent of a single crystal. On heating, the single-crystal deformed martensitic structure transforms into a single crystal of the parent. This "unshearing" of the martensite during the reversion to the parent phase restores the specimen to its original shape. An outline of the shape-memory process, due to Wayman, is shown in Fig. 17.41. Shape-memory alloys are used in a variety of applications, from automotive to aerospace and medical devices (Refs. 43 and 44).

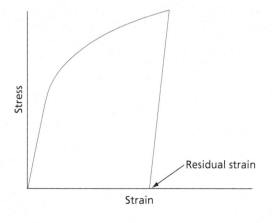

Fig. 17.40 A schematic loading and unloading stress-strain curve for a superelastic alloy deformed in the martensitic state

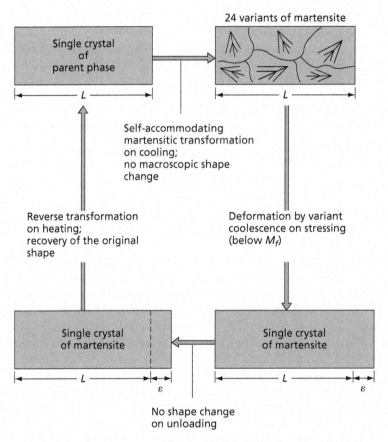

Fig. 17.41 A schematic diagram illustrating the shape-memory processes. (From Wayman, C. M., *Solid → Solid Phase Transformations*, Aaronson, H. I., Loughin, D. E., Sekerka, R. F., and Wayman, C. M., Eds., p. 1138, *Met. Soc. AIME*, Warrendale, Pa., 1981 p. 1138, Figure 12)

Problems

17.1 With the aid of Appendix B and Appendix E, make a sketch similar to that in Fig. 17.13 for $\{10\bar{1}2\}$ twinning in cadmium.

17.2 Now make an equivalent sketch for $\{10\bar{1}1\}$ twinning in titanium.

17.3 Compute the twinning shears associated with $\{10\bar{1}2\}$ twinning in cadmium and $\{10\bar{1}1\}$ twinning in titanium.

17.4 (a) Consider the $\{10\bar{1}1\}$ and $\{10\bar{1}3\}$ twinning systems that have been observed in magnesium.

These are called *reciprocal twins*. Examine Appendix E and determine the significance of this designation for the two twinning systems.

(b) Determine the twinning shear for these two types of twins in magnesium.

17.5 Considering both the $\{10\bar{1}1\}$ and $\{10\bar{1}3\}$ twinning systems in magnesium, would you expect them to form under either a tensile or a compressive stress applied to a magnesium crysal along the direction of its basal plane pole? Explain.

17.6 Appendix E lists the twinning elements for body-centered cubic metals such as iron, as K_1 {112}, $\eta_1 \langle 11\bar{1} \rangle$, K_2 $\{11\bar{2}\}$, and $\eta_2 \langle 111 \rangle$.

(a) How many different {112} twinning planes are there in a body-centered cubic crystal?

(b) Make a list showing the (specific) twinning elements for each of the bcc {112} twinning modes.

17.7 In fcc metals the twinning plane is {111}.

(a) On how many planes of an fcc crystal can twins form?

(b) How many twinning systems are there in an fcc crystal?

(c) List the (specific) twinning elements for each of the fcc {111} twins.

17.8 In the case of badly deformed fcc and bcc crystals, what is the maximum number of different twin traces that one should expect to find?

17.9 Plot the {111} poles on a standard (100) stereographic projection of a cubic crystal. Assume that twinning occurs with (111) as K_1 and with $(11\bar{1})$ as K_2, and draw on the diagram the great circle corresponding to the twinning plane. Next, plot on the stereographic projection the directions corresponding to η_1 and η_2. Label all the plotted data with the proper Miller indices.

(a) On the assumption that the twin forms as a twin of the first kind, rotate the data plotted in the stereographic projection into the orientations that they will assume in the twin.

17.10 Repeat Prob. 17.9 assuming that the twin forms as a twin of the second kind.

17.11 (a) Determine the magnitude of the shear associated with a twin in a face-centered cubic crystal.

(b) Compare this twinning shear with that of the $\{10\bar{1}2\}$ twins in the hcp metals. Which type of twin would be the easiest to nucleate? Explain.

17.12

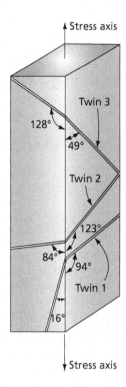

This diagram represents a face-centered cubic crystal with a rectangular cross-section. The crystal has twinned on three planes and the twin traces have been measured with respect to a vertical edge, or the stress axis of the crystal. The angles thus obtained are shown in the figure. Orient this crystal by the two-surface technique following the steps listed below.

(a) Lay out a stereographic projection on a sheet of tracing paper with the front face of the crystal as the basic circle and the top of this circle the stress axis.

(b) Around the basic circle, plot the twin trace orientations corresponding to the front face.

(c) Draw in the great circle corresponding to the side of the crystal and plot on the circle the corresponding twin trace orientations.

(d) Draw in the three great circles representing the three twinning planes. Plot the poles of these three planes.

(e) From the geometry of the fcc crystal structure, determine the orientation of a cube pole; that is, {100}. Plot this on the figure.

(f) Rotate the stereographic projection thus obtained into a standard {100} projection, making sure that the stress axis is also rotated. In order to simplify the result, this last step is best performed on a second sheet of tracing paper.

(g) Draw in the boundaries of the standard stereographic triangle that surrounds the stress axis, thus defining the stress axis orientation.

17.13 Make a rough sketch of the stress-strain curve corresponding to the data in Fig. 17.20.

17.14 (a) Some martensite transformations are completely reversible; however, there may be a large difference in the size of the hysteresis loop that couples a complete temperature-induced cycle. Explain why in some cases the size of the hysteresis is large and in others it is small.

(b) The martensite transformation in steels is normally not reversible. Rationalize this fact.

17.15 (a) What is pseudoelasticity?

(b) What is the shape-memory effect?

(c) What is the meaning of the term *stress-induced martensite*?

References

1. Cahn, R. W., *Acta Met.*, **1** 49 (1953).
2. Blewitt, T. H., Coltman, R. R., and Redman, J. K., in *Dislocations and Mechanical Properties of Crystals*, Wiley, New York, 1957, p. 179.
3. Christian, J. W., and Mahajan, S., *Progress in Materials Science*, 39, 1995, 1–157.
4. Valenzuela, C. G., *TMS-AIME*, **233** 1911 (1965).
5. McLean, D., *Grain Boundaries in Metals*, p. 76, Oxford University Press, London, 1957.
6. Barrett, C. S., ASM Seminar, *Cold Working of Metals* (1949).
7. Haasen, P., and Lawson, A. W. Jr., *Zeits. für Metallkunde*, **49** 280 (1958).
8. Bell, R. L., and Cahn, R. W., *Proc. Roy. Soc.* (London), **239** 494 (1957).
9. He, Y., Li, B., Wang, C. et al., *Nat. Commun.*, **11**, 2483 (2020).
10. Cottrell, A. H., and Bilby, B. A., *Phil. Mag.*, **42** 573 (1951).
11. Christian, J. W., *The Theory of Transformation in Metals and Alloys*, p. 792, Pergamon Press, Oxford, 1965.
12. Venables, J. A., *Deformation Twinning*, AIME Conf. Series, vol. 25, p. 77, Gordon and Breach Science Publishers, New York, 1964.
13. Dickson, J. I., and Robin, C., *J. of Less-Common Metals*, 70, 1980, 1–13.
14. Song, S. G., and Gray III, G. T., *Acta Metall Mater.*, 43, 1995, PR 2325–2337.
15. Kaschner, G. C., Tomé, C. N., McCabe, R. J., Misra, A., Vogel, S. C., and Brown, D. W., *Mat. Sci. Eng. A*, 463, 2007, 122–127.
16. Lee, E. H., Byun, T. S., Hunn, J. D., Yoo, M. H., Farrell, K., and Mansur, L. K., *Acta Mater.*, 49, 2001, 3269–3276.
17. Lee, E. H., Yoo, M. H., Byun, T. S., Hunn, J. D., Farrell, K., and Mansur, L. K., *Acta Mater.*, 49, 2001, 3277–3287.
18. Krishnamurthy, S., Qian, K.-W., and Reed-Hill, R. E., *ASTM-STP* 839, p. 41, American Society for Testing and Materials, Philadelphia, 1984.

19. Bolling, G. F., and Richmond, R. H., *Acta Met.*, **13** 709–57 (1965).
20. Vohringer, O., *Z. für Metallkde.*, **67** 518 (1976).
21. Wayman, C. M., in *Solid-Solid Phase Transformations*, Edited by Aaronson, H. I., Laughlin, D. E., Sekerka, R. F., and Wayman, C. M., *Trans-AIME*, 1981, 1119–1144.
22. Wayman, C. M., *Metallography*, 8, 1975, 105–130.
23. Bain, E. C., *Trans. AIME*, **70** 25 (1924).
24. Burkart, M. W., and Read, T. A., *Trans. AIME*, **197** 1516 (1953).
25. Basinski, Z. S., and Christian, J. W., *Acta Met.*, **2** 148 (1954).
26. Burkart, M. W., and Read, T. A., *op. cit.*
27. Chang, L. C., *Jour. Appl. Phys.*, **23** 725 (1952).
28. Basinski, Z. S., and Christian, J. W., *Acta Met.*, **2** 148 (1954).
29. Wechsler, M. S., Lieberman, D. S., and Read, T. A., *Trans. AIME*, **197** 1503 (1953).
30. Lieberman, D. S., *Acta Met.*, **6** 680 (1958).
31. Klosterman, J. A., *J. of Less-Common Metals*, 28, 1972, 75–94.
32. Muddle, B. C., in *Solid-Solid Phase Transformations*, Edited by Aaronson, H. I., Loughlin, D. E., Sekerka, R. F., and Wayman, C. M., *Met. Soc. AIME*, 1981, 1347–1372.
33. Kaufman, L., and Cohen, M., *Trans. AIME*, **206** 1393 (1956).
34. Machlin E. S., and Cohen, M., *Trans. AIME*, **191** 746 (1951).
35. Kithara, H., Ueji, R., Ueda, M., Tsuji, N., and Minamino, Y., *Materials Characterization*, 54, 2005, 378–386.
36. Machlin, E. S., and Cohen, M., *Trans. AIME*, **194** 489 (1952).
37. Cohen, M., and Wayman, C. M., Proc. Conf: *Treatises in Metallurgy*, Beijing, China, Nov. 13–22, 1981, Tien, J. K., Elliot, J. F., Eds., pp. 445–468, TMS-AIME, Warrendale, Pa., 1981.
38. Kajiwara, S., *Met. Trans. A*, **17A** 1693 (1986).
39. Borgenstam, A., *Mat. Sci. Engin. A*, 273–275, 1999, 425–429.
40. Patel, J. R., and Cohen, Morris, *Acta Met.*, **1** 531 (1953).
41. Maksimova, O. P., *Metal Science and Heat Treatment*, 41, 1999, 322–339.
42. Ortin, J., and Planes, A., *Acta. Met.*, **36** 1873 (1988).
43. Morgan, N. B., *Mats. Sci. Engin. A*, 378, 2004, 16–23.
44. Firstor, G. S., Van Humbeeck, J., and Koval, Y. N., *Mats. Sci. Engin. A*, 2004, 2–10.

Chapter **18**

The Iron-Carbon Alloy System

Learning Objectives

Upon completion of Chapter 18, you will be able to:

1. Identify specific features of iron-carbon diagram with carbon appearing both as graphite and as cementite
2. Compare and contrast proeutectoid nucleation and growth of ferrite during cooling of hypoeutectic austenite and of cementite in hypereutectic compositions
3. Understand the mechanism of 2-phase lamellar pearlite formation upon cooling of eutectoid steel
4. Describe the influences of growth rate and undercooling on pearlitic lamellar spacing
5. Describe the partitioning of alloying elements during growth of pearlite
6. Differentiate between the two basic forms of bainite
7. Utilize TTT diagrams to determine transformation microstructures

18.1 The Iron-Carbon Diagram

We shall now consider alloys of iron and carbon in some detail. There are several reasons for this: first, carbon steels constitute by far the greatest tonnage of metal used by man; second, no other alloy system has been studied in such detail; and third, the solid-state phase changes in steel are varied and interesting. Further, it is becoming increasingly evident that the solid-state reactions of the iron-carbon system are similar in many respects to those that occur in other alloy systems. As such, a study of the iron-carbon system is valuable not only because it helps explain the properties of steels but also because it provides a means of understanding solid-state reactions in general.

The iron-carbon diagram shown in Fig. 18.1 is not a complete diagram in that it is plotted only for concentrations (in weight percent) of less than 12 percent carbon. It does, however, cover that part of the diagram of primary interest in the study of steels and cast irons. It is due to Chipman (Ref. 1) and is currently accepted, with minor changes, as the standard iron-carbon diagram (Ref. 2).

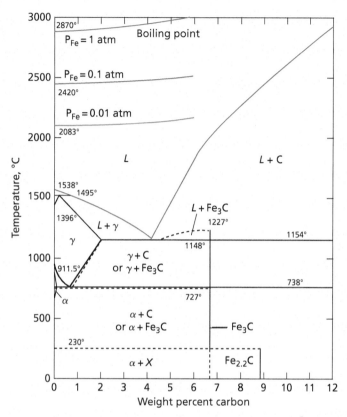

Fig. 18.1 The iron-carbon phase diagram. (From Chipman, J., *Met. Trans.*, **3** 55 [1972] with kind permission from Springer Science and Business Media)

This version of the diagram is complicated because it shows the equilibrium relationships not only between iron and carbon (graphite) but also between iron and two iron-carbides; cementite (Fe_3C) and the Hägg carbide ($Fe_{2.2}C$). Included in this diagram are three nearly horizontal lines in the upper left-hand part of the diagram. These lines represent the boundaries between the liquid and gaseous phases at three pressures; 1.0, 0.1, and 0.01 atmospheres. They thus indicate the dependence of the boiling point temperature on composition at each of the indicated pressures. The two carbides are metastable. In other words, carbon or graphite is more stable than the carbides. However, the decomposition of the carbides does not occur easily and, from a practical point of view, one may consider that, in plain carbon steels, the breakdown of the carbides to form iron and graphite does not occur. Furthermore, when a steel is slowly cooled from the liquid to the solid state, cementite is normally the easiest to nucleate so that when carbon is precipitated from solid solutions of alpha (body-centered cubic) or gamma (face-centered cubic), the resulting precipitate is almost always cementite if the temperature is above 350°C. It should be noted that once cementite has formed, it is very stable. As shown on the phase diagram, the Hägg carbide is actually the more stable carbide below about 350°C. This constituent, however, has been identified in purified iron specimens carburized below 350°C. Hägg carbides have been observed to transform to cementite when steels containing them were heated above 500°C. It is still not clear whether cementite transforms to the

Hägg carbide in a reasonable period of time on cooling. In any event, it will be assumed that in the study of steels one may reasonably use the simplified phase diagram shown in Fig. 18.2.

The phase diagram in Fig. 18.2 is also not a complete diagram in that it is only plotted for concentrations (in weight percent) less than 6.67 percent carbon, the composition of Fe_3C, or cementite. Cementite is an intermetallic compound with negligible solubility limits, and the iron-carbon diagram can be divided at its composition into two independent parts. That part of the diagram containing carbon concentrations higher than 6.67 percent has little commercial significance and is usually ignored. The iron-carbon diagram in Fig. 18.2 is characterized by three invariant points: a peritectic point at 0.17 percent C and 1495°C, an eutectic point at 4.32 percent C and 1154°C, and an eutectoid point at 0.77 percent C and 727°C. The peritectic transformation occurs at very high temperatures and in steels of very low-carbon concentration (the upper left-hand corner of the phase diagram).

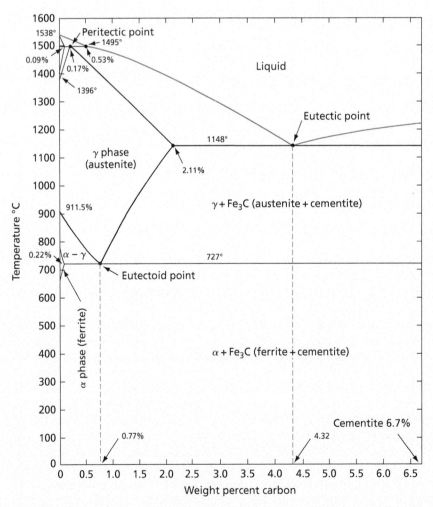

Fig. 18.2 The Fe-Fe_3C metastable phase diagram. (After Chipman, J., *Met. Trans.*, **3** 55 [1972])

The peritectic transformation has only secondary effects on the structures of steels at room temperature. All compositions that, as they freeze, pass through the peritectic transformation region, enter the single-phase face-centered cubic field. Except for coring effects (segregation) caused by the complexities of the peritectic transformation, these alloys are equivalent to higher carbon compositions (above 0.53 percent C) that freeze directly to the gamma or face-centered cubic phase. With the assumption that sufficient time is allowed as the alloys cool to permit diffusion to bring about a homogeneous solid solution, we may disregard the peritectic transformation in considerations of phase transformations that occur at lower temperatures.

The face-centered cubic solid solution, or gamma phase, is given the identifying name *austenite*. A study of the phase diagram shows that all compositions containing less than 2.11 percent C pass through the austenite region on cooling from the liquid state to room temperature. Alloys in this interval are arbitrarily classed as steels. Actually, most carbon steels contain less than 1 percent carbon, with the greatest tonnages produced in the range 0.2 to 0.3 percent C (structural steels used in buildings, bridges, ships, etc.). Only in very rare instances is steel used with more than 1 percent C (razor blades, cutlery, etc.) and then the composition never rises more than a few tenths of one percent above one percent. Compositions above 2 percent are classified as cast irons. However, it is necessary to notice that cast irons of commerce are not simple alloys of iron and carbon, for they contain relatively large quantities of other elements, the most prevalent of which is silicon. In general, cast irons are better considered as ternary alloys of iron, carbon, and silicon. They also differ in another important respect from steels. The presence of silicon promotes the formation of graphite in them. As a result, cast irons may contain carbon in the form of both graphite and cementite. The latter are called white cast irons, whereas the former are called gray cast irons when graphite is in the form of flakes. This fact signifies an important difference between cast irons and steels, for the latter contain only combined carbon in the form Fe_3C.

While the phase transformations associated with the eutectic point at 4.32 percent C are interesting and useful in a detailed study of the structures in cast irons, time does not permit us to consider this section of the phase diagram.

A detailed analysis of the phase transformations at the peritectic and eutectic points of the iron-carbon system are not essential in a study of steels. We shall focus our attention, therefore, on the phase transformations that occur in the eutectoid region of the phase diagram. This study is not simple, because we have to consider not only equilibrium phase transformations, but also those that occur under non-equilibrium conditions. With the assumption of slow-cooling, the phase changes in steels can be predicted with considerable accuracy using the equilibrium diagram. On the other hand, when the transformations do not occur under equilibrium conditons, such as when the metal is cooled rapidly, new and different metastable phases and constituents form. Since these structures are important to the theory of the hardening of steels, the kinetics and principles of their formation are of considerable importance. These will be considered later. For the present, the reactions that occur as austenite is slowly cooled (Ref. 3) will be discussed.

18.2 The Proeutectoid Transformations of Austenite

The microstructures obtained when a steel is slowly cooled from the austenitic region depend on the original carbon concentration of the steel. If this concentration is less than 0.77 percent, the eutectoid composition, then the microstructure will contain two primary constituents: proeutectoid ferrite and pearlite. As discussed later, pearlite is a two-phase lamellar structure consisting of ferrite and cementite lamellas. If the carbon concentration equals that of the eutectoid, the structure will consist only of pearlite, and if it is higher than 0.77 percent, it will contain proeutectoid cementite and pearlite. Both the proeutectoid ferrite and the proeutectoid cementite nucleate preferentially and heterogeneously on the austenite grain boundaries. There are two basic reasons for this: the boundaries contain energetically favorable sites for nucleation and they are regions where the diffusion rates are higher. When a ferrite nucleus forms at an austenitic grain boundary, it normally develops a different orientation relationship with respect to the austenite grains on the two sides of the boundary. Generally, the nucleus will have a simple orientation relationship with one of its two neighboring austenite grains, known as the *Kurdjumov-Sachs relationship* (K-J). This is:

$$\{111\}_\gamma \parallel \{110\}\alpha \text{ (habit plane)}$$
$$\langle 110 \rangle_\gamma \parallel \langle 111 \rangle \alpha$$

18.1

In this relationship, a $\{110\}$ plane of the alpha phase is aligned with a $\{111\}$ plane of the austenite. The boundary between the nucleus and this austenite grain conforms to the Kurdjumov-Sachs relation and is the habit plane. In this habit plane, a $\langle 111 \rangle$ direction in the ferrite nucleus is aligned with a $\langle 110 \rangle$ direction in the austenite. Since this habit plane is a low-energy boundary it normally does not move during the growth of the nucleus. The ferrite grain, thus, grows into only one of the two austenite grains associated with the boundary. In this regard, the boundary that is created between the ferrite nucleus and the austenite grain into which it grows is, in general, a high-energy or noncoherent boundary. However, it has been often observed that when a new phase grows inward from a boundary, this phase may appear as long thin plates (or needles). This produces a morphology usually called a *Widmannstätten plate structure*. Such a structure is commonly observed when austenite transforms into ferrite. It is most readily seen in steel specimens in which the austenite grain size was very large before the transformation took place.

It has been observed that the Widmannstätten plates or laths can both increase in length and thickness as the transformation proceeds. There are two different types of boundaries that actually separate a ferrite lath from the austenite grain into which the lath grows: (1) the parallel sides of the lath, and (2) the edge or end of the lath. The sides of the lath are normally assumed to be low-energy or coherent (or at least semicoherent) boundaries and thus immobile. The boundary at the end of the lath, on the other hand, is expected to be a non-coherent high-energy boundary in agreement with its observed mobility. On this basis, one would normally expect that a lath should only grow in length but not in thickness. However, a solution to the thickening problem of the laths has been proposed by Aaronson (Ref. 4). This proposal, which is supported by experimental observations, assumes that the laths thicken as a result of the lateral movement of small ledges or steps along the coherent faces of the laths. A schematic drawing illustrating this type of growth is given in Fig. 18.3. Part A of this figure shows the serrated interfacial structure of a ferrite grain growing in this manner, while Part B shows the elimination of the serrations as a result of extensive further growth.

In the hypereutectoid steels with greater than 0.77 percent carbon, the proeutectoid cementite also nucleates at the austenite grain boundaries. In this case, the cementite tends to form as a layer that lies only on one side of a boundary and in some cases, under the right conditions, it may actually form a continuous grain boundary network. Thus, it normally grows only into one of the two grains at the boundary. An orientation relationship also exists between the cementite and the austenite grain into which the cementite does not grow. Since cementite is tetragonal, this relationship is complicated and may be found in the literature (Ref. 5).

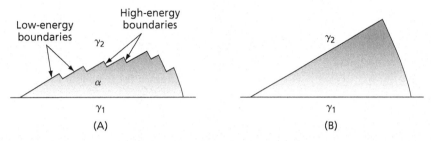

Fig. 18.3 Growth of proeutectic ferrite by the movement of ledges along low-energy gamma-alpha boundaries. (After Aaronson, H. I., *Decomposition of Austenite by Diffusional Processes*, p. 387, Interscience Publishers, New York, 1962)

18.3 The Transformation of Austenite to Pearlite

When the carbon concentration of an iron-carbon alloy is 0.77, it transforms on slow cooling to the eutectoid structure, *pearlite*. For a more detailed historical review of the mechanism and crystallography of pearlite, consult the article by Paul R. Howell (Ref. 6). In practice, steels are almost always cooled so that the temperature falls continuously. Thus, annealed steels are slowly cooled by leaving them in a furnace after the power to the furnace has been shut off, or a normalized steel is cooled by removing the red hot metal from the furnace and allowing it to cool in air. In continuous cooling processes, the nature of the reaction changes with the decreasing temperature. This change is particularly marked when the rate of cooling is appreciable. It goes almost without saying that the resulting microstructures are very difficult to analyze. Much more readily interpretable specimens are obtained when austenite is allowed to transform at constant temperature. True isothermal transformation is possible because of the relatively small heat of transformation involved; normally of the order of 4.2 J per mol, which is about one-fourth of the latent heat of fusion (L_F for pure iron = 15.5 J per mol). In addition, the use of small specimens and the relatively slow reaction rates observed when austenite decomposes allow transformation heat to be removed fast enough to prevent appreciable temperature rises.

Figure 18.4 illustrates a simple experimental method of studying isothermal austenitic transformations. As may be seen by examining Fig. 18.5 (an enlarged section of the iron-carbon diagram containing the eutectoid point), a steel of eutectoid composition is austenitic at temperatures

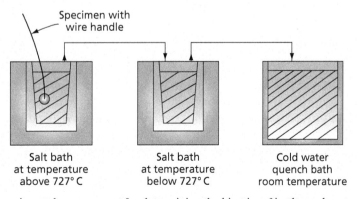

Fig. 18.4 Simple experimental arrangement for determining the kinetics of isothermal austenitic transformations

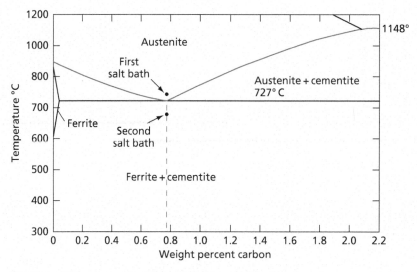

Fig. 18.5 The eutectoid section of the iron-carbon diagram

above 727°C. Thus, if a small furnace containing a crucible filled with a molten salt mixture is held at 730°C, eutectoid-composition steel specimens placed in this bath can be maintained in the austenite phase and at a temperature just above that at which they undergo the eutectoid reaction. This furnace is shown on the left in Fig. 18.4. Just to the right of this unit is a similar furnace that also contains a salt bath but maintains it at a temperature below 727°C. In an experiment of this type, it is convenient to use specimens in the shape of flat discs about the size of a dime. A short length of temperature-resistant wire forms a convenient handle for use in moving the specimens from furnace to furnace. Metal pieces of this size, when placed in liquid salt baths, very quickly attain the temperature of the bath. Consequently, if a specimen originally in the left-hand furnace is quickly removed and inserted into the furnace to its right, we can assume that its temperature changes instantly from just above the eutectoid temperature to just below it. Since austenite is no longer stable below 727°C, it is now able to decompose into other phases. If this solid-state reaction is carried out at temperatures not too far below the eutectoid temperature, the reaction occurs by nucleation and growth and is, therefore, time dependent.

The isothermal decomposition of the austenite is usually followed with the aid of a number of specimens (generally about ten), all of which are quenched simultaneously from the upper temperature to the lower temperature. Individual specimens are then removed from the second bath at increasing time intervals (usually measured on a logarithmic scale) and quickly cooled to room temperature by quenching them in a bath of cold water. This latter fast-cooling effectively stops the isothermal reaction and any austenite still untransformed, at the instant of the quench, undergoes a martensitic (athermal) transformation as the specimen approaches room temperature. The nature of this martensitic transformation will be covered in detail later. It suffices to state here that it is of the same basic type as the iron-nickel transformation considered in Chapter 17. Fortunately, steel martensite has a different appearance under the microscope than the high-temperature reaction products of austenite. After a suitable metallographic polish and etch, the specimens, prepared as outlined above, may be examined metallographically and the relative amount of the isothermal-transformation product determined in each case.

If austenite is allowed to transform isothermally at temperatures just below 727°C, the reaction product is the same as that predicted by the iron-carbon diagram for a very slow, continuous cooling process. As may be seen by examining Fig. 18.5, the stable phases below the eutectoid temperature are ferrite and cementite, and the eutectoid structure is a mixture of these phases. This constituent, called pearlite, consists of alternate plates of Fe_3C and ferrite, with ferrite being the continuous phase. An example of a pearlite structure is shown in Fig. 18.6. Pearlite is not a phase, but a mixture of two phases—cementite and ferrite. It is, nevertheless, a constituent because it has a definite appearance under the microscope and can be clearly identified in a structure composed of several constituents. (See Fig. 18.34.) It should be further recognized that when austenite of eutectoid composition is reacted to form pearlite just below the eutectoid temperature, the two phases appear in a definite ratio. This ratio is easily computed by using the lever rule and assuming that ferrite contains zero percent carbon.

$$\text{Fraction of ferrite} = \frac{6.67 - 0.77}{6.67} \simeq 88\%$$

$$\text{Fraction of cementite} = \frac{0.77}{6.67} \simeq 12\%$$

Since the densities of ferrite and cementite are approximately the same (7.86 and 7.4, respectively), the lamellae of iron and Fe_3C have respective widths of about 7 to 1.

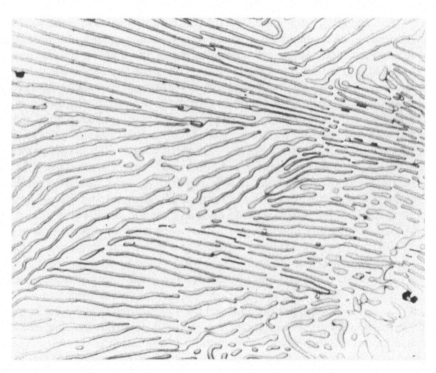

Fig. 18.6 Pearlite consists of plates of Fe_3C in a matrix of ferrite. 2500×. (From Vilella, J. R., *Metallographic Technique for Steel*, ASM Cleveland, 1938. Reprinted with permission of ASM International®. All rights reserved. www.asminternational.org)

The decomposition of austenite to form pearlite occurs by nucleation and growth. Nucleation takes place heterogeneously. If the austenite is homogeneous, nucleation occurs almost exclusively at grain boundaries. A significant observation is that a pearlite colony normally grows into only one of the two austenite grains at the boundary. When it is not homogeneous, but has concentration gradients and contains residual iron-carbide particles, nucleation of pearlite can occur both at the grain boundaries and in the centers of the austenite grains. This would be the case if austenite, once reacted to pearlite, were re-austenitized for too short a time and then decomposed to pearlite a second time.

It is difficult to observe the actual nucleation of a pearlite colony in an unalloyed iron-carbon steel. This is because the reaction occurs at a high temperature and tends to go to completion when the steel is cooled to room temperature, where it can be conveniently examined. To fully understand the nucleation event one needs to understand the orientation relationships between the austenite, the ferrite, and the cementite at a grain-boundary site where a pearlite colony is nucleated. Simply stated, the problem is that cooling to room temperature effectively destroys the austenite structure, thereby making it very difficult to determine with certainty the relations between the three phases once the metal has been returned to room temperature. A number of investigators have attacked this problem with the use of certain high-alloy steels. In these steels it is possible to obtain a partial transformation of austenite to pearlite at higher temperatures and then, on cooling to room temperature, to retain the remaining austenite without it undergoing any further transformation. Dippenaar and Honeycombe (Ref. 5) obtained significant results in this manner by using a steel containing 0.79 percent carbon and 11.9 percent manganese. The specimens were austenitized at 950°C for 1 hour, which gave them an austenitic grain size of 50 μm. Quenching their specimens to room temperature completely suppressed the transformation of the austenite so that specimens possessing a 100 percent austenitic structure were obtained on reaching room temperature. This was possible because the high manganese concentration of the metal made the austenite much more stable. A simple iron-carbon steel with 0.79 percent carbon steel would have been almost completely converted to martensite by the quench. Following the quench, the specimens were reheated to between 400 and 650°C and allowed to transform isothermally. Because of the large manganese concentration in their steel, the transformation to pearlite did not go to completion, but stopped after about 15 to 20 percent transformation. This, however, was sufficient to allow them to study the nucleation and early development of pearlite colonies. It should also be mentioned that their alloy steel results are consistent with observations made on plain-carbon steels.

As a result of this study it was concluded that there were two basic orientation relations between the cementite and ferrite lamellae in a pearlite colony, on the one hand, and the two austenite grains, at the grain boundary where the pearlite colony nucleated, on the other. The deciding factor determining which of these relations existed for a particular pearlite colony was the type of site at which nucleation occurred, that is, whether it had nucleated at a "clean" austenitic grain boundary or on one containing a proeutectoid cementite layer. In the former case, both the ferrite and the cementite lamellae obtain orientations related to that of the austenite grain, into which the pearlite does not grow. This austenite grain will now be identified by the symbol γ_1. The orientation relationship of the pearlite with the austenite in γ_1 has been named the *Pitsch-Petch relation* and in it the ferrite in the pearlite has a relationship to the austenite in γ_1 that is close to the Kurdjumov-Sachs relation. On the other hand, both the ferrite and the cementite in the pearlite were observed to have no simple crystallographic relationship with respect to the grain into which a pearlite colony grows. For convenience, this latter austenite grain will be designated by the symbol γ_2.

In the case where a pearlite colony nucleates at a proeutectoid austenitic grain-boundary cementite layer, the cementite in the pearlite was found to have the same orientation as the cementite in the boundary layer. Since, as pointed out above, the grain-boundary cementite has an orientation related to γ_1, the cementite in the pearlite must also have this same orientation relationship with γ_1. The orientation of the ferrite in the pearlite, on the other hand, was found to be unrelated to the orientations of either γ_1 or γ_2. In this case, where only the cementite of the pearlite has an orientation associated with γ_1, the orientation relationship is known as the *Bagaryatski relation*.

In summary, it is apparent that when a pearlite colony nucleates at an austenitic grain boundary, it grows into only one of the two grains associated with the boundary. If the nucleation occurs on a boundary that does not contain proeutectoid cementite, the orientations of both the ferrite and the cementite of the pearlite are related to that of the grain into which the pearlite does not grow, that is, γ_1, as shown schematically in Fig. 18.7A. The reason for this is that the boundaries formed at the site of nucleation between the austenite of grain γ_1 and the ferrite and cementite lamellae of the pearlite, are low-energy boundaries with a very low mobility. Conversely, the boundaries formed at the other ends of the ferrite and cementite lamellae with the austenite in grain γ_2 are high-energy boundaries with a high mobility. Consequently, the pearlite colony grows only into γ_2. On the other hand, if the pearlite is nucleated at a boundary layer of proeutectoid cementite, it forms on a band of cementite that has a low-energy interface with respect to γ_1 and a high-energy interface with respect to γ_2. Thus, the pearlitic cementite lamellae can grow out into γ_2 as extensions of the proeutectoid cementite layer. This is illustrated schematically in Fig. 18.7B. At the same time, while the ferrite lamellae of the pearlite colony cannot grow into γ_1 because the proeutectoid cementite layer separates them from γ_1, they are able to grow into γ_2 because a simple orientation relationship between this ferrite and the austenite of γ_2 does not exist, so that the boundaries between the ferrite and the austenite, at the ends of the ferrite lamellae away from the layer of proeutectoid cementite, should possess high energies. This work also confirmed the general conclusion that, normally, all the ferrite lamellae in a single pearlite colony have a common orientation, and similarly, all the cementite lamellae in a pearlite colony also possess a common orientation.

It was also deduced that the nucleation of a pearlite colony can occur by sympathetic sideways nucleation of the ferrite and cementite lamellae as indicated schematically in Fig. 18.8. In this regard, the nucleation of a ferrite lamella should increase the carbon concentration of the austenite in its vicinity and favor the nucleation of an adjacent lamella of cementite. Similarly, the nucleation of a cementite lamella should lower the carbon concentration of the austenite next to it and promote the nucleation of another ferrite lamella. This process is repeatable so that as the colony grows, both the lengths of the lamellae and the number of the lamellae tend to increase. There is

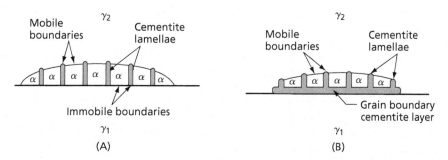

Fig. 18.7 The two primary methods of nucleating pearlite. **(A)** Nucleation of pearlite at an austenite grain boundary (Pitsch-Petch relation). **(B)** Nucleation of pearlite at a grain-boundary layer of cementite (Baryatski relation)

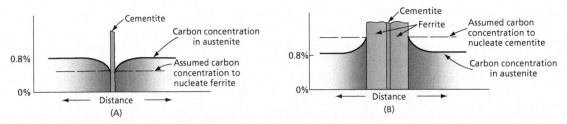

Fig. 18.8 Growing cementite and ferrite lamellae may nucleate each other

also evidence that the number of the lamellae can be increased as a result of branching of both the ferrite and cementite lamellae. This process is indicated schematically in Fig. 18.9.

As pointed out above, Dippenaar and Honeycombe (Ref. 5) used a hyper-eutectoid alloy steel (0.79 percent C, 11.9 percent Mn) in their study of the phase relations that occur during the transformation of austenite to pearlite. This steel had the advantage that the transformation of austenite to martensite could be avoided when the steel was cooled back to ambient temperatures. Some very significant results have also been reported by Hackney and Shiflet (Refs. 7–9), as a consequence of another extensive transmission electron microscope investigation of the pearlite transformation in this steel. Their study included the use of hot stage electron microscopy, which allowed them to follow the growth of the pearlite colonies *in situ*.

These authors present strong evidence for the belief that the growth of the pearlite-austenite interface involves the lateral motion of steps or ledges across the interface (Ref. 10). This type of growth is illustrated for the growth of proeutectic ferrite into austenite in Fig. 18.3. In the growth of the pearlite-austenite interface, the evidence is consistent with an assumption that the ledges run continuously across the ferrite and the cementite lamellae of the pearlite. This strongly implies that the growths of the ferrite and cementite phases of the pearlite (into the austenite) are coupled or interconnected. In other words, the growth of pearlite occurs by an edgewise growth mechanism in which steps sweep laterally across the austenite-pearlite interface as a whole.

18.4 The Growth of Pearlite

As indicated above, the growth of a pearlite colony is accomplished in several ways: (1) by the addition of lamellae, (2) by the branching of the lamellae, and (3) through the extension outward of the ends of the lamellae. Because the pearlite colony has an almost equal rate of growth in

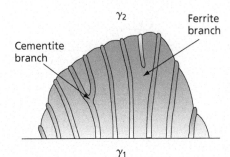

Fig. 18.9 Branching of pearlite lamellae

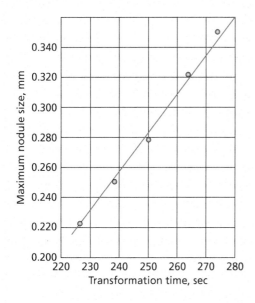

Fig. 18.10 Plot of data showing the linearity of the growth rate of pearlite. Eutectoid high-purity steel transformed at 708°C. (From Frye, J. H., Jr., Stansbury, E. E., and McElroy, D. L., *Trans. AIME*, **197** 219 [1953])

directions parallel and perpendicular to the lamellae, a large developed pearlite nodule is often spherical in shape. Thus, when viewed on a plane surface with a microscope, a fully developed pearlite colony usually has a circular shape.

Pearlite colonies grow unimpeded until they impinge on adjacent colonies. During this period, the rate of growth, as determined experimentally, is constant. This is illustrated in Fig. 18.10. The data for a curve of this type may be obtained by reacting a number of specimens for different lengths of time. The specimens are prepared for metallographic examination and the diameter of the largest pearlite nodule in each specimen is measured. The slope of the curve, obtained when the diameter is plotted against the reaction time, equals the growth rate V. After impingement, pearlite nodules can only grow into austenite remaining between nodules, which consititutes the last stage of pearlite growth.

18.5 The Effect of Temperature on the Pearlite Transformation

The Interlamellar Spacing and the Rate of Growth The interlamellar spacing λ is nearly a constant in pearlite formed from austenite at a fixed temperature and in a given specimen, varying only slightly about a mean. This mean value is generally referred to in speaking of the interlamellar spacing of pearlite. The actual variation about the average is considerably less than one would assume from an examination of a metallographic surface, where the spacing between lamellae may appear to vary widely. This is caused by the fact that the plane of the surface does not intersect all pearlite colonies at the same angle. The true spacing is only observed when the lamellae are perpendicular to the surface. A convenient method for determining this quantity, which has been widely used, is to take the minimum spacing that can be seen on a metallographic specimen surface as a measure of the true spacing. Figure 18.11, from Ridley (Ref. 11), illustrates the nature of the problem of determining λ. In this diagram, λ is the spacing as measured on a surface that is not normal to the pearlite lamellae, while λ_0 is the spacing measured on a surface normal to the lamellae. Actually, the

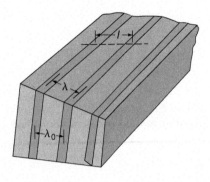

Fig. 18.11 The parameters used for the lamellar spacing of pearlite. (From Ridley, N., *Met. Trans. A*, **15A** 1019 [1984] with the kind permission of Springer Science and Business Media)

true spacing is not the minimum spacing observed in the plane of polish, since the true spacing is not a single fixed value but corresponds to a distribution of spacings about a mean true value λ_0. The line on the upper surface of the crystal in Fig. 18.11 represents a random line laid down on this surface. The distance between the centers of two adjacent lamella along this line is designated l. A mean pearlite intercept $\bar{l}$ may be obtained by superimposing many straight lines of known length on the microstructure in such a manner that they intersect the pearlite lamellae in all directions. The total number of lamellae intercepted by these lines are counted and then the mean intercept is obtained by dividing the total length of these lines by the number of the lamellae intercepted.

It can be shown (Ref. 11), with the aid of several simple quantitative microscopy relations, that:

$$\bar{l} = 2\bar{\lambda}_0 \qquad 18.2$$

where $\bar{l}$ is the mean intercept length and $\bar{\lambda}_0$ is the mean true value of the (pearlite) lamellar spacing.

The temperature at which austenite is transformed has a strong effect on the interlamellar spacing of pearlite (Refs. 12 and 13). The lower the reaction temperature, the smaller $\bar{\lambda}_0$. The spacing of the pearlite lamellae has practical significance because the hardness of the resulting structure depends upon it; the smaller the spacing, the harder the metal. Pearlite formed from austenite at temperatures just below the eutectoid temperature (700°C) has a pearlite spacing of the order of 1.0 μm. The hardness of this structure is about Rockwell C-15, while pearlite formed at 600°C has a spacing of about 0.1 μm and a corresponding greater hardness, Rockwell C-40.

The rate of growth or the growth velocity of pearlite (V) is also a strong function of temperature, as can be seen by examining Fig. 18.12. At temperatures just below the eutectoid, the growth rate increases rapidly with decreasing temperature, reaching a maximum at 600°C, and then decreases again at lower temperatures.

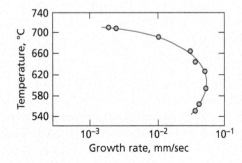

Fig. 18.12 Rate of growth (G) of pearlite as a function of reaction temperature in a high-purity iron-carbon alloy of eutectoid composition. (From Frye, J. H., Jr., Stansbury, E. E., and McElroy, D. L., *Trans. AIME*, **197** 219 [1953])

18.6 Forced-Velocity Growth of Pearlite

The data shown in Figs. 18.10 and 18.12 were obtained some years ago by isothermally transforming austenite to pearlite. In addition to isothermal studies of the transformation of austenite to pearlite, there is another recently introduced technique for this purpose called the *forced-velocity method*. In a typical experiment of this type (Ref. 14), the specimen consists of a high-purity steel rod of eutectoid composition with a diameter of about 6.4 mm. This specimen is placed inside a concentric assembly that can be moved at a controlled fixed rate along the length of the specimen. This unit contains an induction heating coil that heats a short (1 cm) length of the specimen into the austenitic region. It also contains sealed cold water cooling cylinders at each end of the heated zone that produce very steep temperature gradients on both sides of this zone. Finally, it is capable of maintaining a protective vacuum or argon gas atmosphere around the heated section of the specimen. As this assembly is slowly translated along the specimen, the heated zone of the specimen is forced to move along the specimen and, as it does so, the cross-section of the specimen at the rear of the heated zone transforms from austenite into pearlite. In effect, this produces a pearlite front that moves with a fixed velocity, and the steep temperature gradient at this end of the heated zone suppresses pearlite nucleation in advance of the front. The process is similar to that described in Fig. 20.23. Also, as pointed out by Ridley (Ref. 11) this procedure is similar to that used for the production of aligned lamellar structures during the freezing of eutectic alloys.

In the isothermal transformation technique, the growth rate of the pearlite is a function of the temperature. That is, the temperature is the independent variable and the growth velocity is the dependent variable so that if a temperature is assigned to the experiment, the growth velocity is determined. On the other hand, in a forced velocity experiment, one assigns a growth velocity and determines a transformation temperature. A comparison between the forced velocity and the isothermal transformation techniques is given by Pearson and Verhoeven (Ref. 14) in Fig. 18.13. This shows the pearlite growth velocity, V, plotted against ΔT, the difference between the eutectoid and transformation temperatures, on double logarithmic coordinates. All of the data in the figure are from high-purity iron-carbon specimens. Note that the forced velocity data of Pearson and Verhoeven fall on a straight line whose slope is close to $+2$. A slope of $+2$ for the log-log plot in Fig. 18.13 strongly implies that ΔT is proportional to $V^{1/2}$ or that we may write:

$$\Delta T = K_1 V^{1/2} \qquad 18.3$$

where K_1 is a constant. Equation 18.3 is equivalent to Eq. 14.51, which was deduced theoretically for the freezing of an eutectic alloy. The eutectoid transformation of austenite to pearlite is analogous to the freezing of an eutectic alloy, so the theories of both transformations are similar. Thus, it is not surprising that the empirical relation in Eq. 18.3 supports Eq. 14.51. Note that different symbols are used for the variables in these two equations. Thus K_1 in Eq. 18.3 is equivalent to AB in Eq. 14.51 and the growth rate or growth velocity, V, in Eq. 18.3 is the same as R in Eq. 14.51. As mentioned above, Fig. 18.13 also contains data obtained from a number of isothermal transformation studies. These data are in good agreement with the forced velocity data for pearlite growth velocities less than 20 μm/s, which corresponds to a supercooling, ΔT, of about 50 K. For velocities greater then 20 μm, the isothermal data scatter to both sides of the line of slope $+2$. This scatter has been attributed by Pearson and Verhoeven to the fact that at higher growth velocities in an isothermal experiment, problems associated with measuring the growth velocity occur since it is

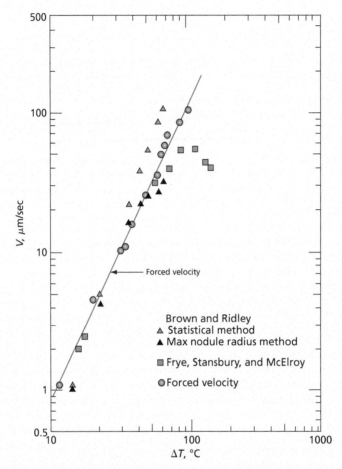

Fig. 18.13 A plot of the logarithm of the growth velocity, V, versus the amount of undercooling, ΔT. The data marked with the symbol "o" were obtained using the forced velocity experimental technique. The other data are from isothermal transformation experiments as reported in Ref. 14. (From Pearson, D. D., and Verhoeven, J. D., *Met. Trans. A.*, **15A** 1037 [1984] with kind permission from Springer Science and Business Media)

very difficult to accurately measure the time variable of a transformation when the total reaction time is very short, as is the case at large growth velocities.

In the eutectic freezing section in Chapter 14 on solidification, it was predicted in Eq. 14.50 that the interlamellar spacing in an eutectic alloy should vary as the inverse square root of the growth velocity. In other words:

$$\lambda = K_2/V^{1/2} \qquad 18.4$$

where K_2 is a constant equivalent to B/A in Eq. 14.50. When Eq. 18.4 is expressed in double logarithmic notation by plotting log V against log λ, one should obtain a straight line with a slope of -2. As may be seen in Fig. 18.14, where forced velocity data from four sources are plotted, the experimental data agree well with Eq. 18.4.

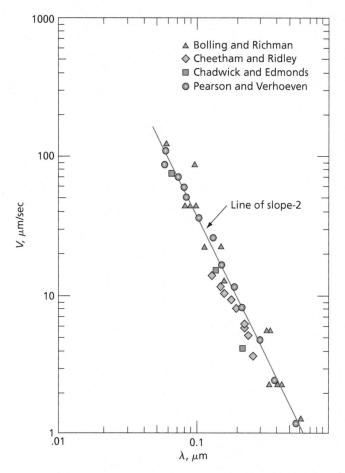

Fig. 18.14 Four sets of pearlite growth velocity versus pearlite interlamellar spacing data. (From Pearson, D. D., and Verhoeven, J. D., *Met. Trans. A.*, **15A** 1037 [1984] with kind permission from Springer Science and Business Media)

Finally, it must be pointed out that there are three primary variables in the pearlite transformation: (1) the undercooling, ΔT; (2) the growth velocity, V; and (3) the interlamellar spacing, λ. These variables are interrelated. The relation between V and ΔT is shown in Fig. 18.13; that between λ and V appears in Fig. 18.14; and that between λ and ΔT appears in Fig. 18.15. Figure 18.15 shows that on logarithmic coordinates, λ varies linearly with ΔT with a slope of -1. This conforms to:

$$1/\lambda = K_3 \Delta T \qquad 18.5$$

Note that Fig. 18.15 contains both isothermal and forced-velocity data. The relationship between the variables in Fig. 18.15 can also be obtained easily from Eqs. 18.3 and 18.4. If this is done, one finds that $K_3 = 1/K_1 K_2$.

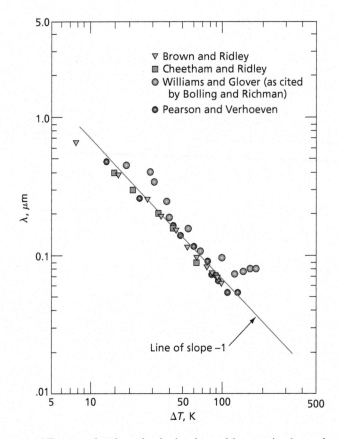

Fig. 18.15 Plot of λ versus ΔT comparing forced velocity data with some isothermal transformation data. (From Pearson, D. D., and Verhoeven, J. D., *Met. Trans A*, **15A** 1037 [1984].) For the source of other data, Ref 14 should be consulted

18.7 The Effects of Alloying Elements on the Growth of Pearlite

Most commercial steels contain alloying elements. Even the class of steels called "plain-carbon steels" may contain between 0.5 and 1.0 percent of manganese and between 0.15 and 0.30 percent of silicon. These elements are either residual impurities from iron ores or scrap, or are intentionally added to deoxidize or remove sulfur. Their addition also improves strength and decreases brittleness. The low-alloy steels, which are also an important class, may contain additional alloying elements, such as cobalt, chromium, molybdenum, and nickel in concentrations that in some cases may equal or exceed several percent. Furthermore, in some low-alloy steels, the manganese concentration may exceed that normally found in the plain-carbon steels and be as high as 5 percent. Since all of these steels are normally capable of undergoing an austenite to pearlite transformation, knowledge about how these alloying elements influence the transformation is important.

18.7 The Effects of Alloying Elements on the Growth of Pearlite

There are several well known effects that the common alloying elements can have on the austenite to pearlite transformation. These are:

1. An alloying element can change the eutectoid temperature. Elements such as Ni and Mn, which tend to make austenite more stable, act to decrease the eutectoid temperature. On the other hand, elements such as Si, Cr, and Mo increase the eutectoid temperature.
2. The presence of an alloying element may change the rate at which the pearlite reaction occurs. In general, the only element that is not known to appreciably retard the pearlite transformation is cobalt. All others act to generally retard the transformation. Since Si, Cr, and Mo raise the eutectoid temperature, the pearlite transformation tends to occur earlier at high temperatures in steels with these elements than it does in plain-carbon steels. However, at lower transformation temperatures, the transformation rates in these steels are usually slower than those in pure iron-carbon steels.
3. The various alloying elements may be partitioned between the ferrite and the cementite in the pearlite as a result of the pearlite transformation. This means that the alloying element that was in solution in the gamma phase does not remain solely in the alpha phase after the transformation. Part of it may be incorporated into the cementite. Thus, the cementite in the pearlite is not simply Fe_3C but may have some iron atoms replaced by atoms of the alloying element.

Tewari and Sharma (Ref. 15) have summarized the significant factors associated with the partitioning of the alloying elements. They point out that:

1. The partitioning of the alloying elements between the ferrite and the cementite of the pearlite is required by thermodynamics when the transformation occurs under approximately equilibrium conditions. In effect, the redistribution of the alloying elements is a growth requirement for temperatures just below the eutectoid temperature. Thus, partitioning is favored by a low degree of supersaturation or when the undercooling, ΔT, is small. Alloying elements that have a strong affinity for carbon and are known as carbide-forming elements should have a strong tendency to partition to the cementite of the pearlite. Elements in this class are manganese, chromium, and molybdenum. On the other hand, elements such as silicon, nickel, and cobalt are expected to concentrate in the ferrite of the pearlite.
2. If the supersaturation of the carbon is high, as with a large degree of undercooling and a correspondingly fast rate of transformation, it may be possible to completely suppress the partitioning.
3. In the temperature interval where partitioning occurs, the pearlite growth rate is controlled by the diffusion of the alloying element along the grain boundaries. On the other hand, at lower temperatures where partitioning does not occur, the growth rate is controlled by the volume diffusion of the carbon atoms.
4. In general, a temperature may exist above which partitioning occurs and below which it does not occur. This partition-no-partition transition temperature is characteristic of the composition of the alloy, and decreases for any given alloying element with an increase in its concentration.

A useful parameter in the study of the partitioning of the alloying elements is known as the *partitioning parameter*. It is defined by the following equation:

$$K_\alpha^{cm} = [C_X^{cm}/C_{Fe}^{cm}]/[C_X^\alpha/C_{Fe}^\alpha] \qquad 18.6$$

where C_X^{cm} and C_{Fe}^{cm} are the respective weight fractions of the alloying element and the iron in the cementite of the pearlite and C_X^{α} and C_{Fe}^{α} are their corresponding weight fractions in the ferrite of the pearlite.

Some of the effects that alloying elements can have on pearlite are illustrated in Figs. 18.16 and 18.17. In particular, Fig. 18.16 shows a set of $1/\lambda$ against T plots corresponding to four eutectoid steels containing from 0.4 to 1.8 percent of chromium and two steels with 1.08 and 1.80 percent of manganese. Note that with increasing chromium the plots are displaced upward and to the right, while with increasing manganese concentration they move in the opposite direction. In addition, the dashed line representing an unalloyed iron-carbon eutectoid steel falls between those of the two alloy classes. Finally, it can be observed that for all of these compositions the data fall reasonably well on straight lines.

Figure 18.17 is interesting since it shows the variation with temperature of three important pearlite transformation parameters in a 1 percent manganese eutectoid steel. On the left is the pearlite growth rate; in the center is the reciprocal pearlite spacing; and on the right is the partitioning parameter. Note that the partitioning parameter is large, that is, about 3.0, just below the eutectoid temperature and that it decreases as the temperature is decreased so that at approximately 670°C its value is close to one. A partitioning parameter of 1.0 signifies that there is no partitioning. Thus, 670°C is approximately the no-partitioning temperature of this steel.

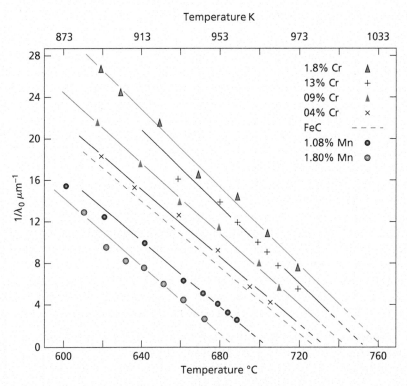

Fig. 18.16 Minimum interlamellar spacing data for eutectoid steel containing additions of manganese and chromium. (From Ridley, N., *Met. Trans. A*, **15A** 1019 [1984] with kind permission from Springer Science and Business Media)

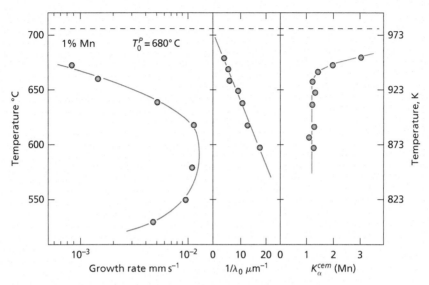

Fig. 18.17 Variation of growth rate, reciprocal minimum spacing, and partition coefficient with temperature for a 1 wt. % Mn eutectoid steel. (From Ridley, N., *Met. Trans. A*, **15A** 1019 [1984] with kind permission from Springer Science and Business Media)

18.8 The Rate of Nucleation of Pearlite

The *rate of nucleation* is the number of nuclei that form in a unit volume (usually a cubic millimeter) per second. As previously mentioned, in austenite of homogeneous composition, nuclei normally form heterogeneously at grain boundaries. Unlike the isothermal growth rate, the isothermal rate of nucleation is a function of time. This is shown in Fig. 18.18. In order to compare nucleation rates at different temperatures, it is necessary to consider the average nucleation rate for each temperature. The variation of the average N with temperature is shown in Fig. 18.19. Plotted on this same figure is the rate of growth as a function of temperature. Much about the changes in pearlitic structure that occur as a function of temperature can be deduced from a study of these curves.

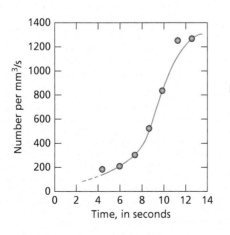

Fig. 18.18 Rate of nucleation (N) of pearlite as a function of time. Eutectoid steel transformed at 680°C. (Mehl, R. F., and Dube, A., *Phase Transformations in Solids*, John Wiley and Sons, Inc., New York, 1951, p. 545)

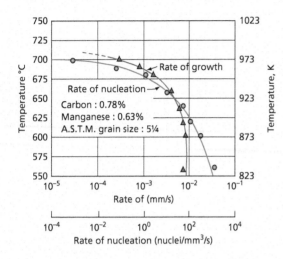

Fig. 18.19 Variation of N and G with temperature in an eutectoid steel. (Mehl, R. F., and Dube, A., *Phase Transformations in Solids*, John Wiley and Sons, Inc., New York, 1951, p. 545)

Let us first consider a temperature only slightly below the eutectoid temperature, 973 K (700°C). At this temperature, the rate of nucleation is very small, approaching zero. The growth rate, on the other hand, has a finite value between 10^{-3} and 10^{-4} mm per sec (mm/s). Only a few pearlite nuclei form and, because of the relatively high growth rate, the nuclei grow into large pearlite nodules. The nodules, in fact, grow to sizes much larger than the original austenite grains; nodule growth proceeds across austenite grain boundaries.

Because of the scarcity of pearlite nuclei and the long distances between nuclei forming at these high temperatures, they may be considered to form randomly throughout the austenite in spite of the fact that they actually form at grain boundaries. If it is now assumed, first, that the rate of nucleation N is constant (the average nucleation rate); second, that the nodules maintain a spherical shape as they grow (until they impinge on each other); and, third, that the growth rate G is constant, then it has been shown by Johnson and Mehl (Ref. 16) that the fraction of austenite transformed to pearlite is given as a function of time by the following reaction equation:

$$f(t) = 1 - e^{(-\pi/3)NG^3 t^4} \qquad 18.7$$

where $f(t)$ is the fraction of austenite transformed to pearlite, N is the nucleation rate, G is the growth rate, and t is the time. The typical sigmoidal curve that results when the reaction curve is plotted is shown in Fig. 18.20.

As the reaction temperature is lowered, the rate of nucleation increases at a much faster rate than the rate of growth, and more and more pearlite colonies are nucleated the lower the reaction temperature. One of the consequences is that at an early stage in the transformation, the austenite grain boundaries become outlined by the large number of pearlite colonies that form along them. Instead of one pearlite nodule growing large enough to consume a number of austenite grains, there are now a number of pearlite nodules growing into a single austenite grain. Under these conditions, it is no longer possible to think in terms of random nucleation. We should think, rather, in terms of grain-boundary nucleation. The analysis of the transformation is now mathematically more difficult, but it has been carried out (Ref. 16). For such cases, the transformation rate depends on the austenite grain size. It has been shown that the transformation rate decreases with increasing the austenite grain size (Ref. 17).

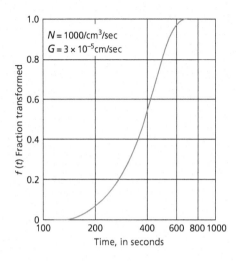

Fig. 18.20 Theoretical reaction curve obtained from Johnson and Mehl equation. (Mehl, R. F., and Dube, A., *Phase Transformations in Solids*, John Wiley and Sons, Inc., New York, 1951, p. 545)

18.9 Time-Temperature-Transformation Curves

Information of a very important and practical nature can be obtained from a series of isothermal reaction curves determined at a number of temperatures. First, consider the theoretical curve of Fig. 18.20. Experimental counterparts of this curve are obtained by isothermally reacting a number of specimens for different lengths of time and determining the fraction of the transformation product in each specimen. A plot of these data as a function of the reaction time gives the desired curve, one of which is shown in the upper diagram of Fig. 18.21.

From this reaction curve the time required to start the transformation and the time required to complete the transformation may be obtained. This is done in practice by observing the time to get a finite amount of transformation product, usually 1 percent, corresponding to the start of the transformation. The end of the transformation is then arbitrarily taken as the time to transform 99 percent of the austenite to pearlite. A plot of these data for a series of temperatures gives Fig. 18.22, which is clearly not an equilibrium diagram, but still a form of phase diagram. It shows the time relationships for the phases during isothermal transformation. The area to the left of the first C-shaped curve corresponds to structures that are austenitic. Any point to the right of the second of these two curves represents a pearlitic structure and is, consequently, a mixture of two phases (cementite and ferrite). Between the two curves is a region of pearlite and austenite with the relative ratios of these two constituents varying from all austenite to all pearlite as one moves from left to right.

One of the significant factors about the pearlite transformation is the very short time required to form pearlite at temperatures around 873 K (600°C). This, of course, could have been deduced from the information of Fig. 18.20, where it can be seen that at this temperature both the rate of nucleation and the rate of growth have a maximum.

The time-temperature-transformation (*T-T-T*) diagram of Fig. 18.22 corresponds only to the reaction of austenite to pearlite. It is not complete in the sense that the transformations of austenite, which occur at temperatures below about 823 K (550°), are not shown. To complete this study, it is necessary to consider two other types of austenitic reactions: austenite to bainite and austenite to martensite.

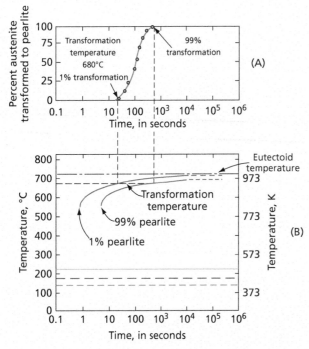
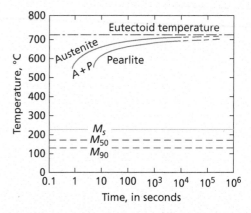

Fig. 18.21 (A) Reaction curve (schematic) for isothermal formation of pearlite. (B) Time-temperature-transformation diagram obtained from reaction curves. (Adapted from *Atlas of Isothermal Transformation Diagrams*, United States Steel Corporation, Pittsburgh, 1951)

Fig. 18.22 The partial isothermal transformation diagram for an eutectoid steel: 0.79 percent carbon 0.76 percent manganese. (Adapted from *Atlas of Isothermal Transformation Diagrams*, United States Steel Corporation, Pittsburgh, 1951)

The bainite reaction will be covered in the next section and the martensite reaction in steels in Chapter 19. With regard to the martensite reaction, it can be stated that in the present eutectoid steel, the reaction is primarily athermal. The horizontal lines drawn in Fig. 18.22 show the M_s (martensite start) temperature and the temperatures at which certain indicated fractions of martensite are obtained. It goes without saying that martensite only forms if the steel, on cooling, does not transform to pearlite at higher temperatures.

18.10 The Bainite Reaction

The bainite reaction is still perhaps the least understood and most controversial of all the austenite reactions (Refs. 18–21). There are a number of reasons for this. Bainite is formed over a rather wide temperature range. The morphology of the bainite microstructure is also very diverse. It has long been recognized that there were significant differences between bainites formed at higher temperatures (roughly between 300 and 500°C) and those formed at lower temperatures (approximately from 200 to 300°C). This has led to the designations *upper bainite* and *lower bainite*, respectively, for the bainites formed inside these two temperature regimes. Furthermore, within each of these temperature intervals the bainite morphology tends to be diverse. Thus, it has been recently shown that in a Cr-Ni alloy steel, upper bainite may assume three basically different morphologies (Ref. 18). Similarly, the morphology of lower bainite may take different forms. The mechanisms by which austenite transforms to bainite are also complicated and remain a source of considerable controversy. It is particularly difficult to investigate the bainite reaction in simple iron-carbon steels because the bainite transformation tends to overlap the austenite to pearlite reaction [at temperatures of the order of 773 K (500°C)]. Steels transformed in this temperature range have structures containing both pearlite and bainite. In certain alloy steels, the presence of the proper alloying elements in substitutional solid solution (in the austenite) has the effect of separating the temperature ranges in which the respective pearlite and bainite reactions occur. The bainite reaction is much more readily studied in these alloys. Some of the conclusions obtained from these investigations are characteristic of bainite reactions in general and can be applied to simple iron-carbon alloys.

The most puzzling feature of the bainite reaction is its dual nature. In a number of respects, it reveals properties typical of a nucleation and growth type of transformation such as the pearlite transformation. At the same time, it shows an equal number that would classify it as a martensitic type of reaction. Like pearlite, the reaction product of the bainite transformation is not a phase but a mixture of ferrite and carbide. In the bainite transformation, the carbon that is uniformly distributed in the austenite is concentrated into localized regions of high-carbon content (the carbide particles), leaving an effectively carbon-free matrix (the ferrite). The bainite reaction thus involves compositional changes and requires the diffusion of carbon. In this it differs markedly from a typical martensitic transformation. Another property of the bainite reaction that differentiates it from martensitic reactions is that it is not athermal. The formation of bainite requires time, and when austenite is reached to bainite isothermally, a typical S-shaped reaction curve is obtained, as can be seen in Fig. 18.23. The similarity of this curve to those obtained in simple nucleation and growth transformations can be seen by comparing it with that previously presented in the discussion dealing with the pearlite reactions. Another set of isothermal reaction curves are shown in Fig. 18.24. These were obtained by Okamoto and Oka (Ref. 22) using a high-purity hypereutectoid iron-carbon steel containing 1.10 percent of carbon that was isothermally reacted at five temperatures between 663 and 422 K (390 to 149°C). The progress of the reaction was followed at each temperature with a dilatometer. Since the M_s temperature for this composition is about 450 K, Curve 5 lies below M_s and the bainite region. The other four curves lie in the lower bainite region. Note that all the curves were fitted to the Johnson-Mehl equation (Eq. 18.7). It is also interesting that these authors identified two basically different lower bainite morphologies in their hypereutectoid steels, adding further evidence of the complexity of the bainite transformation.

While bainite and pearlite are both mixtures of ferrite and carbide, the mechanism of bainite formation differs from that of pearlite. The resulting product also does not have the alternating

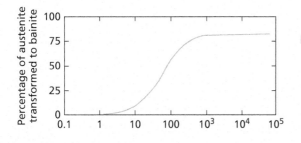

Fig. 18.23 A characteristic of the isothermal bainite transformation is that it may not go to completion. (After Hehemann, R. F., and Troiano, A. R., *Metal Progress*, **70**, No. 2, 97 [1956])

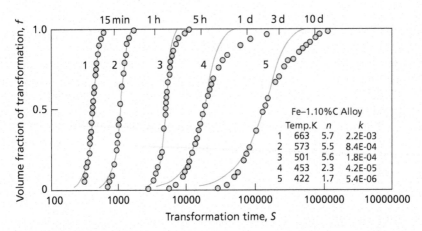

Fig. 18.24 Isothermal reaction curves for a 1.10 percent C steel austempered at various temperatures. The marks O were determined by dilatometry, and the solid lines were derived from Johnson-Mehl equation. (From Okamoto, H., and Oka, M., *Met. Trans. A.*, **17A** 1113 [1986] with kind permission of Springer Science and Media)

arrangement of parallel lamellae of ferrite and cementite found in pearlite. As we have seen, pearlite, due to almost equal growth rates in all directions, tends to develop in the form of spheres. This is not true of bainite, which grows largely as plates or lathes—a typical martensitic characteristic. When observed on a metallographic section, bainite often has a characteristic acicular (needle-like) appearance, which in many respects is similar to that of deformation twins and martensite plates. The formation of bainite plates is also accompanied by surface distortions (surface tilts and accommodation kinks), which has led many authors to conclude that lattice shear is involved in the formation of the plates. A basic difference, however, between the bainite and martensite plates lies in the speed of their formation. Martensite plates are in most cases formed under the condition of a high-driving force and they grow to their final sizes in a small fraction of a second, whereas bainite plates grow slowly and continuously. The growth of bainite plates is apparently retarded by the time required for the diffusion processes that accompany the reaction. However, if bainite forms by a process involving shear displacements, one has to consider the possibility that the relaxation of the accommodation strains, associated with the formation of bainite plates, should be a factor. Relaxation of these strains should be possible by recovery processes at the temperatures where bainite forms. Their relaxation might make either the continued growth of the bainite lamellae or the nucleation of additional lamellae possible.

A significant paper relative to bainite has been published by Bhadeshia and Edmunds (Ref. 23), who used a steel with 0.43 percent C, 3.0 percent Mn, and 2.12 percent Si. In this steel, the bainite transformation occurred slowly and allowed a thorough study of the bainite reaction to be obtained. Bhadeshia and Edmunds were able to produce evidence that the transformation reactions leading to upper and lower bainite, respectively, were different. In proof of this, two distinct "C" curves may be seen in the time-temperature-transformation (*T-T-T*) diagram for their steel in Fig. 18.25. This diagram was obtained by isothermally reacting their alloy specimens for different times at several fixed temperatures. The solid line curves correspond to the time to obtain a transformation of 5 percent at the various temperatures. The dashed lines on either side of the solid line merely show the scatter of their data. They are not boundaries showing other percentages of transformation.

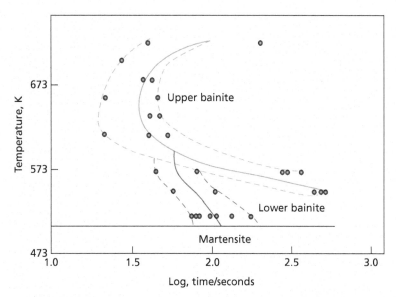

Fig. 18.25 The 5 percent transformation *T-T-T* curve of a silicon alloy steel as determined with a dilatometer. (From Bhadeshia, H. K. D. H., and Edmunds, D. V., *Met. Trans. A*, **10A** 895 [1979] with kind permission of Springer Science and Business Media)

More recent data for isothermal transformation of 1.2 percent C steel also gives two separate curves for the upper and lower bainites (Ref. 20).

Normally, the microstructure of upper bainite is composed of lath- or plate-shaped ferrite elements arranged in packets or sheaves with layers of carbide in between the ferrite laths. Because of the extremely fine structure of this bainite, it is necessary to use an electron microscope to resolve its components. Figure 18.26 shows several upper bainite plates in an electron micrograph made at a magnification of 15,000×. The irregular dark areas are ferrite regions, while the light areas inside them are the carbide particles. The upper bainite shown in this photograph was formed at a relatively high transformation temperature (approximately 733 K). Lower bainite differs in appearance from upper bainite in that the structure is normally coarser and the carbides lie inside the ferrite plates. This is illustrated in Fig. 18.27 for a specimen isothermally transformed at approximately 523 K, where the ferrite plates have a more regular and needlelike shape in the plane of the specimen surface than those in Fig. 18.26. At the same time, the carbide particles, which appear in this latter figure as rather coarse structures paralleling the length of the ferrite plate, assume a different aspect. They become smaller in size and appear as cross-striations (Ref. 24), making an angle of about 55° to the axis of the plate. (See Fig. 18.27.) The net result of this change in the structure of bainite with transformation temperature is a variation in its appearance under the light microscope (that is, at lower magnifications). To illustrate this point, Fig. 18.28 shows the appearance of bainite formed at two different isothermal temperatures.

Something should be said about the nature of the carbides that appear in bainite. It would seem that in carbon steels transformed to bainite at temperatures above 573 K (300°C), the carbides are simply cementite, Fe_3C (Ref. 25). It has been recognized for some time that in bainite formed at lower temperatures [below 573 K (300°C)], carbide other than cementite may form. Investigations have shown that this carbide is often epsilon (ε) carbide (Ref. 25), which has a hexagonal crystal

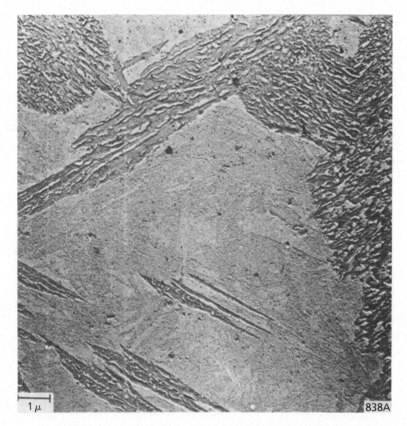

Fig. 18.26 The structure of bainite transformed at 733 K revealed by the electron microscope. Original magnification 15,000× reduced 30 percent in above photograph. (*Trans. ASTM,* **52**, 540, Fig. 20 [1952]. [Second Progress Report of Subcommittee XI of Committee E-4]) Copyright ASTM International. Reprinted with permission

structure instead of the orthorhombic structure of cementite. The carbon concentration in epsilon carbide also differs from that of cementite and is about 8.4 percent instead of 6.7 percent.

With regard to the two basic forms of bainite, that is, upper bainite and lower bainite, the habit plane of upper bainite is close to $\{111\}_\gamma$ while that of lower bainite is irrational. According to Sandvik (Ref. 26), while a number of nonacicular forms of bainite, such as granular bainite, columnar bainite, and grain-boundary bainitic allotriomorphs, have been observed, these are detected only under special conditions. Consequently, we may assume that the two major bainite forms are upper bainite and lower bainite.

An interesting feature of the bainite reaction is that in alloy steels, bainite does not form until the temperature at which austenite is isothermally reacted falls below a definite temperature, designated as B_s (bainite start temperature). Above B_s austenite does not form bainite except in the presence of externally applied stresses. Further, at temperatures just below B_s, austenite does not transform completely to bainite. The amount of bainite formed increases as the isothermal reaction temperature is lowered, as is shown schematically in Fig. 18.29. Below a lower limiting temperature, B_f (bainite finish), it is thus possible to transform austenite completely to bainite. The analogy

Fig. 18.27 The structure of bainite transformed at 523 K (250°C) as revealed by the electron microscope. Original magnification 15,000× reduced 30 percent in above photograph. (*Trans. ASTM,* **50**, 444, Fig. 23 [1952]. [Second Progress Report of Subcommittee XI of Committee E-4].) Copyright ASTM International. Reprinted with permission

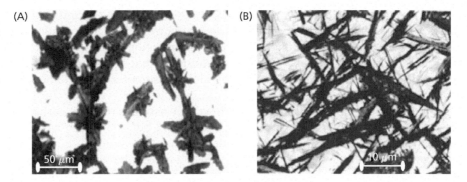

Fig. 18.28 **(A)** Upper bainite in 1.2 percent C steel formed at 723 K (450°C). **(B)** Lower bainite formed at 523 K (250°C). (Reprinted from *Journal of Materials Processing Technology* 189, Sajjadi, S. A., Isothermal transformation of austenite to bainite in high carbon steels, pp. 107–113, Figure 1, Copyright 2007, with permission from Elsevier. https://www.sciencedirect.com/science/article/abs/pii/S0924013607000313)

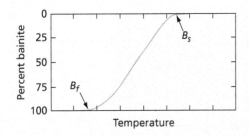

Fig. 18.29 Effect of temperature on the amount of bainite formed in an isothermal transformation (schematic). (From Hehemann, R. F., and Troiano, A. R., *Metal Progress*, **70**, No. 2, 97 [1956]. Reprinted with permission of ASM International®. All rights reserved. www.asminternational.org)

between the temperature dependence of the bainite reaction and the temperature dependence of a martensitic reaction is apparent. The B_s and B_f temperatures are the equivalent of the M_s and M_f temperatures, and the curve of Fig. 18.29 is very similar to those shown in Chapter 17 giving the amount of athermal martensite formed as a function of temperature.

In those alloy steels where the pearlite and bainite reactions do not overlap, the fraction of austenite that is untransformed when the steel is reacted between B_s and B_f is capable of remaining as austenite for indefinitely long periods of time. This is true provided the metal is held at the transformation temperature. If cooled to room temperature, however, the remaining austenite, or some fraction of it, undergoes the martensite transformation. In simple iron-carbon alloys that undergo isothermal transformation in the region where the pearlite-bainite overlap occurs, it can be assumed that the austenite that is not transformed to bainite goes to pearlite.

The fact that the bainite transformation does not go to completion in the temperature interval B_s to B_f implies that in this range of temperatures both nucleation and growth stop before all of the austenite is consumed. Thus, a finite number of nuclei must form, which grow to typical bainite plates. The limitation on the growth of these plates is not difficult to understand. A bainite plate can grow until it intersects another plate, or an austenite grain boundary. Alternatively, the growth of bainite plates can be limited by a loss of coherency between the ferrite in the plate and the parent austenite. On the other hand, the reasons why nucleation stops are not easy to determine.

At this point, it should be mentioned that the dual nature of the bainite transformation, that is, with both nucleation and growth as well as martensitic characteristics, has produced proponents favoring either one or the other as the dominant factor in the bainite reaction. This has resulted in a longstanding controversy which has run for over half a century (Ref. 27). On the one hand, there are those who support the diffusion-controlled reaction model, which views (Ref. 28) the bainitic reaction as a nonlamellar eutectoid reaction and considers that bainite is a nonlamellar counterpart of pearlite. This group considers that bainitic ferrite grows by the diffusion-controlled movement of ledges along the alpha-gamma ferrite plate boundaries. This mechanism is the same as that proposed for the Widmannstätten growth of proeutectoid ferrite, which is described in Sec. 18.2. The mechanism effectively assumes that the carbon concentration in the ferrite is low, implying that the austenite should be enriched in carbon. Thus, the carbides are expected to form on the austenite side of the austenite-ferrite boundary and grow into the austenite. The experimental evidence tends to support the concept that the carbides are precipitated in upper bainite on the austenite side of the boundaries. On the other hand, in lower bainite, the empirical evidence gives some support to the belief that the carbides precipitate in the ferrite.

The alternate view (Ref. 28) of the bainite transformation considers that bainite is formed by a shear mechanism. It is primarily based on the many morphological and crystallographic similarities between martensite and bainite. Some of these were described earlier in this section. These similarities have, in general, not conclusively established the shear or displacive theory as the basic mechanism for bainite. It has also been argued (Ref. 28) that these similarities are consistent with the concept of a nonlamellar eutectoid bainite reaction. More recently, Yasuya Ohmori (Ref. 29) has concluded that bainite formation likely consists of two steps of lattice deformations via individual atomic jump at the interface followed by lattice invariant shear.

18.11 The Complete T-T-T Diagram of an Eutectoid Steel

The time-temperature-transformation (T-T-T) diagram of Fig. 18.21 is shown again in Fig. 18.30. In the latter figure, however, the curves corresponding to the start and finish of transformation are extended into the range of temperatures where austenite transforms to bainite. Because the pearlite and bainite transformations overlap in a simple eutectoid iron-carbon steel, the transition from the pearlite reaction to the bainite reaction is smooth and continuous. Above approximately 800 to 900 K, austenite transforms completely to pearlite. Below this temperature to approximately

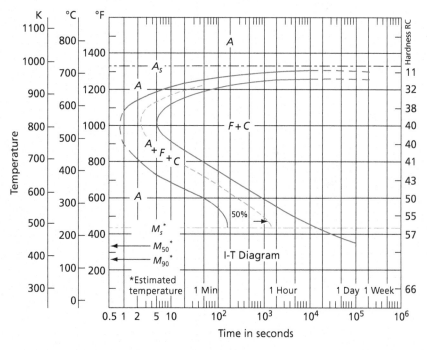

Fig. 18.30 The complete isothermal transformation diagram for an eutectoid steel. Notice that this steel is not a high-purity iron-carbon alloy, but a commercial steel (AISI 1080) containing 0.79 percent carbon and 0.76 percent manganese. The effect of the manganese will be discussed in Chapter 19. Note that the temperature is given in Centigrade, Fahrenheit, and Kelvin on the left and the hardness of the isothermally transformed specimens is shown on the right. In this figure A = austenite, F = ferrite, C = cementite, and M = martensite. (From *Atlas of Isothermal Transformation and Continuous Cooling Diagrams*, American Society for Metals, Metals Park, 1977. Reprinted with permission of AMS International®. All rights reserved. www.asminternational.org)

700 K, both pearlite and bainite are formed. Finally, between 700 and 483 K, the reaction product is bainite only.

An interesting feature of the bainite reaction is that as the reaction temperature is lowered, the rate at which bainite forms decreases. Accordingly, very long times are required to form bainite just above the M_s temperature.

The significance of the dotted line that runs between the two curves, marking the beginning and end of isothermal transformations, should be mentioned. This represents, at any given temperature, the time to transform half the austenite to bainite or austenite to pearlite, as the case may be.

Let us consider some arbitrary time-temperature paths along which it is assumed austenitized specimens are carried to room temperature. These are shown in Fig. 18.31 and represent exercises for showing the principles of the use of time-temperature-transformation (T-T-T) diagrams.

Path 1 The specimen is cooled rapidly to 433 K and left there 20 min. The rate of cooling is too rapid for pearlite to form at higher temperatures; therefore, the steel remains in the austenitic phase until the Ms temperature is passed, where martensite begins to form athermally. Since 433 K (160°C) is the temperature at which half of the austenite transforms to martensite, the direct quench converts

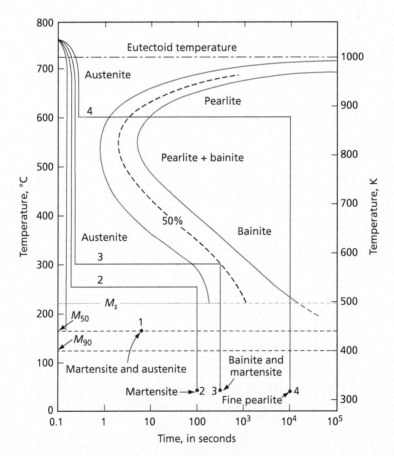

Fig. 18.31 Arbitrary time-temperature paths on the isothermal transformation diagram

50 percent of the structure to martensite. Holding at 433 K forms only a very small quantity of additional martensite because in simple carbon steels, isothermal transformation to martensite occurs only to a very limited extent. At point 1, accordingly, the structure can be assumed to be half martensite and half retained austenite.

Path 2 In this case, the specimen is held at 523 K (250°C) for 100 s. This is not sufficiently long to form bainite, so that the second quench from 523 K to room temperature develops a martensitic structure.

Path 3 An isothermal hold at 573 K about 500 s produces a structure composed of half bainite and half austenite. Cooling quickly from this temperature to room temperature results in a final structure of bainite and martensite.

Path 4 Eight seconds at 873 K (600°C) converts austenite completely (99 percent) to fine pearlite. This constituent is highly stable and will not be altered on holding for a total time of 10^4 s (2.8 hr) at 873 K (600°C). The final structure, when cooled to room temperature, is fine pearlite.

18.12 Slowly Cooled Hypoeutectoid Steels

The steel region of the iron-carbon diagram is reproduced again in Fig. 18.32. Alloys to the left of the eutectoid point are arbitrarily designated as hypoeutectoid, while those to the right are known as hypereutectoid. In the above discussion, attention has been directed primarily to a consideration of the eutectoid composition. The transformations that austenite undergoes in steels, whose compositions fall on the low, or hypoeutectoid, side of the eutectoid point, will now be considered. A typical hypoeutectoid composition is represented by line ac. At point a this alloy is austenitic. Its transformation, on very slow cooling, starts when the temperature reaches point b. At this time the alloy enters a two-phase field of ferrite and austenite. When this happens, ferrite begins to nucleate heterogeneously at the grain boundaries of the austenite, as indicated in Fig. 18.33A. With continued slow-cooling to point c, the ferrite grains grow in size. Since the ferrite is very low in carbon (<0.02 percent C), its growth is associated with a rejection of carbon from the interface back into the austenite, and a corresponding increase in the carbon content of the austenite. The two-phase

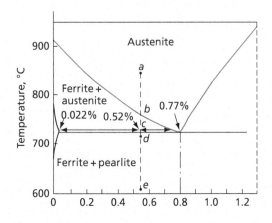

Fig. 18.32 Transformation of a hypoeutectoid steel on slow cooling

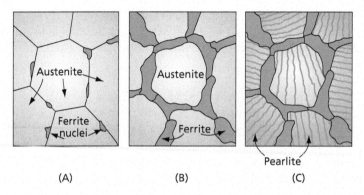

Fig. 18.33 Three stages in the formation of a slowly cooled hypoeutectoid structure corresponding to points *b*, *c*, and *d*, respectively, in Fig. 18.32

mixture that is obtained when the alloy is cooled to point *c* is shown in Fig. 18.33B. Notice that each austenite grain is now surrounded by a network of ferrite crystals. The pertinent data concerning this two-phase mixture are easily deduced from Fig. 18.32. By the lever rule, we have:

$$\text{The amount of ferrite} = \frac{0.77 - 0.52}{0.77 - 0.02} = \frac{1}{3}$$

$$\text{The amount of austenite} = \frac{0.52 - 0.02}{0.77 - 0.02} = \frac{2}{3}$$

and the intersections of the tie lines with the single-phase boundaries show us that the ferrite must contain 0.02 percent C and the austenite 0.77 percent C. The structure thus consists of two-thirds austenite and one-third ferrite in which the austenite has the eutectoid composition and is just above the eutectoid temperature. Slow cooling of this metal through the eutectoid temperature transforms the remaining austenite to pearlite, so that the structure at point *d* consists of a mixture of ferrite and pearlite. (See Fig. 18.33C.) The ratio of ferrite to pearlite is, of course, the same as the ratio of ferrite to austenite which obtains at point *c*, namely, 1 to 3. Continued cooling of the alloy to room temperature causes no visible change in the microstructure. Theoretically, a change should occur since the solubility of carbon in ferrite decreases with temperature, but the quantity of carbon involved is very small, since at the eutectoid temperature ferrite only dissolves 0.022 percent C.

For most practical purposes, in making lever-rule computations one can assume that the carbon content of ferrite is zero. If this assumption is made, then the part of the structure of a hypoeutectoid steel that is pearlite varies directly as the ratio of the carbon content of the steel divided by the eutectoid composition 0.77 percent. Thus, a 0.2 percent C steel should have a structure with approximately one-fourth pearlite, and a steel with a structure composed of half pearlite should have a carbon content of about 0.4 percent C. The microstructures of several slowly cooled hypoeutectoid steels are shown in Fig. 18.34.

The structure of a hypoeutectoid steel serves to show the difference between the phase and constituent concepts. In any one of the photographs of Fig. 18.34, two basic types of structures are clearly evident: the white ferrite areas and the dark pearlitic areas. The constituents of these specimens are, accordingly, pearlite and ferrite. The phases in these structures are, however, ferrite and cementite. In each specimen, cementite is localized in the pearlitic areas, while the ferrite occurs both in the

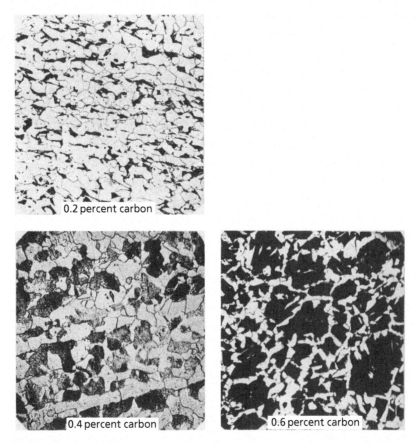

Fig. 18.34 Hypoeutectoid-steel microstructures. Black areas are pearlite, white areas are ferrite. Approximately 300×

pearlite and in the simple ferrite grains. While there is no basic difference between the two forms of ferrite, it is sometimes convenient to differentiate between them. Ferrite that appears in the pearlite is called the *eutectoid ferrite*, while the other is the *proeutectoid ferrite*. The Greek prefix "pro" is used to designate the latter because, on cooling, it forms before the eutectoid-structure (pearlite) forms.

The relative amounts of the two phases in a steel are also found by the lever rule. In this case, the levers extend to the compositons of the phases (ferrite 0 percent and cementite 6.7 percent), instead of to the compositions of the constituents (proeutectoid ferrite 0 percent C and pearlite 0.77 percent C).

18.13 Slowly Cooled Hypereutectoid Steels

Steels having a carbon content that falls above the eutectoid point transform in a fashion similar to those laying below the eutectoid composition. The proeutectoid constituent in this case is cementite instead or ferrite which, for a typical composition (1.2 percent C) (Fig. 18.35), forms between

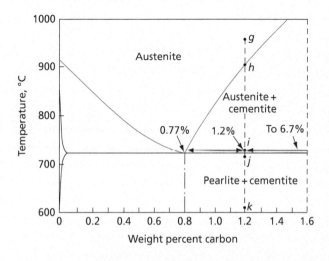

Fig. 18.35 Transformation of a hypereutectoid steel on slow cooling

point h and i as the temperature is lowered along line gk. At point i, the structure is a mixture of cementite (6.7 percent C) and austenite (0.77 percent C). A lever-rule computation shows that 92.7 percent of the structure is austenite and 7.3 percent is cementite. The precipitation of the proeutectoid cementite reduces the carbon content of the austenite to that of the eutectoid point. As a result, slowly lowering the temperature of the specimen through the eutectoid temperature converts the remaining austenite to pearlite. Continued cooling of the specimen to room temperature in this case also causes no visible change in the microstructure, so that the structure obtained at point j may be taken as representative of that visible at room temperature, which is 7.3 percent proeutectoid cementite and 92.7 percent pearlite, which are the constituents of the structure (see Fig. 18.36). The phases are cementite and ferrite. The latter can be determined with the aid of the lever rule and shown to be 18 percent total cementite (eutectoid plus proeutectoid) and 82 percent ferrite (all contained in the pearlite).

A typical hypereutectoid structure is shown in Fig. 18.37. Hypereutectoid microstructures usually differ in appearance from hypoeutectoid ones. The most important difference is in the amount of the proeutectoid constituent. In the above example (1.2 percent C steel that lies 0.43

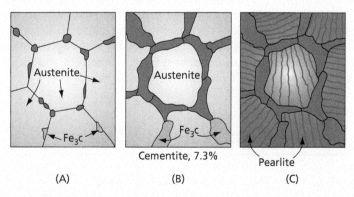

Fig. 18.36 Three stages in the formation of a slowly cooled hypereutectic corresponding to points h, i, j, respectively, in Fig. 18.35

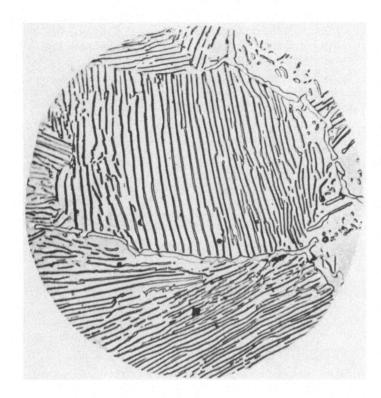

Fig. 18.37 Hypereutectoid-steel microstructure. Notice the band of cementite plates outlining the pearlite colony in the center of the photograph. 1000×

percent above the eutectoid point) there is only 7.3 percent proeutectoid cementite. At an approximately equal distance in weight percent on the other side of the eutectoid point, 0.4 percent C, the structure contains 50 percent proeutectoid ferrite. A particular characteristic of hypereutectoid microstructures is the small amount of the proeutectoid constituent. In the photograph of Fig. 18.37, which is of a 1.1 percent C steel, the proeutectoid cementite appears as a fine network surrounding pearlite areas, as may be seen by studying the boundary around the large pearlite colony occupying the center of the photograph. Each pearlite region was a single austenite grain before the transformation. The proeutectoid cementite thus forms heterogeneously at the austenite grain boundaries.

Another important difference between the structures of low- and high-carbon steels is that in steels ferrite is the continuous phase and cementite the surrounded phase. Thus, the proeutectoid cementite, while outlining the austenite grain boundaries, still consists of a number of disconnected plates. Conversely, proeutectoid ferrite often completely surrounds pearlite areas in the form of a number of contiguous grains.

18.14 Isothermal Transformation Diagrams for Noneutectoid Steels

Isothermal transformation diagrams (T-T-T) for both a hypoeutectoid steel (0.35 percent C) and a hypereutectoid steel (1.13 percent C) are shown in Figs. 18.38 and 18.39, respectively. The similarity between these diagrams and that of the eutectoid steel are at once apparent, but there are,

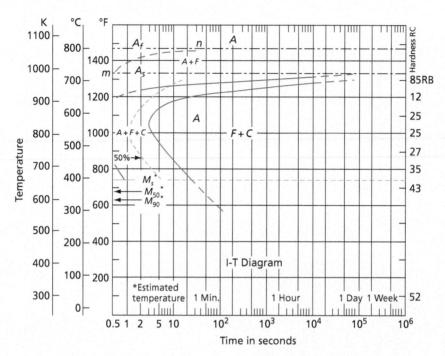

Fig. 18.38 Isothermal transformation diagram for a hypoeutectoid steel: 0.35 percent carbon, 0.37 percent manganese. Note: the A_f line represents the highest temperature at which ferrite can form; A_s is the eutectoid temperature; M_s is the martensite start temperature, A, F, and C have the same significance as in Fig. 18.30. (From *Atlas of Isothermal Transformation and Continuous Cooling Diagrams*, American Society for Metals, Metals Park, Ohio 44073, 1977. Reprinted with permission of ASM International®. All rights reserved. www.asminternational.org)

nevertheless, important differences. Among the latter is the presence in both figures of the curve *mn* corresponding to the start of the isothermal transformation of the proeutectoid constituent. In each case, the designated line lies above and to the left of those defining the isothermal transformation of austenite to pearlite. Notice that these curves are asymptotic (at large times) to constant temperature lines made on each figure to pass through the temperatures at which the two given alloys are first able to form the proeutectoid constituents (ferrite and cementite, respectively) on very slow cooling.

The isothermal-transformation diagram for a hypoeutectoid steel is reproduced again in Fig. 18.40. Several arbitrary cooling paths are made in this diagram in order to illustrate the complete significance of all the lines shown.

In each case, it is assumed that the specimens are austenitized at 1113 K (840°C), which is some 40 K above the temperature at which ferrite is able to first form in this composition. Along Path 1 it is assumed that the specimen is instantly quenched to 1023 K (750°C) and held at this temperature for 1 hr. During the first second of this isothermal treatment, the structure remains entirely austenitic, but at the end of this second the curve designating the start of the ferrite nucleation is crossed and ferrite begins to form. From this point on to the end of 10,000 s (2.8 h) the structure lies in the two-phase austenite-ferrite region. Because of the large length of time at this temperature,

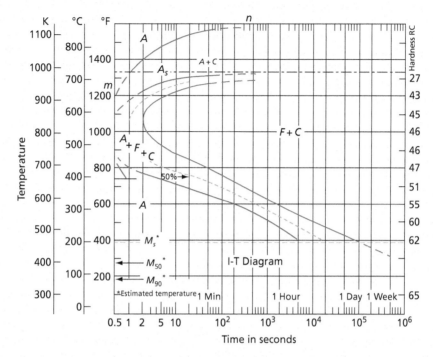

Fig. 18.39 Isothermal transformation diagram for a hypereutectoid steel: 1.13 percent carbon, 0.30 percent manganese. (From *Atlas of Isothermal Transformation and Continuous Cooling Diagrams*, American Society for Metals, Metals Park, Ohio 44073, 1977. Reprinted with permission from ASM International®. All rights reserved. www.asminternational.org)

the amount of ferrite formed should be close to that predicted by the equilibrium diagram for this temperature. No pearlite should form because we are still above the eutectoid temperature [1000 K (727°C)]. In the diagram, Path 1 is completed by a quench to room temperature, which should transform any austenite left at 1023 K (750°C) almost completely to martensite, so that the final structure can be assumed to consist of ferrite and martensite.

Path 2 represents one in which the specimen is assumed to be isothermally transformed at a temperature below the eutectoid temperature, and for this purpose 923 K (650°C) has been selected. Because of the extreme rapidity with which ferrite forms from austenite in this temperature range, even a very rapid quench (cooling time less than 0.5 s) cannot suppress the formation of some ferrite during the quench. As a result, the specimen starts its isothermal transformation as a mixture of ferrite and austenite. The transformation to pearlite in this temperature range is also very rapid and the latter begins to form at once. During this period, from about 0.5 s (the assumed start of the transformation) to the end of 100 s, the austenite is transformed to pearlite. The specimen may be assumed to be completely transformed at the end of 100 s and to consist of a mixture of ferrite and pearlite. Cooling to room temperature at any normal rate of cooling does not change this structure.

Transformations of the type indicated above, in which a hypoeutectoid specimen is quickly cooled to a fixed temperature and allowed to transform at this temperature, represent a nonequilibrium irreversible transformation. One of the results is that the ratio of ferrite to pearlite that

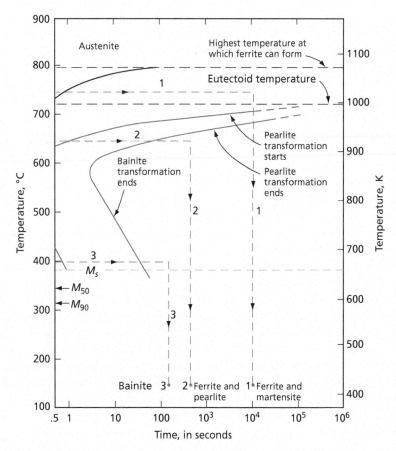

Fig. 18.40 Arbitrary time-temperature paths on the isothermal diagram of a hypoeutectoid steel

is obtained is not the same as that obtained on a very slow continuous cooling cycle approaching an equilibrium transformation. The amount of ferrite in the irreversible process is usually smaller, which means that the amount of pearlite in the final microstructure is greater. This increase cannot occur without a corresponding change in the composition of pearlite: normally, on slow cooling, $\frac{1}{8}$ cementite and $\frac{7}{8}$ ferrite. Transformations of hypoeutectoid steels at temperatures below the eutectoid point, therefore, tend to suppress the amount of proeutectoid ferrite and lower the carbon content of the eutectoid structure (the pearlite). At this point it is well to point out that there is a corresponding effect in the case of hypereutectoid compositions. Transformations that occur below the eutectoid temperature tend to suppress the amount of the proeutectoid cementite and, accordingly, to raise the carbon concentration of the pearlite.

The great rapidity with which ferrite and pearlite form in the particular alloy under consideration precludes the formation of a microstructure that is all bainite. As Fig. 18.39 shows, a quench that takes even as short a period as 0.5 s to reach 673 K (400°C) still passes through the lines designating the start of the ferrite and pearlite transformations. A specimen quickly cooled and held at 673 K (400°C) for about 100 s (Path 3), accordingly, contains bainite mixed with a small amount of ferrite and pearlite. Finally, a direct quench to room temperature should furnish a hardened specimen containing a high percentage of martensite, but with also unavoidable small percentages of ferrite and pearlite.

An analysis similar to the above can be carried out for the hypereutectoid steel shown in Fig. 18.39. The principal difference will be in the nature of the proeutectoid constituent—cementite instead of ferrite.

Problems

18.1 The Johnson-Mehl equation, Eq. 18.7, can in most cases adequately describe the kinetics of the austenite to pearlite transformation. This can be demonstrated by taking several sets of N and G values from Fig. 18.19, which contains data from a 0.78 percent C plain-carbon steel. This figure indicates that at 700°C, $N = 6.31 \times 10^{-4}$ nuclei per mm^3 per sec and $G = 3.16 \times 10^{-4}$ mm per sec. On the other hand, at 550°C, $N = 1000$ nuclei per mm^3 per sec and $G = 8.91 \times 10^{-3}$ mm per sec. Substitute these values of N and G into Eq. 18.7, and with the aid of a computer obtain the pearlite reaction curves for this steel at 700 and 500°C. Plot the fraction transformed, $f(t)$, versus $\log(t)$ as in Fig. 18.20.

18.2 (a) Determine, using the curves obtained in Prob. 18.1, the times required to obtain 1 percent and 99 percent pearlite at 700 and 550°C.

(b) Compare these times with the experimental data for an eutectoid plain-carbon steel shown in the time temperature-transformation diagram (T-T-T) in Fig. 18.21. Is the agreement reasonable?

18.3 After being slowly cooled from the austenite region, a simple iron-carbon steel exhibits a microstructure consisting of 40 percent pearlite and 60 percent ferrite.

(a) Estimate the carbon concentration of the steel.

(b) Describe the equilibrium microstructure that would be obtained if the steel were heated to 730°C and held there for a long period of time.

(c) What would be the equilibrium structure of this steel if it were heated to 850°C?

(d) Make sketches of all of these microstructures.

18.4 In slowly cooled hypoeutectoid steels one normally finds a microstructure in which the pearlite colonies are surrounded by ferrite grains. Explain.

18.5 An iron-carbon steel containing 0.5 percent of carbon has a microstructure consisting of 85 percent pearlite and 15 percent ferrite.

(a) Are these the amounts of these constituents that one would expect to find in the steel if it had been slowly cooled from the austenite region?

(b) In a slowly cooled microstructure, pearlite normally has a 7 to 1 ratio of the widths of the ferrite and cementite lamellae. What would this ratio be in the present case?

18.6 In a slowly cooled hypereutectoid iron-carbon steel, the pearlite colonies are normally separated from each other by a more or less continuous boundary layer of cementite. Explain how this microstructure develops. Use simple sketches to illustrate your answer.

18.7 Consider an iron-carbon alloy containing 1.0 percent carbon.

(a) If this composition is slowly cooled from the austenite region, what would be the respective percentages of the constituents and the phases in its microstructure?

(b) Now assume that it has been cooled at a rate rapid enough to yield a proeutectoid constituent of only 1.2 percent. What would be the percentages of the constituents and the phases in this case?

18.8 With the aid of the data given in Figs. 18.13, 18.14, and 18.15, determine the values of the constants K_1, K_2, K_3 in Eqs. 18.3, 18.4, and 18.5, respectively. Then check your answers by seeing if $K_1 \times K_2 \times K_3 = 1$.

18.9 Substitute the values of K_1, K_2, and K_3 into Eqs. 18.3, 18.4, and 18.5. Now substitute $(727 - T)$ for ΔT in Eqs. 18.3 and 18.5, and then make the following plots:

(a) Plot the temperature T in °C, from 726°C to 526°C, against log V, where V is the growth velocity. This should give a more up-to-date version of the curve in Fig. 18.19. Use μm/s units for V.

(b) Plot T vs log λ where λ is the interlamellar spacing of the pearlite, from $T = 726$°C to 526°C.

(c) Plot λ against V in the interval from $V = 0.1$ to 100 μm/s. Express λ in μm units.

18.10 (a) Explain in detail how bainite differs from martensite and from pearlite.

(b) How do upper and lower bainite differ?

18.11 Answer the following with regard to the T-T-T diagram of a steel of eutectoid composition, shown in Fig. 18.31, and assume that the specimens involved in the various cooling paths were cut from a thin sheet and austenitized at 750°C before cooling. Describe the microstructure resulting from being:

(a) Cooled to room temperature in less that 1 s.

(b) Cooled to 160°C in less than 1 s and then maintained at this temperature for several years.

(c) Quenched to 650°C and held at this temperature for 1 day, then quenched to room temperature.

(d) Quenched to 550°C and held at this temperature for 1 day, then quenched to room temperature.

18.12 A high-carbon steel containing 1.13 percent carbon, with a microstructure similar to that in Fig. 18.36, is heated to 730°C and allowed to come to equilibrium. It is then quenched to room temperature. Make a sketch of the resulting microstructure, identifying the constituents.

18.13 Assume that the 0.52 percent carbon steel of Fig. 18.32 is slowly cooled to the temperature of point c and then quenched to 450°C and held at this latter temperature for a period of a day. Describe the microstructure that one might find. Determine the percentages of the constituents.

18.14 Transpose paths 1, 2, and 3 of Fig. 18.39 to Fig. 18.38 and determine the microstructures corresponding to these paths in the hypereutectoid steel.

18.15 With regard to the data for eutectoid steels in Fig. 18.16:

(a) What is the interlamellar spacing in the pure Fe-C alloy at 660°C?

(b) How much smaller is it at this temperature when the steel contains 1.8 percent chromium?

(c) How much larger is it in the steel with 1.8 pct manganese?

18.16 Describe the meaning of the term *partitioning coefficient*.

References

1. Chipman, J., *Met. Trans.*, **3** 55 (1972).
2. Massalski, T. B., Ed. in Chief, *Binary Alloy Phase Diagrams*, American Society for Metals, Metals Park, Ohio, 1986.
3. Honeycombe, R. W. K., *Met. Trans. A*, **7A** 915 (1976).
4. Aaronson, H. I., *Decomposition of Austenite by Diffusional Processes*, p. 387, Interscience Publishers, New York, 1962.

5. Dippenaar, R. J., and Honeycombe, R. W. K., *Proc. Roy. Soc. Lond. A*, **333** 455 (1973).
6. Howell, P. R., *Materials Characterization*, **40** 1998, 227–260.
7. Hackney, S. A., and Shiflet, G. J., *Scripta Met.*, **19** 757 (1985).
8. Hackney, S. A., and Shiflet, G. J., *Acta Met.*, **35** 1987, 1007–1017.
9. Hackney, S. A., and Shiflet, G. J., *Acta Met.*, **35** 1987, 1019–1028.
10. Zhou, D. S., and Shiflet, G. J., *Met. Trans. A*, **22A** 1991, 1349–1365.
11. Ridley, N., *Met. Trans. A*, **15A** 1019 (1984).
12. Caballero, F. G., García de Andrés, C., and Capdevila, C., *Materials Characterization*, **45** 2000, 111–116.
13. Bolshakov, V. I., and Bobyr, S. V., *Metal Science and Heat Treatment*, **46** 2004, 329–333.
14. Pearson, D. D., and Verhoeven, J. D., *Met. Trans. A*, **15A** 1037 (1984).
15. Tewari, S. K., and Sharma, R. C., *Met. Trans. A.*, **16A** 597 (1985).
16. Johnson, W. A., and Mehl, R. F., *Trans. AIME*, **135** 416 (1939).
17. Hull, F. C., Colton, R. A., and Mehl, R. F., *Trans. AIME*, 50, 1942, 185–207.
18. Mou, Y., and Hsu, T. Y., *Met Trans. A.*, **19A** 1695 (1988).
19. Yang, Z., and Fang, H., *Solid State and Materials Science*, 9, 2005, 277–286.
20. Sajjadi, S. A., and Zebarjad, S. M., *J. Materials Processing Technology*, 189, 2007, 107–113.
21. Quidort, D., and Brechet, Y., *Scripta Mat.*, 47, 2002, 151–156.
22. Okamoto, H., and Oka, M., *Met. Trans. A.*, **17A** 1113 (1986).
23. Bhadeshia H. K. D. H., and Edmunds, D. V., *Met. Trans. A.*, **10A** 895 (1979).
24. *Trans. ASTM*, **52** 543 (1952).
25. Hehemann, R. F., ASM Seminar, *Phase Transformations*, Amer. Soc. for Metals, Metals Park, Ohio, 1970.
26. Sandvik, B. P. J., *Met. Trans, A*, **13A** 777 (1982).
27. Aaronson, H. I., *Met Trans. A*, **17A** 1095 (1986).
28. Hehemann, R. F., Kinsman, K. R., and Aaronson, H. I., *Met Trans.* **3** 1077 (1972).
29. Ohmori, Y., *Scripta Mat.*, 47, 2002, 201–206.

Chapter 19

The Hardening of Steel

> **Learning Objectives**
>
> Upon completion of Chapter 19, you will be able to:
> 1. Utilize continuous cooling transformation diagrams for heat treatment of steels
> 2. Differentiate between hardness and hardenability
> 3. Discuss the effects of carbon, austenite grain size and alloying elements on steel hardenability
> 4. Explain how dimensional changes during martensitic transformation can lead to quench cracks
> 5. Describe the effects of carbon and alloying elements on the martensite start and finish temperatures, and justify double quenching treatment
> 6. Explain basic phenomena that occur during tempering of low and high carbon steels
> 7. Compare and contrast the influence of carbon content on the hardness of quenched martensite with those of tempered martensite, pearlitic, bainitic and spheroidized steels

19.1 Continuous Cooling Transformations (CCT)

The isothermal transformation diagram is a valuable tool for studying the temperature dependence of austenitic transformations. In even a single reaction, such as the austenite to pearlite transformation, the product varies in appearance with the transformation temperature. A specimen that is allowed to transform over a range of temperatures, therefore, has a mixed microstructure that is very difficult to analyze without prior information. Before the original work on isothermal transformations by Davenport and Bain (Ref. 1), the basic austenitic reactions were not clearly defined and our knowledge of them was, to say the least, confused. The time-temperature relationships that are mapped out on an isothermal transformation diagram, however, are strictly applicable only to transformations carried out at constant temperatures. Unfortunately, very few commercial heat treatments occur in this manner. In most cases, the metal is heated into the austenite range and then continuously cooled to room temperature, with the cooling rate varying with the type of treatment and the size and shape of the specimen. The difference between isothermal transformation diagrams and continuous cooling transformation (CCT) diagrams is perhaps most easily understood by comparing these two forms for a steel of eutectoid composition. This particular composition is chosen because of its simplicity, and the pertinent diagrams are given in Fig. 19.1. Two cooling curves, corresponding to different rates of continuous cooling, are also shown in Fig. 19.1. In each case, the cooling curves start above the eutectoid temperature and fall in temperature with increasing times. The inverted shape of these curves is due to plotting the time coordinate (abscissa) according to a logarithmic scale. On a linear time scale the curves would be concave toward the right, signifying a decreasing rate of cooling with increasing time.

Now consider the curve marked 1. At the end of approximately 6 s this curve crosses the line representing the beginning of the pearlite transformation. The intersection is marked on the diagram as point a. The significance of point a is that it represents the time required to nucleate pearlite isothermally at 650°C (the temperature of point a). A specimen carried along line 1, however, only reached the 650°C isothermal at the end of 6 s and may be considered to have been at temperatures above 650°C for the entire 6-s interval. Because the time required to start the pearlite transformation is longer at temperatures above 650°C than it is at 650°C, the continuously

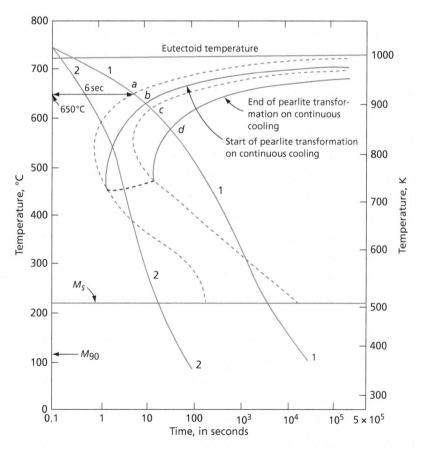

Fig. 19.1 The relationship of the continuous cooling diagram to the isothermal diagram for an eutectoid steel (schematic). The dashed curves represent isothermal transformation

cooled specimen is not ready to form pearlite at the end of 6 s. Approximately, it may be assumed that cooling along path 1 to 650°C has only a slightly greater effect on the pearlite reaction than does an instantaneous quench to this temperature. In other words, more time is needed before transformation can begin. Since in continuous cooling an increase in time is associated with a drop in temperature, the point at which transformation actually starts lies to the right and below point a. (The location of this point may be estimated with the aid of several appropriate assumptions (Ref. 2).) This position is designated by the symbol b. In the same fashion, it can be shown that the finish of the pearlite transformation, point d, is depressed downward and to the right of point c, the point where the continuous cooling curve crosses the line representing the finish of isothermal transformation.

The above reasoning explains qualitatively why the continuous-cooling-transformation (CCT) lines representing the start and finish of the pearlite transformations are shifted with respect to the corresponding isothermal-transformation lines. Why the bainite reaction in this metal does not appear on continuous cooling also needs to be explained. This is not too hard to understand and is related to the fact that the pearlite reaction lines extend over and beyond the bainite-transformation lines. Thus, on slow or moderate rates of cooling (curve 1), austenite in the specimens is converted completely to pearlite before the cooling curve reaches the bainite-transformation range. Since the austenite has already been completely transformed, no bainite can form. Alternatively, as shown by curve 2, the specimen is in the bainite-transformation region for too short a period of time to allow any appreciable amount of bainite to form. An element in this latter conclusion is the fact that the rate at which bainite forms rapidly decreases with falling temperatures. It is generally assumed (as a first approximation) in drawing the continuous-cooling-transformation diagram for this particular alloy, that the transformation along a path such as curve 2 stops in the region where the bainite and pearlite transformations overlap on the isothermal diagram. As a result, the microstructures corresponding to path 2 should consist of a mixture of pearlite and martensite, with perhaps a small amount of bainite, which we shall ignore. The martensite, of course, forms from the austenite that did not transform to pearlite at higher temperatures.

The continuous-cooling-transformation diagram of an eutectoid steel is shown again in Fig. 19.2. A number of cooling curves are also shown on the diagram. The given curves are not quantitative, but rather qualitative representations of how various cooling rates can produce different microstructures. The curve marked "full anneal" represents very slow cooling and is usually obtained by cooling specimens (suitably austenitized) in a furnace that has its power supply shut off. This rate of cooling normally brings the steel to room temperature in about a day. Here the transformation of the austenite takes place at temperatures close to the eutectoid temperature and the final structure is, accordingly, a coarse pearlite and close to that predicted for an equilibrium transformation. The second curve, marked "normalizing," represents a heat treatment in which specimens are cooled at an intermediate rate by pulling them out of the austenitizing furnace and allowing them to cool in air. In this case, cooling is accomplished in a matter of minutes and the specimen transforms in the range of temperatures between 550°C and 600°C. The structure obtained in this manner is again pearlite, but much finer in texture than that obtained in the full annealing treatment. The next cooling curve represents a still faster rate of cooling, such as might be obtained when a red-hot specimen is quenched directly into a bath of oil. Cooling at this rate produces a microstructure that is a mixture of pearlite and martensite. Finally, the curve farthest to the left and marked "water quench" represents a rate of cooling so rapid that no pearlite is able to form and the structure is entirely martensitic.

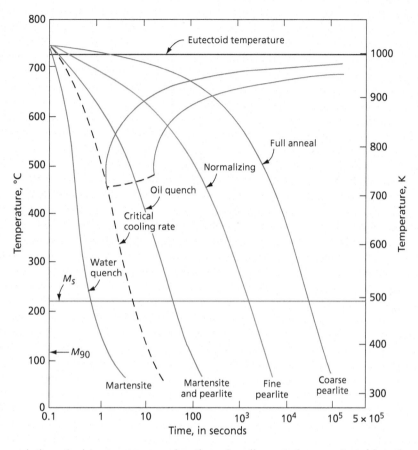

Fig. 19.2 The variation of microstructure as a function of cooling rate for an eutectoid steel

In summary, CCT diagrams provide powerful road maps for predicting microstructural development during normal or interrupted cooling conditions. As such, the diagrams can be used to predict the mechanical properties which will result under a prescribed heat treatment process. Examples of such diagrams for various steels can be found elsewhere (Refs. 3 and 4). As will be discussed in the following sections, CCT diagrams depend on the alloy chemistry and grain size of the parent austenite (Ref. 5). CCT diagrams are determined, similar to the isothermal diagrams, by conventional interrupted quenching and metallographic techniques (Ref. 6), or by in-situ measurement of the magnetic permeability and volume change during transformation (Ref. 7). The latter two techniques make use of the materials property changes that take place when austenite transforms to pearlite, bainite, or martensite. Among these techniques, the volumetric change, which is accomplished by using a dilatometer, is widely used. Computational modeling has also been employed to predict CCT diagrams (Refs. 8 and 9). However, the modeling techniques require verification by other experimental measurements.

19.2 Hardenability

Another curve is also shown in Fig. 19.2 as a dashed line—the critical-cooling-rate curve. Any rate of cooling faster than this produces a martensitic structure, while any slower rate produces a structure containing some pearlite.

In a steel specimen of any appreciable size, the cooling rates at the surface and at the center are not the same. The difference in these rates naturally increases with the severity of the quench, or the speed of the cooling process. Thus, there will be little difference, at any instant, between the temperature at the surface and at the center of a bar of some size when it is furnace-cooled (fully annealed). On the other hand, the same bar, if quenched in a rapid coolant such as iced brine, has markedly different cooling rates at the surface and at the center, which are capable of producing entirely different microstructures at the surface and center of the bar. This effect is demonstrated in Fig. 19.3, where cooling curves are plotted to represent paths followed by the surface and the center of a sizable bar (perhaps 2 in. in diameter) assumed to be quenched in a very rapid cooling

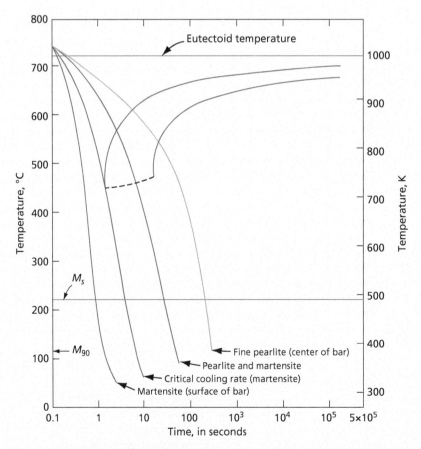

Fig. 19.3 The effect of the difference in the cooling rate at the surface and at the center of a cylindrical bar on the resulting microstructure (schematic)

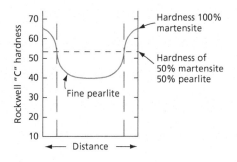

Fig. 19.4 Typical hardness test survey made along a diameter of a quenched cylinder (after sectioning the cylinder)

medium. Two other curves are also plotted on the diagram representing the critical cooling rate, and the rate that gives a structure 50 percent martensite and 50 percent pearlite. It is significant that in the present example these two rates fall within the extremes denoted by the surface and the center of the bar. We may conclude that this specimen will have a structure at the surface that is martensitic, while the center will have a pearlitic structure. The change in the microstructure with distance along a diameter is accompanied by a corresponding variation in the hardness of the metal. This fact can be easily demonstrated by cutting a quenched bar in two on an abrasive cutoff wheel, while taking care to see that the specimen is cooled properly during the sawing so that the metal does not become overheated and the micro-structure altered. The circular cross-section thus exposed is subjected to a number of hardness tests made at equal intervals along a diameter, and the results plotted to give a hardness contour of the type shown in Fig. 19.4. The hardness traverse shows that the martensitic structure near the surface is very hard (Rockwell C-65), while the pearlitic structure near the center is considerably softer (Rockwell C-40). Also plotted on the diagram is a horizontal line corresponding to the hardness of an eutectoid-steel structure containing 50 percent martensite and 50 percent pearlite (C-54). Notice that this intersects the hardness contour in the regions where it rises most steeply. This means that the distance from the surface where the specimen cooled at a rate which produced 50 percent martensite is capable of rather precise experimental measurement. The same position can also be determined by preparing the cross-section as a metallographic specimen and observing it under the microscope. Alternatively, since the pearlite etches darker than martensite, a macroscopic measurement of the position at which the structure effectively changes from martensite to pearlite can be made in terms of a color change. At any rate, the position corresponding to half-martensite and half-pearlite is easy to measure and is used as a criterion for measuring the depth to which a steel hardens with a given type of quench.

The depth at which the 50 percent martensite structure is obtained in a bar of steel is a function of a number of variables that include the composition and grain size of the metal (austenitic), the severity of the quench, and the size of the bar. Let us consider first the effect of changing the diameter of the bar of steel. Suppose that a number of bars of the same steel are given an identical quench in a brine solution and then sectioned so as to obtain hardness contours. The results are shown schematically in Fig. 19.5. An investigation of these curves shows that there is one unique diameter, the value of which is 1 in. The bar with this diameter hardens so that it has the 50 percent pearlite–50 percent martensite structure just at its center. All bars with smaller diameters are effectively hardened throughout, while any bar with a larger diameter has a soft core containing pearlite. This particular diameter is called the *critical diameter*. Its value depends on the steel in question and the means of quenching, and its importance lies in the fact that it gives a measure of the ability of the steel to respond to a hardening heat treatment. The particular steel being discussed

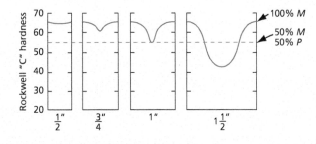

Fig. 19.5 Hardness test traverses similar to that of Fig. 19.4 made on a series of steel bars of the same composition, but with different diameters (schematic)

has a moderate ability to harden, or, as it is more properly stated, it has a moderate hardenability. According to Fig. 19.5, its critical diameter D is 1.0 in. The addition of suitable alloying elements to steels can greatly increase their hardenability and this is shown by corresponding increases in their critical diameters. The critical diameter D of a steel is, consequently, a measure of its hardenability (ability to harden), but it also depends on the rate of cooling (the type of quench). In order to eliminate this latter variable, it is general practice to refer all hardenability measurements to a standard cooling medium. This standard is the so-called *ideal quench*, which uses a hypothetical cooling medium assumed to bring the surface of a piece of steel instantly to the temperature of the quenching bath, and maintain it at this temperature. The critical diameter corresponding to an ideal quench is called the *ideal critical diameter* and is designated D_I.

The ideal quenching medium is assumed to remove the heat from the surface as fast as it can flow out from inside of the bar. While such a quenching medium does not exist, its cooling action on steels is capable of computation and comparison with those of ordinary commercial quenching media, such as water, oil, and brine. Information of this type is frequently presented in the form of curves such as those shown in Fig. 19.6, where the ideal critical diameter D_I is plotted as the abscissa, and the critical

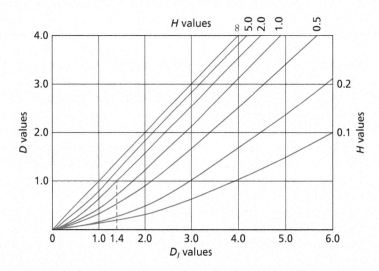

Fig. 19.6 Relationship of the critical diameter D to the ideal critical diameter D_I for several rates of cooling (H values). (After Grossman, M. A., *Elements of Hardenability*, ASM, Cleveland, 1952)

diameter D is plotted as the ordinate. A number of different curves are plotted on this chart, each corresponding to a different rate of cooling. In each case, the rate of cooling is measured by a number known as its H value, or the severity of the quench. This value is also known as the Grossman number (Ref. 10). Some of the values of this number for commercial quenches are given in Table 19.1. The use of the chart is easily illustrated by considering the example given above (Fig. 19.5), where it was determined that the critical diameter D, obtained in a brine quench ($H = 2$), was 1.0 in. The H value for a brine quench with no agitation is 2.0. The intersection of the curve for this severity of quench with the 1.0-in. ordinate occurs at a D_I value (abscissa) 1.4. The ideal critical diameter, or hardenability of the steel in question, is $D_I = 1.4$ in.

Table 19.1 shows the value of agitation during a quench. When hot metal is placed in a liquid cooling medium, gas vapors are formed at the surface of contact between the hot metal and the liquid, which retard the heat flow from metal to liquid. Agitation, or movement of the specimen relative to the liquid, is instrumental in removing the bubbles from the surface and increasing the rate of cooling. The fact that brine, water, and oil are better cooling agents in descending order is closely related to the removal of vapor bubbles from the surface. The lower inherent viscosity of water allows the bubbles to be moved faster in a water quench than in an oil quench. In a brine quench, the presence of salt in the water causes a series of small explosions to occur near the hot surface and therefore violently agitates the cooling solution in the vicinity of the quenched material.

The Grossman method of determining the ideal critical diameter D_I, outlined above, is too time consuming to be of wide practical application. It has been described as a means of introducing the concept of hardenability and the quantity that we use to measure it, namely, D_I. A much more convenient and widely used method of determining hardenability employs the Jominy end quench test.

In the Jominy test, a single specimen takes the place of the series of specimens required in the Grossman method. The standard Jominy specimen consists of a cylindrical rod 4 in. long and 1 in. in diameter. Since Jominy test specimen dimensions are frequently specified in inches, rather than mm, the former units will be used in this section of the text. For conversion, each inch is equal to 25.4 mm. In making a test, the specimen is first heated to a suitable austenitizing temperature and held there long enough to obtain a uniform austenitic structure. It is then placed in a jig and a

Table 19.1 Severity of Quench Values (or Grossman Number) for Some Typical Quenching Conditions

H Value	Quenching Condition
0.20	Poor oil quench—no agitation
0.35	Good oil quench—moderate agitation
0.50	Very good oil quench—good agitation
0.70	Strong oil quench—violent agitation
1.00	Poor water quench—no agitation
1.50	Very good water quench—strong agitation
2.00	Brine quench—no agitation
5.00	Brine quench—violent agitation
∞	Ideal quench

stream of water allowed to strike one end of the specimen. The experimental equipment is shown in Fig. 19.7. The advantage of the Jominy test is that in a single specimen one is able to obtain a range of cooling rates varying from a very rapid water quench at one end to a slow air quench at the other end. After the complete transformation of the austenite in the bar, two shallow flat surfaces are ground on opposite sides of the bar and a hardness-test traverse is made running

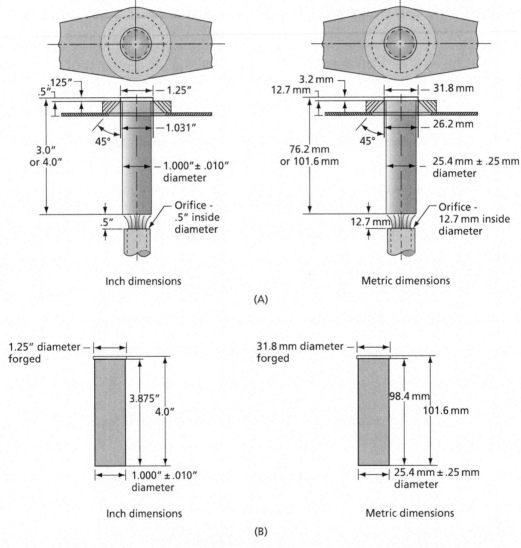

Fig. 19.7 The Jominy hardenability test. **(A)** This shows the arrangement for supporting the specimen in water quenching. Note that below the specimen there is an orifice through which water may be directed on the bottom of the specimen. **(B)** This shows the preferred dimensions for the Jominy test specimen. (From ASTM Standard A255.)

from one end of the bar to the other along one of these prepared surfaces. The data thus obtained are plotted to give the Jominy hardenability curve. A typical example is shown in Fig. 19.8, where it can be seen that the hardness is greatest where the cooling is most rapid—near the quenched end. A great deal of effort has gone into the determination of the cooling rates expected at various distances from the quenched end of a Jominy bar and to correlating these data with cooling rates inside circular bars and other shapes. Of particular importance to our present discussion is the relationship between the size, or diameter, of a steel bar quenched in an ideal quenching medium that has the same cooling rate at its center as a given position along the surface of a Jominy bar. This information is furnished in Table 19.2. The significance of Table 19.2 is that if the position on the Jominy bar where the structure is half martensite is known, this table makes possible the determination of D_I, the ideal critical diameter. For example, consider the Jominy curve in Fig. 19.8. In a 0.65 percent carbon steel, the hardness of 50 percent martensite should be HRC-52, as may be seen in Table 19.3. This table gives the hardness of the initial structure, taken as 100 percent martensite, as well as that at a 50 percent martensitic structure, for steels of different carbon concentrations. According to Fig. 19.8, HRC-52 lies at 3/16 in. from the quenched end of the Jominy bar, and according to Table 19.3, the D_I of the steel is 1.37 in. This is within the experimental error of our previous value, 1.4 in. Additional details on the use of Jominy bar tests to predict hardenability of steels can be found elsewhere (Refs. 10–12).

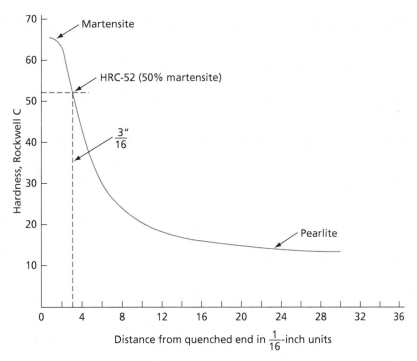

Fig. 19.8 Variation of the hardness along a Jominy bar. (Schematic for steel with 0.65 percent carbon)

Table 19.2 The Hardness of Martensite and of 50 percent Martensite as a Function of the Carbon Concentration of a Steel. (From *1986 Annual Book of ASTM Standards, Sec. 3*, **Standard A 255**, ASTM, Philadelphia, Pa., 1986.)

Carbon Content, Initial Hardness, 50% Martensite Hardness

% Carbon Content	Hardness—HRC Initial 100% Martensite	Hardness—HRC 50% Martensite	% Carbon Content	Hardness—HRC Initial 100% Martensite	Hardness—HRC 50% Martensite	% Carbon Content	Hardness—HRC Initial 100% Martensite	Hardness—HRC 50% Martensite
0.10	38	26	0.30	50	37	0.50	61	47
0.11	39	27	0.31	51	38	0.51	61	47
0.12	40	27	0.32	51	38	0.52	62	48
0.13	40	28	0.33	52	39	0.53	62	48
0.14	41	28	0.34	53	40	0.54	63	48
0.15	41	29	0.35	53	40	0.55	63	49
0.16	42	30	0.36	54	41	0.56	63	49
0.17	42	30	0.37	55	41	0.57	64	50
0.18	43	31	0.38	55	42	0.58	64	50
0.19	44	31	0.39	56	42	0.59	64	51
0.20	44	32	0.40	56	43	0.60	64	51
0.21	45	32	0.41	57	43	0.61	64	51
0.22	45	33	0.42	57	43	0.62	65	51
0.23	46	34	0.43	58	44	0.63	65	52
0.24	46	34	0.44	58	44	0.64	65	52
0.25	47	35	0.45	59	45	0.65	65	52
0.26	48	35	0.46	59	45	0.66	65	52
0.27	49	36	0.47	59	45	0.67	65	53
0.28	49	36	0.48	59	46	0.68	65	53
0.29	50	37	0.49	60	46	0.69	65	53

19.3 The Variables That Determine the Hardenability of a Steel

The hardenability of a steel, which is expressed by its D_I, is a function of its chemical composition and the size of the austenite grains it contains at the instant of quenching. We shall now discuss the effect of these factors on the hardenability, but first a word should be said about the principles of changing hardenability. A metal with a high hardenability is one in which austenite is able to transform to martensite without forming pearlite, even when the rate of cooling is rather slow. Conversely, high rates of cooling are required to form martensite in steels of low hardenability. In either case, the limiting factor is the rate at which pearlite forms at elevated temperatures. Any variable that moves the pearlite transformation lines to the right in a continuous-cooling-transformation diagram, such as Fig. 19.3, makes it possible to obtain a martensite structure at a slower rate of cooling. A movement of the pearlite transformation nose to the right is thus associated with an increase in hardenability. From another viewpoint we can say that anything that slows down the nucleation and growth of pearlite increases the hardenability of a steel.

Table 19.3 This Table Relates the Ideal Critical Diameter, D_I, to the Distance Along a Jominy Bar where the Hardness Corresponds to 50 percent Martensite. (From *1986 Annual Book of ASTM Standards, Sec. 3*, **Standard A 255**, ASTM, Philadelphia, Pa., 1986.)

Jominy Distance for 50% Martensite vs. DI (in.)

"J" $\frac{1}{16}$ in.	DI in.	"J" $\frac{1}{16}$ in.	DI in.	"J" $\frac{1}{16}$ in.	DI in.
0.5	0.27	11.5	3.74	22.5	5.46
1.0	0.50	12.0	3.83	23.0	5.51
1.5	0.73	12.5	3.94	23.5	5.57
2.0	0.95	13.0	4.04	24.0	5.63
2.5	1.16	13.5	4.13	24.5	5.69
3.0	1.37	14.0	4.22	25.0	5.74
3.5	1.57	14.5	4.32	25.5	5.80
4.0	1.75	15.0	4.40	26.0	5.86
4.5	1.93	15.5	4.48	26.5	5.91
5.0	2.12	16.0	4.57	27.0	5.96
5.5	2.29	16.5	4.64	27.5	6.02
6.0	2.45	17.0	4.72	28.0	6.06
6.5	2.58	17.5	4.80	28.5	6.12
7.0	2.72	18.0	4.87	29.0	6.16
7.5	2.86	18.5	4.94	29.5	6.20
8.0	2.97	19.0	5.02	30.0	6.25
8.5	3.07	19.5	5.08	30.5	6.29
9.0	3.20	20.0	5.15	31.0	6.33
9.5	3.32	20.5	5.22	31.5	6.37
10.0	3.43	21.0	5.28	32.0	6.42
10.5	3.54	21.5	5.33		
11.0	3.64	22.0	5.39		

19.4 Austenitic Grain Size

When steel is heated into the austenite region in order to austenitize the metal, the low-temperature structure that is transformed to the gamma phase is, in general, an aggregate of cementite and ferrite (that is, pearlite, or decomposed martensite). In this reversed transformation, the austenite grains form by nucleation and growth; the nuclei form heterogeneously at cementite-ferrite interfaces. Because of the large interfacial area available for nucleation, the number of austenite grains that appear is usually large. The transformation of steel upon heating is, therefore, characterized initially by a small austenitic grain size. However, in the austenite range, thermal movements of the atoms are rapid enough to cause grain growth, so that extended times and high temperatures in the austenite range are capable of greatly increasing the size of the initial austenite grains.

The size of the austenite grain that is attained before a metal is cooled back to room temperature is important in determining a number of the physical properties of the final structure, including the hardening response of the steel. Before describing this latter effect, let us explain

Table 19.4 ASTM Grain-Size Numbers

ASTM Grain-Size Number	Average Number of Grains per Square Inch as Viewed at 100×
1	1
2	2
3	4
4	8
5	16
6	32
7	64
8	128

the most commonly accepted method of designating the austenitic grain size. The ASTM grain-size number is defined by the relationship

$$n = 2^{N-1} \qquad \textbf{19.1}$$

where n is the number of grains per square inch as seen in a specimen viewed at a magnification of 100 times, and N is the ASTM grain-size number. The usual range of austenitic grain sizes in steels lies between 1 and 9. The number of grains per square inch in this interval is given in Table 19.4. Notice that as the grains get smaller (more numerous), the grain-size number increases.

19.5 The Effect of Austenitic Grain Size on Hardenability

The effect of grain size on hardenability has been explained (Ref. 2) on the basis of the heterogeneous manner in which pearlite nucleates at austenitic grain boundaries. While the rate of growth G of pearlite is independent of the austenite grain size, the total number of nuclei that form per second varies directly with the surface available for their formation. Thus, in a fine-grain steel, ASTM No. 7, there is four times as much grain-boundary area as in coarse-grain steel of grain size No. 3. The formation of pearlite in the fine-grain steel is, therefore, more rapid than it is in the coarse-grain steel and, accordingly, the fine-grain steel has a lower hardenability.

The use of a coarse austenitic grain size in order to increase the hardenability of steel is not generally practiced. The desired increase in hardenability is accompanied by undesirable changes in other properties, such as an increase in brittleness and a loss of ductility. Quench cracks, or cracking of the steel due to thermal shock and the stresses incident to the quenching operation, are also more common in large-grain specimens.

19.6 The Influence of Carbon Content on Hardenability

The hardenability of a steel is strongly influenced by its carbon content. This fact is shown in Fig. 19.9, where the variation of the ideal critical diameter D_I with carbon content is plotted for three different grain sizes. In addition to showing that the hardenability increases with increasing carbon content,

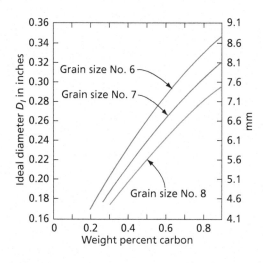

Fig. 19.9 Ideal critical diameter as a function of carbon content and austenite grain size for iron-carbon alloys. (After Grossman, M. A., *Elements of Hardenability*, ASM, Cleveland, 1952)

these curves demonstrate the very low hardenability of simple iron-carbon alloys. For example, an eutectoid steel with about 0.8 percent C and a small grain size (No. 8) possesses an ideal diameter 7.1 mm (0.28 in.). This means that the maximum theoretical diameter of a steel bar of this relatively high-carbon content that can be hardened to its center (in an ideal quench) is about 6.9 mm ($\frac{1}{4}$ in.). Any ordinary quench will, accordingly, not harden even this size bar to its center. Fortunately, so-called commercial carbon steels always contain some manganese and sometimes small amounts of other elements that increase their hardenability. The manganese in these steels is required in order that they can be manufactured economically. In this respect, the isothermal diagrams previously described were for steels containing manganese. The hardenabilities of the latter are considerably higher than if simple iron-carbon alloys had been considered.

Because increasing carbon content is associated with an increase in hardenability, it is evident that the formation of pearlite and proeutectoid constituents becomes more difficult the higher the carbon content of the steel. This statement is true not only for hypoeutectoid steels, but also for those with carbon contents greater than the eutectoid composition (hypereutectoid steels), provided that each steel is completely transformed to austenite before its hardenability is measured. It frequently happens in practice that hypereutectoid steels are austenitized in the two-phase cementite plus austenite region. When this happens, nearly all of the structure becomes austenitic, but a small amount of cementite is stable and does not dissolve. On cooling, the residual carbide particles encourage pearlite nucleation, resulting in lowered hardenability.

19.7 The Influence of Alloying Elements on Hardenability

Each and every one of the chemical elements in a steel has an influence on its hardenability. The degree, of course, varies with the element in question. Of the common alloying elements added to steel, cobalt is the only one known to decrease hardenability. The presence of cobalt in steel increases both the rate of nucleation and the rate of growth of pearlite (Ref. 13), and steels containing this element are more difficult to harden than those that do not contain it.

Other common alloying elements, to the extent that they are soluble in iron, increase the hardenability of steels. There are a number of ways that this effect can be demonstrated. One of the simplest is by empirical hardenability multiplying factors. These useful parameters were first introduced by Grossman (Ref. 14). They make possible a first-order approximation of the hardenability of a steel when only its chemical composition and its austenite grain size are known. A short list of these factors is given in Table 19.5. In this table, column 2 gives essentially the same information as that shown graphically in Fig. 19.9 for the austenitic grain size 7. The ASTM table of the Grossman multiplying factors only considers this austenitic grain size on the basis that most commercial steels, designed for a quench and temper heat treatment, are produced with approximately this grain size. It should also be noted that in this table, the carbon concentration factors are different (smaller) for carbon concentrations below 0.35 percent C than they were in earlier multiplication factor tables. These refinements are based on analyses of experimental data from thousands of heats of steel. A word should be said at this point about the difficulty that has been encountered in determining the carbon multiplying factors when the carbon concentration is below about 0.3 percent (Ref. 15). These factors are normally determined using iron-carbon specimens of high purity. In these steels, the hardenability is very low, which requires that the specimens have small cross-sections in order to develop martensitic structures. At the same time, there is a very critical dependence of the microstructure on the cooling rate in these low hardenability steels. A significant factor also has to be that some of the carbon in the specimens may precipitate to dislocations during the quench, thus lowering the hardness of the martensite.

Another important feature of the ASTM Standard A255 is that it gives a procedure which allows the boron concentration of a steel to be included in the determination of the ideal critical

Table 19.5 Hardenability Multiplying Factors*

Percent	Carbon-Grain Size #7	Mn	Si	Ni	Cr	Mo
0.05	0.026	1.167	1.035	1.018	1.1080	1.15
0.10	0.054	1.333	1.070	1.036	1.2160	1.30
0.15	0.081	1.500	1.105	1.055	1.3240	1.45
0.20	0.108	1.667	1.140	1.073	1.4320	1.60
0.25	0.135	1.833	1.175	1.091	1.54	1.75
0.30	0.162	2.000	1.210	1.109	1.6480	1.90
0.35	0.189	2.167	1.245	1.128	1.7560	2.05
0.40	0.213	2.333	1.280	1.146	1.8640	2.20
0.45	0.226	2.500	1.315	1.164	1.9720	2.35
0.50	0.238	2.667	1.350	1.182	2.0800	2.50
0.55	0.251	2.833	1.385	1.201	2.1880	2.65
0.60	0.262	3.000	1.420	1.219	2.2960	2.80
0.65	0.273	3.167	1.455	1.237	2.4040	2.95
0.70	0.283	3.333	1.490	1.255	2.5120	3.10
0.75	0.293	3.500	1.525	1.273	2.62	3.25
0.80	0.303	3.667	1.560	1.291	2.7280	3.40
0.85	0.312	3.833	1.595	1.309	2.8360	3.55
0.90	0.321	4.000	1.630	1.321	2.9440	3.70
0.95		4.167	1.665	1.345	3.0520	
1.00		4.333	1.700	1.364	3.1600	

*Abstracted from ASTM Standard A255, Table X2.1.

diameter D_I. Boron is an element that can produce a significant increase in hardenability when added to a fully deoxidized steel. Even as small an amount of boron as 0.001 percent can have as strong effect on the hardenability. Its greatest effect is noticed in low-carbon steels. High-carbon steels show a smaller response. The subject of boron will not be considered further here; for further information, consult ASTM A255 (Ref. 16).

As an example of the computation of the ideal critical diameter, D_I, using the Grossman multiplication factors, consider a steel with a carbon content of 0.40 and an austenitic grain size 7. Table 19.5 indicates that for this carbon concentration the carbon multiplying factor, or the *base diameter*, is:

$$D_I = 0.213 \qquad \mathbf{19.2}$$

Suppose now that the steel in question conforms to that classified as an AISI 8640 steel in the American Iron and Steel Institute's system. In this case its composition limits should fall within the following bounds:

carbon	0.38 to 0.43 percent
manganese	0.75 to 1.00 percent
silicon	0.20 to 0.35 percent
nickel	0.40 to 0.70 percent
chromium	0.40 to 0.60 percent
molybdenum	0.15 to 0.25 percent

Let us also assume, for the purpose of a simple illustrative calculation, that the steel under consideration contains 0.40 percent carbon and the maximum concentration of all the other indicated elements. The next step is to find the value of the multiplying factor for each element. This is done in the same manner as is used for finding the base diameter. Thus, opposite 1.00 percent as read in the percent column, there appears in the manganese column the multiplying factor 4.333. The base diameter is multiplied by this factor to give the hardenability of the steel as determined by its austenitic grain size, its carbon content, and its manganese content. In the same manner we find that the multiplying factors for the other elements are

Percentage of Element	Multiplying Factor
1.00% Mn	4.333
0.35% Si	1.245
0.70% Ni	1.255
0.60% Cr	2.296
0.25% Mo	1.75

The total hardenability of the steel in question is obtained by multiplying the base diameter by every one of these factors, or

$$D_I = 0.213 \times 4.333 \times 1.245 \times 1.255 \times 2.296 \times 1.75$$
$$D_I = 5.79 \qquad \mathbf{19.3}$$

The significance of the above numerical value is worth considering. The addition of a total amount of alloying elements equal to less than 3 percent produces a steel with an ideal critical diameter of 5.79 in. Even in a poor water quench ($H = 1.0$), a steel of this hardenability should have a critical diameter of about 5 in. An ordinary carbon steel of the same carbon content (C AISI 1040) contains manganese in the limits 0.60 to 0.90 percent Mn as the only basic alloying element, assuming the maximum manganese content D_I is 0.8 in. and the D (critical diameter) for a poor water quench ($H = 1.0$) is less than $\frac{1}{2}$ in. The importance of the alloying elements in low-alloy steels in developing the hardenability is quite apparent.

Refer again to the allowable compositional limits in the low-alloy steel discussed above. It is interesting to compute its hardenability in terms of the minimum compositional limits rather than the maximum values. When this is done, one obtains a D_I of 2.63 in. It is clear that the hardenability of commercial steels varies within rather wide limits, corresponding to the compositional variations dictated by the problems of manufacture.

The fact that different alloying elements have greatly different effects on hardenability is clearly indicated in Table 19.5. An element that has no effect would have a multiplying factor of unity. For the elements listed in the table, nickel shows the least effect and manganese the greatest. Phosphorus and sulfur, which occur in steels as impurities, are generally considered to have factors of unity.

The hardenability of alloy steels is also visible in isothermal-transformation diagrams. As an example, consider the steel designated AISI 4340. The isothermal transformation diagram for this steel is shown in Fig. 19.10. The hardenability of this steel, for the composition indicated in the figure, is 6.55 in. A significant characteristic of the transformation diagram of this steel is that the pearlite and bainite transformations both exhibit noses. At the upper nose, the diagram shows that the minimum time required to

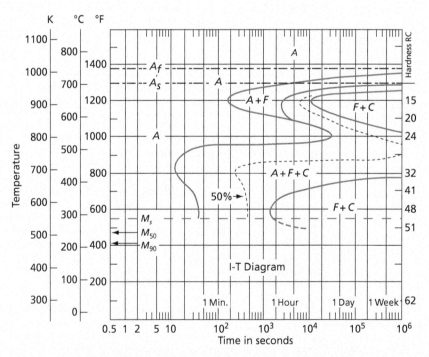

Fig. 19.10 Isothermal transformation diagram of a low-alloy steel (4340): 0.42 percent carbon, 0.78 percent manganese, 1.79 percent nickel, 0.80 percent chromium, and 0.33 percent molybdenum. Grain size 7–8. Austenitized at 1550°F (843°C). (From *Atlas of Isothermal Transformation and Continuous Cooling Diagrams*, American Society for Metals, Metals Park, Ohio 44073, 1977. Reprinted with permission of ASM International®. All rights reserved. www.asminternational.org)

form a visible amount of proeutectoid ferrite is about 200 s (650°C), and just below this temperature the minimum time to form pearlite is somewhat more than 1800 s (30 min). In the same manner, the minimum time for the formation of a visible amount of bainite is slightly over 10 s at 450°C. This transformation diagram should be compared with those for plain-carbon steels given earlier (Figs. 18.30, 18.37, and 18.38 in Chapter 18). The difference in the rapidity with which the transformations occur is quite evident.

The continuous cooling transformation for the AISI 4340 steel is shown in Fig. 19.11. It is apparent in the figure that any cooling rate that brings the steel to room temperature in less than 90 s produces a martensitic structure. The effect of the high hardenability is also clearly shown in the corresponding Jominy curve for this steel, which is shown in Fig. 19.12. At a distance greater than 2 in. from the quenched end of the bar the structure is still 95 percent martensite. Note that Fig. 19.11 does not include pearlitic transformation shown in Fig. 19.10 because pearlitic transformation take place at much slower cooling rates.

The isothermal diagram of Fig. 19.10 is characteristic of those steels in which bainite can be obtained during continuous cooling. In the diagram for a plain carbon steel considered earlier, measurable amounts of bainite do not form on continuous cooling because the pearlite transformation region extends over the corresponding bainite region. In the present alloy steel, the bainite nose extends beyond the pearlite nose, thus making possible the formation of bainite on

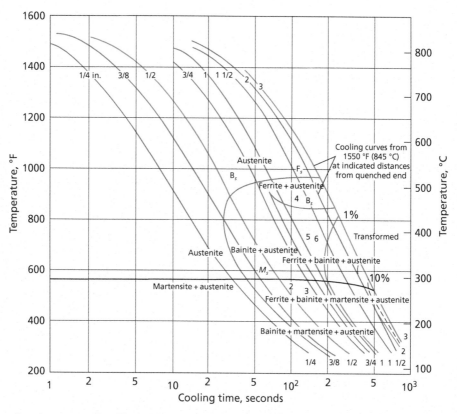

Fig. 19.11 Continuous cooling diagram for 4340 steel. (From *Heat Treaters Guide*, American Society for Metals, Metals Park, Ohio, 1982. Reprinted with permission of ASM International®. All rights reserved. www.asminternational.org)

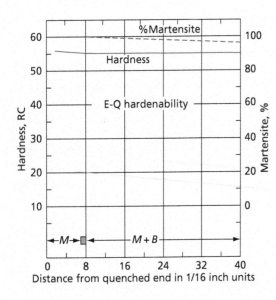

Fig. 19.12 Jominy hardenability curve for 4340 steel. (From *Atlas of Isothermal Transformation and Continuous Cooling Diagrams*, American Society for Metals, Metals Park, Ohio 44073, 1977. Reprinted with permission of ASM International®. All rights reserved. www.asminternational.org)

continuous cooling. The possible structures that can be obtained in this alloy with different cooling rates are shown at the bottom of the diagram in Fig. 19.12.

19.8 The Significance of Hardenability

Is high hardenability desirable in a steel? The answer is that it is not always desirable. This is especially true when steels are to be welded. Steels containing any appreciable amount of added elements are notoriously difficult to weld successfully. In making a weld, two pieces of steel are joined by casting a section of molten metal between them. This operation naturally heats the metal adjacent to the weld and, for some distance to each side of the weld center, steel is raised into the austenitic region. If the hardenability of the metal is high, hard brittle martensite may form on cooling to room temperature. This is accentuated by the quenching effect of the cold steel surrounding the heat-affected zone, for heat flows rapidly from the heated region into the surrounding metal. Because of the above considerations, structural steels, intended for construction of bridges, buildings, and ships, are usually designed to have moderate hardenabilities.

High hardenabilities are desirable in steels that are to be hardened as one of the final steps in their manufacture. It is generally believed that the best combination of physical properties (strength plus ductility) is obtained if the metal is transformed completely to martensite during the hardening treatment. In some cases, if the M_f temperature is below room temperature, this may require that the metal be cooled to subzero temperature in order to remove most of the residual austenite. However, the primary requirement for obtaining a martensitic structure in a given steel object is that the steel have a sufficiently high hardenability to harden under the required quenching conditions. Unfortunately, hardenability is associated with alloying elements in the steel, whose presence raises its cost. As a result, the basic problem is economic and requires the avoidance of the use of steel with too high a hardenability (and too high a cost) for the job at hand. An important factor in determining the required hardenability is the speed of quench that must be used. Of course, the more rapid the quench, the lower the required hardenability, but high cooling rates are also associated with severe thermal shocks. These can cause quench cracks and warping of the finished object, leading to its rejection. Thus, in commercial practice, oil and even air quenches are commonly used in order to reduce damage to steel parts caused by the more rapid quenching processes. The added cost of the high-hardenability steel in these cases is justified by the savings in the number of parts rejected.

19.9 The Martensite Transformation in Steel

The martensitic structure in steels, in the ideal case, is a simple phase that marks it from the aggregates of ferrite and carbides, which we call pearlite and bainite. The martensitic crystal structure is body-centered tetragonal and can be assumed to be an intermediate structure between the normal phases of iron—face-centered cubic and body-centered cubic. The relationship between the three structures has already been discussed in Chapter 17. The Bain distortion in steel is shown in Fig. 19.13. In these drawings, the positions that carbon atoms occupy are shown by black dots. It must be recognized, however, that actually in any given steel specimen only a very small percentage of the possible positions are ever filled. In the face-centered cubic structure there are as many possible positions for carbon atoms as there are iron atoms. This means that if all positions were filled, the alloy would have a composition containing 50 atomic percent carbon. The maximum actually observed is 9.1 atomic percent (2.11 weight percent). Figure 19.13A represents face-centered cubic austenite. In this structure, the carbon atoms occupy the midpoints of cube edges and the cube centers. These are equivalent positions, for in each case a carbon atom finds itself located between two iron atoms along a $\langle 001 \rangle$ direction. The equivalent positions in the austenite, when it is considered a body-centered tetragonal structure, are shown in Fig. 19.13B. Notice that in this cell the carbon positions occur between iron atoms along c-axis edges and in the centers of the square faces at each end of the prismatic cell. Finally, the martensitic structure is shown in Fig. 19.13C. In this last case, the tetragonality of the cell is greatly reduced, but the carbon atoms are still in the same relative positions with respect to their iron-atom neighbors as in the austenite unit cell. The

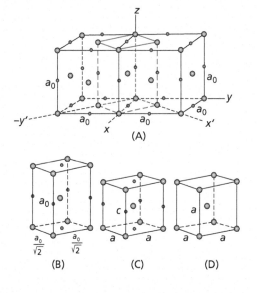

Fig. 19.13 The Bain distortion in the martensite transformation of steels. Black dots represent positions that carbon atoms can occupy. Only a small fraction are ever filled. (A) Face-centered cubic. (B) Tetragonal representation of austenite. (C) Tetragonal martensite. (D) Body-centered cubic

resulting structure is tetragonal only because carbon atoms are inherited from the austenite, and the transformation, which would normally proceed to body-centered cubic, is not able to go to completion. The carbon atoms can be considered to strain the lattice into the tetragonal configuration, and the extent of the tetragonality that occurs may be deduced from Fig. 19.14. Notice that the lattice parameters are plotted as a function of carbon content in both austenite and martensite, and, in each case, the parameters vary linearly with carbon content. In martensite with increasing carbon content the c-axis parameter increases, while the parameter associated with the two a axes decreases. At the same time, the cubic parameter of austenite (a_0) increases with increasing carbon content. These relationships may be expressed in terms of simple equations where x is the carbon concentration.

Martensite parameters (nm) are given in the following equation in nm (Ref. 17).

$$c = 0.2866 + 0.0166x$$
$$a = 0.2866 - 0.0013x$$
19.4

More recent work (Ref. 18) gives slightly different equations as

$$c = 0.28861 + 0.0115x \text{ and}$$
$$a = 0.28661 + 0.00124x$$

Austenite parameter (nm) as a function of carbon concentration is given as (Ref. 19):

$$a_0 = 0.3555 + 0.0044x$$
19.5

A simple computation of the c/a ratio at 1.0 percent carbon yields 1.045. This number should be compared to the corresponding ratio when face-centered cubic austenite is considered as a tetragonal lattice. As mentioned previously, this ratio is 1.414, so that in most steels (less than 1.00 percent C), the martensitic lattice is certainly much closer to body-centered cubic than to face-centered cubic.

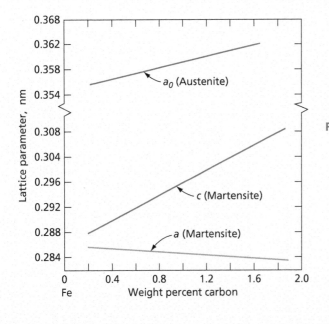

Fig. 19.14 Variation of the lattice parameters of austenite and martensite as a function of carbon content. (Roberts, C. S., *Trans. AIME*, **197** 203 [1953])

Table 19.6 Martensite Transformations in Steel.

Carbon Content	Habit Plane	Orientation Relationship
0–0.4%	$(557)_A$	?
0.5–1.4%	$(225)_A$	$(111)_A \, (101)_M$ $[1\bar{1}0]_A \, [11\bar{1}]_M$
1.5–1.8%	(259)	?

The lattice parameter curves of Fig. 19.14 show that the tetragonality of martensite varies with the carbon content. This should have a bearing on the crystallographic characteristic of the transformations. With reference to the Wechsler, Lieberman, and Read theory of martensite formation, a change in tetragonality of the product phase means a difference in magnitude of the Bain distortion. This, in turn, implies a difference in the size of the required shear and rotation. As a result, it might be expected that both the habit plane and the orientation relationship between the parent and product should vary with carbon content. This is actually observed, as may be seen in Table 19.6.

The best data in Table 19.6 relates to the concentration range between 0.5 and 1.4 percent C, which fortunately includes almost all of the commercially important carbon steels that are hardened to martensite. In this range of carbon contents, the habit plane of the martensite plates is very close to a {225} plane in the austenite. Since there are twelve {225} planes and two possible twin-related orientations of the martensite for each habit plane, there are 24 possible ways that an austenite crystal can form a martensite plate. The corresponding orientation relationship between the lattices in the martensite and in the austenite of these steels is known as the *Kurdjumov-Sachs relationship*. It states that the (101) plane of the martensite is parallel to the (111) plane of the parent austenite, and, at the same time, the [11$\bar{1}$] direction in the martensite is parallel to the [1$\bar{1}$0] direction of the austenite.

The martensite plates that form in the middle carbon region of steels are quite similar in appearance to those that occur in iron-nickel alloys. They are, accordingly, small in size and lenticular in shape. The martensitic transformation in iron-carbon alloys is primarily athermal. However, small amounts of austenite may be transformed to martensite isothermally. A typical curve showing the formation of athermal martensite as a function of temperature is shown in Fig. 19.15. The similarity of this curve to that for the formation of martensite on cooling given for other alloys in Chapter 17 is apparent. However, one important difference should be observed. There is no curve

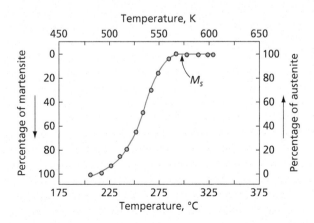

Fig. 19.15 The formation of martensite in a 0.40 percent carbon low-alloy steel (2340) as a function of temperature. (After Grange, R. A., and Stewart, H. M., *Trans. AIME*, **167** 467 [1946])

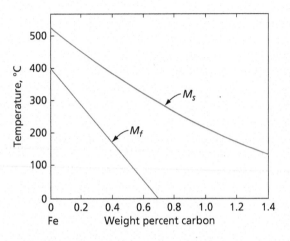

Fig. 19.16 Variation of M_s and M_f with carbon concentration in steel. (After Troiano, A. R., and Greninger, A. B., *Metal Progress*, **50** 303 [1946])

given for the reverse transformation. The reason for this is quite simple: the martensitic reaction in steels is not reversible. Iron-carbon martensite represents an extremely unstable structure with a high free energy relative to the more stable phases—cementite and ferrite. The entrapment of carbon atoms in what should be a body-centered cubic structure, capable of holding at equilibrium an infinitesimal amount of carbon, produces a lattice with a high internal strain. Even moderate reheating promotes its decomposition. The study of the phenomena of martensite decomposition will be considered in the section entitled "Tempering."

Both the M_s (martensite start) and M_f (martensite finish) temperatures in steel are functions of the carbon content, as shown in Fig. 19.16. The M_f temperature, however, is usually not clearly defined. By this it is meant that the martensite reaction can never be theoretically complete even at absolute zero temperature. The transformation of the last traces of austenite becomes more and more difficult the smaller the total amount of austenite remaining. Curves such as the M_f line shown in Fig. 19.16 are based on visual estimations of the structures of metallographic specimens and small amounts of retained austenite are very difficult to measure in a structure composed of many small overlapping martensite plates. The M_f temperature shown in Fig. 19.16 should thus be interpreted as the temperature at which the reaction is completed as far as one can determine by visual means. One technique for the measurement of retained austenite in quenched steels involves quantitative X-ray diffraction measurements and is capable of measuring retained austenite in amounts as small as 0.3 percent. This method has been used to determine the amount of retained austenite in carbon steels quenched to room temperature. The results are shown in the curve at the bottom of Fig. 19.17. This diagram also includes a plot of the M_s temperature versus carbon concentration as well as one showing the volume fraction of the low temperature form of martensite, lath martensite, with carbon concentration. A corresponding curve showing the variation of the volume fraction of the high temperature form, lenticular or twinned martensite, with carbon concentration appears in Fig. 19.22. Notice in Fig. 19.17 that the martensite finish temperature, M_f, occurs at room temperature (20°C) near 0.6 percent C. The amount of retained austenite under these conditions is over 3 percent.

Substitutional alloying elements in steels also affect the martensitic transformation. This may be reflected in the habit plane indices and the orientation relationships between the martensite and the parent austenite. Unfortunately, there is very little data available on these effects. The influence of alloying elements on the M_s temperature is much more easily recognized and measured (Refs. 20 and 21). As a typical example, consider the addition of manganese to steels containing carbon where a very decided lowering of the M_s temperature occurs. This is shown in

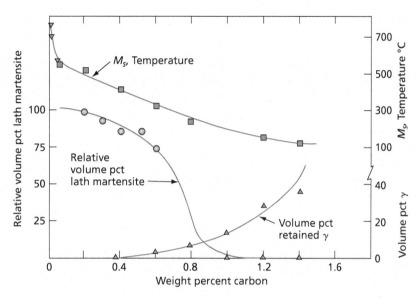

Fig. 19.17 The effect of the carbon concentration on the relative fraction of lath martensite, the M_s temperature, and the volume fraction of retained austenite. (Adapted from Speich, G. R. and Leslie, W. C., *Met. Trans.*, **3** 1043 [1972] with kind permission from Springer Science and Business Media)

Fig. 19.18. One of the interesting features of this set of curves is the very rapid decline of the M_s temperature with increasing manganese content in steels that also contain 1.0 percent carbon. For those compositions containing more than 6 percent manganese, the M_s is so far below room temperature that we can assume that a quench to room temperature produces a structure that remains austenitic indefinitely at room temperature. The famous Hadfield manganese steel (10–14 percent Mn and 1–1.4 percent C) takes advantage of this fact to produce a steel with an austenitic structure that work hardens very rapidly and has an initial very high strength due to the presence of the manganese and carbon in solid solution. This combination of properties makes for a very tough, hard, abrasion-resistant metal. A typical application is in the buckets and teeth of power shovels.

The effect of other solid-solution elements on the M_s temperature varies with the element concerned. Manganese has the strongest effect followed by chromium. All common alloying elements (substitutional) lower the M_s temperature, except for cobalt and aluminum, which raise it.

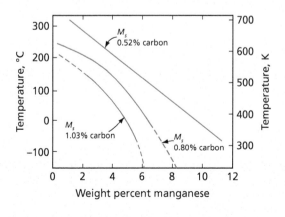

Fig. 19.18 Variation of M_s with manganese content in three series of steels with different carbon concentrations. (After Russell, J. V., and McGuire, F. T., *Trans. ASM*, **33** 103 [1944])

642 Chapter 19: The Hardening of Steel

19.10 The Hardness of Iron-Carbon Martensite

The hardness of high purity iron-carbon martensites is plotted against carbon concentration in Fig. 19.19. This diagram includes data from 10 different sources. The Vickers diamond pyramid hardness scale appears on the left of the figure and is designated (DPH). Recently, it has become the practice to designate this type of hardness number by the symbol HV or hardness Vickers. On the

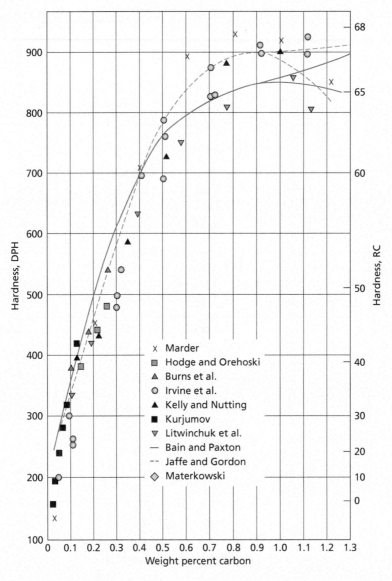

Fig. 19.19 A summary of the experimental data corresponding to the hardness of high purity iron-carbon martensites as a function of their carbon concentrations. (References for the sources of the original data may be obtained from the following source: Krauss, G., *Principles of Heat Treatment of Steels*, American Society for Metals, Metals Park, Ohio, 1980)

right-hand ordinate axis the corresponding Rockwell-C hardness numbers are also given. Note that at lower carbon concentrations, the data from all sources are in reasonable agreement. However, above 0.6 percent carbon an appreciable scatter occurs in the data. Above this concentration the martensite finish temperature falls below room temperature. It is, therefore, reasonable to assume that the high carbon scatter of the data may be associated with differing amounts of retained austenite resulting from a difference in the quenching rates used by the various authors of the data in Fig. 19.19. If this is true, then the upper bound of the data should be more representative of the hardness of the martensite. Note that this boundary tends to rise continuously with increasing carbon concentration. A schematic diagram consistent with Fig. 19.19 but that eliminates the data points is shown in Fig. 19.20. Finally, the effect that retained austenite can have on the hardness of very high carbon steels is clearly shown in Fig. 19.21. This plot shows that for carbon concentrations in excess of 1.0 percent, the hardness of carbon steels, brine-quenched to room temperature, falls continuously with increasing carbon concentration. This is the result of a corresponding increase in the amount of retained austenite. In support of this assumption, as may be seen in Fig. 19.16, the M_s temperature approaches room temperature for carbon concentrations above 1.4 percent.

The hardness of martensite results from the presence of carbon in the steel. In this respect, it is significant that the martensite product in alloys other than steels are not necessarily hard. It is also evident from Fig. 19.19 that an appreciable amount of carbon (about 0.4 percent) is needed in steel to cause a marked degree of hardening. In order to obtain a truly hardened steel, the two

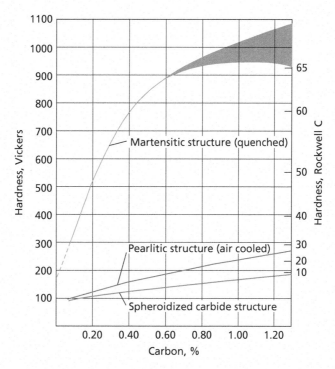

Fig. 19.20 The hardness of martensitic steels as a function of their carbon concentration. The cross-hatched area shows the effect of retained austenite. The hardness of steels with pearlite (plus ferrite) and spheroidized cementite structures are also shown. (From Krauss, G., *Principles of Heat Treatment of Steels*, American Society for Metals, Metals Park, Ohio, 1980. Reprinted with permission of ASM International®. All rights reserved. www.asminternational.org)

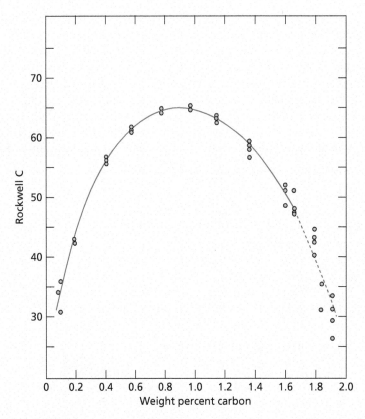

Fig. 19.21 The Rockwell-C hardness versus carbon concentration curve for brine-quenched steels. (With kind permission from Springer Science + Business Media: From Lipwinchuk, A., Kayser, F. X., and Baker, H. H., *Jour. Mat. Sci.*, **11** 1200 [1976])

required factors are: first, an adequate carbon concentration in the metal, and second, rapid cooling to produce a martensitic structure.

Experimental work based on transmission electron microscopy has clearly shown that the martensite in carbon steels can form by two reactions. One gives a structure known as *lath martensite*; the other is a *lenticular martensite* that is internally twinned. The primary factor controlling the relative volume fractions of these two forms is apparently the transformation temperature. A lower transformation temperature favors a higher concentration of the twinned lenticular martensite. Since increasing the carbon concentration generally lowers the M_s temperature, higher carbon steels tend to have large volume fractions of the twinned component. This is demonstrated in Fig. 19.22. On the other hand, in low-carbon steels the martensite is primarily of the lath type.

The lath martensite is characterized by a high internal dislocation density of the order of 10^{15} to 10^{16} m^{-2}. These dislocations are arranged in plate-shaped cells. On the other hand, the twinned martensite normally does not contain a high density of dislocations. The observed two basic forms of martensite are consistent with an assumption that in the lath martensite the macroscopic shear is accomplished by slip, while in the lenticular martensite it occurs by twinning (see Figs. 17.30 and 17.31). The difference in the defect concentration between the two martensite forms has an important bearing on the respective distribution of the carbon atoms inside them. In lath martensite, the carbon atoms tend

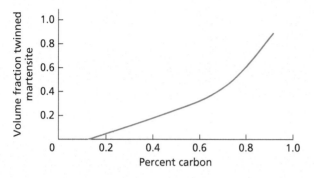

Fig. 19.22 Curve showing the volume fraction of twinned martensite as a function of the carbon concentration. (From Speich, G. R., *TMS-AIME*, **245** 2553 [1969])

to diffuse and segregate around the dislocations. Even in a very rapid quench there is ample time for this diffusion to occur. On the other hand, in the twinned structure where there is a much lower density of dislocations, the carbon atoms are forced to occupy normal interstitial sites.

As may be seen in Fig. 19.19, the hardness of martensite for pure iron is about 200 HV. An annealed pure iron has a hardness of the order of 100 HV. The high hardness of martensitic pure iron is caused by the substructural strengthening effect of the finely spaced cell walls and lath boundaries in the lath martensite. In effect, this is attributing the hardness to the high dislocation density in this structure.

In iron containing carbon there is additional hardening due to the carbon. In this case the carbon is believed to increase the hardness through its interaction with the dislocations. This can occur by segregation of the dislocations in the cell walls and by solid-solution strengthening. With regard to this latter effect, the martensitic transformation can be viewed (Ref. 22) as trapping an abnormally high concentration of carbon atoms in solid solution, thereby producing a large hardening component.

A very important practical fact about the hardness of martensite in steels is that in all of the so-called low-alloy steels (less than about 5 percent total alloying elements) the hardness of martensite can be assumed to depend only on the carbon concentration of the metal. Consequently, if the carbon content of any low-alloy steel is known, the approximate hardness of the steel when it has a martensitic structure can be determined (with the aid of a curve such as Fig. 19.20).

The hardness of a steel containing some fixed fraction of martensite, say 50 percent, is also approximately a function of only its carbon content. In this respect, Fig. 19.23 shows the variation of the Rockwell-C hardness with carbon content in steels containing 50 percent martensite structures. In practice, the hardness of commercial steels usually lies within ± 4 Rockwell-C hardness units of this curve.

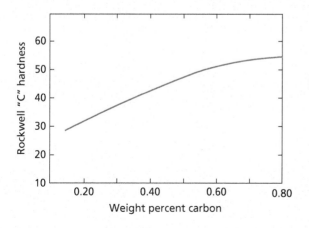

Fig. 19.23 Average hardness of 50 percent martensite structures in low-alloy steels. (After Hodge, J. M., and Orehoski, M. A., *AIME*, TP 1800, 1945)

19.11 Dimensional Changes Associated with Transformation of Martensite

When austenite transforms to martensite, there is a change in volume that can be computed by considering the Bain distortion and the lattice parameters of austenite and martensite. With reference to Eqs. 19.4 and 19.5 and considering a 1 percent carbon steel, in the austenite the lattice parameter is

$$a_0 = 0.3555 + 0.0044(1.0) = 0.3599 \qquad \text{19.6}$$

and the volume of the austenite unit cell (tetragonal form) is

$$V_A = a_0 \cdot \frac{a_0}{\sqrt{2}} \cdot \frac{a_0}{\sqrt{2}} = \frac{(0.3599)^3}{\sqrt{4}} = (0.0233 \text{ nm}^3)$$

In martensite, the lattice parameters are:

$$a = 0.2866 - 0.0013(1.0) = 0.2853$$
$$c = 0.2866 + 0.0116(1.0) = 0.2982 \qquad \text{19.7}$$

The volume of the martensite unit cell is

$$V_M = c \times a \times a = 0.2982(0.2853)^2 = 0.0243 \text{ nm}^3$$

The change in volume is, accordingly,

$$\Delta V = V_M - V_A = 0.0243 - 0.0233 = 0.0010 \text{ nm}^3$$

and the relative change in volume, assuming martensite to form from austenite at room temperature, is

$$\frac{\Delta V}{V_A} = \frac{0.0010}{0.0233} = 4 \text{ percent}$$

When a 1 percent carbon steel transforms to martensite, there is a volume increase of about 4.0 percent that can be considered to be an average value, representative of steels in general, that does not vary widely with the carbon content. This is because we are transforming from austenite with a *c/a* ratio of 1.414 to martensite with *c/a* ratios lying between 1.0 and 1.090, corresponding to the maximum range of carbon contents (0 to 2 percent C).

Because of the many orientations that martensite plates can take in a single austenite crystal, it can be assumed that the volume expansion is isotropic in a specimen of sufficient size. Length changes may, accordingly, be used to measure the deformation associated with the martensitic reaction. In this respect, as is shown in calculus, a small isotropic length change is approximately equal to one-third the corresponding volume change. Therefore,

$$\frac{\Delta l}{l} = \frac{\Delta V}{3V} = \frac{4.0 \text{ percent}}{3} = 1.3 \text{ percent} \qquad \text{19.8}$$

19.12 Quench Cracks

When a piece of steel is cooled in such a way as to form martensite, two basic dimensional changes occur. First, there is the normal thermal contraction due to cooling, but superimposed on this is the expansion of the metal as it transforms from austenite to martensite. Under the right conditions, these volumetric changes can produce very high internal stresses. If these stresses become large enough, they can produce plastic deformation and the steel will be deformed or warped. While plastic deformation tends to reduce the severity of the quenching stresses, the degree to which this is accomplished depends on a number of factors, and it is quite possible to have large enough residual stresses remaining in the metal to actually cause rupture. These localized fractures are called *quench cracks*.

The approximate nature of the residual stress pattern that develops in an actual steel shape on quenching to martensite will now be considered. For this purpose, steel specimens in the form of round cylinders will be considered. When quenched, the surface always cools faster than the center and undergoes the martensitic transformation first, which hardens the surface relative to the center. Supplementing this hardening is the fact that the yield strength, or plastic flow stress of the metal, increases with decreasing temperature. Now, whether or not the surface will be set in tension relative to the center of the bar depends on the sign of the net volumetric change that occurs in the interior of the bar after the surface has hardened. If the expansion in this region, due to the martensitic transformation, is larger than the remaining thermal contraction, the surface will attain a residual state of tension. The center will, of course, remain in a state of compression and the nature of this stress pattern is implied in Fig. 19.24. The reversed stress pattern, in which the surface ends in a state of residual compressive stress and the center in a state of tensile stress, occurs when the thermal contraction in the center of the bar subsequent to the hardening of the surface exceeds the martensitic expansion. Which of these two basic stress distributions eventuates depends on the relative cooling rates at the surface and at the center of the bar. This, of course, is a function of both the size of the bar and the speed of the quench. When the product of these two variables is large (large diameter and fast cooling), the surface hardens although the center is still at a very high temperature. The magnitude of the thermal contraction in this case is large and determines the sign of the volume change in the central regions of the bar. In other words, thermal contraction usually exceeds the expansion of the martensite transformation. When the difference between the cooling rates at the surface and center are only moderate, the center is at a temperature only slightly above the surface when the latter hardens. The thermal contraction in the central areas subsequent to the hardening of the surface is then smaller than the expansion due to the formation of martensite.

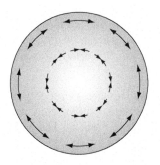

Fig. 19.24 Schematic tangential stress pattern in a quenched cylinder when the surface is left in a state of tension and the center in compression

Several factors help to determine the magnitudes of the residual stresses (Refs. 23 and 24). It should be noted that residual stresses can be transient (developing during cooling) or locked-in stresses remaining within the body (Ref. 25). It was shown in the previous section that the expansion that occurs when steel transforms to martensite at room temperature involves a dilation of about 1.3 percent. Actually, during a quench, austenite does not transform completely to martensite at room temperature, but over a range of temperatures starting at the M_s temperature. It is also true that the higher the temperature at which the transformation occurs, the less the expansion due to the formation of martensite. This follows from a corresponding change in the austenite and martensite lattice parameters. In steels in which the M_s temperature is high, the specific-volume changes are smaller and, as a consequence, there is a reduced tendency to form quench cracks. High-carbon steels and those containing alloying elements that lower M_s are, conversely, more subject to quench cracking. In addition to lowering M_s, high-carbon contents also increase the danger of cracking because of the increased hardness or brittleness that high-carbon concentrations impart to the steel.

The length of time that a steel is maintained at temperatures of the order of room temperature to 100°C after quenching also has an effect on the formation of quench cracks: the longer the time, the more liable the formation of cracks. Several different explanations of this phenomenon have been offered. One of the most attractive is related to the isothermal transformation of retained austenite to martensite. The formation of isothermal martensite adds an additional volumetric strain to an already badly strained metal, thereby increasing the probability of crack formation, distortion, or deformation.

19.13 Tempering

Steels that have undergone a simple hardening quench are usually mixtures of austenite and martensite, with the latter constituent predominating. Both of these structures are unstable and slowly decompose, at least in part, if left at room temperature; the retained austenite transforms to martensite and the martensite undergoes a reaction that will be described shortly. Since specific-volume changes are connected with both reactions, hardened steel objects undergo dimensional changes as a function of time when left at room temperature. Of more general importance is the fact that a structure that is almost completely martensite is extremely brittle and is also very liable to develop quench cracks if aged at room temperature. These factors lead to the conclusion that steels with a simple martensitic structure are of little useful value, and a simple heat treatment called *tempering* is almost always used to improve the physical properties of quenched steels. In this treatment, the temperature of the steel is raised to a value below the eutectoid temperature and held there for a fixed length of time, after which the steel is cooled again to room temperature. The obvious intent of tempering is to allow diffusion processes time to produce both a dimensionally more stable structure and one that is inherently less brittle.

A number of basic phenomena that occur during the decomposition of martensite have been identified. These are as listed next (Ref. 26) and tend to occur in the order presented. However, many of the phenomena tend to overlap to some degree.

- (a) A redistribution of the carbon atoms in the martensite. This occurs roughly between room temperature and 100°C. There are several ways that this redistribution may occur:
    - (i) A segregation of the carbon atoms to lattice defects such as dislocations and twin boundaries.

(ii) A clustering of the carbon atoms, which in turn can occur in several ways (Refs. 27 and 28) and involve spinodal decomposition and ordering.
(b) The precipitation of a transition carbide or carbides. The most generally identified transition carbide is epsilon, ε, carbide. However, in many cases this carbide is not well resolved and it is sometimes referred to as (ε/η) *carbide*, that is, epsilon or eta carbide. The primary difference is that epsilon carbide is hcp and eta carbide is orthorhombic. (See Table 19.7.) The precipitation of the transition carbide leaves a matrix of iron still containing the carbon atoms that were segregated to the dislocations.
(c) The decomposition of retained austenite into a mixture of ferrite and cementite. This structure has often been called *bainite* or *secondary bainite*.
(d) The conversion of the transition carbide and segregated carbon into small rod-shaped cementite particles.
(e) Spheroidization of the rod-shaped cementite to reduce the surface energy of the particles.
(f) Recovery of the ferrite structure.
(g) Recrystallization of the ferrite structure.
(h) Ostwald ripening of the cementite. In this process, the larger cementite particles grow at the expense of the smaller particles. This further reduces the surface energy and is accomplished by the diffusion of carbon atoms through the iron matrix.

Classically, it has been the custom to talk about the three stages of tempering. The first stage was the precipitation of the transition carbide. It is, thus, the same as (b) in the above compilation of phenomena. The second stage was the decomposition of the retained austenite, that is, (c) above. Finally, the third stage corresponded to the formation of cementite, (d) above. The difficulty of the three stages of tempering classification is that as time goes on, additional stages are discovered. Because of this, various authors have tended to use stage notations not

Table 19.7 Crystallographic data for phases of interest in the tempering of martensite. (From Cheng, L., Brakman, C. M., Korevaar, B. M., and Mittemeijer, E. J., *Met. Trans. A.*, **19A** 2415 [1988] with kind permission of Springer Science and Business Media.)

Phase	Structure	Lattice Parameters (Å)	Number Fe Atoms per Unit-Cell
Martensite	bct	$a = 0.28664 - 0.0013$ wt pct C	2
		$c = 0.28664 + 0.0116$ wt pct C	
Ferrite	bcc	$a = 0.28664$	2
Austenite	fcc	$a = 0.3555 + 0.0044$ wt pct C	4
ε-carbide	hex.	$a = 0.2735$	2
		$c = 0.4335$	
η-carbide	orthorhombic	$a = 0.4704; b = 0.4318$	4
		$c = 0.2830$	
Cementite	orthorhombic	$a = 0.45234; b = 0.50883$	12
		$c = 0.67426$	

consistent with the original three-stage concept. This should be recognized when reading papers on the subject of tempering.

To fully grasp the subject of the tempering of martensite, it must be remembered that there are two basically different forms of martensite: the lath martensite that forms primarily in low-carbon steels and the lenticular martensite that is the primary form in high-carbon steels. As may be seen in Fig. 19.17, the lath martensite is the dominant morphological feature in steels with less than about 0.5 percent C. By Fig. 19.23, the lenticular martensite dominates above 0.9 percent C. In the intermediate range between 0.5 and 0.9 percent C one finds mixed structures containing finite fractions of both types of martensite. Because of differences associated with the tempering of the high- and low-carbon containing martensites, it is perhaps best to consider the tempering of high-carbon and low-carbon steels separately. With this in mind, we shall first consider the tempering of a high purity, high-carbon steel containing 1.13 percent carbon. A recent comprehensive analysis of the tempering of such a steel has been made (Ref. 29). As may be seen in Fig. 19.16, the M_f temperature of this steel lies well below room temperature. Consequently, it was observed that brine quenching of this steel to room temperature from a temperature (842°C) in the austenite range resulted in a structure containing 15 percent of retained austenite. However, a subsequent quench into liquid nitrogen (77 K or −196°C) reduced the volume fraction of retained austenite to 6 percent. The data that will now be considered comes from specimens which were given the double (brine and liquid nitrogen) quenching treatment. The tempering reactions were studied in these specimens using both calorimetric and dilatometric (change in length) measurements. These were supplemented by microhardness measurements. A differential scanning calorimeter (DSC) was used to obtain the thermal data and a thermal mechanical analyzer was used in obtaining the change in length data. Recent improvement (Refs. 29 and 30) in these instruments has made them valuable for the study of the kinetics of tempering reactions under isochronal conditions; that is, where the specimens are annealed (heated) at a constant rate. This type of annealing is also known as *non-isothermal annealing*. It should also be mentioned that the non-isothermal investigations were supplemented by some isothermal experiments.

Figure 19.25 shows the isochronal dilation curve obtained using a double quenched specimen heated from −196°C (77 K), the temperature of liquid nitrogen, to 450°C at a rate of 10°C/min. In considering this curve, it should be noted that the normal thermal expansion of the steel, in the absence of a tempering thermal reaction, gives a nearly linear curve that rises steeply as the temperature is increased. That part of the curve shown in Fig. 19.25 between the regions marked Stage 1 and Stage 2 represents a temperature interval in which there is no apparent tempering reaction and thus, its slope is due only to the normal thermal expansion of the steel specimen. Deviations from this slope such as occur in the 5 temperature intervals designated as Stages 1 to 5, represent regions in which the specimen also underwent length changes due to martensitic tempering reactions. The slope of the experimental curve in these regions is accordingly determined by both the thermal expansion of the specimen and the changes in length associated with the thermal reactions. As may be seen in Fig. 19.25, the slope in Stage 1 is steeper than that due to thermal expansion alone. The tempering reaction in this region causes an expansion of the specimen. It was concluded that this expansion was due to the transformation of a part of the austenite that was retained as a result of the double quench to −196°C. It can also be concluded from Fig. 19.25 that relatively large contractions occurred in Stages 2, 3, and 5.

A corresponding differential thermal analysis (DTA) curve for a double quenched 1.13 percent carbon steel is shown in Fig. 19.26. In this case, a similar fully tempered specimen was heated at a rate of 20°C/min, along with the doubly quenched martensitic specimen, and the difference in temperature between the two specimens was recorded. This temperature difference was then plotted

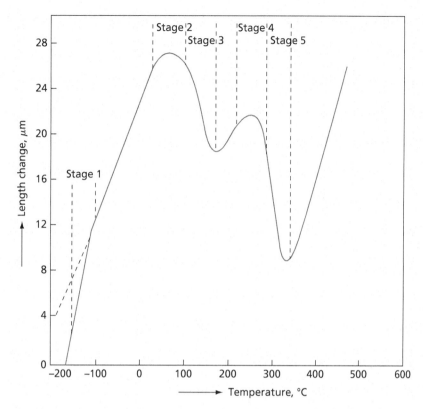

Fig. 19.25 Dilatomer curve of a 1.13 percent carbon steel brine quenched from 842°C to room temperature and then, within 6 min, quenched into liquid nitrogen at −196°C. After the double quench, the change in length of the 10.5 mm long specimen was monitored as it was heated to 450°C at a constant rate of 10°C/min. (Abstracted from Cheng, L., Brakman, C. M., Korevaar, B. M., and Mittemeijer, E. J., *Met. Trans. A*, **19A** 2415 [1988])

against the temperature. In an experiment of this type, the heat released in the martensitic specimen as it undergoes a tempering reaction causes a temperature difference between the two specimens. The undulating solid line in Fig. 19.26 represents the temperature difference between the specimens, while the dashed lines below it show that the experimental curve is resolvable into a set of peaks representing stages of tempering. Four of these stages correlate well with those of the dilatometric curve in Fig. 19.25. Since the data in Fig. 19.26 covered only the interval between room temperature and 450°C, Stage 1 was not represented. An additional peak, marked Stage X in the figure, was observed. The basis for this peak will be discussed presently.

Activation energies associated with the various tempering processes were deduced in terms of the dependence of the data pertaining to the stages upon the rate of heating. This required a number of isochronal annealing experiments carried out at heating rates between 5 and 40°C per minute. The procedure for doing this is covered in the original papers (Refs. 31 and 32). As a result of an analysis of all the dilatometric, calorimetric, and hardness data obtained both isochronally and isothermally, the information displayed in Table 19.8 was deduced. Note that in this table, Stage X in Fig. 19.27 is suggested as being possibly due to the precipitation of the Haggs carbide,

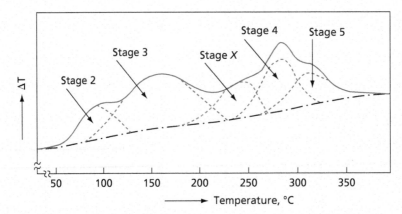

Fig. 19.26 Differential thermal analysis (DTA) curve for a specimen like that in Fig. 19.25. The heating rate in this case was 20°C/min. Data were recorded between room temperature and 450°C. ΔT is the temperature difference between the martensitic specimen and a stable reference specimen. (Abstracted from Cheng, L., Brakman, C. M., Korevaar, B. M., and Mittemeijer, E. J., *Met. Trans. A*, **19A** 2415 [1988])

Table 19.8 Stages in the Tempering of a Martensitic 1.13 Percent Carbon Steel. (Abstracted from Cheng, L., Brakman, C. M., Korevaar, B. M., and Mittemeijer, E. J., *Met. Trans. A.*, **19A** 2415 [1988] with kind permission of Springer Science and Business Media.)

Stage	Temperature Interval, °C	Process	Activation Energy, J/mol	Rate Controlling Mechanism
1	−180 to −100	Transformation of retained austenite to martensite.		Diffusionless transformation.
2	Below +100	Redistribution of carbon atoms by (a) segregation to lattice defects and (b) clustering of carbon atoms.	~80	Volume diffusion of the carbon atoms.
3	80 to 100	Precipitation of (ε/η) transition carbide.	~120	Pipe diffusion of iron atoms along dislocations to accommodate the volume misfit between the transition carbide particles and the iron matrix.
4	240 to 320	Decomposition of the remaining retained austenite to ferrite and cementite.	~130	Volume diffusion of carbon atoms in the austenite.
5	260 to 350	Conversion of the segregated carbon and the transition carbide to cementite.	~200	Combined volume and pipe diffusion of iron atoms.
X	200 to 270	Precipitation of Häggs carbide.		

$Fe_{2.2}C$, which is shown in Fig. 18.1 to be stable at temperatures below 230°C. This would indicate the formation of an additional type of transition carbide that appears in a temperature interval different from that in which the (ε/η) carbide precipitates.

Finally, with regard to this study of the tempering phenomena in a high-carbon steel, consider Fig. 19.27, which shows the hardness of the double quenched specimens as a function of the tempering temperature at which each specimen was given a one hour anneal. Note that the hardness of the specimens increases by about 150 HV 0.3 between room temperature and 100°C. (HV 0.3 signifies a diamond pyramid hardness measurement made with a 0.3 kg load). This increase in hardness with increasing temperature is accredited to the clustering of carbon atoms in the martensite just above room temperature and to the precipitation of (ε/η) carbide just under 100°C. A significant factor with regard to the transition carbides is that they are extremely small in size, as is to be expected considering the low temperature range in which they form. They have, however, been identified with the aid of the electron microscope (Ref. 33) and it appears that the preferred position for their nucleation is the subgrain boundaries inside the martensite. The average diameter of these subgrains is about 1.0 to 0.1 μm and the thickness of the (ε/η) carbide subgrain-boundary network is less than 20 nm. It is not surprising that these carbides can cause an increase in the hardness.

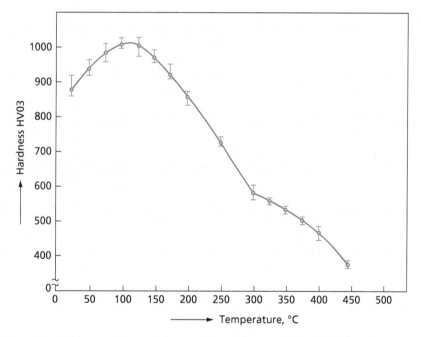

Fig. 19.27 Hardness of a 1.13 percent C steel, quenched in brine and then in liquid nitrogen, as a function of tempering temperature. (From Cheng, L., Brakman, C. M., Korevaar, B. M., and Mittemeijer, E. J., *Met. Trans. A.*, **19A** 2415 [1988] with kind permission of Springer Science and Business Media)

19.14 Tempering of a Low-Carbon Steel

Several processes that occur during the tempering of high-carbon steels do not normally appear in the tempering of low-carbon steels. As may be seen in Fig. 19.16, the martensite finish temperature, M_f, lies above room temperature in steels with less than about 0.6 percent C. This means that the amount of retained austenite in these steels should be small after quenching them from the austenite region, a fact confirmed by Fig. 19.17, which shows that the volume fraction of retained austenite is nearly zero below 0.4 percent C. Thus, Stage 3 in Table 19.8, which is accredited to the breakdown of retained austenite into ferrite plus cementite, should not be observed below approximately 0.4 percent C.

There is also a strong tendency to eliminate or suppress Stage 3 in Table 19.8, where the precipitation of the transition carbide (ε/η) takes place. This is because in a low-carbon steel, most of the carbon atoms are able to segregate to dislocations during the quench, and the segregated carbon atoms are apparently too tightly bound to the dislocations for them to form transition carbide particles. The fact that in low-carbon steels, all or a large fraction of the carbon atoms are able to segregate during the quench, has been clearly demonstrated by Speich (Ref. 34) using resistivity measurements. The electrical resistivity of a steel is a function of the carbon concentration, whether the carbon atoms are in solid solution or segregated at dislocations. However, the increase in resistivity with carbon concentration is greater when the carbon atoms are in solid solution. This is shown by the two dashed lines in Fig. 19.28, where the resistivity is plotted against the carbon content of the steels. Note that the calculated line marked *No segregation* has a much steeper slope than that marked *Complete segregation*. The experimental data are also plotted on this figure as a solid line curve that lies very close to the complete segregation line up to 0.2 percent C. According to Speich's calculations, about 90 percent of the carbon is segregated below the 0.2 percent C concentration. Above 0.2 percent C, the slope of the experimental data curve becomes nearly equal to that of the no segregation curve. Therefore, it is concluded that in steels with less than 0.2 percent C, the carbon of the martensite is almost completely segregated. In addition, because the plot of the experimental data effectively assumes the slope of the no segregation line above 0.2 percent carbon, it is believed that at this concentration the dislocations in a quenched low-carbon steel become saturated with carbon atoms. A consequence of this is that in low-carbon steels with greater than 0.2 percent carbon, the carbon available for precipitation as a transition carbide is about 0.2 percent smaller than the actual carbon content of the steel.

Of the various stages listed in Table 19.8, it would appear that for steels with less than 0.2 percent C, Stage 2, involving the segregation of carbon, actually occurs during the quench, and thus, is not a true stage of tempering. Stage 3, in which the transition carbide is precipitated, is effectively suppressed because the carbon atoms are segregated. Stage 4, where the retained austenite is decomposed to ferrite and cementite, is eliminated because there is no retained austenite. This leaves only Stage 5, in which the segregated carbon forms small rod-shaped cementite particles and the higher temperature phenomena, namely spheroidization of the cementite particles and their growth by Ostwald ripening, and the recovery and recrystallization of the ferrite matrix.

With regard to the tempering phenomena that occur at higher temperatures, Speich has indicated that the rod-shaped particles dissolve near 400°C and are replaced by spheroidal Fe_3C particles. The Fe_3C particles are nucleated at lath boundaries and former austenite grain boundaries, but there may also be general nucleation.

As for the reactions that occur in the ferrite matrix, the first takes place between 500 and 600°C and involves the recovery of the dislocations in the lath boundaries. This produces a

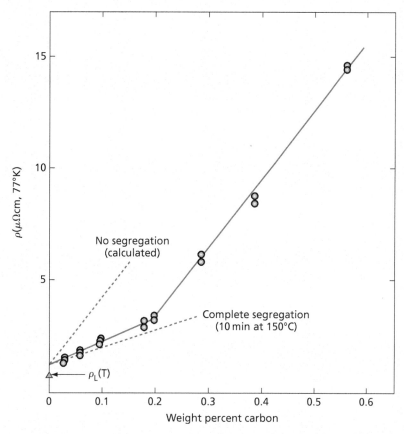

Fig. 19.28 The segregation of carbon during quenching of iron-carbon martensites as detected by electrical resistivity measurements. (From Speich, G. R., *Trans. TMS-AIME*, **245** 2553 [1969])

low-dislocation-density acicular ferrite structure. On further heating to 600 to 700°C, the acicular ferrite grains then recrystallize to form an equiaxed ferrite structure. This recrystallization occurs with greater difficulty at higher carbon cencentrations because of the pinning action of the carbide particles on the ferrite boundaries.

The end result of tempering in a plain-carbon steel is an aggregate of equiaxed ferrite grains containing a large number of spheroidal iron-carbide particles. In the Ostwald ripening stage the spheroidal carbide particles grow. This is accomplished by diffusion processes and results from the fact that larger particles have lower free energies than do smaller particles. A study (Ref. 35) of the growth kinetics of the carbide particles in iron-carbon alloys indicates that the growth rate is diffusion-controlled. However, the problem is complex, as indicated by the fact that the effective diffusion constant for growth lies between the diffusion coefficients for the diffusion of carbon and for the diffusion of iron in the ferrite. Figure 19.29 indicates some of the problems associated with studying the growth of a precipitate. Note that the precipitate particles have a range of particle sizes and that with increasing annealing time, while the average particle grows in size so that the number of particles decreases (Fig. 19.30), the range of particle sizes increases. An excellent series of pictures is shown in Fig. 19.31. These are electron micrographs of the structures of steel specimens tempered in the range where the carbide particles are coalescing.

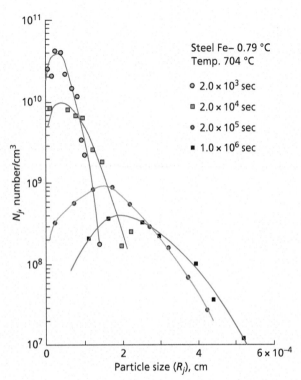

Fig. 19.29 Number of particles per unit volume, N_j, in a given size class as a function of the mean particle size of the class, R_j, for an eutectoid steel spheroidized at 704°C. (From Vedula, K. M., and Heckel, R. W., *Met. Trans.* **1** 9 [1970] with kind permission of Springer Science and Business Media)

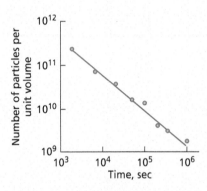

Fig. 19.30 The total number of particles per unit volume as a function of the time of spheroidization for the steel of Fig. 19.29. (From Vedula, K. M., and Heckel, R. W., *Met. Trans.* **1** 9 [1970] with kind permission of Springer Science and Business Media)

19.15 Spheroidized Cementite

Spheroidized cementite is the name given the structure consisting of cementite spheroids embedded in a matrix of ferrite when the particles become large enough to be easily visible under the light microscope. Such a structure is readily attained in a moderate length of time, if the third stage of tempering is carried out at a temperature just below the eutectoid temperature (727°C). A typical photograph (made with a light microscope) of a spheroidized cementite structure is shown in Fig. 19.32. This structure is perhaps the most stable of all the ferrite and cementite aggregates. Martensite, bainite, and even pearlite can be transformed into this microstructure by holding the metal for a sufficiently long time just below the eutectoid temperature. The formation of the spheroidized cementite is, of course, slowest when the starting structure is pearlite, and it is also true that the coarser the pearlite structure the more difficult it is to spheroidize it.

19.15 Spheroidized Cementite

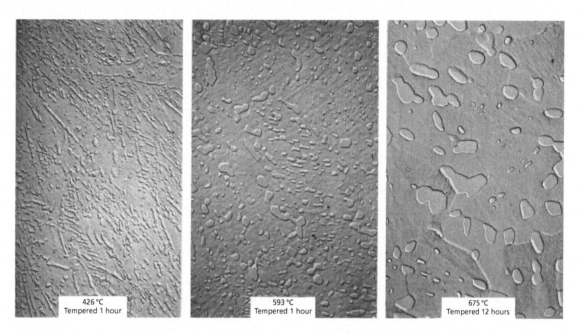

Fig. 19.31 Structure of tempered martensite in a steel with 0.75 percent carbon. Electron microscope. 15,000×. (From Turkalo, A. M., and Low, J. R., Jr., *Trans. AIME,* **212** 750 [1958])

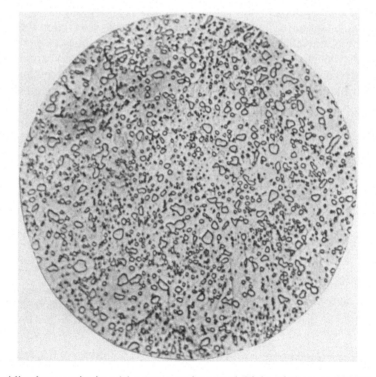

Fig. 19.32 Spheroidized cementite in a 1.1 percent carbon steel. Light microscope. 1000×

The spheroidized cementite structure is especially desirable in softened high-carbon steels, for steels containing this microstructure are more readily machined and give better results when heat-treated. Consequently, high-carbon steel sold under the label "annealed" will almost always possess a spheroidized structure.

19.16 The Effect of Tempering on Physical Properties

The microstructure changes that occur during tempering greatly change the physical properties of steel. The hardness of tempered steels will now be considered. The changes in this property are functions of both tempering time and tempering temperature. Curves such as those of Figs 19.33 and 19.34, where the hardness (as measured at room temperature after the tempering heat treatment) is plotted as a function of tempering temperature, specifically refer to constant tempering times (1 hr) at each tempering temperature. Curves for both medium- and high-carbon steels are shown in Fig. 19.33 and in each case, the specimens employed were refrigerated at −196°C before tempering. The purpose of this treatment was to reduce the retained austenite in the quenched metal to a negligible quantity, so that the results plotted in Fig. 19.33 are truly representative of the effects of tempering martensite. If the specimens had contained a finite amount of retained austenite, an additional hardening component would have been introduced due to the transformation of austenite to martensite or bainite. This is frequently observed in hardness versus tempering temperature curves and appears as a rise in hardness just above room temperature.

A slight increase in hardness (not due to retained austenite) may be observed in the higher carbon steel (1.4 percent) as the tempering temperature is carried to about 93°C. This is undoubtedly associated with the precipitation of (ε/η) carbide. A similar rise is not observed in the lower carbon steel (0.4 percent) because of the much smaller quantity of (ε/η) carbide that can precipitate in this composition. It should also be mentioned that while the precipitation of epsilon carbide undoubtedly contributes a hardening component to the steel, the depletion of carbon from the martensite matrix can be expected to contribute a softening component. The observed hardness, therefore, reflects the result of these two effects. A decided softening of the specimen, however, occurs when reactions associated with the formation of cementite become appreciable. This is shown by the marked drop in hardness that starts at about 200°C. In the early parts of this stage, both the solution of epsilon carbides and the

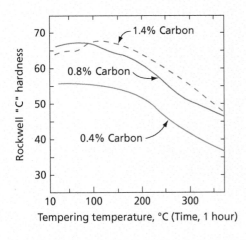

Fig. 19.33 Effect of tempering temperature on the hardness of three steels of different carbon contents. (Lement, B. S., Averbach, B. L. and Cohen, M., *Trans. ASM*, **46** 851 [1954]. Reprinted with permission of ASM International®. All rights reserved. www.asminternational.org)

19.16 The Effect of Tempering on Physical Properties

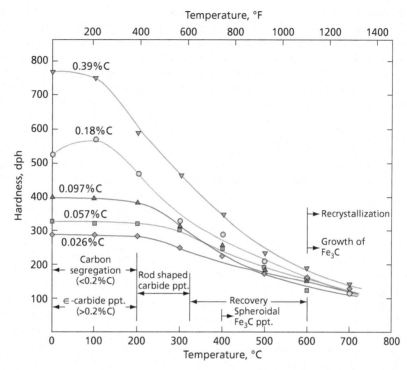

Fig. 19.34 The hardness of low and medium iron-carbon martensite tempered for 1 hour between 100°C and 700°C. (From Speich, G. R., *TMS-AIME*, **245** 2553 [1969])

removal of the carbon from the martensite (low-carbon form) should soften the metal. However, at the same time, cementite precipitates contribute a hardening effect.

When the steel has attained a simple ferrite and cementite structure, further softening results from the growth or the coalescence of cementite particles. This softening, due to growth in size and decrease in number of the cementite particles, continues and becomes more rapid the closer one approaches the eutectoid temperature (727°C). In effect, this means that for fixed tempering times the hardness of tempered martensite will be lower the closer one comes to the eutectoid temperature. The curves of Fig. 19.33 are only drawn for tempering temperatures below 375°C. Above this temperature to 727°C the three given curves can be expected to continue to fall in hardness with approximately the same slope as they exhibit in the temperature range from 200°C to 375°C. Figure 19.34 is from the work of Speich (Ref. 34). It shows the effect of tempering on the hardness (DPH) of steels of low to medium carbon concentration, and complements the data given in Fig. 19.33. In addition, this diagram also outlines the reactions that occur during tempering.

A comprehensive illustration of how the hardness of tempered martensite varies in iron-carbon steels, with both carbon concentration and tempering temperature, appears in Fig. 19.35. The data for this diagram were obtained by using small thin (2.5 mm thick) specimens for the higher carbon steels and similar specimens with half this thickness for the lower carbon steels. The small sizes of the specimens gave them a rapid cooling rate when they were quenched. All of the specimens were austenitized at 927°C for 10 min and then brine quenched. After this, they were tempered at the indicated temperatures for 1 hour. The quench produced 100 percent martensitic

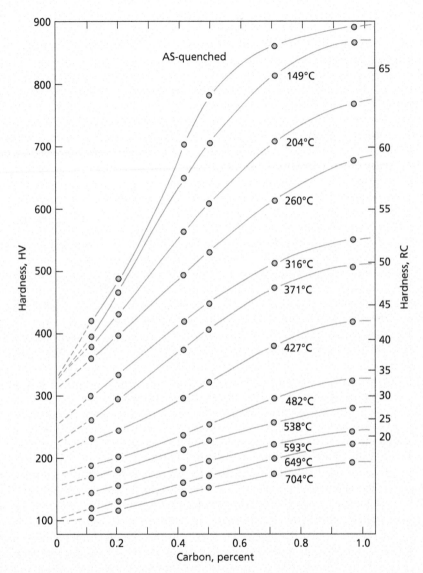

Fig. 19.35 The hardness of tempered martensite in iron-carbon steels. (Abstracted from Grange, R. A., Hribal, C. R., and Porter, L. F., *Met. Trans. A*, **8A** 1775 [1977] with kind permission of Springer Science and Business Media)

structures in all the specimens except for those with the higher carbon concentrations. The volume fraction of retained austenite in the specimens containing 0.5, 0.72, and 0.98 percent carbon were 3, 7, and 13 percent, respectively. The strong effect of carbon on the hardness of quenched steels is clearly shown in Fig. 19.35. Note that the effect of carbon on hardness drops sharply as the tempering temperature is increased.

19.17 The Interrelation Between Time and Temperature in Tempering

The fact that time and temperature have corresponding effects in tempering of steels, especially in the stage where aggregates of cementite and ferrite are involved, has been known for a long time. This is demonstrated in a simple fashion in Fig. 19.36, where a number of curves corresponding to different tempering temperatures are given. In each case, the curve shows the dependence of hardness on the time at the indicated tempering temperature. The same data can also be related by a simple rate equation

$$\frac{1}{t} = Ae^{-Q/RT} \qquad 19.9$$

where t is the time to attain a given hardness, Q is an empirical activation energy for the process, A is a constant, and R and T have their usual significance.

19.18 Secondary Hardening

Most steels are used with either a pearlitic or a quenched and tempered structure consisting of an aggregate of carbide particles embedded in a matrix of ferrite. In either case, a two-phase ferrite and carbide structure is involved. When alloying elements are added to steels, they may enter the ferrite or the carbides in varying amounts depending on the alloying element concerned. Some elements, however, are not found in the carbides. These include: aluminum, copper, silicon, phosphorus, nickel, and zirconium. Other elements are found in both the ferrite and the carbides. A number of these elements in the order of their tendency to form carbides (manganese having the least and titanium the greatest) are: manganese, chromium, tungsten, molybdenum, vanadium, and titanium.

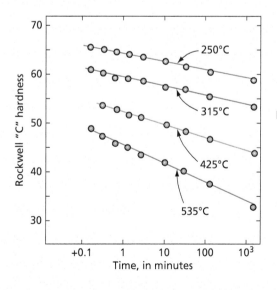

Fig. 19.36 The effect of time and temperature on the hardness of a tempered steel: 0.82 percent carbon, 0.75 percent manganese. (After Bain, E. C., *Functions of the Alloying Elements in Steel*, **ASM**, Cleveland, 1939)

Most alloying elements in steels tend to increase the resistance of the steel to softening when it is heated, which means that for a given time and temperature of tempering, an alloy steel will possess a greater hardness after tempering than a plain-carbon steel of the same carbon content. This effect is especially significant in a steel that contains appreciable amounts of carbide-forming elements. When these elements are tempered at temperatures below 540°C, the tempering reactions tend to form cementite particles based on Fe_3C, or, more accurately, $(Fe, M)_3C$ where M represents any of the substitutional atoms in the steel. In general, the alloying elements are present in the cementite particles only in about the same ratio as they are present in the steel as a whole. However, when the tempering temperature exceeds 540°C, appreciable amounts of alloy carbides are precipitated. The precipitation of these new carbides which, in general, do not conform to the formula $(Fe, M)_3C$, induces a new form of hardening that is believed due to coherency (Refs. 36 and 37). A schematic comparison of the tempering curves of a plain-carbon steel and a steel with large amounts of carbide-forming elements is shown in Fig. 19.37.

That the alloy carbides do not form readily at lower tempering temperatures is undoubtedly related to the fact that at these tempering temperatures the rate of diffusion of substitutional elements is too slow to permit their formation. Cementite can form because the diffusion rate of carbon is still very large at temperatures below 540°C. The formation of cementite depends only on the diffusion of carbon.

The use of carbide-forming elements in order to produce a steel with a high resistance to tempering is best illustrated by the so-called high-speed steels. These are tool steels, whose original purpose was for use as tool bits in lathes and other machines where the cutting edges often get very hot during machining operations. Steels of this type retain their hardness for very long periods of time even at a red heat. A typical high-speed composition is represented by the composition designated 18-4-1, which contains approximately 18 percent tungsten, 4 percent chromium, 1 percent vanadium, in addition to approximately 0.65 percent carbon.

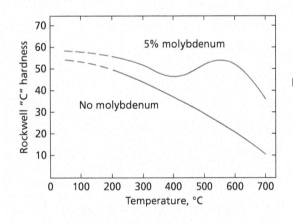

Fig. 19.37 Secondary hardening in a steel containing 0.35 percent carbon. (After Bain, E. C., *Functions of the Alloying Elements in Steels*, ASM, Cleveland, 1939)

Problems

19.1 (a) Determine the ideal critical diameter, D_I, of an unalloyed 0.65 percent C steel with an austenitic grain size 7.

(b) What is the hardness of this steel, when it has (1) a 100 percent martensitic structure and (2) a 50 percent martensitic structure? Give your answers in Rockwell-C units (HRC).

(c) At what distance along a Jominy bar should one find the HRC of the 50 percent martensitic structure?

19.2 Now assume that one has a plain-carbon steel with an austenitic grain size 7 containing 0.65 percent C and 0.75 percent Mn.

(a) Give the HRC values for the 100 percent and 50 percent martensitic structures.

(b) Determine the D_I of this steel.

(c) At what distance along a Jominy bar of this steel would one expect to find the HRC corresponding to 50 percent martensite?

19.3 What diameter of bar of the plain-carbon steel of Prob. 19.2 will have a 50 percent martensitic structure at its center after being properly austenitized and given a very good oil quench?

19.4 The American Iron and Steel Institute (AISI) specifications for steels require only that the concentrations of the alloying elements fall within specified limits. As an example, consider the steel designated AISI 4135. The specifications call for a carbon concentration between 0.32 to 0.38 percent, manganese between 0.60 to 1.00, silicon from 0.20 to 0.35, chromium from 0.75 to 1.20, and molybdenum from 0.15 to 0.25. Compute the D_I for an AISI 4140 steel (grain size 7) containing only the minimum concentration of each alloying element.

19.5 Determine the D_I for AISI 4135 when it contains the maximum concentrations of the various elements. Note: the multiplying factor for 1.20 percent Cr is 3.592.

19.6 The answers to Problems 19.4 and 19.5 are $D_I = 2.2$ and 8.1, respectively. It is interesting to see how this is reflected in their respective Jominy curves. This can be demonstrated with the aid of tabulated distance hardness dividing factors that are listed in ASTM A 255. With these factors it is possible to determine the hardness at various distances along a Jominy bar, given a value of D_I and the initial hardness at 1/16 in. from the quenched end of the bar. This initial hardness, designated as IH, is normally assumed to equal that of a 100 percent martensitic structure and can be obtained from the carbon concentration of the steel. The hardness at a given distance along the Jominy bar is then computed by dividing the value of the IH by the *distance hardness dividing factor* corresponding to the distance in question. The distance hardness dividing factors for steels with $D_I = 2.2$ and $D_I = 7.0$ are listed in the table below.

(a) Determine the two IH values.

(b) Divide the IH values by their respective distance hardness dividing factors to obtain two sets of Jominy hardness values, and plot these as two separate Jominy curves.

19.7 Consider an AISI 4135 steel in which all of the alloying elements have the mean or average value of their ranges.

(a) Determine its ideal critical diameter.

(b) Determine the maximum diameter of a bar of this material that can be quenched so as to obtain a 50 percent martensite–50 percent pearlite structure at its center when given a poor oil quench.

Jominy Distance Hardness Dividing Factors (1/16 in).

D_I, in.	2	3	4	5	6	7	8	9	10	12	14	16	18	20	24	28	32
2.2	1.00	1.07	1.23	1.43	1.66	1.73	1.82	1.90	1.98	2.20	2.30	2.39	2.47	2.56	2.74	2.90	3.03
7.0	1.00	1.00	1.00	1.00	1.00	1.00	1.00	1.00	1.00	1.00	1.01	1.03	1.04	1.05	1.08	1.13	1.15

(c) Make the same computation, but now assume that the bar is given a brine quench.

19.8 In certain cases the grain size of a metal may be of the order of several tenths of a µm. Approximately what is the ASTM grain size number of a metal whose grain size is 0.2 µm?

19.9 (a) Determine the relative volume change when a carbon steel containing 0.6 percent of carbon is reacted to form 100 percent martensite.

(b) What is the corresponding change in a linear dimension?

(c) Assuming that Young's modulus is 206,850 MPa, what tensile stress would have to be applied to this steel in order to obtain a strain equivalent to that found in (b) of this problem?

(d) The compressibility of iron at room temperature is:

$$\frac{\Delta V}{V \Delta P} = 5.773 \times 10^{-6} \text{ per MPa}$$

Determine the magnitude of the pressure increase that would be needed in order to decrease the volume of the 0.6 percent C steel specimen by an amount equal to the expansion that occurs when it transforms from austenite to martensite.

19.10 Suppose that you are given a 2 in. diameter bar of the eutectoid composition plain-carbon steel described in Prob. 19.2 and that it has a very coarse pearlite structure. Describe a procedure that could change the coarse pearlite structure into a relatively fine spheroidized cementite structure.

19.11 (a) Describe the various reactions involved in the decomposition of martensite during tempering.

(b) Describe how the tempering phenomena change with carbon concentration in a steel.

19.12 Equation 19.9 gives the interrelation between time and temperature in tempering. With the aid of the data in Fig. 19.37, determine approximately the activation energy Q in Eq. 19.9. To do this, draw a number of horizontal straight lines on a copy of Fig. 19.36. These lines represent constant hardness values. An intersection of one of these lines with one of those plotted through data obtained at a fixed temperature gives the time, at the indicated temperature, to obtain the hardness in question.

References

1. Davenport, E. S., and Bain, E. C., *Trans. AIME*, **90** 117 (1930).
2. Grange, R. A., and Kiefer, J. M., *Trans. ASM*, **29** 85 (1941).
3. *Atlas of Isothermal and Cooling Transformation Diagrams*, American Society for Metals, 1977.
4. Zhao, M., Yang, K., Xiao, F., and Shan, Y., *Materials Science and Engineering*, A355, 2003, 126–136.
5. Umemoto, N., Komatsubara, N., and Tamura, I., in *Solid-Solid Phase Transformations*, Edited by H. I. Aaronson et al., TMS-AIME, 1111–1115.
6. "Heat Treating", *ASM Handbook*, vol. 4, 1991, 3–19.
7. Eldis, G. T., in *Hardenability Concepts with Applications to Steel*, edited by D. V. Doane and J. S. Kirkaldy, TMS-AIME, 1978, 126–157.
8. Trzaska, J., and Dobrzanski, L. A., *Journal of Materials Processing Technology*, 192–193, 2007, 504–510.
9. Rios, P. R., *Acta Materiallia*, 53, 2005, 4893–4901.
10. "Heat Treating," *ASM Metals Handbook*, Vol. 4, 1991, p. 71.
11. Zehtab, A., Sajjadi, S. A., Zebarjad, S. M., and Moosavi Nezhad, S. M., *Journal of Materials Processing Technology*, **199** 2008, 124–129.
12. Thomas, K., Geary, E. A., Avis, P., and Bishop, D., *Materials of Design*, **13** 1992, 17–22.
13. Mehl, R. F., and Hagle, W. C., *Prog. in Metal Physics.*, **6** 74 (1956).
14. Grossman, M. A., *Trans. AIME*, **150** 242 (1942). (For additional discussion, see "Heat Treating," *ASM Metals Handbook*, vol. 4, 1991, p. 9.)

15. Siebert, C., Doane, D. V., and Breen, D. H., *The Hardenability of Steels*, p. 72, American Society for Metals, Metals Park, Ohio, 1977.
16. *1986 Annual Book of ASTM Standards, Sec. 3*, **Standard A255**, ASTM, Philadelphia, Pa., 1986.
17. Winchell, P. G., and Cohen, M., *Trans. ASM*, **55** 347 (1962).
18. Bernshtein, M. L., Kaputkin, L. M., and Prokoshkin, S. D., in *Solid-Solid Phase Transformations*, Edited by H. I. Aaronson et al., TMS-AIME, 1982, p. 869–874.
19. Roberts, C. S., *Trans. AIME*, **197** 203 (1953).
20. Tsai, M. C., Chiou, C. S., Du, J. S., and Yang, J. R., *Materials Science and Engineering*, A332, 2002, 1–10.
21. Lenel, U. R., and Knott, B. R., *Met. Trans A*, 18A, 1987, 767–775.
22. Hirth, J. P., and Cohen, M., *Met Trans.*, **1** 3 (1970).
23. "Heat Treating," *ASM Handbook*, vol. 4, 1991, 13–18.
24. Hildenwall, B., and Ericsson, T., in *Hardenability Concepts with Applications to Steel*, edited by D. V. Doane and J. S. Kivkaldy, TMS-AIME, 1978, 579–607.
25. "Heat Treating," *ASM Handbook*, vol. 4, 1991, 603–618.
26. Cheng, L., Brakman, C. M., Korevaar, B. M., and Mittemeijer, E. J., *Met. Trans. A*, **19A** 2415 (1988).
27. Olson, G. B., and Cohen, M., *Met. Trans. A*, **14A** 1057 (1983).
28. Ren, S. B., and Wang, S. T., *Met. Trans. A*, **19A** 2427 (1988).
29. Mittemeijer, E. J., van Gent, A., and van der Schaaf, P. J., *Met. Trans. A*, **17A** 1441 (1986).
30. Meisel, L. V., and Cote, P. J., *Acta Metall.*, **31** 1053 (1983).
31. Mittemeijer, E. J., Cheng, L., van der Schaaf, P. J., Brakman, C. M., and Korevaar, B. M., *Met. Trans. A*, **19A** 925 (1988).
32. Cheng, L., Brakman, C. M., Korevaar, B. M., and Mittemeijer, E. J., *Met. Trans. A*, **19A** 2415 (1988).
33. Lement, B. S., Averbach, B. L., and Cohen, M., *Trans. ASM*, **46** 851 (1954).
34. Speich, G. R., *Trans. TMS-AIME*, **245** 2553 (1969).
35. Vedula, K. M., and Heckel, R. W., *Met. Trans.*, **1** 9 (1970).
36. Payson, P., *Trans. ASM*, **51** 60 (1959).
37. Seal, A. K., and Honeycombe, R. W. K. *Jour. of the Iron and Steel Inst.* (London), **188** 9 (1958).

Chapter 20

Selected Nonferrous Alloy Systems

Learning Objectives
Upon completion of Chapter 20, you will be able to:
1. Describe the effects of alloying on the electrical and thermal conductivities of copper
2. Become familiar with the mechanical properties of brasses, bronzes and copper alloys
3. Compare and contrast precipitation hardening in Al-Li and Al-Li-X alloys with those of Al-Cu discussed in Chapter 16
4. Understand beta to alpha transformation in pure titanium
5. Discuss strengthening in dual-phase alpha-beta Ti alloys
6. Explain heat treatment and Widmanstätten plates formation in alpha-beta titanium alloys
7. Describe the influence of grain boundaries on creep failure of turbine blades

20.1 Commercially Pure Copper

Of all the metals, copper has the second highest electrical and thermal conductivities. Only silver has higher abilities to transport electricity and heat. Thus, it is not strange that the most common use for copper is in applications where high electrical or thermal conductivities are required. Table 20.1 lists the room temperature electrical and thermal conductivities of copper, as well as those of a number of other metals. An important consideration is the fact that even small amounts of impurities adversely affect the conductivities, because any irregularity inside the space lattice of a metal causes scattering of electrons and a reduction in their mean free path. Consequently, the purity of copper intended for use in wires and other shapes used in transporting electricity must be carefully controlled. Figure 20.1 shows the effects of a number of alloying elements on the conductivity of copper. Note that almost all of the alloying elements reduce the conductivity of copper, with silver having the least effect. It should be noted that the reduction of the conductivity does not depend on the conductivity of the alloying element, but on the particular effect of the alloying element on the copper lattice. Because of this, when it is desired to harden copper for

20.1 Commercially Pure Copper

Table 20.1 Electrical Resistivity and Thermal Conductivity of Copper and Other Pure Commercial Metals at 293 K

(Metal 100)	Electrical Resistivity at 293 K, μΩcm	Thermal Conductivity Wm^{-1}k^{-1}	Relative Electrical Conductivity (Copper = 100)	Relative Thermal Conductivity (Copper = 100)
Silver	1.63	419	104	106
Copper	1.694	397	100	100
Gold	2.2	316	77	80
Aluminum	2.67	238	63	60
Beryllium	3.3	194	51	49
Magnesium	4.2	155	40	39
Tungsten	5.4	174	31	44
Zinc	5.96	120	28	30
Nickel	6.9	89	24	22
Iron	10.1	78	17	20
Platinum	10.58	73	16	18
Tin	12.6	73	13	18
Lead	20.6	35	8.2	8.8
Titanium	54	22	3.1	5.5
Bismuth	117	9	1.4	2.2

Adapted from Brandes, E. A., Ed., *Smithells Metals Reference Book*, Sixth Edition, Butterworth, Inc., 1983. (Used by permission.)

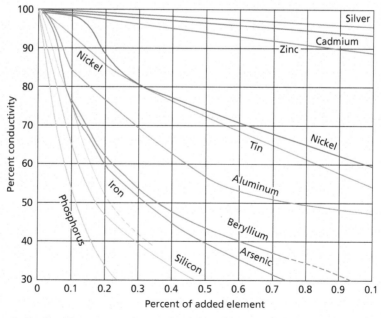

Fig. 20.1 Effect of alloying elements on the conductivity of copper. (From Mendenhall, J. H., Ed., *Understanding Copper Alloys*, Olin Brass Corporation, East Alton, IL, 1977. Used by permission)

electrical applications, silver is normally the best alloying element to use. However, it is usually too expensive. Phosphorus, one of the best deoxidizers used during the melting and refining of copper, unfortunately lowers the conductivity of copper very rapidly.

The three principal forms of commercially pure copper used for conductivity purposes are: (a) electrolytic tough-pitch copper, (b) deoxidized low-phosphorus copper, and (c) oxygen-free electronic copper (OFHC). The first, electrolytic tough pitch, which is also used for automobile radiators, gaskets, kettles, and pressure vessels, contains more than 99 percent copper. It also has from 0.02 to 0.05 percent oxygen, which is added deliberately to remove soluble impurities by forming insoluble oxide particles (for example, Fe in solid solution forms Fe_3O_4 precipitates). Since the maximum solubility of oxygen in copper is approximately 0.004 weight percent at 1339 K, the remaining oxygen forms compound cuprous oxide (Cu_2O) in the interdendritic regions during solidification of the metal. This can be seen in Fig. 20.2A. Upon subsequent working and annealing, these particles tend to strengthen the matrix. However, the alignment of these particles along grain boundaries, as shown in Fig. 20.2B, can also lead to the embrittlement of the copper. Impurities also directly affect electrical conductivity and mechanical properties upon recycling of copper (Ref. 1).

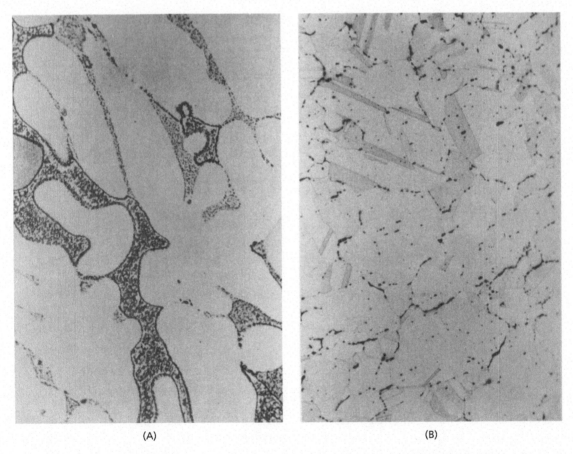

(A) (B)

Fig. 20.2 **(A)** As-cast microstructure of electrolytic tough-pitch copper showing copper dendrites with oxide particles dispersed in the interdendritic regions. **(B)** The same material after hot working and annealing

Deoxidized low-phosphorus copper usually contains from 0.01 to 0.04 percent phosphorus. According to Fig. 20.1, this material has an approximately 15 percent lower conductivity than pure copper. Thus, it is widely used for piping and tubing applications rather than for electrical applications.

Oxygen-free electronic copper, on the other hand, contains more than 99.99 percent copper with a maximum oxygen concentration of only 0.001 weight percent. This material is melted and cast in a carbon monoxide atmosphere and is used in electrical applications. Its conductivity is almost the same as tough-pitch copper, but it has the advantage that it can be cold worked more extensively than tough-pitch copper.

There are other commercially pure coppers available, to which small concentrations of alloying elements are added for specific applications. For example, there are electrical applications that require an increased strength which can be obtained by cold working OFHC. Unfortunately, however, pure copper has a low recrystallization temperature. In order to increase the recrystallization temperature, a small amount of silver, 0.03–0.05 weight percent, can be added. This addition reduces the conductivity by only about 1 percent while increasing the recrystallization temperature to approximately 600 K. The *silver-bearing alloy*, as it is called, is preferred in the manufacture of electric motors and commutators. Other examples include arsenical copper containing about 0.3 percent arsenic which has improved corrosion resistance, and free-cutting copper containing about 0.6 percent tellurium which has excellent machining properties.

20.2 Copper Alloys

The copper-zinc alloys, which are known as brasses, are the most widely used copper-based alloys. Brasses may contain from 3 to 45 percent zinc, with the balance being copper. However, other minor alloying elements, such as lead, tin, or aluminum, may be intentionally added to achieve a desired color, strength, ductility, machinability, corrosion resistance and/or formability. It is because of this variability that brass alloys find a wide variety of applications in items such as coins, jewelry, cartridges, hinges, electrical connectors, terminals, bearings, gears, and springs (Refs. 2 and 3). Table 20.2 lists some of the common brasses and other copper alloys with their compositions. The Cu-Zn phase diagram is shown in Fig. 11.28. It should be noted that alloys containing up to about 36 percent zinc are normally called *alpha brasses* and those containing more than 38 percent zinc are known as *alpha plus beta brasses*. Alpha brasses are frequently single-phase solid solutions, although those containing more zinc may, under some circumstances, contain some of the beta phase.

In general, the cost of a brass is lower, the higher its zinc content. Because of this, "yellow or high brasses" are most commonly used in ordinary applications. "Low brass," on the other hand, is used in cases where superior corrosion resistance is required, such as in water pipes. With respect to corrosion resistance, red brass is even better. Cartridge brass (30 percent zinc) and other deep drawing brasses are usually manufactured with a very low impurity level. Muntz metal, on the other hand, consists of two phases, alpha + beta, and is used mostly in the form of extrusions.

The term *bronze* is commonly used to describe copper-tin-zinc or other complex alloys. Historically, multicomponent alloys of copper and zinc that contain less tin than zinc were called *brasses*, while copper and tin alloys that contain less zinc than tin were called *bronzes*. However, other alloys that contain no tin, or relatively small amounts of tin, are also commonly called *bronzes*. For example, the casting alloys in this category include aluminum bronze (5 to 15 percent aluminum), silicon bronze (over 0.5 percent silicon), nickel bronze (over 10 percent nickel), lead bronze (over 30 percent lead),

Table 20.2 Composition of Some Common Copper-Base Alloys

Copper Alloy No.	Previous Trade Name	Composition, Percent Maximum or Range					
		Cu	Pb	Fe	Sn	Zn	Other Elements
C23000	Red Brass	84.0–86.0	0.05	0.05	—	Rem.	—
C24000	Low Brass	78.5–81.5	0.05	0.05	—	Rem.	—
C26000	Cartridge Brass	68.5–71.5	0.07	0.05	—	Rem.	—
C27000	Yellow Brass	63.0–68.5	0.10	0.07	—	Rem.	—
C28000	Muntz Metal	59.0–63.0	0.30	0.07	—	Rem.	—
C33500	Low-Leaded Brass	62.5–66.5	0.3–0.8	0.10	—	Rem.	—
C34000	Medium-Leaded Brass	62.5–66.5	0.8–1.4	0.10	—	Rem.	—
C34200	High-Leaded Brass	62.5–66.5	1.5–2.5	0.10	—	Rem.	—
C36000	Free-Cutting Brass	60.0–63.0	2.5–3.7	0.35	—	Rem.	—
C46400	Naval Brass	59.0–62.0	0.20	0.10	0.5–1.0	Rem.	—
C51800	Phosphor Bronze	Rem.	0.02	—	4.0–6.0	—	0.1–0.35 P, 0.01 Al
C61300	Aluminum Bronze	86.5–93.8	—	3.50	0.2–5.0	—	6.0–8.0 Al
C65100	Low-Silicon Bronze B	96.0	0.05	0.80	—	1.5	0.8–2.0 Si, 0.7 Mn
C65500	High-Silicon Bronze A	94.8	0.05	0.80	—	1.5	2.3–3.8 Si, 0.5–1.3 Mn

Adapted from CDA's *Standards Handbook* © Copper Development Association (www.copper.org).

and beryllium bronze (around 2 percent beryllium). Some wrought alloys also contain no tin, but are still called bronze. Examples include commercial bronze (alloy designation number C22000), which consists of 90 percent Cu and 10 percent Zn, and jewelry bronze (alloy number C22600), which consists of 87.5 percent Cu and 12.5 percent Zn.

The copper-zinc phase diagram has a retrograde character on its copper-rich side. Thus, the solubility increases from 31.9 percent at the peritectic temperature (903°C) to about 38.3 percent at 450°C and then decreases to about 28 percent around room temperature.

The extensive solubility of zinc in copper is due to the fact that the atomic size difference between these two elements is only about 4 percent. The atomic radius of copper that crystallizes in the fcc structure is 0.1277 nm, while that of zinc with the hcp structure is 0.1332 nm.

The alpha brasses are solid solutions at room temperature. The as-cast structures of these alloys consist of cored solid-solution dendrites with the cores of the dendrite arms containing less zinc than the sides. Upon annealing, the microsegregation due to coring is eliminated. If the alloy has been plastically deformed, or sometimes if it contains residual stresses due to cooling after casting, recrystallization also takes place during annealing. The recrystallized structure commonly contains annealing twins, which appear as narrow bands with straight sides.

The alpha brasses are highly ductile at room temperature and can be easily deformed into intricate shapes by a variety of deformation processes, such as deep drawing, cold rolling, and stamping. In general, brasses have mechanical properties that are superior to those of copper. The solid-solution alloys have higher tensile strengths and elongations at fracture than commercially pure copper. This can be seen in Fig. 20.3, where the effects of zinc additions on the tensile strength, hardness, and elongation of copper are illustrated. This figure also shows the conductivity of the alloy as a function of its zinc concentration. The tensile strength of alpha brass increases with increasing zinc content, whereas the electrical conductivity (and thermal conductivity, not shown in the figure) shows the opposite trend. The tensile strength of the alloy also depends on its grain size, as shown in Fig. 20.4. This behavior, as described earlier, occurs because the grain boundaries inhibit slip and contribute to work hardening and the grain-boundary area per unit

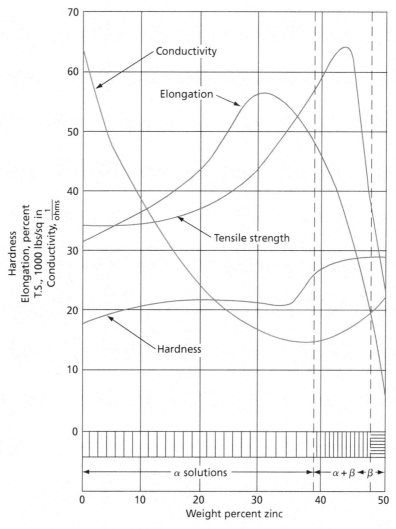

Fig. 20.3 The effect of zinc on the tensile strength, hardness, ductility (tensile elongation), and electrical conductivity of copper. (From Mendenhall, J. H., Ed., *Understanding Copper Alloys*, Olin Brass Company, East Alton, IL, 1977)

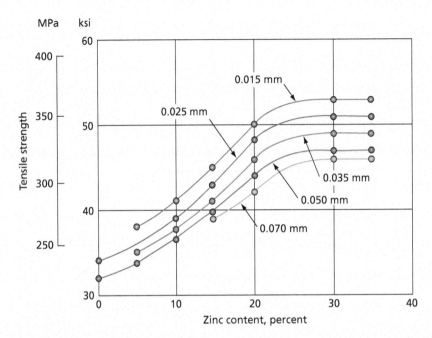

Fig. 20.4 The effect of both the grain size and the zinc concentration on the tensile strength of brass are shown in this figure. (From Mendenhall, J. H., Ed., *Understanding Copper Alloys*, Olin Brass Company, East Alton, IL, 1977)

volume increases as the grain size decreases. The strengthening follows the Hall-Petch relationship (see section 6.11), but the strengthening efficiency seems to decrease as the grain size is reduced to nanoscale (Ref. 4).

The beta (β) phase is a body-centered cubic electron compound that is approximately centered about the composition CuZn. The homogeneity range of the compound decreases with decreasing temperature from about 800°C. At 800°C the homogeneity range is from 39 to 55 percent zinc, while at 500°C it is from 45 to 49 percent. The β phase undergoes ordering when the temperature falls below 470°C. The long-range ordered β' phase has a bcc superlattice in which each zinc atom is surrounded by eight copper atoms, and vice versa.

As discussed in Sec. 11.15, the β and β' phases differ only in that the former has a disordered bcc structure, whereas the latter is ordered.

The disorder-order transformation is so very fast that even rapid quenching cannot suppress ordering, with the result being that it is not possible to retain the disordered phase at room temperature. The ordered phase commonly contains many domains that are separated by antiphase domain boundaries. When a disordered phase transforms to an ordered structure, the transformation involves the nucleation and growth of many separate ordered regions (domains). In the CuZn system, two distinct types of ordered domains are possible. For one type, the bcc structure can be thought of as having a zinc atom at its center with copper atoms at the corners, whereas for the other, a copper atom will be at the center and zinc atoms at the corners. When these two configurations ("phases") come together, a boundary, called an *antiphase domain boundary*, forms between them.

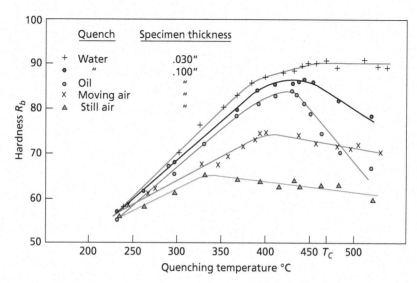

Fig. 20.5 Effect of quenching rate on the yield point immediately after quenching. (Reprinted from "The cause of the strengthening in quenched beta brass," Brown, N., *Acta Metallurgica*, **7** 210 [1959] with permission from Elsevier. https://www.sciencedirect.com/science/article/abs/pii/0001616059900768)

The domain size depends upon the cooling rate. Very slow cooling from the β region allows the development of a relatively coarse domain size, whereas faster cooling results in the formation of a finer domain size. Figure 20.5 shows the effect of the cooling rate on the as-quenched hardness of Cu-48.5 percent Zn. It can be seen that the hardness of the ordered alloy not only depends on the cooling rate during quenching but also on the temperature from which the sample is quenched. The quenching temperature that produces the maximum strength is also a function of the quenching rate. The strength of the quenched ordered alloy is commonly attributed to the size of the antiphase domain; the finer the domain size, the harder the material. On the other hand, Brown (Ref. 5) attributes the strengthening of quenched CuZn to vacancies generated during the rapid disorder-to-order transformation. These vacancies probably originate from interactions between dislocations during long-range ordering.

20.3 Copper Beryllium

In some copper alloy systems, alloying additions are made that exceed the solubility limit on cooling so that second-phase precipitates tend to form during cooling. The copper-beryllium alloys are an example. As shown in Fig. 20.6, the maximum solubility limit of beryllium in copper is 1.6 percent at 620°C (893 K), which decreases to less that 0.5 percent at 300°C (573 K). Consequently, alloys with less beryllium than the maximum (1.6 percent) can have a single phase at high temperatures, but two phases at room temperature; one of these phases will be the alpha phase and the other the intermetallic compound γ with the composition CuBe. The intermetallic phase will appear as precipitate particles within the matrix. These Cu-Be alloys are able to undergo an age hardening reaction similar to those of the Al-Cu or Cu-Al alloys. For example, when the alloy containing 1 percent beryllium is heated to 900 K, it will contain only the alpha

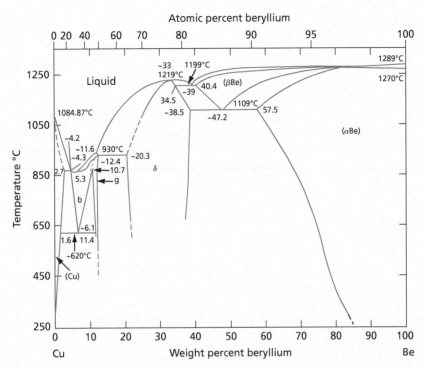

Fig. 20.6 Copper-beryllium phase diagram. (Chakrabarti, D. J., Laughlin, D. E, and Tanner, L. E., *Bulletin of Alloy Phase Diagrams*, **8** 288 [1987]. Reprinted with permission of ASM International®. All rights reserved. www.asminternational.org)

phase. Upon rapid quenching of this material to room temperature, the formation of the γ phase will be suppressed, and the material will consist of a supersaturated solid solution of beryllium in copper. If the alloy is now heated to 600 K, particles of the equilibrium γ phase will eventually form in the alpha phase. However, as in many other precipitation and age hardening processes, the formation of the equilibrium phase is preceded by the formation of GP zones and a metastable γ′ precipitate. As discussed in Chapter 16 on precipitation hardening, it should be noted that in some alloys, such as Al-Cu, the equilibrium θ is preceded by GP zones and two metastable phases (θ″ and θ′). In the Cu-Be system, the precipitation sequence is (1) disk-shaped GP zones, (2) γ′, and (3) γ (CuBe).

As a result of precipitation hardening, a considerable strengthening of the alloy can occur. For example, an alloy containing 1.9 percent Be can attain a hardness of about 42 Rockwell-C by aging at 618 K (345°C) for 4 hours (Ref. 6). This hardness is comparable to those of some hardened steels. As discussed in Sec. 16.5, careful control of the precipitation time and temperature must be exercised to attain the maximum strength level. Aging at higher temperature or longer times may cause the particles to coarsen and soften the alloy. For high strength, small and coherent precipitates, tightly bound to the matrix, are needed. There are also indications that certain copper-Be alloys have another attractive characteristic that needs to be mentioned. These alloys can experience martensitic transformations similar to those that occur in some Cu-Al, Cu-Sn, and Cu-Zn alloys (Ref. 7).

20.4 Other Copper Alloys

Other alloying elements, such as tin, aluminum, or nickel, are also added to copper to obtain higher tensile strength or hardness, wear resistance, special corrosion resistance, or any combination of these. A detailed discussion of these alloys can be found elsewhere (Refs. 6, 8–10). The presence of a second phase in a copper alloy does not always produce an increase in strength and hardness of the copper. The opposite might be the case in some alloys. For example, certain leaded copper alloys contain soft, lead-rich particles. The relatively large size of these particles and their noncoherent boundaries not only do not contribute to hardening of the alloy but also tend to lower its overall strength. However, an advantage of leaded copper alloys is improved machinability, which is an important factor in the manufacture of copper-based alloys.

20.5 Aluminum Alloys

Aluminum and its alloys possess many attractive characteristics including light weight, high thermal and electrical conductivities, a nonmagnetic nature, high reflectivity, nontarnishing nature, high resistance to corrosion, reasonably high strength with good ductility, and easy fabrication. Nevertheless, probably the most important characteristic of aluminum is its low density, which is about one-third that of steels and copper alloys. Because of this, certain aluminum alloys have a better strength-to-weight ratio than high-strength steels.

Among the many alloying elements added to aluminum, the most widely used are copper, silicon, magnesium, zinc, and manganese. These are used in various combinations, and in many cases they are used together with other additions to produce classes of age hardening, casting, and work hardening alloys as shown in Fig. 20.7. All age hardening alloys contain alloying elements

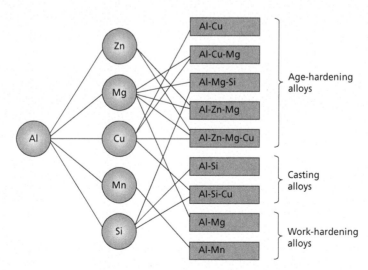

Fig. 20.7 Major aluminum alloy systems. (Hatch, J. E., Ed., *Aluminum: Properties and Physical Metallurgy*, © 1984. Reprinted with permission of ASM International®. All rights reserved. www.asminternational.org)

that dissolve in aluminum at elevated (solutionizing) temperatures and precipitate at lower (aging) temperatures. An example of an age hardening alloy is Al-Cu, which was discussed in Chapter 16. Most casting alloys contain silicon, which improves the fluidity and mold-filling capacity of aluminum alloys and reduces their susceptibility to hot cracking and the formation of shrinkage cavities during solidification. Work hardening alloys frequently contain Mn and Mg, which form a fine dispersion of intermetallic phases and/or impart solid-solution strengthening.

20.6 Aluminum-Lithium Alloys

Metallurgists have long been challenged with the task of developing lighter and stronger alloys. One system that has received particular attention is the Al-Li system. Among the alloying additions, lithium has been identified as the only alloying element (besides beryllium) that increases the elastic modulus of Al while at the same time reduces its density. Each weight percent of lithium decreases the density of aluminum by approximately 3 percent while increasing its elastic modulus by approximately 6 percent. These two characteristics have produced a great deal of interest in research and development work on the Al-Li alloys. Al-Li alloys, however, suffer from both low ductility and fracture toughness. As will be discussed later, the low ductility has been attributed to strain localization caused by precipitate-free zones (PFZ) and/or shearable coherent precipitates. Hence, recent alloy developments have concentrated on modifying the structure of the alloy in order to alter its deformation mechanisms. The most actively pursued method involves the addition of dispersoid forming elements, such as manganese, chromium, iron, and zirconium. Powder metallurgy is the primary technique used in the production of dispersoid forming alloys; it achieves a fine homogeneous distribution of the dispersoids throughout the matrix. The powder metallurgy technique involves the production of alloy powders from a liquid melt by atomization, followed by hot pressing or hot extruding of powder preforms into ingots. Although improvements in ductility have been achieved, the addition of dispersoid elements undoubtedly affects adversely the low-density characteristics of the Al-Li alloys that initially created interest in them.

Lithium has a significant solubility in aluminum, with a maximum of 5.2 weight percent (2.14 at. percent) at the eutectic temperature of 809 K (536°C). Figure 20.8 shows the Al-rich corner of the Al-Li phase diagram. When Al-Li alloys are cooled slowly from the single-phase field, they decompose to form the equilibrium α and δ phases, where the latter is the intermetallic (AlLi) phase. However, when the alloy is quenched to lower temperatures, the decomposition process goes through a continuous precipitation of δ' particles that are finely distributed throughout the matrix. The age hardening sequence of the Al-Li alloys can therefore be written as:

$$\text{Supersaturated } \alpha \rightarrow \delta'(Al_3Li) \rightarrow \delta(AlLi) \qquad \textbf{20.1}$$

The metastable δ' phase, with the composition Al_3Li, is an ordered, coherent precipitate with the Li_2 superlattice crystal structure. This fcc structure is similar to that of Cu_3Au, with the lithium atoms occupying the corners and the aluminum atoms the faces of the fcc unit cell. The particles of δ' form as spheres in a cube-to-cube orientation relationship with the matrix. The misfit strain between the matrix and the precipitate is less than 0.2 percent and the interfacial energy of the boundary between the matrix and the precipitate has been reported (Refs. 11 and 12) to be 2.5×10^2 J/m² and 24×10^2 J/m². Upon continued aging, individual δ' particles have been observed to grow to a size of about 0.3 μm without a breakdown in their interfacial coherency.

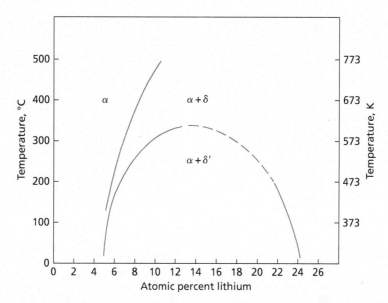

Fig. 20.8 The aluminum-lithium phase diagram showing the apparent metastable miscibility gap for δ'. (From Sanders, T. H, Jr., and Starke, E. A., *Aluminum-Lithium Alloys*, p. 63, Minerals, Metals, and Materials Society, Warrendale, Pennsylvania, 1981)

Attaining such a large size without losing coherency is due to the relatively low misfit strain existing between the δ' particles and the matrix.

Figure 20.9 shows the effect of aging time on the size of the δ' precipitates in a multicomponent Al-Li alloy containing 3 percent lithium, 5.5 percent magnesium, and 0.2 percent zirconium. The behavior of these precipitates in this multicomponent alloy is similar to that in the simple binary alloy. The growth, size, and distribution of the δ' precipitate follow the Lifshitz-Wagner theory of the kinetics of precipitate coarsening and show a classical change in the average particle radius with $t^{2/3}$, where t is the time.

The exact mechanism of the δ' precipitate formation is still undetermined. Probable formation mechanisms include either homogeneous nucleation or spinodal decomposition. Some experiments show that step quenching to eliminate excess vacancies does not influence the δ' precipitation, nor do grain boundaries or dislocations at low aging temperatures. On the other hand, Thompson and Noble (Ref. 13) reported an incubation time for δ' formation in an alloy containing 2 percent lithium, and no incubation time for lithium concentrations greater than 2 percent. This tends to support the spinodal decomposition mechanism, but direct conclusive evidence has not yet been found.

Overaging of Al-Li alloys results in the dissolution of the metastable δ' and the nucleation and growth of the semicoherent equilibrium Al-Li (δ) phase. This phase has been observed to nucleate heterogeneously in both the matrix and at grain boundaries, and independently of δ'.

The ordered coherent δ' precipitate in the Al-Li alloys results in a unique failure mechanism that depends on the microstructural state. In general, strengthening by age hardening is related to the size and distribution of the precipitates and their interaction with dislocations. This is also true for Al-Li alloys. The plastic deformation of aluminum and most of its alloys occurs by the motion of dislocations on {111} close-packed planes along ⟨110⟩ close-packed directions. Three

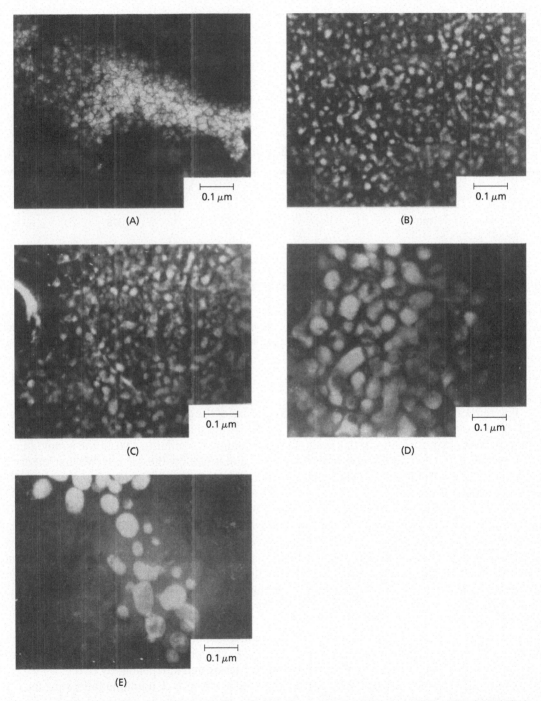

Fig. 20.9 TEM photomicrographs showing the precipitation and aging behavior of δ′ at 200°C (473 K). **(A)** Solutionized and water quenched; **(B)** aged 2 hrs., **(C)** aged 5 hrs., **(D)** aged 30 hrs., and **(E)** aged 50 hrs. (From Abelin, S. P., and Abbaschian, G. J., *Mechanical Behavior of Rapidly Solidified Materials*, Sastry, S. M. L., and MacDonald, B. A., Eds., p. 167, Minerals, Metals, and Materials Society, Warrendale, Ohio, 1985)

failure modes have been identified in Al-Li alloys (Ref. 14), and they are shown schematically in Fig. 20.10. The first mechanism, in Fig. 20.10A, involves the shearing of precipitates. Mechanical loading of a structure containing ordered δ' precipitates that are coherent can result in the passage of dislocations through the precipitates. The passage of a dislocation through an unsheared precipitate lattice creates a disturbance in the order of this lattice and produces what is known as an *antiphase boundary* (APB) in the precipitate. The extra energy associated with the APB normally should impede the motion of other dislocations, and thus have a strengthening effect on the lattice. However, when a precipitate is sheared by the first dislocation, the passage of a second dislocation, lying on the same plane as the first dislocation, restores the order disturbed by the first dislocation. Thus, dislocations in these alloys tend to move in pairs. Such a pair is called (Refs. 13–15) a *superdislocation*. After the passage of a superdislocation, the net cross-sectional area of the precipitate on the slip plane is reduced by an amount proportional to the Burgers vector. As such, the size of the precipitate on the slip plane is effectively reduced and the energy required for the movement of subsequent dislocations is lowered. Thus, once slip starts on a particular plane, further slip on that plane becomes easier. Consequently, intense planar slip can occur leading to dislocation pile-ups at grain boundaries and subsequent grain-boundary failure. This mechanism of failure is called *strain localization*.

The above-mentioned strain localization due to the shearing of precipitates is one reason for the low ductility of certain precipitation-hardened aluminum alloys that include the Al-Li alloys. The second failure mechanism, as shown in Fig. 20.10B, is due to the formation of precipitate-free zones (PFZ) adjacent to the grain boundaries. The precipitate-free zones are softer than the matrix and can become the sites of concentrated deformation that lead to rapid work hardening in these areas and subsequent brittle intergranular failure. Hence, numerous studies have been made concerning the formation of the precipitate free zones (Refs. 16 and 17). From these studies, two different mechanisms of PFZ formation have emerged; one involves vacancy depletion and the other solute depletion.

The formation of a vacancy-depleted PFZ can be explained if the solute concentration is assumed constant throughout the matrix so that there is no segregation of solute at the grain boundaries. In this case, after quenching from the solution temperature, a vacancy concentration gradient will be established in the vicinity of the grain boundaries, as demonstrated in Fig. 20.11A. This occurs because the grain boundaries are sinks for vacancies. The vacancy concentration gradient at a grain boundary will be dependent on both the solution temperature and the severity of the

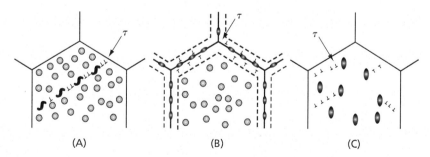

Fig. 20.10 Schematic presentation of tensile deformation and crack nucleation mechanisms: **(A)** slip band fracture, **(B)** grain boundary fracture (PFZ), **(C)** void nucleation and coalescence at inclusions. (From Gysler, A., Crooks, R., and Sanders, E. A., *Aluminum-Lithium Alloys*, p. 264, Minerals, Metals, and Materials Society, Warrendale, Pennsylvania, 1981)

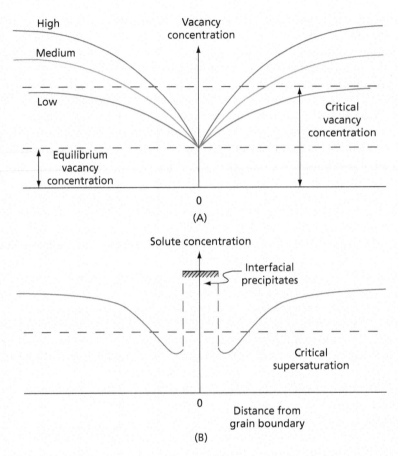

Fig. 20.11 A schematic of the vacancy and solute concentration profiles adjacent to the grain boundary

quench. As the quench rate decreases, the concentration gradient decreases. A similar argument holds for the solution temperature; that is, a higher solution temperature gives a steeper gradient. As discussed in Sec. 16.4, vacancies play a very important role in the formation of GP zones and coherent precipitates. In fact, in many cases, a critical vacancy supersaturation is necessary for the formation of these precipitates. Obviously, for a given set of aging conditions, if the vacancy concentration falls below this critical limit, the nucleation and growth of the precipitates may be so diminished that a precipitate-free zone forms at the grain boundaries.

A precipitate-free zone may also occur when solute is lost near the grain boundaries due to the heterogeneous nucleation of precipitates at the grain boundaries. (See Sec. 16.10.) Heterogeneous nucleation of the equilibrium δ (Al-Li) phase at grain boundaries is possible when during cooling the temperature of the alloy drops below the solvus line. As a result, a solute concentration profile is established adjacent to a grain boundary, as shown in Fig. 20.11B. The size of the solute-depleted zone and its gradient depend on the aging temperature and time (Ref. 18). At low aging temperatures, the width of the PFZ formed is less than that at high aging temperatures. When the vacancy or solute concentration adjacent to the grain boundaries falls below a certain limit at a given aging temperature, homogeneous nucleation of precipitates in these regions may become impossible.

Consequently, a lithium-depleted δ' precipitate-free zone is formed along the grain boundary which contains relatively large particles of the equilibrium δ phase. This yields a microstructure that is sometimes referred to as a *necklace structure*. Since the precipitate-free zones are softer than the matrix that contains precipitate particles, they become preferential sites of greater deformation; that is, strain localization. This develops large stress concentrations at grain-boundary triple points and leads to a lowering of the ductility and thus, to earlier fracture.

The third mechanism of failure in AL-Li alloys appears in Fig. 20.10C. In this case, the presence of hard incoherent dispersoids, added as grain refiners or constituent phases in the form of impurities, leads to microvoid nucleation and coalescence as a result of the passage of dislocations. For example, it has been shown that the addition of magnesium to aluminum-lithium alloys is capable of raising the yield stress, tensile strength, and ductility if the magnesium content does not exceed 4 weight percent (Ref. 19). The particles formed as a result of the magnesium additions provide strengthening by inhibiting dislocation motion, but they also serve as sites for the nucleation of microvoids, which promote the third mechanism of fracture. Higher concentrations of magnesium also result in increasing amounts of intergranular fracture due to the formation of particles at the grain boundaries.

The problems associated with strain localization within the matrix can be minimized by properly alloying and heat treating the alloy. The general purpose is to change the deformation mode of the alloy from dislocation shearing of precipitates to dislocation looping or bypassing of precipitates. Some of the additives also strengthen the matrix. Among the various alloying elements, Mg, Cu, and Zr are the most widely used. Figure 20.12 shows the effect of various alloying elements on the ductility and yield strength of Al-Li-X alloys, where X corresponds to the added alloying element. Table 20.3 compares the properties of a commercial Al-Li alloy with those of selected Al-based alloys that are widely used in the aerospace industry.

Magnesium additions increase the strength of Al-Li alloys by solid solution strengthening of the matrix and by enhancing the δ' precipitations. In aged Al-Li-Mg alloys, magnesium additions produce a 50 MPa increase in yield stress for each weight percent addition if the magnesium concentration is less than 2 percent. Additions beyond 2 percent give a 20 MPa increase for each weight percent addition, which is similar to the effect in Al-Mg alloys. The latter is due to the fact

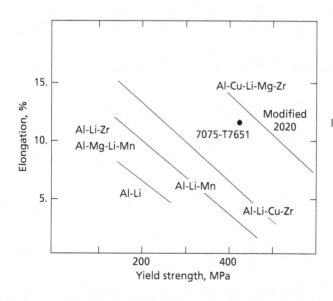

Fig. 20.12 Plot of strength/ductility relationships for various Al-Li-X alloys. (Starke, E. A., Jr., Sanders, T. A., Jr., and Palmer, I. G., *J. of Metals*, Vol. 33, No. 8, 24, 1981)

Table 20.3 Comparison of Typical Mechanical Properties for Aluminum Alloys in Peak Aged Condition

Alloy and Temper	Approximate Nominal Composition	Tensile Strength, MPa	Yield Strength, MPa	Elongation Percent	Average Tensile and Compressive Modulus 10^3, MPa	Shear Strength, MPa	Endurance Limit (500 × 10^6 cycles), MPa
X2020-T6	Al-4.5 Cu-1. Li-0.5 Mn-0.2 Cd	579	531	3	77.2	338	159
2024-T86	Al-4.4 Cu-1.5 Mg-0.6 Mn	517	490	6	73.1	310	124
7075-T6	Al-1.6 Cu-5.6 Zn-2.5 Mg-0.2 Cr	572	503	11	71.7	331	152
7079-T6	Al-0.6 Cu-4.3 Zn-3.3 Mg-0.2 Mg-0.2 Cr	538	469	14	71.7	310	152
7178-T6	Al-2.0 Cu-6.8 Zn-2.7 Mg-0.23 Cr	607	538	11	71.7	359	152

From Balmuth, E. S., and Schmidt, R., in *Aluminum-Lithium Alloys*, edited by T. H. Sanders, Jr. and E. A. Starke, Jr. *TMS-AIME*, Warrendale, Pennsylvania, 1981.

that magnesium reduces the solubility of lithium in the matrix and thus increases the precipitation of the δ' phase. Magnesium additions greater than 2 weight percent promote the formation of the equilibrium Al_2LiMg phase (Ref. 20). Its structure is fcc with a lattice parameter of 1.99 nm. The Al_2LiMg phase precipitates as rods with a $\langle 110 \rangle$ growth direction. The orientation relationship with the alpha matrix is:

$$(110)_{Al_2LiMg} \parallel (110)_\alpha; \quad (110)_{Al_2LiMg} \parallel (111)_\alpha \qquad 20.2$$

Hence, Al_2LiMg is semicoherent and exhibits accommodation dislocations. The formation of Al_2LiMg from solid solution has been reported to proceed as follows:

$$\text{Supersaturated } \alpha \rightarrow \delta' \rightarrow Al_2LiMg \qquad 20.3$$

The Al_2LiMg compound has also been observed to precipitate heterogeneously both within the matrix and at grain boundaries, which implies it forms independently of the δ' intermediate precipitate.

Zirconium is added to Al-Li alloys primarily as a grain refiner. It forms an intermetallic compound, Al_3Zr, which has a tetragonal (DO_{23}) structure. This intermetallic compound is stable at high temperatures and resists coarsening. Hence, it cannot be manipulated or affected by subsequent processing. It has also been reported that zirconium reduces the solubilities of lithium and magnesium in aluminum (Ref. 21).

20.7 Titanium Alloys

Titanium and its alloys are widely used as structural materials in the aerospace and chemical industries because of their lower density combined with high strength and corrosion resistance. The low density of titanium, 4.5 Mg/m^3, as compared with approximately 7.8 Mg/m^3 for steels and superalloys, offers the possibility of substantial reductions in airframe weight and increased thrust/weight ratios of their jet engines. As shown in Fig. 20.13, the tensile strength/density ratios of titanium alloys are considerably higher than those of steels and aluminum alloys at ordinary temperatures. The yield strength/density ratios of titanium alloys are also higher than those of steels and aluminum and magnesium alloys. It is because of this property that titanium alloys have been widely used instead of iron or nickel-based alloys in aerospace applications. In addition, the high melting point of the metal (2093 K) places titanium among the refractory metals, making it attractive for many high temperature applications. However, the chemical behavior of the metal normally limits the application of titanium alloys to low to moderately elevated temperatures. At low temperatures, titanium passivates and becomes almost completely inert to most mineral acids and chlorides. At high temperatures, on the other hand, titanium oxidizes quickly and its properties are greatly altered by the adsorption of oxygen interstitials from the air. It should be noted that heating titanium in air results not only in oxidation but also in dissolution of oxygen and nitrogen in the surface layers of titanium, causing solid-solution hardening of the surface. The thickness of the "air contaminated layer" depends on the temperature as well as the dissolution time. Normally, this layer is removed by machining or other means prior to using the alloy because the presence of the air contaminated layer reduces the fatigue strength and ductility. Pure titanium undergoes an allotropic transformation at about 1158 K; it changes from the close-packed hexagonal crystal structure to the body-centered cubic crystal structure. The high-temperature bcc

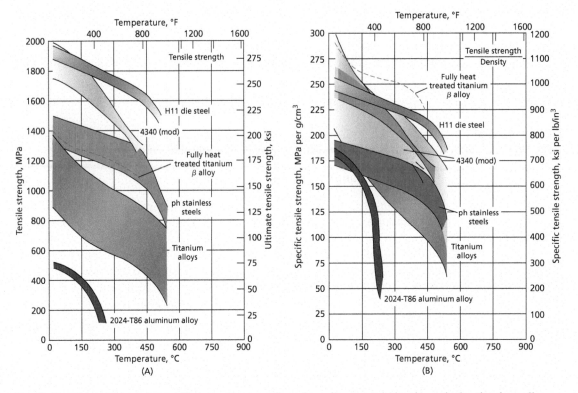

Fig. 20.13 Comparison of short-time tensile strength and tensile strength/density ratio for titanium alloys, three classes of steel, and 2024-T86 aluminum alloys included for annealed alloys with less than 10 percent elongation or heat-treated alloys with less than 5 percent elongation. (From *ASM Handbook*, Volume 2, Properties and Selection: Nonferrous Alloys and Special-Purpose Materials, 1990. Reprinted with permission of ASM International®. All rights reserved. www.asminternational.org)

phase, called the beta phase, has a lattice parameter of 0.332 nm at 1200 K and a density of about 4.35 Mg/m^3. For the low-temperature hcp phase, called the alpha phase, $a = 0.29503$ nm, $c = 0.468312$ nm, and $c/a = 1.5873$. The room temperature density of α is 4.507 Mg/m^3.

The beta to alpha transformation in pure titanium occurs very easily and cannot be suppressed by rapid quenching. Because of this, it is widely believed that the transformation is a diffusionless martensitic transformation, rather than one that is diffusion controlled. Figure 20.14 shows a proposed mechanism by which the bcc structure transforms to hcp. In order to form the hexagonal cell, it is necessary to change the angle *ABC* from 70°32′ to 60°. This can be accomplished by a shear on the (112) plane and in the ⟨111⟩ direction of the bcc structure. Additional contraction and expansion of the cell edges, plus displacement of the center atom, will then result in the formation of the hcp structure.

The allotropic phase transformation temperature of titanium depends strongly on the amount and nature of the alloying elements. Some alloy additions tend to raise the transformation temperature. These are called *alpha* (α) *stabilizers*. Others lower this temperature and are called *beta* (β) *stabilizers*. Among the former elements, carbon, oxygen, nitrogen, and aluminum are the most

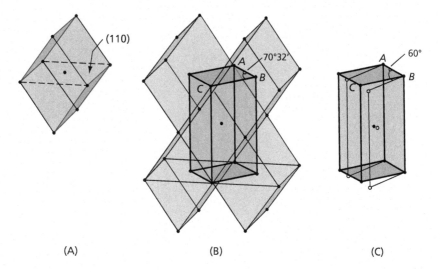

Fig. 20.14 The diffusionless transformation of β titanium to α titanium. **(A)** A bcc unit cell of β titanium; **(B)** an array of the cells shown in (A), with a potential hcp cell outlined; **(C)** shear of the cell in (B) into the hcp cell of α titanium. (From Brick, R. M., Pense, A. W., and Gordon, R. B., *Structure and Properties of Alloys*, 4th Ed., McGraw-Hill, New York, 1977. Used with permission)

important; their phase diagrams are shown in Fig. 20.15. Carbon, oxygen, and nitrogen are rapidly absorbed by titanium when the metal is hot, and consequently some of these elements are expected to be present in all of the commercial grades of titanium. In addition, all of these elements harden and solution strengthen α titanium. Aluminum has significant solubility in both alpha and beta phases, and because of this, it strengthens both phases.

The elements that lower the transformation temperature and stabilize β can arbitrarily be divided into two classes: (1) those that undergo eutectoid transformations, such as chromium, iron, copper, nitrogen, palladium, cobalt, manganese, and hydrogen, and (2) those that are isomorphous with β at high temperatures and form α + β equilibrium phases at ordinary temperatures. Examples of these two classes are shown in Fig. 20.16. Among the latter are molybdenum, tantalum, vanadium and niobium. These elements have limited solubility in the alpha phase. In general, beta stabilizers impart solid-solution hardening to the beta phase but, in contrast with the alpha stabilizers, affect the alpha phase very little. Consequently, a combination of alpha and beta stabilizers can be added to the two-phase alpha and beta alloys, in order to strengthen both phases. Zirconium and hafnium are called sister elements of titanium because they are isomorphous with both the alpha and beta phases.

20.8 Classification of Titanium Alloys

Commercial titanium alloys are classified as "α or near α alloys," "α + β" and "β" alloys. Table 20.4 gives a summary of some of these alloys. The table also lists different grades of commercially pure titanium. It should be noted that the various grades of pure titanium depend on the impurity content, most notably oxygen and iron. Higher purity grades contain lower levels of interstitials and have lower strength, hardness, and transformation temperatures than lower purity grades. As

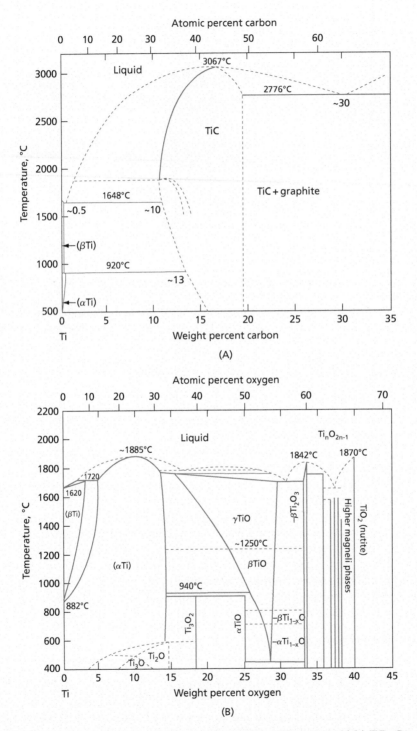

Fig. 20.15 Phase diagrams of titanium with selected α stabilizers. (From Massalski, T.B., *Binary Alloy Phase Diagrams*, ASM, 1990. Reprinted with permission of ASM International®. All rights reserved. www.asminternational.org)

20.8 Classification of Titanium Alloys

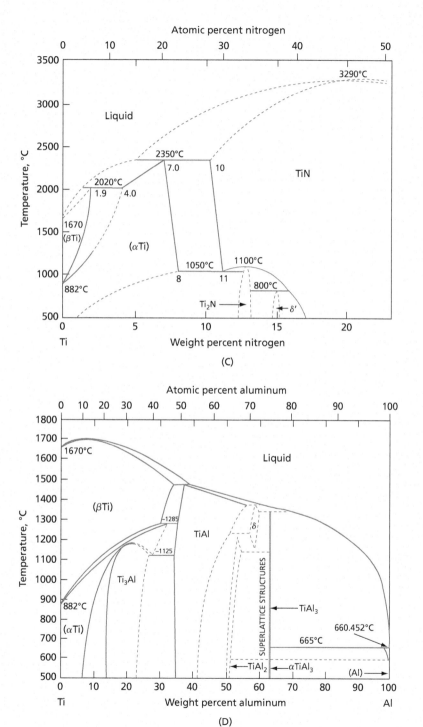

Fig. 20.15 (cont.)

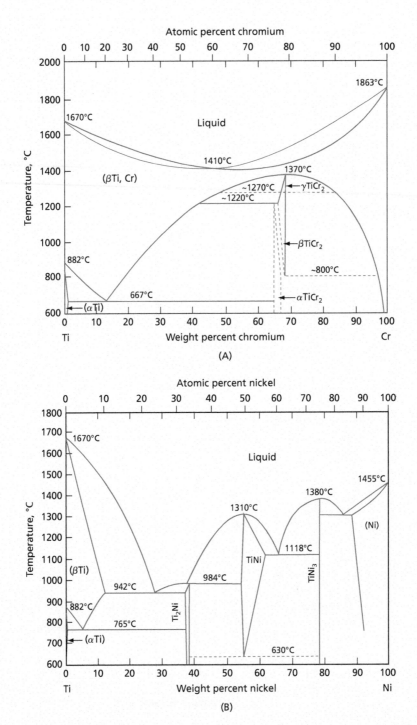

Fig. 20.16 Phase diagrams of titanium with selected beta stabilizers. (From Massalski, T. B., *Binary Alloy Phase Diagrams*, ASM, 1990. Reprinted with permission of ASM International®. All rights reserved. www.asminternational.org)

20.8 Classification of Titanium Alloys 689

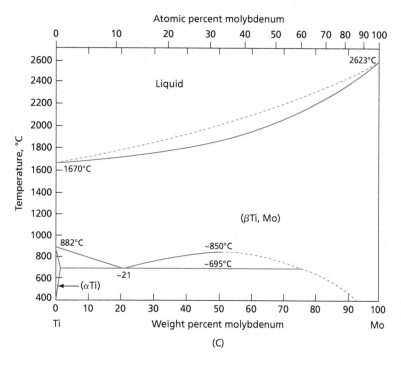

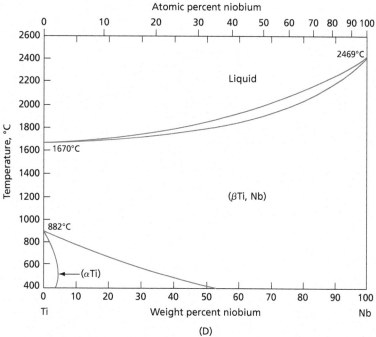

Fig. 20.16 (cont.)

Table 20.4 Summary of Commercial and Semicommercial Grades and Alloys of Titanium

Designation	Tensile Strength (min) MPa	Tensile Strength (min) ksi	0.2% yield Strength (min) MPa	0.2% yield Strength (min) ksi	Impurity Limits, wt% N (max)	Impurity Limits, wt% C (max)	Impurity Limits, wt% H (max)	Impurity Limits, wt% Fe (max)	Impurity Limits, wt% O (max)	Nominal Composition, wt% Al	Nominal Composition, wt% Sn	Nominal Composition, wt% Zr	Nominal Composition, wt% Mo	Nominal Composition, wt% Others
Unalloyed grades														
ASTM Grade 1	240	35	170	25	0.03	0.10	0.015	0.20	0.18	...	...	...	...	...
ASTM Grade 2	340	50	280	40	0.03	0.10	0.015	0.30	0.25	...	...	...	...	...
ASTM Grade 3	450	65	380	55	0.05	0.10	0.015	0.30	0.35	...	...	...	...	...
ASTM Grade 4	550	80	480	70	0.05	0.10	0.015	0.50	0.40	...	...	...	...	...
ASTM Grade 7	340	50	280	40	0.03	0.10	0.015	0.30	0.25	...	...	...	...	0.2Pd
Alpha and near-alpha alloys														
Ti Code 12	480	70	380	55	0.03	0.10	0.015	0.30	0.25	...	...	...	0.3	0.8Ni
Ti-5Al-2.5Sn	790	115	760	110	0.05	0.08	0.02	0.50	0.20	5	2.5	...	...	...
Ti-5Al-2.5Sn-ELI	690	100	620	90	0.07	0.08	0.0125	0.25	0.12	5	2.5	...	...	...
Ti-8Al-1Mo-1V	900	130	830	120	0.05	0.08	0.015	0.30	0.12	8	...	...	1	1V
Ti-6Al-2Sn-4Zr-2Mo	900	130	830	120	0.05	0.05	0.0125	0.25	0.15	6	2	4	2	...
Ti-6Al-2Nb-1Ta-0.8Mo	790	115	690	100	0.02	0.03	0.0125	0.12	0.10	6	...	...	1	2Nb, 1Ta
Ti-2.25Al-11Sn-5Zr-1Mo	1000	145	900	130	0.04	0.04	0.008	0.12	0.17	2.25	11.0	5.0	1.0	0.25i
Ti-5Al-5Sn-2Zr-2Mo(a)	900	130	830	120	0.03	0.05	0.0125	0.15	0.13	5	5	2	2	0.25 Si
Alpha-beta alloys														
Ti-6Al-4V(b)	900	130	830	120	0.05	0.10	0.0125	0.30	0.20	6.0	...	...	...	4.0V
Ti-6Al-4V-ELI(b)	830	120	760	110	0.05	0.08	0.0125	0.25	0.13	6.0	...	...	...	4.0V
Ti-6Al-6V-2Sn(b)	1030	150	970	140	0.04	0.05	0.015	1.0	0.20	6.0	2.0	...	...	0.75Cu, 6.0V
Ti-8Mn(b)	860	125	760	110	0.05	0.08	0.015	0.50	0.20	...	...	...	...	8.0Mn
Ti-7Al-4Mo(b)	1030	150	970	140	0.05	0.10	0.013	0.30	0.20	7.0	...	...	4.0	...
Ti-6Al-2Sn-4Zr-6Mo(c)	1170	170	1100	160	0.04	0.04	0.0125	0.15	0.15	6.0	2.0	4.0	6.0	...
Ti-5Al-2Sn-2Zr-4Mo-4Cr(a)(c)	1125	163	1055	153	0.04	0.05	0.0125	0.30	0.13	5.0	2.0	2.0	4.0	4.0Cr
Ti-6Al-2Sn-2Zr-2Mo-2Cr(a)(b)	1030	150	970	140	0.03	0.05	0.0125	0.25	0.14	5.7	2.0	2.0	2.0	2.0Cr, 0.25Si
Ti-10V-2Fe-3Al(a)(c)	1170	170	1100	160	0.05	0.05	0.015	2.5	0.16	3.0	...	...	...	10.0V
Ti-3Al-2.5V(d)	620	90	520	75	0.015	0.05	0.015	0.30	0.12	3.0	...	...	...	2.5V
Beta alloys														
Ti-13V-11Cr-3Al(c)	1170	170	1100	160	0.05	0.05	0.025	0.35	0.17	3.0	...	...	...	11.0Cr, 13.0V
Ti-8Mo-8V-2Fe-3Al(a)(c)	1170	170	1100	160	0.05	0.05	0.015	2.5	0.17	3.0	...	...	8.0	8.0V
Ti-3Al-8V-6Cr-4Mo-4Zr(a)(b)	900	130	830	120	0.03	0.05	0.020	0.25	0.12	3.0	...	4.0	4.0	6.0Cr, 8.0V
Ti-11.5Mo-6Zr-4.5Sn(b)	690	100	620	90	0.05	0.10	0.020	0.35	0.18	...	4.5	6.0	11.5	...

[a]Semicommercial alloy; mechanical properties and composition limits subject to negotiation with suppliers. [b] Mechanical properties given for annealed condition; may be solution treated and aged to increase strength. [c]Mechanical properties given for solution treated and aged condition; alloy not normally applied in annealed condition. Properties may be sensitive to section size and processing. [d] Primarily a tubing alloy; may be cold drawn to increase strength.

(From *Metals Handbook*, 9th Edition, Vol. 3, American Society for Metals, Metals Park, Ohio, 1980, p. 357. Reprinted with permission of ASM International®. All rights reserved. www.asminternational.org)

shown in Table 20.4, the yield strength of the unalloyed grades vary from 170 MPa to 485 MPa, depending on the impurity and interstitial contents. Unalloyed grades of titanium account for about 30 percent of total titanium production, as compared with 45 percent for the most widely used Ti-6Al-4V, and 25 percent for all other remaining alloys.

20.9 The Alpha Alloys

Alpha alloys contain alpha stabilizers such as Al and Sn, singly or in combination, and are hcp at ordinary temperatures. Aluminum has a strong solid-solution hardening effect on titanium, as may be seen in Fig. 20.17, but it also reduces the ductility. The addition of tin also strengthens α titanium, however, without a significant loss of ductility. In many instances, both alloying elements are added together as in the commercial Ti-5Al-2.5Sn alloy. The α titanium alloys are generally categorized by good strength, toughness, creep resistance, and weldability. However, unlike other titanium alloys, alpha alloys cannot be strengthened by heat treatment. They are generally annealed or recrystallized to remove residual stresses induced during cold working.

Alpha alloys containing extra low levels of interstitials (ELI grades), such as Ti-5Al-2.5Sn-ELI, retain ductility and toughness at cryogenic temperatures. Because of the absence of a ductile-brittle transformation, a property of the bcc structure, these alpha alloys have been extensively used in cryogenic applications. Some alpha alloys contain small additions of beta stabilizers, such as Ti-4Al-1Mo-1V, and contain some retained beta phase. Nevertheless, since the majority of the alloys consists of the alpha phase, they behave more like alpha alloys than alpha-beta alloys.

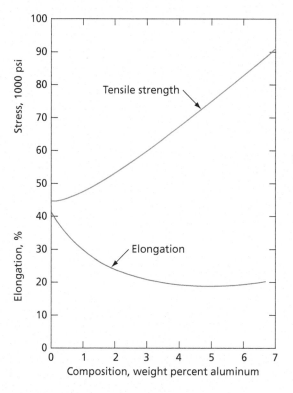

Fig. 20.17 The hardening of α titanium by aluminum in solid solution. (From Brick, R. M., Pense, A. W., and Gordon, R. B., *Structure and Properties of Alloys*, 4th Ed., McGraw-Hill, New York, 1977. Used with permission)

20.10 The Beta Alloys

Beta alloys generally contain one or more of the so-called beta stabilizers, such as molybdenum, vanadium, or chromium. These alloys are characterized by high hardenability and forgeability. They are, however, susceptible to a ductile-brittle transformation and are unsuitable for low-temperature applications.

In the solution-treated conditions, beta alloys have good ductility, toughness, and excellent formability. The most important advantage of beta alloys over alpha alloys is that they can be age hardened to cause partial transformation of the metastable beta phase to alpha. The age hardening process, which is commonly accomplished at 700 to 900 K, results in the formation of finely dispersed alpha particles in the retained beta phase.

20.11 The Alpha-Beta Alloys

Alpha-beta alloys contain mixtures of both the alpha and the beta stabilizers such that the alloys retain a mixture of the alpha and beta phases at room temperature.

The dual-phase structure of these alloys makes them considerably stronger than either the alpha or beta alloys. Moreover, the alpha-beta alloys can be further strengthened by heat treatment. This is accomplished by quenching the alloys from the $\alpha + \beta$ phase field, followed by aging at a moderate temperature. The heat treatment cycle for a typical alloy is schematically shown in Fig. 20.18. It should be noted that the maximum heating temperature is in the two-phase $\alpha + \beta$ region, as marked by point a in the figure, in contrast with the usual age hardening of other alloys such as Al-Cu. For the latter alloys, the maximum temperature is in the one-phase region, and a

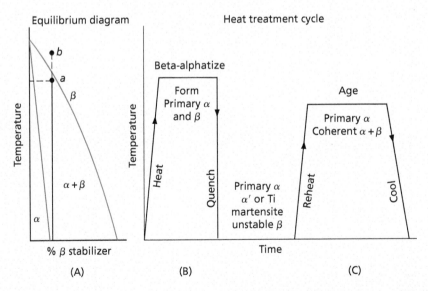

Fig. 20.18 Heat treatment for a typical alpha-beta titanium alloy. **(A)** Partial phase diagram; **(B)** solution treatment; **(C)** subsequent aging treatment. (From Brick, R. M., Gordon, R. B., and Phillips, A., *Structure and Properties of Alloys*, 3rd Ed., McGraw-Hill, Inc., New York, 1965. Used with permission)

single-phase solid solution is formed prior to quenching. For alpha-beta alloys, however, heating to the all beta phase field, to a point such as point *b* in Fig. 20.18A, would result in the formation of the beta phase with excessively large grains. The subsequent age hardening would also be mostly at the beta boundaries, and inhomogeneously distributed. These conditions reduce the ductility of the alloys.

When this alloy is slowly cooled from the beta region, Widmanstätten plates of alpha initially nucleate at the grain boundaries at about 980°C. The hcp plates form with their basal planes parallel to the {110} plane of the bcc matrix. This orientation relationship is similar to that shown earlier for the martensitic transformation of pure Ti. Figure 20.19 schematically shows the formation of Widmanstätten plates in a Ti-6Al-4V alloy. Note that the partial phase diagram shown in the left of this figure is an approximate isopleth for an alloy containing 6 percent aluminum. Upon continued cooling, the plates thicken relatively slowly at their broad sides which are closely matching the matrix, but grow much faster along the edge sides. The final microstructure would consist of alpha plates with the remaining beta in between them.

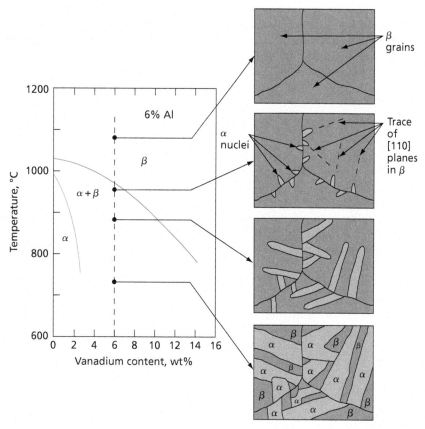

Fig. 20.19 Schematic illustration of the formation of a Widmanstätten structure in a Ti-6Al-4V alloy by cooling slowly from above the beta transus. The final microstructure consists of plates of alpha separated by the beta phase. (From Brooks, C. R., *Nonferrous Alloys*, Amercian Society for Metals, 1984. Reprinted with permission of ASM International®. All rights reserved. www.asminternational.org)

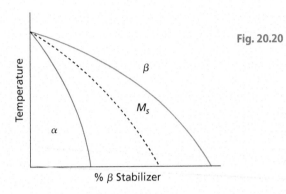

Fig. 20.20 Schematic phase diagram showing the dependence of the martensitic-start (M_s) temperature of beta-stabilized types of titanium alloys on composition. (Reprinted from R. I. Jaffee, "The physical metallurgy of titanium alloys," *Progress in Metal Physics*, Vol 7, ed B. Chamers & R. King, 1958, pp. 65–106, with permission from Elsevier. https://www.sciencedirect.com/science/article/abs/pii/0502820558900042)

On the other hand, when the alloy is heated to the two-phase field, the presence of alpha grains inhibits excessive growth of the beta grains. Quenching of this alloy then suppresses the transformation of the beta phase, until it is reheated to the aging temperature. During aging, alpha precipitate particles may form in a manner similar to that described in the previous section on beta alloys.

It should be noted that upon quenching, beta may decompose by a martensitic type reaction, forming a metastable phase. Depending upon the quenching temperature and alloy composition, the spontaneous transformation product may be α' or α''. The structure of α' has not yet been resolved, but it is indicated to be either face-centered cubic or face-centered tetragonal (Ref. 22) or hcp (Ref. 23). Similarly, α' has been identified as having either a hexagonal (Ref. 22) or orthorhombic (Ref. 23) structure.

Similar to the martensitic transformation in iron base alloys, the formation of martensite begins at the martensite start temperature, M_s, which depends on the alloy composition. This is shown schematically in Fig. 20.20. If the amount of the beta stabilizer is such that its M_s is lower than room temperature, quenching of the alloy will result in the retainment of the metastable beta phase. On the other hand, when the M_s temperature is above room temperature, quenching results in the partial or complete formation of one of the martensites.

The formation of the titanium martensite produces some hardening of the alloy, but greater strength can be attained if the quenched alloy is subsequently "aged." During aging, some of the α' or α decomposes to alpha and beta, but a part of the retained metastable beta phase also decomposes to form coherent alpha precipitates in the beta matrix. Additional details can be found in other references (Refs. 24–26).

It should be recognized that during the aging heat treatment another phase called (ω) may form that has an undesirable embrittling effect on the alloy. The structure and details of the formation of ω are not yet fully understood (Ref. 23). The omega phase precipitation occurs athermally within a narrow composition range in many titanium alloys as a result of rapid quenching from elevated temperatures. Its structures may be either hexagonal or trigonal depending on the solute concentration.

20.12 Superalloys

One of the most demanding applications of materials is in gas turbines of military and commercial aircraft, as well as in other industrial turbines. Gas turbines ingest air from the atmosphere, compress it, add fuel, and burn the mixture to produce hot inlet gases that drive the turbine section. The turbine components exposed to the hot gas are subjected to high-temperature oxidation, hot corrosion, creep, high-cycle fatigue, and thermal fatigue. The alloys being used for these components are nickel-, cobalt-, and iron-base superalloys. In addition to gas turbines, superalloys are also used in space vehicles, nuclear reactors, submarines, and petrochemical equipment, as well as many other high-temperature applications. Among superalloys, the nickel-base alloys are most widely used. In this section, the physical metallurgy of these alloys is briefly discussed. More detailed analysis can be found in other references (Refs. 23, 27–30).

The nominal compositions of typical nickel-base superalloys are shown in Table 20.5. The oxidation and hot corrosion resistance of the alloys depend mostly on the presence of Cr. Chromium also imparts a solid-solution strengthening to the alloy. Molybdenum, aluminum, niobium, titanium, and other alloy additions also contribute to solid-solution hardening, as well as serving other functions. The alloys, however, depend for their strength mostly upon aluminum and titanium, which allow for precipitation hardening of the alloy. The precipitation hardening results from the formation of an ordered compound based on the chemical formula $Ni_3(Al, Ti)$, called γ'. This phase and the matrix, which is called γ, are fcc but they have different lattice parameters. The lattice mismatch between the precipitates and the matrix creates coherency strains that impede dislocation migration, thus resulting in precipitation hardening. Therefore, the mechanical properties of the alloys depend, among other factors, upon the amount, size, and morphology of γ' and the elastic strain induced because of the lattice mismatch. Recent STEM and 3-D atom

Table 20.5 Composition of Superalloys

Alloy	Composition, Percent by Weight										
	C	Cr	Ni	Fe	Co	Ti	Al	B	Mo	Cb	Others
Ni-Base											
Waspaloy[a]	0.05	19.5	balance	...	13.5	3.00	1.30	0.005	4.3	...	0.05Zr
Astroloy	0.05	15.0	balance	...	15.0	3.50	4.40	0.030	5.25	...	...
IN100	0.18	10.0	balance	...	15.0	5.00	5.50	0.015	3.0	...	1.0V, 0.05Zr
René 95[b]	0.15	14.0	balance	...	8.0	2.50	3.50	0.010	3.5	3.50	3.5W, 0.05Zr
Pyromet 31[c]	0.05	22.7	balance	15.0	...	2.30	1.30	0.005	2.0	0.85	...
Ni-Fe Base											
Alloy 901	0.05	13.5	42.7	34.0	...	2.5	0.25	0.015	6.1	...	...
Alloy 718	0.04	19.0	52.5	balance	...	0.90	0.50	0.005	3.05	5.30	...
Alloy 706	0.02	16.0	40.0	balance	...	1.70	0.30	0.004	...	2.75	...
Pyromet CTX-1	0.03	...	37.7	balance	16.0	1.75	1.00	0.008	...	3.00	...
Pyromet CTX-2	0.03	...	37.7	balance	16.0	1.75	1.00	0.008	...	3.00	0.75Hf
Fe Base											
A-286	0.05	15.0	26.0	balance	...	2.00	0.20	0.005	1.25	...	0.30V

[a] Waspaloy is a registered trademark of United Technologies Corp.
[b] René is a registered trademark of General Electric Corp.
[c] Pyromet is a registered trademark of Carpenter Technology Corp.
From *MiCon 78, Optimization of Processing, Properties and Service Performance Through Microstructural Control*, **ASTM STP 672**, American Society for Testing and Materials, Philadelphia, 1979.

probe tomography results (Refs. 31 and 32) show that the characteristics of these interfaces can be defined by a compositional width, δ, and structural width, δ'. The latter width can play an important role in high-temperature mechanical behavior of precipitation-hardened alloys since the slower diffusion in the ordered part of the interface hinders the solid-state diffusion-based phenomena.

Another factor contributing to the strength and high-temperature stability of the alloys is the presence of carbide phases such as $M_{23}C_6$, M_6C, or MC, where M can be chromium, titanium, molybdenum, or tungsten. The formation of the carbides has an important effect on creep properties, but this effect is variable and depends on the alloy composition, their size, and distribution and processing. The relative amounts of the carbides also depend on the temperature and processing, as shown in Fig. 20.21. For example, MC forms upon solidification of the alloy, but disappears during

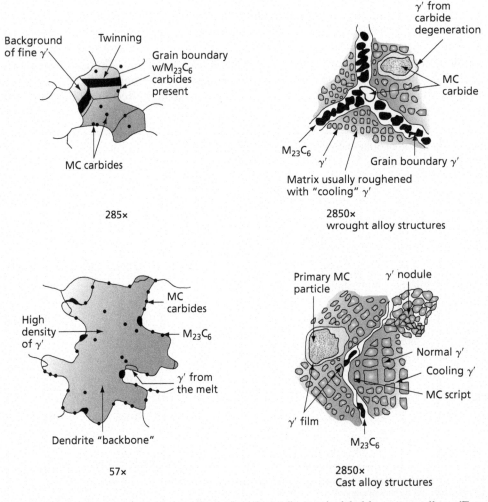

Fig. 20.21 Structures characteristic of wrought (top) and cast (bottom) nickel-base superalloys. (From Sims, C. T., and Hagel, W. C., *The Superalloys*, John Wiley and Sons, Inc., New York, 1972. Reprinted with permission of John Wiley & Sons, Inc.)

aging at lower temperatures. On the other hand, $M_{23}C_6$ is stable at temperatures below 875°C, but not at higher temperatures. The reaction between the two carbides can be written as

$$MC + \gamma \rightarrow M_{23}C_6 + \gamma' \qquad 20.4$$

20.13 Creep Strength

Turbine blades generally experience longitudinal stresses to approximately 140 MN/m² (≈20,000 psi) and temperatures of 650 to 1000°C in their airfoil section. The blade root, on the other hand, is subjected to much higher tensile stresses, around 280 to 560 MN/m², but temperatures not exceeding 760°C. The blades not only have to be strong enough to withstand such high stresses, but they must also have adequate creep resistance.

In order to produce higher creep strengths, the alloy composition can be adjusted to yield larger amounts of γ'. However, such an approach is often accompanied by a reduction in ductility, which may cause premature fracture. Therefore, traditional alloy development is limited. Great improvements in the creep performance can be made by modifying the grain structure of the alloy, however, particularly by eliminating grain boundaries that are normal to the stress direction. In general, grain boundaries contribute to creep and eventual failure of the materials by grain boundary sliding and void formation. It is because of this that large-grain specimens generally have better creep performance than fine-grain ones, provided that the structure remains stable at high temperatures.

For turbine blades, the stress is predominantly is one direction, the axial direction of the blade, and so the creep performance can be improved by eliminating the grain boundaries perpendicular to the stress direction. This has been accomplished by using directional solidification techniques, rather than conventional casting methods, to cast the blades. Directional solidification results in a microstructure that is either columnar, with the long axis of the grains parallel to the stress direction, or a monocrystal. Figure 20.22 shows a comparison of the creep properties of a superalloy (Mar-M200) in the conventionally cast, columnar grain, and monocrystalline microstructures. During conventional casting of an object, the heat of fusion is removed from the casting more or less equally in all three directions and at low temperatures (see Fig. 20.23A). As such, the microstructure consists of equiaxed grains which have nucleated randomly and grown during solidification. It should be noted that each grain normally is formed by growth of a single dendrite colony. On the other hand, if the heat removal is allowed to take place in one direction only, for example, along the blades axis and through the root of the blade, the grains will only grow parallel

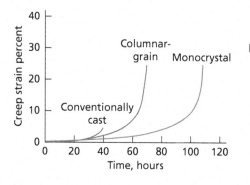

Fig. 20.22 Comparison of the creep properties at 1255 K and 207 MN/m² of MAR-M200 in the conventionally cast, columnar-grain, and monocrystalline forms. (From Sahm, P. R., and Speidel, M. O., *High Temperature Materials in Gas Turbines*, Elsevier Scientific Publishing Company, New York, 1974)

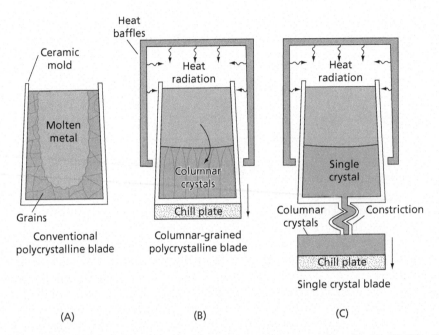

Fig. 20.23 The technique used to obtain monocrystal high-temperature turbine blades. (© Slim Films. From Kear, B. H., *Scientific American*, October, 1986)

to the heat flow direction, as shown in Fig. 20.23B. This will result in the formation of columnar grains along the blade axis. Now, if only one grain existed at the beginning, or if all the grains except one were prevented from growing, the resulting blade would have only one grain, as in Fig. 20.23C. Since this grain contains one dendrite colony, which may have second-phase precipitate in between the dendrite arms, the structure is called a monocrystal.

Problems

20.1 List the most important properties of copper, and explain why copper is a suitable material for automobile radiators, cooking utensils, and electrical wires. For each application, list other competing materials and specify the advantages or disadvantages of using copper over these other materials.

20.2 Using Table 20.1 for values of the resistivity of copper, calculate the resistance of a copper wire 1 m long and 1 mm in diameter. What would the resistivity be if

(a) The wire had $\frac{1}{2}$ mm diameter.

(b) The wire was made of pure iron.

(c) The copper contained 0.5 percent silver.

(d) The copper contained 10 percent zinc.

20.3 Cold worked oxygen-free electronic copper was heated to various temperatures for one hour and quenched to room temperature. Plot schematically the room temperature resistivity and hardness as a function of the annealing temperature, and explain the driving force and mechanisms for the changes in the resistivity and hardness.

20.4 Based on the Cu-Zn phase diagram shown in Fig. 11.23, draw the microstructure of three copper-zinc alloys containing (i) 30 at.%, (ii) 35 at.%, and (iii) 50 at.% Zn at:

(a) 800°C.

(b) 500°C.

(c) 25°C.

Assuming that the alloys were solidified by equilibrium cooling, label the phases and constituents, and draw approximately their correct areal fractions.

20.5 Copper-beryllium alloys containing less than 1.6 weight percent Be exhibit age hardening strengthening. Compare and contrast the age hardening behavior of Cu-1% Be with that of Al-4% Cu.

20.6 The hardness of a supersaturated Al-3% Li-5.5% Mg-0.2% Zr as a function of the aging time and temperature is shown in the following figure.

(a) Describe why at each aging temperature, the hardness goes through a maximum.

(b) Explain why the maximum hardness and the aging time required for the maximum hardness increase as the aging temperature is decreased.

20.7 Compare and contrast the hardness versus aging time and temperature for the Al-Li alloy given in Prob. 20.6 with that of the Al-Cu alloy given in Fig. 16.10. Explain why for the latter alloy, but not for the former, the curves may exhibit two maxima.

20.8 Explain how grain-boundary precipitation can lead to the formation of precipitation-free zones. Prescribe a heat treatment schedule to eliminate or minimize grain-boundary precipitates.

20.9 What are the outstanding properties of titanium and its alloys? Give one application for each property. Describe the allotropic crystal structures of pure titanium.

20.10 Why is it widely believed that the allotropic transformation in titanium is a diffusionless martensitic type?

20.11 Why are some titanium alloying elements called alpha stabilizers while some others are called beta stabilizers? Why are zirconium and hafnium called sister elements of titanium?

20.12 How does the addition of a beta stabilizer strengthen titanium?

20.13 How does the addition of an alpha stabilizer strengthen titanium?

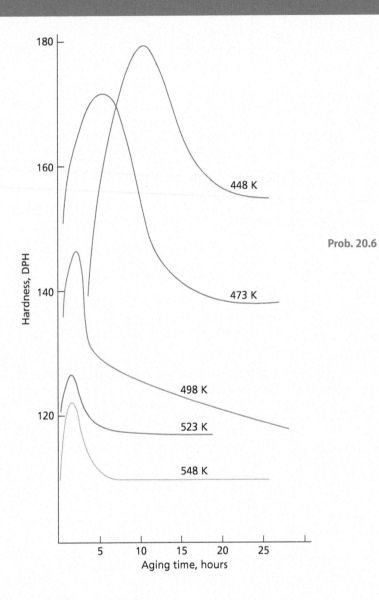

Prob. 20.6 Hardness versus aging time and temperature for an Al-Li-Mg-Zr alloy. (S. Abeln and G. J. Abbaschian, *Mechanical Behavior of Rapidly Solidified Materials*, Sastry, S. M. L., and MacDonald, B. A., Eds., p. 167, Minerals, Metals, and Materials Society, Warrendale, Ohio, 1985)

References

1. Martinez, M., Fernandez, A. I., Segaria, M., Xuriguera, H., Espid, F., and Ferrer, N., *J. Mater. Science*, 42, 2007, pp. 7745–7749.
2. "Introduction to Copper and Copper Alloys," in *ASM Handbook*, volume 2, 1990, pp. 216–240.
3. Shams El Din, A. M., *Desalination*, 93, 1993, pp. 499–516.
4. Wang, L., and Li, D. Y., *Surface and Coatings Technology*, 167, 2003, pp. 188–196.
5. Brown, N., *Acta Met.*, **7** 210 (1959).

6. Brooks, C. R., *Nonferrous Alloys*, p. 321, American Society for Metals, Metals Park, Ohio, 1984.
7. Ganin, E., Weiss, B. Z., and Komen, Y., *Met. Trans. A*, **17A** 1885 (1986).
8. Mendenhall, J. H., Ed., *Understanding Copper Alloys*, Olin Corporation, 1977.
9. Wilkins, R. A., and Bunn, E. S., *Copper and Copper Base Alloys*, McGraw-Hill Book Company, New York, 1943.
10. Sarma, V. S., Sivaprasad, K., Sturm, D., and Heilmaier, M., *Mat. Sci and Engineering A*, doi: 10.1016/J.msea.2007.12.016.
11. Williams, D. B., and Edington, J. W., *Met. Sci. Jour.*, **9** 529 (1975).
12. Cocco, G., Gagherazzi, G., and Schiffini, L., *Jour. Appl. Cryst.*, **10** 325 (1977).
13. Thompson, G. E., and Noble, B., *Met. Sci. Jour.*, **6** 114 (1971).
14. Abeln, S. P., and Abbaschian G. J., *Mechanical Behavior of Rapidly Solidified Materials*, Shastry, S. M. L., and MacDonald, B. A., Eds., p. 167, Minerals and Materials Society, 1985.
15. Tamura, M., Mori, T., and Nakamura, T., *Trans. Jap. Inst. Met.*, **14** 355 (1973).
16. Starke, E. A., Jr., *J. of Met.*, **22** 54 (1970).
17. Jha, S. C., Sanders T. H., Jr., and Dayananda, M. A., *Acta Metall.*, 35, 1987, pp. 473–482.
18. Yao, D. P., Zhang, Y. Z., Hu, Z. Q., Li, Y. Y., and Shi, C. X., *Scripta Metall.*, 23, 1989, pp. 537–541.
19. Dinsdale, K., Harris, S. J., and Noble, B., *Aluminum Lithium Alloys*, p. 101, *TMS-AIME*, Warrendale, PA, 1981.
20. Thompson, G. E., and Noble, B., *Jour. Inst. Met.*, **101** 111 (1973).
21. Mondolfo, C. S., *Aluminum Alloys*, Butterworth, Inc., Boston, Mass., 1976.
22. Brooks, C. R., *Nonferrous Alloys*, p. 363, American Society for Metals, 1982.
23. Collings, E. W., *Titanium Alloys*, p. 90, American Society for Metals, 1984.
24. Sridhar, G., Gopalan, R., and Sarma, D. S., *Metallography*, 20, 1987, pp. 291–310.
25. Zhang, X. D., Evans, D. J., Baeslack, W. A., and Fraser, H. L., *Mat. Sci. Eng. A*, A344, 2003, pp. 300–311.
26. Ahmed, T., and Rach, H. J., *Mat. Sci. Eng. A*, A243, 1998, pp. 206–211.
27. Sims, C. T., and Hagel, W. C., *The Superalloys*, John Wiley and Sons, Inc., New York, 1972.
28. Sahm, P. R., and Speidel, M. O., *High Temperature Materials in Gas Turbines*, Elsevier Scientific Publishing Company, New York, 1974.
29. Wang, W. Z., Jin, T., Liu, J. L., Sun, X. F., Guan, H. R., and Hu, Z. Q., *Mat. Sci. Eng. A*, 479, 2008, pp. 148–156.
30. Chen, W., and Immarigeon, J. P., *Scripta Mat.*, 39, 1998, pp. 167–174.
31. Forghani, F., Han, J. C., J. Moon, J., Abbaschian, R., Park, C. G., Kim, H. S., and Nili-Ahmadabadi, M., *J. Alloys and Compounds*, 777, 2019, 1222–1233.
32. Forghani, F., Moon, J., Han, J. C., Rahimi, R., Abbaschian, R., Park, C. G., Kim, H. S., and Nili-Ahmadabadi, M., *Materials Characterization*, 153, 2019, 284–293.

Chapter 21

Failure of Metals

Learning Objectives

Upon completion of Chapter 21, you will be able to:

1. Compare and contrast materials failure by ductile or brittle fracture
2. Describe nucleation and propagation of cleavage cracks
3. Describe mechanism of cup and cone formation during ductile failure
4. Differentiate between three stages of fatigue failure
5. Explain differences between low cycle fatigue and high cycle fatigue
6. Identify the effects of steel microstructure and amount of martensite on fatigue limit
7. Describe the influence of surface finish and hardening on fatigue failure

Metal failure in structural applications refers to the loss of load carrying capacity of a component. Fracture, or separation of a component into separate pieces, is the most dramatic example of such failure, which can lead not only to the loss of functionality of the component but also to catastrophic damages and personal injuries. Metal may also fail less dramatically when it cannot hold its dimensions through creep deformation in applications that require close dimensional tolerances.

Fracture is commonly categorized as ductile or brittle. The former is associated with a failure that occurs at the end of an extensive plastic deformation. Brittle failure, on the other hand, is associated with a minimum plastic deformation. Alternatively, one may think of ductile fracture as resulting from shear forces that produce deformation along certain crystallographic planes, whereas brittle fracture is due to tensile forces that produce cleavage (splitting) on certain crystallographic planes. However, there is considerable confusion as to how to differentiate explicitly between brittle and ductile fractures. This is mainly because one tends to think in terms of the entire deformation process leading up to the final act of fracture. To this must be coupled the fact that the word *brittle* is associated with a minimum of plastic deformation, while the word *ductile* connotes large plastic deformation. However, metals can fail by cleavage (a basically brittle process) after a relatively large preceding macroscopic deformation. In the same manner, it is quite possible to have a negligible macroscopic strain in a metal that fails by a ductile mechanism. In the last case, the fracture usually occurs in some localized region in which the deformation is very high. In viewing the problem of fracture, it would seem best, therefore, to reserve the terms "ductile fracture" and "brittle fracture" to refer to the actual act of propagating a crack. Thus, a *brittle fracture* is one in which the movement of the crack involves very little plastic deformation of the metal adjacent to the crack. Conversely, a *ductile crack* is one that spreads as a result of intense localized plastic deformation of the metal at the tip of the crack. There can, of course, be no sharp division between ductile and brittle failures, but the extremes of these two methods of failure are quite distinguishable.

Fracture is considered the end result of plastic-deformation processes, and there are many different ways that plastic deformation leads to failure, as can be seen by considering just a few ways that single crystals fracture.

21.1 Failure by Easy Glide

Let us consider the hexagonal metals that deform by easy glide on the basal plane (Zn, Cd, and Mg). In these metals, the single-crystal stress-strain curves depend upon the temperature of the test, as shown in Fig. 21.1A for magnesium. At low temperatures (291 K) the slope of the stress-strain curve rises very rapidly, while at higher temperatures (523 K) the slope of the stress-strain curves is small. The rate at which the metal hardens with strain at lower temperatures is much greater than that at high temperatures. This difference is undoubtedly due to dynamic recovery, with softening occurring more rapidly during a high-temperature test than during a low-temperature test. This conclusion is further emphasized by the fact that increasing the rate of testing has much the same effect on the stress-strain curve as has lowering the temperature (Fig. 21.1B).

From the above it can be concluded that at higher temperatures and at very slow strain rates basal glide involves very little, if any, strain hardening. Once deformation starts along those slip

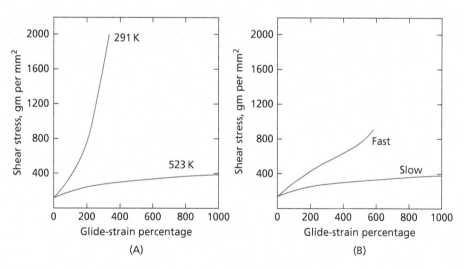

Fig. 21.1 Stress-strain curve of magnesium crystals. **(A)** The effect of temperature on the stress-strain curves. **(B)** The effect of strain rate. The curve marked "fast" represents a rate of straining 100 times faster than that for the curve marked "slow." (From Schmid, E., and Boas, W., *Kristallplastizität*, Julius Springer, Berlin, 1935)

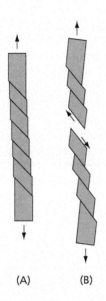

Fig. 21.2 **(A)** High temperatures and low strain rates promote extensive slip on a few slip bands. **(B)** Fracture can occur by shear along these bands

bands in which it is easiest to activate dislocation sources, it can continue on them without appreciable increase in their resistance to slip. This produces coarse, widely separated slip bands, as is shown schematically in Fig. 21.2A. The notches formed where slip bands cut the surface are a factor in their development. The stress concentration at these notches aids the slip process in the bands. Finally, as more and more deformation occurs in the slip bands, the cross-section area of the bands becomes smaller and smaller. This raises the effective shear stress, thereby increasing the tendency for the slip to remain confined to the operating bands.

Provided that another plastic deformation process, such as mechanical twinning or slip on a plane other than the basal plane, does not occur at a late stage in the deformation process, it is quite possible for a crystal, such as that indicated in Fig. 21.2, to shear completely in two along one of the coarse slip bands. Shearing in two along a slip band is one possible mode of single-crystal failure.

21.2 Rupture by Necking (Multiple Glide)

Outside of certain hexagonal metals, single glide is the exception rather than the rule in single crystals. In cubic metals, after only a relatively small deformation, easy glide breaks down into multiple glide on two or more systems. When deformation occurs by slip on several slip systems in a single-crystal tensile test, the rate at which the metal strain hardens can have an important effect on the fracture mechanism. What generally happens is that early in the test, some part of the gage length deforms at a slightly more rapid rate than the rest. The cross-sectional area in this region becomes slightly smaller than in the rest of the specimen, with a corresponding increase in the shear stress. If the rate at which the slip planes harden with increasing strain is small, then continued slip in this reduced section will occur more easily than in the remainder of the specimen. In this manner a neck can form in the gage length. The various steps involved in the development of a neck are outlined schematically in Fig. 21.3 for an assumed case of double glide. The slip planes are indicated in Fig. 21.3A and are assumed perpendicular to the plane of the paper, with the slip directions in the plane of the paper. The neck is shown in Fig. 21.3B, and the ultimate shape of

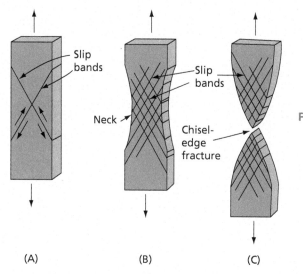

Fig. 21.3 **(A)** Crystal oriented for double slip. **(B)** Development of a neck. **(C)** Chisel edge fracture

the specimen can be observed in Fig 21.3C. Here it can be seen that continued growth of the neck results in a chisel-edge type of failure. Such a failure is frequently observed in metal crystals.

If more than two slip systems operate during the deformation of a metal crystal and a neck is formed, the final fracture can occur when the cross-section at the neck pulls down to a point rather than a chisel edge.

It should not be inferred that the end result of multiple glide is always a simple chisel edge, or point type of fracture. The original orientation of the crystal is an important factor in determining the character of the fracture. It may well be that as the metal is deformed and the lattice of the crystal is reoriented by the deformation, other slip systems will become active in addition to those that were activated at the start of the deformation. It is thus quite possible, for example, for necking, as a result of multiple glide, to give way to a final fracture by shear on a single plane.

21.3 The Effect of Twinning

While the shear associated with mechanical twinning is, in general, small, twinning always involves a lattice reorientation in the twins. This may place new slip systems in a favorable position relative to the stress axis, so that plastic deformation occurs more readily inside the twin than in the parent crystal outside the twin. Thus, when zinc or cadmium crystals are strained in tension, the basal plane of the crystal rotates toward the stress axis, as shown in Fig. 21.4. The effect of this rotation is to decrease the shear-stress component on the basal plane. In order to continue the deformation of the crystal, the tensile force must be increased. A point is eventually reached where the metal undergoes deformation twinning on a $\{10\bar{1}2\}$ plane. When this occurs, the basal plane in the twin is found to be in a position favorable for slip. Fracture then occurs as a result of secondary basal slip inside the twin.

The fracture methods illustrated below, which involve large plastic deformation by slip, differ somewhat from the normal concept of fracture, which usually involves the spreading of a crack. It is perhaps better, therefore, to classify these forms of metal failure as ruptures rather than as fractures.

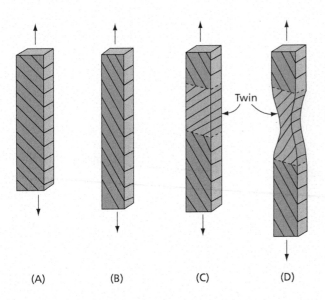

(A)　　(B)　　(C)　　(D)

Fig. 21.4 The lattice reorientation due to twinning may induce failure. **(A)** Zinc or cadmium crystal. **(B)** Rotation of lattice due to slip. **(C)** Formation of twin. **(D)** Deformation leading to fracture in twin

21.4 Cleavage

Under certain conditions, it is possible to split crystals in two pieces along planes of low indices. Let us assume that the block in Fig. 21.5A is a single crystal of zinc. Now suppose that a wedge or knife edge is laid along a basal-plane trace in the manner indicated in the figure and that the knife edge is given a sharp blow with a small hammer. If the temperature at which this operation is carried out is sufficiently low, the crystal will split, or be cleaved, into two parts, the separation following the basal plane. (See Fig. 21.5B.) This operation is called *cleavage* and the plane on which it occurs is known as the *cleavage plane* of the crystal. In the case of zinc, it is, of course, the basal plane (0001).

Zinc crystals are capable of being cleaved at room temperature, but only with some difficulty, and the cleaved surfaces are usually badly distorted. The splitting of zinc crystals becomes progressively easier the lower the temperature at which cleavage is attempted. Very nice cleavages may be readily

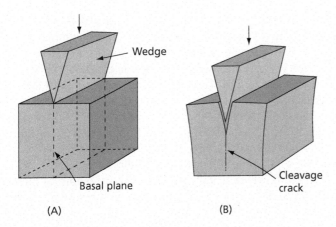

Fig. 21.5 Cleavage of a zinc crystal

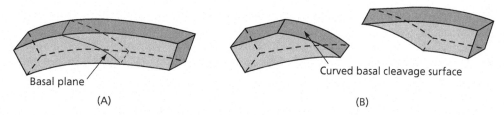

Fig. 21.6 A distorted zinc crystal cleaved. The cleavage follows the curved surface of the basal plane

obtained at the temperature of liquid air (77 K). In fact, if a carefully grown, undistorted zinc crystal is rapidly cleaved at this temperature, the surface may be so perfect that it becomes impossible to focus on the surface with a light microscope.

An interesting fact relative to the cleavage of zinc crystals is that the fracture follows the basal plane even in a bent or a distorted crystal. Thus, if a crystal is first bent and then cleaved, as shown in Fig. 21.6, the fracture surface will show the curvature of the basal plane. This fact has frequently been used in the study of plastic-deformation phenomena (using zinc crystals), to follow the distortion inside crystals. If should be noted that in polycrystalline zinc, grain boundaries may affect the deformation, thus masking brittle fracture behavior (Ref. 1).

Because basal cleavage is such a well-known phenomenon in zinc crystals, it is commonly believed that other hexagonal metals cleave on the basal plane as readily as zinc. However, this is not the case, except possibly for beryllium, which has been observed (Ref. 2) to cleave on (0001) and $\{11\bar{2}0\}$ planes. Magnesium does not cleave easily on the basal plane or on any other plane, nor is there any information in the literature relative to cleavage of cadmium crystals. There is also no evidence that anyone has observed true cleavage in a face-centered cubic metal. The one important class of metals in which cleavage is most frequently observed is the body-centered cubic metals, although the alkali metals (sodium, potassium, etc.) are body-centered cubic, and evidently do not cleave. The cleavage plane in the body-centered lattice form is usually {100}, although there are examples in which it has been indicated that cleavage along {110} is preferred (Refs. 3 and 4).

While most of the commercially important metals are not subject to cleavage, it is still a significant subject because of the fact that iron is a body-centered cubic metal that cleaves. The low-temperature brittleness of steels can be directly attributed to this fact. When fractures occur in polycrystalline iron or steel by transcrystalline cleavage, very little energy is expended in propagating the fractures, so that they closely resemble those that occur in glass or other brittle isotropic elastic solids.

21.5 The Nucleation of Cleavage Cracks

Glass at room temperature can be considered to be a material incapable of plastic deformation. The same is not true of metals that are capable of deforming by slip and twinning even at temperatures approaching absolute zero. It has also been observed that when a metal fails by brittle cleavage, a certain amount of plastic deformation almost always occurs prior to the fracture. This has been interpreted by many as evidence that metals do not fracture as the result of pre-existing Griffith cracks, but that cleavage cracks are probably nucleated by plastic-deformation processes. In confirmation of this point of view is the fact that cleavage can sometimes be made to occur in the interior of annealed polycrystalline specimens; in these specimens, there is little likelihood of the existence of small cracks before the application of stress. In a well-annealed metal, cracks should be self-healing and disappear.

While it is quite possible to nucleate cleavage fractures in metal crystals with a chisel and hammer, this approach tells us nothing about the formation of fracture nuclei in metal specimens under normal conditions of loading. Of more interest is how a crack starts in a single-crystal tensile specimen. A number of these studies have been performed in which zinc or iron crystals were used.

The fracture surfaces of iron single crystals are never perfect cleavages, even though the metal can fail in a completely brittle manner. Thus, while a surface can be a macroscopically flat plane, observation under a microscope at even relatively low magnification shows a considerable amount of surface detail. This was documented by the early work of Briggs and Pratt in 1958, upon investigating deformation and fracture of alpha iron at low temperatures (Ref. 5). They observed that the fracture of alpha iron single crystals of all orientations was completely ductile when tested in tension at temperatures above 173 K($-100°C$). Failure occurs by slip mechanisms of the type outlined at the start of the present chapter. As the temperature is lowered to 90 K($-183°C$), a ductile-brittle transition is observed. At this temperature, the fracture also becomes orientation-dependent. Those near to the (011)–(111) line predominantly fail by brittle cleavage, while samples near to [001] still fail in a ductile fashion by slip. The orientations that cleave are those in which the stress axis of the specimen lies close to a <100> direction. Finally, it should be noted that ductile-brittle transition temperature strongly depends on carbon contents and other alloy additions (Refs. 6 and 7).

In analyzing brittle fracture in iron, it is important to consider the effect of temperature on the critical resolved shear stress (Fig. 21.7). In body-centered cubic metals, such as iron, the temperature-dependence of the critical resolved shear stress for slip is very large, varying by a factor of almost eight in the temperature interval from room temperature to 77 K. Both cleavage and twinning require nucleation, and the high stress levels needed to deform iron plastically at low temperatures favor these nucleation processes.

Over the years a decided change in viewpoint has occurred with regard to how cleavage cracks form in polycrystalline metals. The original view was that nucleation was probably due to interactions between slip dislocations. A number of mechanisms were suggested to rationalize how this could occur. One of the first, due to Zener, was proposed in 1948 (Ref. 8). Another important dislocation

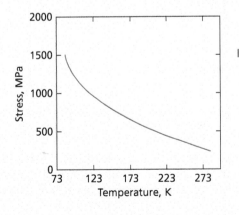

Fig. 21.7 Temperature dependence of the critical resolved shear stress in iron single crystals. Data corresponds to the stress at the upper yield point. (Reprinted from Biggs, W. D., and Pratt, P. L., "The deformation and fracture of alpha-iron at low temperatures," *Acta. Met.*, **6** (11) 694 [1958] with permission from Elsevier)

mechanism for cleavage in iron was proposed by Cottrell in 1958 (Ref. 9). However, at this time it is generally felt that the origin of a cleavage crack in a steel is a carbide that cracked as a result of plastic deformation produced by the load applied to a specimen. Early evidence for this showed (Ref. 10) that cracked grain-boundary carbides could be responsible for the start of cleavage cracks. It has also been shown that for two cracks in the same grain-boundary carbide, only that in the thicker part of the carbide might be able to propagate into the matrix as a cleavage crack. This has been interpreted as showing that the length of the crack in the carbide is critical to whether it can propagate or not, with the larger carbide crack being able to satisfy the conditions for propagation and the smaller crack not. It has also been demonstrated (Ref. 11) that cleavage cracks can result from cracking of the spheroidal carbide particles in a spheroidized cementite structure. On the basis of these and other observations, the concept that cleavage cracks originate at fractured inclusions is now well accepted.

21.6 Propagation of Cleavage Cracks

In elastic solids (cold glass), the strain energy that is released as a crack propagates is converted into the surface energy of the crack surfaces and the kinetic energy of the moving material at the sides of the crack.

An additional energy term must be considered for the propagation of crystalline cleavage cracks. This term is that associated with the plastic deformation which usually accompanies the motion of cracks through crystals. Low temperature and high strain rates tend to raise the yield points of metals and to suppress plastic deformation. The energy term associated with plastic deformation, therefore, becomes less important the lower the temperature of testing and the higher the velocity with which the crack moves. In the following sections certain of the experimental aspects of these effects will be discussed. More detailed analysis can be found in other references (Refs. 12–14).

Plastic deformation associated with a moving crack is most liable to occur just ahead of the crack. The metal in this region is effectively in a state of very high uniaxial tensile stress, with the stress axis normal to the plane of the crack. A simple tensile stress of this type is equivalent to a set of shear stresses on planes at 45° to the tensile-stress axis, as indicated in Fig. 21.8. Because these

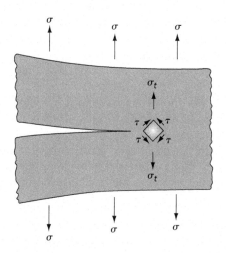

Fig. 21.8 When an external tensile stress (σ) is applied to an elastic body containing a crack, the material just ahead of the crack is subjected to a very large tensile stress (σ_t). This in turn, is equivalent to shear stresses (τ) on planes at 45° to the plane of the crack

shear stresses are large, it is quite possible to nucleate dislocation ahead of the crack on slip planes that are favorably oriented with respect to the shear stress.

Gilman and his associates (Ref. 15) have reported some interesting work dealing with plastic deformation during cleavage fracture, using single crystals of lithium fluoride, which is an ionic cubic solid similar to MgO in which dislocations can be readily revealed by a reliable etching technique. It has been observed that in this material, when a crack moves slowly or stops, dislocations are nucleated in advance of the crack. Actually, there is a limiting velocity below which dislocations are nucleated and above which no evidence of plastic deformation is to be found. An interesting feature of the work on LiF is that the maximum velocity with which a crack can move in this material has actually been measured and is 0.31 the velocity of sound (Ref. 15). This compares quite favorably with the theoretical value for fracture in an elastic material which is 0.38.

Now consider the effect of plastic deformation on cleavage crack propagation. When slip takes place during the movement of a crack, energy is absorbed in nucleating and moving dislocations. This energy comes at the expense of the elastic strain energy that drives the crack. If the work to overcome plastic deformation is too large, the crack may decelerate and stop, which implies that in those crystalline materials which cleave, a minimum velocity must be achieved before the crack can move freely. If the velocity is too slow, then too much energy will be absorbed in the form of slip. The plastic deformation term may well be one of the most important factors that make some metals incapable of cleavage. In this regard, it is interesting to remember that face-centered cubic metals, with their many equivalent slip systems, have not been observed to cleave.

In addition to the energy associated with the formation and growth of dislocations, there is another way that dislocations absorb energy from a moving crack. The cutting of dislocation lines by cracks involves energy losses. This is true particularly when the intersected dislocations are in a screw orientation. When a cleavage fracture passes a screw dislocation, the fracture surface receives a step whose height equals the Burgers vector of the dislocation. The nature of the development of this step is shown in Fig. 21.9. As a crack progresses, the steps that it obtains, due to its intersection with dislocations, tend to run together. If the steps are from dislocations of the same sign, they combine to form larger steps, whereas if they are from dislocations of opposite sign, they cancel.

The step produced on the cleavage plane by a single screw dislocation is, of course, too small to be seen. However, if the crack intersects a large number of screw dislocations of the same sign, then, by combining, steps can be developed large enough to be easily visible. As grown, LiF crystals often contain low-angle boundaries that have a twist component. Such boundaries contain a cross-grid of closely spaced screw dislocations, all of the same sign, and when a cleavage crack crosses one of these boundaries, large-sized steps develop in the cleavage plane. An excellent example of this is shown in Fig. 21.10. Notice how the steps progressively run together to form still steeper steps. The pattern that the steps form on the cleavage surface is known as a *river pattern*. In general, the latter run normal to the crack front, and it is often possible to locate the point of origin of a cleavage crack by following the river pattern back to its source.

The river pattern on brittle fracture surfaces can arise from a number of causes other than the cutting of screw dislocations by a crack. Thus, if a cleavage crack is started, say by a hammer and chisel, in such a manner that the crack is not strictly parallel to the cleavage plane, the fracture must contain steps in order to accommodate the misalignment. River patterns are also observed on the surfaces of glass fractures. Since glass is normally amorphous, the crack does not follow a crystallographic plane and dislocations in the normal sense cannot exist.

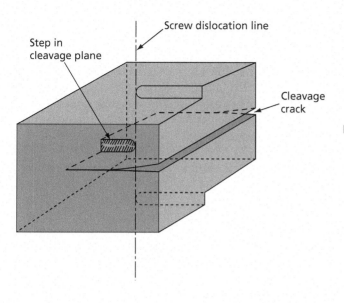

Fig. 21.9 Production of a step in the cleavage plane when a cleavage fracture intersects a screw dislocation

There are several reasons why it is more difficult for a fracture to move along a stepped surface. First, a cleavage plane that contains a large number of steps has a larger surface area and therefore a larger surface-energy term. Second, the advance of the crack not only involves separation of the crystal along cleavage plane segments, but also entails the continued growth of the surfaces of the steps, or small *cliffs*. Unless a secondary cleavage plane, or slip plane, is nearly normal to the surface of the primary cleavage plane, the formation of steps will involve what amounts to plastic tearing of metal in order to form the surfaces of the step. A great deal of energy can be expended in doing this.

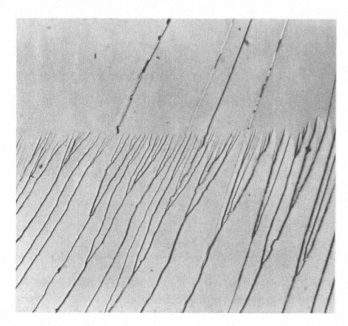

Fig. 21.10 Cleavage steps resulting from the intersection of a cleavage crack with a low-angle boundary containing a grid of screw dislocations of the same sign. 250×. (Gilman, J. J., *Trans. AIME*, **212** 310 [1958])

21.7 The Effect of Grain Boundaries

As shown in the preceding section, a small-angle twist boundary (in a single crystal) adds to the difficulty of motion of a cleavage crack by introducing steps into the fracture plane. Grain boundaries in polycrystalline metals also impede the motion of cracks, and there is sound evidence that the magnitude of the effect is much larger because it is possible to find cleavage cracks in deformed polycrystalline tensile specimens that are no larger than a grain diameter. Such a microcrack in a polycrystalline iron specimen is shown in Fig. 21.11.

Consider first the case where the orientation difference between adjacent crystals is more than a few degrees, but still not large. In this case, the cleavage planes in the two crystals, while approximately aligned, still make a finite angle with each other. It is not possible, under these conditions, for the fracture surface to pass smoothly through the boundary, and what probably happens is that a series of parallel cleavage surfaces are nucleated on different levels. The ultimate result is that the fracture surface develops a series of steps originating at the grain boundary. A typical example can be seen in Fig. 21.14.

In the average polycrystalline specimen, the misalignment between grains is larger than that between the crystals in Fig. 21.12 and, in general, the fracture surface is much more irregular. A study (Ref. 16) of river patterns on the fracture surfaces of polycrystalline specimens shows that the crack also propagates in an erratic fashion, sometimes moving in a direction opposite to the

Fig. 21.11 A cleavage crack that stops at the boundaries of a single grain in a polycrystalline iron specimen. 200×. (Hahn, G. T., Averbach, B. L., Owen, W. S., and Cohen, M., *Fracture*, p. 91, The Technology Press and John Wiley and Sons, Inc., New York, 1959)

Fig. 21.12 Large-cleavage steps that develop when a cleavage crack passes from one crystal to another. Specimen 3 percent silicon-iron alloy, cleaved at 78 K, direction of crack propagation top to bottom. 250×. (Low, J. R., Jr., *Fracture*, p. 68, The Technology Press and John Wiley and Sons, Inc., New York, 1959)

mean direction of movement. Evidence is also found for discontinuous crack propagation, meaning that failure is not merely by the movement of a single-crack front, but that a number of crack segments form and then join together. Since the individual segments may not be on the same level, this usually involves plastic tearing between fracture surface segments.

21.8 The Effect of the State of Stress

Both cleavage-crack nucleation and propagation are favored by high tensile stresses. On the other hand, slip requires shear stress. When deformation occurs by slip, however, the applied stresses tend to be relieved. In other words, it is difficult to achieve large stresses when a metal deforms easily by slip. From these considerations we can conclude that any stress system capable of producing a combination of large tensile stresses and small shear stresses favors cleavage. The nature of the state of stress in a metal specimen is clearly an important consideration in the fracture process.

In simple uniaxial tension, the stress can be viewed as equivalent to a set of shear stresses oriented at 45° to the tensile stress axis. This relationship is shown in Figs. 21.13A and 21.13B, where the tensile stress in one case is assumed horizontal and in the other vertical. If the two tensile stresses (at 90° to each other, as shown in these drawings) are applied simultaneously to the same specimen, the shear-stress components will oppose each other. In the two-dimensional case illustrated, it is clear that under a state of biaxial tension the shear stress in the material is reduced. Also, if a third tensile stress is applied normal to the plane of the previously mentioned

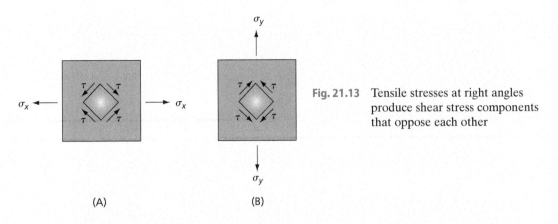

Fig. 21.13 Tensile stresses at right angles produce shear stress components that oppose each other

two stresses, and all tensile stresses are assumed equal, a state of hydrostatic tension will occur in which the material experiences no shear stress whatsoever. The equivalent state of hydrostatic compression is well known—for example, a liquid under pressure.

From the above it can be concluded that whenever a metal specimen is tested under conditions of biaxial or triaxial tension, slip that requires shear stress will tend to be suppressed. Because of this, a high level of tensile stress can be attained and brittle fracture by cleavage can be promoted. Conversely, if a metal specimen is pulled in tension while immersed in a fluid under pressure, the above conditions will be radically altered. In this case, the applied tensile stress will be complemented by two compressive stresses, having shear-stress components in the same direction as those of the applied tensile stress. Deformation by slip in this case is highly favored, and very high ductilities in polycrystalline specimens have actually been attained in this manner.

An easy way of approximating the state of triaxial tension involves the placing of a simple V-notch around the girth of a cylindrical tensile-test specimen. Figure 21.14 represents such a

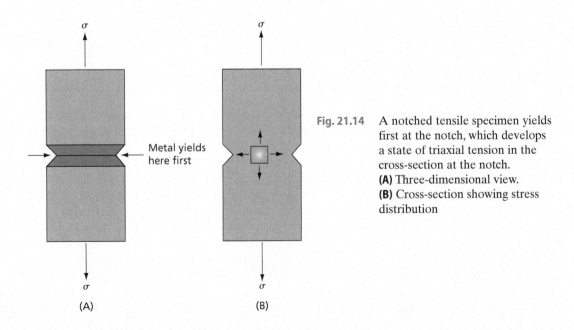

Fig. 21.14 A notched tensile specimen yields first at the notch, which develops a state of triaxial tension in the cross-section at the notch.
(A) Three-dimensional view.
(B) Cross-section showing stress distribution

specimen. When it is loaded in tension, the reduced section at the notch will be the first position to yield. As this region elongates (in the direction of the applied stress), its natural tendency is to shrink in the horizontal plane. This, however, is resisted by the metal lying above and below the notch that has not yet yielded. The metal in the cross-section at the notch is thus placed under three tensile stresses: the vertical applied stress and two induced horizontal stresses at 90° to each other.

21.9 Ductile Fractures

A completely brittle cleavage fracture will show sharp planar facets which reflect light, while a completely ductile fracture presents a rough dirty-gray surface. The reason for the latter is that a ductile fracture surface has a rough, irregular contour where much of the surface is inclined sharply to the average plane of the fracture. Figure 21.15 shows a cross-section through a ductile fracture that illustrates this characteristic of ductile fracture.

The failure of most ductile materials in the polycrystalline form occurs with a *cup-and-cone* fracture. The appearance of a typical specimen that has failed in this fashion is illustrated in Fig. 21.16. This type of fracture is closely associated with the formation of a neck in a tensile specimen. As indicated in Fig. 21.17, fracture begins at the center of the necked region on a plane that is macroscopically normal to the applied tensile-stress axis. As deformation progresses, the crack spreads laterally toward the edges of the specimen. Completion of the fracture occurs very rapidly along a surface that makes an angle of approximately 45° with the tensile-stress axis. In a perfect example, this final stage leaves a circular lip

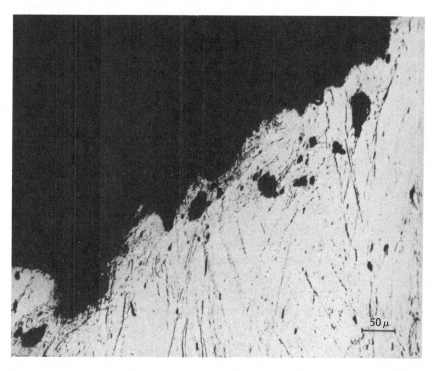

Fig. 21.15 Irregular surface contour of a ductile fracture. (Rogers, H. C., *TMS-AIME*, **218** 498 [1960])

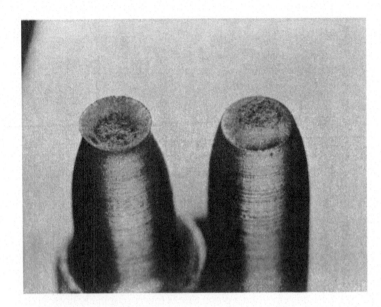

Fig. 21.16 Cup and cone fracture

on one-half of the specimen and a bevel on the surface of the other half. Thus, one half has the appearance of a shallow cup and the other half, a cone with a flattened top.

Rogers (Ref. 17) has shown that in the necked region of a tensile-test specimen, small cavities may form in the metal near the center of the cross-section before a visible crack is found. The density of these pores depends strongly on the amount of deformation, and increases with increasing deformation. Thus, Rogers observed that the number of pores per cm^3 was of the order of 10^3 greater

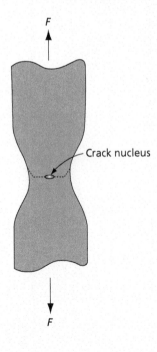

Fig. 21.17 Cup and cone fracture. Crack starts at the center of the specimen and spreads radially, with final completion of the fracture on a cone-shaped surface making an angle of approximately 45° with the tensile axis

21.9 Ductile Fractures

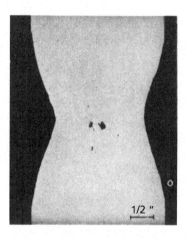

Fig. 21.18 Voids at the center of a copper tensile-test specimen. The two large voids at the center are joined by a crack. (Rogers, H. C., *TMS-AIME*, **218** 498 [1960])

in the region of the neck in his copper specimens than in a position in the gage section well away from the neck. Coalescence of these cavities, as a result of their growth under the applied stress, can be assumed to lead to the formation of a ductile crack. Such a condition is actually shown in Fig. 21.18. There is good evidence that in most (Refs. 18 and 19) commercial metals, these internal cavities probably form at nonmetallic inclusions. The manner in which inclusions nucleate pores may only be surmised, but certainly hard brittle inclusions will impede the natural plastic flow of the matrix surrounding them. The belief that inclusions play a strong role in nucleating ductile fractures is supported by the fact that extremely pure metals are much more ductile than those of slightly lower purity. In a tensile test very pure metals often draw down to almost a point before they fracture.

Once a crack of the type shown in Fig. 21.18 has developed, it can propagate by the *void-sheet mechanism*. This mechanism acts as follows. The stress concentration at the ends of the crack localizes the plastic deformation in these regions into shear bands that make angles of 30° to 40° with the stress axis. The relationship of the shear bands to the crack is shown in Fig. 21.19A. Because the deformation

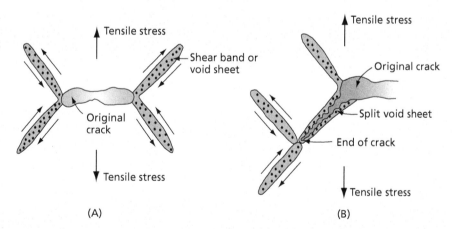

Fig. 21.19 The void sheet mechanism in ductile fracture. **(A)** The stress concentration at the ends of a small crack initiated by void coalescence induces bands of shear strain at its ends. The highly concentrated strain in these bands nucleates pores inside the bands. **(B)** The crack advances when a void sheet splits

inside the bands is very intense, the bands become filled with voids. Rogers has used the term *void sheets* to describe these bands at this point. As the voids in these bands grow, they eventually impinge on each other, with the result that a void sheet splits in two and the crack advances as indicated in Fig. 21.21B. The movement of the crack down the void sheet changes the location of the end of the crack so that new shear bands are formed, as shown in this diagram. At this point, the crack could theoretically move along either of these two new bands, one of which is an extension of the original band that split, while the other is inclined in the opposite direction. However, if the crack continues to move forward in its original direction, it will move away from the specimen cross section at the center of the neck and into a region of decreasing stress. On the other hand, if it moves into the other shear-band orientation, it moves back into the region of maximum stress concentration. This is what normally occurs. Repetition of this process acts to spread the crack across the specimen cross-section. In a ductile metal, the final fracture can occur by several methods: one of these produces a cup and cone, and the other a double cup. When the final fracture occurs so as to form a typical cup-cone failure, as in iron, brass, or duraluminum, it is probable that the shear lip of the cup forms by the void-sheet mechanism. As the central fracture advances toward the surface, the specimen cross-section available to carry the load is progressively decreased. Eventually a condition is reached where the shear bands may extend to the surface. A catastrophic splitting of these bands can then result in the shear lip. This case is illustrated in Fig. 21.20.

An interesting aspect of the fracture that occurs by the void-sheet mechanism is that it produces a fracture surface with a characteristic appearance. At low magnifications the surface has a rough, spongy texture. However, at higher magnifications, when observed by the scanning electron microscope, the true nature of the fracture surface can be seen. Such a photograph is shown in Fig. 21.21. The significant feature of this photograph is the rather uniformly ordered array of cuplets that are found on the surface. Each of these cuplets corresponds to a pore that appeared in the void sheet. The fact that the cuplets are oriented in a single direction is a direct result of the deformation that occurs in the shear band. As indicated in Fig. 21.22A, this tends to elongate the pores in a direction approximately parallel to the tensile stress axis. When the void sheet splits, both of the resulting fracture surfaces will contain cuplets, but they will point in opposite directions on the two sides, as shown in Fig. 21.22B. A close examination of Fig. 21.21 will also reveal that small holes have opened up at the bottom of some cups. These represent a connection with other pores below the surface. These subsurface pores are clearly visible in Fig. 21.15, which is a cross-section cut through a specimen that failed by the void sheet mechanism. The stress axis was in the vertical direction. Notice the inclination of the fracture surface characteristic of a void sheet.

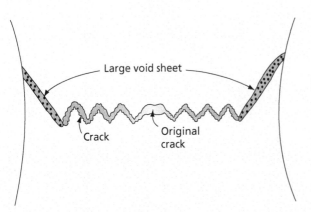

Fig. 21.20 The development of large void sheets that extend to the surface can explain the lip on a cup and cone fracture

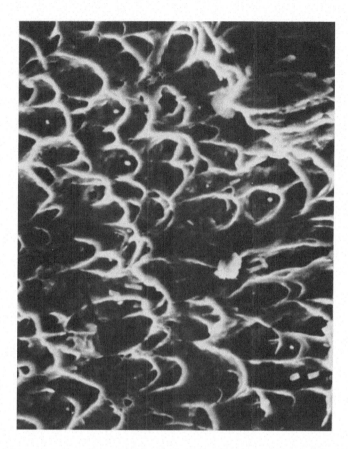

Fig. 21.21 A scanning electron microscope photograph of the fracture surface of an austenitic stainless steel (304) specimen that failed by the void sheet mechanism. Notice the oriented arrangement of cuplets. Magnification 28,000×. (Photograph courtesy of Ellis Verink)

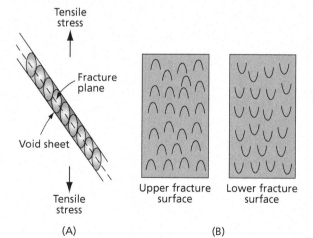

Fig. 21.22 **(A)** The pores in a void sheet are elongated in a direction roughly parallel to the tensile-stress axis. **(B)** When the void sheet splits, this produces oppositely pointing cuplets on the two fracture surfaces. (After Ref. 14)

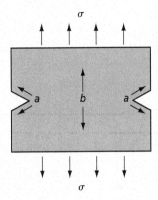

Fig. 21.23 The strain rate at the root of the notch (points a) is larger than at a point such as b in the interior of a tensile specimen

Rogers (Ref. 20) has proposed a mechanism for ductile-crack propagation which is quite interesting and is based on the following concepts. It has been demonstrated that the rate of plastic strain at the surface of a notch is larger than that at a distance inside the metal away from the notch. For example, in Fig. 21.23 the metal should flow more rapidly at points a than at point b. This difference in strain rate may be quite large and depends on the sharpness of the notch. We can conclude that during a tensile test, crystals lying at the surface of the notch are deformed more rapidly than those in the interior of the specimen. Rogers suggests that these crystals should deform like a single crystal that thins down to a chisel edge and ruptures. Ductile-crack movement is thus pictured as the successive pulling apart of crystals that find themselves in succession at a notch surface (Ref. 16). This propagation mechanism for a ductile crack is indicated schematically in Fig. 21.24.

The above method of crack propagation should hold for any ductile-crack extension whether it is the inward movement of an external notch (or neck), or the outward movement of an internal crack that starts at a void. It can also explain the final stages of a tensile fracture in which a double cup-cone fracture is obtained. This type of fracture is shown schematically in Fig. 21.25 and is observed (Ref. 14) in some ductile metals such as copper, aluminum, silver, gold, and nickel. In this type of failure the crack also originates in the center of the specimen and propagates by the void-sheet mechanism until

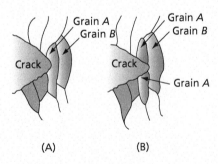

Fig. 21.24 Propagation of a ductile fracture according to Rogers. **(B)** represents a later time than **(A)**. Notice that crystal A has been pulled apart in the indicated time interval. (After Ref. 17)

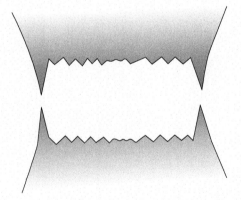

Fig. 21.25 Double cup fracture. This type of failure is observed in many pure face-centered cubic metals. (After Ref. 14)

a thin ring of metal remains. This then fails by the progressive slipping apart of the grains, rather than by the void-sheet mechanism. The result is two cups instead of a cup and cone.

21.10 Intercrystalline Brittle Fracture

Fracture by cleavage is not the only form of brittle fracture that occurs in metals. Brittle fracture can also occur in which the fracture passes along grain boundaries. In some instances this type of fracture can be caused by grain-boundary films of a hard-brittle second phase, like that formed by bismuth in copper. In other cases it may not be due to an actual precipitate, but to a concentration of solute in the metal close to crystal boundaries. Why this concentration of solute should lower the cohesion across the boundaries is not known.

21.11 Blue Brittleness

This is a phenomenon that has been recognized in steels for many years. Its name is a result of the fact that it occurs in the temperature range (several hundred degrees centigrade above room temperature) where a blue oxide film is formed on the surface of a steel specimen. The term "blue brittleness" is somewhat of a misnomer, as the metal does not become brittle in the normal sense. What actually happens is that the elongation in a tensile test undergoes a minimum at the blue brittle temperature. This is shown in Fig. 21.26 for the case of commercial purity titanium (Ref. 21).

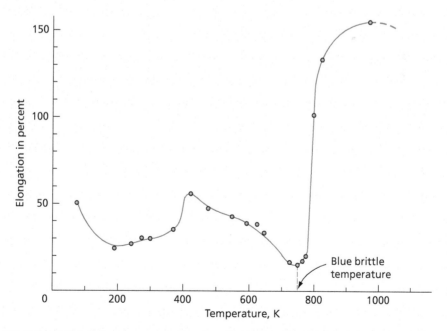

Fig. 21.26 The "blue brittle" phenomenon represents a loss of tensile elongation at the blue brittle temperature. The phenomenon is best known in steels, but occurs in other metals. This curve shows the effect in commercial purity titanium. Strain rate 3×10^{-4} sec^{-1}. (From the data of Ref. 18)

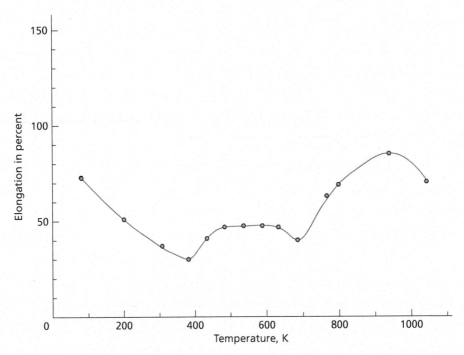

Fig. 21.27 Blue brittleness is associated with dynamic strain aging. In high purity titanium, as shown by this curve, the elongation minimum almost disappears. Strain rate 3×10^{-4} sec^{-1}. (From the data of Ref. 18)

However, the reduction in area of a tensile specimen does not normally show a pronounced minimum at the blue brittle temperature. This signifies that the fracture is not brittle.

The blue brittle phenomenon is associated with dynamic strain aging (see Sec. 9.15). More detailed analysis can be found in other references (Refs. 22 and 23). The interaction between dislocations and impurity atoms acts to bring about the point at which necking starts in a tensile specimen at a relatively small strain and, once necking begins, the strain becomes highly concentrated in the neck. Both effects act to reduce the elongation that is obtained. That this phenomenon is closely associated with the impurity atoms in the lattice is demonstrated in Fig. 21.27. This curve shows the elongation as a function of temperature in high-purity titanium specimens. Notice that the minimum observable in the corresponding curve for commercial purity titanium has nearly disappeared.

21.12 Fatigue Failures

Failures that occur in machinery parts are almost always fatigue fractures. When a metal is subjected to many applications of the same load, fracture occurs at much lower stresses than would be required for failure in a tensile test. The failure of metals under alternating stresses is known as *fatigue*.

21.13 The Macroscopic Character of Fatigue Failure

A fatigue fracture always starts as a small crack which, under repeated applications of the stress, grows in size. As the crack expands, the load-carrying cross-section of the specimen is reduced, with the result that the stress on this section rises. Ultimately the point is reached where the remaining cross-section is no longer strong enough to carry the load, and the spread of fracture becomes catastrophic. Because of the manner in which the fracture develops, the surfaces of a fatigue fracture are divided into three areas with distinctly different appearances, as is shown schematically in Fig. 21.28: crack initiation (stage I), crack propagation (stage II), and final rupture (stage III). In most cases, the stage II surface will have a polished or burnished appearance in the region where the crack grew slowly. This texture arises because the metal surfaces of the crack rub against each other as the specimen is deformed back and forth through each stress cycle. In the last stage, when the specimen finally fractures, there is no rubbing action and the surfaces developed at this time are rough and irregular. Because the latter area usually has a granular appearance, an erroneous conclusion is frequently made relative to fatigue fractures, namely, that the metal crystallized in service and thereby became embrittled.

In machinery components, the amplitudes of the stress cycles are not always of the same magnitude. For example, consider an automobile drive shaft in which the stress cycles are much larger during periods of fast acceleration than when the car moves at a steady speed. Under variable stress amplitudes, the crack may stop spreading when the stress is low and continue to grow when it rises. This alternation of periods of rapid growth with periods of slow or no growth changes the degree of rubbing that the surfaces of the crack undergo, with the result that the surface may attain "beach marks," or a "clamshell" appearance. (See Fig. 21.29.) Usually these ring-shaped markings on the fracture surface are concentric with the origin of fracture and make possible its determination. Their existence on the fracture surface of a metal object is also good evidence that the failed part fractured by a fatigue mechanism. At higher magnifications using SEM, many deformation striations may also be observed between the beach marks. Other evidence in this regard is certainly the presence of the burnished and smooth areas on the fracture surface. Finally, fatigue fractures

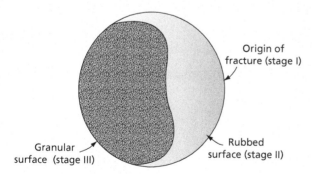

Fig. 21.28 The surface of a fatigue fracture usually shows three regions with characteristically different aspects. The crack initiation site, followed by a smooth or burnished region, corresponding to the slow spreading of the fatigue crack, and the final rough, or granular, belonging to the section that failed as a result of overload.

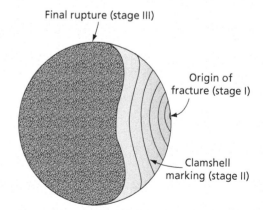

Fig. 21.29 Fatigue fractures in machinery parts subjected to stress cycles that vary in magnitude with time are liable to show a clamshell pattern in the burnished area

are almost always characterized by a lack of macroscopic plastic deformation at small distances from the immediate fracture surface. In this respect, they resemble typical brittle fracture, and if a fractured machinery part indicates that a large degree of plastic flow occurred just prior to fracture, it is usually indicative of the fact that failure occurred as a result of a temporary overload rather than because of a uniform repeating load.

21.14 The Rotating-Beam Fatigue Test

There are as many ways of running a test to measure fatigue as there are ways of applying repeated stresses to a metal. A specimen may be first stretched in simple tension and then the stress direction reversed so that the specimen is placed in compression. Alternating the direction of twist in a torsion specimen is another type of repeated stress. A relatively simple alternating stress is obtained by reversed bending. In some instances fatigue has been studied by using combined stress loadings in which the obvious intent was to subject the metal to conditions approaching those found in machinery parts. Since, in many cases, machine elements and structures are not subjected to a loading that completely reverses the stress magnitude, tests are often run in which the specimen is given a steady load, say, in tension, and then an additional alternating load is superimposed on the steady load.

A rotating beam is the type of fatigue specimen most generally used. One of its greatest advantages is its relative simplicity.

Figure 21.30 shows schematically the basic components of one type of rotating beam fatigue-testing machine. Its main element is a small high-speed electric motor capable of running at a speed of 10,000 rpm. Speeds of this order do not greatly influence the data, while they markedly reduce the time needed to obtain the necessary information. Next to the motor is a large bearing, whose purpose is to relieve the motor of the large bending moment that is applied to the specimen. The specimen proper is mounted in collets that serve as grips. One collet is attached to the shaft driven by the motor while the other is attached to a rotating lever arm. At the end of the latter is a small bearing, used to apply a downward force to the lever arm. The application of this force

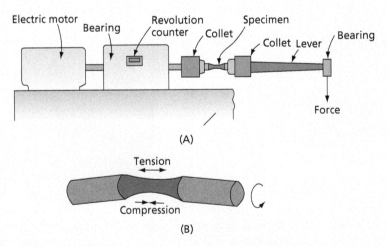

Fig. 21.30 **(A)** One form of rotating-beam fatigue-testing machine. **(B)** Fatigue-test specimen. Specimen is bent while it rotates. Any point in the reduced middle section alternates between states of tensile and compressive stress

places the small circular cross-section specimen in a state of bending, so that its upper surface is in tension while its lower surface is in compression. As the specimen is rotated by the action of the motor, any given position on the surface of the specimen alternates between a state of maximum tensile stress and a state of maximum compressive stress.

In making a test, one measures the number of cycles required to fracture the specimen at a given stress. The stress, of course, is the maximum fiber stress developed on the specimen surface by the bending moment, which is created by hanging a weight on the end of the lever arm. This stress can be easily computed in terms of the magnitude of the applied weight, the length of the applied weight, the length of the lever arm, and the diameter of the specimen at its minimum cross-section. If the maximum tensile stress applied by bending is only slightly less than that which will break the specimen in a simple tensile test, the fatigue tester will run only a few cycles before the specimen fractures. Continued reduction of the stress greatly increases the life of the specimen, so that in plotting fatigue-test results, it is common practice to plot the maximum bending stress against the number of cycles to fracture, using a logarithmic scale for the latter variable. Figure 21.31 shows the type of curve that is usually obtained when the fatigue specimens are steel. This form of curve is significant because there is a stress below which the specimens do not fracture. At this particular stress, called the *fatigue* (or endurance) *limit*, the *SN* (stress-number of cycles) curve turns and runs parallel to the *N* axis. This is an important effect, for it implies that if a steel is only loaded to a stress that is below its fatigue limit, no matter how many times the stress is applied, it will not fail.

There appears to be a reasonably good correlation between the fatigue limit of a steel and its ultimate tensile strength. The ratio of these two quantities, known as the *endurance ratio*, is therefore of some value and usually falls in the range 0.4 to 0.5. Since the yield strength of some steels is often close to one-half the ultimate strength, the endurance limit and the yield stress are often approximately equal. It should not be inferred, however, that the two are equal, because there is no good overall correlation between the two quantities.

Unlike steels, most nonferrous metals do not appear to have an endurance limit. The *SN* curves for these metals generally have the appearance shown in Fig. 21.32 where, with

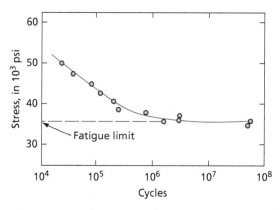

Fig. 21.31 A typical *S-N* curve for a steel. Individual data points, however, usually have a larger scatter than is observed in this figure. (From *Prevention of Fatigue of Metals*, p. 46, The Staff of the Battelle Memorial Institute, John Wiley and Sons, Inc., New York, 1941)

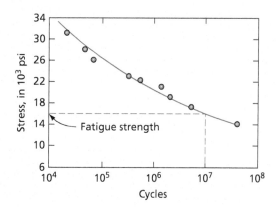

Fig. 21.32 A typical *S-N* curve for a nonferrous metal (aluminum alloy). (From *Prevention of Fatigue of Metals*, p. 48, The Staff of the Battelle Memorial Institute, John Wiley and Sons, Inc., New York, 1941)

decreasing stress, the curve continues to fall steadily, although at a decreasing rate. In speaking of the ability of nonferrous alloys to resist fatigue failure, one generally has to specify how many stress cycles the metal should withstand. The stress that will cause fracture at the end of the specified number of stress alterations, normally 10^7, is known as the *fatigue strength* of the metal, and is illustrated in Fig. 21.32.

21.15 Alternating Stress Parameters

The rotating beam machine in Fig. 21.30 produces a sinusoidal stress on the surface of the fatigue specimen as shown in Fig. 21.33A. Note that the maximum tensile (positive) stress, σ_{max}, equals the maximum compressive (negative) stress, σ_{min} during a cycle. Thus, the average or mean stress, σ_m, is zero. In practice, this is the type of alternating stress that an axle or shaft may be subjected to when it rotates at a constant speed and constant load. However, many engineering components are normally subjected to alternating stresses where the mean stress is not zero. It is convenient to consider these stresses in terms of two components: one is a mean or steady-state stress, σ_m, and the other is an alternating stress, σ_a. Examples may be found in Fig. 21.33, where σ_m is taken as positive so that it lies above the zero stress axis. It is possible for the mean stress to be negative, but this is not normally as significant as when it is positive. Note that the alternating stress, σ_a, is defined by

$$\sigma_a = (\sigma_{max} - \sigma_{min})/2 = \Delta\sigma/2 \qquad 21.1$$

where σ_{max} is the maximum stress in a cycle, σ_{min} the minimum stress, and $\Delta\sigma$ the alternating stress range.

Two algebraic stress ratios are often used in fatigue studies and are defined by the relations

$$R = \sigma_{min}/\sigma_{max} \qquad 21.2$$

and

$$A = \sigma_a/\sigma_m \qquad 21.3$$

where R is the ratio of the minimum stress to the maximum stress in a cycle and A the ratio of the alternating stress to the mean stress. It should be noted that in computing the R ratio, compressive stresses are considered negative.

The endurance limit of a metal tends to rise with an increase in the R ratio. This is clearly shown in Fig. 21.34, where four S-N curves are plotted corresponding to the R ratios $-1.0, -0.3, 0,$ and $+0.3$ of the curves in Fig. 21.33. In these plots a constant stress amplitude is assumed. Because of the interrelation between R and the endurance limit, it is obvious that the R ratio should be seriously considered in the design of components to be used under conditions involving a nonzero mean stress.

It is interesting to note that almost a century ago, Goodman observed that there was a relationship between the limiting stress range that a material can withstand without failing and the mean stress σ_m. Since by definition $\sigma_a = \Delta\sigma/2$, this signifies a relation between the alternating stress, σ_a,

21.15 Alternating Stress Parameters

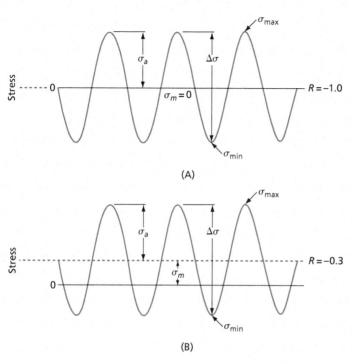

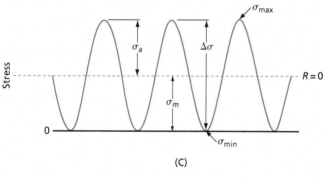

Fig. 21.33 The effect of several values of σ_m upon the cyclic stress pattern for constant value of the stress range

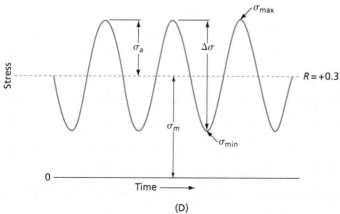

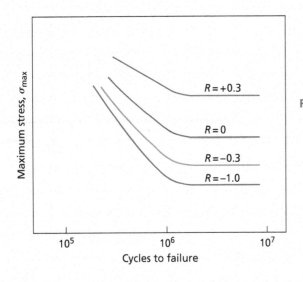

Fig. 21.34 The effect of R on the S-N curves. Note that as R becomes more positive, the endurance limit increases. (From Dieter, G. E., *Mechanical Metallurgy*, 2nd Ed., p. 436, McGraw-Hill, Inc., New York, 1976. Used with permission)

and the mean stress, σ_m. Actually, there is a well-known empirical equation (Refs. 23 and 25) relating these two parameters of a material to its fatigue limit and ultimate stress:

$$\sigma_a = \sigma_e[1 - (\sigma_m/\sigma_u)^\alpha] \qquad 21.4$$

where σ_e is the fatigue limit for the case of completely reversed cyclic loading, as in Fig. 21.33A, σ_u is the ultimate stress, and $\alpha = 1$ for what is called the *Goodman approach*. However, an alternate proposal due to Gerber takes $\alpha = 2$. As may be seen in Fig. 21.35, the Goodman relation is linear and the Gerber parabolic. The Goodman form of Eq. 21.4 is normally preferred (Ref. 24).

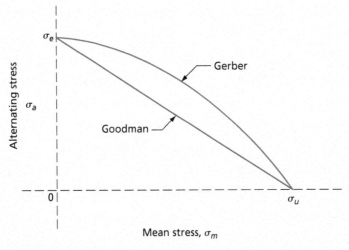

Fig. 21.35 A plot showing the proposed relations of Goodman and Gerber for the dependence of the alternating stress on the mean stress

21.16 The Microscopic Aspects of Fatigue Failure

It is customary to subdivide the fatigue process into three phases: crack initiation, crack growth, and the ultimate catastrophic failure. We shall now consider crack initiation.

It is almost universally agreed that fatigue failures start at the surface of a fatigue specimen. This is true whether the test is made in a rotating-beam machine where the maximum stress is always at the surface, or in a push-pull machine that gives a simple tensile-compressive stress cycle. Furthermore, fatigue fractures start as small microscopic cracks and, accordingly, are very sensitive to even minute stress raisers. It is quite apparent, in light of these considerations, that a fatigue specimen will give results that are representative of the metal tested only if its surface is free of defects. Tool or grinding marks left on the surface make the formation of fatigue cracks easier and may result in low apparent values of the fatigue limit or fatigue strength. A fundamental condition for making a fatigue test is that the surface of the specimen be carefully polished.

In considering the problem of fatigue, the fact that fatigue failures have been observed in specimens tested at 4 K is significant. At this extremely low temperature, thermal energy cannot make any appreciable contribution to the mechanism of fatigue fractures. It can, therefore, be concluded that it is possible to have fatigue failure without thermal activation. This means that while diffusion processes may be involved in some cases of fatigue, they are not necessary to the formation of fatigue cracks.

Slip and twinning are plastic deformation processes that are believed to be able to occur without thermal activation, and it is felt that they are strongly involved in fatigue-failure mechanisms. Because fatigue failures are promoted by the presence of minute stress raisers, the study of mechanisms of fatigue is best done on metals or alloys that have a minimum of nonmetallic inclusions. (The role of nonmetallic inclusions will be briefly discussed presently.) In most metals at temperatures close to room temperature, slip seems to be the predominant factor in fatigue. At very low temperatures, and with steel or iron specimens, mechanical twinning is probably of importance.

A number of dislocation mechanisms have been proposed (Refs. 26 and 27) to explain the experimentally observed phenomena of fatigue. None of these mechanisms, however, is completely satisfactory; therefore, they will not be considered. Rather, a brief discussion will be given of some of the experimental information that is now available.

In polycrystalline metals, slip bands (groups of closely spaced and overlapping slip lines) are observed to form on specimens prior to fracture. The nature of these bands differs somewhat with the metal under consideration and with whether slip takes place by single or multiple glide. Figure 21.36 shows that in the case of a particular low-carbon steel (0.09 percent C), slip lines first become visible at about $\frac{1}{100}$ the number of cycles required to fracture the specimens. It can also be seen that at large values of N, the curve representing the first appearance of slip lies below that for the fracture of the specimens. Slip can thus occur in this metal at stresses well below the fatigue limit. This result has also been observed in other metals, but it is not a universal effect, for there are also metals in which slip lines only appear when the stress is equal to or above the fatigue limit.

During a fatigue test, slip lines appear first in those crystals of a specimen whose slip planes have the highest resolved shear stress. As time goes on and the number of stress cycles increases, the size and number of slip bands rises. The extent and number of the slip bands are also a function of the amplitude of the applied stress; higher stresses give larger values. Microscopic observations in polycrystalline copper show that a periodical array of small, pit-shaped fatigue cracks form along slip bands during the early stages of fatigue (Ref. 28).

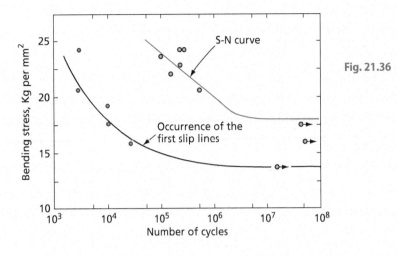

Fig. 21.36 Slip lines may become visible well before a fatigue specimen fractures. Data for flat specimens of low-carbon steel (0.09 percent C) stressed in bending. (After Hempel, M. R., *Fracture*, p. 376, The Technology Press and John Wiley and Sons, Inc., New York, 1959)

In fatigue, the direction of the strain is reversed over and over again. The slip lines that appear on the surface reflect this fact. When the strain is in a single basic direction, the slip steps that appear on a crystal surface have a relatively simple topography. This is particularly true if only a single slip plane is active. The nature of these slip markings is indicated in Fig. 21.37A. On the other hand, under cyclic loading the slip bands tend to group into packets or striations.

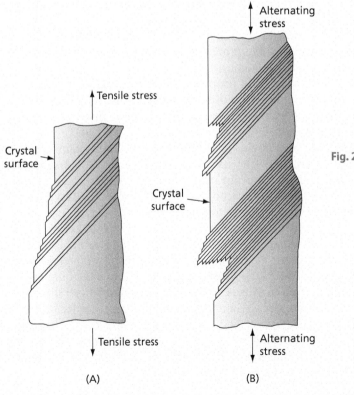

Fig. 21.37 The difference in the surface contours where slip bands intersect the surface. **(A)** One-directional deformation. **(B)** Alternating deformation

The surface topography of these striations is more complex and is indicated schematically in Fig. 21.37B. A basic component of these striations consists of what are now generally called *persistent slip bands* or PSBs. According to Ma and Laird, (Ref. 29) in the copper single crystals that they studied, a typical PSB contained about 5000 slip planes and had the shape of a lamella that traverses an entire crystal. They also state that it is well accepted that PSBs are the only sites where fatigue cracks can form in single crystals cycled at low and intermediate strain amplitudes because they are regions of highly concentrated strain. In addition, while PSBs are not the only possible crack nucleation sites in polycrystals, they are important for furnishing nucleation sites in those strong alloys in which strain localization occurs. It has further been observed (Ref. 30) that the cracks that lead to failure in copper form in the PSBs with the greatest local strain. As a result of their study on copper single crystals oriented for single slip, Ma and Laird also observed that the plastic strain amplitude in a PSB lamella is about 100 times larger than that in the surrounding matrix. Consequently, there should exist a sharp change in the strain level at a PSB-matrix interface. As a result, the longest cracks in their specimens were frequently found at these interfaces. This is not an isolated observation since it has been confirmed by other workers (Ref. 29).

The surface topography of the striations containing the PSBs is more complex than that due to simple tensile loading shown in Fig. 21.37A. In the copper single crystals of Ma and Laird, striations were observed that contained alternating lamellae of the matrix and the PSBs. This implies that the bands shown in Fig. 21.37B should consist of a set of PSBs with matrix lamellae in between them. Such striations are often called *macro-PSBs*. Note that both ridges (*protrusions*) and crevices (*encroachments*) can form and it is possible to nucleate fatigue cracks in either one. Whether or not protrusions or encroachments form in a particular specimen, or a specific grain of a given specimen, is largely a function of crystal orientation. If the shear direction of a striation is nearly normal to the surface, the formation of protrusions or encroachments might be favored. However, if the slip direction is parallel to the surface, they normally do not develop.

One consequence of dislocation movements during fatigue is that small localized deformations called *extrusions* may appear in the PSBs. As may be seen in Fig. 21.38, an extrusion is a small ribbon of metal that appears as if it had been extruded from the surface of the metal. A second photograph showing a number of extrusions appears in Fig. 21.39. Because extrusions are normally accompanied by cracks in the slip packets, they may be of significance in crack initiation. The inverse of extrusions are narrow crevices called *intrusions*. The start of a crack in an intrusion may be seen in Fig. 21.40. These surface disturbances (intrusions and extrusions) have a depth or height of the order of about 10 μm.

21.17 Fatigue Crack Growth

With increasing numbers of cycles, the surface grooves deepen and the crevices and intrusions take on the nature of a crack. When this happens, Stage I of the crack-growth process has begun. Fracture initiation can start at a relatively early point in the fatigue life of a specimen and, under favorable circumstances, Stage I crack growth can continue for a large fraction of the fatigue life. If the specimen has a preferred orientation in which neighboring grains have slip planes with nearly equivalent orientations, it is possible for slip-band cracks to extend across grains. Low applied stresses and

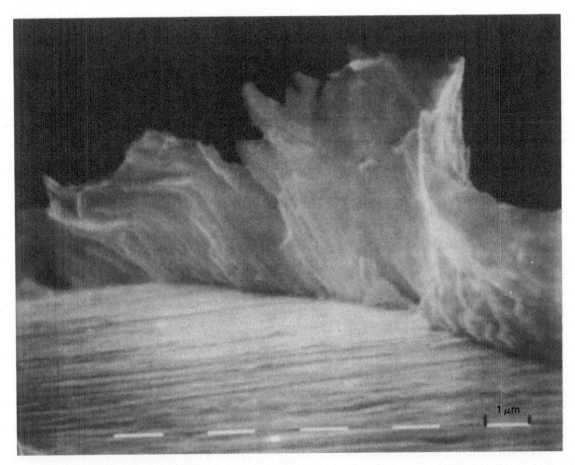

Fig. 21.38 A persistent slip band extrusion in a copper single crystal, tested at a plastic strain amplitude of 2×10^{-3} for 30,000 cycles. (From Ma, B.-T., and Laird, C., *Acta Metall.*, **37** 325 [1989])

deformation by slip on a single slip plane favor Stage I growth. On the other hand, multiple-slip conditions favor Stage II growth. From a practical viewpoint, Stage I is normally of secondary importance to Stage II. It should be noted that in addition to fracture initiation at slip planes, it is also possible for cracks to form at grain boundaries or subgrain boundaries. Slip-band initiation, however, appears to be of the greatest overall importance.

Stage I growth follows a slip plane, whereas Stage II growth does not have this crystallographic character. In this case, the fracture conforms rather closely to fracture mechanics conditions. Thus, if the applied stress favors plane-strain deformation, the fracture surface in Stage II will follow a plane normal to the principal applied tensile stress. On the other hand, in a thin plate with increasing size of the crack, the fracture surface tends to shift into a plane at 45° to the specimen surface. The transition from Stage I to Stage II is often induced when a slip-plane crack meets an obstacle such as a grain boundary. Figure 21.41 is a schematic drawing illustrating the fracture process resulting from Stage I and Stage II crack growth, followed by the final catastrophic failure.

Fig. 21.39 Some PSB protrusions with superimposed extrusions and intrusions in a copper single crystal. A crystal tested at a plastic strain amplitude of 2×10^{-3} for 120,000 cycles. (From Ma, B.-T., and Laird, C., *Acta Metall.*, **37** 325 [1989])

There is a large difference between the crack growth rates in Stages I and II. The growth rate is much slower in Stage I, at about 0.1 nm per cycle, while in Stage II it may be of the order of microns per cycle. This amounts to a difference of about 10,000 times in the two stages. Because Stage I fractures tend to follow crystallographic slip planes, their surfaces are normally much less distorted than those of Stage II fractures. In the latter case, the fracture surfaces may also show ridges due to the fact that in Stage II, fatigue cracks grow by steps; each incremental advance corresponds to one cycle of loading. Basically, the crack moves ahead during the tension part of the cycle and then halts during its compression part. There are several factors involved in this cyclic growth pattern which can be easily understood in the case of a symmetrical tensile and compressive loading pattern of the type shown in Fig. 21.35A, where $\sigma_m = 0$ and $R = -1.0$. This figure shows a sinusoidal wave form; however, it could just as well be one of the other waveforms in Fig. 21.42. In any case, during the tensile part of the cycle, the crack faces are pulled apart from each other and the

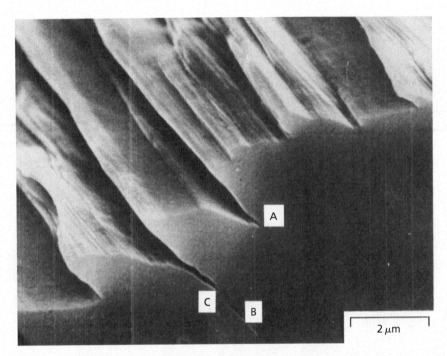

Fig. 21.40 A section through a PSB containing intrusions at *A* and *B*. Note the crack that has started at *B*. (From Neumann, p., *Physical Metallurgy*, P. 1554, Elsevier, Amsterdam, 1983)

crack tip advances. This is accompanied by plastic flow in the region around and in front of the advancing crack tip. This plastic flow tends to blunt the crack tip; that is, it opens up the crack at its extremity, which acts to limit its speed of advance. This process continues until the tensile load reaches its maximum. Following this, the load begins to fall and as it drops, the forward movement of the crack ceases. This is followed by a change in the load from tension to compression. As the compressive load increases, the crack faces are brought together, and the blunted end of the crack

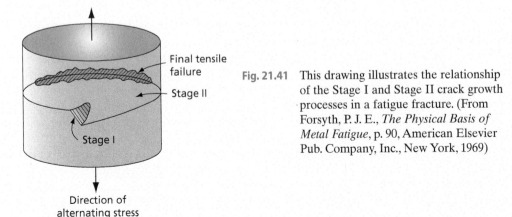

Fig. 21.41 This drawing illustrates the relationship of the Stage I and Stage II crack growth processes in a fatigue fracture. (From Forsyth, P. J. E., *The Physical Basis of Metal Fatigue*, p. 90, American Elsevier Pub. Company, Inc., New York, 1969)

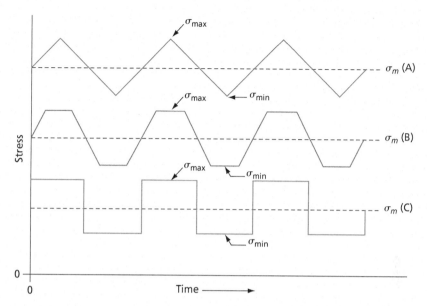

Fig. 21.42 A closed-loop servohydraulic universal testing machine is capable of applying a number of different loading cycles to a fatigue specimen. Three are shown here: **(A)** triangular, **(B)** trapezoidal, and **(C)** square

tends to collapse as a result of reversed plastic flow in the material surrounding the crack tip. This effectively sharpens the crack tip. This sharpening of the crack tip, in turn, aids the renucleation of the crack when the load enters the tensile phase of the next cycle, upon which the entire process is repeated.

A large amount of effort has been devoted to measuring the growth rate of fatigue cracks (Refs. 31–33). The growth of a fatigue crack is relatively easy to follow in a sheet specimen, and there is a great deal of data in the literature that shows a dependence of the growth rate on the applied stress and the square root of the crack length. Normally this takes the form of a power law in which the exponent is close to four. Stated in the form of an empirical equation, this relationship is

$$dc/dN = A(\sigma \sqrt{c})^4 \qquad 21.5$$

where N is the number of cycles, c is the half crack length for a double-ended crack, σ is the peak gross stress, and A is a constant.

21.18 The Effect of Nonmetallic Inclusions

Nonmetallic inclusions reduce the fatigue strength of metals. This fact is illustrated in Fig. 21.45, which gives three *SN* curves for a high-strength medium-carbon alloy steel, AISI (4340). Steel specimens heat-treated to the same strength (ultimate strength 1590 MPa) were used to obtain each of the three curves. The upper and middle curves correspond to vacuum-melted steel, while

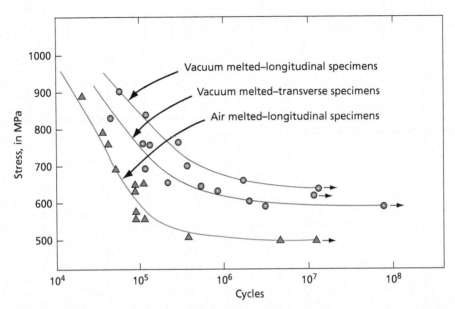

Fig. 21.43 Effect of nonmetallic inclusions on the fatigue strength of a low-alloy steel (4340). All specimens heat-treated to same ultimate strength (1590 MPa). (From Aksoy, A. M., *Trans. ASM.*, **49** 514 (1957). Reprinted with permission of ASM International®. All rights reserved. www.asminternational.org)

the lower curve corresponds to steel that was air-melted. The inclusions in the vacuum-melted metal are smaller in size and much less numerous than in the air-melted steel. This is because gaseous elements are absorbed into the air-melted steel, increasing the number of inclusions. Notice that the air-melted steel has the lower fatigue limit.

The middle curve represents data for vacuum-melted specimens machined from fabricated-steel stock in a direction transverse to the hot-rolling direction. The other two curves are for specimens cut with their long axes parallel to the hot-rolling direction. The middle curve should be compared to that above it, which shows that the fatigue strength or fatigue limit is less when specimens are cut transverse to the rolling direction. The reason for this is probably as follows.

When a metal is fabricated by rolling at a red heat, the inclusions tend to be deformed and elongated in the rolling direction. In a specimen cut transverse to the rolling direction, the long axis of the inclusions lies in a plane perpendicular to the bending stress. On the other hand, in the longitudinal specimens, the long axis of the inclusions is parallel to the bending stress. In the former case, the cross-section of the inclusions normal to the stress is larger than it is in the latter case. This difference in the area of the inclusions normal to the stress is believed to be the biggest factor in the lower fatigue strength of transverse specimens.

Microscopic examinations of metals with high concentrations of nonmetallic inclusions, or second-phase particles, show that small cracks form readily at inclusions (Refs. 34–35). These cracks may appear almost at the start of the test and well before visible slip lines are observed. Not only do inclusions nucleate cracks, but also they aid in their propagation, for the cracks readily jump from one inclusion to the next.

21.19 The Effect of Steel Microstructure on Fatigue

Eighty years ago it was shown (Ref. 36) that the ratio of the fatigue limit to the ultimate strength (endurance ratio) was about 0.60 for steels with tempered martensitic structures, and about 0.40 with austenitic or pearlitic structures. More recent work has shown that the amount of martensite obtained during a quench is very important in determining the fatigue properties of steel. Figure 21.46 shows the result obtained by quenching a number of different alloy steels. All specimens had the same carbon content and were quenched at several different rates of cooling in order to obtain various percentages of martensite in the quenched metals. Each specimen was then tempered to the same hardness (Rockwell C-36) and, following this, its fatigue limit was determined. The relationship between the fatigue limit and the amount of martensite obtained on quenching is shown in Fig. 21.44. The drop in fatigue limit for even as little as 10 percent of structures other than martensite is quite noticeable. This drop constitutes another important reason why steels in machinery parts are almost universally heat-treated before they are tempered, in order to obtain as nearly as possible a 100-percent martensitic structure. Other examples of the pronounced influence of microstructural variables on the crack initiation and fatigue of other steels can be found elsewhere (Refs. 37–39).

21.20 Low-Cycle Fatigue

In recent years, increasing interest has developed in fatigue tests in which failure occurs after a relatively small number of cycles. This has been due in part to the fact that some engineering components, like the spring in an automobile self-starter, will probably never be subjected in its

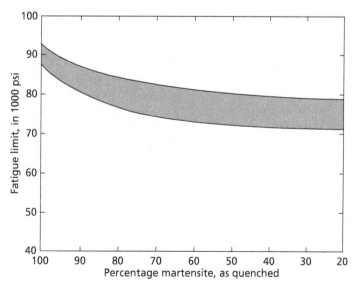

Fig. 21.44   Fatigue limit versus percentage of martensite in "as quenched" specimens tempered to same hardness (Rockwell C-36). Notice that the curve is plotted as a band in order to indicate the scatter of experimental data. (From Borik, F., Chapman, R. D., and Jominy, W. E., *Trans. ASM*, **50** 242 (1958). Reprinted with permission of ASM International®. All rights reserved. www.asminternational.org)

lifetime to more than some tens of thousand of cycles. In other words, they do not need to be designed to undergo as many as 10^7 cycles. Designing such a part in terms of its expected life can reduce both its weight and cost. Another consideration is the large number of important engineering components that are exposed to thermal cycling. This includes nuclear reactor pressure vessels, heat exchanger tubes, and steam and gas turbine rotors and blades. During a heating and cooling cycle, these items undergo large thermal expansions and contractions. In many cases, the thermal expansions and contractions of these items are severely restrained so that they undergo large cyclic strains during a heating and cooling cycle. The number of these cycles is normally relatively small so that the problems associated with thermal cycling can be studied in the region where the total number of cycles to failure is small. Finally, in recent years, a renewed interest in the factors controlling the nucleation growth of fatigue cracks has developed. In this regard, evidence has been found (Refs. 39–41) supporting the view that the same mechanisms are involved in fatigue failures that occur at high stresses and a small number of cycles as occur at low stresses and large numbers of cycles. At the same time, it is generally true that fatigue cracks develop within the first three to ten percent of a fatigue specimen's life. However, in crack growth, after nucleation and initiation (Stage I), there is a basic difference between the fraction of time spent in Stage II growth in a specimen that fails after a few cycles and one that fails after a large number. Normally, a much larger fraction of the fatigue life is spent in Stage II crack growth in low-cycle fatigue. The opposite is true in high-cycle fatigue, where a large fraction of the fatigue life is spent in Stage I. This means that fatigue crack nucleation and growth can be studied more effectively when fatigue failure occurs after a relatively few stress cycles. This has lent additional motivation for the study of low-cycle fatigue (LCF). By general definition, fatigue studies involving fewer than 10^4 cycles are grouped in the *low-cycle fatigue* (LCF) category, while those requiring more than 10^4 cycles fall in the *high-cycle fatigue* (HCF) classification. An additional classification, called *very high cycle fatigue* (VHCF) or *giga cycle fatigue* (GCF), has been introduced in recent years to describe fatigue cycles exceeding 10^7 cycles. Sadananda et al. (Ref. 42) indicate that the fracture is surface crack initiated when the cycle is less than 10^5, whereas in the VHCF or GCF range internal heterogeneous sites nucleate the cracks.

Contributing to the current interest in low-cycle fatigue is the development in recent years of a type of universal testing machine very well suited to this type of testing: the closed-loop servohydraulic testing machine. This device can cyclically load a specimen in a number of different ways, including push-pull, in which a cylindrical specimen is alternately pulled in tension and pushed in compression. Unlike the rotating beam fatigue tester, it is not effectively limited to producing a simple sinusoidal cycle. The wave form may be sinusoidal, triangular, trapezoidal, or square, as shown in Fig. 21.42. In addition, it is easy to change the value of the mean stress, σ_m, with this instrument. A typical closed-loop servohydraulic testing machine may operate at a maximum frequency of the order of 80 Hz, which equals 4800 cycles per min., or close to half the maximum used with a rotating beam instrument. Low-cycle fatigue experiments do not normally require 80 Hz, because at this high a frequency it will produce 10^4 cycles in about 2 minutes. Typical experimental LCF results published in the literature are usually obtained at a frequency of 20 Hz or less.

In a low-cycle fatigue test, the specimens are usually loaded well above their yield stresses. For example, as pointed out in Sec. 21.14, the fatigue limit and the yield stress of a steel are roughly equivalent, while low-cycle fatigue is normally carried out well above the fatigue limit. The high stresses involved in LCF testing signify that finite plastic strains must be developed during an alternating stress cycle. This means that during a single cycle, the stress-strain curve follows a hysteresis loop of finite width, as shown schematically in Fig. 21.45. The loop develops as follows. On the initial loading, the specimen deforms elastically and linearly from point O to point C, the

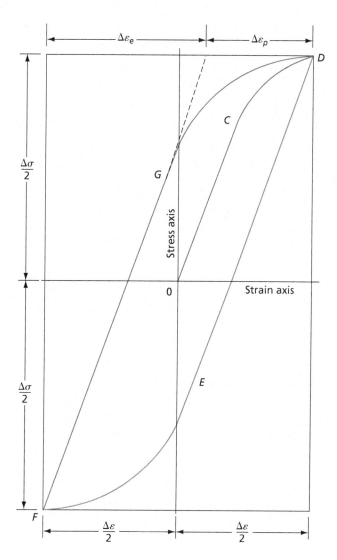

Fig. 21.45 A schematic stress-strain hysteresis loop

yield stress. After this, the stress-strain path is curved as the specimen deforms plastically. At point D, tensile loading of the specimen is stopped and the deformation of the specimen is reversed. At the start of this part of the cycle, the stress falls linearly until point E is reached, corresponding to the yield stress in compression. Note that this compressive yield stress is smaller than the initial yield stress in tension. This is a reflection of what is known as the *Bauschinger effect*; namely, after a specimen has been uniaxially loaded in one sense, the yield stress for loading in the opposite sense is normally smaller than the initial stress. The reversed part of the cycle is completed at point F. On reloading, the stress-strain curve follows path FGD to complete the hysteresis loop. Note that the loop is nearly symmetrical. However, with increasing number of cycles the shape of the loop may change progressively. If the specimen is subjected to loading cycles in which the stress range, $\Delta\sigma$, is held constant, then the strain range, $\Delta\varepsilon$, may either grow or decrease in size. Normally, however, LCF tests are carried out under the condition of a constant strain. This can be easily accomplished with a servohydraulic closed-loop machine. Note that the strain range $\Delta\varepsilon$ consists of two parts: (1) the elastic strain component, $\Delta\varepsilon_e$, and (2) a plastic strain component, $\Delta\varepsilon_p$. These two strain components are shown at the top of the drawing in Fig. 21.45. If the strain range is held fixed, the stress range may either increase or decrease as the number of cycles increases. However, after the specimen has been subjected to approximately 100 cycles, $\Delta\sigma$ usually approaches an approximately constant or steady-state value that depends on the cyclic strain range applied to the specimen. An example of this functional relationship between the steady-state $\Delta\sigma$ and $\Delta\varepsilon_t$ is shown in Fig. 21.46, which shows plots of LCF data obtained (Ref. 43) using specimens of a cast nickel-based superalloy tested in air at three different temperatures. These data correspond to values of $\Delta\sigma$ and $\Delta\varepsilon_p$ observed after each specimen had been subjected to 100 cycles. The cyclic loading involved a controlled total strain amplitude and a triangular completely reversible wave shape. Note that the data plots linearly on double logarithmic coordinates at all three temperatures. This implies the existence of a power law between $\Delta\sigma$ and $\Delta\varepsilon_p$, or that we may write

$$\Delta\sigma = A(\Delta\varepsilon_p)^n \qquad 21.6$$

where A is the strength coefficient and n is the cyclic strain-hardening exponent. In the case of the 1033 K data in Fig. 21.46, it has been determined that $A = 2340$ MPa and $n = 0.46$. If they are substituted into Eq. 21.6, one obtains

$$\Delta\sigma = 2340\Delta\varepsilon_p^{0.46} \qquad 21.7$$

which yields the curve shown in Fig. 21.47. This curve is the cyclic equivalent of a tensile stress-strain curve. It is to be noted, however, that in this case, the strain is the plastic strain and not the total (elastic plus plastic) strain. Furthermore, if one compares a cyclic stress-strain curve, such as that in Fig. 21.47, with a tensile stress-strain curve obtained under similar conditions of temperature and material, significant differences will normally be found between the two curves (Ref. 44).

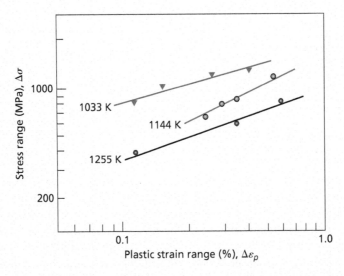

Fig. 21.46 An example showing the variation of the stress range $\Delta\sigma$ (steady state) with the plastic strain range, $\Delta\varepsilon_p$. (From Hwang, S. K., Lee, H. N., and Yoon, B. H., *Met. Trans. A*, **20A** 2793 (1989). Used with kind permission from Springer Science and Business Media)

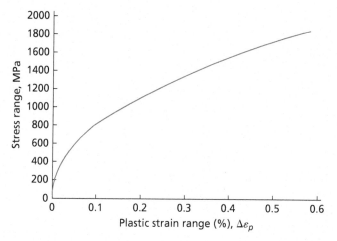

Fig. 21.47 A plot of Eq. 21.6 using the 1033 K $\Delta\sigma$-$\Delta\varepsilon_p$ data in Fig. 21.48

21.21 The Coffin-Manson Equation

The Coffin-Manson equation is a well-known LCF power law relation between the number of cycles to failure, N_f, (called the *fatigue life*) and the plastic strain range $\Delta\varepsilon_p$. It is the result of the fact that these two parameters plot linearly on log-log coordinates. Three examples of such plots are shown in Fig. 21.48. In this case, the data were obtained using an α-β Ti-Mn alloy (Ref. 45). The three curves correspond to three sets of specimens in which the grain size of the alpha phase was varied. Plots such as these were the basis on which Coffin-Manson proposed the following equation:

$$\Delta\varepsilon_p = C(N_f)^\beta \qquad 21.8$$

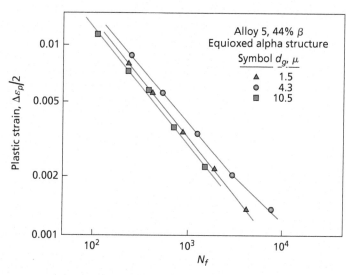

Fig. 21.50 Coffin-Manson plots obtained using three different alpha grain sizes of an α-β titanium alloy. (From Saleh, Y., and Margolin, H., *Met. Trans. A*, **13A** 1275 (1982). Used with kind permission of Springer Science and Business Media)

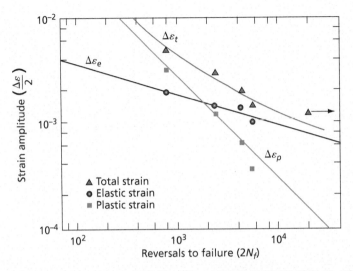

Fig. 21.49 Strain-fatigue life relationships at 1033 K of Inconel-617 (54% Ni, 22% Cr, 12.5% Cr, 9% Mn, 1% Al and 0.07% C). (From Burke, M. A., and Beck, C. G., *Met. Trans A*, **15A** 661 (1984). Used with kind permission of Springer Science and Business Media)

where C and β are material constants usually determined by experiment. Note the constant β is normally a negative number.

A relation similar to Eq. 21.7 has also been observed between the elastic component of the cyclic strain, $\Delta\varepsilon_e$ and the fatigue life, N_f, since a plot involving these two parameters also tends to yield a straight line. This may be seen in Fig. 21.49 showing data obtained at 1033 K using specimens of the nickel superalloy, Inconel-617. Thus, we may write

$$\Delta\varepsilon_e = D(N_f)^\gamma \qquad 21.9$$

where D and γ are also material constants. Observe that in Fig. 21.49, the authors plot the cyclic strain amplitudes (half of the strain ranges) against $2N_f$, the number of load reversals to failure, instead of N_f, the number of cycles to failure. This does not affect the basic functional relations in Eqs. 21.7 and 21.8. In addition, in Fig. 21.49 the elastic strain amplitude is plotted not only against the number of reversals to failure but also against the plastic and total strain amplitudes. Note that in this case, the total strain amplitude does not yield a straight line when plotted against $2N_f$ (or N_f). Since ε_t is the sum of ε_e and ε_p, it follows that

$$\Delta\varepsilon_t = D(N_f)^\gamma + C(N_f)^\beta \qquad 21.10$$

Although the Coffin-Manson equation has been successfully used to predict fatigue life for LCF where the number of cycles to failure is less than 10^4, the equation over predicts life cycles at extremely low cycle fatigue (ELCF) of about hundred (Ref. 46). Additionally, the equation does not adequately describe transition to HCF where alternating stresses are elastic and life cycles are over about 10^5. As such, fatigue life predictions are difficult when a material experiences complex elastic and plastic loading cycles such as in turbine blades in jet engines (Ref. 47).

21.22 Certain Practical Aspects of Fatigue

Most fatigue failures in actual engineering structures are the result of macroscopic stress raisers that were unintentionally incorporated in the object that failed. Any sharp corner or angle on the surface of an object that undergoes repeated stresses is a potential danger point for a fatigue failure. This is especially true if the sharp corner lies in a region in which the stress cycle places the metal in a state of tension during part of the cycle. A well known example of an inadvertent fatigue failure is that of metal airplane propellers that were originally stamped with the manufacturer's name in the middle of the blades. The grooves caused by the stamping served as stress raisers, causing early fatigue failures. Keyways cut into shafts are frequently a source of fatigue damage. Even holes drilled into a metal part can provide stress raisers at the sharp lip of the holes. The V-notches at the bottom of threads cut on bolts are still another source of fatigue cracks.

Since fatigue failures originate at the surface of metallic specimens, strengthening the surface generally improves the fatigue life. Surface hardening for this purpose can be accomplished by a variety of methods. One method often used is shot-peening, where the surface of a metal shaft or other object is bombarded with steel shot. This cold works the surface layers and leaves them in a state of residual compressive stress with a considerable improvement in fatigue properties. Surface hardening by carburizing or nitriding may also be used to harden the surface of metal subjected to repeated stresses.

Problems

21.1 (a) Make a list of the different types of fracture that occur in single crystals.

(b) Now consider a polycrystalline metal. Indicate how grain boundaries add to the fracture possibilities.

21.2 It is possible for zinc crystals to undergo very large plastic strains at temperatures below room temperature. This is not true in polycrystalline zinc, which normally shows a very low ductility at sub-ambient temperatures. Give an explanation for this fact based on the mechanical properties of zinc.

21.3 S. S. Brenner, (*Jour. Appl. Phys.*, **27** 1484 [1956]), has reported that iron whisker single crystals may have a tensile strength of 13,130 MPa. Assume that the iron whiskers have a $\langle 100 \rangle$ orientation; that is, a $\langle 100 \rangle$ direction of the crystal lies along its central axis.

(a) Compute the magnitude of the shear stress on the most highly stressed $\{110\}$ $\langle 111 \rangle$ slip systems when the ultimate tensile stress is attained. How many of these slip systems are there?

(b) The shear modulus, μ, along a $\langle 110 \rangle$ direction of a $\{110\}$ plane is 59,810 MPa. Compute the ratio of τ_{max}/μ.

(c) According to theoretical calculations, the maximum shear strength of a crystal should be of the order of $\mu/10$ to $\mu/30$. How does the value obtained in (b) of this problem compare to the theoretical predictions?

(d) Now assume that one of the slip planes is ideally oriented so as to have both θ and ϕ in the resolved shear stress equation equal to 45 degrees, and compute τ_{max}/μ.

(e) On the basis of these calculations, what can one conclude about the occurrence of slip in the iron whiskers? Does this signify anything about the perfection of the crystal structure in the iron crystal whiskers?

21.4 (a) It is now believed that inclusions in a metal tend to promote both cleavage and ductile fracture. Explain.

(b) How does the shape of the inclusions affect the fracture in fatigue specimens?

(c) How does the multiaxial state of the stress affect fracture in general?

21.5 (a) Both cleavage and twinning in iron are favored by increasing the strain rate. Rationalize this statement.

(b) Describe the development of a cup-and-cone fracture in a tensile specimen.

21.6 A fatigue specimen is cyclic loaded as follows. (1) The specimen is loaded in tension for 1 sec. at 1500 MPa/s, starting from a zero load; (2) it is then held at its maximum load for 1 sec; and (3) it is unloaded for 1 sec. at -1500 MPa/s. This cycle is then repeated over and over.

(a) Draw several cycles showing the cyclic loading pattern.

(b) Determine σ_m, σ_a, R, and A.

21.7 (a) Draw the Goodman and Gerber lines for a steel whose ultimate stress is 1250 MPa and whose fatigue limit (assuming completely reversed loading) is 400 MPa.

(b) If this steel is loaded sinusoidally with an alternating stress $\sigma_a = 150$ MPa, what is the maximum stress that can be applied to the specimen without its failing if the Goodman line is taken as valid?

21.8 According to Hwang, et al., the constants in Eq. 21.6 corresponding to their 1255 K data in Fig. 21.48 are $A = 1920$ MPa and $n = 0.31$. Plot the stress range as a function of the plastic strain range for these data in the manner of Fig. 21.49.

21.9 Determine the constants β and C of the Coffin-Manson equation using the 1.5 μ diameter alpha grain size data of Saleh and Margolin in Fig. 21.50.

21.10 With the aid of the Coffin-Manson equation evaluated in Prob. 21.9, plot a curve of the Saleh and Margolin data on $\Delta\varepsilon_p$ versus log N_f coordinates.

21.11 Figure 21.51 shows the strain amplitudes, $\Delta\varepsilon_e/2$, $\Delta\varepsilon_p$, and $\Delta\varepsilon_t$ plotted against $2N_f$, the number of reversals to failure, on double logarithmic coordinates. Determine equations for the three strain amplitudes as functions of $2N_f$ that are equivalent to Eqs. 21.8, 21.9, and 21.10.

21.12 Plot the three equations obtained in Prob. 21.11 on the same figure using a computer and double logarithmic coordinates. A comparison of your diagram with Fig. 21.51 will then serve as a check on your Prob. 21.11 results.

References

1. Liu, J. H., Huang, C. X., Wu, S. D., and Zhang, Z. F., *Mat. Sci. Eng. A* (2008), doi: 101016/J.msea.2008.01.004.
2. Kaufman, A. R., *The Metal Beryllium*, p. 367. American Society for Metals, Cleveland, Ohio, 1955.
3. Barret, C. S., and Bakish, R., *Trans. AIME*, **212** 122 (1958).
4. Sameljuk, A. V., Vasilev, A. D., and Firstov, S. A., *Int. J. of Refractory Metals and Hard Materials*, 14, 1996, pp. 249–255.
5. Biggs, W. D. and Pratt, P. L., "The deformation and fracture of alpha-iron at low temperatures", *Acta. Met.*, **6** (11) p. 694 (1958).
6. Rinebolt, J.A. and Harris, W. J. Jr., "Effect of alloying elements on notch toughness of pearlitic steels", *Transactions, ASM,* **43**, pp. 1175-1214 (1951).

7. Armstrong, T. and Warner, L. "Low-temperature transition of normalized carbon-manganese steels", *Symposium on Impact Testing*, ASTM International, pp. 40–58 (1956).
8. Zener, C., *Fracturing of Metals*, p. 3, ASM Seminar, 1948.
9. Cottrell, A. H., *Trans. AIME*, **212** 192 (1958).
10. McMahon, C. J., and Cohen, M., *Acta Met.*, **13** 591 (1965).
11. Knott, J. F., *Fracture*, **1** 61 (1967).
12. Ohr, S. M., *Mat. Sci. Eng.*, 72, 1985, pp. 1–35.
13. Cleri, F., Wolf, D., Yip, S., and Phillpot, S. R., *Acta Mater.*, 45, 1997, pp. 4993–5003.
14. Wilsdorf, H. G. F., *Mat. Sci. Eng.*, 59, 1983, pp. 1–39.
15. Gilman, J. J., Knudsen, C., and Walsh, W. P., *Jour. Appl. Phys.*, **29** 601 (1958).
16. Low, J. R., Jr., *Fracture*, p. 68. The Technology Press and John Wiley and Sons, Inc., New York, 1959.
17. Rogers, H. C., *TMS-AIME*, **218** (1960). pp. 498–506.
18. Shabrov, M. N., Sylven, E., Kim, S., Sherman, D. H., Chuzhoy, L., Briant, C. L., and Needleman, A., *Met. & Mat. Trans A*, 35A, 2004, pp. 1745–1755.
19. Brindley, B. J., *Acta Met.*, 18, 1970, pp. 325–329.
20. Rogers, H. C., *Acta Met.*, 7, 1959, pp. 750–752.
21. Garde, A. M., Santhanam, A. T., and Reed-Hill, R. E., *Acta Metall.*, 20, 1972, pp. 215–220.
22. Bergström, Y., and Roberts, W., *Acta Met.*, 19, 1971, pp. 815–823.
23. Salem, A. A., Kalidindi, S. R., and Deherty, R. D., *Scripta Mat.*, 46, 2002, pp. 419–423.
24. Dieter, G. E., *Mechanical Metallurgy*, 2nd Ed., p. 435–438, McGraw-Hill, Inc., New York, 1976.
25. Viswanathan, R., *Damage Mechanisms and Life Assessment of High-Temperature Components*, p. 113, ASM International, Metals Park, Ohio, 1989.
26. Cottrell, A. H., and Hull, D., *Proc. Roy. Soc.* (London), **A242** 211 (1957).
27. Mott, N. F., *Acta Met.*, **6** 195 (1958).
28. Kwon, I. B., Fine, M. E., and Weertman, J., *Acta Metall.*, 37, 1989, pp. 2927–2936.
29. Ma, B.-T., and Laird, C., *Acta Metall.*, **37** 325 (1989).
30. Cheng, A. S., and Laird, C., *Fat. Eng. Mat. Struct.*, **4** 331 (1981).
31. Hahn, C. T., and Simon, R., *Engineering Fracture Mechanics*, 5, 1973, pp. 523–540.
32. Ma, B. T. and Laird, C., *Acta Metall.*, 37, 1989, pp. 369–379.
33. Molent, L., Jones, R., Barter, S., and Pitt, S., *Int. J. Fatigue*, 28, 2006, pp. 1759–1768.
34. Hempel, M. R., *Fracture*, p. 376, The Technology Press and John Wiley and Sons, Inc., New York, 1959.
35. Zhang, J. U., Li, S. X., Yang, Z. G., Li, G. Y., Hui, W. J., and Weng, Y. Q., *Int. J. of Fatigue*, 29, 2007, pp. 765–771.
36. Caz, F., and Persoz, L., *La Fatigue Des Metaux*, 2nd ed., Chap. V, Dunod, Paris, 1943.
37. Gray, G. T., Thompson, A. W., and Williams, J. C., *Met. Trans. A*, 16A, 1985, pp. 753–760.
38. Lin, C-K. and Chang, C-W., *J. of Mat. Science*, 37, 2002, pp. 709–716.
39. Sankaran, S., Subramanya Sarma, V., Padmanabhan, K. A., Jaeger, G., and Koethe, A., *Mat. Sci. Eng.*, A362, 2003, pp. 249–256.
40. Viswanathan, R., *Damage Mechanisms and Life Assessment of High-Temperature Components*, p. 113, ASM International, Metals Park, Ohio, 1989.
41. Skelton, R. P., *Trans. Indian Inst. Metals*, **35** 519 (1982).
42. Sadananda, K., Vasudevan, A. K., and Phan, N., *Int. J. of Fatigue*, 29, 2007, pp. 2060–2071.
43. Hwang, S. K., Lee, H. N., and Yoon, B. H., *Met. Trans. A*, **20A** 2793 (1989).
44. Viswanathan, R., *Damage Mechanisms and Life Assessment of High-Temperature Components*, p. 120, ASM International, Metals Park, Ohio, 1989.
45. Saleh, Y., and Margolin, H., *Met. Trans. A*, **13A** 1275 (1982).
46. Liang Xue, *Int. J. Fatigue*, 30 (2008) 1691–1698.
47. Kai-Shang Li, et al., *Int. J. Fatigue*, 170, 2023, 107512.

Appendices

A Angles Between Crystallographic Planes in the Cubic System* (In Degrees)

HKL	hkl				
100	100	0.00	90.00		
	110	45.00	90.00		
	111	54.74			
	210	26.56	63.43	90.00	
	211	35.26	65.90		
	221	48.19	70.53		
	310	18.43	71.56	90.00	
	311	25.24	72.45		
	320	33.69	56.31	90.00	
	321	36.70	57.69	74.50	
110	110	0.00	60.00	90.00	
	111	35.26	90.00		
	210	18.43	50.77	71.56	
	211	30.00	54.74	73.22	90.00
	221	19.47	45.00	76.37	90.00
	310	26.56	47.87	63.43	77.08
	311	31.48	64.76	90.00	
	320	11.31	53.96	66.91	78.69
	321	19.11	40.89	55.46	67.79	79.11
111	111	0.00	70.53		
	210	39.23	75.04		
	211	19.47	61.87	90.00	
	221	15.79	54.74	78.90	
	310	43.09	68.58		
	311	29.50	58.52	79.98	
	320	36.81	80.78		
	321	22.21	51.89	72.02	90.00

*Abstracted from IMD Special Report Series, No. 8, "Angles Between Planes in Cubic Crystals," R. J. Peavler and J. L. Lenusky, The Metallurgical Society, *AIME*, 29 W. 39 St., New York, N.Y.

A Angles Between Crystallographic Planes in the Cubic System* (In Degrees) (continued)

HKL	hkl								
210	210	0.00	36.87	53.13	66.42	78.46	90.00		
	211	24.09	43.09	56.79	79.48	90.00			
	221	26.56	41.81	53.40	63.43	72.65	90.00		
	310	8.13	31.95	45.00	64.90	73.57	81.87		
	311	19.29	47.61	66.14	82.25				
	320	7.12	29.74	41.91	60.25	68.15	75.64	82.87	
	321	17.02	33.21	53.30	61.44	68.99	83.14	90.00	
211	211	0.00	33.56	48.19	60.00	70.53	80.40		
	221	17.72	35.26	47.12	65.90	74.21	82.18		
	310	25.35	40.21	58.91	75.04	82.58			
	311	10.02	42.39	60.50	75.75	90.00			
	320	25.06	37.57	55.52	63.07	83.50			
	321	10.89	29.20	40.20	49.11	56.94	70.89	77.40	83.74
		90.00							
221	221	0.00	27.27	38.94	63.51	83.62	90.00		
	310	32.51	42.45	58.19	65.06	83.95			
	311	25.24	45.29	59.83	72.45	84.23			
	320	22.41	42.30	49.67	68.30	79.34	84.70		
	321	11.49	27.02	36.70	57.69	63.55	74.50	79.74	84.89
310	310	0.00	25.84	36.87	53.13	72.54	84.26		
	311	17.55	40.29	55.10	67.58	79.01	90.00		
	320	15.26	37.87	52.12	58.25	74.74	79.90		
	321	21.62	32.31	40.48	47.46	53.73	59.53	65.00	75.31
		85.15	90.00						
311	311	0.00	35.10	50.48	62.96	84.78			
	320	23.09	41.18	54.17	65.28	75.47	85.20		
	321	14.76	36.31	49.86	61.09	71.20	80.72		
320	320	0.00	22.62	46.19	62.51	67.38	72.08		
	321	15.50	27.19	35.38	48.15	53.63	58.74	68.24	72.75
		77.15	85.75	90.00					
321	321	0.00	21.79	31.00	38.21	44.41	49.99	64.62	69.07
		73.40	85.90						

B Angles Between Crystallographic Planes for Hexagonal Elements*

HKIL	hkil	Be c/a = 1.5847	Ti 1.5873	Zr 1.5893	Mg 1.6235	Zn 1.8563	Cd 1.8859
0001	10$\bar{1}$8	12.88	12.90	12.92	13.19	15.00	15.23
	10$\bar{1}$7	14.65	14.67	14.69	14.99	17.03	17.28
	10$\bar{1}$6	16.96	16.99	17.01	17.35	19.66	19.95
	10$\bar{1}$5	20.10	20.13	20.15	20.55	23.21	23.53
	10$\bar{1}$4	24.58	24.62	24.65	25.11	28.19	28.56
	20$\bar{2}$7	27.60	27.64	27.67	28.17	31.48	31.89
	10$\bar{1}$3	31.38	31.42	31.45	32.00	35.55	35.98
	20$\bar{2}$5	36.20	36.25	36.29	36.87	40.61	41.06
	10$\bar{1}$2	42.46	42.50	42.54	43.15	46.98	47.43
	20$\bar{2}$3	50.66	50.70	50.74	51.31	55.02	55.44
	10$\bar{1}$1	61.34	61.38	61.41	61.92	64.99	65.33
	20$\bar{2}$1	74.72	74.74	74.76	75.07	76.87	77.07
	10$\bar{1}$0	90.00	90.00	90.00	90.00	90.00	90.00
	21$\bar{3}$2	67.55	67.59	67.61	68.04	70.57	70.86
	21$\bar{3}$1	78.33	78.35	78.36	78.60	80.00	80.15
	21$\bar{3}$0	90.00	90.00	90.00	90.00	90.00	90.00
	11$\bar{2}$8	21.61	21.64	21.71	22.09	24.89	25.24
	11$\bar{2}$6	27.85	27.88	27.91	28.42	31.75	32.16
	11$\bar{2}$4	38.39	38.44	38.47	39.07	42.87	43.32
	11$\bar{2}$2	57.75	57.79	57.82	58.37	61.69	62.07
	11$\bar{2}$1	72.50	72.52	72.54	72.93	72.92	75.18
	11$\bar{2}$0	90.00	90.00	90.00	90.00	90.00	90.00
10$\bar{1}$0	21$\bar{3}$0	19.11	19.11	19.11	19.11	19.11	19.11
	11$\bar{2}$0	30.00	30.00	30.00	30.00	30.00	30.00
	01$\bar{1}$0	60.00	60.00	60.00	60.00	60.00	60.00

*Taylor, A., and Leber, S., *Trans., AIME*, **200**, 190 (1954).

C Indices of the Reflecting Planes for Cubic Structures

Simple Cubic	Body-Centered Cubic	Face-Centered Cubic
{100}	—	—
{110}	{110}	—
{111}	—	{111}
{200}	{200}	{200}
{210}	—	—
{211}	{211}	—
{220}	{220}	{220}
{221}	—	—
{300}	—	—
{310}	{310}	—
{311}	—	{311}
{222}	{222}	{222}
{320}	—	—
{321}	{321}	—
{400}	{400}	{400}
{322}	—	—
{410}	—	—
{330}	{330}	—
{411}	{411}	—
{331}	—	{331}
{420}	{420}	{420}
{421}	—	—
{332}	{332}	—

D Conversion Factors and Constants

Conversion Factors	
Electron volts to ergs	1 eV = 1.60×10^{-12} erg
Electron volts to joules	1 eV = 1.6×10^{-19} J
Calories to joules	1 cal = 4.184 J
Joules to ergs	1 joule = 10^7 erg
Coulombs to statcoulombs	1 C = 3.00×10^9 statcoulombs
Psi to gm/mm²	1 psi = 0.703 gm/mm²
Psi to pascals	1 psi = 6,895 Pa
Psi to MPa	1000 psi = 6.895 MPa
Dynes to newtons	1 dyne = 10^{-5} N

Constants		
Constant	Symbol	Value
Avogadro's number	N_A	6.022×10^{23}/mol
Boltzmann's constant	$k = R/N_A$	1.381×10^{-23} J/°K
Molar gas constant	R	8.314 J/mol °K
		1.987 cal/mol °K
Planck's constant	h	6.626×10^{-34} J/Hz
Elementary charge	e	1.602×10^{-19} C
Rest mass of the electron	m_e	9.11×10^{-31} kg
Speed of light in vacuum	c	2.998×10^8 m/s
Acceleration due to gravity	g	9.81 m/s
		32.17 ft/s

E Twinning Elements of Several of the More Important Twinning Modes

Type of Metal	K_1	η_1	K_2	η_2	Observed in
Body-centered cubic	{112}	⟨11$\bar{1}$⟩	{11$\bar{2}$}	⟨111⟩	
Face-centered cubic	{111}	⟨11$\bar{2}$⟩	{11$\bar{1}$}	⟨112⟩	
Hexagonal close-packed	{10$\bar{1}$1}	⟨10$\bar{1}\bar{2}$⟩	{10$\bar{1}$3}	⟨30$\bar{3}$2⟩	Mg, Ti
	{10$\bar{1}$2}	⟨10$\bar{1}\bar{1}$⟩	{10$\bar{1}\bar{2}$}	⟨10$\bar{1}$1⟩	Be, Cd, Hf, Mg, Ti, Zn, Zr
	{10$\bar{1}$3}	⟨30$\bar{3}\bar{2}$⟩	{10$\bar{1}\bar{1}$}	⟨10$\bar{1}$2⟩	Mg
	{11$\bar{2}$1}	⟨11$\bar{2}\bar{6}$⟩	(0002)	⟨11$\bar{2}$0⟩	Hf, Ti, Zr
	{11$\bar{2}$2}	⟨11$\bar{2}\bar{3}$⟩	{11$\bar{2}\bar{4}$}	⟨22$\bar{4}$3⟩	Ti, Zr

F Selected Values of Intrinsic Stacking-Fault Energy γ_I, Twin-Boundary Energy γ_T, Grain-Boundary Energy γ_G, and Crystal-Vapor Surface Energy γ for Various Materials in ergs/cm² *

Metal	γ_I	γ_T	γ_G	γ
Ag	17[1,*]		790[8]	1,140[4]
Al	~200[2]	120[2]	625[8]	
Au	55[1,*]	~10[10]	364[8]	1,485[4]
Cu	73[1,*]	44[9]	646[5]	1,725[4]
Fe		190[4]	780[8]	1,950[8]
Ni	~400[1,3]		690[8]	1,725[8]
Pd	180[3]			
Pt	~95[3]	196[6]	1,000[6]	3,000[6]
Rh	~750[3]			
Th	115[3]			
W				2,900[7]

1. T. Jøssang and J. P. Hirth, *Phil. Mag.*, **13** 657 (1966).
2. R. L. Fullman, *J. Appl. Phys.*, **22** 448 (1951).
3. I. L. Dillamore and R. E. Smallman, *Phil. Mag.*, **12** 191 (1965).
4. D. McLean, "Grain Boundaries in Metals," Oxford University Press, Fair Lawn, N.J., 1957, p. 76.
5. N. A. Gjostein and F. N. Rhines, *Acta Met.*, **7** 319 (1959).
6. M. McLean and H. Mykura, *Surface Science*, **5** 466 (1966).
7. J. P. Barbour et al., *Phys. Rev.*, **117** 1452 (1960).
8. M. C. Inman and H. R. Tipler, *Met. Reviews*, **8** 105 (1963).
9. C. G. Valenzuela, *Trans. Met. Soc. AIME*, **233** 1911 (1965).
10. T. E. Mitchell, *Prog. Appl. Mat. Res.*, **6** 117 (1964).

* From Hirth, J. P. and Lothe, J., *Theory of Dislocations*, p. 764, McGraw-Hill Book Company, New York, 1968.

Index

A

Accommodation strain, 567
Activation energy, 230, 413
  for grain growth, 255
  for recovery, 231
  for recrystallization, 237, 240
  solidification of metals, 432
  for tempering processes, 651–653
Activation enthalpy for formation of vacancies, 213, 216–217, 238
Activity coefficients, 306
Aging treatment, 514–517
  artificial, 520
  cold, 518
  growth, 514
  at high aging temperature, 517
  incubation period, 514
  at low aging temperature, 517
  overaging, 515
  pre-aging, 517
  precipitation hardening, 514–517
  precipitation rate and morphology, 517
  shape of aging curve, 516
  temperature effect on speed of, 516–517
Alkali halides, cohesive energies of, 80
Allotropic transformation, 299, 302
Alloys, 294. *See also* Nonferrous alloys
  age hardening, 675–676
  allotropic transformation, 299, 302
  aluminium, 675–683
  binary, 294, 430
  compositions, 318
  creep strength, 697–698
  dendritic segregation in, 456
  equilibrium freezing range of, 324
  eutectic, 334–340, 590
  formation, 294
  freezing in, 444, 449–451
  glass-forming, 430
  grain boundaries of, 181
  high entropy (HEA), 346
  hypereutectic copper-silver, 339
  influence on hardenability, 631–636
  intermediate phases, 268–269
  interrelationships between phases, temperature, and composition, 319
  isomorphous alloy systems, 319–320, 323–326
  metallic glasses formed from, 425–431
  monotectic transformation, 343–346
  non-isomorphic, 392–397
  partitioning parameter, 593
  peritectic transformation, 340–343, 347
  phases in, 299–308
  with planar interface, 444–446
  solid-solution, 671
  supercooling, 423–424
  superelastic, 570–571
  superlattices, 329–332, 341
  ternary systems, 294, 315–316
  thermoelastic, elastic deformation of, 569
Alpha alloys, 691
Alpha-beta alloys, 692–694
  martensitic transformation, 694
  Widmanstätten plates of, 693
Alpha brasses, 669–672
Alpha phase, 301–303, 354
Alpha plus beta brasses, 669
Alpha stabilizers, 684
Alternating stress, 726–728
Aluminium alloys, 675–683
  age hardening sequence of, 676
  aluminum-lithium alloys, 676–683
  magnesium additions, 681, 683
  mechanical properties, 682
  overaging of, 677
  precipitation and aging behavior, 677–678
  problems associated with strain localization, 679, 681
  third mechanism of failure in, 681
  zirconium additions, 683
Aluminum bronze, 669
Aluminum precipitation hardening alloys, 522–523
Angle of incidence, 47
Angle of reflection, 47–48
Anisothermal anneal method, 225–226
Anisotropy of crystals, 8–10
Annealing, 670
  anisothermal, 225–226
  cold-worked metal, 223–224
  formation of nuclei, 241–243
  free energy (G) and strain energy (H) relationship, 225
  free-surface effects, 259–260
  full, 620
  geometrical grain coalescence, 249–251
  grain growth, 252–259
  grain size, 260–261
  homogenizing, 457
  isothermal, 225–226
  non-isothermal, 650
  polygonization, 231–236
  preferred orientation, 261–262
  rate of nucleation and rate of nucleus growth, 240–241
  recovery, 226–231
  recrystallization, 226, 236–249, 262–263
  stages of, 227
  stored energy, 223–227
  strain-induced boundary migration, 263
  three-dimensional changes in grain geometry, 251–252
Antiphase boundary (APB), 679
Antiphase domain boundary, 672
Artificial aging, 520
Asterism, 38
ASTM grain-size number, 630
Atmosphere, 275
  definition, 275
  dislocation, 274–276
  drag of, 277–279
  effect on motion of dislocation, 278
Attachment, rate of, 432
Auger electron microscopy (AES), 61–63
  energy of Auger electron, 62
  of silver specimen, 62–63
Auger electrons, 50
Austenitic grain size, 629–630
Austentite, 579
  bain distortion and the lattice parameters of, 646
  eutectoid transformation of, 584
  grain boundary, 585
  isothermal decomposition of, 582
  Kurdjumov-Sachs relationship, 580
  orientation relationships between ferrite, 584
  proeutectoid transformations of, 579–581
  transformation to pearlite, 581–586
  Widmannstätten plate structure, 580
Avogadro's law, 201
Avogadro's number, 69

B

Back-reflection Laue technique, 36–37
Backscattered electron, 51, 54–56
Bagaryatski relation, 585
Bain distortion, 551–533
  Bain cone, 557
  for body-centered cubic transformation, 552
  for face-centered cubic lattice, 551–552
  in indium-thallium alloy, 552–553
  for indium-thallium transformation, 556
  magnitude of atomic movements, 552
  martensite transformations, 555–557
Bainite reaction, 598–605, 649
  in alloy steels, 602, 604
  diffusion-controlled movement of ledges, 604
  dual nature of, 599, 604
  effect of temperatures, 602, 604
  growth of bainite plates, 600
  in iron-carbon steels, 599
  isothermal transformation reactions, 599–600
  Johnson-Mehl equation, 599
  microstructure of upper bainite, 601
  nature of carbides, 601
  time-temperature-transformation ($T\text{-}T\text{-}T$) diagram, 600–601
  upper bainite and lower bainite, 598–599, 601–603

Basal plane, 6, 8, 14
Bauschinger effect, 739
Becker-Döring theory, 483–484
Beryllium bronze, 670
Beta alloy, 692
Beta phase, 295, 301, 354
Beta stabilizers, 684–685
Binary alloys, 294, 430
Binary eutectic temperatures, 354
Binary phase diagrams, 318–355
  congruent points, 328
  copper-zinc-phase diagram, 350–352
  equilibrium heating or cooling, 323–325
  equilibrium phase transformations, 347
  eutectic alloy systems, 334–340
  eutectic transformation, 347
  free-energy-composition, 325–326
  intermediate phases, 348–350
  isomorphous alloy systems, 319–320, 323–326
  maxima curves, 327–329
  minima curves, 327–329
  miscibility gaps, 333–334
  monotectic transformation, 343–346
  peritectic transformation, 340–343, 347
  singular points, 328
  superlattices, 329–332
  tie line, 320, 323, 354
Blowhole, 464
Blue brittle phenomenon, 290, 721–722
Body-centered cubic lattice (BCC), 16
  application of Bragg equation, 33–34
  cleavage plane in, 707
  coordination number (CN) of, 4–5
  distortions (expansions), 275
  ferritic grains, 145
  hard-ball model of, 4
  important directions in, 8–9
  of iron, 8–9
  order-disorder transformation, 352
  slip systems in, 142–143
  Snoek relaxation in, 415
  solubility of carbon in, 269–273
  twins in, 542
  unit cell of, 2–4
Boltzmann equation, 207
Born theory of ionic crystal, 68–72, 80–81
Boyle's law, 201
Bragg angle, 48
Bragg law, 32–36, 45
  angle θ, 34
  application of, 33–34
  expression of relationship, 33
  first-order and second-order reflections, 33–36
Brasses, 669
Bravais lattices, 15
Bright-field image, 46–47
Brittle failure, 702
Brittle fractures, 721
  river pattern on, 710
Bronzes, 669–670
Burgers vector, 144, 148, 158, 194
  characteristics of edge and screw, 99
  of dislocation, 97–100, 106, 117–118
  local, 98
  RHFS local, 98

rules, 98
true, 98
Butterfly martensite, 556

C

Carbon in body-centered cubic iron, solubility of, 269–273
Cartridge brass, 669
Cellulose, 1
Cementite, 272, 577–578, 593
  fraction of, 583
  grain-boundary, 585
  lamellas, 580, 586
  proeutectoid, 580
Cesium chloride lattice, 67
Characterization technique in metallurgy. See Crystal characterization
Chemical interdiffusion coefficient, 382–383
Cleavage plane, 711
Cleavages, 706–711
  basal, 706–707
  cracks, 709–711
  dislocations, 710
  nucleation of, 707–709
  river pattern, 710
  splitting of zinc crystals, 706–707
Cliffs, 711
Climb force on dislocation, 115–118
Close-packed structure, 6, 16
  of body-centered cubic crystal, 4
  coordination number in, 8
  in face-centered cubic lattice, 5, 11
  hexagonal arrangement, 6–8, 14
  stacking sequences, 7
Coffin-Manson equation, 741–742
Coherent particle, 489
Cold aging, 518
Cold-worked metal, 223–224, 286
Components of a system, 294
Composition diagrams, 345
Compressibility, 71–72
Configurationally frozen, 428
Congruent points, 328
Conjugate slip system, 152
Considere's criterion for necking, 155–156
Constitution diagrams, 318
Continual mechanical twinning, 547–548
Continuous cooling transformations (CCT), 618–621, 628
Continuous growth, 436–437
Coordination number (CN) of crystal structure, 4–5, 8
  of diamond structure, 81
Copper alloys, 669–673
  with aluminum, 675
  composition, 670
  copper-beryllium alloys, 673–674
  deoxidized low-phosphorus copper, 669
  effect of alloying elements on, 667
  electrical resistivity and thermal conductivity, 667
  electrolytic tough-pitch copper, 668
  leaded, 675
  oxygen-free electronic copper, 669
  pure, 666–669
  with tin, 675

Copper-lead phase diagram, 343
Copper-nickel phase diagram, 325–327
  free-energy-composition curves for, 324
Copper-silver phase diagram, 334
Copper-zinc-phase diagram, 350–352
  intermediate phases, 352
  terminal phases, 352
Coring, 456, 461
Cottrell-Bilby theory of strain aging, 283–287
Cottrell-Bilby twinning mechanism, 545
Cottrell theory, 281
Coulomb potential energy, 68
Covalent bonding, 81–84
Crevices (encroachments), 731
Critical diameter, 623
Critical plane, 151
Critical resolved shear stress, 134–138
Cross-slip of an extended dislocation, 143–144, 147–149
  in magnesium, 144
Crystal, 2
Crystal binding, 66–67, 70, 72–78, 81–84
  Born equation, 70
  covalent bonding, 81–84
  Debye frequency, 76–78
  dipole-quadrupole term, 79
  dipoles, 72–74
  induced dipoles, 74–76
  in inert-gas atoms, 74
  internal energy of crystal, 66–67
  interrelation between space and time, 82
  ionic crystal, 67–72
  lattice energy (U), 66–67, 70, 76
  metallic bonding, 81–84
  molecular crystals, 80
  quadrupole-quadrupole term, 79
  van der Waals forces, 72
  zero-point energy, 78–79
Crystal characterization, 32–36, 38–42, 49–50, 57–58, 61–63
  Auger electron microscopy (AES), 61–63
  Bragg law, 32–36
  Debye-Scherrer (power) method, 38–42
  depth of focus, 57–58
  elastic scattering, 49
  electron interactions, 49
  electron probe X-ray microanalysis, 58–59
  electron spectrum, 51
  inelastic scattering, 49–50
  Laue techniques, 36–38
  microanalysis of specimens, 58
  picture point (element size), 56–57
  rotating-crystal method, 38
  scanning electron microscope (SEM), 51–53
  scanning transmission electron microscope (STEM), 63–64
  topographic contrast, 53–56
  transmission electron microscope (TEM), 43–48
  X-ray diffractometer, 42–43
  X-ray spectrum, 59–61
Crystalline structures of metallic elements, 15–16
Crystallographic directions, 18, 747–749
Crystal structure, 2

Cubic indices for planes, 12–13
Cu-Co phase diagram, 346

D

Darken's equation, 383, 386
Dark-field image, 47
Debye frequency, 76–78
Debye-Scherrer (power) method, 38–42
    determination of crystal structures, 42
    diffraction line, 42
    interplanar spacings and Bragg angles, 42
    principles, 39–42
    schematic representation, 41
Debye theory, 202
Deformation, 9, 20
Deformation twinning, 534–536, 547–548, 556, 560–561
    difference between slip and, 534
    interatomic spacing, 534
    shear associated with, 534–535
Dendrite arm spacing (DAS), 458, 461
Dendrites, 338, 442
Dendritic freezing, 444, 449–451, 461
Dendritic growth, in metals, 441–444
    branched growth, 442
    in cubic crystal, 443
    dendrite tip and side branching, 444
    direction of, 442
    recalescence, 444
    schematic representation, 442
    secondary, 443
    tertiary branches, 443
Density of coincidence sites, 191
Deoxidized low-phosphorus copper, 669
Depth of focus, 57–58
Detachment, rate of, 432
Dewar flask, 563
Diamond, 205
Diamond pyramid hardness (DPH), 520
Differential scanning calorimeter (DSC), 650
Diffraction patterns, 45–46
    beam directed along cubic crystal, 47
    Bragg's law, 48
    selected area, 48
    spacing of spots in, 48
Diffusion, 358–397
    Arrhenius plots for measurement of, 401–402
    basal plane, 366
    chemical, 383–384
    controlled growth, 496–500
    Darken's equations, 367–371
    Fick's first law, 391–392
    Fick's second law, 371–373
    free-surface, 388–391
    grain-boundaries, 388–391
    interstitial, 400–417
    intrinsic diffusivities, determination of, 377–378
    Kirkendall effect, 362–366
    low-solute concentrations, 383–384
    Matano method, 373–376
    in non-isomorphic alloys, 392–397
    pore formation, 366–367
    pure metals, self-diffusion in, 378–380
    substitutional, 415
    in substitutional solid solutions, 359–362
    using radioactive tracers, 384–387
    Zener ring mechanism, 365
    of zirconium and uranium, 376
Diffusion coefficient, 397, 459
    chemical interdiffusion coefficient, 382–383
    Fick's first law, 391
    grain-boundary, 388
    interdiffusion, 379, 385–387
    intrinsic, 385–386, 392
    liquid, 467
    self, 378–382, 384–387
    temperature dependence of, 380–383
    tracer-, 385–387
    tracer-diffusion coefficients, 386
Diffusion-controlled growth, 504
Diffusion-controlled velocity equation, 504
Diffusionless phase transformations, 550
Diffusivity-concentration curve, 376
Dipole-quadrupole term, 79
Dipoles, 72–74
Direction indices in cubic lattice, 11–12
Dislocations, 46–47
    in AISI 316L stainless steel, 91
    atmosphere, 274–276
    basic orientations of, 95–96
    bending-stress distribution and, 128–130
    boundary between sheared and un-sheared parts, 92
    Burgers vector, 97–100
    climb, 107–108
    climb force on, 115–118
    creep deformation, 90–91
    cross-slip of an extended, 147–149
    definition, 89–90
    density, 138, 281–282
    drag-stress dislocation velocity relationship, 278–279
    edge, 91–92, 94, 96, 99, 107–108, 112–114, 144
    extended, 106–107
    extrinsic and intrinsic stacking faults, 105–106
    in face-centered cubic lattice, 101–105
    Frank-Read source, 124, 147
    in hexagonal metals, 106–107
    interaction between impurity atoms and, 256
    interaction energy between solute atom and, 275–276
    intersections, 108–111
    jogs, 109–111
    kinks, 109
    loop, 96–97, 116
    movement of a single, 93
    negative edge, 94, 96
    nucleation of, 125–127
    parallel edge, 159
    piled-up, 90
    plastic deformation and, 123–159
    positive edge, 94, 96
    screw, 92–95, 99, 111–112, 144–145
    shear stress and, 93
    on single slip plane, 93
    in slip planes, 234
    in soft annealed crystal, 282
    stair-rod, 148–149
    strain energy of, 118–119
    stress field of, 111–114
    theory, 93
    in transmission electron microscope, 90–91
    vector notation for, 100
    in yield stresses, 86–89
Disorder-order transformation, 672
Divacancy, 217
Divorced eutectic, 460
Domain, 330
Double cross-slip mechanism, 145–147
Drag stress, 277–279
Ductile fractures, 715–721
Ductile metal, 702
Dynamic strain aging, 287–290
    in metals containing interstitial solutes, 290
    physical manifestations of, 288
    Portevin-LeChatelier effect, 289
    strain rate sensitivity, 288
    temperature, 289
    yield stress, 288

E

Edge dislocation, 144
    climb force on, 115–118
    interaction energy between interstitial solute atom and, 277
    interstitial atoms and, 274
    in polar coordinates, 114
    positive, 116
    strain energy of, 119
    stress field of, 112–114
Einstein equation, 284
Elastic after-effect (zero stress), 409, 416
Elastic scattering, 49
Electrolytic tough-pitch copper, 668
Electron interactions, 49
Electron probe X-ray microanalysis, 58–59
Electron spectrum, 51
Electron theory of metals, 84
Endurance ratio, 725
Energy, 66–68, 70, 76, 78–79
    of Auger electron, 62
    coulomb potential, 68
    Gibbs free, 203–205
    internal, 66–67
    lattice, 66–67, 70, 76
    strain, 225
    zero-point, 78–79
Energy Dispersive Spectrometers (EDS), 59
Energy-dispersive X-ray spectrometer, 59
Enthalpies, 204, 300
    formation of a single vacancy, 214
Entropy, 201–203
    change due to vacancies, 210–211
    of crystal, 211
    formation of a single vacancy, 214
    homogeneous mixture, 207
    intrinsic, 270–272
    measuring, 202–203
    of mixing, 206, 211, 308
    of mixing associated with carbon atoms, 271

Entropy (*Continued*)
  statistical mechanical definition, 205–209
  statistical mechanics, 205–209
  thermodynamics, 202
  vibrational, 211
Equation of state, 201
Equilibrium diagrams, 318
Equilibrium freezing range of alloy, 324
Equilibrium vacancy concentration, 214
Etching, 2
Eutectic alloy systems, 334–340
  composition, 335–340
  dendrites, 338
  effect of cooling, 340
  eutectic point, 335, 340
  eutectic temperature, 338–340
  hypereutectic structure, 339–340
  microstructure, 335–340
  proeutectic constituent, 338
Eutectic freezing, 466–470, 590
Eutectic point, 335, 340
Eutectic transformation, 347
Eutectoid ferrite, 609
Eutectoid steel, 598, 605–607
  carbon contents, 631
  continuous-cooling-transformation diagram of, 620–621
  isothermal transformation diagram for, 605
  time-temperature paths, 605–607
Eutectoid temperature, 596
Extremely low cycle fatigue (ELCF), 742
Extrinsic stacking fault, 105–106
Extrusions, 731

F

Face-centered cubic lattice (FCC), 5, 16
  atomic arrangement in octahedral plane of, 5–6
  austenitic grains, 145
  Bain distortion, 551–553
  comparison with HCP, 7–8
  low stacking fault energy, 548
  peritectic transformation, 340
  slip systems in, 138–140
  twin boundary in, 542–543
  twins in, 541–542
  uniqueness of, 5
  unit cell, 3
Fatigue failures, 722–724
  crack growth, 731–735
  effect of steel microstructure, 737
  extremely low cycle fatigue (ELCF), 742
  fatigue (or endurance) limit, 725
  fatigue strength, 726
  giga cycle fatigue (GCF), 738
  high-cycle fatigue (HCF), 738
  low-cycle fatigue (LCF), 737–741
  persistent slip bands (PSBs), 731
  practical aspects of, 743
  rotating-beam fatigue test, 724–726
  strain-fatigue life relationships, 742
  very high cycle fatigue (VHCF), 738
Ferrite, 272, 593, 608
  alpha-gamma, 604
  fraction of, 583
  lamellas, 579, 586
  proeutectoid, 580
  solubility of carbon in, 272

Ferromagnetism, 303
Fick's laws, 371–373, 391–392
  first law, 391–392
  Grube method, 372
  Matano method, 373–376
  mobility of an effective force, 391–392
  partial-molal free energy, 391–392, 394–396
  second law, 371–373
First law of thermodynamics, 204, 206
Focused ion beam (FIB) instrument, 44
Forced-velocity growth of pearlite, 589–592
Formal crystallographic theory of twinning, 536–542
  lattice rotations, 539–541
  in magnesium, 541
  nomenclature relating to, 540
  noncoplanar lattice vectors, 538
  plane of shear, 538
  rational directions, 538
  relation between shear and second undistorted plane, 541
  second undistorted plane, 537
  shearing action, 536–537
  size and shape of unit cell, 536
  spatial relationships, 538
  symmetry conditions, 541
  in zinc, 541
Fourier transform of regular lattice, 48
Free-energy equation, 205–206, 212
  of deformed metal, 225
  of ideal solution, 306–307
  for irreversible reaction, 205–206
  relation to stored energy, 225
  of single solution, 296–297
  for vacancies, 212
Freezing, 423, 432, 437, 439–441, 445–449, 452, 457, 466
  in alloys, 444, 449–451
  dendritic, 444, 449–451, 461
  effect of solidification velocity, 449
  enthalpy, 437
  equilibrium, 423, 432, 437, 447, 452, 457, 466
  eutectic, 466–470, 590
  free-energy change for, 466
  of ingots, 451–454
  at local equilibrium, 445
  non-equilibrium, 447–448
  Scheil equation, 446–449
  stable interface, 439–441
  temperature inversion during, 441
Frequency factor, 381
Fusion, heat of, 432–434

G

Gamma phase, 579
Gas-bubble formation, 463–464
Gay-Lussac law, 201
Geiger counter, 42
Gerber parabolic, 728
Gibbs free energy, 203–205, 270, 345
  of a pure substance, 299–300
  thermoelastic martensitic transformation, 568
Gibbs phase rule, 313–315
Giga cycle fatigue (GCF), 738
Glass transition point, 427–428
Glicksman, M. E., 444

Gold-nickel phase diagram, 329
Goodman approach, 728
Grain boundaries, 163–164
  of alloys, 181
  of alpha brass, 671–672
  coincidence sites, 190–191
  configuration, 260–261
  between crystals of different phases, 180–183
  density of coincidence sites, 191
  driving force per unit area for grain-boundary movement, 261
  dynamic recovery, 177–179
  energy, 169–172
  five degrees of freedom of, 167–168
  grain size, 183–185, 187–190
  grooves, 260
  Hall-Petch relation, 185–190
  intergranular fractures, 164
  low-energy dislocations (LADs), 172–177
  at low temperatures, 164
  on mechanical properties, 185–187
  metal failure and, 712–713
  motion, 256
  in nanocrystalline materials, 187–190
  at nucleation sites, 565
  polycrystalline materials, 163–164
  Ranganathan relations, 191–192
  second-phase inclusion, 257–258
  separating crystals, 181
  single-phase, 181–182
  small-angle, dislocation model of, 164–167
  in solid metals, 180
  stress field of, 168–169
  surface tension of, 179–180
  surface tensions, 179–183
  tilt, 168, 194–197
  transgranular fractures, 164
  twist, 168, 192–194
  two-phase, 181–182
Grain growth, 180, 226, 252–259
  activation energy, 255
  effect of impurities on, 256–259
  effect of solutes, 256
  free-surface effects, 259–260
  geometrical coalescence of grains, 249–251
  growth-inhibiting factors, 263
  law, 252–255
  limiting size, 260–261
  preferred orientation, 261–262
  secondary recrystallization and, 262
  thermal grooving and, 260
Graphene, 15
Graphite, 205
Gray cast irons, 579
Griffith cracks, 707
Grossman number, 625
Growth kinetics, 493–496
Grube method, 372
Guinier and Preston (GP) Zones, 2
Guinier-Preston or GP zones, 517

H

Habit plane or plane, 550
  in martensite transformations, 561–562
Hägg carbide, 577–578
Hall-Petch relationship, 185–190, 672

Index   757

Hard-ball model of crystalline lattice, 4
Hardening of steel, 618–664
  austenitic grain size, 629–630
  carbon content, 630–631
  high, 636–637
  influence of alloying elements, 631–636
  interrelation between time and temperature, 661
  iron-carbon martensites, 642–645
  Jominy hardenability test, 625–627
  of low-carbon steel, 654–656
  martensite transformations, 622–623, 628, 637–641, 646
  quench cracks, 647–648
  secondary, 661–662
  spheroidized cementite, 656–658
  of tempered martensite in iron-carbon steels, 660
  tempering, 648–656, 658, 661
  variables determining, 628–629
Heat treatment, 9
Heterogeneous nucleation, 476, 490–493
Heterophase fluctuation, 479
Hexagonal close-packed lattice (HCP), 6
  basal plane of, 6
  close-packed structure, 6
  comparison with FCC, 7–8
  crystallographic features, 6
  octahedral plane of, 6
  slip systems in, 140–142
Hexagonal indices, 14–15
High-cycle fatigue (HCF), 738
High Entropy Alloys (HEA), 346
High resolution electron backscatter diffraction (HR-EBSD), 145
Homogeneous mixture, 207
Homogeneous nucleation, 476–477
Homogenization, 457–461
Homopolar linkages, 81
Hot short, 182
Hume-Rothery rules, 273–274
Hydrogen molecule, 82
Hypereutectic copper-silver alloys, 339
Hypereutectoid steel, 599, 609–615
  carbon contents, 631
  isothermal diagram of, 611–615
  microstructure, 611
  slowly cooled, 609–611
  stages in formation, 610
  transformations of, 613–614
Hypoeutectic compositions, 340
Hypoeutectoid steel, 607–609
  amount of austenite, 608
  amount of ferrite, 608
  forms of ferrite, 609
  slowly cooled, 607–609
  structure, 608–609
  transformation of, 607–608
Hysteresis, 411
  loss, 10

I
Ideal critical diameter, 624
  Grossman method of determining, 625
Ideal gas, equation of state for, 201
Ideal quench, 624
Impurities
  atoms, 256
  effect on grain growth, 256–259

Indium-thallium alloy, 552–554
  Bain distortion in, 552–553
  martensite transformations in, 553–554
Indium-thallium martensitic transformation, 553–554, 556–561
  Bain cone in, 557
  Bain distortion for, 556–557
  habit plane in, 562
  plane of zero distortion, 561
  rational plane in, 562
  shearing process, 558–559
  stabilization phenomenon, 564–565
  tetragonal twinning, 558–561
Induced dipoles, 74–76
Inelastic scattering, 49–50
Ingots, 419
  casting of, 465
  killed-steel, 465
Interaction constant, 276–277
Intercrystalline brittle fracture, 721
Interdiffusion coefficient, 379, 385–387
Interface controlled growth, 501–504
Interface-controlled velocity equation, 504
Interface growth rates, 504
Intergranular fractures, 164
Intermetallic compounds, 349
Internal energy, 202
Internal energy of perfect crystal, 66–67
Internal friction, 412, 523
  interstitial diffusion, 401–402
  relaxation time, 409–415
  relaxed modulus, 411
  specific damping capacity, 412
  of torsion pendulum, 412
  unrelaxed modulus, 410
Interphase precipitation, 525–527
Interstitial atoms, 274
Interstitial diffusion, 400–417
  Arrhenius plots for measurement of, 401–402
  in body-centered cubic lattices, 409–410
  in body-centered cubic metals, 416
  of carbon in alpha iron, 412
  of carbon in body-centered cubic iron (alpha-iron), 415
  experimentally determined, 415–416
  expression for diffusivity, 400–401
  free-energy term, 401
  internal-friction measurements, 401–402
  relaxation time, 409–415
  Snoek effect, 402–409
Interstitial solid solution, 268
  atom formation of, 269
  carbon in body-centered cubic iron, solubility of, 269–273
  equilibrium in, 270–273
  intrinsic entropy, 270–272
  solute atoms in, 269
Intrinsic diffusion coefficient, 385–387, 392
Intrinsic entropy, 270–272
Intrinsic stacking fault, 105–106
Intrusions, 731
Invariant plane strain, 550
Inverse segregation, 461–462
Ionic crystals, 68–72, 80–81
  alkali halides, 80–81
  Born theory of, 68–72, 80–81

  formation of, 67–72
  interactions in, 68
  Madelung number, 71
  plastic deformation mechanisms, 68
  van der Waals terms, 80–81
Iron-carbon alloys, 576–616
  bainite reaction, 598–605
  boundaries between liquid and gaseous phases, 577
  cementite, 577–578
  coring effects, 579
  diagram, 576–579
  equilibrium phase transformations, 579
  eutectic point, 578
  eutectoid point, 578
  eutectoid section of, 582
  eutectoid steel, 598, 605–607, 611
  Hägg carbide, 577–578
  hypereutectoid steel, 614–615
  hypoeutectoid steel, 607–609
  isothermal transformation diagrams, 611–615
  noneutectoid steel, 611–615
  orientation relations, 584
  peritectic point, 578
  peritectic transformation, 578–579
  phase diagram, 577
  phase transformations, 579
  proeutectoid transformations of, 579–581
  time-temperature-transformation curves, 597–598
  time-temperature-transformation (T-T-T) diagram, 605–607
Iron-carbon eutectoid steel, 594
Iron-carbon martensites, 642–645
Iron crystal, 8–9
Iron-nickel martensitic transformation, 562–564
  athermal transformation, 563
  austenite transformation, 563
  deformation associated with, 562
  formation of martensite plates, 563
  reverse transformation, 563
  stabilization phenomenon, 562, 565
Iron-nickel peritectic transformation, 342
Iron sulfide, 182
Isomorphous alloy systems, 319–320
  equilibrium heating or cooling, 323–325
  free-energy-composition, 325–326
Isothermal anneal, 225–226
Isothermal austenitic transformations, 581–582
Isothermal martensite, 648
Isothermal transformation diagrams, 611–615, 618–619
  of iron-carbon alloys, 611–615
  vs continuous cooling transformation (CCT) diagrams, 618–619
Isothermal-transformation technique, 581–583, 589, 597–600, 605
  austenitic transformations, 581–583
  bainite reaction, 599–600
  for eutectoid steel, 605
  growth rate of pearlite, 589, 598
  for hypereutectoid steel, 611–615
  for hypoeutectoid steel, 612
  time relationships for phases, 597

758 Index

J
Jogs, 109–111
Johnston-Gilman theory, 281
Jominy hardenability test, 625–627

K
Killed-steel ingots, 465
Kinetic theory, 200–202
Kinks, 109
Kirkendall effect, 362–366
Kronberg and Wilson boundary, 190, 194
Kurdjumov-Sachs relationship, 580
Kurdjumov's martensitic theory and laws, 568

L
Lamellar eutectics, 467–470
Lamellar spacing, 469–470
lateral crystal growth, 438–439
Lateral growth, 438–439
Lath martensite, 644
Lattice energy (U), 76
Laue techniques, 36–38, 227
   arrangement of Laue back-reflection camera, 36
   back-reflection, 36–37
   Laue photograph, 37
   transmission, 36–37
Law of conservation of energy, 206
Lead bronze, 669
Leaded copper alloys, 675
Lenticular martensite, 644
Lever rule, 608
Lifshitz-Wagner theory of precipitation, 677
Lignin, 1
Liquid miscibility gap, 344
Liquid Phase Separation (LPS), 344, 346
Lithium, 673–683
Local equilibrium, 445
Logarithmic decrement, 412
Low-cycle fatigue (LCF), 737–741
Low-energy dislocations (LADs), 172–177
Lüders bands, 279–281
Lüders lines, 280

M
Macro-PSBs, 731
Macroscopically homogeneous body of matter, 293
Macrostructure, 2
Madelung energy, 80
Madelung number, 71
Magnesium-lithium phase diagram, 330
Martensite transformations, 533
   athermal transformation, 554–555
   Bain distortion, 551–553, 555–557
   butterfly martensite, 556
   crystallographic theory, 555–561
   deformation twin interfaces and, 554
   diffusionless phase transformations, 550
   dimensional changes associated with, 646
   growth of martensite plates, 566
   habit plane in, 561–562
   habit plane or plane, 550
   in indium-thallium alloy, 553–554, 556–561
   invariant plane strain, 550
   iron-nickel, 562–564
   isothermal formation, 564
   Kurdjumov's martensitic theory and laws, 568
   nucleation of martensite plates, 565–566
   phase change, 550
   phenomenological crystallographic theory, 555–561
   plane strain associated with, 552
   plastic deformation, effect of, 567
   reversibility of, 554
   rotation of transformed lattice, 555
   at screw-dislocation-intersection, 554
   shape-memory effect, 571–572
   shear deformation, 555
   stabilization, 564–565
   stabilization in, 555
   of steel, 622–623, 628, 637–641, 646
   stress effects, 566–567
   stress-induced martensite (SIM), 569–571
   surface martensite, 556
   thermoelastic, 567–569
   thin plate martensite, 556
   titanium martensite, 694
   volume relaxation, 554
   Wechsler-Lieberman-Read theory, 555–556
Martensitic reactions, 533
   deformation associated with, 646
Matano-Boltzmann method for diffusivity, 374
Materials engineer, 1
Mechanical twinning, 534–535, 541–542
Metal failure, 702–743
   blue brittleness, 721–722
   brittle failure, 702–703, 710
   brittle fractures, 721
   cleavages, 706–711
   Coffin-Manson equation, 741–742
   cup and cone fracture, 715–716
   ductile crack, 702–703
   ductile-crack movement, 720
   ductile fractures, 715–721
   by easy glide, 703–704
   fatigue failures, 722–735, 737–741, 743
   fracture, 702–703
   grain boundaries and, 712–713
   nonmetallic inclusions, effect of, 735–736
   rupture by necking (multiple glide), 704–705
   stresses and, 713–715
   by twinning, 705–706
   void sheet mechanism, 717–718
Metallic bonding, 81–84
Metallic glasses, 425–431
   deformation of, 431
   glass transition point, 427–428
   physical and mechanical properties, 431
   production of, 430
   by quenching of pure metal, 429
   sinking point, 427
   softening point, 427
   supercooled liquid and, 425–428
   from supercooled liquids, 431
Metallography, 31–32. *See also* Crystal characterization
   three-dimensional diffraction gratings, 32
   using X-rays, 32
Metals, 200–202, 215, 441–444, 533, 535–542
   dendritic growth, 441–444
   enthalpy and entropy of formation of vacancy in, 215
   formal crystallographic theory of twinning, 536–542
   thermal behavior of, 200–202
   twinning, 533, 535, 541–542
Metal structures, 2
   body-centered cubic lattice (BCC), 2–5, 8–9
   coordination number (CN), 4–5, 8
   face-centered cubic lattice (FCC), 3, 5–8
   hexagonal close-packed lattice (HCP), 6–8
   Miller indices, 10, 12–15
   polycrystalline metals, 2, 9
   polymorphic, 16
   preferred orientations of crystals, 9–10
   stereographic projection, 16–24
   topographic features of a metal surface, 53
   unit cells, 2–4
Microanalysis of specimens, 58
   Auger electron microscopy (AES), 61–63
   electron probe X-ray microanalysis, 58–59
   scanning transmission electron microscope (STEM), 63
Microsegregation, 456
Microstructures, 2
Miller indices, 10, 23, 39, 47, 100, 192
   of cubic crystals, 12–13
   determination of, 15
   of directions, 11–12
   for hexagonal crystals, 14–15
   of second cube diagonal, 11
   in square brackets, 11
Miscibility gaps, 333–335
Mixing-entropy equation, 211–212
Mole fractions, 296, 304–305, 309
Molybdenum, 51, 59
Monotectic transformation, 343–346
Mosaic structure, 232

N
Nanoprecipitates, 2
Necking, 155–156
Necklace structure, 681
Negative edge dislocations, 94, 96
Neon, 69
Newton's Second Law, 257
Nickel bronze, 669
Noncubic crystals, 13
Noneutectoid steel, 611–615
Nonferrous alloys, 666–694
   alpha, 691
   alpha-beta, 692–694
   aluminium, 675–683
   beta, 692
   commercially pure copper, 666–669
   copper, 669–674
   creep strength, 697–698

Index   759

superalloys, 694–697
titanium, 683–691
Non-isomorphic alloys, diffusion in, 392–397
Non-isothermal annealing, 650
Nonpolar molecules, 80
Nucleating pearlite, 585–586
Nucleation, 475–476, 490–493, 529–530
  aging treatment and, 515
  Becker-Döring theory, 483–484
  of cleavages, 707–709
  in commercial metals, 424
  critical radius, 477
  diffusion controlled growth, 496–500
  embryos, 476, 479, 481–485, 488, 490, 493
  of ferrite lamella, 585
  during freezing, 423, 485–487
  gas-bubble formation and, 463
  growth kinetics, 493–496
  heating transformations, 504–505
  heterogeneous, 476, 490–493, 529–530
  heterophase fluctuation, 479
  homogeneous, 476–477
  interface controlled growth, 501–504
  liquid from vapor phase, 476–483
  locations for, 529
  of martensite plates, 565–566
  during melting, 424
  of pearlite, 583–584, 595–597
  precipitate particles and, 500–501, 505–507, 523–524
  during recrystallization, 475
  on slip planes, 529
  solidification of metals, 423–425
  solid-state reactions, 487–490
  supercooling, 423–424
  superheating, 424–425
  twin, 543–545
  Volmer-Weber theory, 483
  Zener's theory, 496–497, 499

O

Octahedral planes, 5–6, 8, 138
One-component system, 293
Orowan equation, 158–159, 279
Orowan's mechanism, 527–528
Oxley, Mark P., 63
Oxygen-free electronic copper, 669

P

Partial-molal free energies, 391–392, 394–396, 503
Partial-molar free energies, 304
  graphical determinations of, 310–312
  solutions and, 297, 304
Partitioning parameter, 593
Pauli exclusion principle, 82
Pcal second, 426
Peach–Kohler equation, 118
Pearlite, 579, 608
  Bagaryatski relation, 585
  bainite reaction, 598–605
  bainite transformation, 599
  effect of temperature on, 587–588
  effects of alloying elements, 592–595
  eutectoid transformation of austenite to, 589
  forced-velocity growth of, 589–592
  growth of, 586–587
  interlamellar spacing, 591
  interlamellar spacing of, 587–588
  isothermal growth rate, 595
  isothermal transformation technique, 589
  Johnson and Mehl equation, 596–597
  methods of nucleating, 585
  nucleation of, 595–597
  orientation of ferrite in, 585
  Pitsch-Petch relation, 584
  rate of growth, 587–589, 594
  time-temperature-transformation curves, 597–598
  transformation parameters, 594
  undercooling and, 591
Peritectic transformation, 340–343, 347
Permanent dipoles, 80
Persistent slip bands (PSB), 731
Phase diagrams, 480
  aluminum-copper, 505–506
  aluminum-lithium, 677
  antiphase boundaries, 330
  antiphase domains, 330
  binary, 318–355
  congruent points, 328
  copper-beryllium, 674
  copper-lead, 343
  copper-silver, 334
  copper-zinc, 350–352, 670
  Cu-Co, 346
  equilibrium heating or cooling, 323–325
  equilibrium phase transformations, 347
  eutectic alloy systems, 334–340
  eutectic transformation, 347
  Fe-Fe3C metastable, 578
  free-energy-composition, 325–326
  gold-nickel, 329
  intermediate phases, 348–350
  isomorphous alloy systems, 319–320
  lever rule, 320–323
  magnesium-lithium, 330
  maxima curves, 327–329
  minima curves, 327–329
  miscibility gaps, 333–334
  monotectic transformation, 343–346
  nickel-magnesium diagram, 350–351
  peritectic transformation, 340–343, 347
  singular points, 328
  superlattices, 329–332
  ternary, 353–355
  tie line, 320, 323, 354
  of titanium, 686–689
Phase(s), 180, 293, 295, 299–308
  in alloy system, 299–308
  alpha, 301–303, 354
  beta, 295, 301, 354, 672
  boundaries between crystals of different, 180–190
  continuous or matrix, 295
  definition, 180, 293
  delta, 342
  discontinuous or dispersed, 295
  equilibrium between two, 298–299
  free energy of, 301, 306–307
  gamma, 294, 303, 579, 674
  Gibbs phase rule, 313–315
  of graphite, 205
  Greek symbols, 294
  of ideal solution, 304–305
  intermediate, 295
  liquid, 294
  mixtures, 295
  of nonideal solution, 305–308
  of one-component system, 299–303
  partial-molar free energies, 304, 310–312
  physical nature of mixtures, 295
  of pure iron, 294
  solid, 294–295
  solutions, 295–298
  ternary systems, 315–316
  three phases in equilibrium, 312–313
  of two-component systems, 304, 308–309, 312–313
Picture point (element size), 56–57
  ratio of, 57
Pipe, 465
Pitsch-Petch relation, 584
Plain-carbon steels, 592
Plane of shear, 538
Planes of zone, 18–20
Plasmon excitation, 49
Plastic deformation, 128–134, 149–151, 155–159, 224, 548
  bend gliding, 128–130
  on cleavage crack propagation, 709–710
  compression deformation, 149–151
  Considere's criterion for necking, 155–156
  creation of point defects during, 224
  critical resolved shear stress, 134–138
  cross-slip, 147–149
  crystal structure rotation, 149–151
  double cross-slip, 145–147
  face-centered cubic (FCC), 151–153
  martensitic transformation, 567–568
  Orowan equation, 158–159
  in polycrystalline material, 142
  relation between dislocation density and stress, 156–157
  rotational slip, 130–132
  by slip, 548
  slip bands, 145
  on slip planes, 529
  slip planes and directions, 133–134
  slip systems, 134, 138–143
  stored energy of, 224
  strain aging, 282–290
  Taylor's relation, 157–158
  twinning in, 547–548
  work hardening, 153–155
Polarizability, 74
Pole mechanism, 545
Polycrystalline metals, 2, 9
Polygonization, 231–236
  in bent and annealed iron-silicon single crystals, 233–235
  dislocation movements in, 232–236
  orientations of planes, 234
  rate of, 233
  recovery process, 231–232
  subboundaries, 232, 234, 236
  subgrains, 232, 234
Polyphase systems, 298
Porosity, 462–465
  gas, 465
  interdendritic, 465
  wormhole, 464

Portevin-LeChatelier effect, 289
Positive edge dislocations, 94, 96
Powder diffraction files (PDFs), 43
Precipitate-free zones (PFZ), 676, 679–680
Precipitate particles, 500–507
  dissolution of, 505–507
  interference of growing, 500–501
  nucleation of, 500–501, 505–507
Precipitation hardening, 511–512
  aging temperatures and, 517
  aging treatment, 514–517, 519–522
  alloys and, 519–522
  aluminum alloys, 522–523
  cold aging, 518
  energy per unit volume, 529
  growth of, 528
  Guinier-Preston or GP zones, 517
  hardening theories, 527–529
  homogeneous and heterogeneous nucleation of particles, 523–524
  interaction of dislocations, 527–528
  interphase precipitation, 525–527
  naturally aged, 518
  Orowan's mechanism, 527–528
  pre-aging influence, 517
  precipitate particles, 517–519
  quenching, 517, 523–524, 526, 529
  recrystallization and, 530
  reversion, 518
  sequences in aluminum-silver system, 522–523
  solution treatment, 513–514
  solvus curve, 512–513
  spinoidal transformation, 524
  strain energy, 528
  strengthening of alloy, 674
  Widmanstatten structure, 530
Preferred orientations of crystals, 9–10
Preformed nuclei, 241
Primary slip system, 151
Prism planes, 14
Proeutectic constituent, 338
Proeutectoid ferrite, 579, 609
Proeutectoid transformations of austenite, 579–581
Pseudoelasticity, 570
Pure coppers, 666–669

Q

Quadrupole-quadrupole term, 79
Quantum, or wave mechanics, 81
Quantum theory, 67
Quench cracks, 647–648
Quenched specimen, 319
Quenching
  agitation during, 625
  brine quench, 625
  cooling rate during, 673
  precipitation hardening, 517, 523–524, 526, 529
  of pure metal, 429

R

Ranganathan relations, 191–192
Raster, 52
Recalescence, 444
Reciprocal lattice, 48
Recovery stage of annealing, 226–236
  activation energy, 231

geometrical coalescence of grains, 249–251
at high and low temperatures, 236
isothermal behavior of, 236
polygonization, 231–236
in single crystals, 228–231
subboundaries, 232, 234, 236
subgrains, 232
Recrystallization, 226, 670
  activation energy, 237, 240
  driving force for, 243
  formation of nuclei, 241–243
  geometrical coalescence of grains, 249–251
  grain size, 243–245, 247
  impurities, effects of, 245–246
  nucleation rate, 240–241
  nucleus growth rate, 240–241
  precipitation phenomena and, 530
  secondary, 262–263
  solute, effects of, 245–246
  strain, effect of, 239–240
  strain-induced boundary migration, 263
  temperature, 239
  time and temperature, effect of, 237–238
  variables in, 245
  of zirconium, 239–240
Relaxation time, 429, 459
  experimental determination of, 409–415
  internal friction, 409–415
Relaxed coincidence boundary, 196
Relaxed modulus, 411
Resonance effect, 82
Restrictive equations, 309
Ridges (protrusions), 731
Rimming, 465
Rockwell-C hardness, 643–645
Rose's channel, 546
Rotating-beam fatigue test, 724–726
Rotating-crystal method, 38

S

Scanning electron microscope (SEM), 51–53
  depth of field of, 51
  depth of focus characteristic of, 57
  field of view, 51
  schematic drawing of, 52
  specimen surface in, 51–53
Scanning transmission electron microscope (STEM), 63–64
Scheil equation, 446–449
Schmid orientation factor, 159
Schottky defect, 210
Screw dislocation, 144, 224
  climb force on, 115–118
  in long crystal, 115
  strain energy of, 118–119
  stress field of, 111–112
Seam, 464
Secondary bainite, 649
Secondary electrons, 49
  energy distribution of, 50
Secondary hardening of steel, 661–662
Secondary recrystallization, 190, 262–263
Second-order transformation, 303
Second-order twin, 540
Segregation, 455–456
  complete, 654

of gaseous solutes, 463
inverse, 461–462
no, 654
Selected area diffraction patterns, 48
Self-diffusion coefficient, 378–382, 384–387
Shape-memory effect, 571–572
Sharp yield point, 279–282
  Cottrell theory, 281–282
Shear modulus, 89
Sievert's law, 462
Silicon bronze, 669
Silver-bearing alloy, 669
Silver-magnesium system, 352
  intermediate phases in, 349
  phase diagram, 348–349
Simpson's Rule, 376
Single-phase binary systems, 304
Size factor, 273
Slip bands, 145
Slip lines, or slip traces, 87
Slip systems, 134
  in body-centered cubic crystal, 142–143
  deformation and, 548
  difference between twinning and, 534–535, 547
  dislocations, 544–545, 548–549
  in face-centered cubic structure, 138–140
  in hexagonal crystals, 140–142
  metals deforming by, 548
  plane, 87–88, 548–549, 559
  shear stress for, 544, 548
  slip on equivalent, 138
  twinning shear by, 546
Small-angle, dislocation model of, 164–167
Snoek relaxation in bct martensite, 415
Sodium chloride lattice, 67
  Born exponent for, 72
  cohesive energy for, 72
  compressibility, 71–72
Solidification of metals, 419–423, 431–434, 436–439, 454–455, 457–461, 466–470
  activation energy, 432
  atomic movements, 431–432
  castings, 419, 454–455
  continuous growth, 436–437
  dendritic growth, 441–444
  eutectic freezing, 466–470
  freezing, 444–446, 449–454
  fusion, heat of, 432–434
  homogenization, 457–461
  ingots, 419
  inverse segregation, 461–462
  lateral crystal growth, 438–439
  liquid phase, 420–423
  liquid-solid interface, 434–436
  metallic glasses, 425–431
  nucleation, 423–425
  porosity, 462–465
  rates of attachment and detachment, 432
  Scheil equation, 446–449
  segregation, 455–456
  stable interface freezing, 439–441
  vaporization, heat of, 432–434
Solid solutions, 267–268
  dislocation interactions, 274–276
  drag stress, 277–279
  electromotive series positions, 274
  Hume-Rothery rules, 273–274
  interaction constant, 276–277

Index 761

interaction energy, 276
intermediate phases, 268–273
interstitial, 268–269
Lüders bands, 279–281
sharp yield point, 279–282
size factor, 273–274
solute atoms, 274, 276
strain aging, 282–290
substitutional, 267–268, 273–274
Solid-state reactions, 487–490
Solubility of carbon in body-centered cubic iron, 269–273
Solute atoms, 273, 275
in body-centered cubic metals, 410
interaction of dislocations and, 274
interstitial solid solution, 269
substitutional solid solutions, 274, 276
Solution treatment, 514
Solvent, 267
Southern yellow pine, 1
Specific damping capacity, 412
Sphalerite, 67
Spherical triangles, 26
Spheroidized cementite, 656–658
Spinoidal transformation, 524
Spontaneous reactions, 203, 205
Stable interface freezing, 439–441
Stair-rod dislocation, 148–149
Standard stereographic projection, 23–24
Standard stereographic triangle for cubic crystals, 24–27, 153
crystal axes, 27
crystallographic directions, 26
plotting of crystallographic data, 26–27
standard 100 projection, 23–25
Statistical mechanics, 200–201
Steels, 294
austenitic grain size, 629–630
carbon content, 630–631
continuous cooling transformations (CCT), 618–621
eutectoid, 598, 605–607, 620–621, 631
hardenability, 622–629, 631–637
high-speed, 662
hypereutectoid steel, 599, 611–615
hypoeutectoid steel, 607–609
iron-carbon eutectoid, 594
low-carbon, 654–656
martensite transformation in, 637–641
quench cracks, 647–648
spheroidized cementite, 656–658
tempering, 648–656
use of carbide-forming elements, 662
Stereographic projection, 16–18, 153
of a cubic crystal, 47
effect of rotation, 21–23
standard, 23–24, 151
Wulff net, 20–23
Stirling's approximation, 271
Stored energy, 223–225
cold-worked metals (annealing), 223–224
free-energy equation and, 225
maximum value of, 224
of plastic deformation, 224
release of, 225–227
Strain, 158–159, 416–417, 679, 681
anelastic, 416–417
dislocations, rate of, 158–159
localization, 679, 681

in polycrystalline metal, 159
shear, 158–159
Strain aging, 282–290
aging period, 282–283
blue brittleness phenomenon, 290
Cottrell-Bilby strain aging equation, 283–287
definition, 283
dynamic, 287–290
Harper equation, 286
loads and, 282–283
metals deforming by slip in, 548
Portevin-LeChatelier effect, 289
yield point, 283
Strain energy, 225
in precipitation phenomena, 528
Strain-fatigue life relationships, 742
Strain-induced boundary migration, 263
Strain rate sensitivity, 288
Stress-induced martensite (SIM), 569–571
Stretcher strains, 280
Structural periodicity, 195
Subboundaries, 232, 234, 236
coalescence of, 234
Subgrains, 232, 234
angle of rotation of, 234
Substitutional atoms, 274
lattice distortion associated with, 274
Substitutional solid solutions, 267–268, 273–274
diffusion in, 359–362
electromotive series positions, 274
Hume-Rothery rules, 273–274
Kirkendall effect, 362–366
penetration curves, 363
pore formation, 366–367
size factor, 273–274
solute atoms in, 400
Superalloys, 694–697
composition of, 695
effect of carbides, 696
nickel-base, 695
strength and stability of, 696
Superdislocation, 679
Superlattices, 329–332, 341
Surface martensite, 556
Surface tension, 179–182
of a bismuth-copper interface, 182
consequences of grain-boundary movement, 180
of grain boundary, 179–180
induced boundary motion, 258
between liquid and solid, 182
System, 294

T

Temperature inversion, 441
Tempering, 648–656
effect on physical properties, 658–660
interrelation between time and temperature in, 661
of low-carbon steel, 654–656
Temper roll, 281
Terminal solid solutions, 295
Ternary eutectic temperature, 354
Ternary phase diagram, 353–355
isothermal section, 354–355
solidus points in, 354
utilization of, 354
Ternary systems, 294, 315–316

Tetragonal twinning planes, 558–560
Textures (preferred orientation), 9–10
Thermal grooving, 259–260
Thermodynamics, 200
entropy, 202–203
entropy change in, 202
Gibbs free energy, 203–205
Gibbs free-energy change, 205
internal energy, 202
of solutions, 295–298
spontaneous reactions, 203
Thermoelastic martensitic transformation, 567–569
accommodation strain, 567
elastic and frictional terms, 568
elastic deformation of thermoelastic alloys, 569
of nonferrous alloys, 569
plastic deformation, 567–568
reversed transformation, 569
during structural phase transformation, 568
thermodynamic equilibrium equation, 568–569
Thin plate martensite, 556
Three-dimensional diffraction gratings, 32
Tie line, 320, 323, 354
Tilt boundaries, 168, 194–197
Time-temperature-transformation curves of pearlite, 597–598
Tin sweat, 462
Titanium alloys, 683–691
allotropic phase transformation, 684–685
allotropic transformation, 683
alpha and beta stabilizers, 684–685
beta to alpha transformation, 684
chemical behavior of, 683
classification, 685, 691
commercial and semicommercial grades, 690
at different temperature, 683
melting point of, 683
oxidation effect, 683
phase diagrams, 686–689
tensile strength/density ratios, 683–684
yield strength/density ratios, 683
Topographic contrast, 53–56
Torsion pendulum, 409
damped vibration of, 410
energy loss per cycle in, 411
fractional loss of energy per cycle, 411–413
frequencies, 414
measure of internal friction, 412
stress-strain curves for, 410–411
vibration amplitude, 411
Tracer-diffusion coefficients, 385–387
Transformation temperature, 504–505
Transmission electron microscope (TEM), 43–48
bright-field images and dark-field images, 46–47
diffraction of electrons, 45
diffraction patterns, 46–48
lenses, 44
as a microscope, 46
sample preparation, 44
schematic drawing of, 44

Transmission Laue technique, 36–37
True isothermal transformation, 581
Twin boundaries, 542–543
    in body-centered cubic metals, 542
    coherent boundary and incoherent boundary, 543
    of face-centered cubic metals, 542–543
    in hexagonal metals, 542
    interfacial energy of, 542
    presence of dislocations in, 543
    stacking sequence of close-packed planes, 542
Twinning, 533, 752
    Bain distortion, 551–533
    continual mechanical, 547–548
    deformation, 534–536, 547–548, 550, 556, 560–561
    on face-centered cubic stress-strain curves, 548–549
    formal crystallographic theory of, 536–542
    growth, 543–545
    in indium-thallium alloy, 553–554
    on individual lattice planes, 534
    interfacial energy, 542
    lattices of, 536, 539–541
    martensite reaction, 550–551
    martensitic plates, 534
    mechanical, 534–535, 541–542
    metal, 533, 535, 541–542, 548–549
    metal failure by, 705–706
    plane, 536
    in plastic deformation, 547–548
    in recrystallization phenomena, 535
    Rose's channel, 546
    rotation of atoms, 536
    second-order twin, 540
    shear, 541, 546–547
    work hardening rate, 548–549
    of zinc, 533, 541–542
Twin nucleation, 543–545
    in body-centered cubic crystal, 544
    deformations and, 545
    dislocations, 544–545
    growth and, 543–545

in HCP crystals, 544
homogeneous nucleate deformation, 544
hydrostatic pressure and, 543–544
shear stress for, 544
stress required for, 545
Twist boundaries, 168, 192–194
    example involving, 192–194

U

Unit cells, 2–4
    in body-centered cubic crystal, 2–3
    of face-centered cubic lattice, 3
    of hexagonal close-packed lattice (HCP), 6
Unrelaxed modulus, 410

V

Vacancies, 224, 481
    concentration of, 211, 214–215
    divacancy, 217
    entropy, 201–203
    entropy change due to, 210–211
    equation of state, 201
    free-energy equation for, 210, 212
    Gibbs free energy, 203–205
    internal energy, 202
    interstitial atoms, 217–220
    jump rate of, 216–217
    kinetic theory, 200–202
    motion of, 209–210, 215–217
    principle of diffusion, 209
    Schottky defect, 210
    spontaneous reactions, 203
    statistical mechanics, 200–201, 205–209
    thermal behavior of metals, 200–202
    vacancy, definition, 209
    vacancy-depleted PFZ, 679–680
    vibrational entropy, 211
Valence electrons, 83
Van der Waals crystals, 72–80
    attractive energy, 79
    dipole-quadrupole term, 79
    dipoles, 72–74

forces, 72, 80
of inert-gas solids, 74, 76
in molecular crystals, 80
quadrupole-quadrupole term, 79
repulsive force, 76
second-order nature of interaction, 76
Vaporization, heat of, 432–434
Very high cycle fatigue (VHCF), 738
Vibrational entropy, 211
Vickers diamond pyramid hardness scale, 642
Vogel-Futcher-Tammann equation (VFT equation), 426
Void sheet mechanism, 717–718
Volmer-Weber theory, 483

W

Water quench, 620
Wavelength Dispersive Spectrometers (WDS), 59
Wechsler-Lieberman-Read theory, 555–556
White X-ray beam, 34
Widmannstätten plate structure, 580
Wood structure, 1
Work hardening, 153–155
    plastic deformation and, 153–155
    rate, 548–549
Wulff net, 20–23

X

X-ray diffraction techniques, 38
X-ray diffractometer, 42–43
X-ray microanalyzer, 63
X-ray photons, 60–61
X-ray spectrum, 59–61

Z

Zener ring mechanism, 365
Zener's relationship, 257, 261
Zero-point energy, 67, 78–79
Zincblende lattice, 67–68
Zirconium, 16
Zone axis, 18–19

Physical Metallurgy Principles

...dition

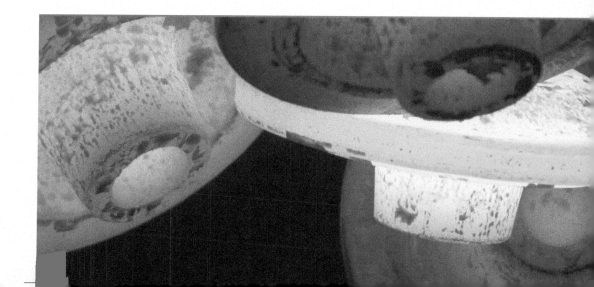

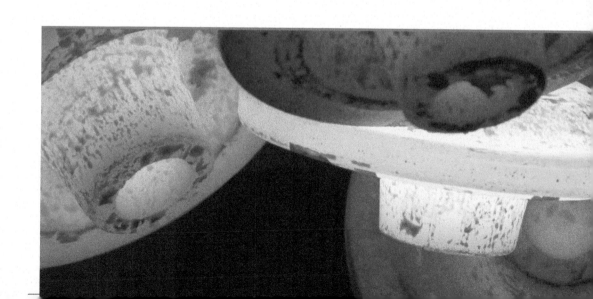

Physical Metallurgy Principles, Fifth Edition

Reza Abbaschian, Lara Abbaschian

SVP, Product: Cheryl Costantini

VP, Product: Thais Alencar

Portfolio Product Director: Rita Lombard

Senior Portfolio Product Manager: Timothy Anderson

Product Assistant: Emily Smith

Learning Designer: MariCarmen Constable

Content Manager: Samantha Enders

Digital Project Manager: Nikkita Kendrick

VP, Product Marketing: Jason Sakos

Senior Director, Product Marketing: Danae April

Product Marketing Manager: Mackenzie Paine

Content Acquisition Analyst: Deanna Ettinger

Production Service: MPS Limited, RPK Editorial Services

Designer: Chris Doughman

Cover Image Source: Jackfoto/ Shutterstock.com

Notice to the Reader

Publisher does not warrant or guarantee any of the products described herein or perform any independe connection with any of the product information contained herein. Publisher does not assume, and expres any obligation to obtain and include information other than that provided to it by the manufacturer. expressly warned to consider and adopt all safety precautions that might be indicated by the activities des and to avoid all potential hazards. By following the instructions contained herein, the reader willingly ass in connection with such instructions. The publisher makes no representations or warranties of any kind, not limited to, the warranties of fitness for particular purpose or merchantability, nor are any such rep implied with respect to the material set forth herein, and the publisher takes no responsibility with res material. The publisher shall not be liable for any special, consequential, or exemplary damages resultin part, from the readers' use of, or reliance upon, this material.

Printed in the United States of America
Print Number: 01   Print Year: 2024

Copyright © 2025 Cengage Learning, Inc. ALL RIGHTS

WCN: 01-100-371

No part of this work covered by the copyright herein reproduced or distributed in any form or by any mea as permitted by U.S. copyright law, without the prior permission of the copyright owner.

The names of all products mentioned herein are use identification purposes only and may be trademarks registered trademarks of their respective owners. Ce Learning disclaims any affiliation, association, connec sponsorship, or endorsement by such owners.

Previous Editions: © 2009, © 1992

For product information and technology assistance, co **Cengage Customer & Sales Support, 1-800-354- or support.cengage.com.**

For permission to use material from this text or produc requests online at **www.copyright.com.**

Library of Congress Control Number: 2024931732

ISBN: 979-8-214-00166-1

Cengage
5191 Natorp Boulevard
Mason, OH 45040
USA

Cengage is a leading provider of customized learni Our employees reside in nearly 40 different countr digital learners in 165 countries around the world. local representative at **www.cengage.com.**

To learn more about Cengage platforms and servic or access your online learning solution, or purchase for your course, visit **www.cengage.com.**

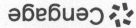